XII International Conference Radiation-Thermal Effects and Processes in Inorganic Materials 2016

IOP Conference Series: Materials Science and Engineering
Volume 168

Tomsk, Russia
4 - 12 September 2016

ISBN: 978-1-5108-3672-3
ISSN: 1757-8981

Printed from e-media with permission by:

Curran Associates, Inc.
57 Morehouse Lane
Red Hook, NY 12571

Some format issues inherent in the e-media version may also appear in this print version.

Printed by Curran Associates, Inc. (2017)

For permission requests, please contact the Institute of Physics
at the address below.

Institute of Physics
Dirac House, Temple Back
Bristol BS1 6BE UK

Phone: 44 1 17 929 7481
Fax: 44 1 17 920 0979

techtracking@iop.org

Additional copies of this publication are available from:

Curran Associates, Inc.
57 Morehouse Lane
Red Hook, NY 12571 USA
Phone: 845-758-0400
Fax: 845-758-2633
Email: curran@proceedings.com
Web: www.proceedings.com

Table of contents

Volume 168

XII International Conference Radiation-thermal Effects and Processes in Inorganic Materials

4–12 September 2016, Tomsk, Russian Federation

Accepted papers received: 19 January 2017
Published online: 7 February 2017

Preface

Physical and chemical phenomena in inorganic materials in radiation, electrical and thermal fields

Technologies and equipment for their implementation

The use of radiation-thermal processes in nanotechnology

Adjacent to the main theme of the conference issues

IOP Conf. Series: Materials Science and Engineering **168** (2017) 011001 doi:10.1088/1757-899X/168/1/011001

XII International Scientific Conference "Radiation-Thermal Effects and Processes in Inorganic Materials"

International Scientific Conference "Radiation-Thermal Effects and Processes in Inorganic Materials" is a traditional representative forum devoted to the discussion of fundamental problems of radiation physics and its technical applications. The conference was held in Tomsk, Russia.

The program of the Conference covers a wide range of technical areas and modern aspects of radiation physics, its applications and related matters.
Topics of interest include, but are not limited to:
- Physical and chemical phenomena in inorganic materials in radiation, electrical and thermal fields;
- Research methods and equipment modification states and properties of materials;
- Technologies and equipment for their implementation;
- The use of radiation-thermal processes in nanotechnology;
- Adjacent to the main theme of the conference issues

The conference was attended by leading scientists from countries near and far abroad who work in the field of radiation physics of solid state and of radiation material science.

During the conference School-Conference of Young Scientists was held.

Peer review statement

All papers published in this volume of *IOP Conference Series: Materials Science and Engineering* have been peer reviewed through processes administered by the proceedings Editors. Reviews were conducted by expert referees to the professional and scientific standards expected of a proceedings journal published by IOP Publishing.

IOP Conf. Series: Materials Science and Engineering **168** (2017) 012001 doi:10.1088/1757-899X/168/1/012001

Evolution of Defective State of Aluminum Oxide Irradiated with Chromium Ions after Annealing in Oxygen Environment

A V Kabyshev[1], **F V Konusov**[2]

[1] Professor, National Research Tomsk Polytechnic University, 30, Lenin Avenue, 634050, Tomsk, Russia

[2] Senior researcher, National Research Tomsk Polytechnic University, 30, Lenin Avenue, 634050, Tomsk, Russia

E-mail: konusov@hvd.tpu.ru, kabyshev@tpu.ru

Abstract. The characteristics of interband and exponential optical absorption of leucosapphire and polycrystalline corundum (polycor) after irradiation with chromium ions and subsequent annealing in vacuum at 300–1800 K and in air at 300–800 K are studied. Contributions of defects with different thermal and chemical stability into optical parameters were established. The effect of intrinsic radiation defects, of substitutional defects and of complexes on base of oxygen and defects on formation the focal point in absorption spectra owing to fulfilment of the Urbach rule was determined. Heating in air of strongly defective material synthesized in surface layers of alumina by the ion–heat modification influences the characteristics of defects and the electronic structure of band gap negligibly.

1. Introduction

The application of irradiation with ions for the modification of properties of dielectric materials stimulates studying the radiation damage processes and the changes in their electronic structure [1–6]. Postimplantation thermal annealing of materials is a necessary stage of modification of their properties [2, 4, 5]. Thermal annealing in vacuum and in reducing atmosphere stabilizes the properties of material owing to coagulation of implanted elements into clusters and formation of new phases, having characteristics different from those for pristine matrix [2, 4]. Thermal annealing in oxygen–containing environment influences the electronic structure of material owing to realization oxidative and restoration reactions with participation of defects and oxygen and formation of oxygen–containing complexes (OCCs) [3, 4]. Localized states (LSs) induced by radiation defects (RDs) well manifest in leucosapphire and polycrystalline corundum (polycor), which contain a low concentration of impurities [1–3, 5]. The implantation of ions with different ability to substitute for lattice atoms of Al_2O_3 and to form the solid solutions allows evaluating of their role in changing of properties [1–6].

RDs and complexes on their base induce in the band gap (BG) of Al_2O_3 the complicated spectrum of LSs [1–3, 5]. The study of the produced nonequilibrium state in dielectrics requires approaches using generalized microscopic parameters characterizing structural disordering of the system of a whole [7, 8]. In the analysis of radiation–induced disorder is taken into account the principle of equivalence of static and dynamic components of its [7, 8]. That allows determining quantitatively the contribution of the atomic disordering into properties change. The parameters of induced disordering, calculated from absorption spectra with using of Urbach rule, depend on concentration of RDs [5–7] and stoichiometric and phase–composition of compounds [8].

The aim of this paper is to study the characteristics of the optical absorption of leucosapphire and polycor after irradiation with chromium ions Cr^{n+} and subsequent annealing in vacuum and in air, to determine the thermal and chemical stability of the defects of different nature and to determine the degree of their influence on the optical properties.

2. Experimental

Leucosapphire and polycor plates were irradiated with Cr^{n+} ions in the frequency pulse mode (the ion energy was 50–150 keV, the fluence Φ was 10^{16}–$2{\cdot}10^{17}$ cm^{-2}, and the current density in the pulse was 10^{-2}–10^{-3} A/cm^2). Annealing after irradiation was performed in the temperature range T_{ann1}=300–1800 K at the residual pressure P≤10 Pa and in air in the temperature range T_{ann2}=300–800 K at atmospheric pressure P=10^5 Pa. Choice of the Cr^{n+} ions for irradiation was caused by high probability for synthesis of hard solutions on base of these ions in the surface layers of alumina [2, 4–6]. The high concentration of RDs allows considering this solution as disordered.

The spectral dependence of absorption coefficient α(hν) calculated from the diffusion reflection spectra according to [2, 5, 6] was approximated by the Urbach rule of general form

$$\alpha(h\nu,T) = \alpha_{00} \times e^{((h\nu - h\nu_0)/E_U(T))}, \qquad (1)$$

where E_U is the Urbach energy which depend on the total dynamic and static disorder in the lattice [7, 8]. The pre–exponential factor α_{00} and the characteristic energy $h\nu_0$ are the parameters of the Urbach focus [7, 8]. For the energy ranges Δ(hν) in which the Urbach rule is satisfied we calculated E_U and the pre–exponential factor α_0 from the simplified equation

$$\alpha(h\nu) = \alpha_0 \times e^{(h\nu/E_U)}. \qquad (2)$$

To evaluate the applicability of the concept of Urbach focus we calculated the dependence $\alpha_0(E_U)$ taking into account (1) and (2) according to equation

$$\alpha_0 = \alpha_{00} \times e^{(-h\nu_0/E_U)}, \qquad (3)$$

where parameters α_{00} and $h\nu_0$ were determined from the spectra α(hν, Φ, $T_{ann1,\ 2}$). Isoabsorption optical gap $E_{g\alpha'}$ was calculated from spectra for a number fixed values of α'. Value $E_{g\alpha'}$ corresponds to energy of transitions hν for execution of equality α(hν)=α' [6, 7, 8]. Values $E_{g\alpha'}$ and E_U were compared with dependence $E_{g\alpha'}(E_U)$, calculated with using of parameters $h\nu_0$ and α_{00} according to equation [7, 8]

$$E_{g\alpha'} = h\nu_0 - [\ln(\alpha_{00}/\alpha']\times E_U. \qquad (4)$$

The interrelation between the absorption by the LSs produced by RDs and the interband absorption was verified using the approximation of the α(hν) in the ranges Δ'(hν) by the power–like law

$$(\alpha \times h\nu) \propto (h\nu - E'_g(E''_g))^m, \qquad (5)$$

where m=1/2 and 2 correspond to the direct and indirect interband allowed transitions across the optical gaps E'_g and E''_g [5]. We determined the contribution of local levels of RDs into the absorption by using the expansion of the α(hν) in terms of elementary components in the Gaussian form.

3. Result and discussion

The charge state of the substitutional defects Cr_{Al}^{n+} (n=2–4) and intrinsic RDs (anion and cation vacancies V_O and V_{Al}, as well as interstitial defect Al_i) is determined by Φ of ions and vary after annealing in vacuum, that reflects on the absorption spectra (figures 1 and 2). The interband absorption in the ranges Δ'(hν)=2.4–4.2 and 3.4–4.8 eV is caused by electronic transitions in clusters Cr_{Al}^{n+}...Cr_{Al}^{n+} which were formed after annealing at T_{ann1}=300–1300 K. The complexes on the base Cr_{Al}^{n+} and intrinsic RDs (T_{ann1}=1300–1600 K) determine the exponential absorption with participation of their LSs in ranges Δ(hν)=1.5–3.4 eV and 3.1–3.9 eV and the interband absorption in ranges

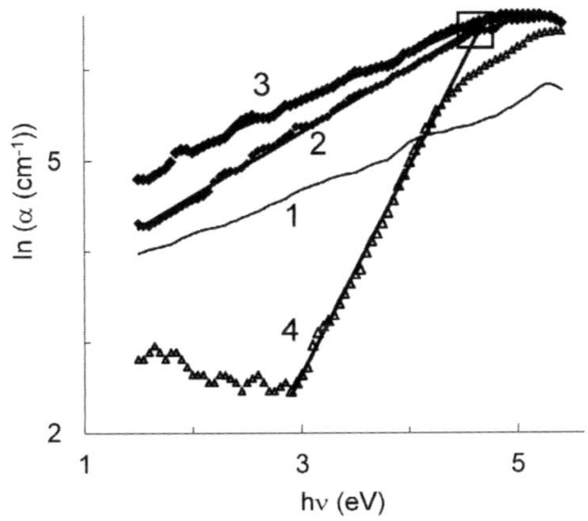

Figure 1. Absorption spectra $\alpha(h\nu)$ of leucosapphire before (1) and after irradiation with Cr^{n+} ions and after annealing in vacuum at T_{ann1}=1800 K (2) and in air at T_{ann2}=500 K (3) and 800 K (4). The fluence is $\Phi=10^{17}$ cm^{-2}.

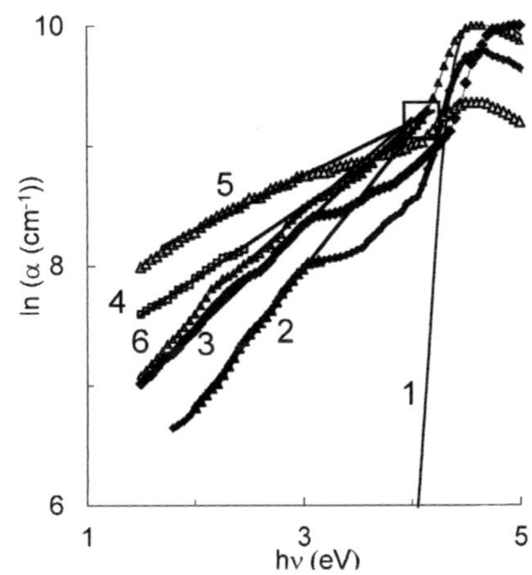

Figure 2. Absorption spectra $\alpha(h\nu)$ of polycor before (1) and after irradiation with Cr^{n+} ions and after annealing in vacuum at T_{ann1}=1470 K (2, 3) and in air at T_{ann2}=500 K (4, 5) and 800 K (6). The fluences are $\Phi=10^{16}$ (2, 4) and 10^{17} cm^{-2} (3, 5 and 6).

$\Delta'(h\nu)$=2.8–4.1 eV and 3.5–4.5 eV (figure 2) [2, 5, 6]. Absorption after $T_{ann1}\geq$1700 K was caused by single defects Cr_{Al}^{n+} and residual defect complexes (figures 1 and 2). Annealing in air stimulates the dissociation of defect complexes and change of the charge state of RDs. At those conditions, OCCs with participation of RDs are formed. The changing in the spectra $\alpha(h\nu, \Phi, T_{ann1})$ and in the Urbach parameters α_0, E_U, α_{00} in single crystals and in polycrystals is subject to the general laws (figures 1 and 2). Two stages of annealing in air were obtained at T_{ann2}=300–500 and 500–800 K. The change of optical parameters depend on building of materials and on concentration of RDs, which determined by Φ and T_{ann1} (figures 1 and 2). The redistribution of LSs density from range 4.0–5.4 eV to range 1.4–4.0 eV and from LSs with 3.0–4.0 eV to levels with 1.5–3.0 eV occurring after annealing at T_{ann2}=300–500 K and accompanies by the expansion of ranges $\Delta(h\nu)$ and by growth of values E_U in these ranges (figure 1, curve 3; figure 2, curves 4 and 5). Between the value of Urbach energy after annealing in vacuum $E_U(T_{ann1})$ and its value after annealing in air $E_U(T_{ann2})$ the relation in form of the equation $E_U(T_{ann2})=a\cdot E_U(T_{ann1})+b$ (a=1.1–2.2, b=0.2–0.5 eV) was obtained (figure 3). The increase in E_U occurs in 1.5–2.2 times after irradiation of Al_2O_3 by fluences $\Phi\leq5\cdot10^{16}$ cm^{-2} and increase in E_U occurs only in 1.1–1.5 times after $\Phi>10^{17}$ cm^{-2} (figure 3). The optical gap is narrowing on the value 0.2–0.4 eV (figure 4). The growth in quantity of disorder in lattice of Al_2O_3, which is caused by influence of oxygen environment, occurs proportional to quantity of disorder, which is induced by irradiation of ions and annealing in vacuum. Relation $E_U(T_{ann2})/E_U(T_{ann1})$ is increased with depth of level localization in BG and it is achieved of maximal values at energy $h\nu\sim$3.6 eV (figures 1 and 2). Smoothing and disappearance of the local bands with centers at energies ε=2.6 eV and 3.2 eV shows, that contribution of continuous component of spectrum of LSs is increased owing to absorption of oxygen on RDs and formation of OCCs (figures 1 and 2).

The change of absorption parameters in range $\Delta(h\nu)$=1.5–4.0 eV after annealing at T_{ann2}=500–800 K is caused by exchanging by electrons between the energetic levels of substitutional defects, the levels of divacancies $F_2^{0\ldots n+}$ (their LSs were identified in [1, 2, 9]) and acceptor levels created by OCCs ($\varepsilon\geq$4.0 eV). The recharge of divacancies $F_2^{n+}\rightarrow F^0\rightarrow F^+$, which stimulated by the sequential acts

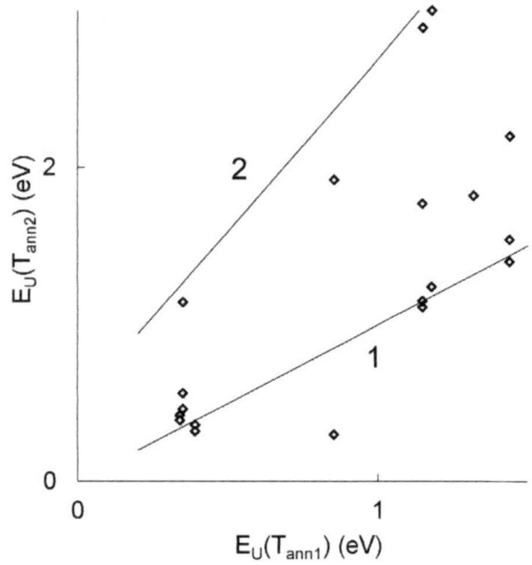

Figure 3. Dependency of the Urbach energy after annealing in air $E_U(T_{ann2})$ from its value $E_U(T_{ann1})$ after prior irradiation of Al_2O_3 with Cr^{n+} ions and subsequent annealing in vacuum.

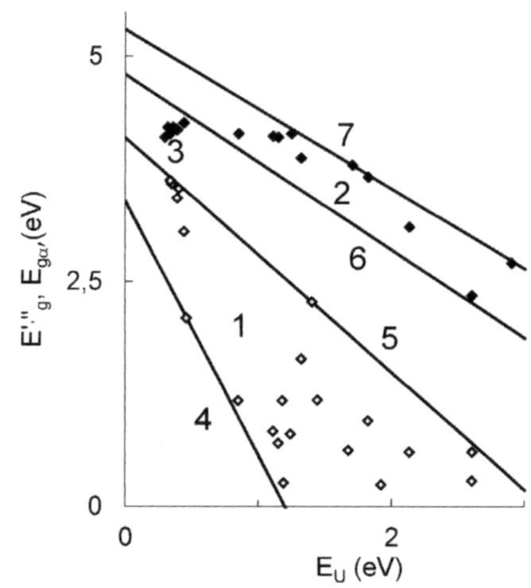

Figure 4. $E'{,}''_g$ vs. E_U (1–3) and $E_{g\alpha'}$ vs. E_U (4–7) for indirect (1) and direct transitions (2 and 3) after irradiation and annealing in vacuum at T_{ann1}=1470 K and 1800 K and in air at T_{ann2}=500 K and 800 K.

of adsorption and dissociation of molecules O_2 and formation of stable chemical bonds, influences on optical parameters also (figures 1 and 2). Shift of range of localization $\Delta(h\nu)$ to high energy region is manifested in leucosapphire in compare with polycor (figures 1 and 2). Values $E_U(T_{ann2})$ are concentrated near the values of $E_U(T_{ann1})$ that are typical for annealing in vacuum in case of weak influence adsorption of oxygen on parameters of RDs (figure 3, line 1) and values $E_U(T_{ann2})$ are distributed between the lines 1 and 2 in case of strong influence of adsorption on properties. Change in structure and in characteristics of LSs induced by RDs and OCCs is reflected in the parameters of interband absorption, calculated from spectra in ranges $\Delta'(h\nu)$=1.5–4.0 and 3.0–5.3 eV (1) (figure 4).

The change in the structure of LSs causes the intersection of the linear sections of the spectra $\alpha(h\nu)$, where relation (2) is valid in the high energy range. The convergence region (is denoted by a rectangle in figures 1, 2) is located at $h\nu_0$=4.8±0.08 eV (α_{00}=800±90 cm^{-1}) in leucosapphire (figure 1) and at $h\nu_0$=3.95±0.15 eV(α_{00}=8500±500 cm^{-1}) in polycor (figure 2) and can be interpreted as a focal point of generalized Urbach rule, just as in [6–8]. Annealing in air supplements the criteria of Urbach rule (1) for irradiated Al_2O_3 (similar points were fixed at $h\nu_0$=4.4 eV for Cr^{n+} and Ti^{n+} ions and at 4.2 eV for Si^{n+} ions) [5, 6]. The similarity of focal points for ions with different ability to substitution of cation of lattice is caused by existence of interrelation between the kinetic of RDs accumulation and parameters of the exponential absorption [1–5, 6]. The fan–like character of curves $\alpha(h\nu, \Phi)$ persists after adsorption of oxygen in the surface layers of alumina (figures 1 and 2). The separation of focal points for single and polycrystals occurs (figures 1 and 2). At that, the low–energy shift of focal point on ~ 0.3 eV was fixed in polycor. The focal point in polycor at $h\nu_0$=3.95 eV is situated between the levels induced by the interstitials defect $Al_i^{0(+)}$ (ε=4.1 eV [2, 9]) and divacancy F_2^{0} (ε=3.6 eV [2, 5, 9]). This point is connected with boundary of deep LSs in BG of irradiated alumina [2, 8]. Focal point in leucosapphire at $h\nu_0$=4.8 eV is neared to energies of F^+–center [9, 10] and of defect Cr_{Al}^{3+} (ε=4.7–4.8 eV [5, 6, 10]). Besides, value $h\nu_0$=3.95 eV is neared to value E_g=4.3 eV in films Al_2O_3 containing the amorphous and ordered phases [11–13]. Biographical defects have a stabilized effect on parameters of RDs (figure 2). Contribution of composition disorder into total disorder of alumina is 15–20 % according [6]. Influence of distortion of stoichiometry is negligible. The basic component of disorder in alumina lattice is the RDs formation, as show the optical parameters (figures 1–4).

Other criteria of this approach is accordance of dependence between parameters α_0 and E_U, which calculated from the spectra, with dependence $\alpha_0=f(E_U)$, which calculated according equation (3) with using of focal point coordinates $h\nu_0=3.95$ eV, $\alpha_{00}=8500$ cm^{-1} (figure 5). The values E_U and α_0 until and after impact of oxygen are united into common array for leucosapphire and polycor (figure 5). The values α_0 situated lower the curve 2 correspond to spectra of alumina after $\Phi\leq5\cdot10^{16}$ cm^{-2}, and values α_0 situated higher the curve 2 – to spectra of polycor after $\Phi\geq10^{17}$ cm^{-2}. The basic contribution to this dependency give the clusters $Cr_{Al}^{n+}...Cr_{Al}^{n+}$ and complexes on the base of RDs. Isoabsorption optical gap $E_{g\alpha'}$ and optical gaps $E'^{,''}_g$ for interband direct (ranges $\Delta'(h\nu)$=3.3–4.0 and 4.0–5.3 eV) and indirect transitions ($\Delta'(h\nu)$=1.5–4.0 and 3.0–5.1 eV) are changed in correlation with E_U changing (figure 4). This relationship has a general character for materials containing the high concentration of defects [6–8]. The straight lines 4–7 in figure 4 are calculated according to equation (4) (at value $\alpha_{00}=8500$ cm^{-1} and value pairs $h\nu_0=3.40$ eV, $\alpha'=500$ cm^{-1} (4); 4.1 eV, 2300 cm^{-1} (5); 4.8 eV, 3200 cm^{-1} (6); 5.3 eV, 3500 cm^{-1} (7)) limit the arrays 1–3 of the experimental values $E'^{,''}_g$ and E_U. At that, values E_g and E_U obtained before and after impact of the oxygen on alumina surface are united in the common array (figure 4). Values of gap for direct transitions which were obtained after T_{ann2}=500–800 K are excluded from this picture (figure 4, array 3). This caused by influence of the OCCs on narrowing of the BG of alumina. At that, value $h\nu_0=3.4$ eV is neared to energies of F_2^0 (ε=3.6 eV [2, 9]), and $h\nu_0=4.1$ eV is neared to focal points 3.95 eV in spectra of polycor (figures 2 and 4). Values $h\nu_0=4.8$ and 5.3 eV coincide with energy of bands of F^+–center (ε=4.8 and 5.4 eV [9, 10]) and of defects Cr_{Al}^{3+} (ε=4.7–4.8 eV [5, 6, 9, 10]). Relations between the parameters of the exponential and interband absorption allowed us to conclude, that accumulation of OCCs has a weak effect on the electronic structure of new strongly defective material which was formed by irradiation with ions and by subsequent annealing.

Analysis of spectra $\alpha(h\nu)$ by expansion on the elementary components of Gaussian form with the centers at energies ε_i allowed us to estimate the influence of interaction between the oxygen and RDs, which have local energetic levels, on the total disorder induced by ions in the lattice of alumina (figure 6). The intensity of bands with centers at ε_i=1.7 eV in leucosapphire and at 1.5–1.6 eV in polycor caused by the substitutional defects Cr_{Al}^{3+} is decreased in 1.5–4 times owing to adsorption of oxygen.

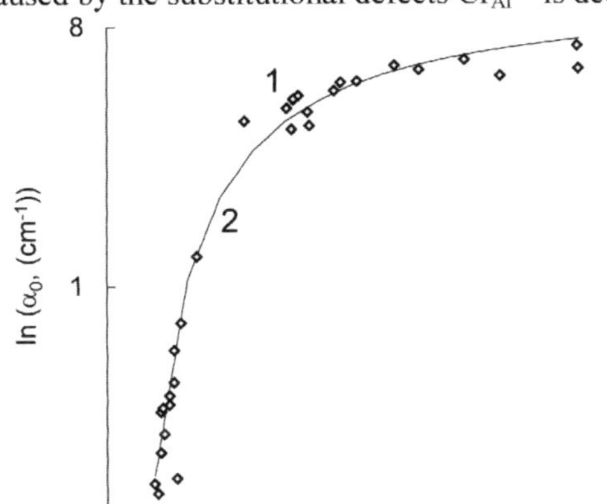

Figure 5. α_0 vs. E_U in Al_2O_3 after irradiation at fluence $\Phi=10^{16}$–10^{17} cm^{-2} and annealings in vacuum and in air. Parameter α_0 (2) was calculated from Eq. (2) and Eq. (3) on array (1) at $\alpha_{00}=8500$ cm^{-1}, $h\nu_0=3.95$ eV.

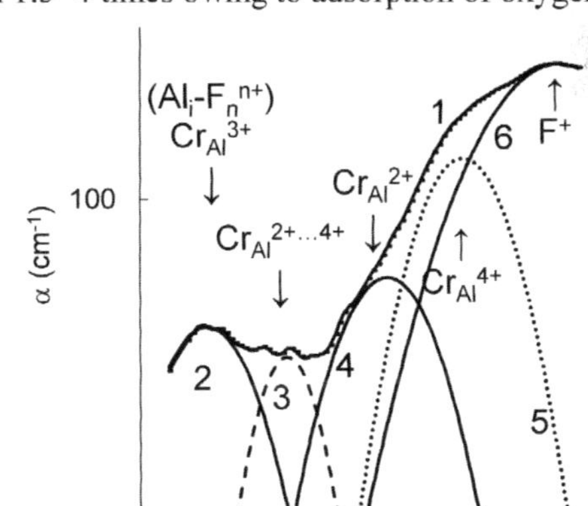

Figure 6. Absorption spectrum $\alpha(h\nu)$ of leucosapphire after irradiation with Cr^{n+} ions at fluence $\Phi=10^{17}$ cm^{-2} and subsequent annealings in vacuum at T_{ann1}=1800 K and in air at T_{ann2}=800 K and its decomposition.

Stability of centers to oxidation is increased as growth of ions fluence and temperature of vacuum annealing from $T_{ann I}$=1470 to 1800 K (figure 2). The weakening of bands induced by electronic transitions in the defects Cr_{Al}^{3+} was caused by recharging Cr_{Al}^{3+}→$Cr_{Al}^{2+,\ 4+}$, as shows the relations between concentration N_i of defects Cr_{Al}^{n+} with different n. Divacancies F_2^0 and clusters of the intrinsic defects $Al_i^{(0,+)}$...$F_{2,3}^{n+}$ (n=2–4) (were identified in [2, 5, 9]) have influence on the recharging of defects also. The processes of recharging of defects are predominate according to scheme F^+→F^0, $Cr_{Al}^{2,4+}$→Cr_{Al}^{3+} and F_2^+→F_2^0, as show the absorption spectra. The stability of the bands induced by RDs to oxidation is higher in polycor in compare with leucosapphire owing to influence on optical absorption of boundaries between the structural fragments in the polycrystals (figures 1, 2 and 6). The band at ε_i=2.15 eV induced by substitutional defects $Cr_{Al}^{2+...4+}$ and band at ε_i=3.0 eV connected with the cation vacancy $V_{Al}^{(0,-)}$ in leucosapphire disappear after annealing in air. Annealing at T_{ann2}=500–800 K promotes the redistribution of absorption in polycor in the bands of F^+–centers from the band with center at ε_i=4.8 eV to ε_i=5.3 eV (figure 2). The new band with ε_i=4.4 eV appears in leucosapphire and this band is caused by forming of OCCs on the base of V_O, Al_i and O_2 (figure 6). At that, bands induced by F^+– and Al_i–centers disappear. We cannot exclude the influence of the electronic transitions (at ε=3.7–3.9 eV) between the LSs of compounds $(Cr_2O_3)_x(Al_2O_3)_y$ [5] on the absorption processes in substitutional defects and divacancies.

4. Conclusion

Annealing in air of alumina after irradiation with chromium ions and subsequent annealing in vacuum changes characteristics of the energetic levels induced by the RDs and complexes on their base. This is reflected in parameters of the exponential and interband absorption and in their interconnection. The charge state of substitutional defects and intrinsic RDs and degree of their influence on properties are determined by fluence of ions and by annealing temperature in vacuum and in air. Annealing in air of irradiated alumina stabilizes the structure of defects and their energetic characteristics. Fulfillment of criteria of generalized Urbach rule is supplemented owing to realization of oxidation and reduction reactions between the defects and oxygen molecules. Accumulation of OCCs with participation of RDs changes negligibly the electronic structure of irradiated alumina with small decreasing of maximal optical gap from 4.5 to 4.0 eV.

Acknowledgments

The research was particularly supported by grant from Russian funds for basic research RFBR No. 13-08-98061 and megaproject of National Research Tomsk Polytechnic University (VIU_IHTPh_85_2014) «Materials for Extreme Conditions».

References

[1] Kabyshev A V, Konusov F V and Remnev G E 2014 *J. Surf. Invest.* **8** 158
[2] Kabyshev A V and Konusov F V 2006 *Inorganic Materials* **42** 756
[3] Kozakiewicz A G, Davidson A T and Derry T E 2002 *Nucl. Instr. Meth. Phys. Res.* **B 191** 582
[4] Marques C, Alves E and McHargue C 2002 *Nucl. Instrum. Meth. Phys. Res.* **B 191** 644
[5] Kabyshev A V and Konusov F V 2006 *J. Surf. Invest. (in Russian)* **3** 58
[6] Kabyshev A V and Konusov F V 2008 *J. Surf. Invest. (in Russian)* **4** 103
[7] Orapunt F and O`Leary S K 2004 *Appl. Phys. Lett.* **84** 523
[8] Da Silva J H D, Campomanes R R and Leite D M G 2004 *J. Appl. Phys.* **96** 7052
[9] Evans B D 1995 *J. Nucl. Mater.* **219** 202
[10] Ogasawara K, Ishii T, Tanaka I and Adachi H 2000 *Phys. Rev.* **B 61** 143
[11] Costina I and Franchy R 2001 *Appl. Phys. Lett.* **78** 4139
[12] Novikov Yu N, Gritsenko V A and Nasyrov K A 2009 *Appl. Phys. Lett.* **94** 222904
[13] Xu W L, Chen H and Zheng M J 2006 *Optical Materials* **28** 1160

Assessing energy efficiency of electric car bottom furnaces intended for thermal energization of minerals

A I Nizhegorodov
Irkutsk National Research Technical University, 83, Lermontov Street, 664074, Irkutsk, Russia

E-mail: nastromo_irkutsk@mail.ru

Abstract. The paper deals with a new concept of electric furnaces for roasting and thermal energization of vermiculite and other minerals with vibrational transportation of a single-layer mass under constant thermal field. The paper presents performance calculation and comparative assessment of energy data for furnaces of different modifications: flame and electric furnaces with three units, furnaces with six units and ones with series-parallel connection of units, and furnaces of new concept.

1. Introduction

In the course of upgrading of electric furnaces with unit-type releasing [1], several engineering solutions were applied that have changed their energy efficiency [2]. Creation of furnaces with series-parallel connection of units enabled the increase of the efficiency coefficient [2] while use of "zero" nonelectrical unit resulted in additional decrease of energy consumption. Thereby minimal specific energy intensity of roasting at the rate of 173–175 MJ/m^3 was reached [1, 2], but it turned out to be not enough since the electricity cost is still high. In order ecological, fire-safe and explosion-proof electric furnaces find their level in industry, they should become competitive. Car bottom furnaces [3] are a decisive step to achieve the goal.

2. Experimental

Furnaces with unit-type releasing operate as gravity chutes: material motion occurs due to gravity forces, grains move with acceleration, and average interval δ between them increases: in the upper part of the unit $\delta = 0$ while in the lower part $\delta = \delta_{max}$ (figure 1). It is important to know how intervals δ will change. Table 1 shows the experimental values of vermiculite movement time along unit's sections of 0.96 m long.

Table 1. Time of vermiculite movement along unit's sections.

Section #	Section 1	Section 2	Section 3	Section 4	Section 5
Movement time t, sec	0.37	0.17	0.13	0.1	0.09

It is obvious that grain concentration in the upper part of the unit is maximal; there are almost no hollow spaces between them ($\delta \approx D$) therefore the full power of incident heat radiation flux falls on work material. As the material moves, the time of sections' passing declines, intervals grow, concentration decreases and becomes minimal in the lower part of the unit and average interval for vermiculite concentrate with weighted-mean grain diameter D becomes equal to δ = (6–6.5) D.

Relation of total area of grain projection on unit's surface to the area of 0.2 m^2 occupied by them at the first section, by regular structure of a single-layer mass is 0.785. The same relation at the fifth section goes down to 0.046 and the concentration decreases by a factor of 17. Therefore only small part of radiant energy is absorbed by vermiculite here.

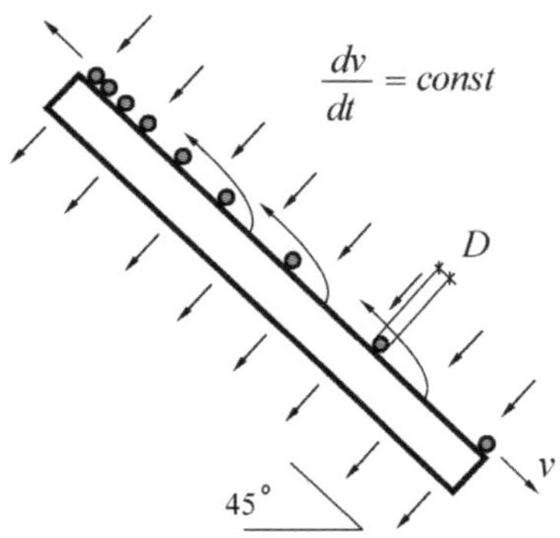

Figure 1. Movement of vermiculite in an electric furnace with unit-type releasing.

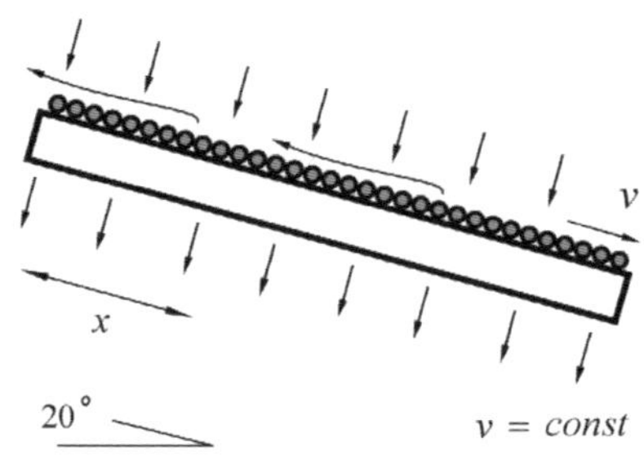

Figure 2. Movement of vermiculite in an electric car bottom furnace.

It is possible to drastically enhance heat transfer conditions by changing the way of material movement so that over the whole surface of the bottom the relation of total area of grain projection to the area of 0.2 m^2 at any section is maximum possible for the model of regular flow and makes up ~ 0.785. This could be reached by vibratory way of raw material feed, figure 2 (input and output heat fluxes are symbolically shown with arrows). By 15-20^0 inclination, granular material moves due to vibrations applied to the base. Asymmetric vibrations are particularly efficient since they cause one-way movement without inclination [4]. It is possible to organize such a movement on the account of installment of conical springs at the side where the flux should move to.

3. Result and discussion

Figure 3 demonstrates a single-phase block of a furnace with its own bottom. A complete furnace should contain three or six such blocks.

It consists of a frame 1, a bin 2, a drive 3 and a dosing drum 4 delivering raw material through a trough 5. The surface of a bottom 6 is produced from separate combined plates 7 made of heat-resistant steel to exclude its warping by heating. The plates are laid with an overlap to avoid retrograde movement of raw material. The bottom is installed on rollers 8 in frame guides. On the panel 9 there are heads 10 holding strip heaters 11 over the bottom surface with a clearance ranging from 0.002 up to 0.008 m. The bottom is covered with a removable heat-insulated lid 12 forming slit-like roasting space through which granular material moves forward for 2.5-5 seconds.

To deliver heat-treatable material, the bottom is secured by conical 13 and coil 14 springs. By rotation of a supplementary shaft with a bearing 15, a plunger 16 excites via a spring 17 asymmetric vibrations of the bottom inducing one-way movement of raw material. The whole block can be additionally inclined towards its movement.

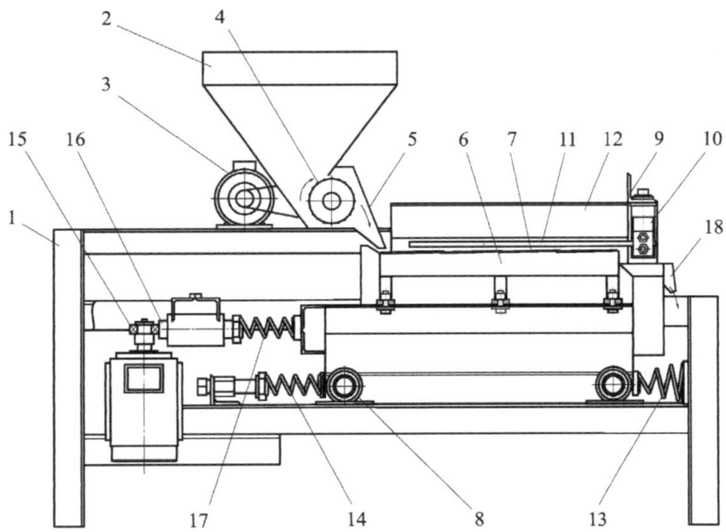

Figure 3. A single-phase block of a furnace with its own bottom.

The bottom makes resonance low-frequency vibrations (5…10 Hz) therefore it is possible to adjust the average speed of a single-layer mass movement on the account of the change of excitation frequency (vibration velocity control) and inclination angle.

Finished product is shot into a receiving bin (not depicted on the figure) through a trough 18.

Let us compare the capacities of a new furnace and of a pilot three-unit furnace with unit-type releasing [5] by vermiculite roasting. Its hourly expanded product capacity Π was ~ 1.75 m^3 with roasting duration t ~ 2.84 sec and unit dimensions as follows: $B = 0.96$ m (width), $l - 0.89$ m (length of a work section). In equivalent to expanded grains from all three furnace's units, their total volume V_Σ was equal to:

$$V_\Sigma = \Pi \times t \times (1 - k) / 3600, \qquad (1)$$

where k is the porosity index of expanded vermiculite mass (~ 0.365).

The total volume is $876600 \cdot 10^{-9}$ m^3 while the volume of a single grain, expanded from the particle with an average nominal diameter $D = 0.004$ m, equals $33.5 \cdot 10^{-9}$ m^3.

The number of all grains in the furnace m equals $876600 \cdot 10^{-9} / 33.5 \cdot 10^{-9} = 26167$ while the number of grains in one unit is $m_1 = 26167 / 3 = 8722$. By regular structure of a single-layer mass on the bottom surface, the average interval between grains equals $\delta = 0.0107$ mm, the clearance between them is ~ 0.0067 mm whereas vermiculite concentration in the unit will be $8722 / (0.96 \times 0.89) = 10208$ grains/m^2.

In the new furnace vermiculite mass moves in such a way that the equality $\delta \approx D$ is fulfilled and therefore grains occupy the whole surface of the bottom. With the same dimensions of the bottom (0.96×0.89 m), the number of grains along bottom width and length will be $0.96 / 0.004 = 240$ and $0.89 / 0.004 = 222$ respectively. The total number of grains will be 53280 pieces while vermiculite concentration on the bottom equals $53280 / (0.96 \times 0.89) = 62359$ grains/m^2 which is 6.1 times more.

Let the time of vermiculite stay on the bottom be $t = 2.84$ sec by equivalent temperature conditions (on nichrome surface it is ~ 750° C). Vibrations of the bottom should secure average speed of expanding grains movement at the rate $v = 0.31$ m/sec. Based on the formula (1), it is possible to determine the unit's production capacity per working hour, m^3:

$$\Pi = \frac{V_{\Sigma} \times 3600}{t \times (1-k)} = 3.55.$$

Hourly capacity of a three-phase furnace with three such blocks makes 10.65 m^3; it is 6 times as much as the capacity of a pilot unit-type furnace by equal electrical power consumption. Therefore specific energy intensity of vermiculite roasting will considerably decrease in the new car bottom furnace and be equal to 250.1× (1.75 / 10.65) = 41.1 MJ/m^3 (compare to ~250.1 MJ/m^3 for the pilot furnace [5]).

Figure 4 demonstrates a strip diagram depicting change of specific energy intensity of various furnaces by vermiculite roasting: 1 – flame furnace, 3 – three-unit furnace, 3 – six-unit furnace with series-parallel connection of units, 4 – car bottom furnace.

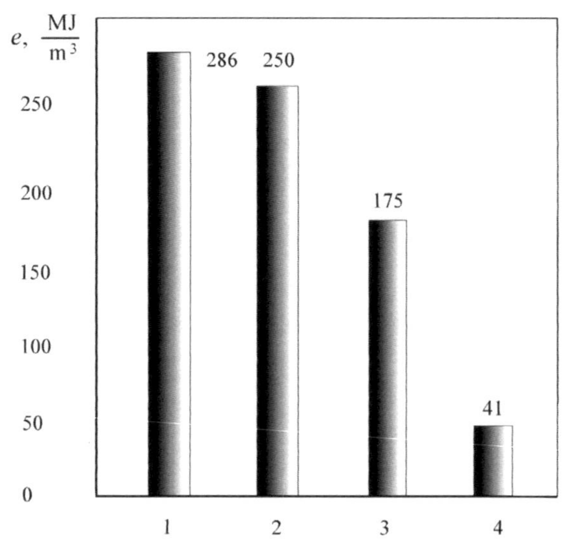

Figure 4. Strip diagram of specific energy intensity of various furnaces.

Forecast of specific energy intensity of roasting in a three-block furnace of new concept is 41.1 MJ/m^3 which make only 14.4 % of energy intensity of one of the most advanced flame furnaces (286 MJ/m^3) working on hydrocarbon fuel [6].

4. Conclusion

Study of capacities of such energotechnological equipment for vermiculite roasting and thermal energization of other materials has been recently started. It is necessary to model the process of control of granular material movement on the furnace's bottom, the processes of heat absorption and heat radiation transfer to working environment taking into consideration its optical properties, to determine energy relations by heat absorption of minerals under energization as well as to carry out the whole number of experimental investigations. But the forecast made encourages therefore a pilot sample of a car bottom furnace is being produced.

References

[1] Nizhegorodov A I 2016 Using und assessing energy efficiency of electrical ovens with unit-type releasing intended for thermal energization of sungulite-vermiculite conglomerates *IOP Conf. Series: Materials Science and Engineering* **110** doi 10.1088/1757-899X/110/1/012014

[2] Nizhegorodov A I 2015 Energotechnological equipment for vermiculite concentrate processing (Irkutsk: INRTU) p 250

[3] Nizhegorodov A I 2015 On investigation of analytical models of systems for granular material delivery to furnaces with vibratory bottom *J. Vestnik of ISTU* **12** 223-229

[4] *Vibrations in Engineering Handbook. Vol. 4 Vibration Processes and Machines* 1981 ed E E Lavendel (Moscow: Mashinostroenie) pp 509

[5] Nizhegorodov A I 2015 Technologies and equipment for vermiculite processing: optimal fractionation, electrical roasting, additional concentration (Irkutsk: INRTU) p 172

[6] Akhtyamov R Ya 2009 Vermiculite as a raw material for production of fire-proof heat-insulating materials *J. Refractories and Technical Ceramics* 1/2 59-64

IOP Conf. Series: Materials Science and Engineering **168** (2017) 012003 doi:10.1088/1757-899X/168/1/012003

Considering adaptation of electrical ovens with unit-type releasing to peculiarities of thermal energization of mineral raw materials

A V Zvezdin and T B Bryanskikh
Irkutsk National Research Technical University, 83, Lermontov Street, 664074, Irkutsk, Russia

E-mail: alexzvez@gmail.com

Abstract. The paper gives a short overview of technologies of mineral raw material thermal treatment where application of electrical ovens with unit-type releasing is possible. Efficiency of such ovens for vermiculite concentrate and conglomerate roasting is proved by more than 13-years experience of their industrial operation. The paper furthermore considers alternative connections of energotechnological blocks of an oven in order to determine its efficient design for specific technology related to one or another mineral raw material.

1. Introduction

The first electrical ovens with unit-type releasing appeared in Russia in the beginning of 2000s [1, 2] and were used for expansion of vermiculite concentrates and conglomerates of Kovdorski, Tatarski and Koksharovski deposits [2, 3, 4]. Over the years ovens have been improved; their reliability increased while roasting energy intensity went down. The ovens proved to be usable for obtaining expanded vermiculite from vermiculite conglomerates [5], raw material containing vermiculite in weight fractions ranging from 30% up to 70% with associated sands and granulated materials of various mineral composition.

In 2015 there was tested an oven with an additional nonelectrical unit which decreased energy consumption on the account of inner energy of not finally expanded vermiculite [6]. Tests were carried out on vermiculite-sungulite conglomerates.

Waste of Kovdorski deposit contains millions of tons of vermiculite and sungulite [7]; the latter turns into an effective chemical for detoxification of technogeneousely polluted soils and water objects [8]. The technology of their dividing was developed [9] but now it incorporates the processing step on thermal energization of the parent raw material.

Such properties of electrical ovens as compactness and light weight, low energy intensity, simplicity of design and use allow extending the field of their application in diverse technologies. Let's consider some of them.

2. Experimental

Experimental researches established [10] that thermal activation of quartz sand and plagiogranite by temperature range of 500-650°C promotes decrease of their radioactivity on the account of diffusion processes of radionuclides. Radon maximal emanation occurs under temperatures of endothermic effects induced by dehydration and structural phase transformations in minerals. Thermal treatment considerably decreases the level of natural radioactivity of different minerals used in building that ranges from 5-6 Bq/kg for sands up to 420-440 Bq/kg for plagiogranite and improves ecological properties of building products.

Potassium sulfate is an important fertilizer. Main source of potassium is *KCl* extracted from sylvinite or other potassium-containing ores. It is produced as a powder with further granulation through pressing. As the results, the so-called fertilizer mixtures are obtained [11]. Authors have suggested and experimentally investigated the method of hardening of such granules. Thermal treatment within temperature range of 300-350°C is an efficient hardening method preventing them from further wear. This is one more promising field of application of electrical ovens with unit-type releasing.

Experiments described in the work [12] show that it is possible to increase the volume of hydraulic binders with the help of composite substances with application of waste products. It will allow reducing consumption of clinker in cement and obtaining materials with considerably new properties. Waste of Korkinski coal deposit hosts large amounts of opal-cristobalite rocks. Their thermal energization within the temperature range of 500-900°C results in considerable increase of binders' strength.

The authors' activity is aimed at the solution of one of urgent tasks such as obtaining of binding materials from industrial waste. Our ovens can serve the purpose as the basis of the considered technology.

The research team of the Institute of Chemistry and Technology of Rare Elements and Mineral Raw Materials of the Kola Science Center of the Russian Academy of Sciences has developed the method of water treatment from acidulation and ions of heavy metals [13]. The character of the invention lies in the fact that chemical for water treating is the product with the content of serpentine minerals of 80-95% by mass, particle size of 0.00002–0.0002 m by the consumption of chemical within the range of 0.1–15.0 kg/m^3. Before water treatment, the chemical is subject to thermal activation within the temperature range of 650-820°C.

Figure 1. An electrical oven with unit-type releasing with removed lids and incandescent heaters.

Figure 2. An electrical oven with unit-type releasing with thermal lids on units.

An electrical oven with unit-type releasing constructed as a sequence of inclined electrical and nonelectrical units enables thermal processing of serpentine minerals. Within the framework of the agreement with the above-mentioned institution, the team of the Technopark of Irkutsk National Research Technical University under the supervision of doctor of technical sciences Nizhegorodov A.I. has developed, produced, tested and carried out start-up and adjustment operations for the pilot oven at the laboratory of the institute (figure 1). This field seems to be one of the most promising for the application of ovens with unit-type releasing.

We have made a short overview of technologies for application of not only ovens with unit-type releasing but ones with moving bottom platform [14].

3. Result and discussion

The thermal activation time for different minerals considerably varies: it ranges from 2 to 3 sec by vermiculite roasting while thermal activation of serpentine minerals takes from 15 to 20 sec. If the size of processed granulated solids lies in the range of 0.001–0.005 m, the time interval stays as a rule within the mentioned limits.

The authors have developed a concept of an alternative electrical oven with unit-type releasing composed of a set of energotechnological blocks, one of which is demonstrated on the figure 2.

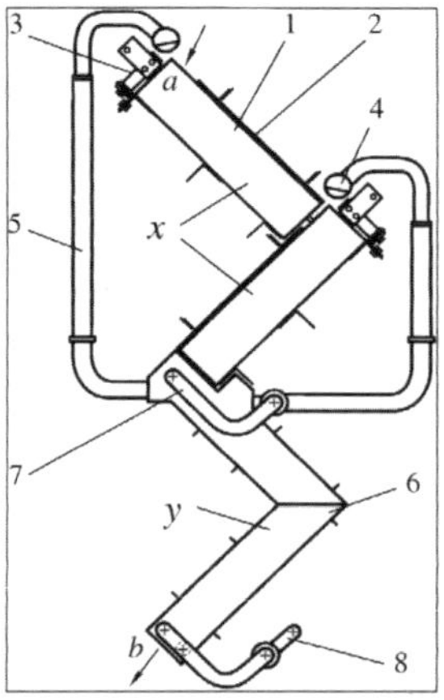

Figure 3. Energotechnological block of an oven with alternative structure

As shown on the figure 3, the block consists of electrical units 1 (x) with thermal lids 2 and fixing heads 3. Perforated suction tubes 4 through pipelines 5 are joined to heat chambers of an additional unit 6 (y) by pipes 7 while pipes 8 connect these chambers to an exhaust fan (not depicted on the figure 3). Points a and $в$ correspond to the input of raw material and the output of half-finished or finished product.

Hot air passes through heat chambers creating vacuum flask effect due to which high temperature is kept in slit-like space of the unit 6. Here the exchange of heat flows takes place: the energy of small grains in the form of radiant fluxes is being transmitted to greater grains that haven't been fully heated. At the same time the heat of surface layers of greater grains conductively passes to their deeper layers.

So, for the vermiculite by weighted-mean nominal diameter of a not finally expanded grain equal to $D = 0.0042$ mm, the time of temperature equalization on the periphery and in the center will be 1.2 sec in accordance with the following expression:

$$\tau = 0.677\,\delta^2 / \chi$$

($\delta = 0{,}5D$, χ is the index of thermal diffusivity along flakes).

But for other minerals this time can be considerably longer taking into consideration particle fineness.

The combination of electrical and additional (nonelectrified) units for the mentioned block can be expressed with the formula $2x + y$. For the oven depicted on the figure 1 the formula takes the form $3x + y$, while the time of material movement through the oven is equal to 3.2 sec.

The time can be increased up to required values via the efficient combination of elements of the basic block (y and x) so that work material is "fed" with heat energy from outer source as it moves; for instance:

$$(2x + y) + (1x + y) + (1x + y) \;-\; 7.8 \text{ sec},$$

or:

$$(2x + y) + (x + 2y) + (x + 2y) \;-\; 11.1 \text{ sec}.$$

Some materials [8] require soaking for several minutes after thermal activation. Such processing chain should include a heat-insulated bin with a heating "jacket".

4. Conclusion

Study of capacities of such energotechnological equipment for thermal energization of different materials has been recently started. It is necessary to carry out experiments on thermal activation of above-mentioned and many other minerals as well as materials of technogenic origin.

Ovens with moving bottom platform [14] developed by the research group have similar adaptive capacities. But it is another alternative requiring particular consideration.

References

[1] Nizhegorodov A I 2007 New concept of ovens for roasting of vermiculite concentrates *J. Construction and road building machinery* **10** 19-20

[2] Nizhegorodov A I 2007 Some issues of preparation and roasting of vermiculite concentrates in electrical ovens *J. Stroitel'nye materialy* **11** 16-17

[3] Nizhegorodov A I 2009 Technological complex for roasting of vermiculite concentrates *J. Construction and road building machinery* **2** 30-32

[4] Nizhegorodov A I 2009 Narrow-band fractionation as a quality factor of vermiculite concentrates *J. Stroitel'nye materialy* **9** 68-69

[5] Nizhegorodov A I 2009 Adapted technological complex for preparation and roasting of vermiculite concentrates with high content of inert material *J. Construction and road building machinery* **12** 28-31

[6] Nizhegorodov A I 2016 Using und assessing energy efficiency of electrical ovens with unit-type releasing intended for thermal energization of sungulite-vermiculite conglomerates *IOP Conf. Series: Materials Science and Engineering* **110** doi: 10.1088/1757-899X/110/1/012014

[7] Gershenkop A Sh 2011 Vermiculite concentration disengagement from vermiculite ores mill tailing and its use in heat resistant concrete *Mining J.* **11** 57-59

[8] Kremenetskaya I P 2008 Reagent used to immobilize heavy metals of overburden rocks containing serpentine *J. Vodo'ochistka. Vodopodgotovka. Vodosnabzheniye* **4** 33-40

[9] Tereshchenko S V, Kremenetskaya I P 2014 Areas of complex use of phlogopite production waste *Proc. All-Russian Conf. On Ecological Strategy of Mining Branch Development (Apatity)* vol 1 (Saint-Petersburg: Renome) pp 272–279

[10] Vetrova Yu V, Pavlenko V I 2006 Analysis of dependence of radioactivity on different types of quartz sand processing *Proc. Int. Conf. on Engineering and Technological Research for Sustainable Development (Moscow)* (Moscow: Moscow State University of Engineering Ecology) p. 112

[11] Berdichevskiy A I, Zinyuk R Yu et al 2000 On the issue of potassium sulfate granulation *Proc. 3rd Scient. and Techn. Conf. of Postgraduate Students (Saint-Petersburg)* part 1 (Saint-Petersburg: Saint-Petersburg State Technological Institute (Technical University)) p. 42

[12] Nosova A N, Fomina E V 2013 Thermal activation of opal-cristobalite rocks, waste of Korkinskiy coal deposit *Proc. Int. Conf. "Technical sciences – from theory to practice" (Novosibirsk)* No 24 (Novosibirsk: SibAK) pp 106-112

[13] Makarov V N, Kremenetskaya I P et al 1999 A method for water treatment in open water reservoirs from acidulation and heavy-metal ions *RF patent 2136608 IPC CO2F1/62 publ. 10.09.99*

[14] Nizhegorodov A I 2015 On investigation of analytical models of systems for granular material delivery to ovens with vibratory bottom platform *J. Vestnik of ISTU* **12** 223-229

Disclosure of the oscillations in kinetics of the reactor pressure vessel steel damage at fast neutron intensity decreasing

E Krasikov[1] and V Nikolaenko[2]
[1]Laboratory Head, National Research Centre «Kurchatov Institute»,
1, Kurchatov Sq., Moscow, 123182, Russia.
[2]Group Head, National Research Centre «Kurchatov Institute»,
1, Kurchatov Sq., Moscow, 123182, Russia.

E-mail: ekrasikov@mail.ru

Abstract. Fast neutron intensity influence on reactor materials radiation damage is a critically important question in the problem of the correct use of the accelerated irradiation tests data for substantiation of the materials workability in real irradiation conditions that is low neutron intensity. Investigations of the fast neutron intensity (flux) influence on radiation damage and experimental data scattering reveal the existence of non-monotonous sections in kinetics of the reactor pressure vessels (RPV) steel damage. Discovery of the oscillations as indicator of the self-organization processes presence give reasons for new ways searching on reactor pressure vessel (RPV) steel radiation stability increasing and attempt of the self-restoring metal elaboration. Revealing of the wavelike process in the form of non monotonous parts of the kinetics of radiation embrittlement testifies that periodic transformation of the structure take place. This fact actualizes the problem of more precise definition of the RPV materials radiation embrittlement mechanisms and gives reasons for search of the ways to manage the radiation stability (nanostructuring and so on to stimulate the radiation defects annihilation), development of the means for creating of more stableness self recovering smart materials.

1. Introduction.

Modern nuclear power plants based on PWR type reactors. As the main safety barrier RPV have to be capable to working at any operating regimes including in an emergency.

Radiation stability is a dominating factor of the RPV life cycle. In turn substantiation of the normative dependencies of the RPV materials radiation embrittlement was carried out in research reactors at neutron fluxes that were two and more orders higher than at RPV operation. Surveillance specimens (SS) programmes are more adequate for RPV embrittlement control but at this case lead factor may be also high (up to 20). So, scale of the flux effect influence touch upon hypersensitive questions of the RPV real radiation embrittlement. This context became the main motivation of the present work carrying-out and publication.

Operating RPVs are subject to multi-factor influence. It is practically impossible to reproduce some of this factors («long-time bias», e.g.) in the framework of experimental investigations including RPV surveillance specimens tests. Detailed information that can be obtained by means of taking through RPV wall samples immediate from the decommissioned RPVs is more representative than received by any another ways and therefore has a highest value.

2. Experimental procedures and test results

Along with routine investigations in Russia systematic research on actual radiation embrittlement of the decommissioned pressure vessels via through samples (trepans) has been carried out.

The earliest commercial PWR prototype unit Novovoronezh-1 (210 MWe) RPV after 20 years (1964-1984) of operation was trepanned in 1987. Then Novovoronezh-2 (365 MWe, 1969-1990), the oldest PWR type experimental prototype reactor EPR (1956-1986) and, finely, nuclear icebreaker «Lenin»" RPVs also were trepanned.

Through-thickness RPV wall samples had been cut out using trepanning tools such as boring machine with annular drill, anodic-mechanical and electric resistance cut machines. Typical dimensions of trepans were 120-140 mm in diameter and 120-140 mm lengthwise. Test pieces and specimens were fabricated by means of electro-discharge wire cutting machine.

The general dependence of the Transition Temperature Shift (TTS) on radiation embrittlement coefficient (REC) A_F is as follows:

$$TTS=A_F\times F \qquad (1),$$

where A_F is the radiation embrittlement coefficient (REC - chemical factor) and F is the fast (E>0,5MeV) neutron fluence in units of $10^{18}cm^{-2}$(fluence factor), n – coefficient (~1/3).

For Russian RPV Cr-Mo-V base steel

$$A_F=A\times(P+0{,}07Cu) \qquad (2),$$

where A_F – coefficient (800 at 270°C), P and Cu – are the mass concentrations of these elements.

One of the Novovoronezh-1 RPV trepans for parallel study was delivered to IVO (Finland). Teamwork obtained serrated results (Fig.1) force to lost in thoughts. No less unexpected became the results of a special research of the SS Kolskaya-3 NPP [1] that confirm the subsistence of the flux effect at the initial section of the neutron exposition as well as data on post irradiation tests of the trepans cut out of nuclear icebreaker "Lenin" RPV [2].

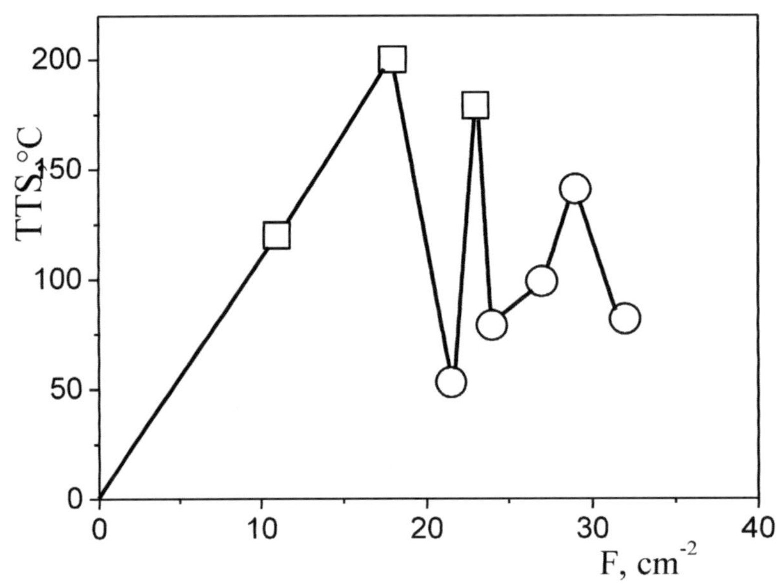

Figure 1. Dependence of the Novovoronezh-1 RPV weld metal TTS on the fast neutron fluence (□– NRC "Kurchatov Institute", ○– IVO Finland).

Particular curiosity to provoke paradox results of the SS zero embrittlement obtained in Russia (Fig.2)[3] and abroad (Fig.3)[4-7]. Stress of incomprehension diminished when finding analogous data on non monotonous SS degradation with commentary "quite atypical" [4].

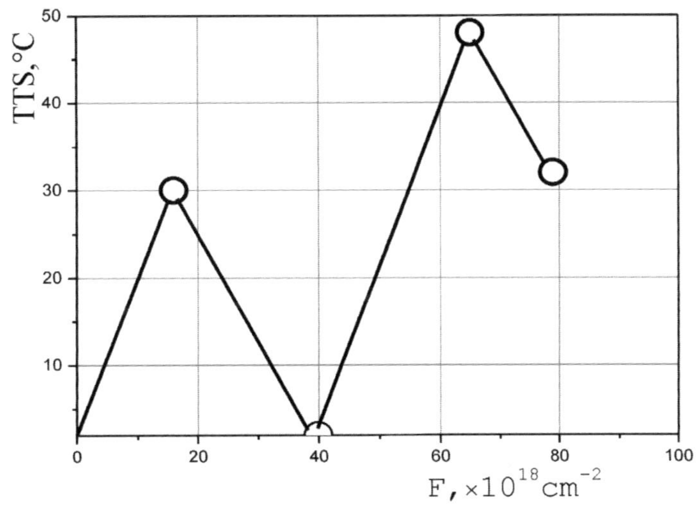

Figure 2. Dependence of the Rovno-1 RPV base metal TTS on the fast neutron fluence. Neutron flux~$6\times10^{10}cm^{-2}c^{-1}$ (specimens from upper containers).

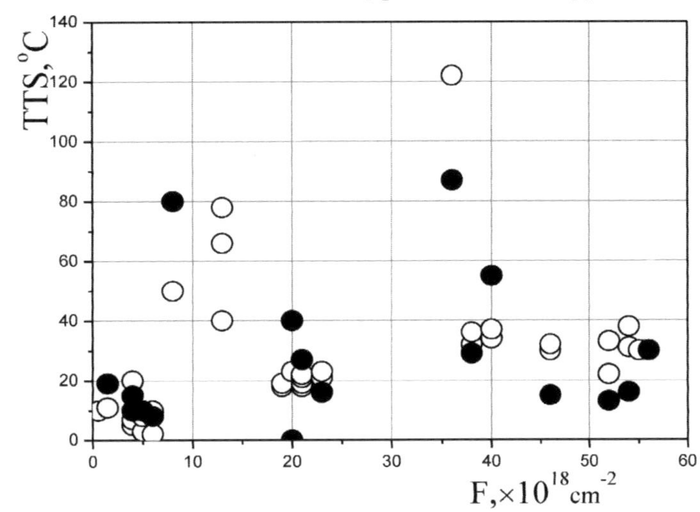

Figure 3. Dependence of the TTS on the fast neutron fluence *from surveillance data for* Spanish nuclear power plants base metal(○) and welds(●)

In addition to this circumstances attention is drawn to the fact that the considerable scatter of the experimental data take place. Fig.3 presents data scattering for SS of the Spanish nuclear power plants. One can note unexpected low embrittlement at the region of high fluence and zero embrittlement at the neutron fluence area of $2\times10^{19}cm^{-2}$.

3. Data analysis and discussion

Revealed non monotony has to be inspired by the peculiarity of the copper impurities containing metal degradation process at low flux when after forced embrittlement at the initial stage mitigation follows [8]. At the same time refinement of the steel up to threshold level of 0,05% Cu that slows down flux effect changes the scatter slightly, SS data scatter exceeds this parameter for data obtained in test reactors (high flux) [9]. One can assume, that factor exists which generate oscillating process in mechanism of embrittlement and become more apparent with neutron flux lowering. Conversely, at high level of the flux it is hardly noticeable because amplitude of a oscillation decrease while frequency increase. As an analogue of temporal self-organization in solids under irradiation one can cite professor V.S. Chmelevskaya data on periodic change in length of the plutonium specimens with different content of ^{238}Pu isotope (fig. 4[10]).

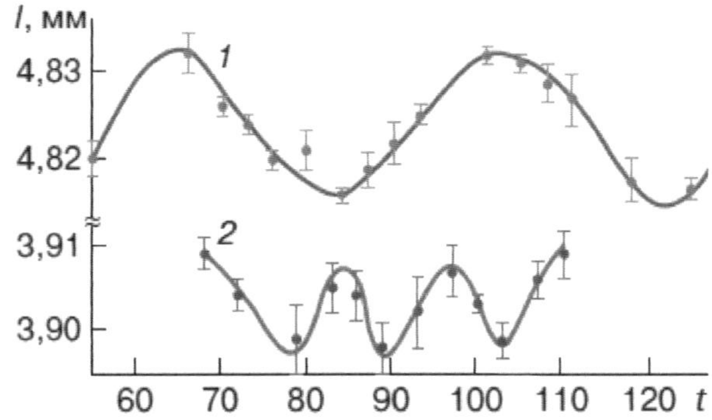

Figure 4. Change in length (mm of the plutonium specimens with different content of ^{238}Pu isotope against time (hours): 1 - 7,3% ^{238}Pu, 2 - 80% ^{238}Pu

Hypothetically we assume the presence of the oscillation component in kinetics of the neutron embrittlement process that arise from radiation defects periodic beginning and annealing, which in its turn is a result of the metal structure temporal self-organization.

Similar to indicated in Fig.2 effect we find out the mention concerning structural and phase transitions in the amorphous and nanocrystalline $Ti_{50}Ni_{25}Cu_{25}$ alloy upon high-pressure torsion [11]. Fig.5 demonstrates volume fraction of crystalline phase **V** as a function of the number of the mobile anvil rotations **n** for the amorphous Ti50Ni25Cu25 alloy.

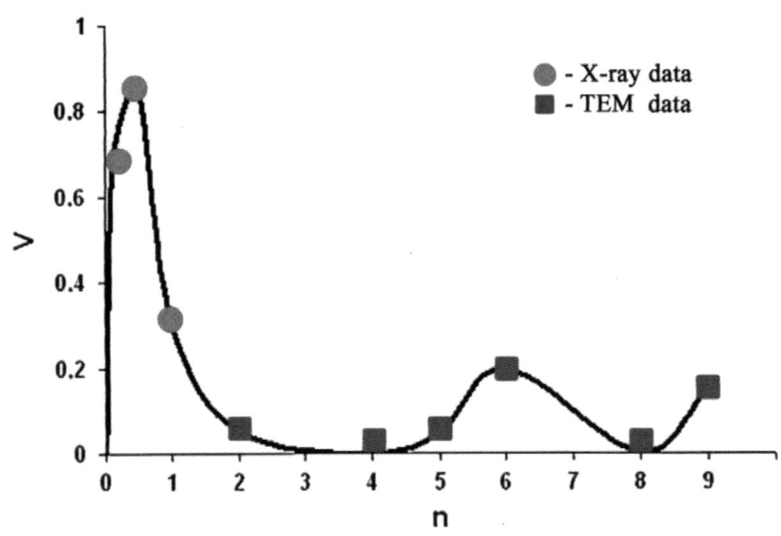

Figure 5. Volume fraction of crystalline phase **V** as a function of the number of the mobile anvil rotations **n** for the amorphous Ti50Ni25Cu25 alloy.

Assumed periodicity of the embrittlement confirms the analysis of the dependence of the transition temperature shift against neutron fluence from surveillance data for WWER-440/213 base metal and welds (fig.6).

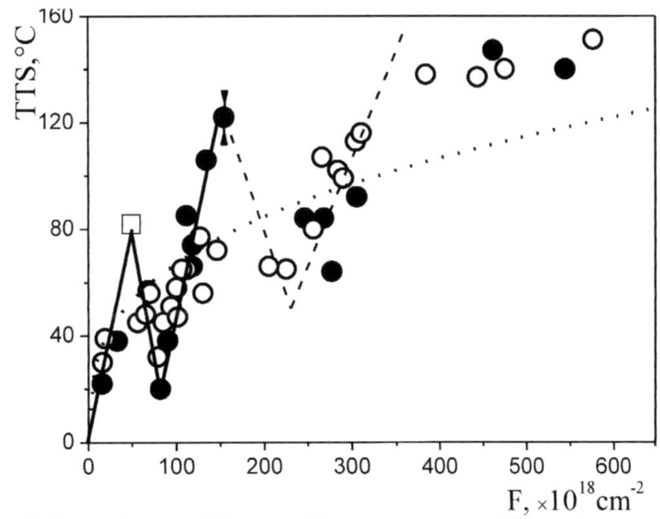

Figure 6. Dependence of the transition temperature shift against neutron fluence from surveillance data for WWER-440/213 base metal and welds. ○ – flux~30×10^{11}cm^{-2}c^{-1}, ● – flux ~4×10^{11}cm^{-2}c^{-1}, □ – flux ~ $1{,}3\times10^{11}$cm^{-2}c^{-1}.

Possibly linear approximation is simplification, but one can notice the serrated character of the dependence and high amplitude of the oscillations at the initial stage of neutron exposition. In addition an attempt was undertake to consider the dependence of the transition temperature shift against radiation embrittlement coefficient (chemical factor) $\mathbf{A_F}$ from surveillance data for WWER-440/213 weld and base metal (fig.7). In this case initial data were used for neutron fluences at the range of $(40\div60)\times10^{18}$cm^{-2}. Again oscillating process become apparent. It is necessary note that data for different intensities and materials (weld and base metals) were used. It seems that oscillation factor may advance to the forefront.

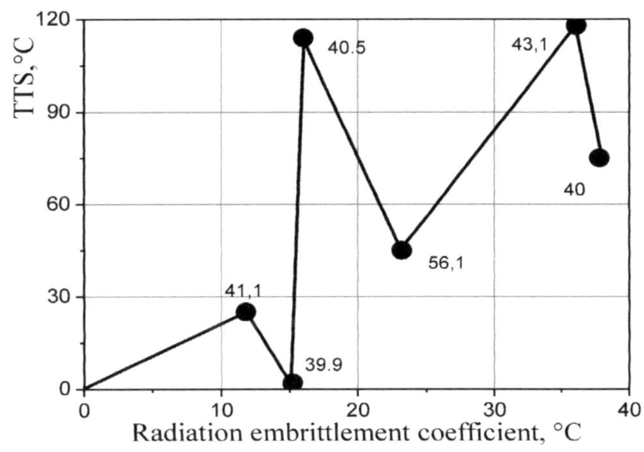

Figure 7. Dependence of the transition temperature shift against radiation embrittlement coefficient (chemical factor) A_F from surveillance data for WWER-440/213base metal and welds.

Conclusion

Disclosure of the oscillations - sign of the self-organization - in kinetics of the reactor pressure vessels (RPV) steel damage at fast neutron intensity decreasing testifies the periodic transformations of the metal structure and underlines the actuality of the thorough knowledge radiation damage mechanisms deepening.

As a hypothesis we suppose that at some stages of irradiation damaged metal have to be partially restored by irradiation i.e. neutron bombardment. Nascent during irradiation structure undergo occurring once or periodically transformation in a direction both degradation and recovery of the initial properties. According to our hypothesis at some stage(s) of metal structure degradation neutron bombardment became recovering factor. As a result oscillation arise that in tern lead to enhanced data scatter.

So, one can assume, that factor exists which generate oscillating process in mechanism of embrittlement and become more apparent with neutron flux lowering. Conversely, at high level of the flux it is hardly noticeable because amplitude of a oscillation decrease while frequency increase. Hypothetically we assume the presence of the oscillation component in kinetics of the neutron embrittlement process that arise from radiation defects periodic beginning and annealing, which in its turn is a result of the metal structure temporal self-organization.

Revealing of the wavelike process in the form of non monotonous parts of the kinetics of radiation embrittlement process – sign of self-organization, that testify to periodic transformation of the structure – actualize the problem of more precise definition of the RPV materials radiation embrittlement mechanisms and give reasons for search of the ways to manage the radiation stability (nanostructuring [12] and so on to stimulate the radiation defects annihilation), development of the means for creating of more stableness self recovering smart materials.

References

[1] Nikolaenko V and Krasikov E 2004 *Atomic energy* **97** 82 (in Russian)

[2] Krasikov E and Nikolaenko V 2008 *Atomic energy* **105** 5 (in Russian)

[3] Gerard R and Fabry A 1996 *Effects of radiation on material* Eds. D Gelles, R Nanstad West Conshohocken, PA, Ann Arbor ASTM STP ***1270*** 294

[4] Krasikov E and Nikolaenko V 2015 *Atomic energy* **119** 46 (in Russian)

[5] Debarberis L and Ballesteros A 2004 *IAEA Specialists Meeting on Irradiation Effects and Mitigation in Reactor Pressure Vessels and Internals* Davis M Gus Khrustalny-Moscow, Kurchatov Institute **1** 82

[6] William L and Server M 2004 *IAEA Specialists Meeting on Irradiation Effects and Mitigation in Reactor Pressure Vessels and Internals* Davis M Gus Khrustalny-Moscow, Kurchatov Institute **1** 112

[7] Williams T 2013 *IAEA Technical Meeting* on *Knowledge Gaps in Toughness and Irradiation: Lessons of the Past, Prospects for the Future* Davis M Vienna IAEA **2** 44

[8] Krasikov E and Nikolaenko V 2016 *IOP Conf. Series: Materials Science and Engineering* IOP Publishing **110** 012001 doi:10.1088/1757-899X/110/1/012001 1-6.

[9] Debarberis L and Ballesteros A 2002 *Review of Dose Rate Effects on RPV Embrittlement: Joint EPRI-CRIEPI RPV Embrittlement Studies* Nanstad R Palo Alto-Tokyo EPRI-CRIEPI **1** 75

[10] Chmelevskaya V 2000 *Soros educational journal* **6** 91.

[11] Sundeev R, Glezer A and Shalimova A 2014 *Materials Letters* **133** 32

[12] Bai X, Voter A, Hoagland R and Nastasi M 2010 *Science* **327** 1631

IOP Conf. Series: Materials Science and Engineering **168** (2017) 012005 doi:10.1088/1757-899X/168/1/012005

The effect of electric exposure on the gas cleaning systems filter efficiency

V V Makarov, O P Kuznetsova and M Yu Nikolaev
Omsk State Technical University, Omsk, Russia.

E-mail: makarov@omgtu.ru

Abstract. This article discusses different characteristics of dust-gas flow and electrical action algorithms of perceiving fields providing better gas cleaning in gas-cleaning equipment. We consider the most promising for the gas purification maximum degree to combine these algorithms into one system, and it can significantly increase the gas cleaning equipment efficiency.

1. Introduction

February 9, 2014 Omsk, Russia was covered with black snowfall. It is not the first time in Omsk. There were recorded 2 more cases of black snowfall in 2012. Rosprirodnadzor of the Russian Federation stated that it was HES emissions. The contamination area was 7.7 thousand hectares. An unscheduled inspection revealed that the precipitation was Ekibastuz coal ash, Kazakhstan. The audit found out that the HES filters were ineffective, and the station authorities were given the responsibility to bring the degree of electrostatic gas cleaning system to the design values. The electrostatic precipitator as a gas purification machine operates through two main electrophysical processes - cleaned flue gas particles charging with corona discharge ions and the charged particles motion to the collecting electrode under the electrostatic forces influence. The electrical gas cleaning process is that at a certain voltage value applied to the interelectrode gap, the field strength near the corona electrode becomes sufficient for its appearance; its consequence is to fill the electrode gap outer part mainly with negatively charged ions. Under the influence of electric field forces they move from the corona electrode to the collecting one. Ash or dust particles meeting on their way ions adsorb them, charge, and under the force field influence they also move towards the collecting electrodes, where they are collected. The electrodes are periodically shaken, a layer of collected dust is destroyed, and the dust falls into the electrostatic precipitator hoppers, where it is intermittently or continuously removed. A feature of the electrostatic precipitator is that it provides the maximum purification degree of the electric field strength in the interelectrode space, when there are sparks in the electrostatic precipitator, but there are no transition arcs. The purpose of the article is to study the processes and to develop recommendations on optimization of electric power modes of electrostatic gas cleaning system.

2. Objects and experimental methods

The voltage level with sparks varies non-linearly over a wide range and depends on many factors: the properties of the exhaust gases, the dust content in the gas, humidity, temperature, dust particles size, chemical composition and their electrical conductivity, the layer formation conditions on the collecting electrodes and other factors. Thus, the first problem to be solved in the course of work is the development of interelectrode space intensity algorithms change in electrostatic precipitator considering the dynamics of the filtered gases properties. One of the most difficult problems in the electrical gas purification is cleaning gas from dust with high resistivity, which is Ekibastuz coal ash, Kazakhstan. Such gases give rise to back-corona on the collecting electrodes, decrease breakdown voltage, electric field voltage, increase the adhesive layer properties on the collecting electrode, its poor cleaning, and as a result, reduce of the current load due to the dust layer locking action. These processes result in decrease in electrostatic gas purification degree. Thus, the second problem is to reduce back-corona effects. The third problem is the secondary dust loss associated with the precipitator shaking mechanism operation. The dust loss that exists due to continuous collecting electrodes shaking can be significantly reduced if there is an optimum shaking mode implementation. The amount of ash loss reduction through shaking optimization depends on the physico-chemical properties of dust-gas atmosphere. The optimal mode of corona electrodes shaking must support the dust layer on the discharge tools where the corona current provides the most effective dust collection in an electrostatic precipitator. Thus, the next issue to be solved is the development and optimization algorithms of shaking electrodes control mechanisms. [1].

Electrooptical monitoring sensors (EOS-2) of the electrostatic precipitator dust and ash emissions were developed in FSBEI HPE "Omsk State Technical University" to solve the mode configuration tasks of dust and ash clearing equipment operation [5], the equipment operation processes control and qualitative assessment of the gas cleaning process in general, as well as for environmental monitoring purposes. The sensors are installed directly on the gas ducts and are electrooptical devices, carrying out continuous automatic monitoring of dust and ash emissions consisting of three blocks: a radiation source, radiation detector and control unit.

Analysis of the DVR-4 operation (digital voltage regulator) and determining the mode effectiveness with through periods form of feeding voltage at the electrostatic precipitator substations were carried out according to the following procedures:

1. Set the sensitivity and maximum dust meter scale.
2. Continuous shaking of all semifields during 6 minutes.
3. Installation of test mode for 4-8 minutes.
4. Trend analysis of the dust loss at the precipitator exit.
5. Repeat steps 3-5 for other supply modes of electrostatic precipitator fields.

3. Results and discussion

All electrostatic precipitator substations use the power supply units ATF-1000 Series. They are the third generation units; the main difference is that there is no a power magnetic amplifier in the unit circuit, it is replaced with a thyristor power regulator. The advantage of the ATF unit series is speed, compactness and the basic units' durability, the size and weight of the unit are reduced. These power units are single-phase version units which used the silicon bridge rectifiers with symmetries shielding. The block of power thyristors (two thyristors connected in anti-parallel) combines the functions of smooth voltage regulation on the primary winding of the power transformer and the unit switching on and off. The survey proved that all power units are in satisfactory condition. Analysis of DVR-4 (digital voltage regulator) operation in electrostatic precipitation substations showed the overall operability of power supply units and automatic control modes maintenance of electrostatic precipitation voltage [3]. During the analysis of work and settings following shortcomings were identified:

a) spark selectors excessive sensitivity;

b) unstable operation mode with pulse skipping of supply voltages;

c) visual mismatch of current sensor readings on the digital display DVR and milliammeter pointer of the regular voltage controller of converter-rectifier panel (CRP) type.

Air suction reduction is known to lead to improved gas purification degree. In addition, gas air-conditioning with water reduces the high dust resistance typical for Ekibastuz coals burning, and to exclude reverse crown formation. In this regard, the examination program included work on the examination of housing lining state of the electrostatic precipitator and flue gas conditioning scrubber. As a result of thermal imaging examination [2], we can draw the following conclusions: in general, the state of scrubbers and electrostatic precipitator flues linings is good (Fig. 1). The false air is found out at the sampling platform and the electrostatic precipitator inlet. The false air to detect on the electrostatic precipitator housing by means of the electrostatic imager is not possible due to the shielding effect of the coating around the electrostatic precipitator corrugated casing.

The digital control unit (DCU) analysis with shaking electrodes mechanisms showed that all DCU-1 installed in electrostatic precipitator substations are operating in the normal mode [4].

Experimental deviations indications studies of CPR regulator currents and voltages with respect to the CRP control readings have been completed. When reading the current-voltage characteristics the experimental data shown in Fig. 2, 3, 4 were obtained.

Figure 1. Lining of the scrubber and electrostatic precipitator walls

The experimental data led to the following conclusions:

1. CPR current regulator readings in comparison with the CRP control readings are practically the same (Fig. 2):

- the minimum deviation error is 0.47%;
- the maximum deviation error is 56% (at the beginning of CRP scale up to 25 mA);
- mean error is 6.9% (CRP milliampermeter accuracy class is 2.5 giving an instrumental measurement error of 5%, moreover at the beginning of the scale it is almost impossible to make an accurate visual counting. Besides the analogue-digital conversion error in DVR may be 100% at the beginning of the scale and 5% for 100 mA).

2. DVR regulator indications voltage in comparison with the CRP control readings are practically the same (Fig. 3):

- the minimum deviation error is 0.29 %;
- the maximum deviation error is 5.4%;
- mean error is 2.6 % (accuracy class miliampermetra PVP is 2.5, which makes it an instrumental measurement error of 2.5%, the first samples were made with 18-22 kV).

3. The random error proves that DVR current regulator reading measurement is significantly less than CRP regulator due to malfunctioning of information visual assessment with pointer instruments. It was indicated due to a curve dependence comparison Icomp., measured with an oscilloscope, and currents CRP and DVP.

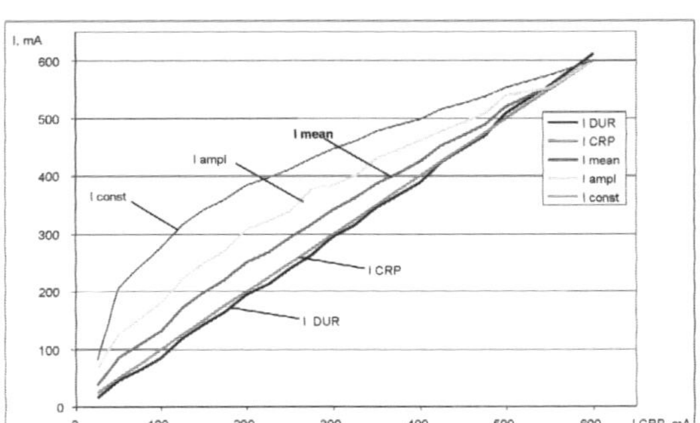

Figure 2. Comparative experimental characteristics of current values for different measuring systems with respect to CRP milliammeter readings

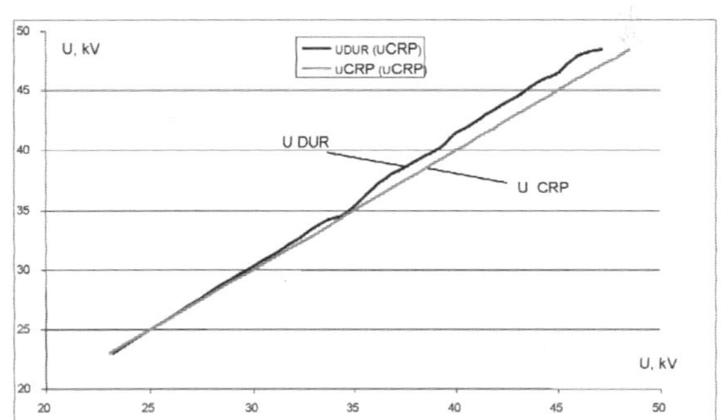

Figure 3. Comparative experimental characteristics of DVR voltage values relative to the CRP voltmeter readings

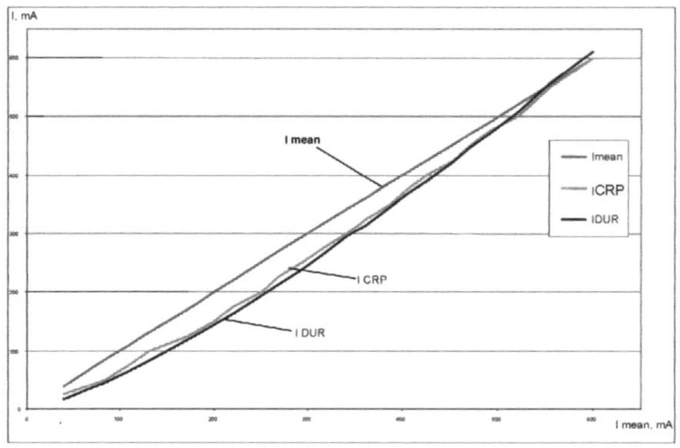

Figure 4. Comparative experimental characteristics of DVR and CRP currents values with respect to FLUKE oscilloscope readings (the mean value of current I mean)

The current-voltage characteristics taken in the flue gases (Fig. 5) allow estimating the fields' state and the back corona presence.

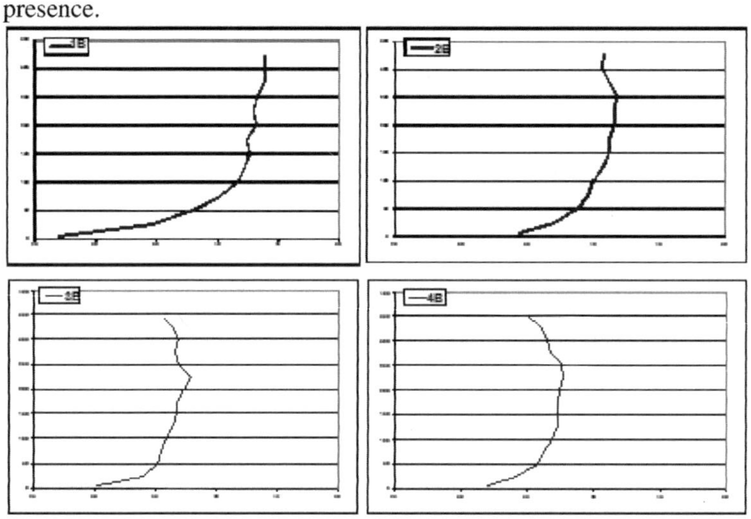

Figure 5. The current-voltage characteristics of the electrostatic precipitator semifields operating on gas

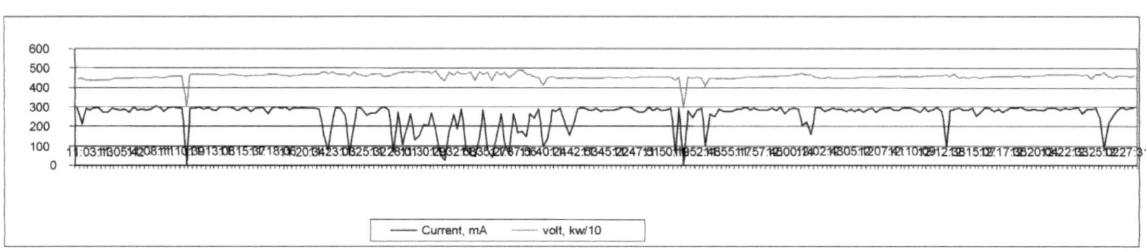

Figure 6. Trends in the dust meter readings

The nature of the curves for electrostatic precipitator semifields allows concluding the back corona existence on these electrostatic precipitators and the need to apply for these semifields through periods supply and maintaining the voltage on the electrodes using electrostatic precipitators through DVR algorithms close to the current-voltage characteristics inflection point.

We show trends graphs of dust meter readings, current sensors and voltage fields in the transition from the skip pulses mode (2/8, 2/6) into a continuous flow voltage pulses mode at 100 Hz frequency in Fig.6. Relative dust and ash losses are decreases in 2 times at the supply voltage with through periods form.

The analysis revealed a dust meter trend to increase dust discharge from the electrostatic precipitator during normal shaking modes. In the period from 9 a.m. to 14 p.m. pauses were reduced and a palpitations electrodes shaking was carried out. It is clear from Fig. 7 that after the pause time reducing dustiness is also reduced.

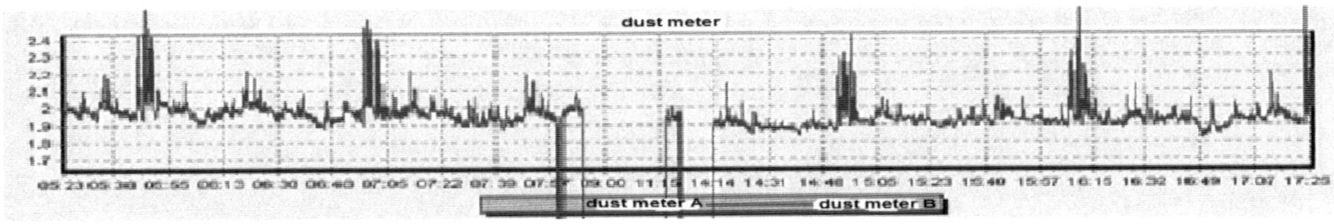

Figure 7. Graph of output dustiness at the dust meter readings

There was made are graphs confirming the link between shaking mechanisms operation pause times and the voltage change on the electrostatic precipitator electrodes allowing justifying 3 minutes pauses (Fig.8).

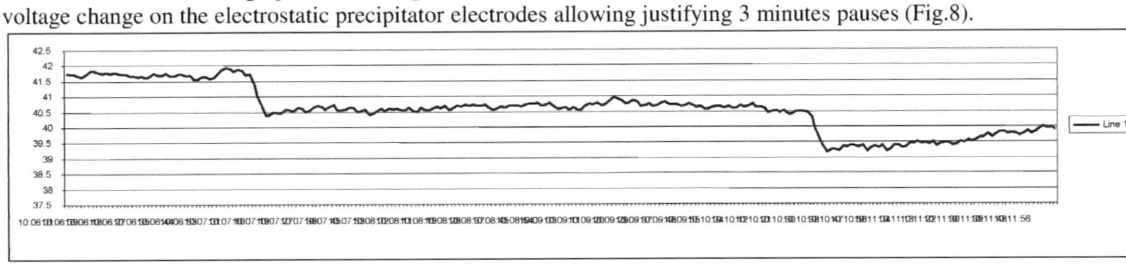

The voltage on the electrostatic precipitator field

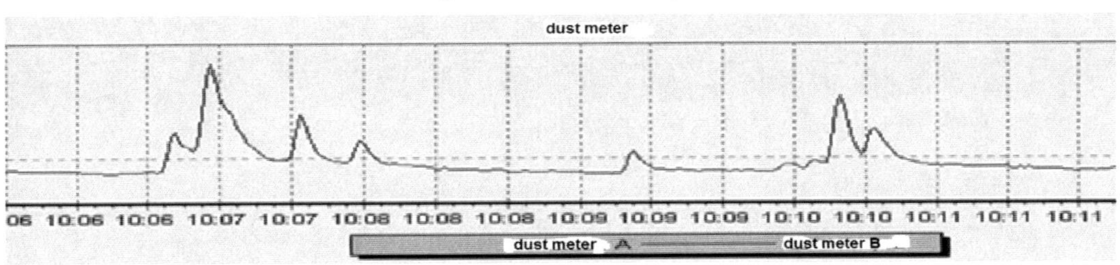

Figure 8. Output dustiness on dust meters

The feeding through periods form voltage was applied and it reduced power consumption in half, DVR operation algorithms were changed, the new software firmware was completed on all DVR controllers. As a result ash losses were reduced by 10-15%.

4. Conclusion

The examination and adjustment of DVR and DCU of electrostatic precipitators provide the potential possibility for increasing the cleaning from ash efficiency. The quantitative results of the proposed solutions effectiveness can be obtained with controlling comparative measurements of ash and dust losses in normal and recommended modes using GRAVIMAT device.

Acknowledgements

The authors express their sincere gratitude to Doctor of Technical Sciences, Professor of Omsk State Technical University Zakharenko Vladimir Andreyevich for his disproportionately significant contribution to the improvement of ecological situation in Omsk, the Russian Federation and the actual reduce of ash and dust losses as a result of HES gas-cleaning equipment.

References

[1] Zakharenko V A, Shkaev A G 2015 Improving the gas treatment industry efficiency. *Young Russia: advanced technology - to the industry!* **2** 32-35

[2] Zakharenko V A 2014 Methods and tools for contactless telemetry for thermal control problems in the industry Omsk

[3] Valke A A, Zakharenko V A, Murashko D N, Ponomarev D B, Ponkrashin V V, Shkaev A G 2012 Software Digital Controller CPR-4 strain *Navigator in the world of science and education* **4-7 (20-23)** 37

[4] Zakharenko V A, Shkaev A G, Saifutdinov R N 2003 Dust and ash losses optoelectronic sensor *Sensors and systems* **12** 27-29

[5] Shilnikov P Y, Zhikhareva A V, Zakharenko V A 2015 Autonomous shock pulse setting *Young Russia: advanced technology - to the industry!* **2** 83-85

IOP Conf. Series: Materials Science and Engineering **168** (2017) 012006 doi:10.1088/1757-899X/168/1/012006

Intensification of electrochemical properties of the molten chloride electrolytes of the cerium subgroup lanthanides

O M Shabanov, S I Suleymanov[1] and A O Magomedova
Dagestan State University, 367000, Russia, Republic of Dagestan, Makhachkala, M. Gadzhieva str., 43-a

E-mail: s.sagim.i@ya.ru

Abstract. The electrical conductivity of molten chloride electrolytes of the cerium subgroup lanthanides increases with rising electric field strength and strive to achieve the limiting high voltage values (Wien effect). On exposure of the high-voltage microsecond pulsed fields, the melts are transited into a prolonged non-equilibrium state with increased electrical conductivity and electrolyze current density. During the relaxation processes in non-equilibrium melts, increased electrical conductivity tends to restore the values that are specific to equilibrium systems.

1. Introduction

In various modern industries there is an increasingly widespread use of rare earth metals (REM). In an industrial scale mainly REM alloys and relatively pure lanthanum, cerium and neodymium are produced. REM and their alloys are advantageously prepared by electrolysis of anhydrous molten chloride mixtures Ln-MCl (Ln = REM and M = Na, K). The structures of the molten electrolytes, as well as the nature and distribution their structural species, determine their observed physicochemical properties, the mechanisms and kinetic pathways that decrease energetic efficiency of metals production technologies [1]. For better understanding the structure of equilibrium molten electrolytes and improving the chemical and electrochemical technologies, their properties should be studied in a non-equilibrium state, which can be achieved by various external influences, for example, by the action of strong electric pulses [2]. We have found that electrical conductivity of molten alkaline-earth chlorides and their mixtures with alkaline chlorides increase with increasing external electric field strength (EFS) and strive to achieve the limiting high-voltage values in the fields of the order of 1 MV/m [3], in the same way as in the Wien effect in the aqueous electrolyte solutions.

After the high-voltage pulsed discharges in the electrolytes having been completed, their low voltage conductivity (measured by usual AC Bridge) turns out to be increased [4] i.e., in them the phenomenon of activation (the "memory effect") is observed. The activation degree of electrical conductivity reaches up to 20 % in the case of individual alkali earth metals chlorides and to 55 % in the case of their mixtures with potassium chloride. These observations were interpreted as a consequence of the stimulated dissociation of complex formations. The activation of the melts is followed by the recombination of complex ions during prolonged relaxation processes in the non-equilibrium melts.

[1] Author to whom all correspondence should be addressed.

IOP Conf. Series: Materials Science and Engineering **168** (2017) 012006 doi:10.1088/1757-899X/168/1/012006

As far as we know, Wien effect and the phenomena of the activation of the molten chloride electrolytes of lanthanides do not have analogues in the literature. In this review we present some results of studies on the dependence of the electrical conductivities of molten individual lanthanides chlorides and their binary mixtures with potassium chloride on the EFS and their activation followed by a prolonged relaxation in a non-equilibrium state.

2. Experimental

The measurements of the electrical conductivity dependence on the EFS for any electrolyte must be carried out exclusively in the regime of short time electric pulses in order to avoid a change in the temperature of a sample, introduction into it of possible electrolysis products, which may change the electrolytic nature of the conductance, and other side phenomena. To this end, we have constructed high-voltage setup for pulsed conductometry. The electric circuit of our experimental setup, the measuring cell, and the technique of determining the electrical conductivity of an electrolyte in strong electric fields were described in our earlier works (see, e.g., [3, 4]). The setup involves the digital pulsed memory oscilloscope ASK-8 which records waveforms of attenuated current and voltage during the pulse discharge in a studied sample. The representative oscillograms (waveforms) without electric breakdown in the molten $LaCl_3$ are outlined in figure 1. The waveforms do not contain the signs of electric breakdown (the voltage disruption and current jump), the current and voltage of discharge pass through maximums reached in about 1 μs; after reaching the maximum, they fall exponentially. Full pulse duration of several microseconds excludes introduction of the electrolysis products into the electrolyte and it's any significant overheating. The electric resistance of the electrolytes is determined as U/I at the moment of maximum of the current, when the quasi-steady state condition $dI/dt=0$ (σ=const) is fulfilled and Ohm's law is obeyed again. The system oscilloscope-computer records and saves the electric and temporal parameters of the high-voltage pulse discharges (HVPD) for any time in its continuation. Knowing the attenuation factor of the voltage and current, one can evaluate their absolute values at any instant while discharge goes on. The above arguments suggest that observed behavior of the molten electrolytes in the used EFS take place with conservation of the ionic conductivity nature.

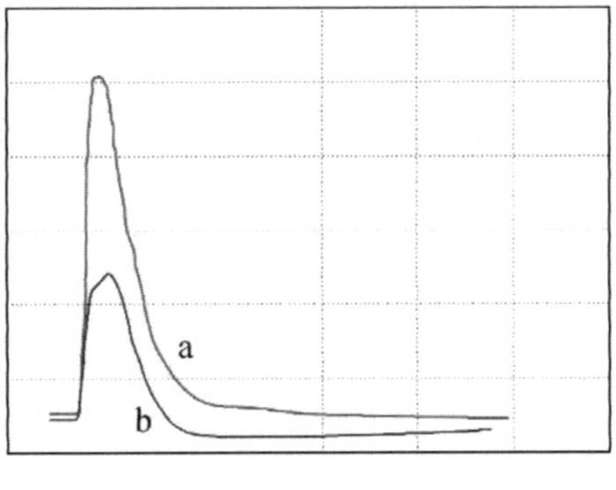

Figure 1. Characteristic oscillograms of voltage (curve a) and current (curve b) high voltage discharge in molten $LaCl_3$. The sweep rate is 0.2 μs/graduation. The voltage sensitivity is 2 V/graduation, the current sensitivity is 5 A/graduation.

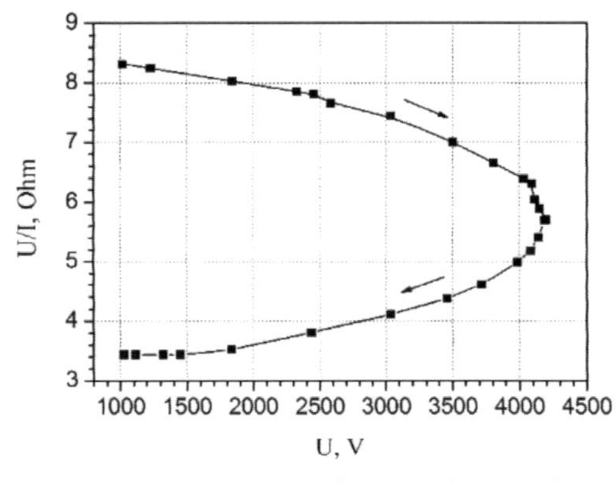

Figure 2. Dependence of ratio U/I on voltage U before (curve →) and after (curve ←) reaching the maximum current.

To study the phenomenon of the activation of the electrolytes we measured their conductivity by usual Ac Bridge RLC - before the application of high-voltage pulses to the equilibrium melts and after

IOP Conf. Series: Materials Science and Engineering **168** (2017) 012006 doi:10.1088/1757-899X/168/1/012006

the HVPD in them having been completed. The relaxation dynamics of the non-equilibrium systems was defined by observing the changes in the conductivity in time after HVPD having been completed.

Anhydrous metal chlorides were prepared by dehydration of the hydrates by slowly heating to 400 °C with ammonium chloride under high vacuum. The anhydrous salts were molted and experiments were carried out in the dry argon atmosphere.

3. Wien effect in molten lanthanides chlorides and their mixtures with potassium chloride

We have established the dependence of electrical conductivity on EFS for molten chlorides $LnCl_3$ (Ln = La, Ce, Nd) and their mixtures with KCl at different temperatures. The conductivities of individual molten lanthanides chlorides and their mixtures with potassium chlorides rise with EFS end tend to the limiting values in the fields of the order of 1 MV/m. This dependence is presented in figures 3-8 at the temperatures indicated in the captions.

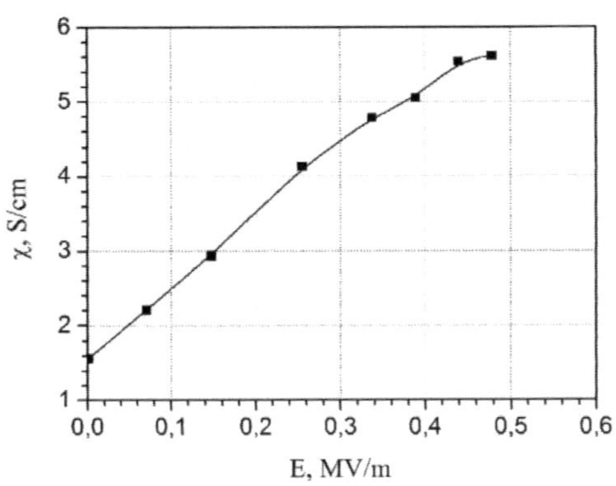

Figure 3. Dependence of specific conductivity of $LaCl_3$ melt on EFS at 1000 °C.

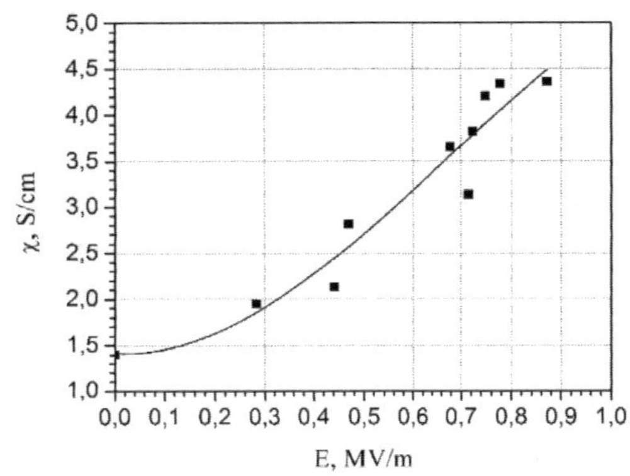

Figure 4. Dependence of specific conductivity of $0.2LaCl_3$-KCl melt on EFS at 1000 °C.

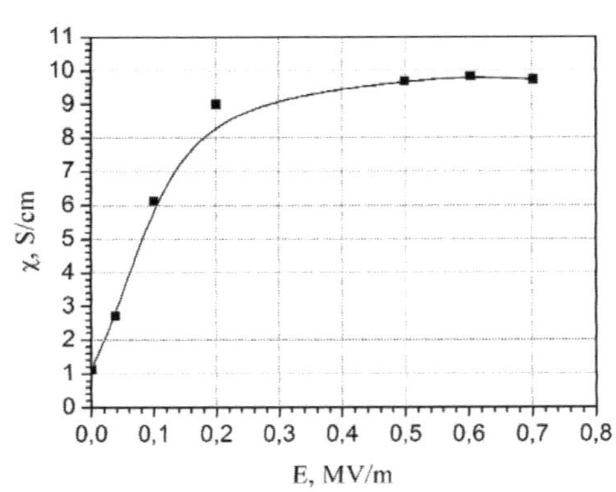

Figure 5. Dependence of specific conductivity of $CeCl_3$ melt on EFS at 840 °C.

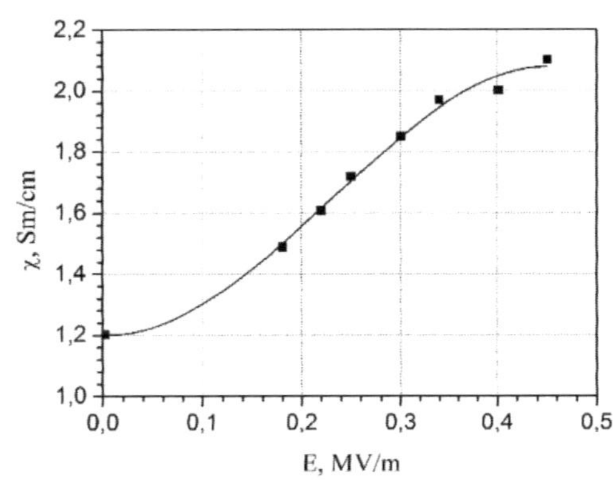

Figure 6. Dependence of specific conductivity of $0.2CeCl_3$-KCl melt on EFS at 840 °C.

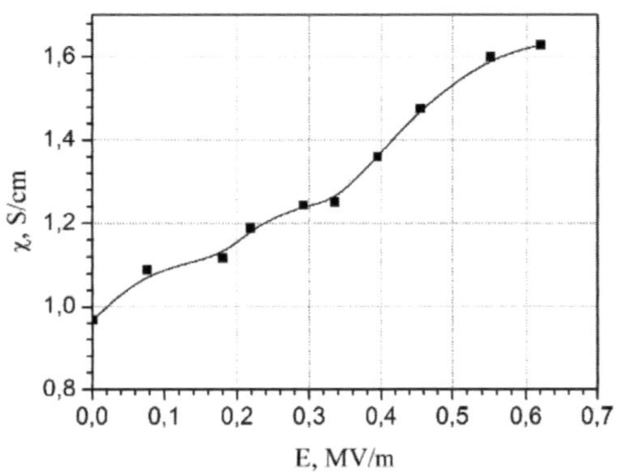

Figure 7. Dependence of specific conductivity of $NdCl_3$ melt on EFS at 1000 °C.

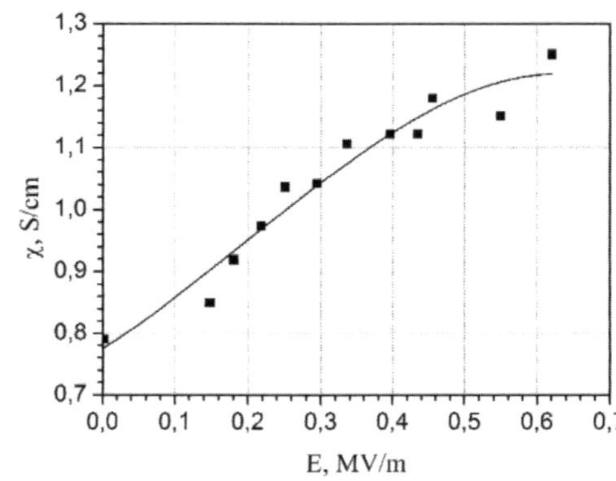

Figure 8. Dependence of specific conductivity of $0.8NdCl_3$-KCl melt on EFS at 1000 °C.

Limiting high-voltage conductivity χ of molten $LaCl_3$, $CeCl_3$, and $NdCl_3$ surpasses their usual ones $\chi(0)$ by 290%, 600%, and 80%, respectively. The effect of a high field in their binary mixtures with KCl is 150%, 120%, and 230%, respectively. These results are comparable to those obtained previously for $MgCl_2$ 260% [5], $KAlCl_4$ 700% [6], and $MgCl_2$-KCl 470% [5, 7].

An analysis of the HVPD oscillograms in figure 2 shows that U/I for full discharge sweep at $U \rightarrow 0$ in the lower curve (after reaching the maximum current) is lower than the initial value (at $\tau \rightarrow 0$) determined from the upper curve. In our experiments on Wien effect, every next voltage pulse was applied to an electrolyte only after the initial low-voltage conductivity was restored. This means that the electrical conductivity of a sample measured by a conventional ac bridge increases due to the passage of a high-voltage pulse through the sample; that is, the "memory" effect is observed. Electrolyte activation also manifests itself in increasing the electrical conductivity of the electrolyte. The dependence of the activation degree $\Delta\chi/\chi(0)$ on the voltage amplitude is presented in figures 9 and 10 for the mixtures of lanthanum and cerium chlorides as examples.

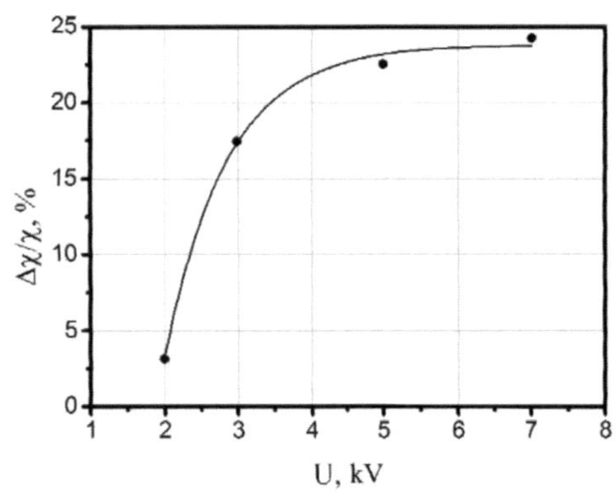

Figure 9. Variation of the activation degree of $LaCl_3$-0.6KCl melt as function of voltage amplitude of activating pulses at 810 °C.

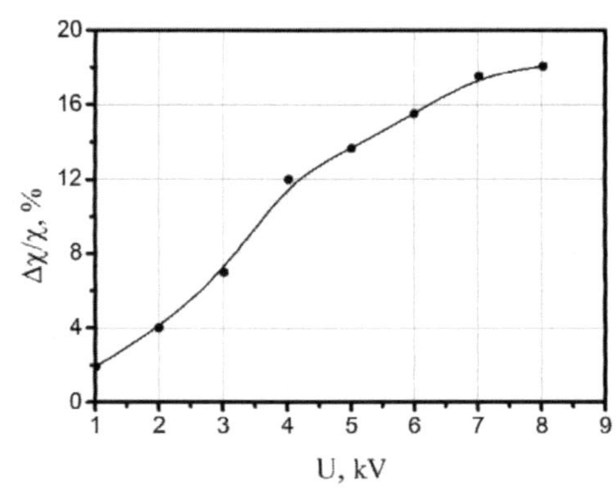

Figure 10. Variation of the activation degree of $NdCl_3$-0.2KCl melt as function of voltage amplitude of activating pulses at 810 °C.

The graphs of figures 9 and 10 show that the value of the activation degree increases with increasing voltage amplitude of single activating pulse and tends toward the saturation with increasing voltage amplitude. The increased electrical conductivity of activated molten electrolytes decreases with the time and tends to the initial (equilibrium) values, the relaxation of conductivity occurs in oscillatory regime. It can be seen from the following figures:

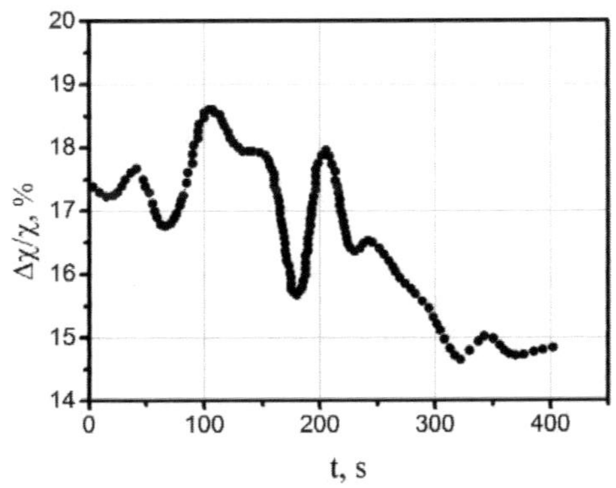

Figure 11. Change of the relative values $\Delta\chi/\chi(0)$, % in time in molten mixtures $LaCl_3$-0.6KCl activated by single pulses of the voltage amplitude 6 kV at 810 °C.

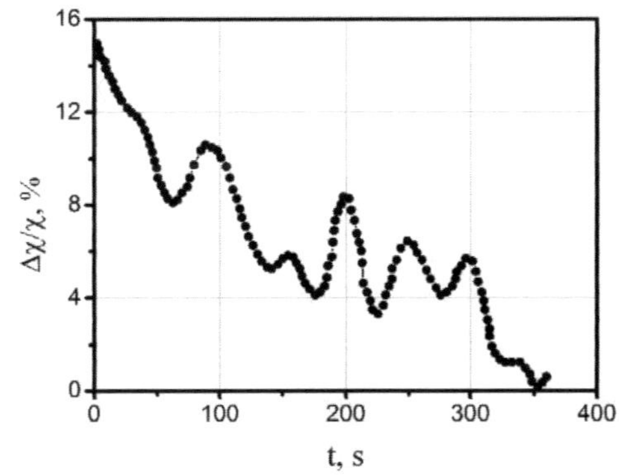

Figure 12. Change of the relative values $\Delta\chi/\chi(0)$, % in time in molten mixtures $NdCl_3$-0.2KCl activated by single pulses of the voltage amplitude 6 kV at 810 °C.

The relative change in the electrical conductivity is $\Delta\chi = [\chi(\tau)-\chi(0)]/\chi(0)$, where $\chi(0)$ is the initial low-voltage electrical conductivity and $\chi(\tau)$ is the low-voltage conductivity at time τ after the end of a high-voltage pulse. It is seen from figures 11 and 12 that the increased electrical conductivity of the electrolyte in the nonequilibrium state decreases in time. The relaxation has the anomalous duration reaching up to five minutes.

The activated molten electrolytes show also the intensification of the electrolysis process. Figure 13 presents the current-potential curves of electrolysis in the systems $CeCl_3$-KCl-NaCl.

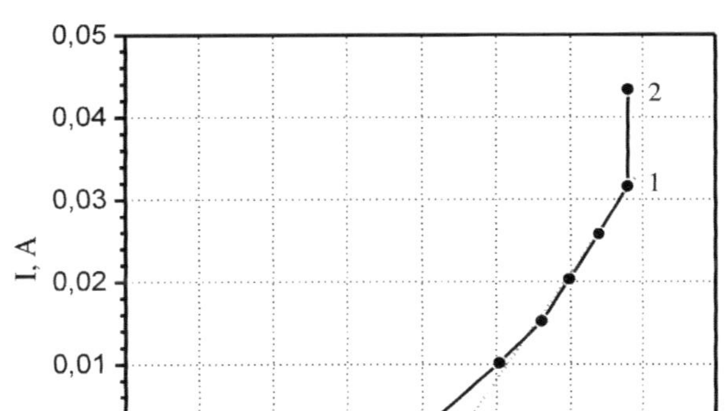

Figure 13. Current-potential curve of the equimolar system $CeCl_3$-KCl-NaCl at 660 °C, 1 - beafore activation, 2 - after activation.

The value of the decomposition potential of $CeCl_3$ was found equal to 2.2 V, in good agreement with the literature values [8]. As can be seen from figure 13, at potential 3.4 V the current was equal to

0.032 A (point 1); after exposition to the electrolysis cell of the activating pulse with the voltage amplitude of 6 kV, the current becomes equal to 0.044 A while maintaining the original voltage 3.4 V (point 2). This means that, as a result of the melt transfer to a non-equilibrium state, the electrolysis process is intensified by ~ 38%.

4. Discussion

The electrochemical production of REM from the chloride melts has a number of well-known advantages compared with others electrolytes, this method is promising one. Physicochemical, structural and electrochemical properties of the molten electrolytes usually are been investigated by substantially modern techniques. These results clearly indicate that the structure of the molten REM trichlorides and their mixtures with alkali chlorides are known to be thermodynamically nonideal, and there has been a good deal of work to understand the structural effects factors that contribute to the nonideality. Their Raman spectra were identified as the totally symmetric stretching vibration of the octahedral $LnCl_6^{3-}$ complexes forming loose network-clusters composed of edge bridged octahedrons connected to each other [9].

The presented above behavior of the melts can be understood on the basis of their equilibrium structure and its changes under the influence of external perturbations action. Stimulated dissociation of the complexes and the destruction of the grid structure can improve the conductivity and other electrochemical properties of these melts. The increase in the electrical conductivity of the molten chloride electrolytes with increasing EFS and the achievement of the limiting of high-voltage values are do both an increase of current carries concentrations and their motilities. These factors are major reasons which cause observed phenomena, and they are the consequence of simplifying the structure of the melts $LnCl_3$ and mixtures $LnCl_3$-KCl as consequence of stimulated dissociation of complex structural species.

The relaxation has the anomalous duration reaching up to five minutes. The conductivity failures from the steps of figures 11 and 12 can be attributed to a recombination of complex ions, they are formed and decomposed again, releasing simple ions and causing a jump of conductivity, and so on: the relaxation of conductivity occurs in oscillatory regime obviously expressed. The results above can be considered as the experimental confirmation of the dissociation of the complex ions stimulated by strong pulses. The prolonged relaxation of the molten electrolytes can serve the intensification of electrochemical technologies of RSM production.

Acknowledgements
This work was supported by Russian Foundation for Basic Research, project № 14-08-00033-a and Ministry of Education and Science of Russian Federation, state task № 2560.

References

[1] Yoon S Y, Flint J H, Kipouros G J, and Sadowy D R 2001 *J. Light Metals* **1** 111
[2] Diller I M 1969 *Nature* **224** 877
[3] Shabanov O M, Gadzhiev S M, Magomedova A O, Dzhamalova S A 2003 *Chem. Phys. Letters.* **380** 352
[4] Shabanov O M 2013 *Activation of molten and solid electrolytes* (Lambert Academic Publishing)
[5] Shabanov O M, Gadzhiev S M, Iskakova A A, Kachaev R T, Magomedova A O and Suleymanov S I 2011 *Russian J. Electrochem.* **47** 221
[6] Shabanov O M, Kazieva L A, Kachaev R T, Magomedova A O, and Suleimanov S I 2015 *Russian Metallurgy (Metally)* **8** 654
[7] Gadzhiev S M, Shabanov O M, Magomedova A O, and Dzhamalova S A 2003 *Russian J. Electrochem.* **39** 425
[8] Virgil C M, Popescu A M, and Zuca S 2003 *Z. Naturforsch* **58a** 57
[9] Takeshi U, Toshiyuki F, Yasuhiko I, Akihiro U, and Hajimu Y 2013 *Z. Anorgan. Allgem. Chem.* **639** 765

IOP Conf. Series: Materials Science and Engineering **168** (2017) 012007 doi:10.1088/1757-899X/168/1/012007

Influence irradiation argon ion SnO_2 on optical and electrical characteristics

O Asainov, S Umnov and V Temenkov
Physical Technical Institute, National Research Tomsk Polytechnic University, Tomsk, Russia

E-mail: usp@tpu.ru

Abstract. Tin oxide in the form of films has been deposited by reactive magnetron sputtering on glass substrates a room temperature .Process was carried out in such mode when the deposited films were conductive. The deposited films were irradiated with argon ions. Have been studied happening at that the changes optical and electric properties of films. Have been investigated optical properties of films in the range of 300-1100 nanometers by means of photometry. For research structure of films was used the x-ray diffractometry. Diffractometric researches have shown that the films deposited on a substrate have crystal structure from shares of a quasicrystal phase and after influence of argon ions she completely became quasicrystal. It is established that change transmission of a film correlates with change her electric resistance. Average value transmission in the range of 380-1100 nanometers as well as the electric resistance of a film with growth of irradiation time increases to the values exceeding initial. At the same time at irradiation time ~ 13,2 sec. are observed their slight decrease. To this value of irradiation time there corresponds the minimum value of electric resistance and transmission films. Change of transmission coefficient correlates with change of surface resistance.

1. Introduction

The transparent conductive oxide (TCO) films are widely used in various industries [1-4]. These optically transparent electrodes are for displays, solar panels, photoelectric devices, touch panels, etc. One of them is tin oxide, which has found wide practical application. In some practical applications, there is no need to simultaneously achieve high value of transmittance and low resistance. In such cases it is possible to use an undoped tin oxide. Pure tin oxide has conductivity and transparency. This is due to the existence of intrinsic point defects that act as vacancies [5,6]. Annealing deposited films leads to a modificated in the optical and electrical properties. Annealing films at an air, the between O and Sn ratio changed that leads to a decrease in their resistance [7,8]. However, the transmission and resistance depends on the duration of the annealing. Annealing the tin oxide films in a vacuum results in a significant reduction of resistance [9]. In work [10] it is shown that the optimization of oxygen content and the substrate temperature one can receive films with improved transmittance and resistance values. Tin oxide can be in the form of SnO2 and SnO. Under heating, these phases interact with one another that leads to the modification of films properties [9,11,12]. The effect of films annealing in various atmospheres (air, vacuum or hydrogen atmosphere) is studied in work [9]. The correlation was found between the structural and optical properties of SnO2 films. In the work [11] tin oxide films were annealed at different time period, however the correlation between the electrical specific resistivity and optical and structural properties was not found. In works [13], it was studied the effect of oxygen content in a reactive gas mixture on the properties of the deposited tin oxide films. It is shown that films consist of an amorphous and crystal phase [14]. The formation of textured films in the course of the sedimentation of tin oxides and its transformation into polycrystalline SnO2 at a low temperature (~ 200 °C) is investigated in [15]. Stoichiometry of films depending on the oxygen content and its influence on electric properties were investigated in [16]. The influence of sedimentation rate of a film on specific electric resistance has been studied in [17]. The mechanism of conductivity of pure oxide tin films deposited at various temperatures of a substrate has been investigated in work [18]. In work [19] the influence of the application of an external constant magnetic field on resistance and an optical transmission of the deposited films is investigated. Oxygen negative ions of the magnetron discharge [20] also have an impact on the films properties. The influence of ionic stimulation in the course of films sedimentation on optical properties of a film is considered in [21]. The properties modification of already deposited films happens at the irradiation by an ionic beam of already deposited films [22-24]. In this research, the tin oxide films were deposited by a reactive magnetron dispersion, and the films after sedimentation were irradiated by a bunch of argon ions.

2. Experimental details

Thin films of tin oxide were deposited on the glass substrates using reactive magnetron sputtering on the target in size 700x100x5 mm from tin (purity of 99,5%). The sputtering was carried out at a working pressure of 0.3 Pa in the gas mix O2 and Ar (O2 – 60%, Ar – 40%). The voltage of the magnetron discharge was ~ 275 V. Under these conditions the deposited films were dielectric. The power the discharge at sedimentation was stabilized at the level 1.2 kW. The heating of substrates was not carry out. The substrates were placed on a carrier that was able to carry out a reciprocating motion under the magnetron and the ion source in a vacuum chamber. The distance between a sample and a target was 8 cm. Speed movement of the carriage was 3.62 cm/sec. After the deposition, tin oxide films were irradiated with argon ions. The ions beam was located at an angle 90o to a substrate surface. Ions were generated by a source of ions with the closed drift of electrons at a voltage 2800 V and discharge current 0.28 A. The current density on a surface of samples was 2.4 *10-3 A/cm2. However, the pressure in a chamber was 0.166 Pa. The thickness of the deposited films was measured using profilometer MicroMeasure 3D Station. The measurement a spectrum of a transmission in the range of 300-1100 nanometers the was used spectrophotometer SF-256. The structure of films was studied on the x-ray diffractometer XRD6000. The average size of crystallites was estimated by half of width of x-ray diffraction lines.

3. Results and discussion

Conductive films of tin oxide were deposited on a glass substrate. For research of a structural films SnO2 was deposited thickness ~800 nm, and for optical and electrical – 264 nm.

Figure 1 shows the diffractogram films of tin oxide before and after irradiation by Ar ions. It is found that the films SnO2 after magnetron deposition have crystal structure with shares of amorphous (quasicrystal). The analysis of XRD data showed that the deposited films consist of pure SnO2 tetragonal structure without the presence of other phases – SnO and Sn. Similar results were observed.

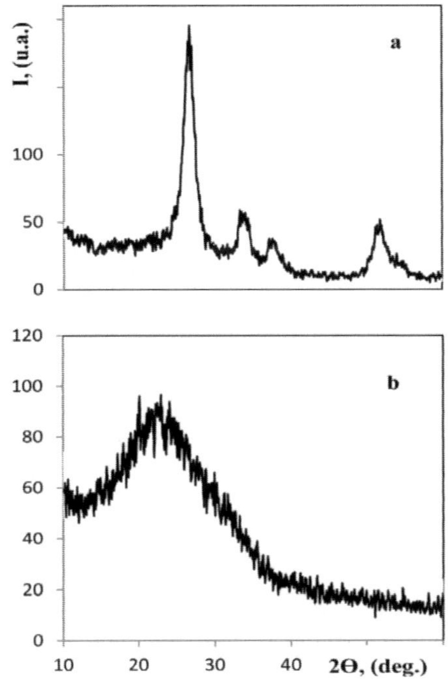

Figure 1. XRD diagram samples of SnO_2: a- before irradiation: b-after irradiation

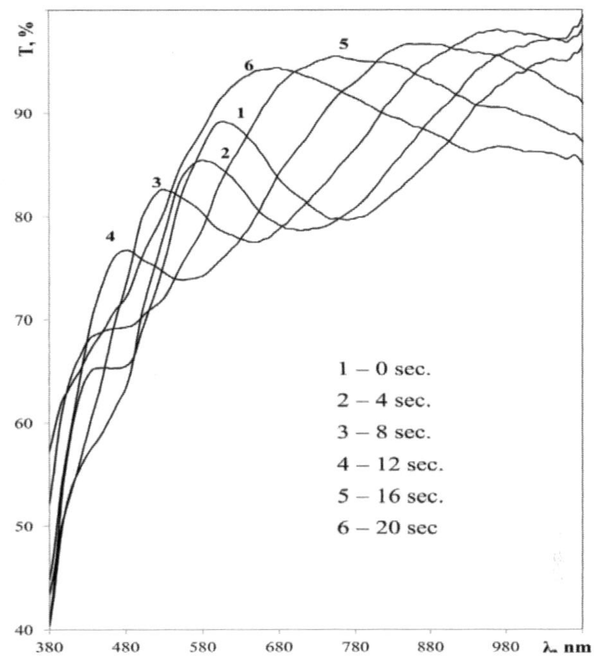

Figure 2. Dependence of the transmission on the irradiation time

in work [17]. Moreover, presence of other phases of tin oxide in films it was not revealed. The size of crystallites makes ~ 10 nanometers. After irradiation of structure became quasicrystal (amorphous). Figure 2 shows the results measurement transmission of films SnO2 at various of irradiation times. It is found that the transmission T (λ) in depending on irradiation time it has different character. On the impact at 0 to 13.2 sec., the dependences T (λ) have similar qualitative nature. For these times is characteristic existence of area of a maximum transmission.

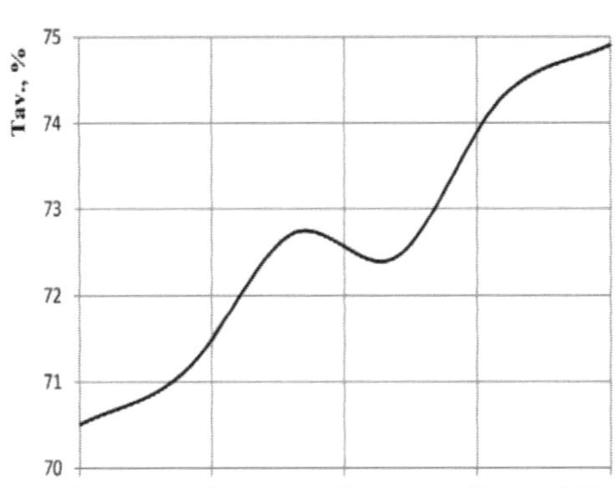

Figure 3. Dependence of the average value of the transmission on the irradiation time

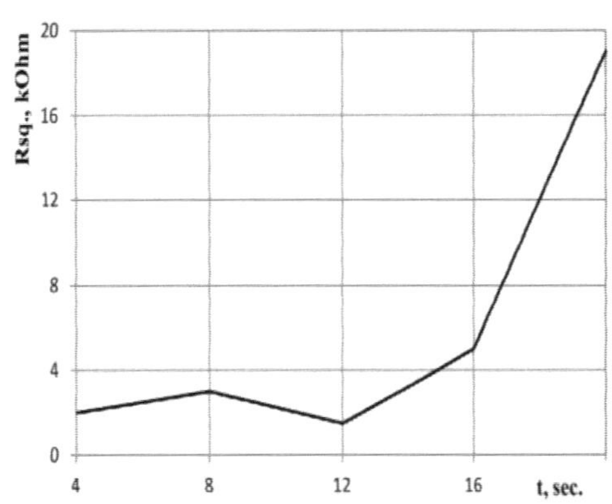

Figure 4. Dependence of surface resistance on the irradiation time

Shift of a maximum transmission to the area short lengths waves at time of exposure an ionic beam less than 13.2 sec. are possible, is connected with formation of polycrystalline structure SnO2 of [12] nanodisperse sizes. Decrease in the maximum value transmission at influence time of an argon ions beam from 0 to 13.2 sec. is possible, connected with destruction chemical bond of Sn-O. However, in the surface layer is formed metal tin and free oxygen which is sorbed. The formation in the surface layer of metal particles of Sn leads to growth of reflection a light stream and as a result – to decrease in a transmission. At further influence of an ionic beam there is a desorption of the formed oxygen and the composition of a film becomes closer to stoichiometric that provides to improvement it transparency of a film.

Figure 3 shows the change of size average value transmission of samples with the deposited SnO2 films in the range of 380-1100 nm depending on irradiation time ions Ar. It if found that the transmission with increase in influence time of ions first decreases, and then increases. The minimum transmission is observed at influence time of a beam ~ 13.2 sec. Generally, increase value average transmission happens improvements transmission in the field of lengths waves of 500-900 nm. It is connected with improvement stoichiometry composition of a films as a result of a desorption of tin metal. Small decrease transmission is connected with destruction of the existing communications and education new under the influence of the bombarding argon ions.

Fig. 4 demonstrates the measurements of a film surface resistance depending on the influence time of an ion beam. It correlates with the change of average transmission. On an initial the plot resistance of a films increases. It on all visibilities is connected with a desorption of metal tin from a surface. At further impact of an ionic beam on a films happens some decrease of resistance at the expense of oxygen desorption. At big times of impact of an ionic beam the composition of a film approaches stoichiometric and resistance of a film increases up to a condition of dielectric.

4. Conclusions

Conductive polycrystalline films of pure SnO2 were deposited at the room temperature using reactive magnetron sputtering. After deposition of a films SnO2 were irradiated the argon ions in the result what films became quasicrystal. With increase in of irradiation time average value transmission of a film increases from 70% up to 75%. However, at irradiation time ~ 13.2 sec. are observed insignificant decrease (to 72. 45%). This time there corresponds the minimum value of superficial resistance. At other irradiation times resistance of a films grows with growth of irradiation time. Is observed the correlation between optical and electrical properties of films SnO2. Results research show that optical and electrical properties of films can be changed after deposition.

References

[1] Harrison P G, M J Willett M J 1988 *Nature* **332** 337
[2] Batzill M, Diebold U 2005 *Surf. Sci.* **79** 47
[3] Zhao W M, Zhang M, Ai Z et al 2014 *J. Phys. Chem. C* **118** 23117
[4] Qian J, Liu P, Xiao Y et al 2009 *Adv. Mater.* 21 3663
[5] Jarzebski Z M, Marton J P 1976 *J. Electrochem. Soc.: REVIEWS AND NEWS* **123** 199
[6] Kılıç C, Zunger A 2002 *Phys. Rev. Lett.* 2002 **88** 095501
[7] Alterkop B, Parkansky N, Goldsmith S and Boxman R L 2003 *J. Phys. D: Appl. Phys.* **36** 552
[8] Kim I H, Ko J H, Kim D et al 2006 *Thin Solid Films* **515** 2475
[9] Mukashev B M, Tokmoldin S Zh, Beisenkhanov N B et al 2005 *Mater. Sci. Eng: B* **118** 164
[10] Abhijit De and Swati Ray 1991 *J. Phys. D: Appl. Phys.* **24** 719
[11] Mukhamedshina D M, Beisenkhanov N B, Mit K A 2006 *Thin Solid Films* **495** 316
[12] Karapatnitski I A, Mit K A, Mukhamedshina D M et al 2002 *Surf. Coat. Technol.* **151-152** 76
[13] Goodchild R G, Webb J B and Williams D F 1985 *J. Appl. Phys.* **57** 2308
[14] Gubbins M A, Casey V, Newcomb S B 2002 *Thin Solid Film* **405** 270
[15] Mukashev B N, Aimagambetov A B et al 2007 *Superlatives and Microstruct.* 42 103
[16] Lee W H, Son H C, Moon H S et al 2000 *J. Power Sources* 89 102

[17] Shamala K S, Murthy L C S, Narasimha Rao K 2004 *Bull. Mater. Sci.* **27** 295
[18] Shikha Bansal, Dinesh K. Pandya , Subhash C. Kashyap 2012 *Thin Solid Film* **524** 30
[19] Tadatsugu Minami, Hidehito Nanto and Shinzo Takata 1988 *Japan. J. Appl. Phys.* **27** L287
[20] Klaus Ellmer and Thomas Welzel 2012 *J. Materials Research* **27** 765
[21] Won-Kook Choi, Hyung-Jin Jung and Seok-Keun Koh 1996 *J. Vac. Sci. Technol.* A **14**, 359
[22]Asainov O, Umnov S, Shinin A 2015 *Journal of Physics: Conference Series* **652** 012046
[23]Asainov O, Umnov S, Temenkov V 2016 *Journal of Physics: Conference Series* **669** 012059
[24]Umnov S, Asainov O, Temenkov V 2016 *Journal of Physics: Conference Series* **124** 012148

Calculation of the ion current distribution along the cathode in glow discharge under the existence of a periodic relief and an insulating film of varying thickness on its surface

G G Bondarenko[1] and V I Kristya[2]
[1] National Research University Higher School of Economics (HSE), Myasnitskaya st. 20, Moscow, 101000, Russia
[2] Bauman Moscow State Technical University, Kaluga Branch, Bazhenov st. 2, Kaluga, 248000, Russia

E-mail: kristya@bmstu-kaluga.ru

Abstract. An approximate analytical expression for the ion current density near the cathode in glow discharge is obtained in the presence of a periodic relief of small amplitude and an insulating oxide film of varying thickness on its surface. It is found that ion focusing at the cathode sections with the minimum film thickness, located on any parts of the surface relief, takes place, resulting in an increase of the film thickness non-uniformity with time. Therefore, under the existence of an oxide film on the cathode, its sputtering in glow discharge is determined mainly by the film thickness non-uniformity and not by the surface relief.

1. Introduction

Glow discharge in gases is used in many types of electronic devices (gas lasers, plasma display panels, discharge illuminating lamps) [1-3]. Its feature is a thin layer near the cathode (the cathode sheath (CS)) with the voltage drop of the order of 10^2 V and a high value of the electric field strength [1]. Therefore, details of the process of glow discharge plasma interaction with the cathode surface are determined by the CS characteristics, such as the CS length and the cathode voltage drop. A reduction of the cathode voltage drop, leading to a decrease of the ion energies at the cathode and intensity of its sputtering in the discharge, can be achieved by formation of a thin insulating oxide film on the cathode surface, which ensures a higher value of its effective ion-electron secondary emission rate γ_{eff} [4, 5]. Because usually a relief exists on the cathode and the oxide film thickness is varied along its surface, the electric field line bending in a near-cathode discharge layer and the focusing of ions at some sections of the cathode surface can take place [6, 7]. This should result in a non-uniformity of its sputtering in the discharge and in a decrease of the cathode service time. In particular, as it was shown in [7-10], the ion current density has a higher value at the metal cathode relief tops, whereas in the case of the cathode with an insulating film of varying thickness the ion flow is focused at its more thin sections. But the combined effect of both these factors on the ion current density distribution and the cathode relief time evolution was not studied before.

In this work, the ion current distribution along the cathode in glow discharge under the presence of a periodic relief of small amplitude and an insulating oxide film of varying thickness on its surface is calculated, and their influence on the cathode sputtering is investigated.

2. Mathematical model

Let the discharge CS be bounded by plane $z=0$ and the metal cathode, on which there is a thin dielectric film, which thickness is periodically varying along the cathode in the direction of x-axis (see figure 1). The cathode surface is described by equation $z_s = d_c + h_s \cos kx$, equation of the boundary between the film and the cathode metal substrate has the form $z_b = d_c + H_{f0} + h_b \cos kx$, and the film thickness is determined by the dependence $H_f(x) = z_b - z_s = H_{f0} + h_f \cos kx$, where $h_f = h_b - h_s$, d_c is the CS length, $k = 2\pi / l_s$, h_s and l_s are the amplitude and period of the cathode surface relief, h_b is the amplitude of variation of the film-substrate boundary z-coordinate.

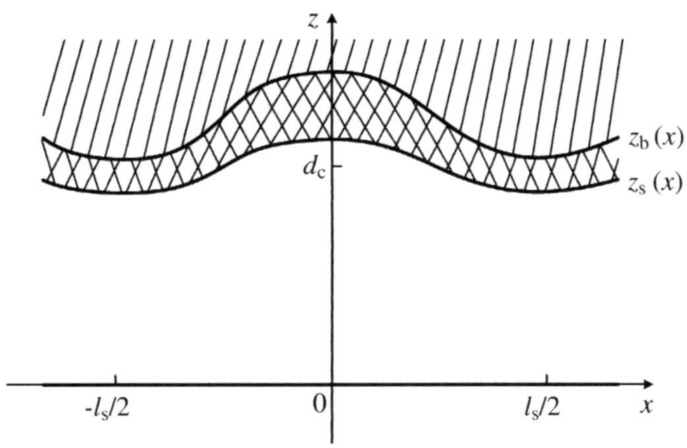

Figure 1. A schematic of the CS geometry near the cathode with a periodic relief and a surface insulating film of varying thickness.

We will suppose that the lateral dimension l_s of the cathode surface relief elements greatly exceeds the average characteristic length λ of the ion and electron path between collisions with gas atoms and the amplitude of the surface relief elements is rather small, i.e.

$$h_b \ll h_s \ll \lambda \ll l_s \ll d_c. \quad (1)$$

In the discharge, due to bombardment of the cathode by ions, positive charges are accumulated on the film surface. As a result, a strong electric field arises in the film, and when it reaches value of the order of 10^8-10^9 V/m, the field electron emission from the metal cathode substrate into the film starts [11, 12]. The emitted electrons move in the film to its outer boundary and neutralize positive surface charges, preventing further increase of the cathode surface charging, i.e. the discharge becomes stationary [13, 14].

Under condition (1) the macroscopic description of charged particles motion in the gas can be used. Therefore, taking into account that the space charge of electrons in the CS and in the film is small [1, 15], the CS equations can be written in the form [1, 7]:

$$\nabla \vec{j}_e = \alpha |\vec{j}_e|, \ \nabla \vec{j}_i = \alpha |\vec{j}_e|, \ \vec{j}_e = -e\mu_e n_e \vec{V}_e, \ \vec{j}_i = e\mu_i n_i \vec{V}_i, \ \Delta\varphi = -en_i/\varepsilon_0, \ \Delta\psi = 0 \quad (2)$$

with boundary conditions

$$\vec{E}(x,0) = 0, \ \varphi(x,0) = 0, \ \varphi(x,z_s) = \psi(x,z_s), \ j_e(x,z_s) = -\gamma_{eff} j_i(x,z_s),$$
$$(1+\gamma_{eff}) j_i(x,z_s) = j_{FN}(x), \ \psi(x,z_b) = -U_c, \quad (3)$$

where $\vec{j}_e$ and $\vec{j}_i$ are the electron and ion current densities in the discharge, α is the coefficient of gas ionization by electrons, φ, $\vec{E}=-\nabla\varphi$ and ψ, $\vec{E}_f=-\nabla\psi$ are the potentials and the electric field strengths in the CS and in the film, n_e and μ_e are the electron density and mobility, n_i и $\mu_i = k_i/\sqrt{E}$ are the ion density and mobility, k_i is a the constant for a given type of the gas, U_c is the cathode voltage drop, ε_0 is the dielectric constant. The field electron emission current density from the metal cathode substrate into the film is defined by the Fowler-Nordheim formula [16, 17] $j_{FN}(x) = j_F\left(E_f(x)/E_F\right)^2\exp\left(-E_F/E_f(x)\right)$, in which j_F and E_F are parameters dependent on the substrate and film materials.

3. Ion current at the cathode

Solution of the boundary value problem (2), (3) under condition (1) can be presented as the sum of unperturbed values and their perturbations caused by the cathode surface non-uniformity:

$$\begin{gathered} n_e(x,z) = n_{e0}(z) + n_{e1}(z)\cos kx,\ n_i(x,z) = n_{i0}(z) + n_{i1}(z)\cos kx, \\ j_i(x,z) = j_{i0}(z) + j_{i1}(z)\cos kx,\ E_f(x) = E_{f0} + E_{f1}\cos kx, \\ \varphi(x,z) = \varphi_0(z) + \varphi_1(z)\cos kx,\ \psi(x,z) = \psi_0(z) + \psi_1(z)\cos kx, \end{gathered} \tag{4}$$

where

$$n_{e1} << n_{e0},\ n_{i1} << n_{i0},\ j_{i1} << j_{i0},\ E_{f1} << E_{f0},\ \varphi_1 << \varphi_0,\ \psi_1 << \psi_0 .$$

Substitution of expressions (4) in (2), (3) results in two systems of equations for the zero-order quantities and the first-order quantities with the corresponding boundary conditions. From the first system it follows [7]:

$$\varphi_0(z) = -(U_c - E_{f0}H_{f0})(z/d_c)^2,\ E_{c0} = -\varphi_0'(d_c) = 2(U_c - E_{f0}H_{f0})/d_c,\ j_{i0}(d_c) = k_i\varepsilon_0 E_c^{3/2}/ed_c, \tag{5}$$

where E_{f0} is determined by equation

$$(1+\gamma_{eff})\, j_{i0}(d_c) = j_F\left(E_{f0}/E_F\right)^2\exp\left(-E_F/E_{f0}\right). \tag{6}$$

Solving the second system, which is rather cumbersome and is therefore not presented here, with the asymptotic method [7,18] under condition $kd_c \gg 1$, following from (1), it can be found that

$$\begin{gathered} \varphi_1(z) = \left(E_{f1}H_{f0} + E_{f0}(h_b - h_s) + E_{c0}h_s\right)\exp\left(k(z-d_c)\right), \\ j_{i1}(z) = -\left(kj_{i0}(d_c)/E_{c0}\right)\left(E_{f1}H_{f0} + E_{f0}(h_b - h_s) + E_{c0}h_s\right)\exp\left(k(z-d_c)\right), \end{gathered} \tag{7}$$

where

$$E_{f1} = -\left(E_{f0}(h_b - h_s) + E_{c0}h_s\right)/\left(H_{f0} + E_{c0}\left(2 + E_F/E_{f0}\right)/kE_{f0}\right).$$

It is seen in expressions (7) that perturbations of the electric field strength and the ion current density, caused by the cathode surface non-uniformity, are concentrated in a thin near-cathode layer with the thickness of the order of $l_s \ll d_c$ and have negligible values outside of it, i.e. they do not influence the CS macroscopic characteristics noticeably.

It can be found from (2), (4) and (7) that the ion current density, bombarding the cathode, is described by expression

$$j_i(x,d_c) = j_{i0}(d_c) + j_{i1}(d_c)\cos kx, \tag{8}$$

where

$$j_{i1}(d_c) = -\left(k\,j_{i0}(d_c)/E_{c0}\right)(1-\chi)\left(E_{f0}(h_b - h_s) + E_{c0}h_s\right),\ \chi = \left(1 + \left(2 + E_F/E_{f0}\right)E_{c0}/kE_{f0}H_{f0}\right)^{-1}.$$

Therefore, the cathode is sputtered non-uniformly and numerical estimations show that for the oxide films with $H_{f0} \approx 10^{-8}$ m and $l_s \approx 10^{-5} - 10^{-4}$ m magnitude of the coefficient χ is much less than unity.

4. Time evolution of the cathode surface relief

Variation of the cathode surface z - coordinate with time due to ion sputtering is described by expression [7]:

$$\frac{dz_s(x)}{dt} = \frac{M_a}{N}\left(RJ_i(x,d_c) - J_{sb}(x,d_c)\right), \quad (9)$$

where N and M_a are the cathode material density and the mass of sputtered atom, R is the cathode sputtering rate equal to its sputtering yield averaged over ion energies, $J_i(x,d_c) = j_i(x,d_c)/e$ and $J_{sb}(x,d_c) = \beta R j_{i0}(1 - kh_s \cos kx)$ are the flow densities of ions bombarding the cathode and atoms sputtered from the cathode, which return back due to scattering on background gas atoms [8,9,19], $\beta < 1$ is a fraction of sputtered atoms re-deposited on the cathode, e is the ion charge.

Substitution of expressions for $J_i(x,d_c)$ and $J_{sb}(x,d_c)$ in (9) gives an equation for the cathode surface relief amplitude:

$$\frac{dh_s}{dt} = -\frac{1}{\tau_s}\left[\frac{E_{f0}}{E_{c0}}(h_b - h_s) + (1-\beta)h_s\right], \quad (10)$$

where $\tau_s = Ne/M_a Rk\, j_{i0}(d_c)$ is the characteristic time of relief variation in the discharge.

It follows from (10) that in case of the metal cathode without an oxide film on the surface, i.e. at $h_s = h_b$, a decrease of the relief amplitude proceeds in the discharge due to focusing of the ion flow on the relief tops (because of $j_{i1}(d_c) < 0$). But when an insulating film of varying thickness exists on the cathode, then in the case of its lower thickness at the relief tops (at $h_s < h_b$) their more intensive sputtering, followed by a decrease of h_s, goes on, i.e. a reduction of the film thickness on the relief tops, accelerating with time, takes place as a result of an increase of the magnitude in square brackets in the right-hand side of equation (10). However, under a higher film thickness on the relief tops (at $h_s > h_b$), the ion focusing at the relief bottoms exists already at

$$(h_s - h_b)/h_s > (E_{c0}/E_{f0})(1-\beta) \quad (11)$$

because of $E_{f0} \gg E_{c0}$, which results in an h_s increase and the film thickness decrease on the relief bottoms.

For instance, in case of the aluminum cathode with an oxide film in the argon normal glow discharge ($p = 1330$ Pa, $U_c = 100$ V, $d_c = 3\times10^{-4}$ m, $j_{i0}(d_c) = 2\times10^2$ A/m^2, $\lambda \approx 5\times10^{-6}$ m, $E_{c0} = 6.6\times10^5$ V/m, $E_F = 1.4\times10^{10}$ V/m, $E_{f0} = 4\times10^8$ V/m, [1, 15]) under $H_{f0} = 2\times10^{-8}$ m, $h_s = 2\times10^{-9}$ m, $l_s = 2\times10^{-5}$ m, $R = 0.01$ condition (1) is satisfied, $\chi = 0.04$, $E_{c0}/E_{f0} = 1.6\times10^{-3}$ and $\tau_s = 6\times10^4$ s. As it follows from (11), ions are focused at the cathode sections with lower film thickness at $|h_f|/h_b > 10^{-3}$, i.e. under a very small film thickness non-uniformity, due to the fact that the electric field in the film is three orders of magnitude greater than in the discharge volume.

5. Conclusions

Under the existence of a relief and a thin dielectric film on the cathode surface in glow discharge, ion focusing on its sections with the minimum film thickness, located at any parts of the surface relief, takes place, followed by an increase of the film thickness non-uniformity with time and formation of pores in the film. Therefore, in the presence of an oxide film on the cathode, its sputtering in glow discharge is determined mainly by the film thickness non-uniformity and not by the surface relief.

Acknowledgements
This work was performed in frameworks of the program "Organization of scientific researches" of the Russian Federation Ministry of Education and Science in the Bauman Moscow State Technical University.

Support from the Basic Research Program of the National Research University Higher School of Economics is gratefully acknowledged.

References

[1] Raizer Y P 1991 *Gas discharge physics* (Berlin: Springer)
[2] Lister G G, Lawler J E, Lapatovich W P and Godyak VA 2004 *Rev. Mod. Phys.* **76** 541
[3] Samukawa S. et al 2012 *J. Phys. D: Appl. Phys.* **45** 253001
[4] Boeuf J P 2003 *J. Phys. D: Appl. Phys.* **36** R53
[5] Bondarenko G G and Korzhavyi A P 2007 *Russ. Phys. J.* **50** 125
[6] Kobayashi K and Shimuzu K 1988 *J. Electrochem. Soc.* **135** 908
[7] Kristya V I 2010 *Glow Discharges and Tokamaks* ed S A Murphy (NY: Nova Science Publishers) chapter 7 pp 329-368
[8] Bondarenko G G and Kristya V I 2008 *J. Phys.: Conf. Series* **100** 062009
[9] Kristya V I 2008 *J. Surf. Investig.* **2** 203
[10] Bondarenko G G, Kristya V I and Tun J N 2016 *Russ. Phys. J.* **58** 1313
[11] Moon K S, Lee J and Whang K-W 1999 *J. Appl. Phys.* **86** 4049
[12] Stamenković S N, Marković V Lj, Gocić S R and Jovanović A P 2013 *Vacuum* **89** 62
[13] Riccardi P, Ishimoto M, Barone P and Baragiola R A 2004 *Surf. Sci.* **571** L305
[14] Kristya V I and Tun Y N 2014 *Bull. Russ. Acad. Sci. Phys.* **78** 549
[15] Kanter H and Feibelman W A 1962 *J. Appl. Phys.* **33** 3580
[16] Forbes R G 1999 *J. Vac. Sci. Tech.* B **17** 534
[17] Hickmott T W 2000 *J. Appl. Phys.* **87** 7903
[18] Nayfeh A H 1981 *Introduction to Perturbation Techniques* (NY: Wiley-Interscience)
[19] Valles-Abarca J A and Gras-Marti A 1984 *J. Appl. Phys.* **55** 1370

The formation of nanopores in metal materials after irradiation by beams of Ar^+ with energy of 30 keV

V A Ivchenko
Institute of Electrophysics, Ural Branch, Russian Academy of Sciences, ul. Amundsena 106, Ekaterinburg, 620016, Russia
Yeltsin Ural Federal University, Yekaterinburg, 620002 Russia

e-mail: ivchenko2008@mail.ru

Abstract. In this paper are the results of direction observations of nanopores in the subsurface volume of metals materials Pt and Pd(CuAg) using field-ion microscopy (FIM). Radiation of tip specimens was carried out with ions having an energy ~ 25-30 keV in the fluency range of $10^{16} – 10^{18}$ ions/cm^2, the current density lying within 150– 340 μA/cm^2. Nanopores have been observed immediately after removal of the first atomic layers from the irradiated surface. It was established that, the threshold for ion-implanted platinum corresponds to fluence F = 10^{17} ions/cm^2. For Pd(CuAg) it was revealed that nanopores have been down to 80 nm deep with current density 340 μA/cm^2. Their dimensions and volume fractions were determined. The obtained results can be used for prediction of radiation stability of materials based on fcc metals.

1. Introduction

Intensely developing studies of changes in the structure of metals and alloys after interaction between particle beams and the surface show that ionic treatment results in the formation of specific condensed states, as well as unique strength and physical proper- ties in materials that cannot be obtained using conventional methods [1–6]. It was known that pore swelling in materials under irradiation causes shortening of the operational life of products and drastic deterioration of physical and mechanic properties of materials. Therefore, studies of the appearance of nanopores on the surface and in the subsurface layer of the studied objects under exposure to radiation on an atomic scale appear to be of current interest.

In this work, the results of modification of the surface and subsurface region of metals materials after irradiation by medium-energy charged particles (up to 30 keV) within a nanometer-deep surface region are shown. The method of field-ion microscopy (FIM) allows studying experimentally structural states appearing under such exposure. FIM allows directly visualizing and registering an atomically pure surface, analyzing the object of study in the subsurface region with the spatial resolution of the crystal lattice atoms and thus determining structural changes on the surface and in the subsurface region and various new states of the materials appearing under radiation exposure.

The aim of this work was to study experimentally (with atomic resolution) the phenomena of nanopore formation in metals materials after their interaction with charged beams of Ar^+ ions. In particular, it was intended to analyze the nanopore size distribution in the subsurface region as dependent on the distance from the irradiated surface and radiation fluence, determine the nanopore formation threshold, and thus determine the optimum modes of radiation exposure to obtain nanostructured subsurface regions in ion-implanted metals and alloys.

2. Experimental

The objects of interaction with accelerated gas ion beams were pure metals (Pt) and constructional materials Pd (CuAg), used in special instrumentation for low-current contacts. Samples intended for study were prepared in the form of needle emitters with a tip curvature radius of 30–50 nm made of metal billets by electro polishing. The sample tips were evaluated using a field-ion microscope; i.e., an ion surface image of the corresponding materials in the initial state was obtained before their irradiation. Field emitters evaluated for ion implantation had an atomically smooth surface of the tip sample, close to a semispherical one. Such a surface was prepared by in situ field evaporation of the surface atoms.

Irradiation of needle samples (Pt) previously attested in a field ion microscope was carried out using gas ion (Ar^+) beams accelerated to 30 keV, with fluences F = 3 10^{16} to 10^{18} ion/cm^2 and ion current density j = 150– 200 μA/cm^2, for Pd50Cu30Ag20 (wt%) - ion current density j = 200 – 340 μA/cm^2 , with energy 25 keV. Bombardment was carried out in the direction parallel to the sample tip axis.

Implanted tip samples were again placed into an FIM, field ion surface micro patterns were registered using a photo or video camera with controlled removal of atomic layers, and changes in the structural state of bulk materials were analysed.

A field ion microscope was equipped with a micro channel ion–electron converter enhancing the brightness of surface micro images by 10^4 times. The refrigerant agent was generally liquid nitrogen (T = 78 K); the imaging gas was spectrally pure neon.

3. Results and discussion

Ion images of evaluated field emitters registered a regular ring pattern of the pure metal single crystal surface pointing to the virtual absence of structural defects (Fig. 1).

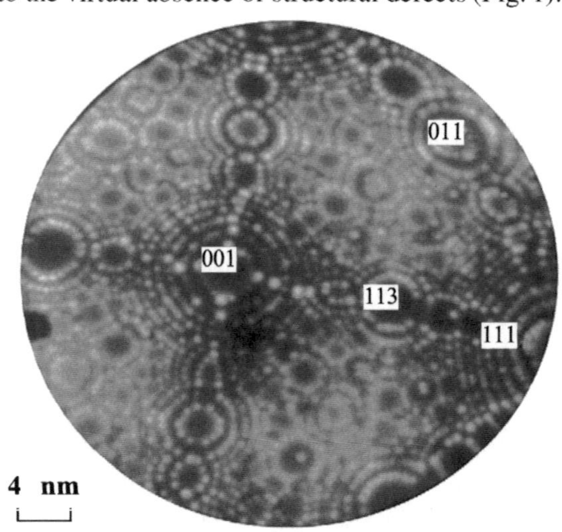

Figure 1. Neon image of validated Pt single crystal prepared for ionic irradiation.

In the microphotograph, faces {111}, {110}, and {113} are also registered, which are characteristic for crystals with an fcc lattice, in addition to the {001} face. The ring contour lines in ionic images are edges of the corresponding families of crystallographic planes of certain directions. Neighboring rings (from any family of concentric rings) are images of parallel atomic

layers. The rings themselves in an ionic micro image generally consist of separate bright points corresponding to images of surface atoms located at the sites of atoms in the step kinks.
In the course of studies of the crystalline structure of platinum irradiated to $F = 10^{16}$ to 10^{18} ion/cm^2 with an energy of 30 keV and j = 200 μA/cm^2, ion contrast of nanopores was observed in the subsurface layer, apart from formation of nanosize blocks [4]. Such a contrast was observed at a fluence of 10^{17} ion/cm^2.

The contrast of nanopores in Pt was registered simultaneously with the appearance of an ionic image of the surface studied by FIM and was preserved after removal of several atomic layers of the implanted metal (Fig. 2a). Ion contrast of vacancy nanopores at the moment of field evaporation of the latter before the appearance of a nanopore of the atomic layer was registered in the form of a contrast of "craters." Then, as the atomic layers around the defect evaporated, a vacancy cluster cross section with a size below that before the nanopore opening was observed. Finally, the nanopore outlet from the material in the course of the further evaporation of atomic layers from the surface generally ended in a dislocation loop (Fig. 2b).

The size of nanopores was estimated on the basis of the ionic contrast both using the area of the sample surface section by the defect measured on the basis of a calculation of the local surface radius on the face where the defect had appeared and by its depth via calculation of the number of evaporated atomic layers from the beginning of appearance of the nanopore contrast to its full disappearance. Results of quantitative analysis showed that nanopores had both spherical and cylindrical shapes. According to our estimates, their diameters were 1–5 nm. The size of nanopores across the depth was 1–9 nm. Successive evaporation of the surface atomic layers by electric field to the depth of 60 nm from the irradiated surface allowed elucidating specific regularities of nanopore formation. In particular, information could be obtained on the nanopore concentration and their distribution in the subsurface layer of the irradiated material. As a result, it was found that up to 40% of the pores were concentrated in the subsurface layer with a thickness of 10 nm. Later, the bulk fraction of nanopores decreased logarithmically (Fig. 3). As is known from [7], argon atoms in the case of applied beam energies have a projective path in platinum not above 10 nm.

It is, hence, obvious that threshold swelling in the material was not related to the presence of implanted argon ions in the defects, as nanopores were observed in the metal up to the depth of 60 nm.

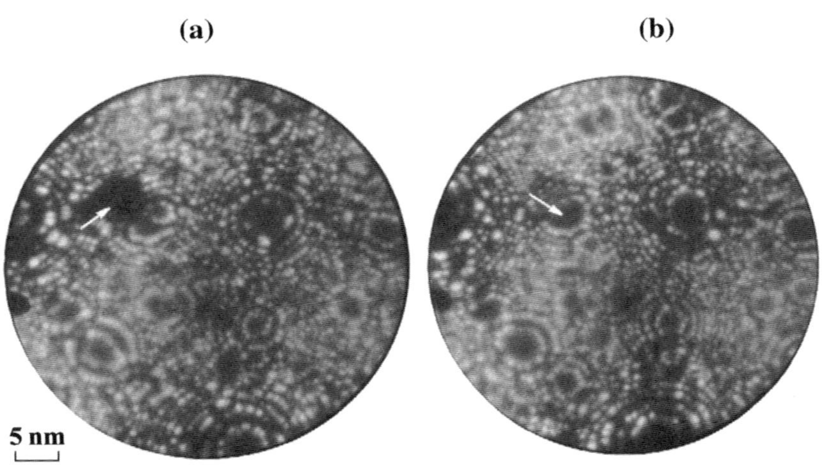

Figure 2. Neon images of pure Pt after irradiation by Ar^+ ions (E = 30 keV, F =10^{17} ion/cm^2): (a) typical ionic contrast of a vacancy nanopore (shown by an arrow); (b) contrast of a dislocation loop at the nanopore outlet from the material after field evaporation of 8 atomic layers with respect to the (001) face (~1.6 mm) (shown by an arrow).

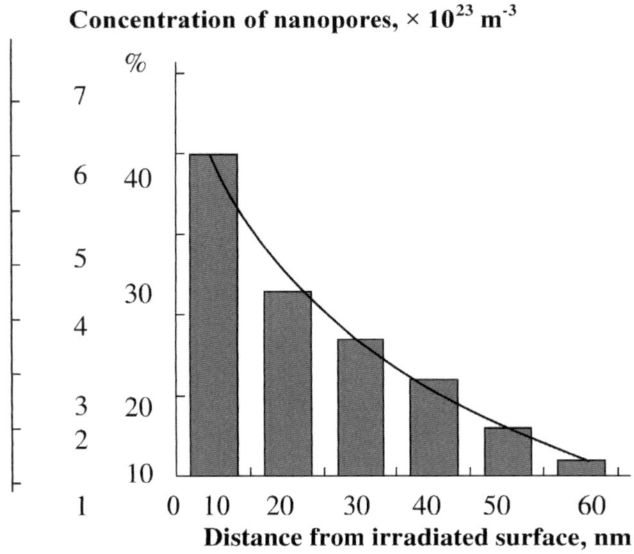

Figure 3. Concentration of nanopores in platinum irradiated by Ar^+ ions ($F = 10^{18}$ ion/cm^2).

Detailed analysis of experimental data allowed assuming that ion implantation resulted in continuous appearance and migration of implanted atoms and vacancies to drains and coalescence of individual vacancies into vacancy nanoclusters (nanopores).

As a result of argon ion implantation (E = 25 keV, j = 340 μA/cm^2, F = 10^{18} ions/cm^2) the structural state of ternary solid solution Pd(CuAg), Fig. 4a [8], changed. This being clearly evident from Fig. 4b, which revels an atomically clean surface of the irradiated alloy. The micrograph shows a superstructural (001) pole of a particle of the PdCu ordered B2 phase [8].

The examination of the subsurface volume of the said alloy in the course of the field evaporation of atoms of the material itself revealed an ion contrast similar to that due to nanopores and found [9].The said contrast appeared as crater after the first surface layer of the irradiated solid solution had evaporated. Fig. 5 shows a typical example of such contrast. Quantitative analysis has shown that the nanopores were ellipsoidal in shape. According to our estimates their diameters were within 3-12 nm and their depth within 4-25 nm. The ratio of nanopores volume to that of the materials analysed, wherein the pores were found, was assessed on the basis of quantitative measurements of the size and the number of vacancy clusters in the subsurface volume of the irradiated alloy. Fig. 5 illustrates the spatial disposition of nanopores in the subsurface volume. Analysis has shown that the volume fraction of vacancy clusters decreases monotonically with the depth of occurrence of defects (Fig. 6).

According to our estimates the maximum depth of the occurrence of nanopores after irradiation with j = 340 μA/cm^2, based on the calculation of the evaporated atomic layers, was found to be ca 80 nm and this is referred to the irradiated surface. With the decrease of the irradiation current density down 300 μA/cm^2 the depth of occurrence of nanopores dropped to 17 nm. At the same time it is known [7] that argon ions at the energies used here. Are implanted into pure components comprising of the alloy down to a depth ≤ 15 nm.

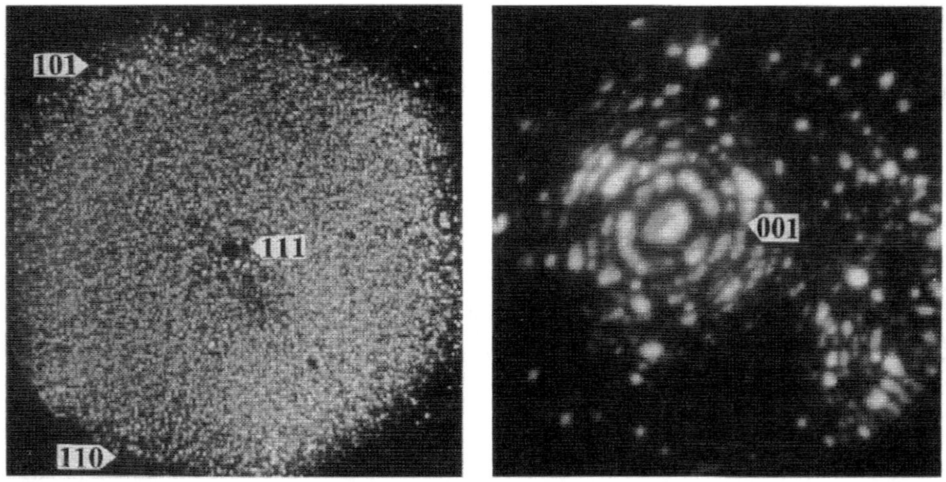

(a) **(b)**

Figure 4. (a) Micrograph of Pd(CuAg)solid solution surface certified in FIM just before irradiation (annealing at 1123 K for 1 h and water quenching); (b) micrograph of the Pd(Cu) ordered phase particle following ion implantation.

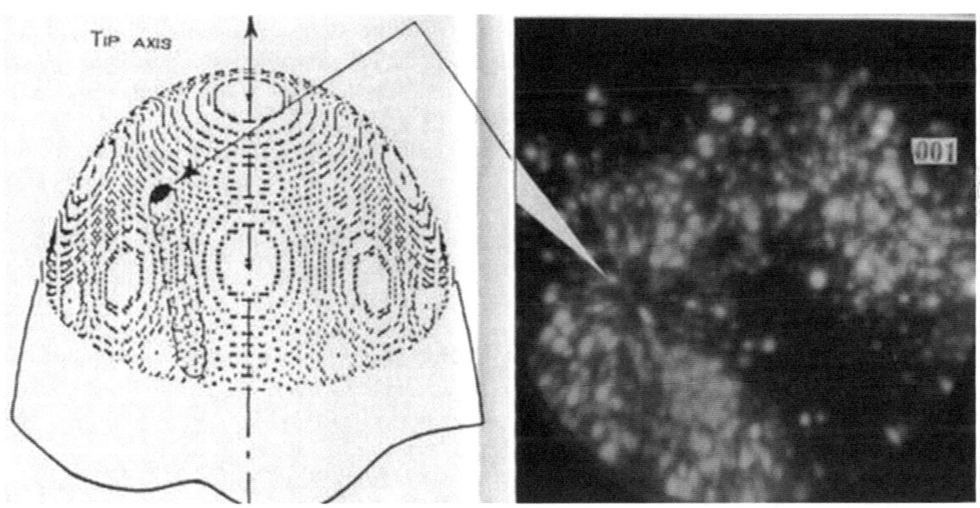

Figure 5. Schematic representation of the spatial disposition of the nanopores in the subsurface volume of the Pd(CuAg) specimen fnd typical ion contrast of nanopores in Pd(CuAg) after irradiation.

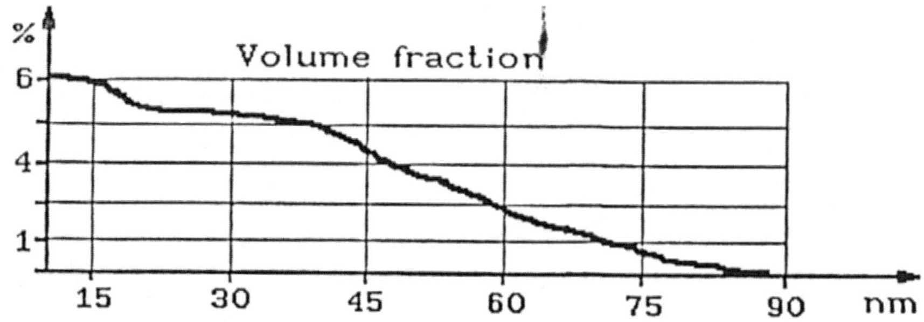

Figure 6. The variation of the volume fraction of nanopores with the depth of an irradiated Pd(CuAg) specimen in the subsurface layers (E = 25 keV, j = 340 μA/cm^2, F = 10^{18} ions/cm^2).

4. Conclusion

Thus, an experimental study of nanopore formation both on the surface and in the subsurface metal materials bulk within a nanometer range from the irradiated surface was carried out on the atomic scale.

Ion implantation modes at which nanopore formation started were determined. For pure metals (Pt), nanopores appeared at fluences of 10^{17} –10^{18} ion/cm^2 in the energy range of 20–30 keV, starting from j = 200 μA/cm^2. The distribution and bulk fraction of nanopores in the subsurface material bulk were studied. As a result, it was found that up to 40% of the pores were concentrated in the subsurface layer with a thickness of 10 nm. Later, the bulk fraction of nanopores decreased logarithmically.

The FIM technique has been used to detect porous swelling in the subsurface volume of Pd(CuAg) alloy irradiated with low-energy (E = 25 keV) argon ions. This swelling is due to radiation-induced precipitation break-up taking place in the solid solution. It was established that nanopores are elliptical in shape with diameters within 3 – 12 and height within 4 – 25 nm.

The obtained results can be used for prediction of radiation stability of materials based on fcc metals.

Acknowledgements

This work was performed as part of State Task No. 0389-2014-0002 and it was supported in part by RFBR Grant No. 16-08-00615.

References

[1] Guseva M I 1989 *Ion Implantation into Semiconductor Materials, in Itogi Nauki I Tekhniki. Ser. Physical Basics of Laser and Beam Technology* (VINITI, Moscow) [in Russian]
[2] *Surface Modification and Alloying by Laser, Ion and Electron Beams* 1983 Ed. by Poate J M, Foti G, and Jacobson D C (Plenum, New York)
[3] Ivchenko V A and Medvedeva E V 2009 *Izv. Vysch. Uchebn. Zaved. Fiz.* **8/2** 410
[4] Ivchenko V A, Syutkin N N 1999 *Pis'ma Zh. Tekh. Fiz.* **25** 6 410
[5] Ivchenko V A, Ovchinnikov V V, Goloborodsky B Yu and Sytkin N N 1997 *Surf. Sci.* **384** 46
[6] Ivchenko V A, Syutkin N N, and Kuznetsova L Yu 2000 *Pis'ma Zh. Tekh. Fiz.* **26** 13 5
[7] Burenkov A F, Komarov F F, Kumakhov M A, and Temkin M M 1985 *Spatial Distributions of Energy Evolved in a Cascade of Atomic Collisions in Solids* (Energoatomizdat, Moscow) [in Russian]
[8] Ivchenko V A, Syutkin N N 1995 *Surf. Sci.* **87/88** 257
[9] Boowkett K M and Smith D A 1970 *Field Ion Microscopy Defects in Crystalline Solids* **2** (Amsterdam, London: North Holland Publishing Company)

About mechanisms of radiation-induced effect of nanostructurization of near-surface volumes of metals

V A Ivchenko
Institute of Electrophysics, Ural Branch, Russian Academy of Sciences, ul. Amundsena 106, Ekaterinburg, 620016, Russia
Yeltsin Ural Federal University, Yekaterinburg, 620002 Russia

E-mail: ivchenko2008@mail.ru

Abstract. Mechanisms of the radiation-induced development of nanostructures in subsurface metal regions have been analyzed based on field-ion microscopy data. It is concluded that the modification of near-surface metal regions on a nanometer scale as a result of the interaction with Ar^+ ion beams proceeds by several mechanisms. In particular, for a fluence of $F = 10^{16}$ ion/cm^2 (at an ion energy of $E = 30$ keV), the main contribution is due to the ion channeling. A tenfold increase in the ion fluence leads to prevailing deformation mechanism in nanostructure formation in the subsurface metal regions.

1. Introduction

As is known, the interaction of accelerated ions with substances leads to the formation of special condensed states in irradiated targets and, hence, unique strength and other physical properties of ion-modified materials. Experimental studies of changes in the structure of metals and alloys upon exposure [1-4] showed that ion irradiation initiates the formation of amorphous, nano-, and submicrocrystalline structures in the subsurface regions of targets. These significant changes in the structural state can lead to improvement of the physical (in particular, mechanical) properties of substances as compared to those achieved by traditional processing methods.

The results of our previous investigations [5, 6] of the surface structure of metals, carried out using the field-ion microscopy (FIM) method, showed that the irradiation of metal targets by charged particles (argon ions) with moderate energies (up to 40 keV) led to nanostructurization of a near-surface region of pure metals. Using the FIM technique, it is possible to study the subsurface layers of irradiated materials by controlled removal of atoms from the sample surface and analyze its structure in the course of layer-by-layer field evaporation of atoms. Sequential imaging of the atomically clean surface of a sample at cryogenic temperatures provides quantitative information on the positions of individual atoms and atomic groups and their changes in as a result of the implantation of positive argon ions. Thus, it is possible to study the atomic structure of radiation-induced defects in the crystalline lattice, determine their distribution in the volume, estimate the thickness of a modified subsurface layer, etc.

This work presents an attempt to analyze the effect of ion implantation on the atomic structure of platinum depending on the variable parameters (including the ion energy, ion current density, and dose) of irradiation with a beams of accelerated argon ions. The aim was to elucidate the mechanisms of formation of nanostructural states in the ion-modified subsurface region of a target. Investigation of the laws of these phenomena is necessary for developing methods of controlled modification of the physical properties of materials, which is an important area in advanced nanotechnologies. In addition, the interaction of ions with substances provides a basis for methods of diagnostics of the initial materials and changes to them caused by ion irradiation.

2. Experimental

The targets for irradiation were made of platinum (99.99% Pt) and tungsten. The samples had the shape of point emitters with a tip curvature radius of 30-50 nm, which were made from metal wire blanks by electropolishing. Then, the samples were attested in a 1 field ion microscope to obtain images of Pt in the initial state. The attested Pt points were irradiated by a beam of 30-keV Ar^+ ions at a beam current density of j = 150 μA/cm^2 (T = 70°C) or j = 200 μA/cm^2 (T = 200°C) to a total fluence of $F = 10^{16}$–10^{17} ion/cm^2. The 2 ion beam was incident in the direction parallel to the axis of the point. The ion-implanted points were again placed into the field ion microscope, and sequential FIM images were recorded (by photo or video camera) during controlled removal of atomic layers. This experimental material was used for analysis of the defect structure of point samples. The field ion micro- scope was equipped with a micro channel ion-electron converter that enhanced the brightness of FIM patterns by 10^4 times. The cooling agent was liquid nitrogen (T = 78 K), and the imaging gas was neon of spectral purity grade.

3. Results and discussion

The initial (FIM-attested) samples prior to irradiation had an atomically smooth surface of the emitter tip with a nearly hemispherical shape. This surface was obtained in situ due to field evaporation of the surface atoms. Ion images of the field emitters exhibited an almost perfect ring pattern characteristic of pure metal single crystals, which was indicative of the absence of structural defects (Figure 1a).

Figure 1b presents a typical neon image of the atomically clean surface of pure platinum irradiated with Ar^+ ions at E = 30 keV to a total fluence of $F = 10^{16}$ ion/cm^2. The field-ion micrographs exhibited distortions in the contrast of ring patterns of crystal face images. These distortions in the ring pattern of ion contrast revealed the appearance of defects in the perfect structure and determined a particular contrast due to defects of different types induced in the emitter by external actions. In the given case, changes in the ion contrast of irradiated platinum as compared to that of the initial (FIM-attested) metal, observed in a 1.5-nm-thick layer under the irradiated surface, are indicative of the presence of a block nano dimensional structure in the subsurface region.

Results of quantitative analysis of the distribution of block sizes in the modified subsurface region (with a volume of $V \approx 250$ nm^3) of platinum irradiated to a fluence of 10^{16} ions/cm^2 have been reported previously [4]. The volume fractions of nanoblocks of various sizes were determined using the Rosiwal linear method based on the Cavalieri-Acker principle.

Analysis of the ion contrast of atomic arrangement in nanoblocks (Figure 1b) showed that atoms occurred virtually in their crystal lattice sites, while blocks were misoriented relative to each other. Taking into account that the target used in experiments represented a platinum single crystal (a field ion microscope point with a tip radius of 30-50 nm is almost always single-crystalline), it can be suggested that the main mechanism of nanostructural state formation at a fluence of 10^{16} ions/cm^2 is related to the phenomenon of channeling [7].

As a result of the irradiation to a higher fluence of ~10^{17} ions/cm^2 (Figure 1c), the formation of a block nanocrystalline structure (with block sizes within 1-5 nm) is observed in a near-surface layer with a thickness of no less than 20 nm under the irradiated surface.

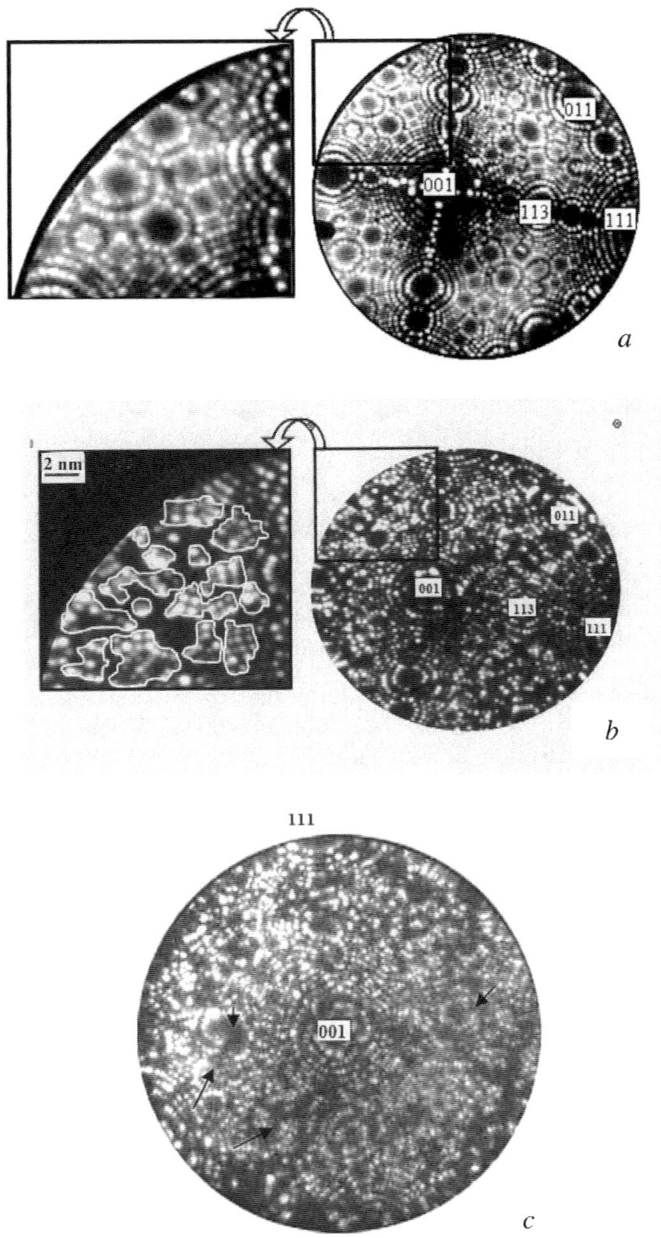

Figure 1. Neon FIM mages of Pt points (a) in the initial (attested) state, (b) upon irradiation with 30-keV Ar^+ ions to $F = 10^{16}$ ion/cm^2 ($T = 70°$ C), and (c) upon irradiation with 30-keV Ar^+ ions to $F = 10^{17}$ ion/cm^2 ($T = 70°$ C), arrows indicate the typical ion contrast of nanoparticle boundaries and stacking faults.

Analysis of the corresponding experimental data allowed the transverse and longitudinal dimensions of nanocrystalline blocks (Figure 2) and the widths of boundaries between blocks to be determined.

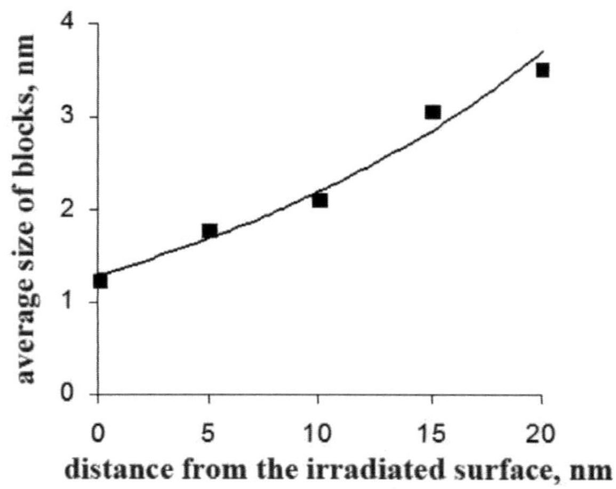

Figure 2. Variation of the average block size in depth of the transverse cross section of a sample irradiation with 30-keV Ar^+ ions to $F = 10^{17}$ ions/cm^2.

According to the obtained estimations, the width of these boundaries varied within 0.4-0.8 nm in various regions of nanodimensional blocks in ion-irradiated platinum points. The field-ion contrast observed on the irradiated platinum surface reveals a pattern that is typical of the grain boundaries and stacking faults [8]. This pattern is observed on FIM images of almost all faces of nanocrystals (Figure 2). This result implies that, at $F = 10^{17}$ ions/cm^2, the mechanism of the nanoblock structure formation in the body of the material changes.

It was shown previously [9, 10] that analogous nanocrystalline block structures could also be formed as a result of intense plastic deformation. Based on these experimental data, it can be suggested that the nanoblock structure under consideration, which appears at the stage of irradiation to $F = 10^{17}$ ions/cm^2 is formed as a result of deformation processes that develop in the material in the course of ion irradiation and during the subsequent period of relaxation. This hypothesis of the formation of nanostructural subsurface layer is probably confirmed by the experimental results of Didenko et al [11]. It was ascertained [11] that a surface layer modified by ion implantation is characterized by the generation of point defects, dislocation loops, and dislocations. In addition, dislocation substructures (DSSs) can form in this subsurface zone. According to the classification of DSSs [12], the ion irradiation can also lead to nanostructural states as a result of complicated evolution of dislocations.

Therefore, another possible mechanism of the formation of a nanostructural state in a subsurface layer of platinum irradiated to $F = 10^{17}$ ions/cm^2 is based upon the deformation model.

Thus, the effect of ion implantation on the crystal- line structure of platinum was experimentally studied on atomic spatial scale for variable parameters (including the ion energy, ion current density, and dose) of ion irradiation with a beam of accelerated argon ions. The obtained FIM data were used to analyze the possible mechanisms of nanostructural state formation in a subsurface layer within nanometer range.

It was established that modification of the crystalline lattice in a subsurface layer of ion-irradiated platinum depended on the regime of metal point irradiation with 30-keV Ar^+ ions. At a fluence of $F = 10^{16}$ ion/cm^2, the observed nanostructurization of the subsurface layer is most probably related to the phenomenon of channeling.

A tenfold increase in the ion fluence led to a change in the mechanism of nanostructural state formation in the irradiated region. The deformation character of the ion contrast observed on the metal surface during controlled sequential removal of atomic layers, followed by an analysis of the state of the near-surface layer using the experimental FIM data, led to the conclusion that the deformation mechanism plays a dominant role in the formation of nanostructure in the subsurface layer. It can also be expected that the formation of a nanostructural state in a near-surface layer of ion-irradiated materials can lead to a significant increase in their physical (in particular, mechanical) properties.

Acknowledgments
This work was performed as part of State Task No. 0389-2014-0002, was supported in part by RFBR Grant No. 15-08-06744.

References

[1] Guseva M I and Gordeeva G V 1986 *Phys. Status Solidi A* **95** 385
[2] Ovchinnikov V V 1996 *Izv. Ross. Akad. Nauk: Metally* **6** 104
[3] *Surface Modification and Alloying by Laser, Ion and Electron Beams* 1983 (Ed. by Poate J M, Foti G and Jacobson D C; New York: Plenum)
[4] Ivchenko V A and Medvedeva E V 2009 *Izv. Vysch. Uchebn. Zaved. Fiz.* **8/2** 410
[5] Ivchenko V A and Medvedeva E V 2009 *Perspekt. Mater* **7** 119
[6] Ivchenko V A and Medvedeva E V 2010 *Bull. Russ. Acad. Sci.: Phys.* **74** 2 217
[7] Nastasi M, Mayer J W and Hirvonen J K 1996 *Ion-Solid Interactions: Fundamentals and Applications* Cambridge Solid State Science Series, **27** (Cambridge: Cambridge University Press)
[8] Boowkett K M and Smith D A 1970 *Field Ion Microscopy Defects in Crystalline Solids* (Amsterdam, London: North Holland Publishing Company)
[9] Ivchenko V A and Syutkin N N 1999 *Pis'ma Zh. Tekh. Fiz.* **25** 6 410
[10] Ivchenko V A, Efros B M, Popova E V, Efros N B and Loladze L V 2003 Fiz. *Tekh. Vys. Davlen.* **13** 3 109
[11] Didenko A N, Sharkeev Yu R, Kozlov E V and Ryabchikov I 2004 *Long-Range Interaction Effects in Ion-Implanted Metal Materials* (Tomsk: Izdat. NTL)
[12] Koneva N A, Kozlov E V and Trishkina L I 1991 *Metallofizika* **12** 149

IOP Conf. Series: Materials Science and Engineering **168** (2017) 012011 doi:10.1088/1757-899X/168/1/012011

Laser Initiation of PETN containing Nickel Inclusions

B P Aduev, A A Zvekov, D R Nurmukhametov and A P Nikitin

Federal Research Center of Coal and Coal Chemistry
650000, 18 Sovetsky av., Kemerovo, Russia. Type the author addresses here

lesinko-iuxm@yandex.ru

Abstract. The spectral and kinetic characteristics of pentaerythritol tetranitrate (PETN) containing nickel nanoparticles glow initiated by laser pulses was studied with high temporal resolution. It was shown that glow which is chemiluminescence arises as a result of chemical reaction initiation. We suggest that the glow is concerned on excited nitrogen dioxide NO_2 luminescence. The reaction propagation leads to the explosion in the microsecond time range that is accompanied by thermal glow of the reaction products with temperature T=4300 K.

1. Introduction

Utilization of composite materials based on pentaerythritol tetranitrate (PETN) containing inclusions of metal nanoparticles in low concentrations allows one to diminish drastically the critical energy density of laser initiation of explosion [1-4]. This gives a principal way to application of such materials as explosives in detonators' cups initiated with laser irradiation.

The dependencies of critical energy density of these explosives initiation with YAG-Nd^{3+} laser (λ=1064 nm) operating in the Q-switch mode on the inclusions' concentration with one typical size and on the size of Ni inclusion with definite concentration were studied in our works [3,5]. The minimal critical energy density corresponding to 50% explosion probability H_{cr}=0.7 J/cm^2 was obtained for the samples containing 0.1 %wt nickel inclusions with average radius 160 nm.

It was experimentally and theoretically shown in [6,7] that the laser radiation is absorbed by inclusions. That leads to their heating up to evaluated temperatures. As many researchers believe the consequence is hot points formation where the self-sustaining chemical reaction propagates making the entire sample explode.

In order to give an insight into the primal chemical reactions, it is essential to apply the methods allowing one to observe the appearing products of PETN molecules decomposition with high temporal resolution starting from the laser pulse.

The present paper is devoted to the spectral and kinetic study of PETN polycrystals containing ultrafine Ni particles glow initiated with neodymium laser pulses.

2. Experimental section

The experimental samples were prepared from PETN powder with narrow size distribution. The typical grains size at the distribution maximum was 1 – 2 μm. The dopant was the Ni powder with the size 160 nm at the distribution maximum. The Ni powder was synthetized by Roman Kolmykov at Kemerovo State University. The EDX analysis with spectrometer JEOL JED 2400 was used to estimate the nickel oxide content in the metal powder. Thus the nickel oxide weight percent was determined as 12±1%. The procedure of powder mixing and samples pressing was thoroughly described in [8].

The samples were pressed in the hole with diameter 3 mm at the center of a copper plate with thickness 1 mm using a hydraulic press. This way the samples with typical density 1.7 g/cm^3 were obtained.

We used a YAG:Nd^{3+} laser produced by Lotis that was operated in the Q-switch mode as an initiation source. The pulse duration fwhm was 14 ns, wavelength was 1064 nm and the energy density at the sample surface was 2 J/cm^2.

The main elements of registration systems were a polychromator and a streak camera based on an image intensifier tube (IIT). The sample's glow arising during its irradiation with a laser pulse and explosion was directed to the spatial-temporal slit of the polychromator. The slit width was 0.1×0.2 mm determined the spectral and temporal resolution of the registering system. The spectral resolution achieved was 10 ns. The temporal resolution depends on the streak-camera scanning parameters. The polychromator decomposes the glow in the spectral range 350-750 nm. The spectral band width 0.2 mm falls onto the photocathode of the streak-camera. The streak-camera do the temporal scan of the spectrum. The glow on the output screen of the streak-camera is a matrix that vertical elements allows one to obtain the spectral dependencies of samples glow intensity at different time moments while horizontal ones give kinetics of glow intensity at a definite wavelength. This light matrix is read from the IIT screen with a PSD-matrix and transferred to a computer where the image numerical processing is carried out.

The spectral sensitivity of the optical registration system was calibrated with the standard lamp TRSh 2850-3000. The standard lamp has a definite color temperature with known emission coefficient thus its spectrum is readily calculable with the Plank equation. The calibration included the lamp spectrum registration and ration of measured and calculated intensity determination at the measurement conditions.

The spectrum restoration was performed by division of measured spectrum of the glow intensity with the sensitivity values at every wavelength at a definite time moment or at one wavelength at different time moments.

3. Results and discussion

The explosion glow intensity kinetics that is presented on the fig. 1 is almost identical in entire studied spectral range. The measurement with streak-camera using higher temporal resolution showed that the glow intensity rising in the first peak area (fig. 1) ceased after the laser pulse ending. Then one observes the intensity decreasing almost to background noise level during about 1 µs with following flash that takes 1.5 – 2.5 µs. The last flash of explosion glow is bimodal typically (fig. 1).

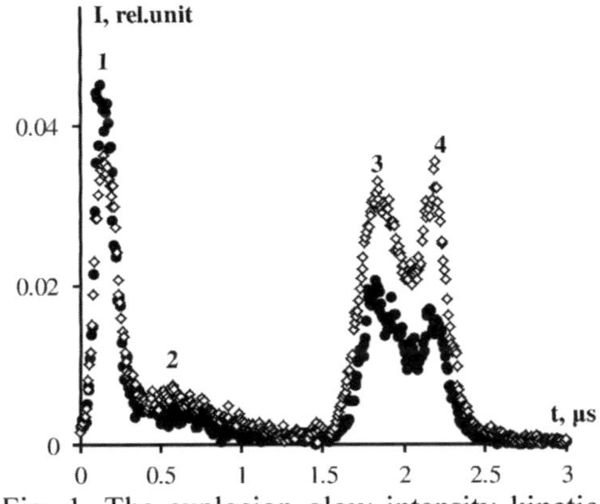

Fig. 1. The explosion glow intensity kinetics of PETN+Ni. 1 – λ = 420 nm; 2 – λ = 600 nm.

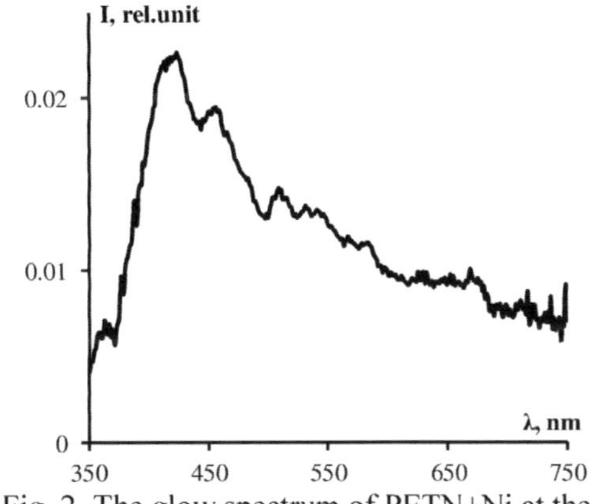

Fig. 2. The glow spectrum of PETN+Ni at the end of the laser pulse.

The glow spectrum at the end of the laser pulse is shown on the fig. 2. A broad band with vibrational structure and maximum glow intensity at the wavelength 420 nm is observed on the spectrum. This glow could be associated with laser induced fluorescence, but it would demand the three-photon absorption at least, which is a highly improbable process. Moreover, at the laser pulse intensity used we have not notice any glow when the samples without inclusions were irradiated.

As it was shown in [6, 7], the laser irradiation is absorbed by metal inclusions mostly causing their heating. The estimated temperatures are higher than typical melting points of metals. At the same time a PETN layer is heated to the inclusion's temperature. The layer with thickness could be estimated as

$$d=(\chi\cdot\tau_i)^{1/2} \quad (1)$$

where χ is thermal diffusivity of PETN, τ_i is pulse duration. Thus the PETN layer with thickness d=40 nm is heated to the temperatures above its melting point (T_m=414.4 K) or even PETN evaporation point on the pulse front. The high temperatures at the hot-spot makes the PETN molecules decompose in the heated layer surrounding the inclusion. The forming small molecules and radicals react chemically with each other and the reaction propagates. The measured glow spectrum shown on the fig. 2 could be assigned to the emission of am excited PETN fragment. We assume that this fragment is nitrogen dioxide NO_2.

The reaction development on the consequent stages causes the deep fragmentation of intermediates, heating of the reaction products and reaction propagation over entire volume of the sample ending with explosion. The spectra measured at the time moments 2, 3, and 4 (the time one may see on the fig. 1) are presented on the fig. 3. It is seen that at t values about 0.5 μs the glow arises from the hot-spots still. At t > 1.5 the reaction propagates over entire sample that causes the explosion.

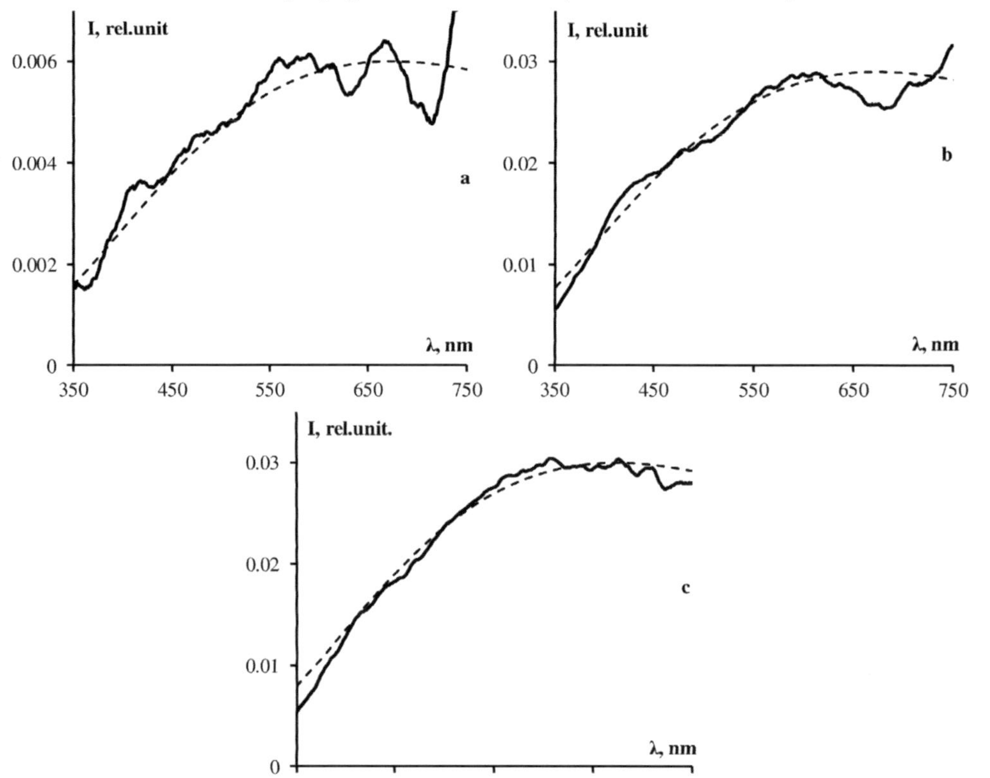

Fig. 3. The glow spectrum of PETN+Ni at . a – at the time moment 2 on the fig. 1; b - at the time moment 3 on the fig. 1; c – at the time moment 4 on the fig. 1.

The spectra shown on the fig. 3 are described satisfactory in terms of Plank formula at the temperature 4300 K. The approximation results are shown as dotted lines on the fig. 3. Thus on the reaction propagation stage the thermal glow is predominant. We interpret this result as an argument in favour of thermal explosion mechanism.

References

[1] Ioffe V B, Dolgolaptev A V, Aleksandrov V E, Obraztsov A P 1985 *Combustion, Explosion and Shock Waves* **21** 316

[2] Aleksandrov V E, Dolgolaptev A V, Ioffe V B, Koval'chuk V M, Levin B V, Obraztsov A P 1983 *Combustion, Explosion and Shock Waves* **19** 384

[3] Aduev B P, Nurmukhametov D R, Zvekov A A, Liskov I Yu 2016 *IOP Conference Series: Materials Science and Engineering* **110** 012010

[4] Aduev B P, Belokurov G M, Nurmukhametov D R, Nelyubina N V 2012 *Combustion, Explosion, and Shock Waves* **48** 361

[5] Aduev B P, Nurmukhametov D R, Furega R I, Zvekov A A 2014 *Russian Journal of Physical Chemistry B* **8** 352

[6] Aduev B P, Nurmukhametov D R, Zvekov A A, Nikitin A P, Nelyubina N V, Belokurov G M, Kalenskii A V 2015 *Instruments and Experimental Techniques* **58** 765

[7] Aduev B P, Anan'eva M V, Zvekov A A, Kalenskii A V, Kriger V G, Nikitin A P 2014 *Combustion, Explosion, and Shock Waves* **50** 704

[8] Aduev B P, Nurmukhametov D R, Tsipilev V P, Furega R I 2013 *Combustion, Explosion, and Shock Waves* **49** 215

Destruction of humic substances by pulsed electrical discharge

G L Lobanova[1], T A Yurmazova[2], L N Shiyan[2], K I Machekhina[2*] and M A Davidenko[2]

[1]Laboratory 12, Institute of High Technology Physics, Tomsk Polytechnic University, 634050, Tomsk, Russia.

[2]Department of General and Inorganic Chemistry, Institute of High Technology Physics, Tomsk Polytechnic University, 634050, Tomsk, Russia.

*E-mail: mauthksu@yandex.ru

Abstract. Currently, the water recourses in the territory of Tomsk region are groundwater which is limited to the high concentration of iron and manganese ions and organic substances. These impurities present in water in different forms such as soluble salts ant the colloid forms. Therefore, the present work is a part of a continuations researcher of the processes in natural waters containing humic substances at the influence of pulsed electrical discharges in a layer of iron pellets. It is shown that the main stage of water purification process of humic substances during treatment by pulsed electric discharge in the layer of iron granules is a difficult process including several stages such as formation of iron oxyhydroxide colloid particles, sorption and coagulation with humic macromolecules substances, growth of particle dispersed phase and precipitation. The reason for the formation and coagulation of the dispersed phase is a different state of charge of the colloid particles (zeta potentials of (Fe (OH)3) is +8 mV, zeta potentials of (Humic substances) is -70 mV.The most intense permanganate oxidation reduction to the maximum permissible concentration occurs at the processing time equal to 10 seconds. The contact time of active erosion products with sodium humate is established and it equals to 1 hour. The value of permanganate oxidation achieves maximum permissible concentration during this time and iron concentration in solution achieves maximum permissible concentration after filtration.

1. Introduction

The formation of stable colloid systems occurs in groundwater containing dissolved humic substances, silicates and heavy metal ions. In the papers [1-3] the formation of colloid solution is investigated. It is shown that the stable colloid solution forms at the molar ratio of iron ions, the silicates and humic substances equal 1: 7: 2, respectively.

Currently, there are various water treatment technologies to clean groundwater such as aeration, ozonation, treatment of pulsed electric discharge, membrane filtration [4-6]. Along with that [7-9] the existing methods do not allow to reduce the humic substances concentration by more than 30-40% of the initial concentration (20 mg/L) and the humic substances concentration exceeds the maximum permissible concentration after treatment. .However, the removal of iron from natural waters is a complex scientific and technical problem, even when the concentration of humic substances does not exceed the limits of standard values. Therefore, the development of technology with integrated physical and chemical effects of the water are required. In view of this the use of spark electric discharges in combination with electro metal erosion in water purification technology is promising due to redox reactions, stimulation of sorption processes, destruction of organic matter, water disinfection.

The present work is a part of a continuations study of the physical and chemical processes complex in natural waters containing humic substances at the influence of pulsed electrical discharges in a layer of iron pellets.

The objective of the work is to investigate the influence of phenomena initiated by the discharge channels and secondary redox processes on the stability of humic substances in groundwater.

2. Experimental approach

Object of the research is the groundwater containing humic substances from the village Vertikos Tomsk region. The experimental work was carried out in an accredited laboratory of Radiation spectroscopy in Tomsk polytechnic university using attorney equipment. For the experiments concerning the effect of the discharge on the stability of humic substances, the installation was used according to the paper [10]. This installation was constructed in the 12th laboratory of High Technologies Physics Institute in TPU.

Russian standard methods were used to determine the chemical composition of natural waters in accordance with the sanitary standard 2.1.4.1074-01 and GOST 51232-98.

The organic substances concentration was determined by the analysis of the chemical oxygen demand (COD), permanganate oxidation (PO) and the total organic carbon (TOC) content.

The iron and silicon concentrations in the solution were determined by means of ICP-OES plasma optical emission spectrometer (Varian, USA). The concentration of iron (II) was determined according technique State Russia Standard 4011-72 with o-phenanthroline. The pH value was measured by means of a WTW Multiline P4 multifunctional device (WTW GmbH, Germany).

The colloid particle size distribution and the ζ-potential in the model solutions were determined using a Zetasizer Nano ZS analyzer (Malvern Instruments, UK), which allows measuring the particle size in the range 0.6…6000 nm. The particle size was determined by dynamic light dispersion (at a dispersion angle of 173°). The measurement result was in the form of numerical particle distribution s $\varphi(r)=dN/dr$. The mode δ_m was found from the maximum of this distribution and was taken as an average particle size because the distribution functions approximated normal log functions.

The chemical composition of the surface formed by electro metal granules in water and aqueous solutions was studied using infrared diffuse reflectance spectroscopy FTIR using a spectrometer Nicolet 5700 (USA).

3. Results and discussion

Indicators of the quality of groundwater for different wells Vertikos village are shown in Table 1. Table 1 demonstrates that the main indicators exceeding the maximum permissible concentration values are color, iron, manganese and permanganate oxidation.

Table 1. Quality indicators of groundwater v. Vertikos Tomsk region.

Determined parameters	The groundwater of v. Vertikos	
	№ 1	№ 2
pH, ед. pH	6.9	6.9
Color, grad	39	31
Turbidity, mg/L	29.3	27.8
Hardness, ^{0}H	4.8	4.6
PO, mgO/L	7.2	7.3
HCO_3^-, mg/L	329.5	328.5
Iron, mg/L	6.3	5.9
Manganese, mg/L	0.55	0.53
Silicon, mg/L	10.8	11.1

Processing groundwater in a pulsed electric discharge (PED) v. Vertikos brought to change parameters such as pH, permanganate oxidation the total iron concentration (Fig. 1).

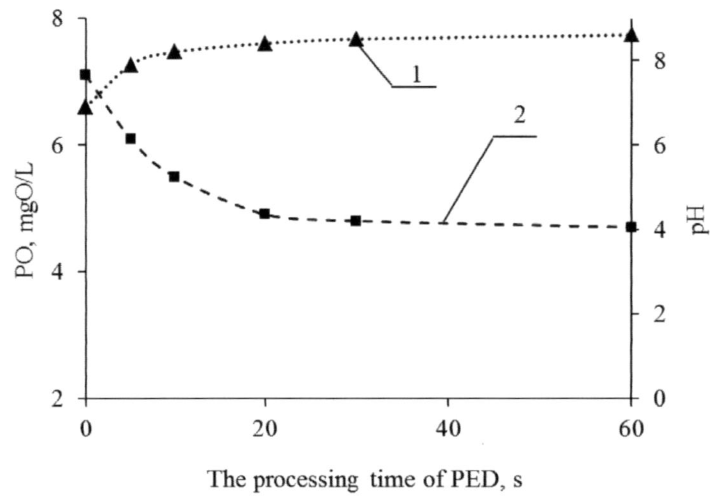

Figure 1. Dependence of the processing time of PEDon pH (1) and value PO (2) (holding time before analysis - 1 hour)

It is seen in Fig. 1 that pH of the solution increases due to the interaction erosion products with water and decomposition of water by the action of the pulse electric discharge on iron pellets. The reduction of permanganate oxidation (PO) indicates that the impurities oxidize by the action of discharge on the water.

The formation and growth of colloidal particles occurs under the influence of electrostatic interaction forces. Value of ζ-potential (mV) for particle erosion products has a positive value and it equals +8 mV. After the interaction of iron compounds hydrolysis products having a positive ζ-potential with the humic substances (ζ-potential = -70 mV) occurs recharge of surface erosion products and value ζ-potential becomes negative and equals -27 mV.

The qualitative composition of deposits obtained after coagulation of erosion products with humic substances from groundwater n. Vertikos determined by IR - spectra presented is shown in the Fig. 2 and Table 2. The IR–specters of sodium humate and erosion products present on the fig. 2 for comparison.

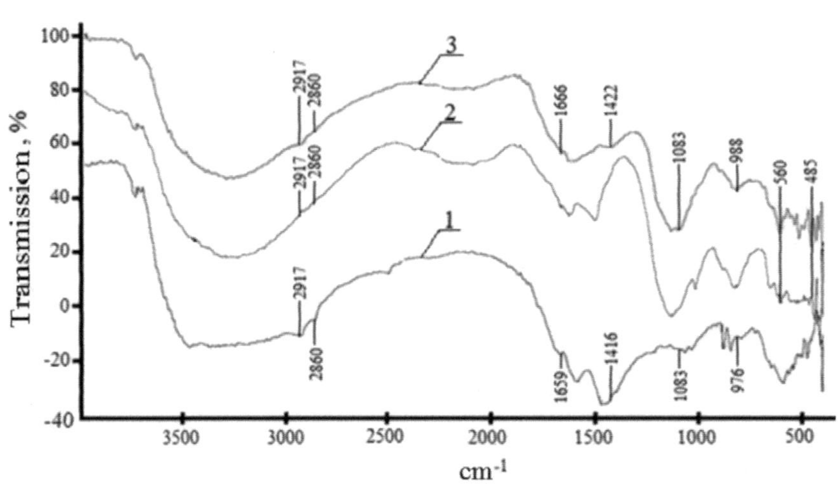

Figure 2. IR-spectra: 1 – sodium humate; 2 – erosion products; 3 – sodium humate with erosion products.

Table 2. Characteristic vibrational frequencies bands in the IR-specters.

Humic substances	Sodium humate	Erosion products of iron pellets	Sodium humate, besieged on erosion products	Functional groups
2920	2917	2917	2917	–OH
2860	2867	2867	2853	–OH
1650	1659	отс.	1666	–C–H
1400	1416	отс.	1422	–C–O
1063	1083	отс.	1083	COO^-, HOH, $^+NH_3$
980	976	отс.	988	C_6H_5–

The analysis of IR–spectra is shown that there are three spectrum bands: 560 cm^{-1}, attributable to stretching vibrations and Fe_2O_3 strip 485 cm^{-1} attributable to the stretching vibrations FeOOH. There are 3 bands of stretching vibrations 1420 cm^{-1} related to the symmetric stretching vibrations of carboxylate ions (C-O), characterized by the presence of humic substances in the sediment, and stretching vibrations at a frequency of 1650 cm^{-1}, is characterized by fluctuations due (C-H) in their aromatic fragments at the same curve. The band 1083 cm^{-1} refers to fluctuations -COO- bonds, HOH, $^+NH_3$, and the band 988 cm^{-1} characterize vibrations of the aromatic ring fragments.

Ground water from v Vertikos was treated by pulsed electric discharge in a layer of iron pellets and it had been settled during 1 hour. The upper layer was analyzed for total iron and the total organic carbon (Fig. 3).

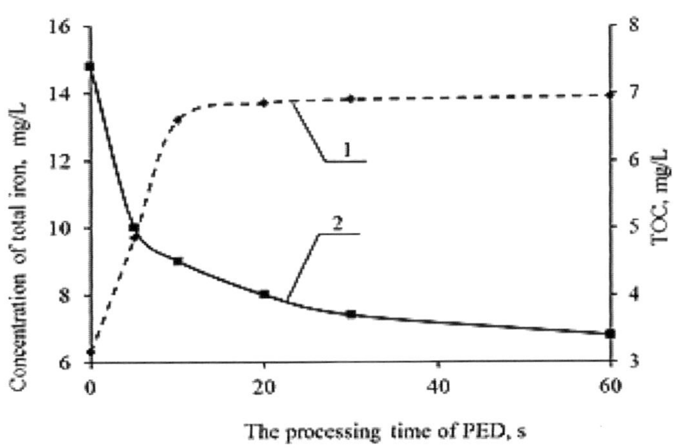

Figure 3. Dependence of the processing time on total iron concentration (1) and TOC (2)

Figure 3 demonstrates that the concentration of humic substances decreases according to the processing time od PED , whereas the iron ions (II) concentration in solution corresponds to 14 mg/L. Iron remaining after interaction with the humic substances as hydroxides can be removed by filtration. Table 3 shows that the main indicators of water quality after all treatment steps do not exceed the maximum permissible concentration for drinking water.

The two stage of process have been established based on the results obtained by the action of a pulsed electrical discharge in a layer of iron pellets in groundwater. The first stage - is electro-erosion stage and destruction of organic substances which takes place at the discharge action is short time (τ = 10 c), the second stage is a long and it occurs after discharge shutdown (τ = 1 hour).

Table 3. Quality indicators of groundwater v. Vertikos Tomsk region before and after treatment

Parameter	Units	Ground water v. Vertikos				MPC
		source of water	PED (10 s)	Contact (1 h)	Filtration	
pH	units pH	6.9	7.5	7.5	7.5	6–9
Color	grad	39	7	5	5	20
Turbidity	mg/L	29.3	56.3	23.5	1.0	1.5
hardness	⁰Ж	4.8	4.8	4.8	4.8	7.0
PO	mgO/L	7.2	5.2	1.0	1.0	5.0
Iron		6.3	35.1	14.0	0.2	0.3
Manganese	mg/L	0.55	0.55	0.1	0.05	0.1
Silicon		10.8	10.7	6.9	6.9	10.0

4. Conclusion

1. It is found that the limiting stage water purification process of humic substances during spark treatment in the layer of iron granules is a complex of processes including formation of iron oxyhydroxide colloid particles, sorption and coagulation together with humic macromolecules substances, growth of particle dispersed phase and precipitation. The reason for the formation and coagulation of the dispersed phase is a different state of charge of the colloid particles (zeta potentials ζ (Fe $(OH)_3$) = +8 mV, ζ (Humic substances) = -70 mV.
2. It is shown that the most intense permanganate oxidation reduction to the maximum permissible concentration occurs at the processing time equal to 10 seconds.
3. It is shown the role of secondary processes. The contact time of active erosion products with sodium humate is established and it equals to 1 hour. The value of permanganate oxidation achieves maximum permissible concentration during this time and iron concentration in solution achieves maximum permissible concentration after filtration.

Acknowledgements
The study was supported by National Research Tomsk Polytechnic University, the project No. 7.1504.2015.

References

[1] Shiyan L N, Tropina E A, Machekhina K I, Gryaznova E N, An V V 2014 *SpringerPlus* DOI: 10.1186/2193-1801-3-260, URL: http://www.springerplus.com/content/3/1/260
[2] Machekhina K I, Shiyan L N 2014 *Advanced Materials Research* **1040** 342-346
[3] Voyno D A, Machekhina K I, Shiyan L N 2014 *Advanced Materials Research* **971-973** 266-269
[4] Serikov L V, Tropina E A, Shiyan L N, Frimmel F H, Metreveli G, Delay M 2009 *Journal for Soils and Sediments* **9** 103-110
[5] Svetleishaya E M, Mitchenko T E, Astrelin I M 2014 *Journal of Water Chemistry and Technology* **36** 25-30
[6] Kornev J, Yavorovsky N, Preis S, Khaskelberg M, Isaev U, Chen B-N 2006 *Ozone: Sci. and Engineering* **28** 207-215
[7] Zherlitsyn A G, Shiyan V P, Shiyan L N, Magomadova S O 2015 *Journal of Physics: Conference Series* **652**
[8] Machekhina K I, Titova E N, Kornev Y I, Shiyan L N, Preys S, Magomadova S O 2016 *Key Engineering Materials* **685** 648-652
[9] Kornev I, Preis S, Gryaznova E, Saprykin F, Khryapov P, Khaskelberg M, Yavorovskiy N 2014 *Industrial and Engineering Chem. Res.* **53** 7263-7267
[10] Danilenko N B, Savelev G G, Yavorovsky N A Yyrmazova T A *2008 Russian Journal of Applied Chemistry* **81** 803-810

Radioprotective materials with tungsten nanopowder additives

V Gavrish, G Baranov, T Chayka and N Derbasova
Laboratory of biotechnologies and environmental monitoring, Sevastopol State University, 33 Universitetskaya str., 299053, Sevastopol, Russian Federation

E-mail: 5brachman5@gmail.com, infbox2001@mail.ru, KimChayka@yandex.ru

Abstract. There's been studied influence of submicron powder additives obtained by processing hardmetal waste TTK (TiC-WC-TaC-Co), on strength properties of cement. This modified cement is used as a structural material for containers at transportation and storage of radioactive waste.

1. Introduction

Safe handling of radioactive waste includes secure insulation which gives safety both to environment and human health during the whole period of radiation hazard. Safe handling is achieved by a system of physical barriers on the way of ionizing radiation and radioactive substance in the environment. One of these barriers is package for radioactive waste [1].

Design and constructional material of a container being an element of the packaging must ensure its structural integrity and operability while handling the package for radioactive waste [2].

In recent years various concrete containers have been used for this purpose. These containers have become an alternative to metal containers widely used previously. Concrete is famous for its good radioprotective properties and resistance to external factors [3].

But over time, changes occurring during storage of the waste have an influence on their migration properties and defense capability of physical barriers [4]. Therefore in the present moment increase in strength and therefore durability of concrete materials used for containers' production in compliance with necessary technical regulations remains a topical problem, especially in radioactive waste handling. This problem could be successfully solved by using various additives at the manufacturing stage.

One of the promising directions in improving properties of construction materials based on concrete to produce radioactive waste storage containers is nanomodifiers which allow increasing significantly strength properties of the material. However, now their large-scale application in the construction industry is minimal. This is due to the cost of nanopowder and its low volume of production that doesn't meet the industry demand.

This paper continues with the papers [5-9]. Previous studies have shown that production of nanopowder by the proposed method can significantly reduce its cost, as well as to increase production to quantity required for the industry due to modularity of the technological cycle.

The aim of this paper is to study ability to modify concrete by submicron size powder.

To succeed in it we've tried to make the following:

– to analyze elemental and granulometric composition of the powder being an additive to concrete;

– to study strength properties of concrete modified with the powder;

– to determine an optimal concentration of modifiers for concrete.

2. Experimental part

Concrete is a complex composite material consisting of sand, coarse aggregate (crushed stone and other materials), cement and water [10]. Cement paste (cement and water) fills all the voids in the sand, covering its particles with a thin layer. Cement-sand mortar fills in its turn all the voids in the coarse aggregate, which forms a rigid spatial structure inside the product.

Strength of concrete depends in many ways on strength and quality of a coating solution. Therefore, this paper deals with changes of ultimate strength in modified concrete without coarse aggregate added.

To study changes in concrete strength properties we applied materials widely used in construction (table 1).

Table 1. Source materials.

Material	Brand	Specifications and technical documentation
Portland-cement	PC II/A–SH–400	Specification BV.2.7–46-96
Bank sand	1,9 – 2,2 mm fraction	State Standard 8736–93

A mixture of titanium carbide, tungsten carbide and tantalum carbide, obtained by waste processing, was used as a concrete modifier.

Qualitative and quantitative composition analysis of the powder being a modifier was made using ElvaX Light spectrometer (see table 2).

Analysis of granulometric composition of modifiers with the laser particle-size analyzer "Analisette 22 NanoTec" is shown in figure 1.

The samples obtained were subjected to destruction by the standard procedure using UM mashine – 5.

Table 2. The powder – modifier's elemental composition.

At. number	Element	Content, %
42	Mo	0.036
26	Fe	0.071
20	Ca	0.211
41	Nb	1.587
14	Si	2.009
27	Co	4.560
73	Ta	5.076
22	Ti	7.821
74	W	78.629

It is found that percentage of the particles depending on size is as follows (figure 1): up to 500 nm - about 30 %, 500 nm –1 micrometer – 70%. This calculation model shows a ratio of length to diameter of the particles in the range of 1,441 to 2,495.

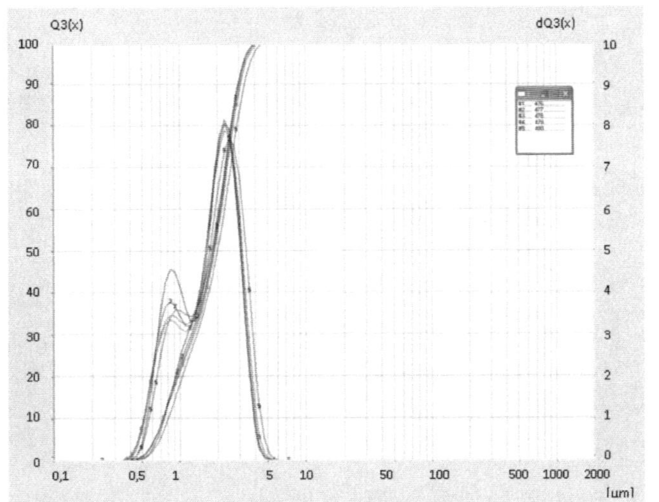

Figure 1. Distribution of powder particles depending on size

The sample of sanded concrete was a mixture of portland-cement PC II / A–SH–400 (table 1) and sifted river sand in 1:3 and 1:2 ratio. The powder was used as a modifier in various concentrations (1, 3, 5, 7, 9, 10 wt.%). The resulting mixture was stirred in a tumbling drum for 15 minutes, and then water was gradually added while stirring until the required solution consistency was reached. The resulting cement-sand mortar was poured into a mold (18x18x16 mm) and allowed to cure at the temperature of 20–25 ° C for 28 days.

The obtained samples were subjected to destruction using standard procedures. The results are presented in figure 2.

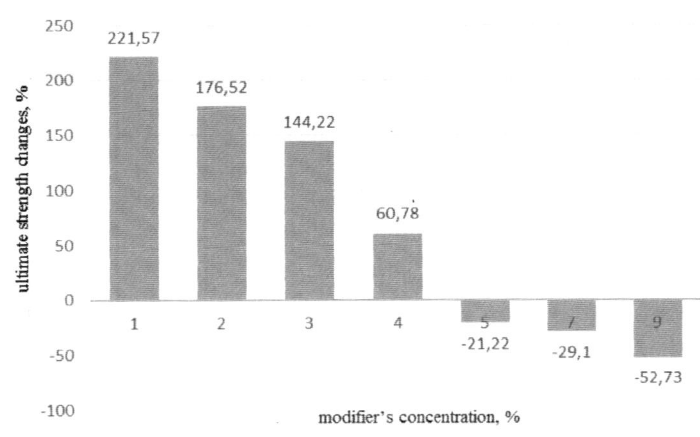

Figure 2. Changes of ultimate strength in sanded concrete samples at different concentrations of powder (1:2 cement/sand ratio).

As it is known, increasing cement fraction in the solution results in increased strength. In fact, modifier's influence on cement matrix strength (figure 2) is most pronounced at low ratio of cement to sand (1:3). The optimal powder concentration is 1% that increases material's ultimate strength by 222%.

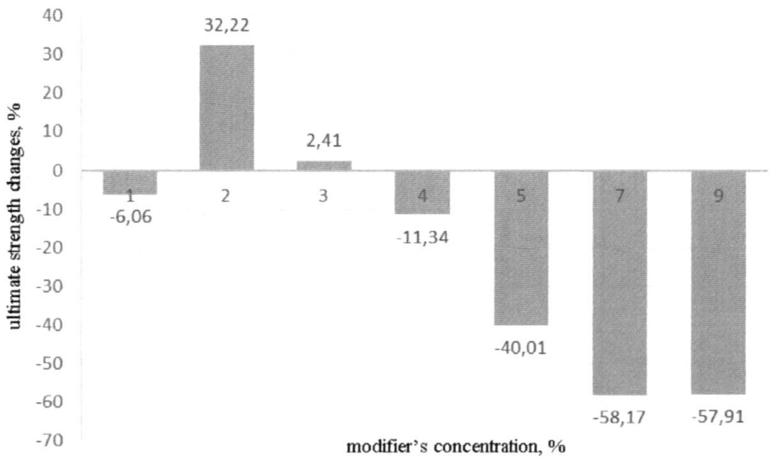

Figure 3. Changes of ultimate strength in sanded concrete samples at different concentrations of powder (1:3 cement/sand ratio)

The more the cement fraction the less influence of the modifier (figure 3). The optimal cement fraction is 2 % that gives an increase in ultimate strength by 32 %.

In our opinion, increase in strength properties of sanded concrete by adding modifiers of submicron size is related to formation of microcracks and residual microporosity. Powder additives form structure types which are certain traps for developing cracks under load. This assumption requires further research.

3.Conclusion

There's been analyzed elemental and granulometric composition of the powder used as modifiers for sanded concrete.

There's been proven modifier's influence on increase of strength properties of sanded concrete. The optimal concentration of modifiers is found to be 1 %, with increase in strength in more than 2 times.

The observed effects can be used in creation of new structural materials for radioprotective material production (storage containers or radioactive waste transporting equipment).

In the future we expect to study the obtained samples for increase of their radioprotective properties and durability.

References

[1] Disposal of Radioactive Waste. Principles, Criteria and General Safety Requirements. **NP-055-14**. Approved by Order of the Federal Environmental, Industrial and Nuclear Supervision Service No. 379, dated of August 22, 2014. Registered in the Ministry of Justice of the Russian Federation, No. 35819, dated of February 02, 2015. Entered into force on February 14, 2015

[2] Gataullin R M, Medelyaev I A and Sharafutdinov R B 2008 The use of advanced technologies for solving problems of safe radioactive waste management *Nuclear and radiation safety* **4** 68-75

[3] Sorokin V T, Demin A V, Kascheev VV and Iroshnikov V V 2013 Containers for radioactive waste low and medium level *Nuclear and radiation safety* **2(68)** 1-8

[4] Pronkin N S, Sharafutdinov RB and Geraskin N I *2011 Safety regulation of radioactive waste management* (Moscow: National Research Nuclear University Moscow Engineering Physics Institute Press) p 264

[5] Baranov G A, Gavrish M V and Smirnov S B 2011 Obtaining nanosized powders in the processing of waste on the basis of tungsten-containing alloys, study of their granulometric composition *J. of Mechanical Engineering the NTUU "KPI"* **63** 42 – 46

[6] Burya A I, Gubskaya M I and Baranov G A 2011 The study of the influence of the superdispersed tungsten on the properties of composites based on the ultra-high molecular weight polyethylene. *J. East Ukrainian National University named after Dahl.* **8 (162)** 56 – 61

[7] Baranov G A, Gavrish M V, Smirnov S B and Derbasova N M 2011 Processing of tungsten-containing waste to produce nano-sized powders, investigation of their granulometric composition Collection of scientific works of international scientific-tech. *conf.* (Kharkov UkrVODGEO) 432 – 442

[8] Gavrish V M, Mikhailov V V, Shapovalov V A, Chayka T V, Baranov G A and Derbasova N M 2013 Possible applications of nanopowders for the modification of lubricants, polyethylene and alloys. *Collection of studies, SNUNEI* **3 (47)** 49 – 55

[9] Chayka T V, Gavrish V M, Baranov G A, Mikhailov V V and Derbasova N M **2013** Research of possibility of improvement of lubricating materials with nanomodifiers obtained by microbiological method. *Collection of scientific works SWorld* **4 (15)** C. 3 – 9

[10] Vasilchenko V N, Zhigalov Ya A, Nosovskiy A V and Sandul G A 2013 Protective barriers in nuclear power: some scenarios to degradation of the system "container + radioactive waste", *Nuclear and radiation safety* **4(60)** 26-32

IOP Conf. Series: Materials Science and Engineering **168** (2017) 012014 doi:10.1088/1757-899X/168/1/012014

Dynamic model of the threshold displacement energy

A I Kupchishin[1,2] and A A Kupchishin[1]
[1] Abay Kazakh National Pedagogical University, 13 Dostyk, Almaty, Kazakhstan
[2] Al-Farabi Kazakh National University, 71 al-Farabi ave., Almaty, Kazakhstan

E-mail: ankupchishin@mail.ru

Abstract. A dynamic (cascade-probability) model for calculating the threshold displacement energy of knocked-out atoms (E_d) was proposed taking into account the influence of the instability zone (spontaneous recombination). General expression was recorded for Ed depending on the formation energy of interstitial atoms E_f and vacancies E_i, on the energy transfer coefficient α and the number of interactions i needed to move the atom out of the instability zone. The parameters of primary particles were calculated. Comparison of calculations with experimental data gives a satisfactory agreement.

1. Introduction

As it is known, the interaction of the neutral and charged particles (electrons, ions, gamma rays, neutrons, etc.) with materials forms the radiative defects of vacancy and interstitial types of various sizes and configurations [1 – 4]. This significantly changes the structure and properties of the substance [5 – 10].

For calculation of the concentration of radiative defects in the materials the minimum energy required for the formation of a displaced atom (from the node-crystalline lattice), or so-called primary knocked-out atom (PKA) should be known. In radiation physics, this energy is called the threshold displacement energy and expressed as E_d or T_d. The value of E_d was qualitatively assessed for the first time by Seitz based on the following considerations. The energy needed for separation of the atom or ion (the sublimation energy) from the solid surface in common solid crystal $E_c \sim 5$ eV. Sublimation (evaporation) takes place on the surface of the crystal, where it is enough for separation to overcome only the half of interatomic bonds compared with an atom located inside the crystal. The binding energy of the inner atom therefore is approximately equal to 10 eV. If the atom moves slowly from the lattice node along the path of least resistance to some interstitial position, thereby allowing the neighboring atoms to return to the original position or if there is an adiabatic motion, this process requires energy, which is approximately equal to $2E_c$. In fact, knocked-out atom receives a sharp blow and enters the interstitial position almost unable to return. Generally, the atom moves not along the line of least resistance, and the neighboring atoms do not have time to relax. A reasonable estimation for the value of consumed energy (according to [1, 2]) is ~ 4 – 5 E_c, i.e. 20 – 25 eV, and, therefore, E_d is equal to 25 eV. It should be noted that this estimation is quite crude, since it does not account the real picture of the formation of interstitial atoms and vacancies (Frenkel couples). It is clear that such non-adiabatic formation of the displaced atom must depend both on the PKA energy E_2 and the material type and the direction of particles emission. The higher the PKA energy, the faster it moves, and the better such motions are. At the same time, the movement time should be much less than the atoms relaxation time. If these times are comparable, the interstitial atom formation process is close to adiabatic and the liminal formation energy must be less than the above mentioned $4E_c$ and is close to the E_c. This leads to the fact that we have a distribution on the threshold energy, i.e. in the energy range of 0 ~ 200 eV the formation of stable Frenkel couple (vacancy + interstitial atom) is less than one.

We have proposed a dynamic (cascade-probability) model of calculation of the threshold displacement energy.

2. Fundamentals of the model

The average energy of the atom displacement can be obtained from the athermal approach more strictly than in [1, 2]. Since the instability zone (spontaneous recombination) exists around the vacancy [3], kicked-out atom is necessary to undergo several collision (i) to get beyond it. In this case, basically, the atom will lose energy by elastic collisions with surrounding atoms. In the approximation of solid bodies, energy spectrum $\omega(E_2)$ of the secondly kicked-out atoms (from PKA) is described by uniform distribution ($\omega(E_2)dE_2 = AdE_2$). Moreover, the energy transfer coefficient α (i.e., the fraction of energy transmitted by primary atom to secondary) can be determined by the average energy of kicked-out atoms:

$$\overline{E_2} = \frac{\int_0^{E_{2\max}} E_2 \omega(E_2) dE_2}{\int_0^{E_{2\max}} \omega(E_2) dE_2} = \frac{E_{2\max}}{2}. \quad (1)$$

Here the good approximation gives $\alpha = 0.5$.

The average collisions number $\bar{i}$ can be determined using cascade-probability function $\psi_i(h)$:

$$\bar{i} = \sum i \cdot \psi_i(h), \quad (2)$$

where [11]

$$\psi_i(h) = \frac{1}{i!}\left(\frac{h}{\lambda}\right)^i \exp\left(-\frac{h}{\lambda}\right).$$

Here h is the depth of observation, λ is interaction path, $\lambda = 1/\sigma n$, σ is total cross section, n – the number of atoms per cubic centimeter.

Since

$$\sum_{i=0}^{\infty} \psi_i(h) = 1, \quad \bar{i} = h/\lambda. \quad (3)$$

x, y, z and r coordinates may be considered instead of h:

It follows that the i value depends on the motion direction of the atom in the instability zone. Typical view of such zone is shown in Figure 1.

In general

$$\bar{i} = \frac{1}{N}\sum_1^N i(x, y, z). \quad (4)$$

Where N is the number of points, i (x, y, z) is the i values in x, y, z points on the outer surface of spontaneous recombination zone. Ultimately, the value of E_d is determined as follows:

$$E_d = \frac{E_V + E_f}{(1-\alpha)^i}. \tag{5}$$

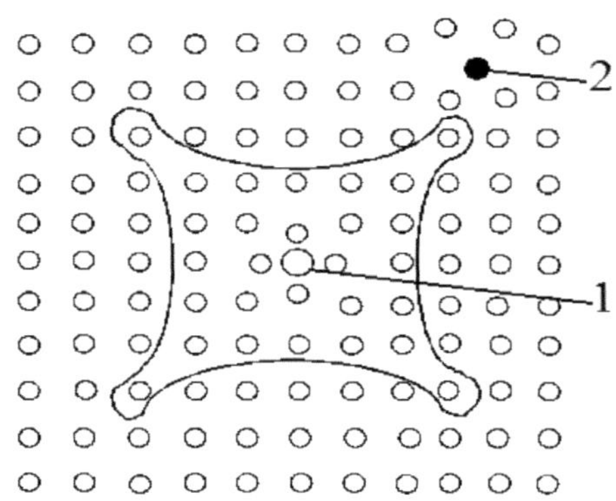

1 – vacancy; 2 – interstitial atom

Figure 1. The instability zone for knocked-out atom (flat picture).

3. Results and discussion

Regarding (1)-(5) and $E_f + E_V = 5$ eV and $\alpha = 0.5$ the values of i and E_d are to be calculated. The results are given in Table 1. It can be seen that the experimental data are in agree with calculations

Table 1. Threshold energy of the atoms and the incoming particles, and the average number of interactions.

Threshold energy	Al	Ti	Cu	Mo	Ag	Au
E_d',(eV) [12], experiment	32	29	22	37	28	35
E_d, (eV, calculated)	30.3	28.3	2.4	37.3	26.4	32.5
E_{th}^e of the electron, (MeV)	0.3	0.44	0.44	0.87	0.78	1.365
E_{th}^p of the proton, (eV)	231	360	360	900	760	1730
E_{th}^α of the α-particle, (eV)	71	100	100	240	200	450
$\bar{i}$	2.6	2.5	2.1	2.9	2.4	2.7

Thus, in our view, it is possible to introduce the following definition: "The threshold displacement energy E_d is the minimum energy that should be given to an atom of the material so that it is displaced from the lattice site and get out of the instability zone after i interactions, forming a stable Frenkel pair".

Radiation physics is characterized by the existence of the minimum energy of damaging particles, in which the defect formation is possible. In this regard, let's consider the elementary act of interaction of primary particles with atoms, leading to the formation of PKA.

For the test process, the threshold kinetic energy E_{th} of the incoming particles has been calculated. Let's consider the process of elastic collision of particles possessing kinetic energy E_1, with fixed atom. If the primary particle is unrelativistic and heavy, the maximum kinetic energy E_{th}^i, that it should have for knocking out an atom is equal to:

$$E_{th}^i = (m_1 + m_2)^2 / 4m_1 \cdot m_2 \cdot E_d. \tag{6}$$

Where m_1 and m_2 are the mass of the incoming particle (e.g. ion) and fixed atom, respectively. Or effective separation of an atom from the crystal lattice the inequality should be saved: $E_{2max} \geq E_d$ [3, 11].

During the electron-atom collisions $m_1 << m_2$. For example, for E_d = 20 eV [5], $E_1 \geq 0.1$ MeV, which indicates the inapplicability of the non-relativistic mechanics for electrons [12]. Consequently, for the electron irradiation in the problems related to the formation of displaced atoms, relativistic kinematics should be used so the maximal value of E^e_{th}, which should have a primary particle-conductive identified the following equation:

$$E^e_{th} = \frac{\sqrt{4m_1^2c^4 + 2\mathrm{E}_d \cdot m_2c^2} - 2m_1c^2}{2}. \qquad (7)$$

Table 2 shows the results of E_{th} calculations for different elements in electron, proton and alpha radiation. As the table shows, with increasing atomic number the value of E_{th} is generally increased.

However, there are deviations from this relationship, for example, molybdenum, which is associated with the difference in the threshold displacement energy for this item

4. Conclusions

1. With the use of cascade-probability method the dynamic (cascade-probability) model developed for calculating the threshold displacement energy of atoms a stable Frenkel couples.
2. The energy of transfer rate, the number of interactions, the threshold displacement energy E_d and the maximum energy for electrons, protons, alpha particles, reaching the used in defects in aluminum, titanium, copper, molybdenum, silver and gold was calculated. The available experimental data are satisfactorily described the proposed model.

Acknowledgments
The study was supported by the Grant of the Ministry of Education and Science of the Republic of Kazakhstan (a.№ 274, 229).

References

[1] Dins D, Vinyard J, 1960 *Radiation effects in solids* (M.: IL)
[2] Kelly B, 1970 *Radiation damage of solids* (M.: Atomizdat)
[3] Vinyard J, 1961 *Achievements in Phys. Sciences* The dynamics of radiation damage **74** 3 435 – 459
[4] Kupchishin A I, Kupchishin A A, Voronova N A, Kirdyashkin V I, Gyngazov V A 2016 *IOP Conf. Series: Material Science and Engineering* **110** 012039
[5] Thompson M V, 1971 *Defects and radiation damage in metals* (M.: Mir)
[6] Krasikov E, Nikolaenko V, 2016 *IOP Conf. Series: Materials Science and Engineering* **110** 012001
[7] Surzhikov A P, Galtseva O V, Vasendina E A, Vlasov V A, Nikolaev E V 2016 *IOP Conf. Series: Materials Science and Engineering* **110** 012002
[8] Ivchenko V A 2016 *IOP Conf. Series: Materials Science and Engineering* **110** 012003
[9] Surzhikov A P, Chernyavskii A V, Ghyngazov S A, Frangulyan T S 2002 *Russian Physics Journal* **45** 12 1190-94
[10] Malyshev A V, Lysenko E N, Vlasov V A and Nikolaeva S A 2016 *Ceramics International* **42** 16180-16183
[11] Boos E G, Kupchishin A I 1988 *Solution of physical problems and cascaded probabilistic method* (Alma-Ata: Nauka) **1 2**
[12] Orlov A N, Trushin A N 1983 *Energy Point defects Ying Metals* (M: Energoatomisdat)

IOP Conf. Series: Materials Science and Engineering **168** (2017) 012015 doi:10.1088/1757-899X/168/1/012015

Study of the influence of filler on the mechanical properties of composites based on polyimide

A I Kupchishin[1,2], B G Taipova[1] and N A Voronova[1]
[1]Abay Kazakh National Pedagogical University, 13 Dostyk, Almaty, Kazakhstan
[2] Al-Farabi Kazakh National University, 71 al-Farabi ave., Almaty, Kazakhstan

E-mail: ankupchishin@mail.ru

Abstract. Physical and mechanical properties of polymer composite films on the basis of polyimide filled with high-temperature superconductors (yttrium ceramics) were studied. Calculations under the catastrophic models were carried out. Exponential and parabolic models show the best agreement with the experiment.

1. Introduction

Creation of polymer ceramic and composite materials with improved or novel properties and characteristics is an essential prerequisite and condition for further development of scientific and technological progress. The constant growth of requirements for modern technology leads to the appointment of the polymer, its nature, operation conditions and other factors [1, 2]. Advances in the field of superconductivity are key to energy, electronics, high-energy physics, air, land and sea transport, space exploration, medicine and many other areas [3]. Increased applications of superconducting materials and increase of the efficiency of their work require further improvement of performance, in particular mechanical properties [4]. Products of flexible HTS (high temperature superconductors) are of practical interest. Since the ceramic HTS materials have a high brittleness and low strength, important materials to be studied are film systems of polymeric composite materials (PCM) on the basis of their compounds having structural perfection polycrystallites and plastic properties of polymers [5]. Preparation of YBCO-films with desired properties can be carried out when understanding features of the formation of the cationic composition of the films during their growth [6]. Establishing a connection between the structure and properties of defects makes it possible to obtain information about the microscopic parameters that are important for understanding the mechanism of the processes in these materials.

The synthesis of new materials is based on the variation, integration and combination of objects of different nature, in particular, organized molecular structures, as well as on the optimization of their spatial organization [7]. Directional surface modification of HTS materials lies in the change of the chemical composition of sample surface by modifying the electrolyte composition and properties, adsorption of various surfactant particles, etc. [8-10].

To date, a huge amount of experimental data requiring theoretical interpretation has been collected, but existing models, including those of mechanical properties can not explain some of these data.

2. Methodology

Physical and mechanical properties of polymer composite films of polyimide base filled with HTSn of two concentrations were studied: C = 0.1 % wt., 0.5 % wt. Fine-grained powder of $YBa_2Cu_3O_{6,7}$

was used as filler. Test samples were prepared by mechanical mixing of the polyimide-based varnish and fine crystalline powder of $YBa_2Cu_3O_{6+x}$, when x = 0.7. Mechanical tests were conducted on the tensile machine of C-TAP 2.0 type with uniaxial tension of special reverser at the constant load. The load had been applied until the film was disrupted at the unclamping rate of 1.53 mm/min and with increasing of load for 0.06 dynes/min. The work area was 40-mm long and 5-mm wide. The experimental results were processed using a special computer software that is installed as a standard Windows application [11].

3. Discussion

Figure 1 shows the dependence of the relative elongation ε on the applied stress σ unirradiated composite material for two values of the concentration of the second component. Introduction of the filler in the form of fine crystalline yttrium ceramic powder in polyimide causes a significant change in the plastic and strength properties of the polymer. Thus, when changing the filler content from C = 0.1 wt.% to C = 0.5 wt.%, a decrease of tensile strength (σ) from ~ 20 MPa to about 2 MPa in comparison with pure polyimide film is observed, and deformations of these materials are significantly reduced from Δε = 41% to Δε = 43%, respectively. In the region of 40 - 47 MPa, for samples of PCM there is a sharp jump in the relative elongation, shifted to lower stresses with increasing filler concentration (YBCO). Increasing the concentration leads to a stabilization of the plastic properties of PCM and deterioration of its elastic properties (σ reduced from 90 MPa to 72 MPa). This effect of filler concentration on the mechanical properties of PCM is due to a change of the porosity of material and structural activity of the filler related to polymer matrix. Change in the plastic properties of PCM samples is due to a high content of O_2 in YBCO crystals [11].

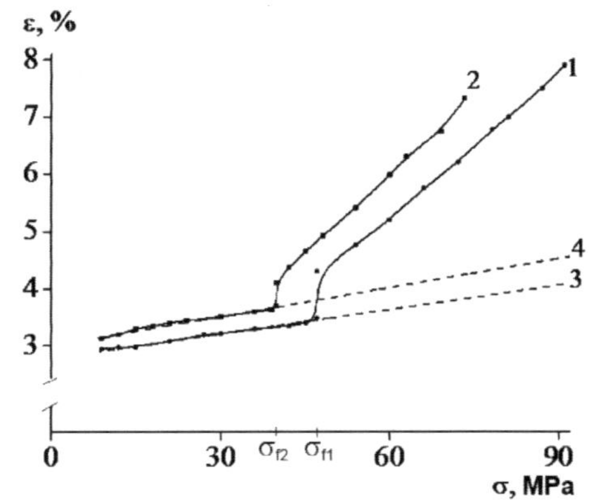

1 – C = 0.1 wt.%; 2 – C = 0.5 wt.%
solid line is the experiment, dashed line is calculation by the linear model

Figure 1. Dependence of elongation ε on the applied load σ for polyimide composite materials for two concentrations of the second component.

In the stress interval of 0 – σ_{f1} (curve 3) and 0 – σ_{f2} (curve 4) dependences of ε from σ are satisfactorily described by a linear model.

Establishment of the dependence of ε from σ (instead of σ from ε, as is generally accepted) is the most correct from the physical point of view [11].

The formula for the reduced strain ε_p on σ was used to describe the dependence of ε from σ.

In this case $$\varepsilon_p = \begin{cases} f(\sigma) - \varepsilon_0, when 0 \le \sigma \le \sigma_f, \\ f(\sigma) - \varepsilon_0 - \Delta\varepsilon_f, when \sigma > \sigma_f. \end{cases} \quad (1)$$

here σ_f is equal to σ_{f1}, or σ_{f2} depending on the curve (1 or 2), ε_0 is the shifting value, $\Delta\varepsilon_f$ is changing of deformation during the phase transition.

Let's consider three types of catastrophic models: exponential, quadratic exponential and parabolic [12]:

$$f(\varepsilon) = \exp\left(\frac{\sigma}{\sigma_0}\right) - 1, \tag{2}$$

$$f(\varepsilon) = \exp\left(\frac{\sigma^2}{E_1^2}\right) - 1, \tag{3}$$

$$f(\varepsilon) = \frac{\sigma^2}{E_2^2}. \tag{4}$$

In (2) σ_0 is the value of stress when ε_p increases *e* time.

The results of calculations of ε_p from σ for different models are shown in Figure 2 (a, b). Analysis of dependencies of ε_p from σ shows that the experimental data are described best by parabolic and exponential models. It should be noted that for the concentration of 0.1 wt% σ_{f1} is 47 MPa, $\varepsilon_0 = 2.78$ and $\Delta\varepsilon_{f1} = 0.82\%$ and for 0.5 wt.% σ_{f2} is 39.8 MPa, $\varepsilon_0 = 3$ and $\Delta\varepsilon_{f2} = 0.4\%$, respectively.

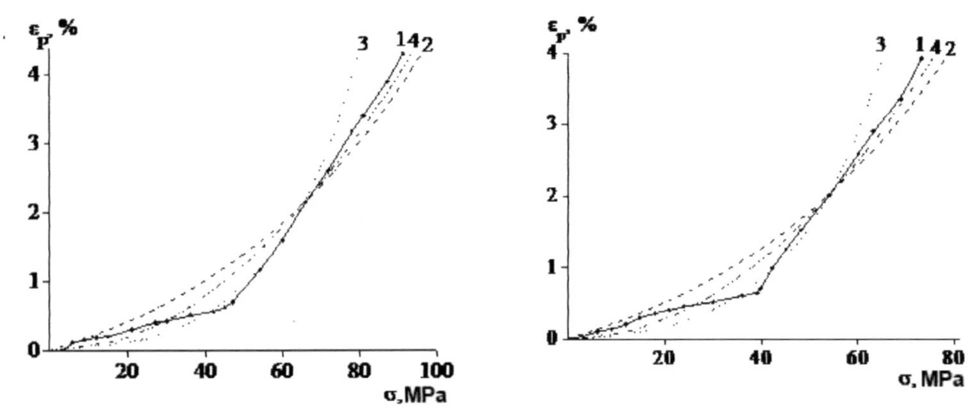

a b

a – 0.1 wt.%; b – 0.5 wt.% of HTS

1 – experiment, 2 – exponential model; 3 – exponential-quadratic model; 4 – parabolic model

Figure 2. Dependence of ε_p from σ for composite material based on polyimide for various models.

At the same time the largest difference between the calculations and the experimental data is observed in the areas of stress σ_{f1} and σ_{f2}. For a quantitative explanation of this effect it is necessary to use the appropriate models of phase transitions.

4. Conclusion

1. Introduction of yttrium ceramic filler into the polyimide causes a significant change in plastic and strength properties of polymer matrix. In the region of small deformations, dependence of ε from σ is

satisfactorily described by Hooke's law. When $\sigma = \sigma_f$ there is a sharp jump. When $\sigma > \sigma_f$ the dependence is more complex.
2. The appropriate calculations were made using the exponential, exponential-quadratic and parabolic models. Parabolic and exponential models have the best agreement with the experimental data (curves shown). The largest difference is observed in the region of σ_f that is due to the influence of phase transition on $\varepsilon(\sigma)$ dependence.

Acknowledgments
The study was supported by the Grant of the Ministry of Education and Science of the Republic of Kazakhstan (a.№ 274, 229).

References

[1] Kudaykulova S K, Iskakov R M, Kravtsova V D, Umerzakova M B, Abadie M, Zhubanov B A et al. 2006 *Polymers for special purposes* (Almaty)
[2] Kabyshev A V, Konusov F V, Pavlov S K, Remnev G E 2016 *IOP Conf. Series: Material Science and Engineering* **110** 012006
[3] Chaud X, Bourgault D, Chateigner D et al. 2006 *Supercond. Sci. Technol.* **19** S590 – S600
[4] Prikhna T A, Vlassenko A V, Dub S N, Melnikov V S, Savchuk Y M, Moschil V E, Sergienko N V, Nagornyi P A, Sverdun V B, Sho Ks, Gavalek V, Litskendorf D, Habisreuter T, J Raber, Jol A 2008 *Superhard materials* **4** 3 – 24
[5] Drobyshev A S, Kupchishin A I, Muradov A D, Sarsembayev G B, Taipova B G 2008 *Proceedings of NAS RK. Ser. phys.-mat.* The effect of low temperatures on the change in the composition of "Polymer-high temperature superconductor" system Almaty **6** 3 – 6
[6] Vorobyev A K, Gaponov S V, Drozdov M N, Klyuenkov E B, Masters D V 2000 *Solid State Physics* **42** 4 589 – 594
[7] Solovyov A L, Dmitriev V M 2009 *Physics of low temperatures* **35** 3 227 – 264
[8] Kopaev Yu V, Sofronov V M 2005 *Phys. Lett. A.* **344** 297 – 302
[9] Gyngazov S A, Surzhikov A P, Frangul'yan T S, Chernyavskii A V 2002 *Russian Physics Journal* **45** 8 753-58
[10] Malyshev A V, Lysenko E N and Vlasov V A 2015 *Ceramics International* **41** 10 13671-13675
[11] Muradov A D, Sarsembayev G B 2014 *Proceedings of "Solid State Physics" International Scientific Conference* Effect of YBCO filler on the mechanical properties of the polyimide composite Astana 142 – 143
[12] Kupchishin A I, Taipova B G, Kupchishin A A, Voronova N A, Kirdyashkin V I, Fursa T V 2016 *IOP Conf. Series: Material Science and Engineering* **110** 012037

The effect of temperature, static load and electron beam irradiation on the deformation of linear polymers

A I Kupchishin[1,2], M N Niyazov[1], N A Voronova[1], V I Kirdiashkin[1] and A T Abdukhairova[1]
[1]Abay Kazakh National Pedagogical University, 13 Dostyk, Almaty, Kazakhstan
[2] Al-Farabi Kazakh National University, 71 al-Farabi ave., Almaty, Kazakhstan

E-mail: ankupchishin@mail.ru

Abstract. A comprehensive experimental study on the effect of temperature, static load and electron beam irradiation on thermomechanical properties of polytetrafluoroethylene material was conducted. Four regions having different nature were detected on the curves of dependences of deformation (ε) from temperature (t). The obtained curves of ε from t prove the presence of phase transitions in this film. Irradiation significantly affects the dependence of $\varepsilon(t)$ and $\varepsilon(\sigma)$.

1. Introduction

Extensive use of polymeric materials and polymer-based composites in different fields of science and technology is due to particular combination of properties not possessed by a significant portion of traditional materials. One of the most important characteristics is the maintenance of high mechanical strength under external actions [1 – 8]. The use of polymers in industry contributes the identification of contradictions between price and properties, as the competition for the production of quality products is quite high. After mechanics of composite materials had found wide application and rapid development, its main goal was the modification of materials and improvement of their mechanical properties. The use of polymers in demanding industrial applications, when the mechanical load and temperature vary at the same time, determines the presence of the required thermomechanical properties in them [9, 11]. Among all PTFE materials have very good properties. It has a high melting point and densely packed linear chain gives it stiffness at increased temperatures [12, 13].

In this work, experimental studies on the effect of temperature, static load and electron beam irradiation on the deformation of non-irradiated and irradiated polytetrafluoroethylene were carried out.

2. Methodology

The appropriate experimental setup have been developed and manufactured for the experiments on the combined effect of temperature, pressure and irradiation. Its scheme is shown in Figure 1. The setup consists of the following blocks: rack, base, experimental chamber, holders, clamps, sample, thermo-electronic heater (TEH), temperature sensor, statistical load, measuring sensor for sample length. The heater is a U-shaped stainless steel tube with low heat capacity, inside of which the spiral wire with high resistivity is located. TEH is separated from the frame by insulator. Needle-electronic thermometer was applied to measure the temperature (t). The dependence of t from time (τ) is shown in Figure 2. As follows from the figure when $\tau > 5$ sec, this dependence is almost linear. To measure the length of the sample a special body displacement sensor developed by ScienceCube was used. In addition, the experiments were recorded on video.

Industrial polytetrafluoroethylene with a thickness of 100 μm was selected as the test material. Film samples were cut using a special device. Length of the test material was 7 cm, the working part was 5 cm (2 cm of polymer sample were fixed in the clamps), and width was 0.5 cm. This film was subjected to uniaxial stretching at the constant statistic load.

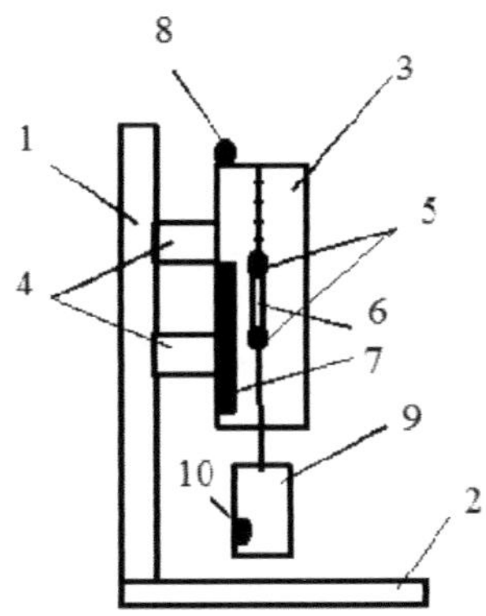

1 – rack; 2 – base; 3 – experimental chamber; 4 – holders; 5 – clamps; 6 – sample; 7 – heater; 8 – temperature sensor; 9 – statistical load; 10 – measuring sensor for sample length

Figure 1. The experimental setup.

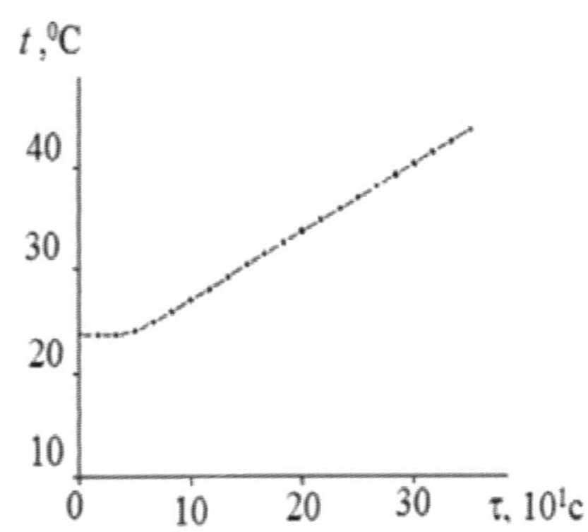

Figure 2. Dependence of temperature on time in the experimental chamber.

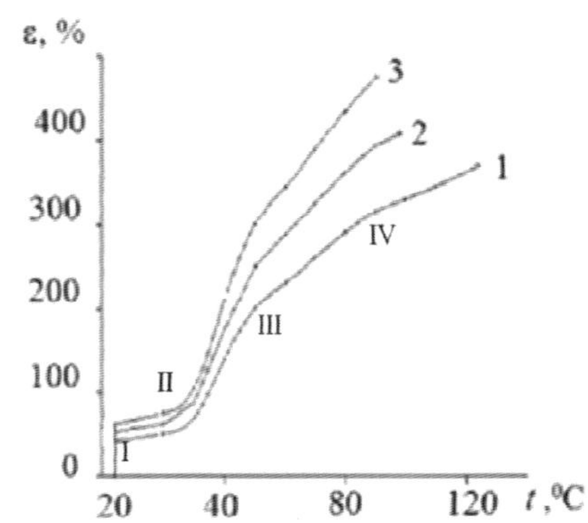

1 – σ = 9; 2 – σ = 11; 3 – σ = 13 MPa

Figure 3. Thermomechanical curves for polytetrafluoroethylene under different constant loads.

Previously the maximum load at which rupture of the tested polymer occurs had been determined using special cargo. For this experiment the stress was 21 MPa. Then the stress was calculated as a part of the limit stress. The next step was fixing the deformation at room temperature for a given load. During the experiment the load was not varied. After fixing, the sample was placed in the chamber and thermionic heater was turned on. Then all the necessary parameters were determined.

Irradiation of samples was carried out using ELU-6 linear electron accelerator with energy of 2 MeV in air. The dose was 5 kGy. Film samples were irradiated for three minutes at the distance of 40 cm from the output window of the accelerator. The height of the fixation of the samples during irradiation coincides with the center of the output window and was equal to 30 cm. Working temperature of the samples was equal to 23 °C and relative humidity was 55 %.

3. Discussion

Experimental work on study of deformation dependence on temperature at the different stresses was conducted: $\sigma_1 = 9$; $\sigma_2 = 11$ and $\sigma_3 = 13$ MPa. Tests were conducted at the temperatures ranged from 23 to 120 °C. As a result of a series of experiments (Figure 3), it was found that PTFE behaves differently at different stages of heating. Application of too much stress results in destructive processes taking place in several stages: I. Straightening of free polymer chains (23 °C); II. Straightening of bonded chains (23-35 °C); III. Catastrophic destruction (35 – 45 °C); IV. Destruction of chains with strong bonds (45 – 120 °C) (Figure 3). When load is applied at room temperature it causes a sharp elongation of the samples. The reason for this behavior of material is the straightening of macromolecules, which constitute the twisted complexes. Further, in the interval of 23 – 35 °C there is a slight increase in related elongation, which is associated with the straightening of hard bonded chains. In the interval of 35 – 45 °C there is a catastrophic destruction of the weak chains, including those bonded with the presence of phase transitions. At t > 40 °C ε(t) curve has a trend of a gradual recovery in saturation. This is due to the disruption of the hard, more durable chains. The maximum deformation at t = 85 °C and σ = 13 MPa reaches 500%.

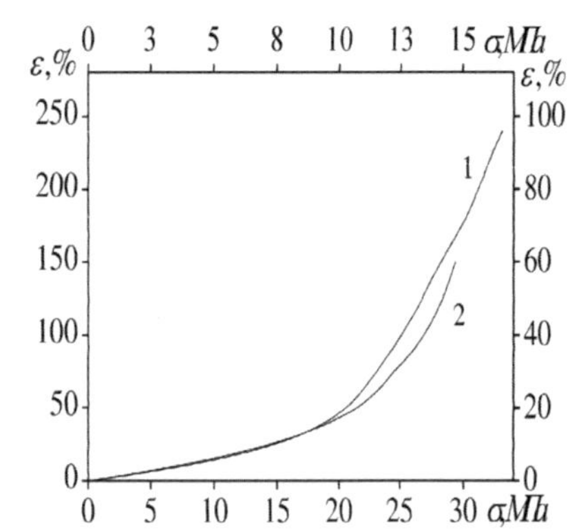

Figure 4. Dependence of deformation from stress for unirradiated (left and bottom axis) and irradiated (right and upper axis) polytetrafluoroethylene material.

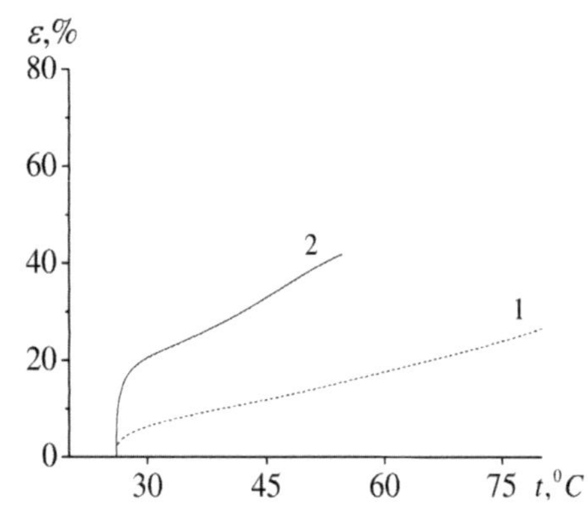

1 – σ = 5; 2– σ = 7 MPa

Figure 5. Dependence of deformation from temperature for polytetrafluoroethylene irradiated with a dose of 5 kGy under different constant loads.

Then the experimental dependence of ε from σ were determined both for non-irradiated (dots) and irradiated (crosses) polytetrafluoroethylene with a dose of 5 kGy. As it follows from the figure, with increasing stress the deformation slowly increases up to σ ~ 18 MPa and then sharply grows exponentially. It also displays the curve calculated using cascade-probabilistic models [10] of ε(σ) dependence according to the formula: $\varepsilon = \exp(\sigma/\sigma_0) - 1$, for this experiment σ_0 = 28 MPa. A satisfactory agreement of experimental and calculated data is seen. And σ_0 = 20 MPa. A similar dependence is shown for irradiated PTFE.

After irradiation by 5 kGy, samples of the material lose flexibility and begin to break at a lower deformation than before irradiation. While the decrease of elongation in 7 times is observed in comparison with unirradiated material (Figure 4), and the strength does not change significantly.

4. Conclusion

1. Experimental studies of the dependence of deformation ε from the temperature at various stresses $\sigma_1 = 9$ MPa, $\sigma_2 = 11$ MPa, $\sigma_3 = 13$ MPa in the polytetrafluoroethylene were carried out. It was found that in the temperature range of 23 – 30 °C ε varies slightly. Further, at t = 30 – 55 °C there is a significant elongation of the samples. At t > 55 °C elongation growth is slowing. The maximal deformation is ~500 % at t = 85 °C and σ = 13 MPa.
2. Irradiation of polytetrafluoroethylene samples leads to loss of plasticity, a significant decrease in deformation (in comparison with non-irradiated material) ~ 240 %, which is due to the destruction of all polymer chains. In this case, the strength is practically reduced to zero.
3. Experimental dependence of ε from σ for both unirradiated and irradiated materials is satisfactorily described by cascade-probabilistic model.

Acknowledgments
The study was supported by the Grant of the Ministry of Education and Science of the Republic of Kazakhstan (a.№ 274, 229).

References

[1] Panshin Y A, Malkevich S G, Dunaevskaya T S 1978 *Fluoropolymers* (L.: Khimiya)
[2] Ed. By LWall, transl. from English. Ed. by IL Knunyants, VA Ponomarenko 1975 *Fluoropolymers* (M.: Mir)
[3] Pikayev A K 1987 *Modern radiation chemistry. Solid and polymers. Applied aspects* (M.: Nauka)
[4] Ivchenko V A 2016 *IOP Conf. Series: Materials Science and Engineering* **110** 012003
[5] Surzhikov A P, Galtseva O V, Vasendina E A, Vlasov V A, Nikolaev E V 2016 *IOP Conf. Series: Materials Science and Engineering* **110** 012002
[6] Surzhikov A P, Frangylyan T S, Ghyngazov S A 2012 *Russian Physics Journal* **55** 4 345-52
[7] Surzikov A P, Lysenko E N, Ghyngazov S A, Frangulyan T S 2002 *Russian Physics Journal* **45** 10 989-96
[8] Lysenko E N, Surzhikov A P, Zhuravkov S P, Vlasov V A, Pustovalov A V and Yavorovskii N A *2014* J. Therm. Anal. Calor. **115** 1447–52
[9] Ed. by VK Milinchuk, VI Tupikov. 1986 *Radiation resistance of organic materials. Reference* (M.: Energoatomizdat)
[10] Kupchishin A I, Taipova B G, Kupchishin A A, Voronova N A, Kirdyashkin V I, Fursa T V 2016 *IOP Conf. Series: Material Science and Engineering* **110** 012037
[11] Kupchishin A I, Kupchishin A A, Voronova N A, Kirdyashkin V I, Gyngazov V A 2016 *IOP Conf. Series: Material Science and Engineering* **110** 012039
[12] Istomin N P, Semenov A P 1981 *Antifriction properties of composite materials based on fluoropolymers* (M.: Nauka)
[13] Mashkov Y K, Ovchar Z N, Surikov V I, Kalistratova L F 2005 *Composite materials based on polytetrafluoroethylene* (M.: Mashinostroenie)

Study of the electron irradiation effect on the structure of treatment systems biomaterials

O V Yesyrev[1], A I Kupchishin[1,2], A T Abdukhairova[1], M K Nauryzbayev[2], N N Khodarina[1] and V C Cherednichenko[1,2]
[1] Abay Kazakh National Pedagogical University, 13 Dostyk ave., Almaty, Kazakhstan
[2] Al-Farabi Kazakh National University. 71 al-Farabi ave., Almaty, Kazakhstan

E-mail: ankupchishin@mail.ru

Abstract. Experimental studies on the effects of pollution and electron irradiation on the structure of a number of treatment systems biomaterials of Sorbulak lake-storage were carried out. It was found that contamination with heavy metal and electron irradiation, respectively, affect the structure of biomaterials.

1. Introduction

Currently, there are various methods of wastewater treatment. The most commons are: mechanical, physical-chemical, chemical and biological methods. Depending on the hazard degree and nature of contamination wastewater treatment can be performed by any method or complex methods (combined method). Treatment process includes sludge treatment (or excess biomass) and disinfection of wastewater before being discharged into the pond [1 – 3]. Solution of the issues on the protection and rational use of water resources is related to the activities for the prevention of water sources pollution as a result of the dumping of industrial and domestic wastewater. In these measures, construction and improvement of sewage treatment plants are very important, as part of which the main role in terms of construction and value is assigned to biological treatment facilities. In wastewater treatment systems biological method is the final and after it is applied the waste water can be used in water recycling or discharge into surface waters. Taking into account the large volumes of domestic, industrial and rain sewage the prospect of an increasingly wide spread of biological, radiation-biological and biochemical methods becomes more clear as the most effective and cheap methods for cleaning or additional cleaning of drains. These methods provide the preservation of flora and fauna ability to use a variety of dissolved minerals and organic compounds contained in wastewater [4 – 6]. Degradation of organic compounds may take place until the complete mineralization. Biological and biochemical methods of purification of wastewater are close to the natural self-purification bioprocess of aqueous media, including using radiation. Here, the basic laws of biochemical transformation of pollutants are observed in the concentrated form [7, 8]. A dilatometric study of similar materials are presented in [9] Today, when the deterioration of environmental conditions has particularly deep resonance because of the high population density and production, bioremediation biotechnology contributes significantly to improving the environmental situation and living conditions of the people, as a reserve for environmental well-being for years to come. The biggest negative influence on the environment comes from heavy metals [10 – 12]. In this paper, experimental studies on the effect of contamination and electron irradiation on the structure of biomaterials (reed, sedge) of Sorbulak lake treatment systems.

2. Experimental

Stems of reeds and sedges uncontaminated, contaminated (with wastewater substances) and irradiated with electrons were used as a biomaterial. The expedition was organized for the collection of biological material. Uncontaminated materials were collected near clean waters, and polluted ones at the shore near the top of Sorbulak lake dam. Irradiation was carried out in air using ELU-6 linear electron accelerator. The electron energy was 2 MeV, beam current - 0.5 uA x cm^{-2}, the temperature of treated materials was 23 K, relative humidity was 55%. Irradiation of the biomaterials was controlled using the remote of high-energy electrons accelerator.

Study of the structure was carried out using an automated digital optical microscope (ACM) LeicaDM6000 M. This microscope is a precision system for research purposes with high-resolution digital cameras and software for analyzing and storing images. The device is unique and has the following features and options: 1. Automatic mode of formation and control of the light (reflection) for the experimental methods of light and dark field. 2. Automated current mode of the transmitted light. 3. Automated dimmers for regulation of change of aperture and illuminator aperture. 4. Motorized z focus gear and motorized sample platform corresponding to x, y and z displacement in the experiment. 5. X / Y panel motorized stepper motor, removable sample platform with appropriate external dimensions: 234 mm x 157 mm. 6. In this case, the range of motion is 76 mm x 50 mm, the minimum step – 0.3 μm. 7. Z panel corresponds to the removable panel: range of movement – 25 mm, minimum step: 0.015μm, maximum speed – 5 mm/s, minimum speed – 1 mm/s, maximum load – 4 kg. 8. Motorized encoded turret has a 6-lens with magnification of 5x, 10x, 20x, 50x, 100x, 150x times respectively. 9. Unique memory function allows to shift lens simultaneously and to use contrast method. Selection of different colors of the image is used (to detect small or subtle elements in contrast to the sample).

3. Results and discussion

Figure 1 a and b shows the micrographs of uncontaminated stems of reed and sedge. On the background of the uniform surface, a large number of natural micropores with the size of about 50 – 70 microns are observed in the structure of reed stem, whereas no such pores are observed in sedges. At the same time, there are various incorporatings with the sizes of from 30 to 150 microns, associated with the presence of various minerals in the soil. Structure of the contaminated materials (Figure 2) both reed and sedge changes significantly, the pores become larger. The largest changes are observed in the sedge, which is associated with the accumulation of heavy metals in it. Table 1 shows the content of heavy metals in reeds and sedges biomaterials. It is seen that reed adsorbs iron strongly, while sedge adsorbs iron, manganese and strontium. Moreover, there is an increase in the concentration of manganese in contaminated sedge compared with blank material more than 50 times.

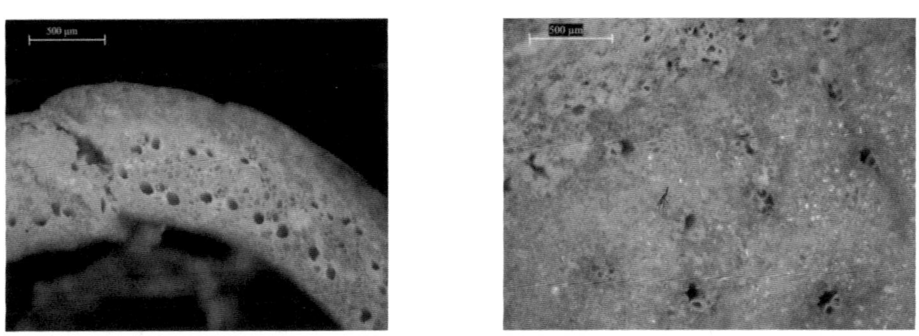

(a) (b)

Figure 1. Optical micrographs of stems of the uncontaminated reed (a) and sedge (b).

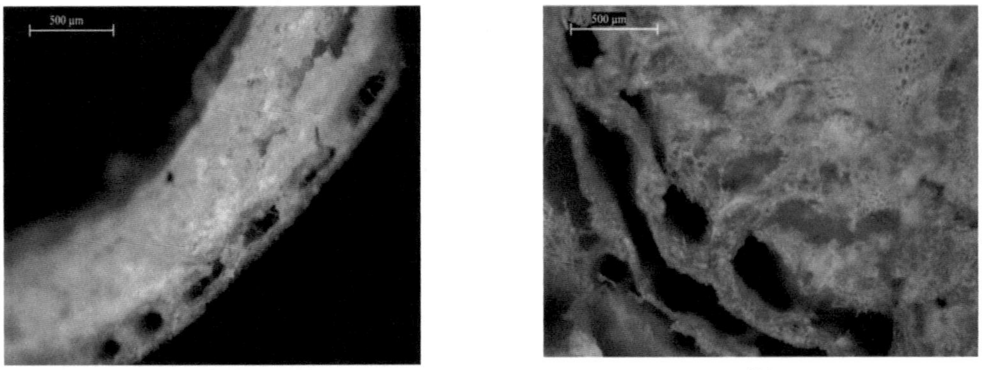

(a) (b)

Figure 2. Optical micrographs of stems of the contaminated reed (a) and sedge (b).

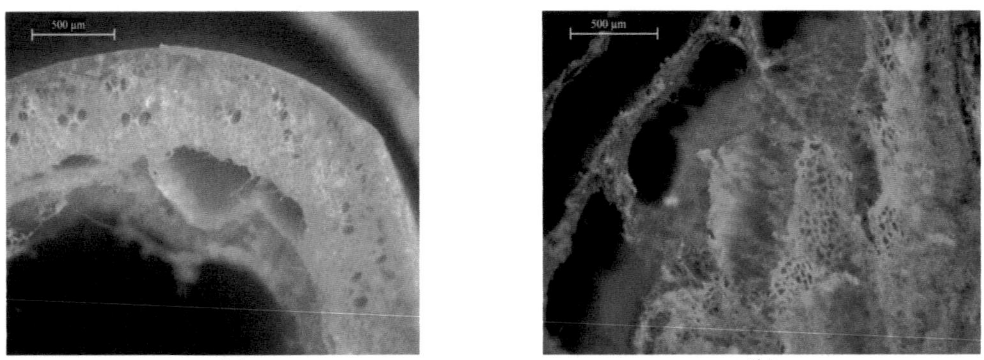

(a) (b)

Figure 3. Optical micrographs of stems of the contaminated and irradiated reed (a) and sedge (b).

Absorbed dose was used to determine the degree of electronic radiation exposure to biomaterials. In the case of the accelerated electrons fixed-energy, the current density, the distance from the accelerator output window and the irradiation time, the absorbed dose equals to:

$$D=\frac{dE}{d\xi} j\tau \times 10^3, \text{ [Gy]},$$

Here, E is the energy of high-energy electrons in MeV, ξ is reduced thickness, kg/m^2 (ξ =xd; d is for density, kg/m^3); x is sample thickness, m; $\frac{dE}{d\xi}$ is the total loss of the electron energy; j is particle current density, μA/m^2; τ is sample irradiation time, s.

Electron irradiation with the dose of 10 kGy leads to the increase in the micropore sizes and cracking of the material (Figure 3).

Table 1. The content of heavy metals in biomaterials of reeds and sedges.

mg/kg	Pb	Cd	Zn	Cu	Fe	Ni	Co	Mn	Cr	Sr
Blank reed	5.00	2.40	67.20	12.70	360.00	24.60	0.24	367.0	19.70	48.00
Cont. reed	5.00	1.00	48.00	4.80	612.00	4.20	0.32	390.0	7.20	43.20
Blank sedge	6.40	0.80	32.00	9.60	648.0	20.40	0.40	70.0	25.20	49.60
Cont. sedge	7.00	1.00	32.00	6.40	720.0	12.60	0.48	3800.0	8.40	89.60

4. Conclusions

1. The method of optical microscopy was used for experimental study of the influence of contamination and electron irradiation with 2 MeV on the structure of biomaterials of reeds and sedges.
2. It was established that the contamination affects significantly the structure of the stems of both reed and sedge. In sedge there is a significant accumulation of heavy metals, which are grouped into larger cluster. Irradiation with high-energy electrons in some cases leads to cracking of materials and the formation of large pores in the biomaterials.

References

[1] Solovchenko A Y 2011 *Biotechnology* Environmental and photobiotechnology for sewage treatment **6** 70 - 88
[2] Yakhin O I, Lubyanov A A, Seregin I V, Yakhin I L 2014 *Agrochemistry* **12** 61 - 78
[3] Poklonov V A, Kotelevtsev S A, Shestakova T V 2012 *Water: chemistry and ecology* **5** 66 - 69
[4] Pelevina I I, Aleshchenko A V, Antoshchina M M, Biryukov V A, Reva Y V, Minayeva N G 2015 *Radiation Biology. Radioecology* **55** 1 57 - 62
[5] Zakirov R K 2001 Abstract of cand. dis. for techn. sciences. Kazan, Intensification of biological wastewater treatment processes using the reagent and ultrasonic treatment methods 19
[6] Akhmadullina F Y, Zakirov Y V, Pronina Y V, Sirotkin A S 2008 *Proceedings of the Samara Scientific Center of the Russian Academy of Sciences. Special issue* Prospects for the use of low-frequency ultrasound for intensification of the processes of extended aeration of industrial waste of Kazakhstan 35 - 39
[7] Brahmi M, Belhadi N H, Hamdi H, Hassen A 2010 *J. Environ. Sci. (China)* **22** 1218 – 1224
[8] Buonanno M, Randers-Pehrson G, Bigelow AW, Trivedi S, Lowy FD, et al. 2013 *In Vitro Studies. PLoS ONE* 207-nm UV Light -A Promising Tool for Safe Low-Cost Reduction of Surgical Site Infections **8**(10)
[9] Surzhikov A P, Frangylyan T S, Ghyngazov S A 2012 Russian Physics Journal **55** 4 345-52
[10] Kupchishin A I, Kupchishin A A, Voronova N A, Kirdyashkin V I, Gyngazov V A 2016 *IOP Conf. Series: Material Science and Engineering* **110** 012039
[11] Esyrev O V, Voronova N A at all. 2016 *IOP Conf. Series: Material Science and Engineering* **110** 012036
[12] Yessyrev O V, Khodarina N N, Aryutkin K N, Tyumenev S. et al 2013 *Proceedings of the higher educational institutions. Physics* Study of the composition, structure and activity of materials of biosystems of "Sorbulak" collector **56** 1/2 122 - 126

IOP Conf. Series: Materials Science and Engineering **168** (2017) 012018 doi:10.1088/1757-899X/168/1/012018

The use of ultrasonic exposure for the modification of synthesized composite materials

D A Negrov, E N Eremin and V V Kuznetsov
Omsk State Technical University, Omsk, Russia

E-mail: negrov_d_a@mail.ru

Abstract. To provide more effective process of the ultrasonic influence while pressing powder materials, it's necessary to increase the strength of natural vibration frequency of the acoustic system transfer devices to the change of a technological load. Solving this task allows one to design a new ultrasonic technology of solid-phase synthesis of polymer composite materials. The technological load value on waveguide system was determined as the research result and the dependence of the frequency stability index of amplification coefficient was found. The rational parameters of ultrasonic exposure during the synthesis of PCM were defined.

1. Introduction

The service life of modern machines and mechanisms is largely depends on the mechanical and tribological materials properties of friction units. The details of friction units made of different metals and alloys have been gradually replaced by polymers and polymeric composite materials, particularly on the basis of polytetrafluorethylene for the last few years [1-3]. The main properties of the polytetrafluoroethylene (PTFE) allow one to use it in the friction units without lubrication. However, the lack of durability of PTFE in operation at high specific loads and sliding velocity requires the development of new ways and methods of increasing the mechanical and tribological properties [4, 5].

To achieve the essential increase of mechanical and tribological properties of polymer composites, it is possible to create improved technologies for their production, in the direction of raising the level of the external energy deposition impact and activating the components directly during their synthesis [6]. Modification of the polymers released by the industry can be carried out by physical methods, both at the stage of their receiving and in the course of their processing. Physical methods of modification can be divided into a number of forms:

- thermal influence (low-temperature or thermochemical processing);
- vacuum and compressor processing (pressure, vacuum, explosive loading);
- periodic deformation (vibration or ultrasonic processing).

From the above mentioned methods of external energetic influence on polymeric materials the most attractive method is appending the energy of ultrasonic vibrations directly while pressing the product, therefore in polymers a number of the physical and chemical phenomena leading to an intensification of a manufacturing process, decrease in the equipment energy intensity, quality improvement of ultimate products are observed. [7, 8].

However, the intensification of pressing process of the polymeric composite materials (PCM) of the ultrasound energy restrains by low effectiveness of ultrasonic sending devices (wave guides). So, the waveguide system design considering the influence of technological loading is an actual task. Its decision allows us to create a new solid-phase technology of obtaining (synthesis) polymeric composite materials. In this connection, acoustic parameters of the waveguide system and their influence on modification of PCM structure were researched.

2. Study subject and methods of research

The study subject was modified polytetrafluoroethylene: 8% - cryptocrystalline graphite, 6% - carbon fiber, 2% - molybdenum disulfide (F4SKG8 UV6M2).

A special installation on the basis of a hydraulic press was developed to research the influence of ultrasonic vibrations while pressing powders of polymeric composite materials [8]. The ultrasonic installation consists of two parts: a power part and a vibratory part. The power part of installation includes a hydraulic press MT – 50, with the maximum pressing effort of 500 kN.

The samples were exposed to heat treatment (agglomeration) at the temperature of 360 ° C after ultrasonic pressing. To observe the same modes of agglomeration, for comparison, samples without impact of ultrasonic vibrations on the pressed composite material were made.

The assessment of the tensile strength limit of samples was performed with the tensile testing machine R 0.5 at the strain rate of 20 mm / min in accordance with State Standard GOST 11262-80. The tribological testing of samples was carried out on a special stand [9], the sliding velocity was V = 0,75 m / s, at the pressure of P = 2 MPa without lubrication.

3. Results and discussion

For increasing the pressing process efficiency it is essential to fulfill the following requirements which are imposed on the waveguide system:

1) the ultrasonic tool should provide necessary vibration amplitude for a given output tool area;
2) the dynamic voltage should not exceed an endurance limit of the system material;
3) the ultrasonic system should work in the mode of a steady resonance under the technological loading change in the process of ultrasonic pressing.

The ultrasonic waveguide tool represents a resonant half-wave core (fig. 1) or set of cores with the adjusted law of changing cross-sectional area on its axis making longitudinal vibrations.

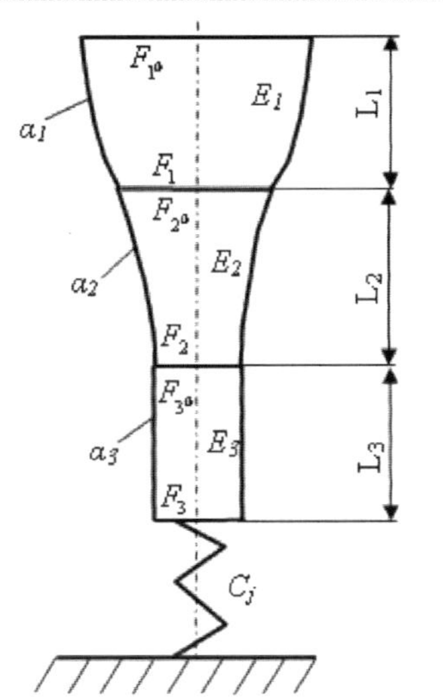

Figure 1. Scheme of the ultrasonic tool.

Calculation of this waveguide system is the complex task being solved relative to two opposite options: the maximum frequency stability at the specified amplification coefficient and the maximum amplification coefficient at the specified frequency stability.

The index of sustainable natural frequency of vibrations to loading change was determined by a formula:

$$\Omega = \left\{ \frac{\left[\frac{(EF)_{i-1}}{E_i F_i}\Delta\Psi_{i-1} + \alpha_i\right]\left[\left(\frac{\alpha_i}{\beta_i} tg\beta_i L_i + 1\right) - \left(\alpha_i - \beta_i tg\beta_i L_i\right)\right]}{1 - \frac{l}{\beta_i}\left[\frac{E_{i-1}F_{i-1}}{E_i F_i}\Psi_{i-1} + \alpha_i\right] tg\beta_i \lambda_i} \right\} \frac{f}{\Delta f}, \qquad (1)$$

where $\lambda_i = 2\pi f/c$, $\beta_i = \sqrt{1 - \frac{\alpha^2}{\lambda^2}}$, *E*- Yung module, *F*- cross-sectional area of the waveguide, *α*- indicator of exponent, *L*- sector length, *i* – sector number .

The offset frequency of Δf was defined experimentally on the ultrasonic installation, while the frequency of vibrations of the acoustic system changes in the range 16,8 – 18,8 kHz. The vibration amplitude of the ultrasonic tool (A) and pressing time (t) were fixed (A=14 microns, t=90 c). Besides, ultrasonic pressing of powder composition materials was carried out under different pressures (P) within the range of 43–87 MPa. The obtained dependences of strength limit on the frequency of ultrasonic tool (waveguide) vibrations are shown in a Fig. 2.

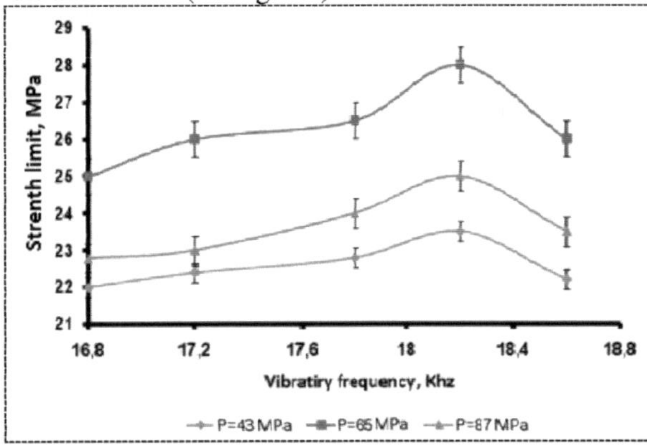

Figure 2. Dependence of strength limits on vibration frequency wave guide while ultrasonic pressing of PKM.

Follows from the achieved results that influence of technological loading in the process of ultrasonic pressing causes essential change in resonance vibration frequency of the acoustic system (=400 Hz), and the most optimal pressure of ultrasonic pressing makes 65 MPa. With molding pressure increase over this value strength limit while extending decreases by 10%. This is due to the fact that pressure grows with increasing load acting on the acoustic assembly, which in turn causes a reduction in the vibrations amplitude of the waveguide in the ultrasonic pressing process.

As a result of the mechanical tests it was found that the optimal pressing time is 90 seconds (Fig. 3). At the same time, the samples produced by ultrasonic pressing the tensile strength higher by 16% and the elastic modulus by 23% in comparison with samples made by the technology without ultrasound use. Further increase of pressing time does not alter the tensile strength and modulus.

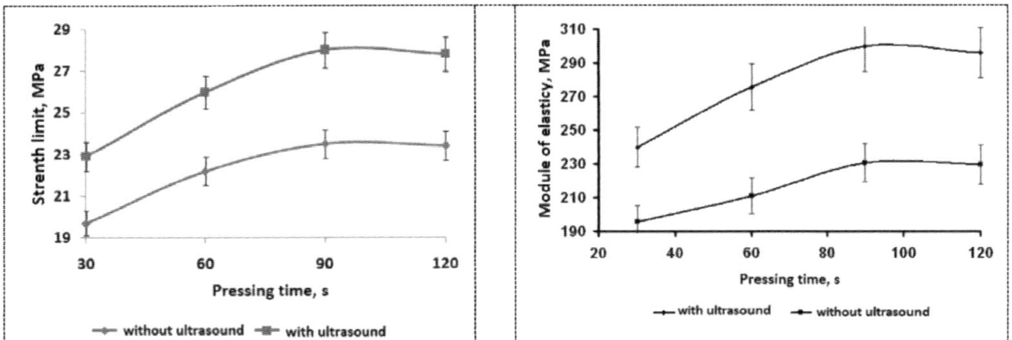

(a) (b)

Figure 3. Dependence of strength limit (a) and module of elasticity (b) on pressing time.

Influence of vibration amplitude of a wave guide punch on strength and elastic modulus of composition material has pronounced extremal character with a maximum of 14 microns (fig. 4). At the same time molding pressure was 65 MPa, while pressing time of 90 seconds.

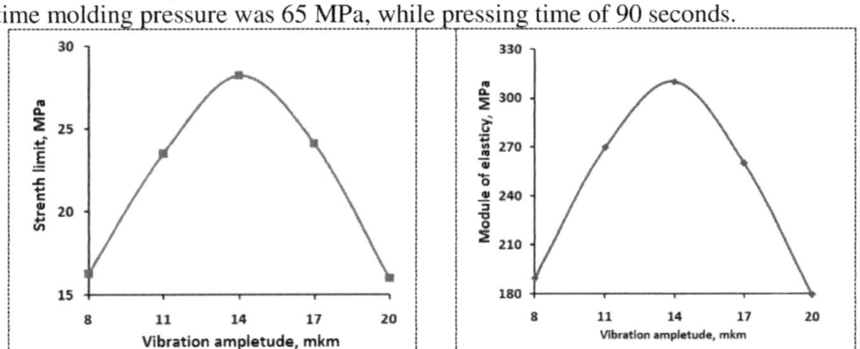

(a) (b)

Figure 4. Dependence of strength limit (a) and module of elasticity (b) from vibration amplitude of a wave guide punch.

On the basis of the above explained researches results rational parameters for the mode of ultrasonic influence while pressing (vibration amplitude of a wave guide punch of 14 microns, extrusion pressure of 65 MPa, time of pressing of 90 seconds) are set. On these parameters were made samples for tribotechnical tests which results are given in a figure 5.

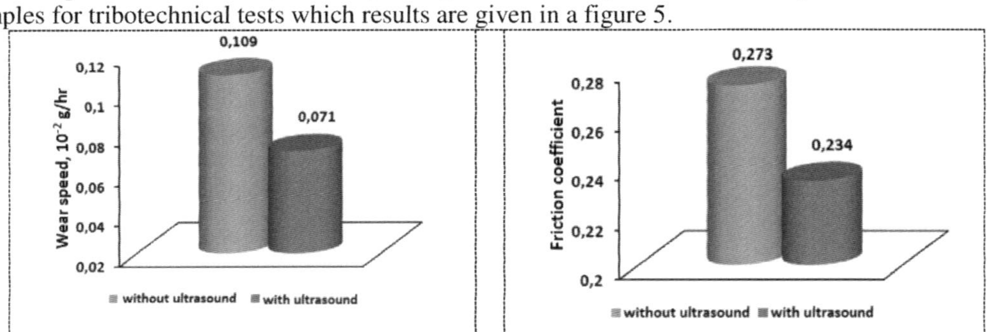

(a) (b)

Figure 5. Tribotechnical characteristics of the received materials:

a – wear speed; – friction coefficient.

As a result of the carried-out tribotechnical tests it is established that the use of ultrasonic pressing allows to decrease wear speed on 34,8%, and friction coefficient on 14,3%.

During ultrasonic processing polymers essentially facilitated occurrence and plastic development deformation of the powder particles, is carried out the activation of dislocations occurring as a result of the absorption of acoustic energy in the areas of crystal lattice defects and other structural imperfections. Consequently, in a short time, local heating occurs around these absorption material sources, the increase in molecular mobility that determines the improvement of mechanical and tribological properties of the produced composite material.

4. Conclusion

As a result of the carried out researches the size of technological load of the waveguide (acoustic) system is determined. Dependence of an indicator of frequency stability on strengthening coefficient was established.

Rational parameters of ultrasonic influence at synthesis of PKM are determined.

The conducted researches have shown that the use of ultrasonic pressing allows to increase strength by 16%, the elasticity module by 23%, at the same time the speed of wear decreases by 34,8%, and friction coefficient by 14,3%.

References

[1] Kobzev D E, Baronin G S, Dmitriev V M and others 2012 Intensification of solid-phase plunger extrusion of the nanomodified polyethylene of high density by the ultrasonic exposure *Materials Science* **4** 37-40

[2] Adamenko N A, Agafonova G V, Arisova V N and others 2009 Structural changes during the explosive composition processing of PTFE-4 with polyoxy benzoyl *Materials Science* **3** 18-25

[3] Petrova P N, Fedorov A L 2010 Development of polymeric composites on the basis of polytetrafluoroethylene with the increased wear resistance for the units of dry friction *The herald of machine building* **9** 50–53

[4] Zhang D, Dong G, Chen Y and Zeng Q 2014 Electrophoretic deposition of PTFE particles on porous anodic aluminum oxide film and its tribological properties *Applied Surface Science* 466-474 DOI: 10.1016/j.apsusc.2013.11.114

[5] Mashkov Yu K, Ovchar Z N, Surikov V I and others 2005 Modification of the structure and composite properties of materials based on polytetrafluoroethylene *Structural modification Machine building* 168

[6] Negrov D A and Eremin E N 2011 Investigation of the effect of ultrasonic vibrations on the structure and mechanical properties of the polymer composite material *Omsk Scientific Herald* **2** 17–20

[7] Eremin E N and Negrov D A 2014 Development of a technology for the fabrication of articles made of complex-modifiend polytetrafluoroethylene for dry friction assemblies *Chemical and Petroleum Engineering* **49** DOI: 10.1007/s10556-014-9822-0

[8] Negrov D A and Eremin E N 2014 Effect of ultrasonic vibrations on changing the mechanical and tribological properties of polytetrafluoroethylene modified with boron nitride *Dynamics of Systems, Mechanisms and Machines (Dynamics)* 1–4 DOI: 10.1109/Dynamics.2014.7005686

[9] Mashkov Yu K, Eremin E N and Negrov D A 2013 The influence of the ultrasonic vibrations energy on the properties of polymer composites *Materials Science* **3** 42–45

IOP Conf. Series: Materials Science and Engineering **168** (2017) 012019 doi:10.1088/1757-899X/168/1/012019

Studies of the generation mechanisms of steady vortex formations in the channels of nuclear-power installations for purposes of improving the reliability and safety of their work

O Mitrofanova

Professor, *National Research Nuclear University "Moscow Engineering Physics Institute"*,
31 Kashirskoe ave., Moscow 115409, Russia

E-mail: omitr@yandex.ru

Abstract. The analysis of the results of experimental researches on revealing the mechanisms of vortex formation in channels of complex geometry in the neutral and conductive media is carried out. The directions of researches related to the study of mechanisms of vortex generation and accumulation of energy by large-scale vortex structures are considered for the possibility of predictions of the man-made accidents and catastrophic natural phenomena. The main goal of ongoing investigations is the solution of the task aimed at improving the safety of nuclear power installations and, in particular, of the fast neutron reactors with liquid-metal coolants, and the prevention of emergency modes arising from acoustic, magnetic and hydrodynamic resonance effects.

1. Introduction

The use of the swirl flows in the engineering is associated in the first place with the solution of the problem of enhancement of heat and mass transfer processes in order to improve the current and create new power plants, propulsion systems and heat exchangers. The positive effects derived from the swirling of a flow in technical devices, include also the equalization of temperature irregularities, the stabilization of the flows and various physical and chemical processes in fluid media [1, 2]. The swirl flows has been successfully used in the separation devices, for thermal protection of the walls of the channels, provide an efficient and cleaner burning of fuel in combustion chambers, etc [3]. Thus, the generation of organized vortex or swirl flow in technical devices, usually is carried out purposefully, and the mechanism of vortex formation is connected with the effect on the flow of different arrangements creating a mechanical torque.

However, it should be noted that in some cases the spontaneous vortex formation and swirling of the flow can play a negative role. For example, appearance of large-scale vortexes in the collectors of nuclear reactors, leads to the decrease of mass flow rate in the most power-stressed channels of the core, owing to pressure fall in the central region of collector.

It was found that the formation of large-scale vortex structures in the pressure chambers of nuclear reactors is caused by the design features and conditions of coolant inlet from the loops. In this regard, a comprehensive study of the influence of hydrodynamics effects on the operating parameters and conditions of heat removal in nuclear reactors is one of the most important aspects of ensuring the safety nuclear power installations.

In [4] it was shown that vortex structures arising both in the external flow over bodies and in internal flows can be the sources of acoustic oscillations. Swirl and curved flows, the presence of which leads to appearance of steady vortex structures, may occur in various elements of thermal-hydraulic circuit of NPI. The reliability and safety of operation of nuclear power systems essentially depend on intensity of vibrations, resulting from hydrodynamic effects of the flow of coolant on the internal elements of the core. Vibrations are usually accompanied by pervasive acoustic noise. Due to the fact that the spectrum of the generated oscillations is determined by the elastic characteristics of the elements of the channels and the internal vortex structure of the flow, it carries great information about the state of the hydromechanical system. It is known that if the power spectrum at low frequencies corresponds to the inversely proportional dependence of the oscillation frequency, the system undergoes the development of so-called flicker-noise process, when the power spectrum of pulsations unlimited grows with lower frequencies of oscillations, and may lead to catastrophic consequences and the destruction of the system itself [5].

In this regard, in the ongoing scientific research a special place is given to the development of the acoustic method of research and the establishment of experimental techniques for the simulation of resonant effects related to the emergence of stable vortex structures.

The use of liquid metal as a coolant requires a study of the effects of the emergence of self-excitation of magnetic fields in the turbulent process of movement of a conductive medium. It is known that turbulence is not always is the cause of the destruction of large-scale vortex structures, but under certain conditions is able to create them, bringing the instability in the flow of conductive coolant [6]. Even more dangerous phenomenon, which leads to violations of flow of liquid-metal coolant, may be the emergence of large-scale vortex motion with non-zero helicity [7, 8].

This article presents some results of experimental researches of mechanisms of the emergence of stable vortex structures leading to acoustic, resonance and electromagnetic effects in complicated hydro mechanical systems.

2. Generation of acoustic oscillations

The mechanism of generation of acoustic oscillations associated with the formation of stable vortex structures in fluid flows is considered on the example the impact swirl flow.

The results of the experiments revealed a correlation between the existence of large-scale vortex flow structure and excitation of acoustic oscillations, leading to vibration processes in hydro mechanical systems.

Detailed description of the experimental setup and the results of the experiments which helped to identify the self-regulatory effect of acoustic oscillations in the impact swirl flow were presented in [9]. The experiments were conducted using the vortex chamber (figure 1a), which is a rigid metal hollow cylinder 1 with a tangential air supply 4 in its bottom part and a Plexiglas plate 2 glued in the upper end of the vortex chamber, through a small hole in which swirl stream flows out. Above the outlet of the vortex chamber was located loose obstacle 3, made in the form of a flat disk. The hydrodynamic scheme of the flow is shown in figure 1 b. The geometric dimensions of the obstacles, the vortex chamber and velocity parameters of a swirl flow were varied. It was found that the excitation of oscillations under condition of the critical flow out of the swirl air jet from the vortex chamber, its upper end surface is subjected to the action of the asymmetrical force that causes bending oscillations.

It was considered two conditions of the outflow of the impact swirl jet: subcritical, in the absence of matching the frequency of the vortex flow with the eigenfrequencies of the hydro mechanical system, and critical, corresponding to the condition of resonance of vortex frequency with the eigenfrequency of the elastic upper surface of the vortex chamber. It was shown that in the both outflow conditions the flow moves under the obstacle at the complicated closed trajectories, imageries of which correspond to the Lissajous figures with the certain ratios of frequency of harmonic oscillations in mutually perpendicular directions.

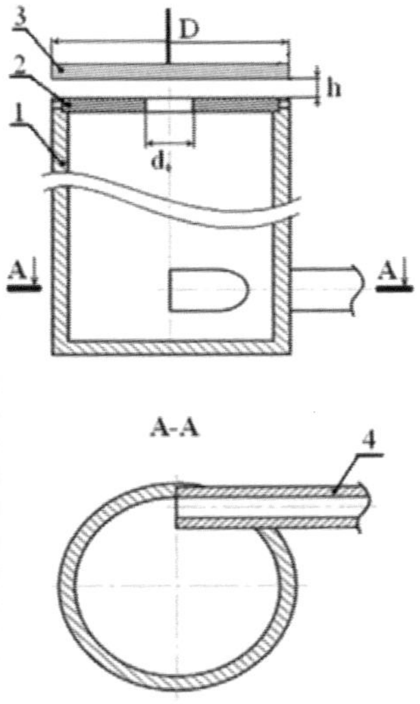

a)

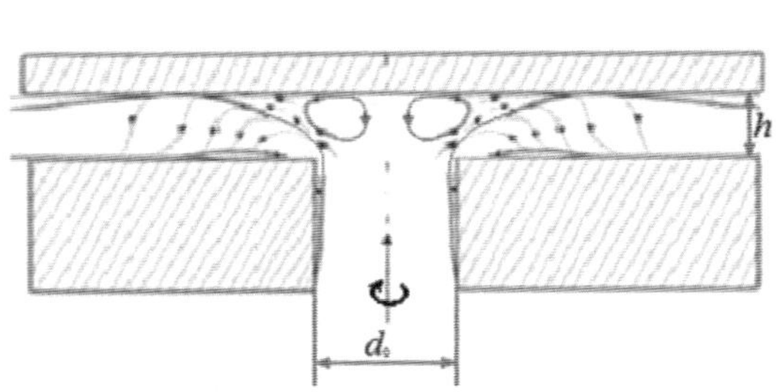

b)

Figure 1. Experimental study of the impact swirl flow: a) scheme of the work area of the experimental set: 1 - vortex chamber, 2 – a Plexiglas plate glued in the upper end of the vortex chamber, 3 – obstacle, 4 - tangential supply of air, b) hydrodynamic scheme of the flow

As it was shown in [9], at coincidence of the flow velocity with the limited critical velocity of flow rate in the swirl jet, the flow acquires a well-defined vortex structure, separating into a thin longitudinal spiral vortices (figure 2), the frequency of rotation which coincides with the closest natural frequency oscillation of the vortex chamber. In considered case this frequency is the frequency of the bending oscillations of the upper Plexiglas plate 2.

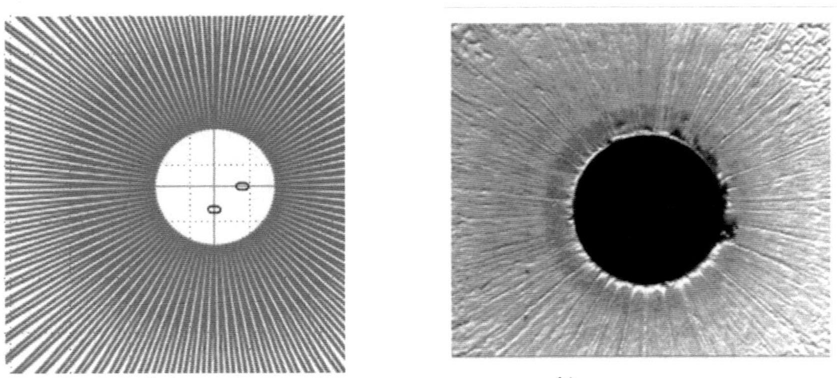

a) b)

Figure 2. The comparison of calculation results with visualization of the flow pattern in the resonance mode: (a) the result of the addition of harmonic oscillations with frequencies f_1 = 2796 Hz and f_2 = 27 Hz in the form of Lissajous figures in cylindrical coordinates; b) – a picture of the vortex wake on the bottom surface of the gap.

The effect of self-regulation of acoustic oscillations is expressed in a resonant amplification of the amplitude of the natural frequencies of the hydro mechanical system due to the capture of energy of acoustic oscillations generated by vortex structure of the flow.

On the basis of this study it can be concluded that the formation of the vortex flow structure can lead to resonant amplification of the amplitude of oscillation of the hydro mechanical system and consequently to provoke the beginning of the vibration process.

3. Cavitations process in a channel of variable cross-section

Experimental studies of the formation of the vortex flow structure in a channel of variable cross-section were carried out on a hydrodynamic bench using the method of registration of acoustic oscillations and optical visualization of the vortex structure of the flow. The working section of experimental setup was made to vary the geometry of the flow part of the channel and swirl flow arrangements that serve as the generators of vortexes or creating a swirling of the flow in the scale of all channel. The scheme of the work area is presented in figure 3a.

Using the developed system for registration of acoustic signals, experiments were conducted to identify natural frequencies of the working outline and to detect the field of local vorticity corresponding to the visualized vortex structure of the flow in the channel of variable cross-section. Thermal measurements were carried out using thermocouples installed along the length of the work area. The amplitude-frequency characteristics of the detected acoustic waves were measured at various points in the work area for various flow rates of water and conformed to the sound spectrum. The Reynolds number Re ranged from $7.5 \cdot 10^3$ to $1.5 \cdot 10^5$.

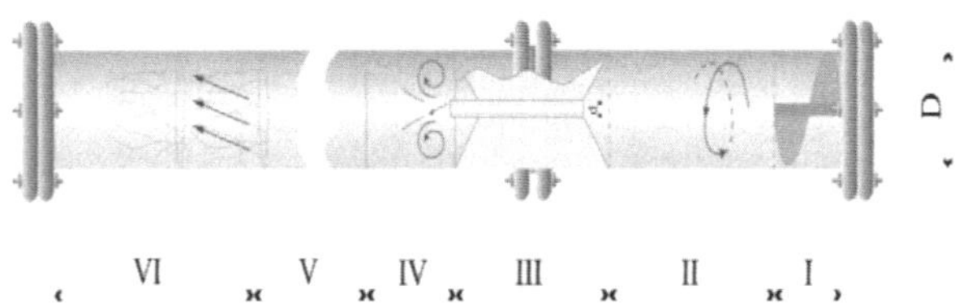

a)

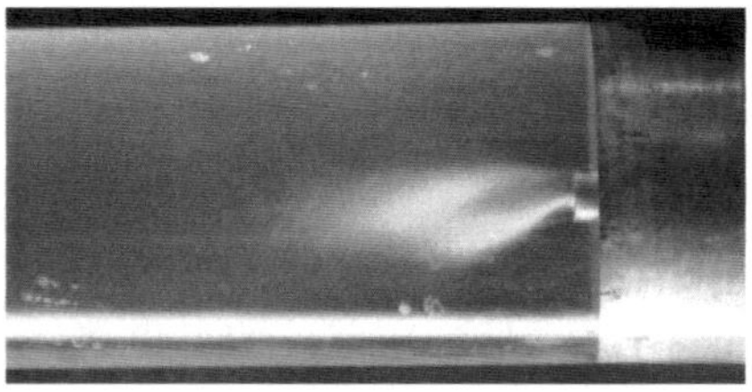

b)

Figure 3. The scheme of the working section of experimental setup (a); the flow pattern at the exit of the narrow part of the channel in the area of acoustic cavitations IV (b).

Numerical calculations of the velocity fields, pressure and temperature showed that in the area IV (see figure 3a) the effect of throttling has occurred. It followed by a temperature drop to the saturation temperature corresponding to the cold boiling of liquid (i.e. the cavitations process). Analysis of measured amplitude-frequency characteristics showed that at the entrance of the flow into the cavitations zone after its exit from the narrow part of the channel, the spectrum of acoustic oscillations changes sharply. The maximum amplitude of acoustic oscillations corresponds to the frequencies in range 10000 - 20000 Hz.

Studies have shown that the flow regimes of the water coolant, in which can be realized cavitations processes in channels of variable cross-section, is highly undesirable, as can lead to destruction of elements of heat power equipment of nuclear power installations.

4. Investigation of the mechanisms of vortex formation in conductive environments. The interaction of vortex and electromagnetic fields

The studies aimed at identifying the mechanisms of vortex generation and accumulation of energy in conductive media are very essential nowadays. Theoretical analysis and experiments show that the effect of these mechanisms is determined by the interdependence of electric, magnetic and vortex fields in conductive media and their influence on the processes of heat and mass transfer in a conductive environment.

An experimental setup was designed to explore the relationship of electromagnetic forces and inertial forces in the liquid conductive media and to identify the mechanism of generation of large-scale vortex motion in the conductive fluid, placed in a magnetic field [10]. The main components of the setup were: a container with an electrolyte, embedded permanent magnet in the form of a ball, the movable electrodes of various configurations, power supply and recording equipment for photo and video.

In experiments managed to obtain the reproducible and well-rendered pictures of the spiral vortex structures, leading to the large-scale swirling of the flow. With the aim of obtaining quantitative data on the distribution of the velocity field of a moving electrolyte there were produced photographs of the swirl flow. The distribution of flow velocity was determined due to obtained parameters of the tracks left by gas bubbles that are indicators of vortex motion in the liquid. Images were obtained at different polarity of the power source and change of the amperage flowing through the circuit. Tangential and angular velocity of rotation were determined on base of measuring the lengths of tracks and their distances to the center of rotation.

As an illustration figure 3 shows the dynamics of development of vortex flow obtained for acetic acid at the position of a positively charged electrode at an angle 55^0 relative to the horizontal axis. On the video shots are represented successive stages of the formation of spiral-vortex structures of the flow while passing an electric current through the immobile electrolyte solution. Gas bubbles formed in the process of electrolytic reactions are the indicators of vortex motion.

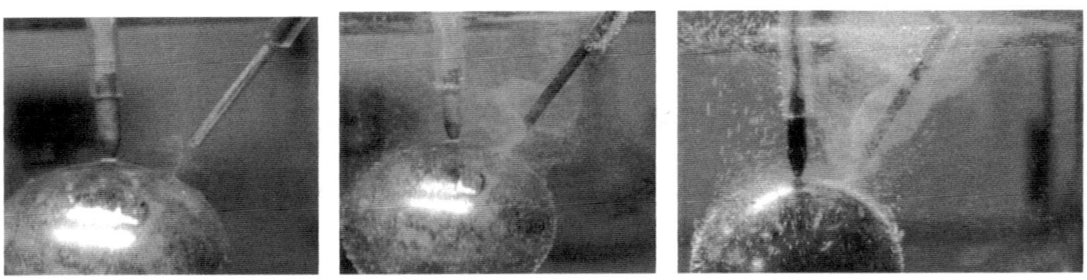

a) b) c)

Figure 3. Visualization of the occurrence of vortex motion in the electrolyte solution: (a) the stage of generation of vortex motion, (b) the formation of a stable spiral-vortex structure, (c) the two-layer helical vortex with downward and upward movement.

It was proposed a physical model of vortex flow generated in a conductive medium placed in a constant magnetic field while creating a potential difference between the electrodes and at their various orientations relative to the magnetic field lines.

Operating experience of fast neutron reactors, as well as the results of experimental studies conducted at the sodium fast reactor BN-600 [7], show that in the implementation of the integrated circuit arrangement of the structural elements of the active zone and collector system of fast reactors in specific areas of a thermal-hydraulic path of the primary circuit can arise large-scale vortices with a spiral-helix motion of liquid metal coolant. In the monograph [1], it was noted that thermal physical properties of liquid metals may contribute to the formation of stable helical vortex structures.

Non-zero helicity of the velocity field $H \neq 0$ ($H = \mathbf{u} \cdot \boldsymbol{\omega} \neq 0$, where $\boldsymbol{\omega} = \nabla \times \mathbf{u}$ is a curl of local flow velocity $\mathbf{u}$) may lead to appearance of "abnormal" electromotive force (EMF). This effect described in [6, 7] as the hydro dynamical alpha-effect is reduced to the emergence of large-scale hydro dynamical instabilities in fluid flows.

Numerical computations for various cases of turbulent flow of liquid metal coolant in complicated channels when applying an external transverse magnetic field have shown that the presence of large-scale vortices leads to the generation of induced magnetic field on the same scale. The configuration of the induced vortex magnetic field in the moving conducting medium depends on the mutual orientation of spiral-vortex motion and direction of an external magnetic field.

The results of the calculations point to the crucial role of convective transfer magnetic field, which reveals "freezing" magnetic field, which is transported due to spiral-vortex structures.

The fulfilled studies are devoted the solution of problems of reliability and safety of operation of nuclear reactors of vessel type with water and liquid metal coolants.

Acknowledgments
Studies are supported by the Russian Foundation for Basic Research (project no. 16-08-00687-a), and the Research and Educational Human Resources of an Innovative Russia Federal Target Program.

References

[1] Mitrofanova O 2010 Hydrodynamics and Heat Transfer of Swirling Flows in Channels of Nuclear Power Facilities (Gidrodinamika i teploobmen zakruchennykh potokov v kanalakh yaderno-energeticheskikh ustanovok) Fizmatlit Moscow 288

[2] Mitrofanova O 2003 Hydrodynamics and Heat Transfer in Swirling Flows in Channels with Swirlers (Analytical Review) *High Temperature* **41** 4 518-559

[3] Gupta A K, Lilley D G and Syred N 1984 Swirl Flows Alacus Press 588

[4] Mitrofanova O V, Egortsov P P, Kokorev L S, Kruglov V B and Chernov A I 2010 An investigation of the mechanism of acoustic vibration in swirl flows *High Temperature* **48** 2 222-230

[5] Koverda V P, Skokov V N 1998 *Journal of Experimental and Theoretical Physics* 1998 **86** 5 953–958

[6] Moffat G 1981 Special Issue Celebrating the 25th Anniversary of the Journal *J Fluid Mech* **106** Batchalor G and Moffat H Eds Cambridge: Cambridge University Press 49

[7] Kirko I M and Kirko G E *Magnetic Hydrodynamics: Modern View on Problems (Magnitnaya gidrodinamika. Sovremennoe videnie problem*) Moscow: Regular and Chaotic Dynamics 632

[8] Mitrofanova O V, Podzorov G D and Tokarev Yu N 2015 Simulation of Magnetohydrodynamic Effects on Generation of Large Scale Vortices in a Liquid Metal Coolant *High Temperature* **53** 3 414–424

[9] Mitrofanova O V and Pozdeeva I G 2015 Self-Adjustment Mechanism in Impinging Swirling Flows *Fluid Dynamics* **50** 5 646–654.

[10] Mitrofanova O V Study of the generation mechanism of large-scale vortex motion in electrically conductive media *High Temperature* **53** 6 848–855

IOP Conf. Series: Materials Science and Engineering **168** (2017) 012020 doi:10.1088/1757-899X/168/1/012020

The quasicrystal model of cluster systems in condensed matter

G Melnikov
South-West State University (SWSU) Kursk, 305040, Russia

E-mail: melnikovga@mail.ru

Abstract. The paper proposes a quasicrystal model of the structure of clusters. The model is based on the similarity of the structure of clusters and macroscopic structure of quasicrystals. It offers a formula to calculate the radii of successive coordination spheres in quasicrystalline films. The formula is based on the properties of Fibonacci sequence and characteristics of the power potential of interaction between particles.

1. Introduction

The structures of cluster systems and nanoscale objects are usually modeled by means of face-centered cubic (fcc), body-centered cubic (bcc) or hexagonal close packings of particles, with atoms in the cluster configuration corresponding to the volume of the sample structure. Even in an ideal crystal lattice of the cubic structure the model of close packing of hard spheres cannot describe the sequential arrangement of the coordination spheres in real crystals. The replacement of solid balls by soft spheres in the model has no notable effect calculation results. Currently there are attempts to develop alternative models of cluster system structures

Quasicrystal models of cluster structure are based on structural properties of quasicrystals. Quasicrystals are a large class of ordered materials occupying an intermediate position between classical crystals and amorphous substances. The first quasicrystal was discovered by Shechtman and colleagues in 1984 in the situation with the rapid cooling of the melt $Al_{86}Mn_{14}$ [1]. To date we know more than one hundred systems based on aluminum, gallium, copper, cadmium, nickel, titanium, tantalum, and other elements in which quasicrystals can form [2, 3].

Mathematical model of an ideal quasicrystal had been independently developed in two theories, that subsequently turned out to be equivalent. Levin and Steyngardt [4] built a model of a quasicrystal based on two unit cells with an irrational ratio of their quantities. Certain construction rules, that are based Fibonacci series, lead to a structure without any periodicity in the arrangement of atoms, but with some properties that are typical of the crystal.

A common model of the structure of quasicrystalline objects is a two fragment model based on quasi-periodic covering of line, plane or space with two basic structural units. For one-dimensional quasicrystal this model leads to a Fibonacci series of short S and long L segments with S = 1 and $L = \Phi = 1.6180339\ldots$. In the two-dimensional case the two fragment model is a Penrose tiling [5] composed of two types of rhombus with acute angles at the vertices $\pi/5$ and $2\pi/5$, in three-dimensional case it is the generalization of the Penrose tiling formed by rhombohedrons of two types and is called Amman-McKay network [2-5].

The model made by Kalugin, Kitaev and Levitova [6] is based on the "cubic" crystal in 6-dimensional space, which results in icosahedral quasicrystal. This crystal is used to cut out so-called "pipe", i.e. a layer of atoms that has the size of interatomic order and is concluded between three-dimensional hyperplanes. Then this pipe is projected onto physical space. Thanks to such arrangement it is possible to get continuum theory of the quasicrystal, to describe dislocations and clarify the nature of low-frequency collective modes.

Most of the existing theories of quasicrystal structuring are based on the idea of cluster approach to constructing aperiodic partitions. Within the framework of the cluster approach [7, 8] it is proved that quasicrystals are units of clusters. Their structure cannot be represented as a package of identical elementary cells. They present a unique packaging way of overlapping clusters, which allows us to realize the most stable and energetically favored local atomic configurations. Long-range order and extraordinary properties are caused by essentially localized fluctuations called phasons [7- 9].

In the Penrose partition it is possible to distinguish two types of clusters made from three rhombs shaped that are shaped like a cube projection. These are either clusters made from two wide and one narrow rhombs, or an elongated clusters made of one wide and two narrow rhombs. Each rhomb is always a part of either one or two mutually overlapping clusters of these types [9]. Locally equivalent units of the same type are grouped together along the respective circumferences, the diameters of which differ by Φ^n times. This feature allows us to draw an analogy with the circular property of linear fractional transformations (Möbius transformations) in the theory of functions of complex variables. Linear fractional transformation maps an arbitrary circle, that lies in the complex plane, into an equivalent circle, with direct lines regarded as circles of infinite radius. [9].

Dimensional model of the quasicrystal and Fibonacci series associated with the golden section are used to build a new type of optical elements: aperiodic gratings and multilayer structures [10-13]. Because of their unique physical properties the peculiarities of light diffraction on one-dimensional quasi-periodic structures (quasicrystals) have become the subject of various investigations in physics [10-13].

2. Quasicrystal model of cluster systems

In this paper we have adopted two fragment model based on the quasi-periodic covering of the space by two elementary structural units to be the basic model of the structure of cluster systems. In Penrose tiling model such elements are thin and thick rhombs with equal sides a with atoms of matter in their vertices. Thin rhomb has acute angle $\frac{\pi}{5}=36^0$, thick rhomb has acute angle $\frac{2\pi}{5}=72^0$. Atoms in the vertices of the thin rhomb form two dimers (see. Figure 1), similarly the atoms in the vertices of the thick rhomb form two other dimers

The first dimer is formed by the atoms of the thin rhomb with the centers of the masses of the atoms located at a distance equal to the rhomb small diagonal:

$$R_{\dim}^{/}=2aSin\frac{\pi}{10}=a(\Phi-1)=\frac{a}{\Phi}=0.618...a \quad (1)$$

the second dimer is formed by atoms lying at a distance of the long diagonal of the thin rhomb:

$$R_{\dim}^{//}=2aCos\frac{\pi}{10}=a\sqrt{\Phi+2}=1.902...a \quad (2)$$

and also

$$R_{\dim}^{//}=R_{\dim}^{/}Ctg\frac{\pi}{10}=R_{\dim}^{/}\Phi\sqrt{\Phi+2}=3.078...R_{\dim}^{/} \quad (3)$$

where $\Phi=1.6180339$ - golden section constant.

The atoms in the vertices of the thick rhomb, like in the thin rhomb, form two more dimers with distances between the atoms $R_{\dim}^{III}$ and $R_{\dim}^{IV}$ respectively

$$R_{\dim}^{III} = 2aSin\frac{\pi}{5} = a\sqrt{3-\Phi} = 1.175...a \tag{4}$$

$$R_{\dim}^{IV} = 2aCos\frac{\pi}{5} = a\Phi = 1.618...a \tag{5}$$

$$R_{\dim}^{IV} = R_{\dim}^{III}Ctg\frac{\pi}{5} = R_{\dim}^{III}\frac{\Phi}{\sqrt{3-\Phi}} = 1.377...R_{\dim}^{III} \tag{6}$$

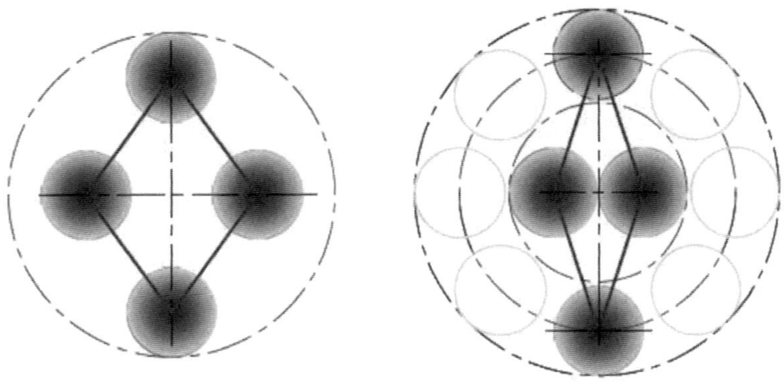

Figure 1. Golden rhombs pattern of the formation of atoms dimers in the structure of the cluster.

Each of four dimers makes the core of a forming cluster, so four types of clusters should be distinguished in this model and in X-ray electron diffraction analysis each of them demonstrates a diffraction pattern. General X-ray pattern is the result of superpositioning the four types of reflections that represent specific features of four types of clusters in the structure of the quasicrystalline system.

In the series of papers [14-17] the author et.al. have proposed a model of cluster formation in a disordered condensed media. According to this model the core of the cluster is a dimer around which a system of successive coordination spheres builds up. The result is the "shell structure" of a cluster. The appearance of defects in the shell structure of the cluster leads to unification and redistribution of the mutual arrangement of coordination spheres compared with the crystalline material phase and determines the temperature dependence of the coordination numbers in the cluster structure.

For a system of successive coordination spheres of spherical particles the radii of such spheres are represented by [16, 17]

$$R_n = R_{\dim}\sqrt{pF_n} \tag{7}$$

where $R_{\dim}$ - the distance between the particles in each of the four dimers (1), (2), (4) and (5), F_n - Fibonacci numbers, $p = 1, 2, 3, ...$ - natural numbers.

The rhomb side a can be found by the formula (1), which $R_{\dim}^{I}$ should be understood as the radius of the first coordination sphere. It is determined by the diameter of the matter particle and characteristics of the interaction potential [18]

$$R_1 = R'_{\dim} = \left(\frac{n}{m}\right)^{\frac{1}{n-m}} \sigma_0 \tag{8}$$

where R_1 - the radius of the first coordination sphere, σ_0 - the diameter of the sphere which simulates the atom, $n \succ m$ - integers for Mie potential.

Table 1. The radii of successive coordination spheres in the structure of a quasicrystalline system measured in the units of a .

	Fibonacci Numbers						
	1	2	3	5	8	13	21
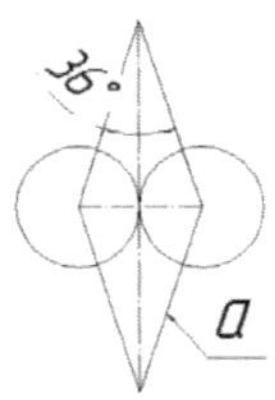	0.618	0.874	1.070	1.382	1.748	2.228	2.832
	1.175	1.662	2.035	2.627	3.323	4.236	5.385
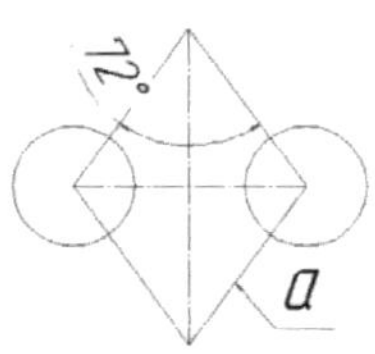	1.618	2.288	2.802	3.618	4.576	5.834	7.415
	1.902	2.690	3.294	4.253	5.380	6.858	8.716

Table 2. The radii of successive coordination spheres measured for the diamond film and macroscopic diamond.

Number KS	R_n / a	R_n, A	R_n, A [19], Diamond film	R_n, A [20], Bulk diamond
1	0.618	1.73	1.44	1.544 (111)
2	0.874	2.45	2.48	2.522 (220)
3	1.070	3.00	2.96	2.957 (311)
4	1.175	3.29	3.30	3.566 (400)
5	1.382	3.88	3.70	3.886(331)
6	1.618	4.54	–	4.368 (422)
7	1.748	4.90	5.11	4.633 (333) (511)
8	1.902	5.33	–	5.044 (440)
9	2.035	5.71	5.96	5.847 (533) 11 KS
10	2.228	6.25	–	6.367 (551) 13 KS
15	2.815	7.89	7.80	7.975 (840) 20 KS
16	3.315	9.30	8.80	9.223 (773) 27 KS
17	3.615	10.14	10.2	10.205 (955) 33 KS

Taking into account the said assumptions, the rhomb side will be determined by the formula

$$a = \Phi\left(\frac{n}{m}\right)^{\frac{1}{n-m}} \sigma_0 \tag{9}$$

Now let's calculate the radii of coordination spheres for carbon films. The diameter of carbon atoms is $\sigma_0 = 1.544A$, so for the Lennard-Jones potential (m = 6, n = 12) the rhomb side will be $a = 2^{\frac{1}{6}}\Phi\sigma_0 = 2.804A$. Considering Table 1 and placing the radii of coordination spheres in monotonic increase order we'll obtain absolute values of the radii of coordination spheres with the rhomb side size $a = 2.804A$.

Table 2 calculation data on the radii of successive coordination spheres for the diamond film consisting of two-dimensional Fibonacci clusters. These results, obtained by the proposed calculation method, are compared with X-ray diffraction data for diamond films [19] and the results of theoretical calculations for bulk diamonds [20].

As the proposed cluster model resulted in obtaining the values of the radii of the first six coordination spheres that are in good agreement with the results of X-ray diffraction patterns in amorphous carbon films [19] and have just a small offset from the radii of coordination spheres in the crystal macroscopic diamond [20], the short-range ordering (within five coordination spheres) in tightly packed carbon systems can be simulated by means of quasicrystal models.

3. Conclusion

Quasicrystal models of the structure of cluster systems in the structure of carbon films can be used to predict the position of coordination spheres consistently with the results of X-ray diffraction measurements. Successive arrangement of coordination is explained by peculiarities of Fibonacci series and the golden section. Carbon films form up from quasicrystalline clusters according to the principle of the close packing of golden rhombs, which is mathematically expressed by the following rule: the radius of the n-th coordination sphere is proportional to the square root of the corresponding Fibonacci number.

The determining parameter of the model is the side of golden rhombs; it depends on the diameter of the matter particles and the nature of the interaction between particles. In case of Mie power potential it depends on the exponent in the repulsive and attractive parts of the potential.

Acknowledgments
The work financially supported the Ministry of Education and Science of Russia in the basic tasks of the State.

References

[1] Shechtman D, Blech I, Gratias D, Cahn J W 1984 *Phys. Rev. Lett.* **53(20)** 1951
[2] Vekilov Y K, Chernikov M A 2010 *Physics-Uspekhi* **53(6)** 537
[3] Bendersky L 1985 *Phys. Rev. Lett.* **54** 2422
[4] Levine D, Steinhardt P J 1984 *Phys. Rev. Lett.* **53(26)** 2477
[5] Penrose R 1974 J. *Inst. Math. Its Appl.* **10** 266
[6] Kalugin P A, Kitaev A Yu, Levitov L S 1985 *JETP Lett.* **41 (3)** 145
[7] Abe E, Yan Y, Pennycook S J R 2004 *Nat. Mater.* **3(11)** 759
[8] Abe E, Tsai A 2004 *J. of Non-Crystalline Solids* **334–335** 198
[9] Madison A E 2013 *Physics of the solid state* **55(4)** 855
[10] Socolar J E S, Steinhardt P J 1986 *Phys. Rev.* **B34** 596
[11] Yuvaraj Sah, Ranganath G S 1995 *Optics Communications* **114** 18
[12] Grushina N V, Korolenko P V, Markov S 2008 *Bulletin of Moscow University. Physics. Astronomy* **2** 40

[13] Grushina N.V., Zotov A.M., Korolenko P.V., Mishin A.Y. 2009 *Bulletin of Moscow University. Physics. Astronomy* **4** 47
[14] Melnikov G.A., Ignatenko N.M., Melnikov V.G., et al. 2014 *Journal of Nano- and Electronic Physics* **6(**3) 03016(4pp)
[15] Melnikov G.A., Ignatenko N.M., Melnikov V.G., et al. 2015 *Journal of Nano- and Electronic Physics* **7,4(2)** 03016(4pp)
[16] Melnikov G., Ignatenko N., Melnikov V., Cherkasov E. 2015 *IOP Conf. Ser.: Mater. Sci. and Engineering* **81** 012032
[17] Melnikov G., Ignatenko N., Krasnych P., et al. 2016 *IOP Conference Series: Materials Science and Engineering* **110** 012064
[18] Galliero G., Latte T., Bessieres D., Boned Ch. 2007 *J. Chem. Phys.* **127** 184506
[19] Robertson J. 2002 *Materials Science and Engineering R* **37** 129
[20] Andreev V.D., Malik V.R., Podzyarey G.A. 2001 *J. Struct. Chem.* **42(2)** 233

IOP Conf. Series: Materials Science and Engineering **168** (2017) 012021 doi:10.1088/1757-899X/168/1/012021

Cluster melting in effective potential model

G Melnikov[1], S Yemelianov[1], N Ignatenko[1], E Cherkasov[1] and O Manzhos[1]
[1]*Kursk, 305040, Russia,* South-West State University (SWSU)

E-mail: melnikovga@mail.ru, rio_kursk@mail.ru, inmkstu@bk.ru

Abstract. The paper shows that the melting temperature of clusters and nanoparticles is determined by the depth of potential well of pair interaction potential, the number of particles in a cluster, and the particle packing pattern. A cluster is in solid (crystalline or quasicrystalline) state, if the potential energy of interaction between cluster particles corresponds to global minimum, in this case the geometric configuration of particle spatial arrangement is considered to be the structure of a solid cluster. A formula is proposed to calculate cluster melting temperature depending on the number of particles in a cluster.

1. Introduction

Since recently phase transitions in cluster systems have been a matter of focal theoretical scientific interest and experiments. Some examples of such studies are the papers by B.M. Smirnov, [1–4]; R.S. Berry et. al. [5–10]; D.J. Wales [11, 12] and others. A comprehensive bibliography on melting processes running in cluster systems and nanoparticles can be found in G. Makarov's review [13]. Research results have revealed main features of phase transitions in cluster systems. One of them is the phenomenon of two co-existing phases: a cluster system has a certain temperature region that ranges from the temperature of transition into solid state T_{sol} to the temperature of transition into liquid state T_{liq}. The region of two phase coexistence is determined by the difference between these two temperatures, with this difference tending to approach zero as the number of the particles in the cluster tends to infinity. Within this co-existence region there is such temperature at which free energies of solid and liquid states get equal. In the condition of thermodynamic equilibrium this temperature is called *cluster melting temperature*.

Solid-liquid state phase transition in clusters is understood as configuration excitation in the cluster structure. The complexity of its description can be explained by the difficulty to find the global minimum of cluster energy and an optimal configuration of cluster atoms spatial arrangements that will agree therewith. Potential energy surface (PES) of a cluster has multiple local minimums and saddle nodes. For a Lennard-Jones cluster with particle number 13 there are 988 PES local minimums, and these local minimums are exponentially increasing in number as the quantity of atoms in the cluster increases [13].

In lattice models the melting of a cluster is described as first order "order-disorder" phase transition which is accompanied by spasmodic internal energy change and entropy of the system at a melting temperature when free energies are equal sufficiently to distribute atoms in the cluster system. Here the function of elementary configuration excitation belongs to void as an exited vacancy. For example, in cluster X_{13} , that is structured as a icosahedron, phase transition into liquid state takes place when only one atom moves out of the occupied shell onto its surface. Lattice models utilize traditional Lindemann criterion of melting and more accurate criteria for Etters-Kaelberer clusters [14,15] that can jump in the melting point.

Phase transitions in cluster systems are accompanied by an abnormal heat capacity behavior close to the melting point. Negative heat capacity of a cluster, which was predicted in [16, 17], was experimentally found for sodium ion cluster [18]. Size effect for melting temperature, that was tentatively described in [19], today has been verified by many experiments, including diffraction methods (by means of transmission electron microscopy it has been found that diffraction pattern changes in the melting point, which is explained by cluster structure disordering under melting effect [20]); reflection methods, that are based on the changes in the reflectivity of the clusters implanted into a transparent template when they transit through the melting range [21]); calorimetric methods (when the cluster, that has been deposited onto the surface of a thin-film calorimeter, is melting there is a sharp increase in the heater's power consumption rate [22]); optical techniques (used to identify spectral changes in the clusters that consist of a giant molecule and clusters of inert gases [7, 23]). Cluster melting temperature determines the disintegration threshold of nanostructure elements and thin films.

Let's assume that global conformation of a cluster is a geometric configuration in the spatial arrangement of atoms (or centers of masses of molecules) in a cluster and this configuration corresponds to the global minimum of the total energy of interaction between the cluster particles. In the approximation of the first-order interaction of two particles in the cluster when their interaction energy can be described through pair interaction potential as the function of one argument, which is the distance between the particles, the energy of the interaction of all cluster particles can be defined as the sum-total of the energies of first-order interactions N of the particles in its structure:

$$E\left(r_{ij}\right)=\sum_{i=1}^{N-1}\sum_{j=i+1}^{N}V\left(r_{ij}\right) \tag{1}$$

To find the global minimum of function (1) is a serious problem in the physics of cluster systems. There are numerous theoretical works devoted to its solution and informative databases that contain the values of cluster minimal energy with relative conformations for different numbers of atoms and different interaction potentials [24]. In order to find the minimum of function (1) different mathematical methods can be applied and different interaction potentials can be used [25, 26].

In this paper we refer to the power potential of interaction of particles with potential well effective depth that depends on the number of particles in a cluster system as it helps us to determine the dependence of cluster melting temperature on the number of particles in the cluster structure.

2. The depth of potential well of pair interaction potential

In the most general case we have defined the depth of potential well of pair interaction potential by the formula [27]

$$\varepsilon_{eff}=\varepsilon_0+\Delta\varepsilon \tag{2}$$

where ε_0 - the depth of potential well of pair interaction potential, $\Delta\varepsilon$ - actual energy increment predetermined by the interaction with the nearest neighbors. The calculation of is the main objective in describing inter-molecular interaction of particles within effective field.

With When studying the interaction of a cluster with a free-moving particle having a kinetic energy that is proportional to absolute temperature T with the help of cluster model the author of this paper obtained the following relationship for $\Delta\varepsilon$ [27 – 29]

$$\Delta\varepsilon = \left(\frac{Z-1}{Z+1}\right)^2 kT \tag{3}$$

where Z - the number of particles in a cluster at a temperature T.

According to formulas (2) and (3) efficient depth of the potential well of interaction potential will be defined by

$$\varepsilon_{eff} = \varepsilon_0 + \left(\frac{Z-1}{Z+1}\right)^2 kT = \varepsilon_0\left[1+\left(\frac{Z-1}{Z+1}\right)^2 \frac{kT}{\varepsilon_0}\right] \tag{4}$$

The depth of the potential well of pair interaction potential ε_0 is determined by the critical temperature of a matter. For inert gases and simple fluids with Lennard-Jones potential $\varepsilon_0 = 0,769T_c$, with Buckingham potential (6-exp) $\varepsilon_0 = 0,82T_c$, with Morse potential $\varepsilon_0 = 0,947T_c$.

Today more than twenty potentials are used to describe interactions between atoms in crystals and fluids. Quite often power potentials are used, for example, Mie potential [2]

$$\varphi(r_{ij}) = \frac{\varepsilon_0}{n-m}\left(\frac{n^n}{m^m}\right)^{\frac{1}{n-m}}\left[\left(\frac{\sigma}{r_{ij}}\right)^n - \left(\frac{\sigma}{r_{ij}}\right)^m\right] \tag{5}$$

where ε_0 - the depth of potential well, σ – effective diameter of interacting particle, $n \geq m$ – integer exponents, with $m = 6$, $n = 9....24$.

In the proposed model Mie's power potential shall be changed by replacing the value $\varepsilon_0 = 0,769T_c$ by the effective depth of potential well that is found by formula (4) and writing it as

$$V(r_{ij}) = C_{ij}\varepsilon_0\left[1+\left(\frac{Z-1}{Z+1}\right)^2 \frac{kT}{\varepsilon_0}\right]\left[\left(\frac{\sigma_0}{r_{ij}}\right)^n - \left(\frac{\sigma_0}{r_{ij}}\right)^m\right]$$

$$C_{ij} = \frac{n}{(n-m)}\left(\frac{n}{m}\right)^{\frac{m}{n-m}}, \sigma_0 = \left(\frac{m}{n}\right)^{\frac{1}{n-m}} R_1 \tag{6}$$

With exponent values $m = 6$, $n = 12$ Mie potential (5) is simplified and is called Lennard-Jones (LJ) potential. Then in accordance with formula (6) the radius of the first coordination sphere will be: $R_1(LJ) = 2^{\frac{1}{6}}\sigma = 1.1225\sigma$, for other values of parameters (m,n) we'll get: $R_1(6,9) = 1.1447\sigma$, $R_1(6,18) = 1.0959\sigma$, $R_1(6,24) = 1.08006\sigma$ etc. It is worth noting that in Lennard-Jones potential the coefficient of proportionally 1.1225 of the first coordination sphere radius and the effective diameter of a particle correspond to fcc lattice, while with $m = 6$, and $n = 18$ it corresponds to bcc lattice.

3. Cluster melting temperature

For macroscopic crystals there is empirical rule according to which the crystal melting temperature is proportional to the potential well of pair interaction potential. In the first approximation this rule applies both to cluster systems and nanoparticles [30]. The influence of the surrounding particles on the energy of first-order interaction results in the dependence of cluster melting temperature on the quantitative composition of a cluster or its geometrical dimensions. To evaluate the depth of effective potential well we shall use formula (4). For the cluster system $T_m = c\varepsilon_{eff}$, $c = const$, and considering the formula (4) melting temperature will be defined by the following relationship.

$$T_m(Z)=c\varepsilon_0\left[1-c\left(\frac{Z-1}{Z+1}\right)^2\right]^{-1} \tag{7}$$

Emprical constant c will be defined through passage to the limit. For a macroscopic sample with $Z\to\infty, T_m(Z)\to T_m(\infty)$ i.e. cluster melting temperature will tend to the melting temperature of the macroscopic sample, hence by applying approved data on inert gases and formula (7) we'll get the empirical constant value:

$$c=\left[1+\frac{\varepsilon_0}{T_m(\infty)}\right]^{-1}=\frac{2}{3}(\Phi-1)=0.412... \tag{8}$$

Considering the obtained value of the empirical constant the relationship (7) can be written as

$$T_m(Z)=0.56(56)...(\Phi-1)\varepsilon_0\left[1+\left(\frac{Z-1}{Z+1}\right)^2\right] \tag{9}$$

where Φ=1.6180339... – golden section ratio that defines Lindemann criterion for cluster melting [31, 32].

In cluster systems with a finite number of particles the cluster melting temperature reduced to pair interaction energy will be the universal function of the number of particles in the cluster irrespective of the cluster nature.

In formula (9) the depth of potential well of pair interaction potential will depend on the potential type, thus any cluster melting temperature predictions made by means of this formula may be uncertain. So it might be proper to write formula (9) as follows

$$T_m(Z)=\frac{1}{2}T_m(\infty)\left[1+\left(\frac{Z-1}{Z+1}\right)^2\right] \tag{10}$$

where $T_m(\infty)$ - bulk sample melting temperature that can be experimentally found with sufficient accuracy.

The obtained relationship to be used to calculated cluster melting temperature (10) does not contain any empirical constants, satisfies limiting processes and agrees with empirical data received by means of different methods.

Figure1 shows reduced melting temperature of inert gas quasicrystal clusters - cluster qualitative composition dependences that have been found by formula (10) (b), and similar diagrams obtained for argon by molecular dynamics methods (a) [33]. According to theoretical model proposed in [34] for organic nanocrystals the dependence of cluster melting temperature from cluster size is essentially similar to our calculations and simulations made in [33].

The paper [34] was devoted to the size effect of melting temperature in organic nanocrystals that were produced by filling porous materials with appropriate organic liquids. Its authors showed that the dependence of the melting temperature of such object on its size can be described by the following formula:

$$\frac{T_m(r)}{T_m(\infty)}=\frac{\sigma^2(\infty)}{\sigma^2(r)}=\exp\left[\frac{-(\alpha-1)}{(r/r_0)-1}\right] \tag{11}$$

where r - the radius of a crystal; $T_m(\infty)$ - bulk crystal melting temperature; r_0 - critical radius at which all atoms of a nanoparticles are located on its surface; $\sigma^2(r)$ - mean square displacement

(MSD) of atoms of a particle with a radius r; $\sigma^2(\infty)$ - is MSD of atoms of the relevant bulk crystal; α - the ratio of MSDs of surface and internal atoms of the crystal.

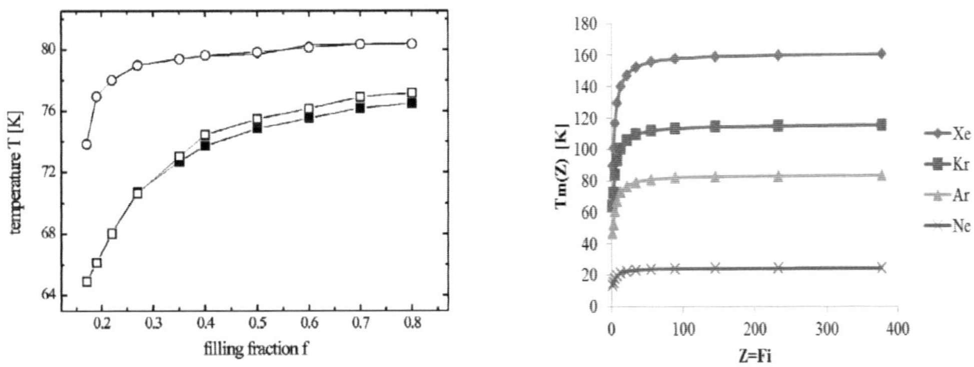

a) b)

Figure 1. a) Crystallization and melting temperatures of argon clusters in triple point T=83,3 K, b) melting temperature of noble gases clusters that has been found by formula (10).

In [35–37] it was demonstrated that the dependence of a metal cluster melting temperature on the cluster radius can be described by simple relationship

$$T_m(R) = T_m(\infty)\left[1 - \frac{\alpha}{R}\right] \quad (12)$$

where α - empirical constant that is approximately equal for all metals with fcc structure.

In such systems the relationships (11), (12) and our formula (10) yield such cluster melting temperature - number of particles relations that have good qualitative agreement (calculation disagreement is about 10-15%), so simultaneous solution of these equations allows us to determine the relationship between the radius of a cluster system and the number of particles in this system.

4. Conclusion

Melting temperature of clusters and nanoparticles is determined by the depth of potential well of pair interaction potential, the number of particles in a cluster and particle packing pattern. A cluster is in solid (crystalline or quasicrystalline) state, if the potential energy of interaction between cluster particles corresponds to global minimum, in this case the geometric configuration of particle spatial arrangement is considered to be the structure of a solid cluster.

A certain energy imparted to a cluster system destructs the original particle arrangement that agrees with global minimum, and a new particle configuration is formed. This particle rearrangement can be interpreted as the configuration of melting onset;this melting process ends when the initial global configuration of particles is lost completely. Cluster melting process runs with a constant number of particles in the cluster in contrast to the process of particle emission either out of cluster space or from the surface of a cluster or a nanoparticle.

Acknowledgments

The work financially supported the Ministry of Education and Science of Russia in the basic tasks of the State.

References

[1] Smirnov B M 1994 *Uspekhi Fizicheskikh Nauk (UFN)* **37** 1079
[2] Berry R S, Smirnov B M 2009 *Uspekhi Fizicheskikh Nauk (UFN)* **179(2)** 147
[3] Berry R S, Smirnov B M 2005 *Uspekhi Fizicheskikh Nauk (UFN)* **175(4)** 367
[4] Berry R S, Smirnov B M 2013 *Uspekhi Fizicheskikh Nauk (UFN)* **183(10)** 1029
[5] Berry R S 1993 *Chem. Rev.* **93** 2379
[6] Berry R S 1998 *Nature* **393** 212
[7] Berry R S 1999 *Theory of Atomic and Molecular Clusters* (Berlin: Springer) 1
[8] Berry R S, Jellinek J, Natanson G 1984 *Phys. Rev.* **A 30** 919
[9] Berry R S, Smirnov B M 2004 *JETP* **98(2)** 366
[10] Jellinek J, Beck T L, Berry R S 1986 *J. Chem. Phys.* **84** 2783
[11] Wales D J, Berry R S 1990 *J. Chem. Phys.* **92** 4283
[12] Wales D J, Berry R S 1994 *Phys. Rev. Lett.* **73** 2875
[13] Makarov G N 2010 *Uspekhi Fizicheskikh Nauk (UFN)* **53** 179
[14] Etters R D, Kaelberer J B 1975 *Phys. Rev. A.* **11** 1068
[15] Etters R D, Kaelberer J J 1977 *Chem. Phys.* **66** 5112
[16] Bixon M, Jortner J J 1989 *Chem. Phys.* **91** 1631
[17] Labastie P, Whetten R L 1990 *Phys. Rev. Lett.* **65** 1567
[18] Schmidt M et al. (2001) *Phys. Rev. Lett.* **86** 1191
[19] Pawlow P 1909 *Z. Phys. Chem.* **65(1)** 545
[20] Takagi M J 1954 *Phys. Soc. Jpn.* **9** 359
[21] Garrigos R, Cheyssac P, Kofman R 1989 *Z. Phys. D* **12** 497
[22] Lai S L et al. 1996 *Phys. Rev. Lett.* **77** 99
[23] Jortner J Z 1994 *Physikalische Chemie* **184** 283
[24] Wales D J, Doye J P K, Dullweber A., et al. 2012. *The Cambridge Cluster Database* http://www-wales.ch.cam.ac.uk/CCD.html
[25] Posypkin M A 2010 *Bulletin of the Nizhny Novgorod University. N.I. Lobachevsky* **1** 210
[26] Evtushenko Y, Posypkin M A 2013 *Optimization Letters* **7(4)** 819
[27] Melnikov G A, Verveyko V N, Melichov Ju F, et al. 2011 *Bulletin of Moscow State Technical University. N.E. Bauman. Series: Natural sciences* **3** 108
[28] Melnikov G A, Ignatenko N M, Melnikov V G, et al. 2015 *Journal of Nano- and Electronic Physics* **7** 4(2) 04087
[29] Melnikov G, Ignatenko N, Melnikov V, Cherkasov 2015 *IOP Conference Series: Materials Science and Engineering* **81** 012032
[30] Davydov S Yu 2015 *Semiconductor Physics and Technology* **49** 1683
[31] Melnikov G A, Ignatenko N M, Melnikov V G et al. 2016 *The collection of materials of the III International Scientific Conference. Tula: Izd. of TGPU them. L.N. Tolstoy* 49
[32] Melnikov G A, Ignatenko N M, Melnikov V G et al. 2016 *The collection of materials of the III International Scientific Conference. Tula: Izd. of TGPU them. L.N. Tolstoy* 43
[33] Soprunyuk V P, Wallacher D, Huber P et al. 2003 *Phys. Rev. B.* **67** 144105-1
[34] Jiang Q, Shi H X, Zhao M 1999 *J. Chem. Phys.* **111(5)** 2176
[35] Peters K F, Cohen J B, Chung Y W 1998 *Phys. Rev. B* **57** 13430
[36] B Castro T, Reifenberger R, Choi E, Andres R P.1990 *Phys. Rev. B* **42** 8548
[37] Ben David T, Lereah Y, Deutscher G et al. 1995 *Philos. Mag. A* **71** 1135

Effect of ion irradiation on the surface energy of deposited coatings

E N Eremin[1], S A Guchenko[2], S S Kasymov[2], V M Yurov[2] and M V Vedyashkin[3]
[1]Omsk State Technical University, Omsk, Russia
[2]E.A. Buketov Karaganda State University, Karaganda, Kazakhstan
[3]National Research Tomsk Polytechnic University, Tomsk, Russia

E-mail: exciton@list.ru

Abstract. We investigated multi-element coatings exposed to argon ion bombardment. The coatings were irradiated using a multi-ampere hollow-cathode ion source. The arc current was 1 A, and the potential of the substrate was maintained equal to 300 V. The surface tension (surface energy) of the coatings was measured before and after irradiation through the size-dependence of the microhardness and electrical resistivity of coatings on their thickness. Ion irradiation was found to affect the surface energy of the coatings in different ways. This is due to both the structure of the coating and its elemental composition.

1. Introduction

In recent years, the concept of high entropy multi-element alloys and coatings based on these alloys has been widely developed [1–5]. Various properties of these coatings are being investigated, including radiation resistance under irradiation by ions at different energies [6-8].

The properties of the coatings are primarily affected by ion bombardment at the stage of their origination due to relaxation of stresses in the area subjected to ion irradiation and crystal structure rearrangement. Point defects, active adsorption centers, are formed on the surface. The mobility of adatoms on the surface (surface diffusion) enhanced by low-energy inert gas ion bombardment of the growing film also plays an important role. The number of point defects can be increased by increasing either the ion flux energy or the ion current density. Defect formation is followed by the reverse process of defect recombination, "annealing", which reduces defect concentration. These two processes result in an equilibrium number of nucleation centers, which can be varied by changing the ion irradiation parameters.

This paper presents the results of the research in the ion bombardment induced effect on the surface energy of some multi-element coatings.

2. Materials and methods

In the experiments, we used Zn-Cu-Al, Mn-Fe-Cu-Al and Cr-Mn-Si-Cu-Fe-Al cathodes manufactured through induction melting. These cathodes were used to deposit coatings on a steel substrate using the ion-plasma unit HHB-6.6-I1 at different process conditions.

The atomic force microscope NT-206 was used to perform a nanoscale study of the coating surface.
Two methods developed and reported by the authors in [9, 10] were used to determine the surface tension (surface energy) of the deposited coatings. The first method involves measurement of the surface tension by determining the dependence of the microhardness on the deposited coating thickness. The dependence of the microhardness of the deposited coating on its thickness is described by the formula:

$$\mu = \mu_0 \cdot \left(1 - \frac{d}{h}\right), \qquad (1)$$

IOP Conf. Series: Materials Science and Engineering **168** (2017) 012022 doi:10.1088/1757-899X/168/1/012022

where μ is the microhardness of the deposited coatings; μ_0 is the microhardness of the "thick" sample; h is the thickness of the deposited coating. The parameter d is related to the surface tension σ by the formula:

$$d = \frac{2\sigma\upsilon}{RT}. \tag{2}$$

Here, σ is the surface tension of the bulk sample; u is molar volume; R is gas constant; T is temperature. The coordinates $\mu \sim 1/h$ ($1/h$ is the inverse thickness of the coating) indicate a straight line, the slope determined by d, and the surface tension of the deposited coating (σ) is calculated by formula (2).

The second method is used to measure the dependence of electrical conductivity Ω of the deposited coating on its thickness h that is described by the formula (1):

Argon ion irradiation of the coatings was performed using a hollow cathode multiampere ion source. The arc current was 1 A, and the substrate potential was maintained equal to 300 V.

3. Results and discussion

Figure 1 shows the AFM images, and Table 1 presents the surface energy values of the investigated coatings.

Table 1. - Surface energy of the deposited coatings.

Coating	Before irradiation	After irradiation
	σ, J/m^2	σ, J/m^2
Zn–Cu–Al	0.243	0.241
Cr–Mn–Si–Cu–Fe–Al	0.711	1.422
Mn–Fe–Cu–Al	0.367	0.122

As shown in Table 1, all three investigated coatings behave differently under ion irradiation: the Zn-Cu-Al coating exhibits radiation resistance, and the surface tension value remains practically unchanged; the surface tension of the Cr-Mn-Si-Cu-Fe-Al coating increases by a factor of two, and the tension of the Mn-Fe-Cu-Al coating decreases by a factor of three. These changes are associated with changes in the surface coating structure during ion bombardment (Fig. 1).

Radiation resistance of the Zn-Cu-Al coating is attributed to its distinct globular structure (Fig. 1). The presence of the system of "balls" leads to elastic scattering of argon ions, so that the local deformation is negligible.

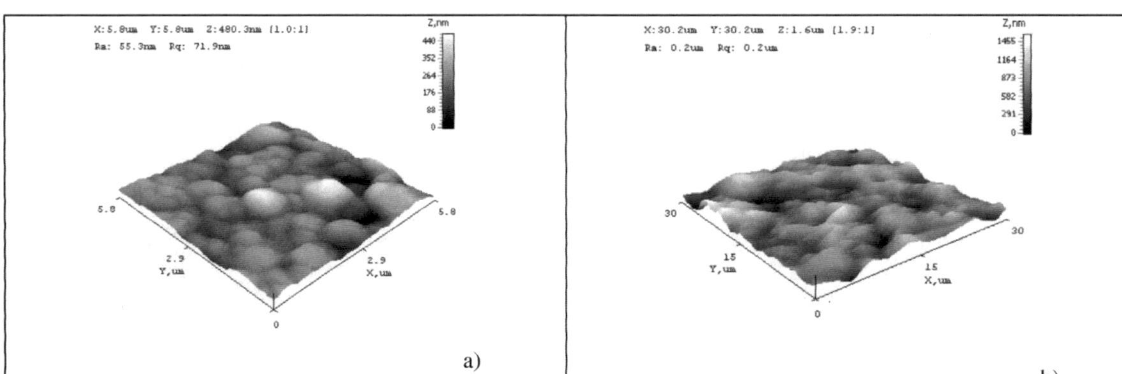

IOP Conf. Series: Materials Science and Engineering **168** (2017) 012022 doi:10.1088/1757-899X/168/1/012022

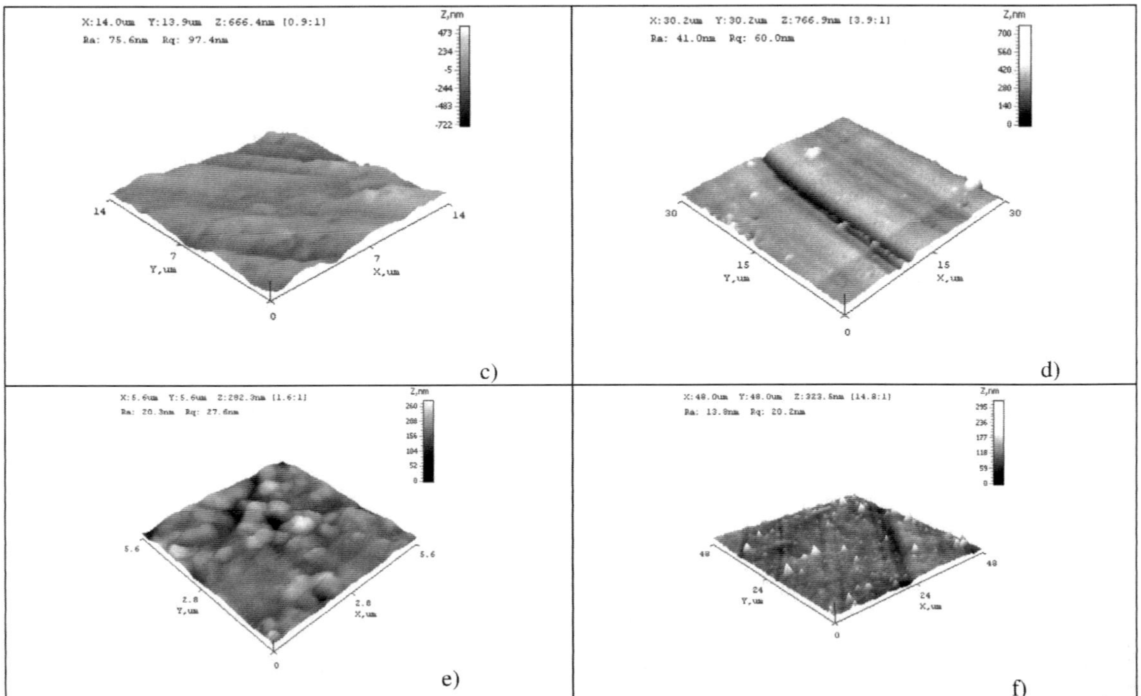

Figure 1. AFM images of the Zn–Cu–Al, Cr–Mn–Si–Cu–Fe–Al and Mn–Fe–Cu–Al coatings before (a, c, e) and after argon ion irradiation (b, d, f).

The wear resistance of the coating depends on the work of its destruction, which is equal to:

$$A = \sigma \cdot S, \tag{3}$$

where S is the sample surface area.

Table 2. Surface tension of pure metals (M) at 300 K [10].

M	σ_{300}, J/m^2	M	σ_{300}, J/m^2	M	σ_{300}, J/m^2	M	σ_{300}, J/m^2	M	σ_{300}, J/m^2	M	σ_{300}, J/m^2	M	σ_{300}, J/m^2
Li	0.152	Sr	0.730	Sn	0.205	Cd	0.294	Fe	1.508	Gd	1.285	Ac	1.023
Na	0.071	Ba	0.683	Pb	0.300	Hg	0.07	Co	1.463	Tb	1.331	Th	1.723
K	0.037	Al	0.633	Se	0.193	Cr	1.873	Ni	1.426	Dy	1.380	U	1.105
Rb	0.012	Ga	0.003	Te	0.425	Mo	2.573	Ce	0.777	Ho	1.434	Np	0.613
Cs	0.002	In	0.129	Cu	1.056	W	3.373	Pr	0.908	Er	1.470	Pu	0.610
Be	1.258	Tl	0.276	Ag	0.934	Mn	1.217	Nd	0.998	Tm	1.518	Am	0.973
Mg	0.623	Si	1.386	Au	1.036	Tc	2.173	Sm	1.025	Yb	0.797	Bk	0.998
Ca	0.818	Ge	0.931	Zn	0.399	Re	3.123	Eu	0.875	Lu	1.625	-	-

Formula (3) demonstrates that increased surface energy will lead to the increase in wear and radiation resistance. The multi-element Cr-Mn-Si-Cu-Fe-Al coating presented in Table 1 meets this condition. Similar results were obtained in [6, 7].

Since the surface energy value for solid solutions is additive, the coating should be selected based on the surface energy of each component of the alloy in order to increase the radiation resistance of solid solutions (single-phase alloys). Table 2 shows the surface tension (surface energy) values for the majority of the metal elements of the periodic table.

Within the framework of the molecular theory, the friction force is equal to:

$$F_{tr} = \int \sigma dL \approx \sigma \cdot L, \tag{4}$$

where L is the path length.

Formula (4) indicates that growth in the surface energy causes decrease in friction. This case is valid for the Mn-Fe-Cu-Al coating.

4. Conclusion

The presented results show that ion irradiation (or any other irradiation) can be used to change the basic properties of coatings in either direction through the change in the surface energy. In practice, it is crucial for both aircrafts in the ionosphere and components of nuclear reactors.

References

[1] Otto F, Yang Y, Bei H, George E P 2013 *Acta Materialia* **61** 2628–38

[2] Tsai M H, Yeh J W 2014 *Mater. Res. Lett.* **2** 107–23

[3] Schuh B, Mendez-Martin F, Vulker B et. al. 2015 *Acta Materialia* **96** 258–68

[4] Pogrebnyak A D, Bagdasaryan A A, Yakuschenko I V, Beresnev V M 2014 *Advances in Chemistry* **83(11)** 1027–61

[5] Surzhikov A P, Lysenko E N, Malyshev A V, Vlasov V A, Suslyaev V I, Zhuravlev V A, Korovin E Y and Dotsenko O A 2014 *Russian Physics Journal* **57** 5 621-626

[6] Komarov F F, Pogrebnyak A D, Konstantinov S V 2015 *Journal of Applied Physics* **85(10)** 106–11

[7] Yurov V M, Laurinas V Ch, Guchenko S A, Zavatskaya O N 2014 *Hardening techniques and coatings* **1** 33–36

[8] Ghyngazov S A, Vasil'ev I P, Surzhikov A P, Frangulyan T S, Chernyavskii A V 2015 *Technical Physics* **60** 1 128-32

[9] Eremin E N, Syzdykova A Sh, Guchenko S A et. al. 2016 *IOP Conf. Series: Materials Science and Engineering* **110** 012017

[10] Yurov V M 2011 *Eurasian Physical Technical journal* **8** 1(15) 10–14

Effect of ultraviolet and x-ray radiation on optical properties of epoxy polymers dyed with organic phosphors

V CH Laurinas[1], S S Kasymov[1], V M Yurov[1], E N Eremin[2] and M V Vedyashkin[3]
[1]E.A. Buketov Karaganda State University, Karaganda, Kazakhstan
[2]Omsk State Technical University, Omsk, Russia
[3]National Research Tomsk Polytechnic University, Tomsk, Russia

E-mail: exciton@list.ru

Abstract. Highly purified industrial bisphenol and cycloaliphatic epoxy oligomers of ED-24 and UP-612 brands were used to produce optically transparent products. UV radiation of a low-pressure mercury lamp with 80% of the light energy at 254 nm was used to study photodegradation. X-ray apparatus with 0.7BSV-Ag tube was used as an ionizing radiation source to investigate the effect of X-rays on the spectra of organic dyes in epoxy polymer. The threshold value of the energy generated by ruby laser which indicated the degradation in the test samples recorded by light scattering method was determined to study radiation resistance of epoxy polymers. Basically, all the dyes exhibited high resistance to UV light. The observation of the absorption spectra showed that on average, a third of the dye molecules in the matrix experienced photobleaching within 200 hour exposure. The exception was coumarin 1, which was completely decolourized after 40 hours of exposure. X-ray irradiation of the samples for two hours results in the change in the optical density equivalent to that caused by 40 hour exposure to UV irradiation. However, in the first case, the matrix optical density is proportional to the irradiation time, and in the second case, it remains stable upon further UV irradiation. The comparison of photoaging processes in dyed and undyed epoxy polymers showed that the investigated organic dyes do not have a sensitizing effect on the matrix. The stability of the optical properties of the epoxy matrices exposed to the effects of different factors was found to depend on the nature of epoxy polymer and the technique of its production. The results of these effects are significantly different in the character of the change in the optical density and mechanisms of chemical transformations in polymer.

1. Introduction

One of the promising areas for application of transparent polymeric materials is laser technology, where these materials can be used to make active elements with lasing dyes, bleaching filters for Q-switching, and others. A dye laser with a dye-doped solid matrix combines the advantages of solid-state lasers, which can generate radiation in a wide spectrum range completely overlapping the entire visible spectrum [1–3].

Practical use of laser elements largely depends on their ability to provide the operational life and to efficiently convert pump radiation of different power levels. These parameters depend on both the solid medium and the dye incorporated [4–6].

This paper presents the results obtained from the experimental study of photo- and X-ray degradation of dyed epoxy polymers.

2. Materials and methods

Epoxy oligomers are typically produced through the reaction of epichlorohydrin with phenol. The structure and properties of cross-linked epoxy polymers and the kinetics and thermodynamics of their formation are well studied [7]. The basic reactions occurring during epoxy polymer curing using amines, most commonly used curing agents, can be represented by the following scheme (Fig. 1).

The reacting system components, starting materials and reaction products, are hetero- and auto complexes that can be formed both inter- and intramolecularly. At high temperatures, the formed epoxy groups are condensed with unreacted hydroxyl and with each other that induces formation of additional lattice points.

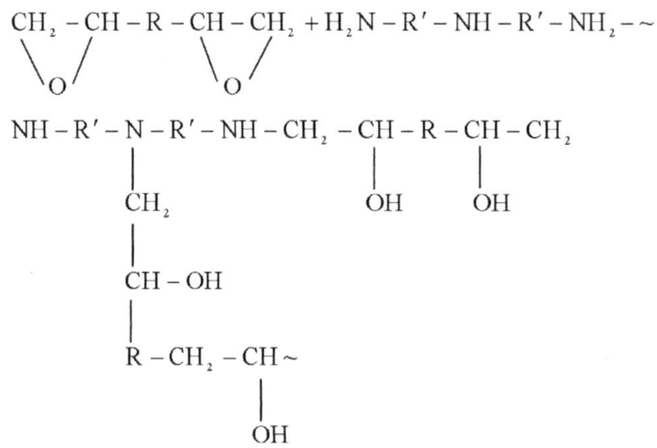

Figure 1. Scheme for producing epoxy polymers.

The synthesis of epoxy polymers was performed using the technology developed in the Ukrainian Research Institute of Plastics (Donetsk) in the 80-ies of the last century.

To obtain optically transparent products, we used highly purified industrial bisphenol and cycloaliphatic epoxy oligomers of ED-24 and UP-612 brands. The absorption spectra of these oligomers have two maxima at 227 and 282 nm. The corresponding absorption coefficients are significantly different and equal to 112200 and 17800 for ED-24, and to 500 and 25 cm^{-1} for UP-612. The transmittance of the curing agents in the visible spectral region is also high. The hardeners include a large number of curing agents of both basic and acid series.

An important advantage of epoxy polymers is the ability to vary their properties over a wide range through the selection of epoxy oligomers and curing agents. To study photodegradation, we used UV radiation generated by low-pressure mercury lamp with 80% of the light energy at 254 nm. The samples were placed in a chamber at 0.25 m from the source. The resultant ozone was blown off. The absorption and emission spectra of the samples were measured after every 20 hours of exposure.

The X-ray effect on the organic dye spectra of epoxy polymer was studied using an X-ray apparatus URS-1.0 with 0.7BSV- tube as an ionizing radiation source. The samples were mounted close to the windows of the X-ray tube. The voltage of the X-ray tube was 45 kV, and its current was 10 mA. The samples were exposed to radiation for 8 hours, and the optical spectra were measured every 2 hours.

The threshold value of the ruby laser energy which indicated the start of the sample degradation recorded using the light-scattering method was determined to study radiation resistance of epoxy polymers. An aperture with a diameter of 3.8 mm was mounted in the path of the ruby laser radiation, and the sample was placed behind the aperture. The energy level of laser radiation was measured using the IKT-1H device and averaged over the whole spot area, $S = 0.113\ cm^2$. The laser operated in a passive Q-switching mode. Phthalocyanine vanadium solution in chlorobenzene was used as a photo-optical shutter.

3. Experimental results and discussion

Fig. 2 shows the absorption and emission spectra of dyed epoxy polymers after UV irradiation.
All the tested dyes largely exhibited good resistance to UV light. The observation of the absorption spectra showed that around a third of the dye molecules in the matrix were photobleached during 200 hour exposure. The exception was coumarin 1, which was completely decolorized after 40 hours of exposure. The changed luminescence intensity was proportional to the changed dye concentration in epoxy polymer.X-ray irradiation lasted for 8 hours. The absorption and luminescence spectra were measured every 2 hours.

No changes observed in the absorption spectra indicate that the tested dyes are radiation resistant. The luminescence intensity slightly decreased after exposure due to the overlapping of the dye and epoxy polymer absorption bands that leads to non-radiative losses of the absorbed radiation in the matrix.

Exposure of the samples to X-ray irradiation for 2 hours causes the changes in the optical density equivalent to those occurring during 40-hour UV irradiation. However, in the first case, the optical density of the matrix is proportional to the irradiation time, and in the second case, the density remains constant under further UV irradiation.

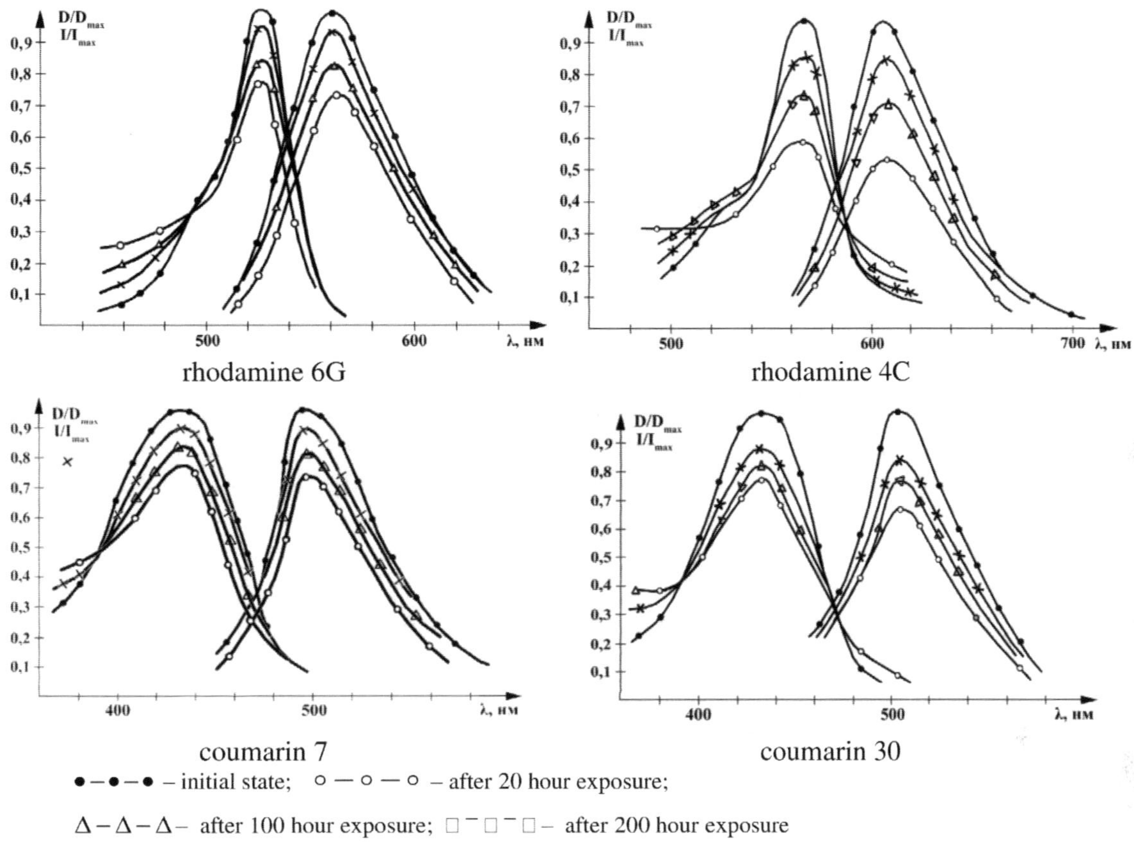

Figure 2. Effect of UV irradiation on absorption and luminescence spectra of dyed epoxy polymers.

Tables 1–2 show the position of the absorption bands after exposure to UV light and X-rays, respectively. The polymers in the tables are arranged in order of decreasing optical clarity. Sample 1919 is amine-cured epoxy polymer, and other samples are anhydride-cured epoxy polymer treated at different curing temperatures.

Table 1. Short-wavelength transmission cutoff of the epoxy polymer clarity after 200 hour UV irradiation.

Sample no.	1914	1904	1913	1901	1915	1900	1903	1919
Transmission cutoff T = 32%, λ, nm	302	325	332	316	336	346	347	370
Δλ	2	15	17	26	-28	31	35	95

As can be seen from the data presented in Tables 1–2, samples 1914, 1904 and 1901 produced at low curing temperature with different polymer-curing agent ratio exhibit the greatest stability of optical properties. In Table 1, the sign "-" before the value λΔ for sample 1915 indicates that the transmission cutoff under UV exposure shifted toward the short wavelength region. This indicates the photopolymerization processes in epoxy polymer caused by the crosslinking agents due to photodissociation of the molecules in the hardener, which possesses a high absorption capacity in a free state in the blue spectral region.

Table 2. Short-wavelength transmission cutoff of the epoxy polymer clarity after 6 hour X-ray ageing.

Sample no.	1914	1904	1913	1901	1903	1900	1915	1919
Transmission cutoff T = 32%, λ, nm	340	358	365	360	370	385	405	completely blackens after 1 hour exposure
Δλ	40	48	50	70	58	70	30	

The study of the effect of UV radiation on optical clarity of the epoxy polymer revealed cavities on the sample surface visible to the naked eye. These cavities are attributed to the impact of ozone. The size and number of defects indicated the polymer resistance to ozone. The surface of sample 1904 remained virtually clean throughout the irradiation time. Thus, epoxy polymer 1904 exhibited the most complete set of performance characteristics required for creation of active laser media for extreme conditions.

The comparison of photoaging of dyed and decolorized epoxy polymers revealed that the investigated organic dyes do not have a sensitizing effect on the matrix. Under X-ray irradiation, the optical density of the samples was proportional to the exposure time. The comparison of the data in Tables 1 and 2 revealed a complete correlation between the resistance of epoxy polymers to UV and X-ray irradiation. However, the effects of UV and X-ray irradiation on the samples are significantly different. The radiation-induced absorption occurs in the absorption spectra of epoxy polymers in the region of 400 to 440 nm: in samples 1901–1914, it occurs after 6 hours of X-ray irradiation, in sample 1915, it can be observed after 4 hour irradiation, and in sample 1919, it is recorded after 10 minute irradiation.

The studies showed that the resistance of the epoxy matrix optical properties to the effects of various factors depends on the nature of epoxy polymer and the technique of its production. The results of these effects are significantly different in the character of the optical density changes and mechanisms of chemical transformations in polymer.

We determined the threshold value of the ruby laser irradiation energy density which indicated the sample degradation visually observed in the diffused light of the helium-neon laser.

Epoxy samples of different chemical composition were studied:

- Sample 1924 dyed with rhodamine 6G ($c=4.68 \cdot 10^{-5}$ g/g);
- Sample 1925 dyed with rhodamine C ($c=4.68 \cdot 10^{-5}$ g/g);
- Samples 1924 and 1927 bleached and modified with diaplast;
- Sample 1929 dyed with rhodamine 6G ($c=10^{-4}$ mol/l) unmodified;
- Sample 1927, polymethylmethacrylate dyed with rhodamine 6G.

The data on the damage threshold are presented in Table 3.

Table 3. Radiation strength of the polymer samples.

Sample no.	PMMA	1927	1929	1924	1925
E, J/cm^2	6.5	5.3	4.8	4.4	3.5

The results of the study show that the damage threshold of the modified polymethyl methacrylate is higher than that of epoxy polymers.

Incorporation of diaplast in the epoxy polymer matrix improved its viscoelastic properties and increased the damage threshold. This fact is in full accordance with the data reported in numerous studies that established the dependence of the polymer damage threshold on its viscoelastic properties.

Thus, the damage threshold of both epoxy polymers and PMMA can be increased through matrix modification. Moreover, the above stated results show that low-molecular plasticizers are more suitable for this purpose.

4. Conclusion

Several tens of polymers and their compositions have been used as active laser media for more than 50 years of the laser technology development. These studies identified the materials, which are most promising from a practical standpoint. It was shown that with respect to the requirements for thermal stability, mechanical hardness, solvent resistance and optical clarity over a wide wavelength range, 17 polymers are found to be suitable for their use in high power lasers.

However, 5–6 polymers (PMMA, PS, PU, PC, and epoxy polymers and their modifications) have been commonly used as active laser media so far. Epoxy polymers exhibit a wide range of differences in their properties depending on the dye, modifier and production technologies. This is the aspect we aimed to consider in the current paper.

References

[1] Korobkin Yu V, Sidorov O I, Studentov V B Active medium of lasers based on the colored epoxy and triazine-containing polymers 1997 *Journal of Technical Physics* **67(10)** 71–76

[2] Solodova T A Creation of solid active media for lasers based on organic and hybrid polymers 2015 *Thesis for the degree of Doctor of Physical and Mathematical Sciences, Tomsk* 109

[3] Laureto J J, Dessiatoun S V, Ohadi M M and Pearce J M Open Source Laser Polymer Welding System: Design and Characterization of Linear Low-Density Polyethylene Multilayer Welds 2016 *Machines* **4** 14

[4] Pascault J-P *Thermosetting Polymers. Marcel Dekker* 2002 477

[5] Belfiore L A *Physical Properties of Macromolecules. Wiley* 2010 803

[6] Yurov V M, Laurinas V Ch, Eremin E N, Gyngazova M S Determination of glass transition temperature for polymers by methods of thermoactivation spectroscopy 2016 IOP Conf. Series: Materials Science and Engineering **110** 012018.

[7] Chernin I Z, Smekhov F M, Zherdev Yu V *Epoxy polymers and compositions M.: Chemistry* 1982 232

Thermodynamic models of radiation-induced processes in solids

V M Yurov[1], E N Eremin[2], S S Kasymov[1], V CH Laurinas[1] and A V Chernyavskii[3]
[1]E A Buketov Karaganda State University, Karaganda, Kazakhstan
[2]Omsk State Technical University, Omsk, Russia
[3]National Research Tomsk Polytechnic University, Tomsk, Russia

E-mail: exciton@list.ru

Abstract. A thermodynamic model is proposed to qualitatively describe the radiation-induced processes in solids: temperature dependence of the X-ray radio luminescence output, dependence of these processes on the excitation density, energy accumulating in a solid under exposure to ionizing radiation and its temperature dependence. The proposed model and the formula derived can be used to develop radiation-resistant and radiation-sensitive materials.

1. Introduction

Mathematical and computer modeling of radiation defects in solids has been widely researched in a number of papers, among which we particularly mention the papers [1–3] which provide an extensive bibliography. Nevertheless, the development of high-quality analytical models of radiation-induced defect formation is advantageous for both theoretical considerations and practical applications.

In [4], we addressed the issue of the response of the defect subsystem in solids to external effects (temperature, pressure, light, etc.) using a nonequilibrium statistical thermodynamics approach.

Crystal defects (color centers, etc.) were considered as a non-interacting particle system immersed in a thermostat. Quantum transitions caused by the interaction of defects with the thermostat are dissipative (with probability of P_0) in contrast to those occurring during the interaction with an external field (with probability of P_1). The dissipative processes always make the secondary field (system feedback) smaller than the primary one that causes defects formation.

For the response function $F=P_0/P_0+P_1$, the obtained expression is

$$F = \frac{1}{1+\frac{\tau_p}{\tau}\frac{2\Delta S}{k}\exp\left(-\frac{E_m - G^0/N}{kT}\right)}, \tag{1}$$

where τ is radiative lifetime τ_r is relaxation time, ΔS is the entropy change during the radiationless transition, k is the Boltzmann constant, E_m is the energy of the defect ground state, G^0 is the thermostat Gibbs energy, N is the number of defects and T is temperature.

In this paper, we attempt to show how to use thermodynamic model (1) to analyze the radiation-induced processes in solids.

2. Temperature dependence of the radio luminescence yield

The simplest pattern of the temperature dependence of the luminescence yield can be observed for sufficiently pure crystals (Fig. 1).

The nature of this luminescence is well-established: the emission appears during radiative annihilation of the self-trapped exciton caused by the electron recombination with the V_K-center. This recombination may occur at different points of the lattice without migration losses. The shape of the dependence I(T) is characteristic of the intra-center processes, and it can be well described by the Mott formula:

$$I(T) = \frac{I_0}{1 + A\exp(-q/kT)}, \tag{2}$$

where q is the activation energy of the nonradiative transition, A is a constant, T is temperature and k is the Boltzmann constant.

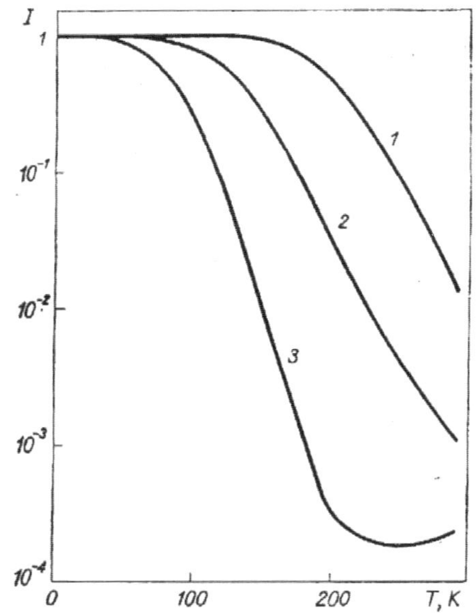

Figure 1. Temperature dependence of the luminescence intensity for pure NaBr (1), LiCl (2) and KI (3) crystals excited by X-irradiation [5].

The Mott formula coincides in its form with formula (1), where the response function $F = I/I_0$. However, in this case, the activation energy

$$q = \frac{E_m - G^0/N}{kT} \tag{3}$$

is related to both the luminescence center parameters (E_m and N), and the crystal properties, the matrix (G^0 and T).

Equations (1) and (3) enable interpretation of the difference in the X-ray luminescence quantum yield and its temperature behavior for different crystals that cannot be directly derived from (2).

Equation (3) shows the dependence of the activation energy q for temperature quenching on the number of luminescence centers N even in the absence of coagulation processes. This does not correlate with the conventional model of the luminescence center potential curves.

We found out this effect experimentally long time ago [6], and it is illustrated in Fig. 2.

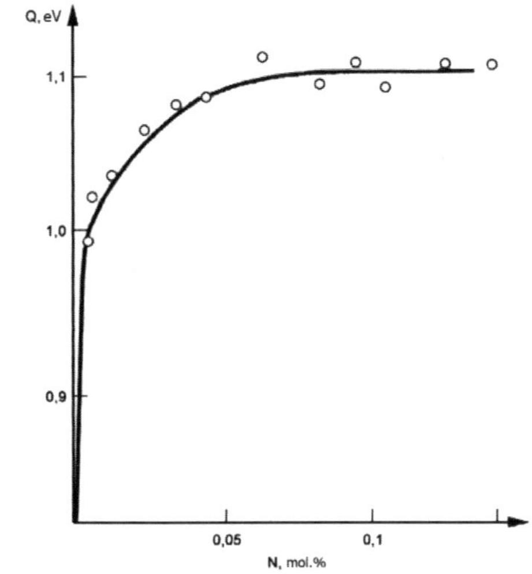

Figure 2. Dependence of the activation energy Q (eV) on the luminescence center concentration N(mol.%) for KBr-Tl [6].

3. Dependence of the crystal x-ray luminescence quantum yield on excitation density

In case of one type of defects present in the solid, the defect concentration N is proportional to the excitation intensity, and the dependence for the recombination luminescence yield obtained for the AHC series by (1) is shown in Fig. 3.

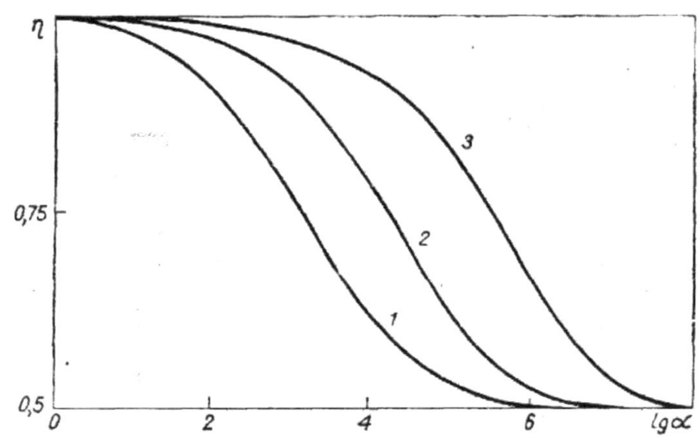

Figure 3. Dependence of the luminescence yield on the excitation density for KCl (1), KBr (2) and KI (3) [5].

The assumptions are rough, and in practice the dependence η(e) is much more complex, however, the pattern presented in Fig. 3 indicates that the greater the Gibbs energy of the host crystal, the greater the excitation intensity required for recombination luminescence quenching.

4. Storage of energy in dielectrics under the exposure of ionizing radiation

Equation (1) can be reduced to a linear form under certain assumptions. Then, taking the stored energy S as a response function, we get:

$$S = b\frac{kT}{G^0} \cdot N, \tag{4}$$

where b is the constant.

At low irradiation doses, the energy storage is a linear function of the irradiation time (Fig. 4).

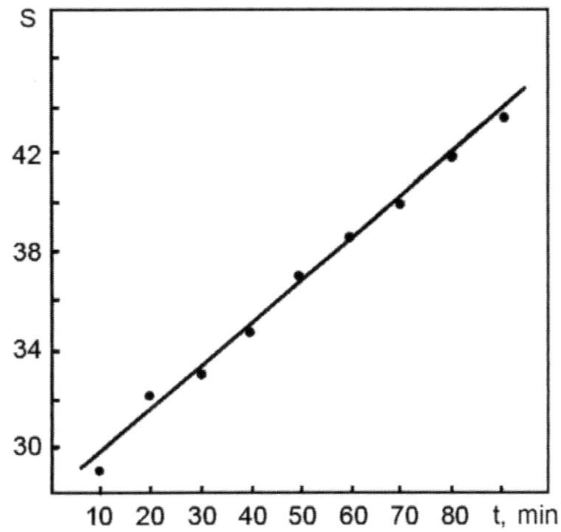

Figure 4. Dependence of the stored light sum in the TSL peak T=200 K for the crystal BiOCl on the X-raying time.

The rate of the defect accumulation is determined by the slope of the curve (4):

$$v_d = tg\alpha = a\frac{kT}{G^0}. \quad (5)$$

Table 1 shows the values v_d for the AHC series obtained from the experimental data.

Table 1. – The defect accumulation rate and Gibbs energy for the AHC series.

Crystal	Defect accumulation rate, v_d, $cm^{-3}\ c^{-1}$	ΔG^0 Gibbs energy, kJ/mol
NaCl	$0.7\cdot10^{-13}$	-356.2
KCl	$3\cdot10^{-13}$	-409.05
KBr	$2\cdot10^{-13}$	-380.7
KI	$0.3\cdot10^{-13}$	-323.24

Table 1 shows that the greater absolute value of Gibbs energy results in the greater rate of defect accumulation due to the negative Gibbs energy. Since the defect accumulation rate defines the crystal radiation resistance, the crystals with greater absolute values of Gibbs energy will be radiation resistant. For example, oxides of most of the elements (MgO, Al_2O_3, etc.) have greater absolute values of Gibbs energy and, hence, these are radiation resistant. Note that $G^0=\mu$ (chemical potential) for metals.

The temperature dependence of the F-center generation can be divided into three stages [5]: the first stage is linear dependence, the second stage is temperature independence, and the third stage is the decrease with the increased defect concentration. The complexity of the dependence of the F-centers generation is considered to be caused by the change in the lifetime of various halogen defects complementary to F-centers. We offer another interpretation of this effect based on formula (4). In fact, the dependence of Gibbs energy on the temperature is of the form:

$$G^0 = A + BT + CT^2, \quad (6)$$

where A, B, C are constants.

According to formula (4), we have a linear dependence of energy storage on temperature at insignificant T – $A>BT+CT^2$. If $G^0\approx BT$, S does not depend on temperature. At greater T, Gibbs energy $G^0\approx CT^2$ and S~1/T, that is the energy decreases hyperbolically with the increased temperature grows. A qualitative pattern of the temperature dependence S is shown in Fig. 5.

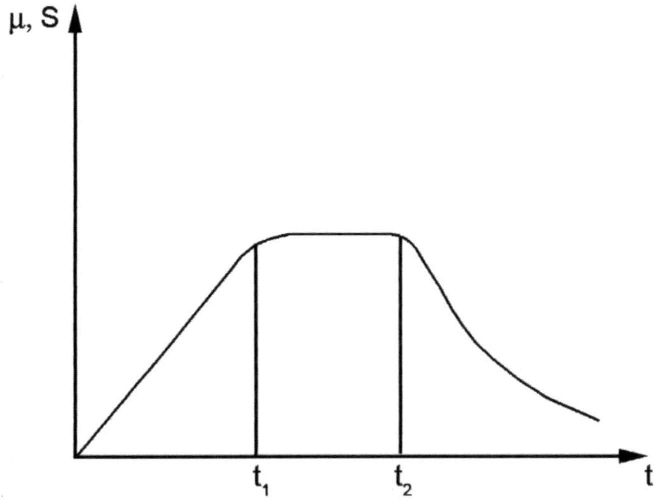

Figure 5. Three stages of temperature dependence of the optical absorption coefficient (μ) of F-centers and stored energy (S) on temperature (t°C).

5. Conclusion

The study results show that the proposed thermodynamic model enables qualitative description of numerous radiation-induced processes in solids under exposure to ionizing irradiation. Considering that Gibbs energy G^0=U-TS+PV (U is internal energy, T is temperature, S is entropy, P is pressure and V is volume), formulas (4) and (5) can be used to perform the analysis of the effect of pressure (P), phase transitions (S) [7-9] and other effects in solids exposed to ionizing irradiation.

Acknowledgements

This work was supported by the Ministry of Education and Science of the Russian Federation within the framework of the government "Science" program.

References

[1] Karkina L E, Yakovenkova L I 2011 Modeling the atomic structure of defects in crystals. Ekaterinburg, Ural Branch of the RAS 463
[2] Ostapchuk P N 2012 The kinetics of radiation-induced processes in crystals. Thesis for the degree of Doctor of Physical and Mathematical Sciences Kharkov 318
[3] Kharchenko D O, Lysenko I O, Kharchenko V O Modelling of microstructural transformations in the systems exposed to irradiation 2012 *Advances in Metal Physics* **13** 1 101–87
[4] Yurov V M Thermodynamics of luminescent systems 2005 *Bulletin of University of Karaganda. Physics* **3** (39) 13–25
[5] Schwartz K K, Ekmanis Yu A 1989 Dielectric materials: Radiation processes and radiation resistance. Riga: Zinatne 187
[6] Yurov V M, Murzakhmetov M K, Kuketaev T A Concentration effects in temperature quenching of the luminescence of impurity centers 1989 *Optics and Spectroscopy* **67** 6 1396–99
[7] Voevodin V N, Nekludov I M 2006 Evolution of the structural-phase state and radiation resistance of structural materials. Kiev: Naukova Dumka 376
[8] Surzhikov A P, Frangylyan T S, Ghyngazov S A 2012 *Russian Physics Journal* **55** 4 345-52
[9] Gyngazov S A, Surzhikov A P, Frangul'yan T S, Chernyavskii A V 2002 *Russian Physics Journal* **45** 8 753-58

IOP Conf. Series: Materials Science and Engineering **168** (2017) 012025 doi:10.1088/1757-899X/168/1/012025

Effect of Thermal Exposure on Structure of the Ultrafine-Grained Zr-1Nb Alloy

E N Stepanova[1], **G P Grabovetskaya**[2], **I P Mishin**[2], **D Yu Bulynko**[2] **and A S Dubrovskaya**[1]
[1] Tomsk Polytechnic University, 30 Lenin Avenue, Tomsk, Russia
[2] Institute of Strength Physics and Materials Science of Siberian Branch of Russian Academy of Sciences, 2/4, Akademicheskii Avenue, Tomsk, Russia

E-mail: enstepanova@tpu.ru

Abstract. Effect of annealing at temperature range of 573–823 K on stability of the ultrafine-grained structure of the Zr-1wt.%Nb alloy was studied by methods of transmission electron microscopy. Growth kinetics of grain–subgrain structure elements of alloy was investigated.

1. Introduction

An effective method of improving the strength and operational characteristics of metal polycrystals at low homologous temperatures is grain refinement to submicron dimensions (grain size d < 1 μm) [1–3]. All most widely spread methods of the ultrafine-grained (UFG) state formation in metallic materials are based on Severe Plastic Deformation (SPD). SPD methods allow obtaining the UFG state in volume billets, which is a structure of grain type (size of grain–subgrain structure elements 0.05–0.5 μm) and contains predominantly high-angle grain boundaries [1, 4]. However, UFG state formed in metallic materials by SPD methods has a high density of lattice and grain boundary dislocations, elastic lattice distortions and long-distance stress fields and, as a consequence, is metastable. One of the possible ways of increase in stability of the UFG state and its mechanical properties is formation of internal structure in alloys. Such structure combined two and/or some structural elements (or phases) with different dispersion is obtained by means of directing use of structural and phase transformations initiated by SPD, as well as subsequent thermal treatments [5, 6]. In this connection, study of development features of such process, as recrystallization, phase transformations, solid solution decay, and phase coalescence during UFG state forming by SPD methods and subsequent thermal treatments is of great relevance.

On the basis of the above reasoning, a purpose of this work was experimental investigation of development features of structural and phase transformations at the UFG Zr-1wt.%Nb alloy (further Zr-1Nb) during annealing and electron beam irradiation at the temperature range of 573–873 K.

2. Experimental procedure

Commercial zirconium Zr-1Nb alloy (E110 grade) was used as an initial material for study in this work. Previously, it was found [3] that used Zr-1Nb alloy in as-received condition has polycrystalline structure with the grain size of 3–5 μm. X-ray structural analysis indicates that the alloy contains not only the main α-Zr phase, but also small amount of β-Zr phase (total volume fraction ~2.5 vol. %), β-Nb (solid solution of Zr on the base of Nb) and solid solution of Nb in β-Zr phase. Secondary phases in the form of particles, which dimensions vary from a few tens of nanometers to several microns, are present in the bulk and at grain boundaries of alloy.

UFG structure in alloy was formed using one of the SPD methods: pressing with the change of deformation axis and gradual temperature decrease at the interval of 973–623 K.

Investigation of thermal stability of the Zr-1Nb alloy UFG structure was carried out by means of annealing at the temperature range of 573–873 K, including annealing under constant electron beam exposure. Energy of electron beam was 35 keV, current density was equal to 75 and 110 μA, and time of irradiation was 1 hour.

Alloy structure was investigated with the use of JEM-2100 transmission electron microscope. The size of structure elements of the UFG alloy was measured from the relevant dark-field image micrographs by means of the secant method. Data sample was not less than 200 elements.

The volume fractions and lattice parameters of the phases were determined with accuracies of ±1% and 0.0001 nm, respectively, using a Shimadzu XRD7000 diffractometer with Cu–K_{α} radiation source. PowderCell program was used during identification of diffraction patterns.

3. Results and discussion

The typical electron micrographs of the UFG structure of the Zr-1Nb alloy after SPD is shown in Fig. 1. It is evident (Fig. 1, a) that an entangled deformational contrast is observed and some structure elements are poorly distinguished in the bright-field image of the structure. Electron diffraction patterns of such structure obtained for an area of 1.4 μm^2 show diffraction rings formed by the reflections from individual crystallites (Fig. 1, a). At the same time almost all of the reflections exhibit azimuthal smearing. Such kind of electron diffraction patterns is characteristic for the UFG state obtained by SPD method and indicates high-angle misorientations of the structure elements and presence of elastic stresses at the individual grains [1, 4]. Grain–subgrain structure elements, which dimensions vary between 0.1 and 0.5 μm, are clearly distinguished in the dark-field image (Fig. 1, b). The average size of grain–subgrain structure elements determined from the dark-field image is (0.3±0.12) μm. Obtained UFG structure contains only α-Zr and β-Nb phases. Volume fraction of the β-Nb phase does not exceed 0.5 vol. %. This phase is observed in the form of particles at the grain boundaries and volume of main α-Zr phase. This is indicative that under mentioned conditions of SPD process partial dissolution of secondary phases and formation of supersaturated solid solution occur in the Zr-1Nb alloy.

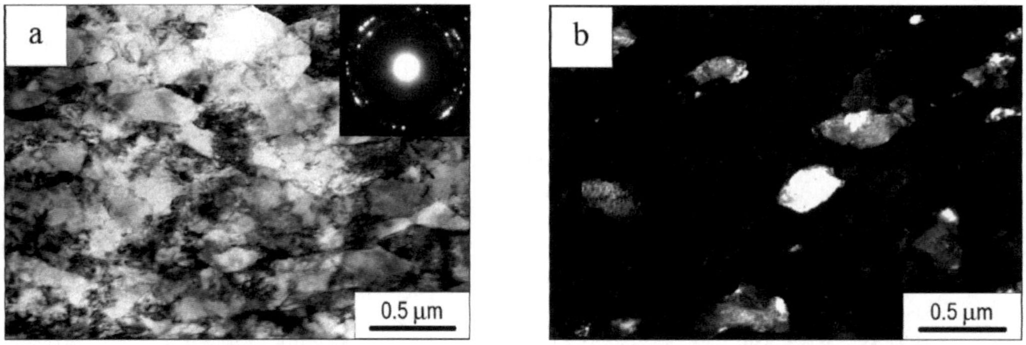

Figure 1. Electron microscope image of the UFG structure of Zr-1Nb alloy: (a) bright-field image and microdiffraction pattern, (b) dark-field image.

Electron microscopic studies of Zr-1Nb alloy microstructure after hour annealing at the temperature range of 573–873 K showed that UFG structure of alloy is stable up to a temperature of 723 K. However during annealing at the temperature range of 573–673 K relieving of defect UFG structure is observed: dislocation density at the grain volume decreases and boundaries of individual grains reveal. At the same time the fringe contrast is detected at some grain boundaries that is inherent in the equilibrium state of grain boundaries (Fig. 2, a). After annealing at temperature of 723 K individual recrystallized grains with dimensions of 1–2 μm appear in structure (Fig. 2, b). Density of such grains is low, so the average size of the structural elements of the alloy found from the dark-field image increases slightly.

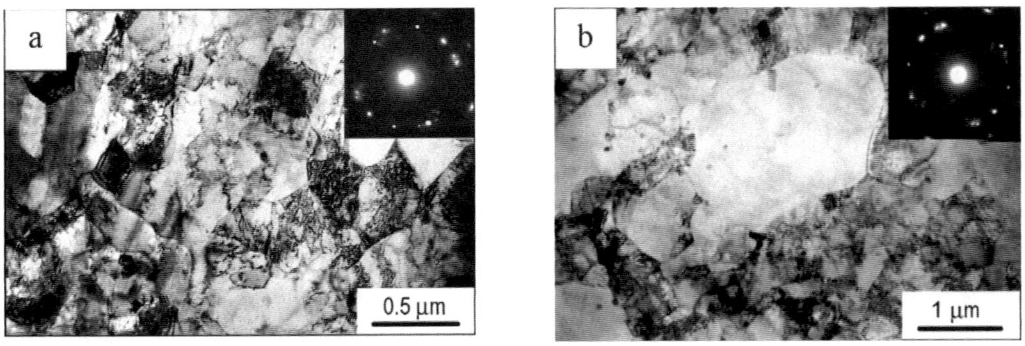

Figure 2. Electron microscope image and microdiffraction pattern of the UFG structure of Zr-1Nb alloy: (a) structure after annealing at 673 K for 1 h, (b) structure after annealing at 723 K for 1 h.

After annealing at temperature of 773 K recrystallization and grain growth are observed in the whole volume of samples. The average grain size increases up to 1.2 μm. At the same time during annealing at temperature of 873 K for 1 h collective recrystallization is actively developing and average grain size becomes equal to 2.5 μm (Fig. 3, b).

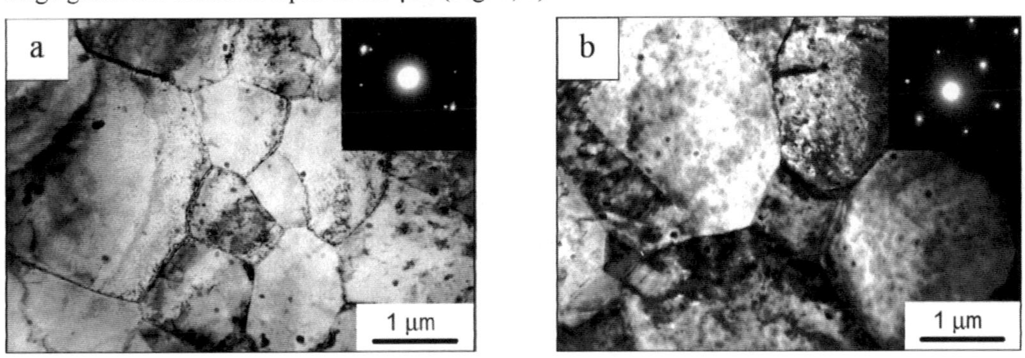

Figure 3. Electron microscope image and microdiffraction pattern of the UFG structure of Zr-1Nb alloy: (a) structure after annealing at 873 K for 1 h, (b) structure after annealing at 723 K for 1 h under the electron beam irradiation

Stability of the UFG Zr-1Nb alloy reduces when annealing is carried out under conditions of electron beam irradiation. In this case growth of structure elements is observed throughout the sample volume already at 723 K. At the same time average element size has increased from 0.3 to 1.5 μm for 1 hour of annealing (Fig. 2, b). Recrystallization and grain growth are accompanied by precipitation of nano-sized particles of the β-Nb phase at the UFG Zr-1Nb alloy at temperatures of 723 K and above (Fig. 2, b and 3). Volume fraction of the β-Nb phase increases up to 1–1.5 %.

Annealing at temperature of 753 K for 2, 3, 5, 7 and 9 hours were performed to study growth kinetics of average size of grain–subgrain structure elements during annealing of investigated alloy. Kinetic curve of growth of grain–subgrain UFG structure elements of Zr-1Nb alloy at temperature of 753 K is represented at the Fig. 4. It is seen that there are two stages on this curve: with high and low rate of the structure element growth. High rate of structure element growth is observed at the first hours of annealing (2–4 hours). Then the rate of the structure element growth significantly slows down.

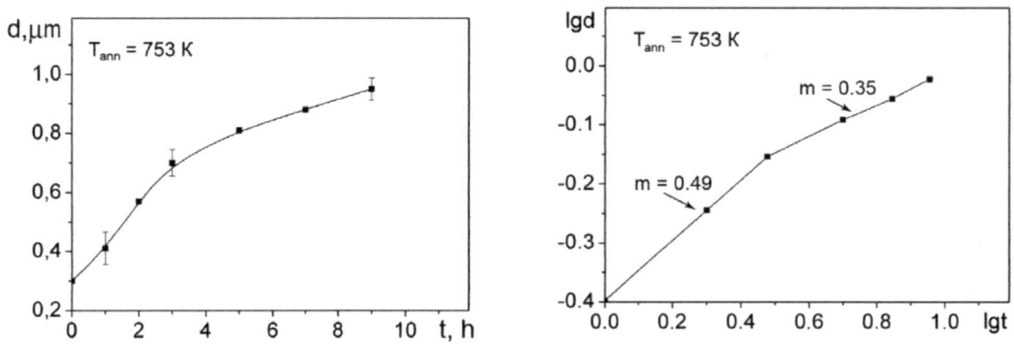

Figure 4. Kinetic curve (a) and logarithmic dependence (b) of change in the average size of the grain–subgrain UFG structure elements of the Zr-1Nb alloy during annealing at temperature of 753 K.

It is known [7] that depending on the development of the recrystallization process grain growth kinetics in materials with the UFG structure can be described by one of the following equations:

$$d = Kt^m \quad (1)$$

$$d = d_0 + Kt \quad (2)$$

where d – average grain size, K – constant factor, m – exponent of grain growth characterized dependence of grain growth on time at given temperature.

For the majority of fine metals grain growth during recrystallization is described by the equation of type (1) with m equal to 0.5. Rearrangement of d–t dependence for investigated zirconium alloy in logarithmical coordinates showed that kinetics of UFG structure element growth is described by an equation of type (1) on the both stages (Fig. 4, b). However, on the first stage the m value in the equation (1) is equal to 0.49, and on the second one – 0.35 (Fig. 4, b). Value $m = 0.49$ on the first stage of the UFG structure element growth for investigated heterophase Zr-1Nb alloy is closed to the m value for the fine metals. Apparently, this is due to a small volume fraction of the β-Nb phase particles, which are able to inhibit migration of grain boundaries. (It was noted above that content of secondary phases decreases to 0.5 vol. % in the process of SPD). With increasing of annealing time, solid solution decay and precipitation of nano-sized particles of the β-Nb phase occur in the Zr-1Nb alloy. These processes prevent grain boundary migration, so the rate of recrystallization process slows down.

4. Conclusion

Partial dissolution of secondary phases and formation of supersaturated solid solution are observed during UFG structure obtaining at the heterophase Zr-1Nb alloy with the use of pressing with the change of deformation axis and gradual temperature decrease at the range of 973–623 K. Hour annealing at the temperature range of 723–873 K result in growth of UFG structure elements and decay of supersaturated solid solution with the precipitation of nano-sized particles of the β-Nb phase. Growth kinetics of the UFG structure elements has double-stage character and is described by power dependence on the both stages. Slowing down the rate of element growth of the UFG grain-subgrain structure at the second stage is connected with precipitation of nano-sized particles of the β-Nb phase prevented grain boundary migration.

Acknowledgements
This work was supported by the Russian Foundation for Basic Research (project No. 15-08-03823).

References

[1] Kolobov Yu R, Valiev R Z, Grabovetskaya G P, Zhilyaev A P, Dudarev E F, Ivanov K V, Ivanov M B, Kashin O A, and Naidenkin E V 2007 *Grain Boundary Diffusion and Properties of Nanostructured Materials* (Cambridge: Cambridge International Science Publishing) 58-122
[2] Stepanova E N, Grabovetskaya G P, and Mishin I P 2015 *J. Alloys Compd* **645** S271
[3] Noskova N I, Mulyukov R R 2003 *Submicrocrystalline and Nanocrystalline Metals and Alloys* (UrO RAN, Ekaterinburg,) 119–190
[4] Grabovetskaya G P, Mishin I P, Stepanova E N, Chernov I P, Bulynko D Yu 2015 *Steel in Translation* **45** (2) 111
[5] Bachmaier A, Pippan R, Kerber M, and Setma D 2012 *Acta Mater.* **60** (3) 860
[6] Tian Y Z, Zhang Z F, Freuderger J, and Pippan R 2013 *Mater. Sci. Eng. A* **568** 184
[7] Ovid'ko I A 2009 *Materials physics and mechanics* **8** 174

The Investigation of Ion Implantation as a Technique for Manufacturing GaAs Magneto-Sensitive Detectors

G F Karlova[1] and B I Avdochenko[2]
[1]Leading Researcher at NIIPP, JSC, Tomsk, Russia
[2]Professor at TUSUR, Tomsk, Russia

E-mail: karlovagf@yandex.ru

Abstract. This paper studies thin active layers of n-n_i and n^+-n-n_i -types produced by means of silicon ion implantation into a semi-insulating GaAs substrate. The results of these structures' physical parameters investigation are presented. Based on the structures the Hall-effect sensors are designed that have the linearity of Hall voltage dependency on magnetic density $U_H(B)$ of at least 1% in the range of up to B<1.2 T.

1. Introduction

The interest of hardware engineers in sensitive and reliable magnetic-field detectors is dictated by both widespread development of automation and microelectronics and necessity of field density measurement. That is because detectors are the primary items for data acquisition from the object under measurement. The sensor operating threshold is determined by the ratio of the Hall-effect sensor probes voltage in the absence of magnetic field U_{res} to the output voltage in the presence of magnetic field U_H. In [1], the GaAs magnetic field detector based on epitaxial structure has been designed. The further step is to develop the magnetically controlled integrated circuit based on Schottky-barrier FETs. This represents a great challenge as the material with submicron thickness is required to create such devices. Furthermore, the net cost of epitaxial structures is relatively high. The ion implantation technique enables overcoming these difficulties and producing submicron structures for Hall-effect sensors.

This paper describes the technique for producing ion-doped GaAs structures, and reports the results of physical parameters investigation for these structures and Hall-effect sensors based on them.

2. Structure manufacturing technique

In the course of the research, the two types of structures have been obtained. They are n-n_i and n^+-n-n_i, where n is a carrier density in active layer, n^+ is a carrier density in contact layer, and n_i is a carrier density in a substrate. Silicon doping was carried out by means of the installation "Vezuvy-5". Semi-insulating undoped GaAs substrate had the orientation of (100) and specific electrical resistance of more than 10^7 Ohm·cm. Post-implantation annealing was conducted in arsenic atmosphere at the temperature of 1123 K.

The preliminary estimation of the doped Si ion concentration profiles is conducted using the Gaussian distribution [20]:

$$N(x) = (D/ (2\pi)^{1/2} \cdot \Delta R_p) \cdot \exp[-(x-R_p)^2 / 2 \cdot \Delta R_p^2], \qquad (1)$$

where D is an ion-implantation dosage; x is a current position; R_p is a projective range for ions, ΔR_p is an rms deviation of the range for ions (what is called struggling). The range and the struggling were determined by means of Lindhard-Scharff-Schiott theory [2]. To find experimental dependencies of distribution profiles on different implantation parameters, the measurements were conducted under the following conditions: a set of electron energies of 20–130 keV at the dosage of $3.75 \cdot 10^{12}$ cm^{-2}; a set of implantations dosages of $2.5 \cdot 10^{12}$–$2 \cdot 10^{13}$ cm^{-2} at the electron energy of 130 keV; and the dual implantation.

For produced layers several measurements of carrier density profile and mobility (μ) were conducted: measurement by Van-der-Pauw method with layer-by-layer etching under the room temperature; measurement by means of the profilometer EDK-6817 with mercury probe of small diameter; temperature measurement in the range of 85–400 K.

3. Key experimental results

1. The structures obtained have the following parameters: the structures of the first type — carrier density $n = (1.3–3.5)\cdot 10^{17}$ cm^{-3}, layer thickness $d = 250–340$ nm, carrier mobility $\mu = 3000–3500$ cm^2/V·s, layer resistance $r = 600–650$ Ω/sq; the structures of the second type — $n^+ = (1–2)\cdot 10^{18}$ cm^{-3}, $d = 70–80$ nm, $n = (1.3–3)\cdot 10^{17}$ cm^{-3}, $d = 250–300$ nm, $\mu = 2600–3200$ cm^2/V·s, $r = 400$ Ω/sq.
2. The parameters of Hall-effect sensor (input and output resistance R_{in} and R_{out}, sensitivity γ, residual voltage U_{res}, operating current I) are dependent on Si ion implantation mode (the energy and the dosage), presence of n^+-layer, and condition of the interface between the n-layer and the substrate. These data are shown in the following table.

Table 1. Types and parameters of the structures.

#	Type	n, cm^{-3}	R_{in}, kΩ	R_{out}, kΩ	I_{max}, mA	U_{res}, mV	γ, V/A·T
1	n-n_I	$2.0\cdot 10^{17}$	4	4	1.5	30–50	600–800
2	n-n_I	$3.0\cdot 10^{17}$	1.75	1.8	10	1–3	200
3	n-n_I	$2.5\cdot 10^{17}$	1.5–3	1.5–3	7	32–45	400
4	n^+-n-n_I	$2.0\cdot 10^{17}$ (n^+=$1\cdot 10^{18}$)	1.5–3	1.6–3	8	4–12	400
5	n^+-n-n_I	$3.5\cdot 10^{17}$ (n^+=$1{,}5\cdot 10^{18}$)	1,3–2	1.3–2	10	1–4	300–400

3. With the growth of density in n-layer, thickness of a space-charge region driven by surface potential gets lower. This leads to decrease of input resistance R_{in}, widening of operating current range, and reduction of residual voltage U_{res} and sensitivity γ (see structures 2 and 3 in Table 1).
4. The presence of n^+-layer in Hall-effect sensor structure results in contact resistance drop and screening the surface potential off. In general, it improves Hall-effect sensor parameters (see structures 4 and 5 in Table 1).
5. The presence of trapping sites at the n-n_i interface results in Hall-effect sensor sensitivity decrease as exemplified by the analysis of the relaxation curves obtained for the produced structures by means of GRAN-6 system.
6. The ohmic contacts' resistance asymmetry makes the main contribution to the U_{res} value.
7. The saturation area that is seen in a current-voltage curve can be explained by a backward control through the substrate and the presence of traps at the active layer surface [3].

4. Discussion of the experimental results

Figure 1 shows the profiles of Si ions doped with the dosage D of $3.75\cdot 10^{12}$ cm^{-2} and the energy of 130 keV (the solid curve is the one calculated using (1); the dots represent distribution of the electrically active carrier density after annealing). The difference between experimental and calculated curves is explained by the diffusion of doped ions during the annealing at the temperature of 1123 K. From the dependency of layer density on temperature it can be seen that density stays constant and equals to $1.5 \cdot 10^{12}$ cm^{-2} up to 400 K. The variation of mobility with the temperature in the range of up to 160 K is explained by ionized impurity scattering. In the range of T>160 K, it is explained by acoustic phonon scattering. These dependencies agree with calculated temperature dependencies for GaAs presented in [3]. However, the absolute value of the mobility is lower and at the room temperature equals $3.5 \cdot 10^3$ cm^2/V $\cdot$s. It points to the presence of significant number of impurities in GaAs layers obtained. To explore the influence of deep trapping sites at the interface between the thin layer and the substrate the samples under investigation were put into the reflection-type microwave resonator operating at the frequency of 38 GHz. The conductivity was estimated through the proportional change of microwave power reflected from the resonator. The measurement was performed alongside with the application of the illumination by a visible-spectrum LED to the side of thin layer (or substrate) and the bias to n^+-n or n-n_i interfaces. The typical picture is shown in Figure 2.

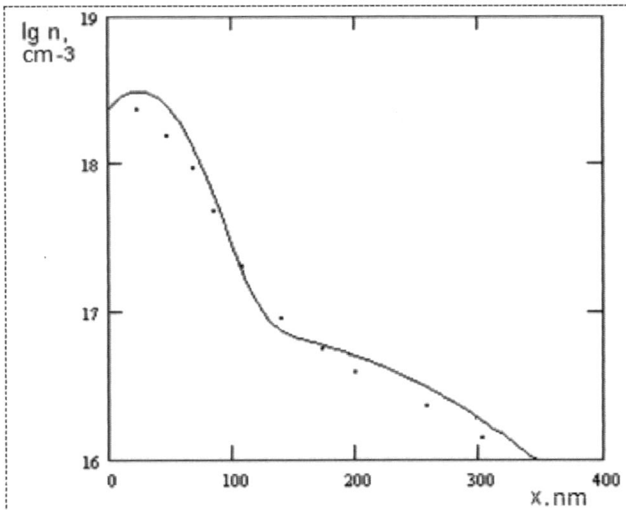

Figure 1. Distribution of carrier density across the depth (solid curve corresponds to calculated dependency, dots correspond to experimental data).

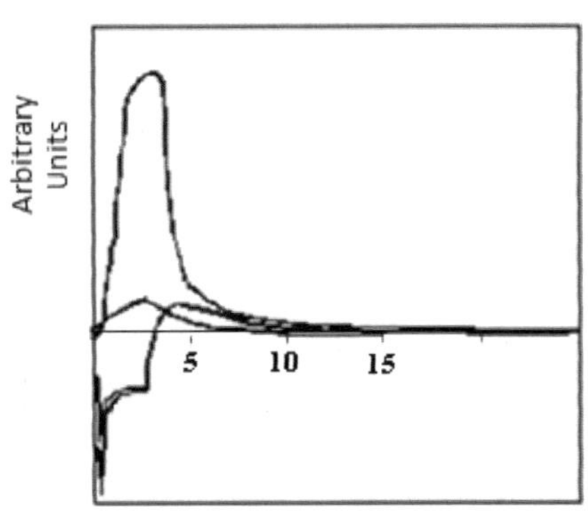

Figure 2. Photoresponse of the structure at different voltages under the forward (upper curves) and back (lower curves) biases.

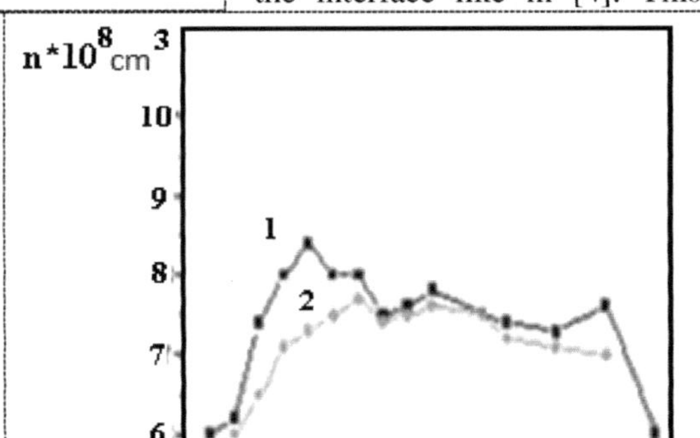

Figure 3. Distribution of carrier density across the depth depending on implantation parameters: curve 1 – 130 keV, $4{\cdot}10^{12}$ cm^{-2}; curve 2 – 130 keV, $3{\cdot}10^{12}$ cm^{-2}.

Figure 2 shows that the long-term relaxation of the light-induced conductivity comes from the recharge of deep level located inside the n-n_i-junction. The recharge is resulted from the illumination (i. e. injection of holes into the n-n_i-junction). For the samples under investigation, the charge carrier density profile in a bulk of epitaxial layer is estimated by means of voltage-capacitance curve obtained using the EDK 68-07 system. There is a peak at the interface between the thin layer and the substrate that can be explained by presence of deep trapping sites localized at the interface like in [4]. This peak corresponds to the peak of light-induced conductivity when the reverse bias is applied to the structure (see Figure 2). The analysis of the relaxation conductivity curves for the interface between the thin layer and the substrate shows that these defects can be explained by the quality of semi-insulating substrates.

The most important parameter of Hall-effect sensors is a residual voltage. It is associated with the resistance of planar ohmic contacts. For planar ohmic contacts with the length d_l and width w it can be found using the long line model [3]. The current I distribution under the contact with the length d is described by the following equation:

$$\frac{dI}{dx} = -J(x)w \qquad (2),$$

where x is a position, w is a width of the contact, J is a current density that equals to

$$J(x) = \frac{U(x)}{r_{конт}} \qquad (3),$$

where r_{cont} is a contact reduced resistance [$\Omega \cdot cm^2$], $U(x)$ is a channel potential related to a metal contact.

$$\frac{dU(x)}{dx} = \frac{-I \cdot R_{пов}}{w} \qquad (4),$$

where R_{surf} is a surface resistance [Ω/sq] of a semiconductive thin layer under the contact. The equations (1) and (3) resolve into the long line equation:

$$\frac{d^2U(x)}{dx^2} = \frac{U}{L_t} \qquad (5),$$

where $L_t = (r_{cont}/R_{surf})^{1/2}$ is a so called attenuation distance.

The solution to the equation (5) in respect to sensors is presented in [5]. The expression for the residual voltage has the following form:

$$U_{ocm} = I_0 \cdot r_{конт} \left(\mathrm{ch} \frac{d}{L_t} \right)^{-1}. \qquad (6)$$

The r_{cont} and subsequently $U_{res.k}$ depend on depth of alloying. The value of contact resistance is defined both by the condition of the interface between alloyed and non-alloyed regions and by the condition of metal-semiconductor junction.

In the course of the research, investigations of r_{cont} dependency on evaporated contact alloying parameters (maximal annealing temperature, annealing time, annealing temperature gradient), epitaxial substrate surface treatment, and contact annealing atmosphere were conducted. Changes of contact alloying parameters and subsequent measurements of r_{cont} for tailored samples using Cox method [3] were carried out to obtain the dependency of r_{cont} on the maximal temperature with the minimum value shift depending on annealing atmosphere. In the hydrogen atmosphere the minimal reduced contact resistance was lower than the one in the nitrogen atmosphere. The reduced contact resistance is shown to be dependent on the carrier density in the active region, although the contact layer of all structures under investigation had low resistance. The minimal value of r_{cont} at the carrier density $n = 2 \cdot 10^{17}$ cm^{-3} and d = 0.4 um is $4 \cdot 10^{-6}$ $\Omega \cdot cm^2$ while $U_{res} = 0.4$ mV, $U_H = 40$ mV, and the ratio $U_{res}/U_H = 0.01$. The asymmetry of r_{cont} in sensor chips result in appearance of U_{res} in a magneto-sensitive detector.

It is also found out that the low residual voltage in Hall-effect sensor can be obtained only when the carrier density profile at the interface between the thin layer and the substrate is sharp. Figure 3 demonstrates concentration profiles for two implantation parameter sets (the curve 1: 130 keV, $4 \cdot 10^{12}$ cm^{-2} and 50 keV, $5 \cdot 10^{12}$ cm^{-2}; the curve 2: 130 keV, $3 \cdot 10^{12}$ cm^{-2} and 50 keV, $3 \cdot 10^{12}$ cm^{-2}). For the first parameter set the residual voltage value is no more than 0.1 mV.

The current-voltage characteristics for all samples under investigation were measured. All of them were linear in the range up to the value U_L that corresponds to the limiting current I_{lim}. At $U > U_L$ the current-voltage characteristic becomes saturated wherein the lower $(n \cdot d)$ value, the earlier saturation comes. This phenomenon can be explained by a backward control through the substrate, and the presence of traps at the active layer surface [3].

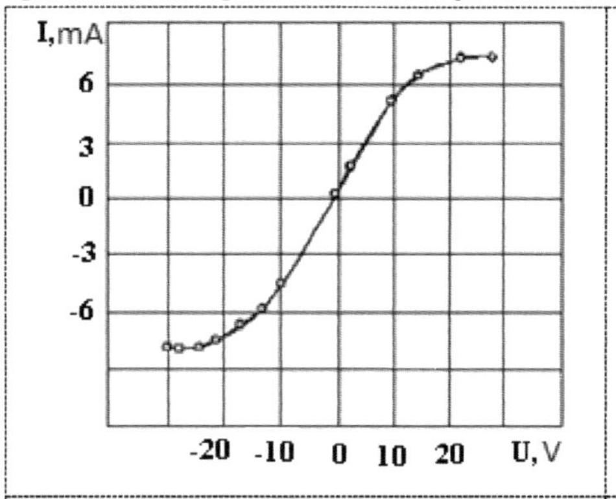

Figure 4. Current-voltage characteristic of Hall-effect sensor.

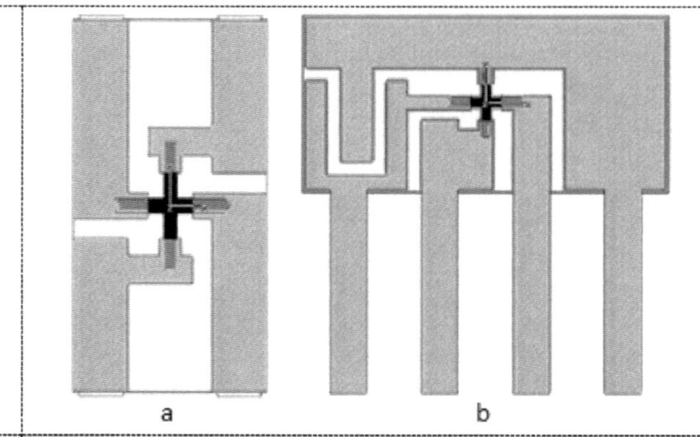

Figure 5. The detector structure: a – contacts at the different sides, b – contacts at the same side.

At the interface between the active layer and the substrate there is a charge-depletion layer of the definite thickness with the maximal width

$$d_{\max} \approx \left(\frac{1}{N_d} \ln \frac{N_d}{n_i} \right)^{\frac{1}{2}} \qquad (7),$$

where N_d is a dopant density at the homogeneous doping, n_i is an intrinsic carrier density. In this case, the saturation current based on the model of field effect transistor without a gate [3] can be found using the following expression:

$$I_{нас} = N_d \cdot q \cdot A_{акт} \cdot U_н \qquad (8),$$

where A_{act} is a thickness of active region of a channel. The expressions (7) and (8) show that as the dopant density in the active region decreases, the value of d_{max} increases and the saturation current gets lower.

The Hall-effect sensors are built on the obtained structures by means of the conventional GaAs technology used for manufacturing integrated circuits based on Schottky-barrier-gate FETs. Active region isolation of the magneto-sensitive detectors is performed by means of H, He and N ion doping. The Figure 5 shows the Hall-effect sensor structure. The die can be encapsulated in a compound or a SOT-8 package. In the first case, the total thickness will be no more than 1 mm, in the second one it will be 1.8 mm. The construction of the magneto-sensitive detector based on Hall-effect sensor with a ferrite ring helps to increase significantly the sensitivity of the current detectors designed today. The produced magneto-sensitive detectors have sensitivity γ=200–560 V/A·T. The dependencies of output voltage on magnetic density are linear and have nonlinearity factor no more than 1% at B<1.2 T. The residual voltage for the structures with sharp profile is no more than $0.1U_H$.

5. Conclusions

Based on the presented investigation, the following conclusions can be made:

1. The conducted estimation of carrier density profiles for ion doped structures is consistent with the experimental results obtained.
2. It is shown that to ensure low residual voltage carrier density profile at the interface between the thin layer and the substrate should be sharp.
3. The doped ions annealing parameters that ensure low r_{cont} are found. These parameters also ensure residual voltage value corresponding to the calculated one that is defined on the basis of the resistance of planar ohmic contacts.
4. The current-voltage curves of the structures obtained are linear only in a definite range, and then they go into saturation. The saturation can be explained by the backward control through the substrate, and the presence of traps at the active layer surface. The saturation current estimation agrees completely with the calculation.

The conducted investigation shows that it is possible to design Hall-effect sensors with high sensitivity ($\gamma \geq$=560 V/A T), low residual voltage ($U_{res} \leq U_H$), and nonlinearity factor of no more than 1% at B<1.2 T by means of thin submicron layers ion implantation.

Acknowledgments
The research was conducted with financial support from the Ministry of Education and Science of the Russian Federation within the base part of the federal assignment № 2014/225 (project code 3643).

References
[1] Karlova G, Porochovnichenko L and Umbras L Patent N2262777. 20.10.2005. *Bulletin* N29
[2] Rissel X and Ruge I1983 *Ion Implantation* (Moscow: Science)
[3] Shur M 1991 *Modern devices based on gallium arsenide* (Moscow: World)
[4] Karlova G, Porochovnichenko L The influence of interface between a thin layer and a substrate on the parameters of epitaxial semiconductor GaAs structures and Hall-effects sensors on their base. *Proceedings of the international conference "Physico-Chemical Processes in Inorganic Materials"* (Physical Chemical Problems-10), October 10–12 (2007) **2** 94–86
[5] Karlova G F, Belopolova T J *Theses of the IXth Russian conference on semiconductor physics.* September 28 – October 3 (2009). Novosibirsk-Tomsk 332

Research of creep deformation in amorphous and nanocrystalline alloys at variable temperature field

V A Fedorov[1], A D Berezner[1] and T N Pluzhnikova[1]
The Tambov state university n.a. G.R. Derzhavin,
392000, Internatsionalnaya str.,33, Tambov city, Russia

E-mail: qwert1009@mail.ru

Abstract. Investigations of Co-based amorphous alloys and Fe-based nanocrystalline alloy failure strain at creep tests within a temperature range being from 300 to 1023 K have been performed. A creep curve analytical form for all investigated alloys has been established.

1. Introduction

Investigation of processes running in amorphous and nanocrystalline metal samples is of interest considering that the given type of materials has unique magnetic and mechanical properties in relation to base metals. It is common knowledge that in the process of amorphous and nanocrystalline metals heating their structure can be rebuilt and, consequently, the above-said properties can change.

In spite of various scientific works [1-3] aimed at search of mechanical creep strain state reference parameters, explicit establishment of these parameters' relation with a material structure is complicated due to non-uniqueness of models in the framework of which these parameters were suggested. The three models being applied for qualitative description of processes running in amorphous and nanocrystalline metal alloys are most often considered. Among these are: a free volume model [4], a structure relaxation model [5] and a fluid model [6].

Within the framework of these models there are values related to creep strain parameters: ductility and a relative strain value at creep. However, due to elongation function explicit form absence data applying within the framework of the given models is complicated. In this context the aim of the work hereof is the following one: the experimental and analytical investigation of amorphous and nanocrystalline metal alloys creep strain behavior at the combined effect of a temperature field and fixed load.

2. Materials and research methods

Co-based amorphous and Fe-based nanocrystalline metal alloys, prepared by rolling at cooling rate 10^6 K/s and were used in the experiments, their elemental composition being shown in Table 1.

Table 1. Alloys elemental composition.

Alloy (AMAG)	Co%	Fe%	Ni%	Si%	Mn%	B%	Cr%	Cu%	Nb%
170	70.42	4.7	10.46	9	2.1	2	1.3	-	-
179	78.1	3.31	8.19	5.48	1.61	2	1.31	-	-
180	78.65	4.03	4.73	7.22	1.88	2	1.49	-	-
183	82.69	2.21	-	7.77	4.19	2	1.14	-	-
186	85.41	2.27	-	5.15	4.07	2	1.1	-	-
200	-	80.22	-	8.25	-	-	-	1.44	10.09

Band samples of 50 mm (length), 3.5mm (width), 0.02 mm (thickness) dimensions served as subjects of research. Heating rate was maintained constant as 1-3 K/s at creep tests. Sample temperature was measured with a Testo-845 laser pyrometer. Samples elongation was measured with a laser triangulating sensor.

3. Experimental results and their discussion

It has been shown that creep curves for amorphous and nanocrystalline alloys have characteristic view at different heating rates (see Fig.1).

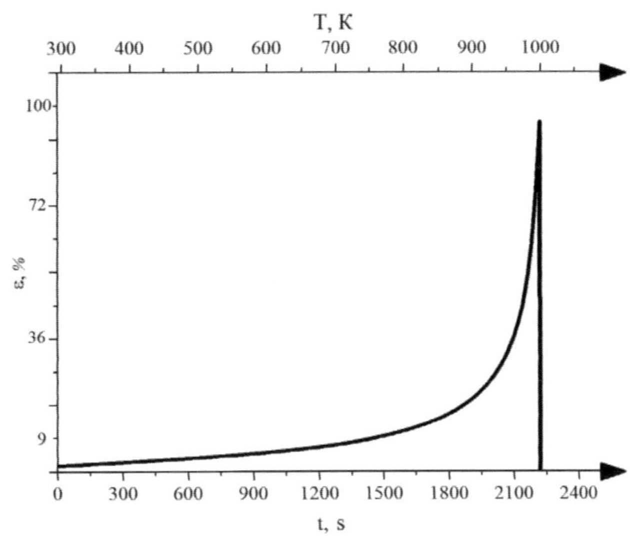

Figure 1. Typical creep curve for amorphous and nanocrystalline alloys.

It is found that for all experimental curves the selection of the following hyperbolic dependence form is possible:

$$x(t) = x_0 + \frac{Ct}{B^2 - Bt} \quad (1)$$

Where x_0 means sensor initial reading; C, B mean analytical parameters with fixed values.

Creep strain time t can be determined through experiment initial conditions (i.e. heating rate and temperature) in the form of the following linear relation:

$$t = \frac{T - T_0}{V} \tag{2}$$

where T_0 means the initial temperature of the experiment (~ 293 K), T is sample heating temperature. The (1) and (2) relations allow establishing between heating and elongation of the bands:

$$x(T) = x_0 + \frac{C(T - T_0)}{B^2 V - BT + BT_0} \tag{3}$$

Correlation between analytically suggested function (1) and the statistical data obtained in the course of the experiment was checked by means of covariation and a correlation coefficient. A $\{r\}$ data correlation coefficient value was calculated and the value is as follows: $0.9< \{r\} <1$ ($\{r_{max}\} = 0.9972$). The $x(t)$ values calculated separately from the (1) formula with a maximal accuracy coincide with the experimental (tabular) data by 99.97 per cent (accurate to triangulating sensor indicated error being ± 0.001 mm, average-absolute error being ±0.0000049 mm and relative error being $\delta_{min}= 0.000244$). Consequently, correlation method applying to the (1) equation and the experimental data is adequate and the suggested formula (1) actually describes a creep process to the extent that makes this process worthwhile (the scope of continuum mechanics).

The physical meaning of the B value was determined at the (1) dependence analyzing; the B value takes on time values at which band rupture occurs, this depending on experiment initial conditions (i.e. sample quality, etc.) and also on a heating mode. A dimensional analysis of the C value showed that the given parameter is a coefficient with a m·s quantity dimension, and its values depend only on experiment initial conditions. This means the (1) analytical form persistence accurate to the variation of sensor reading initial value x_0.

For the achievement of the object in view in the work hereof an additional experimental check was performed allowing assessing the adequacy of C and B values applying. Sensitivity of the given parameters to the initial heat treatment effect and heating rate variability was assessed. The measured curves were interpolated with the (1) analytical dependence together with applicable coefficient values selection so that the interpolation accuracy calculated by means of a correlation coefficient would have the prescribed value of 99.97 percent. The averaged magnitudes of the C and B values were then obtained for each set of experiments. The averaged analytical curves plotted by the C and B values correspond to the averaged experimental curves plotted by the experiment tabular data. Prediction calculation of resistivity constant average quantity was performed in order to additionally check the (1) and (3) analytical equations in accordance with Ohm's law:

$$R = \int \frac{\rho dl}{S} = \frac{U}{I} = \frac{\bar{\rho}\Delta l}{\bar{S}} \tag{4}$$

The calculation showed that samples resistivity at creep (temperature) decreases and varies within the interval of values being from 10^{-4} to 10^{-5} Ohm/m. The (1) derived functions passed the check for conformity to dimensionality and predict the magnitude of the resistance value. This result is confirmed by methods of four- point probe measurement and bridge tests within the interval of preservation of sample's cross-section area measured in mm^2, this interval being equivalent to the temperature interval before crystallization.

Consequently, the (1) and (3) equations and their derivatives together with the C and B parameters included in them not only describe the creep process but also allow predicting the behavior of not initially considered values.

From the [7] literary source it is known that amorphous samples creep strain is a non-homogeneous process; and this fact can indicate the absence of Newtonian flow in the creep process. This fact permits performing a check to prove practicality of the (1) equation by means of alternative proof of Newtonian flow absence in the amorphous alloys structure at variable temperature field effect.

The calculation was performed in accordance with the assumption of pure shift presence in the creep process. This hypothesis is put forward in the [6] work. It is known that viscosity of Newtonian and non-Newtonian liquids differs in functional view [8]. This condition permits viscosity function applicability assessing in order to investigate internal processes proceeding in the alloys being under consideration in the present work.

According to the classic approach, Newtonian liquid viscosity with the presence of internal friction can be derived from the following equation:

$$\eta = \frac{\tau}{\dot{\gamma}} \tag{5}$$

η is laminar flow viscosity, τ is shearing mechanical stress, $\dot{\gamma}$ is sample's relative strain rate.

If the (5) equation is not satisfied then the alloys under consideration cannot be considered Newtonian substances. As the creep process investigated by the present work methods is a thermally-activated one, then the strain rate is a temperature function and, consequently, is related to viscosity; and this is an infeasible condition for Newtonian flow presence [8].

The (5) viscosity is merely a temperature function and cannot be calculated according to the creep test data due to curve slope continuous changing of the $\tau(\dot{\gamma})$ function graph; this function should be a straight line with a fixed slope of curve.

However, creep process special features assume the presence of pseudoplastic or thixotropic flow in the samples under consideration. Such assumptions being true, creep strain can proceed according to the power rheology law of pseudoplastic flow:

$$\tau(\dot{\gamma}) = \mu \cdot (\dot{\gamma})^n \tag{6}$$

where μ is a value of alloy consistency related with viscosity by direct dependence.

Non-Newtonian viscosity of the pseudoplastic flow can be found out in the form of the following equation:

$$\eta = \mu \cdot (\dot{\gamma})^{n-1} \tag{7}$$

as a temperature function decreasing in the process of creep strain. In case of thixotropic flow presence shear load should simultaneously depend on strain durability and rate (as a function of two variables); however, material viscosity at thixotropic strain cannot be explicitly found due to complex dependence between basic rheological values. For purposes of the above-made assumptions analytical checking direct substitutions of the derived function of the (1) formula, were performed in the (5) and (6) equations, the derived function being in relative, dimensionless form, by means of the following additional constraint equation:

$$\varepsilon = \frac{x - x_0}{x_0} \tag{8}$$

and τ value increment was found out separately (the increment was obtained from the shearing flow equation). On account of shear stress determination

$$\tau = \frac{F}{A} \tag{9}$$

the sample shearing flow area was found, where A is a function of elongation and mean value of the $\bar{a}$ thickness (taking into account bands through-thickness rupture in the process of creep):

$$A = \bar{a} x_0 \,. \tag{10}$$

As a result, in accordance with the equation of pure shift (taking into account the presence of the G shear modulus at strain) the dependence of the following form was obtained:

$$\tau = \frac{F}{A} = G \cdot arctg(\frac{\Delta x}{x_0}); \tag{11}$$

The dependence can be transformed into the following form through substitution of $\Delta x = (x - x_0)$ for ε:

$$\tau = G \cdot arctg(\varepsilon); \tag{12}$$

Taking into account (1) creep function preservation at differentiation the recurrence relationship between elongation and the rate of bands relative elongation was found:

$$\varepsilon = \sqrt{\frac{C\dot{\gamma}}{x_0} - \frac{C}{Bx_0}} \tag{13}$$

By way of substitution of the (13) in the (12) a final condition of non-Newtonian flow of amorphous metal bands was obtained:

$$\tau(\dot{\gamma}) = G \cdot arctg(\sqrt{\frac{C\dot{\gamma}}{x_0} - \frac{C}{Bx_0}}). \tag{14}$$

The above equation can be transformed into the following form:

$$\tau(\dot{\gamma}) \approx G \cdot (\sqrt{\frac{C\dot{\gamma}}{x_0} - \frac{C}{Bx_0}}) \tag{15}$$

under condition of small shearing angles presence.

From the (14) equation it is evident that the dependence of mechanical shearing creep stress on relative elongation rate is of nonlinear character and cannot be transformed into the (5) form; the dependence also do not have the $\tau(\dot{\gamma}, t)$ form due to the possibility of the time component elimination from the (13) equation.

However, the (14) equation shows the evident presence of material pseudoplastic behavior; consequently, viscosity can be found from the (7) equation at small strains:

$$\eta(T) \approx G \cdot (B - \frac{T - T_0}{V}) \tag{16}$$

From the strength of materials course it is known that the G value is numerically equal to shear load at reference strain value, at the load of the following form:

$$.\tau = \frac{F}{A} = \frac{mg}{\bar{a}x_0} = \frac{mg}{\bar{a}\Delta x} \Leftrightarrow x = 2x_0 \tag{17}$$

This permits numerical assessment of test samples viscosity value at room temperatures as follows:

$$\eta(T) \approx \frac{mg}{\bar{a}\Delta x} \cdot (B - \frac{T - T_0}{V}) \approx 10^{11} Pa \cdot s \tag{18}$$

The obtained viscosity value coincides with the value measured according to other methods.

Thus the (14) equation analytically proves the absence of Newtonian and thixotropic flows within the temperature interval up to crystallization temperature values in the conditions of thermally-activated creep of amorphous alloys.

Similar arguments can also be given for a noncrystalline alloy at initial stages of strain when intercrystalline amorphous structure retains its phase. The rheological approach as a whole can take place only at the preservation of cross-section area value of the quantity and small strains.

System passing through the crystallization temperature threshold, the alloys creep more and more corresponds to metals creep curve and, consequently, the suggested (1) dependence permits deducing a single analytic expression from metal glass phase transition into metal state.

4. Conclusions

The obtained dependence of creep rate in a variable temperature field can be used for an analytical investigation of the given type of strain within the framework of various models; assuming description of phase transition through crystallization temperature.

Acknowledgements

Financial support from RFBR (grant № 15-01-04553-a).

References

[1] Strombom J , Henderson P, Savage S 1992 *Script.Met.et Mat.* **27** 1179-1184
[2] Kozikowski P et al. 2012 *Script.Mat.* **67** 763-766
[3] Tsao S, Spaepen F 1985 *Acta Metall.* **33** 891-895
[4] Granato A 1992 *Phys. Rew. Lett.* **68** 974-977
[5] Vlasák G, Švec P, Duhaj P 2001 *Mat. Sci. and Eng.* A. **304** 472-475
[6] Luborsky F 1983 *Amorphous metallic alloys* (*London: Butterworths*) p 543
[7] Wilkinson W 1960 *Non-newtonian fluids* (*London: Pergamon Press*) p 17-63

Fractal parameterization analysis of ferroelectric domain structure evolution induced by electron beam irradiation

A G Maslovskaya and T K Barabash

Physics Engineering Department, Amur State University, 21 Ignatyevskoe Shosse, Blagoveshchensk 675000, Russia

E-mail: maslovskayaag@mail.ru

Abstract. The article presents some results of fractal analysis of ferroelectric domain structure images visualized with scanning electron microscope (SEM) techniques. The fractal and multifractal characteristics were estimated to demonstrate self-similar organization of ferroelectric domain structure registered with static and dynamic contrast modes of SEM. Fractal methods as sensitive analytical tools were used to indicate degree of domain structure and domain boundary imperfections. The electron irradiation-induced erosion effect of ferroelectric domain boundaries in electron beam-stimulated polarization current mode of SEM is characterized by considerable raising of fractal dimension. For dynamic contrast mode of SEM there was revealed that complication of domain structure during its dynamics is specified by increase in fractal dimension of images and slight raising of boundary fractal dimension.

1. Introduction

Nowadays SEM techniques have become increasingly important in many areas of scientific work and in practice. The unique SEM analytical abilities of registration of specific responses electron beam-induced allow thoroughly exploring materials which possess high sensitivity with respect to external exposure. The urgent applications of SEM consist in surface topography visualization as well as study of potential relief of polar materials [1]. In this way as from middle of the last century SEM capabilities were employed to obtain images of domain structures, to study and to modify electric properties of ferroelectric materials. In this case one of the key problems consists in studying of sample modification which can be potentially stimulated by electron beam irradiation. In the series of studies some unconventional SEM modes have been described to analyze and to modify ferroelectrics. In particular, electron beam polarization current mode [2] enables domain structure to be visualized and simultaneously polarization switching current to be registered.

Notice that fractal geometrical features of ferroelectric domain structure have been previously reported in [3-13]. The formation of fractal ferroelectric domain structures results from self-organization process of physical system. So as to identify self-similar character of ferroelectric domain structure concepts of fractal theory can be used [14-15]. The fractal analysis yields one of the directions of mathematical description of wide class of objects which have complex and nonregular structures. The fractal analysis of geometry of ferroelectric domain structure images obtained with SEM can be characterized by specific features because of potential contrast pattern in SEM is due to electron beam interaction with samples. Using secondary electrons emitted from surfaces of domains with inverse polarizations allows us to register potential contrast of a specimen. Videosignal and as a consequence observed contrast indicate polarization processes in ferroelectrics in some active probing modes specifically in electron beam induced polarization current mode [2]. Therefore inhomogeneities of the images correspond to variation of polarization distribution. This can be attributed to accumulation injected charge, distributions of defects and stochastic nucleation process (Barkhausen effect). The fractal analysis has become sensitive technique to study of domain boundary irregularity and it can be used for quantitative characterization of observed modification of domain structures. The present study is therefore aimed to analysis of fractal characteristics of ferroelectric domain structure images visualized with static and dynamic modes of the SEM.

2. Research methodology outline

First in order to estimate the scaling characteristics of SEM-images we can apply the algorithms of fractal and multifractal analysis designed for two-dimensional raster images. The fractal dimension of object is a major quantity of fractal theory. The fractal dimension yields numerical estimation of inhomogeneity and nonregularity of self-similar structure. Here covering method based on ratio of "number of particles N – measure" can be applied [14]:

$$N \sim (\delta/\delta_0)^D, \tag{1}$$

where δ or δ_0 – variable covering parameters, D – fractal dimension of the object.

The equation (1) enables one to obtain the technique for calculation of fractal dimension D of physical object providing a variation of one of characteristics δ or δ_0. In particular, box-counting method assumes a variation of the sizes of particles of the covering set δ_0. A binary image is divided into 2^{2n} parts covering the raster image. Then the number of element N is counted which contains even if the one point of the image. After that relation between number of element N covering the analyzed image and linear size of element δ_0 is defined both in logarithmic scale using the following equation [14]:

$$\ln N \sim -D \cdot \ln \delta_0 . \tag{2}$$

As a consequence the fractal dimension D can be calculated from the approximation of the function $N(\delta_0)$ by means of least square procedure.

In order to estimate the fractal dimensions of cluster structures we can use the dependence between perimeter P and area S of a cluster provided clusters or "islands" can be assigned to image [15]:

$$P = C \cdot \delta^{1-D} \cdot S(\delta)^{D/2} . \tag{3}$$

Assuming that the unit is equal to measure of length δ (it is the one pixel for a raster image). In this case the relation between perimeter and square (3) can be specified with the following equation:

$$\ln P = (D/2) \cdot \ln S . \tag{4}$$

Whereas complexity and irregularity of analyzed domain structures can take place the authoring modification of algorithm was designed to calculate fractal dimension of bitmapped images of domain boundaries. Modification is based on equation (1) and the following principles are assumed. The process of cluster marking is realized by means of cluster identification Hoshen-Kopelman algorithm [16] with the recursive procedure excluding incorrect cluster marking in case of repeated viewing.

Moreover the method of multifractal parametrization [14-15] and the method of wavelet transformation modulus maxima [17] were also applied to analyze more complex images of domain structures which can be characterized as self-affine objects. These methods permit one to estimate multifractal scaling properties considering in details some images which representing nonuniform fractal objects. Multifractals due to having statistical geometrical properties require fractal dimensions spectrum to describe them. Compared with fractal analysis, the initial image is divided into clusters with the subsequent reduce in cluster size l. Then a matrix C is created provided each element of C is equal to number of the filled cells. In this case image consisting of $N \times N$ points has in a (i, j) cluster $\left(i, j = \overline{1,k}, k = N/l\right)$ a determined quantity of points. In multifractal parametrization not only the quantity of the clusters comprising points of self-similar structure is counted, but also their specific weight [15]:

$$P_{i,j} = C_{i,j} \Big/ \sum_{i,j} C_{i,j} , \tag{5}$$

where $\sum_{i,j} C_{i,j}$ is total number of units in a matrix of clusters.

Then we use the measure $M_d(q,l) = \sum_{i,j=1,k} P_{i,j}^q \cdot l^d = N(q,l) \cdot l^d$, where l is the cell size, q is a moment order. Change in P_{ij} depending on l according to power law is defined by sequence of the factors $\tau(q)$ characterizing a measure. Further the weighed number of cells $N(q,l)$ can be expressed in a form: $N(q,l) = \sum_{i,j=1,k} P_{i,j}^q \propto l^{-\tau(q)}$ and the indicator of weight is defined by formula $\tau(q) = -\lim_{l \to 0} (\ln N(q,l)/\ln l)$.

Therefore calculation of the $N(q,l)$ and $\tau(q)$ is performed at a variation of parameter of deformation of q. The Renyi's multifractal spectrum can be evaluated by the equation

$$D(q) = \tau(q)/(q-1). \tag{6}$$

Algorithmization and program implementation of examined methods were performed to study scaling characteristics of SEM domain structure images. All of the initial images of ferroelectric domain structures were initialized as raster images represented in a bitmap format.

3. Application of fractal analysis methods to ferroelectric domain structure images obtained with static contrast mode of the SEM

Previous studies have indicated that analysis of geometrical properties of ferroelectric domain structure images visualized with SEM techniques possesses a number of specific features owing to the fact that the potential contrast image is result of electron beam interaction with the sample. In this terms tonality of domains as well as domain boundaries observed with SEM is defined by features of the used mode as well as mechanisms of contrast formation. In particular, at first the electron probe initiates specific effects in the irradiated sample, and subsequently registers them in the "active" scanning modes. One of priority factors of fractal analysis of ferroelectric domain structures consists in correct interpreting the images of the analyzed element of domain structures and correct submitting initial data.

The present study confirms that SEM images of static domain configurations are characterized by a complex scaling. The fractal dimensions of ferroelectric domain structure images were calculated by means of box-counting method (2). The results suggest fractality of images and distribution of fractal dimensions over the range from 1.6 to 1.9 for typical ferroelectrics. The fractal dimensions were evaluated for set of images of each objects. Whence we obtained the following results: D=1.72–1.88 for Rochelle salt, D=1.71–1.92 for triglycine sulfate (TGS), D=1.60–1.77 for barium titanate ($BaTiO_3$), D=1.84–1.89 for bismuth titanate ($Bi_4Ti_3O_{12}$), D=1.64–1.78 for lead titanate ($PbTiO_3$), D=1.86–1.95 for sodium nitrite ($NaNO_2$), D=1.52–1.56 for PZT ferroelectric ceramics. However the classical box-counting method enables us to estimate only the quantitative characteristics of space filling of fractal and it does not allow describing form of structures of which the image consists.

The irregularity and roughness degree of domain boundaries for typical ferroelectrics was identified by slit island method (4) and it is characterized by fractal dimension from the almost regular forms with D=1.05 to strongly curved with pronounced fractal properties (D=1.66). For instance, the fractal dimension of the domain boundaries separating 90° a-components of spontaneous polarization and boundaries separating antiparallel 180° c-components in monocrystals of $Bi_4Ti_3O_{12}$ was the quite considerable and it is distributed over the range of D=1.509–1.780. The example of result of fractal dimension calculation for the line restricting clusters (a- and c-domains) is shown in figure 1 a.

The performed multifractal analysis for domain structure images has allowed us to calculate the spectral characteristics specifying dispersion measure of fractal dimensions. The figure 1 b illustrates the Renyi's multifractal spectrum (6) computed for typical domain structure of TGS crystal (positive domains or dark contrast were subject to the analysis).

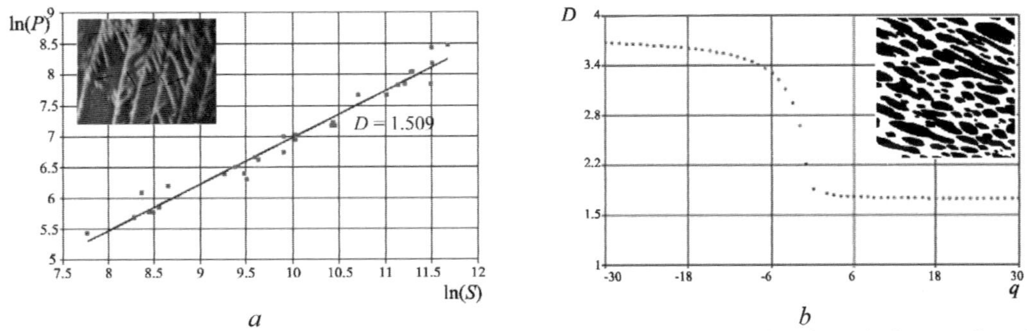

Figure 1. The result of fractal analysis performed by slit island method for bismuth titanate domain structure image shown in the insert figure – *a*; the Renyi's multifractal spectrum for TGS crystal – *b*.

The computed scaling factor $\tau(q)$ is nonlinear which is relative to multifractal structures. The analysis of Renyi's multifractal spectrum permits us to define a set of the dimensions forming the structure and also to estimate width of a fractal spectrum ($\Delta\alpha \approx 2$). The value of $D(q=0)=1.86$ corresponds to the fractal dimension calculated by covering method. The analyzed image of TGS domain structure has a self-similar structure with distribution of fractal dimension over the range of 1.7 – 3.7. The wavelet transformation modulus maxima method allows visualizing a picture of wavelet-coefficients and extremum skeleton which show hierarchical structure of a profile of the image. Also this method leads to the spectral characteristics of $\tau(q)$ and $f(q)$ which are in good agreement with the data obtained by multifractal parametrization method.

The fractal analysis of image of lens-like domain boundary of TGS crystal observed with electron beam induced polarization current mode [2] at probe current $I=10^{-9}$ A and beam accelerating voltages U=3-5 kV so long as irradiation time was varied. Microphotographs and corresponding fractal dimensions calculated by slit island method (4) are demonstrated in figure 2.

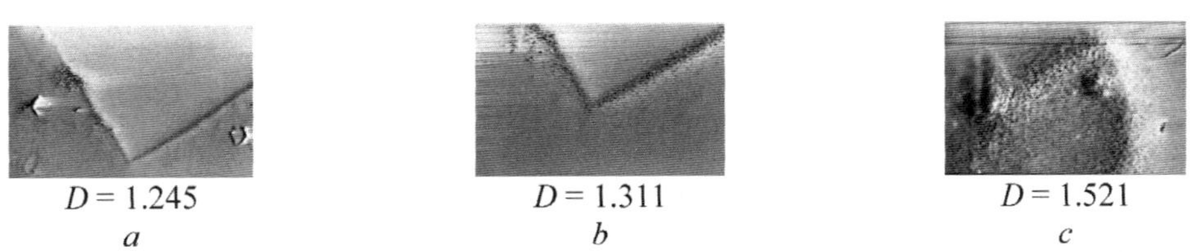

Figure 2. The TGS domain boundaries observed with electron beam stimulated polarization current mode and corresponding values of computed fractal dimensions: *a* – the image of boundary of lens-shaped domain, *b* – erosion of boundary under prolonged irradiation, *c* – the image of domain fragment after 5 minute of irradiation.

The fractal dimensions were estimated to express roughness and irregularity domain degree. The durational irradiation of sample leads to erosion of image: line of boundary disintegrates on series of the points and becomes wider as it is shown in figure 1 c. The erosion could be caused by modification of boundary area structure by means of creating ramified antiparallel structure with width up to 10 μm. Note also that increase in fractal dimension indicates complication of ferroelectric domain geometry.

4. Estimation of fractal characteristics of ferroelectric domain structure images visualized with dynamic contrast mode of the SEM

Taking into consideration performance capabilities of dynamic contrast imaging modes, there is a practical interest examining evolution of fractal dimension during sequential stages of domain structure dynamics induced by an electron beam of SEM. Let us perform analysis of experimental data obtained with potential electron contact mode [2].

The fractal dimensions of sequential stages of domain structure dynamics in barium titanate ($BaTiO_3$) were calculated by box-counting method (2). The result can be attributed to complication of domain structure with fractal dimension distribution of the following range: D=[1.608,1.613,1.729,1.734]. Meanwhile, the degree of irregularity and roughness of domain boundaries changes slightly: D^*=[1.261,1.263,1.269,1.277]. Providing $Bi_4Ti_3O_{12}$ is subjected to field directed perpendicular to the sample surface in potential electron contact mode, we can conclude that electron beam exposure results in movement of boundaries dividing anti-parallel c-components of spontaneous polarization.

The domain growth with parallel to field c-component of spontaneous polarization is realized by expansion of the wedges which are formed on 90° domain wall. This process is specified by increase in the fractal dimension from D=1.784 to D=1.807 which is calculated by covering method. On the other hand, fractality of domain boundaries remains in rather narrow interval $D\in(1.61,1.66)$. Characteristic feature of polarization switching process in TGS is reproducibility of process of time cycle. The analysis of the scaling characteristics of SEM-contrast of TGS domains during domain structure dynamics observed in the potential contact mode was carried out using of the videofile with visualization of a full cycle of polarization switching in TGS [2], which is illustrated in figure 3. In order to obtain data suitable for the fractal analysis quantization of video with an interval of 1 s has been performed.

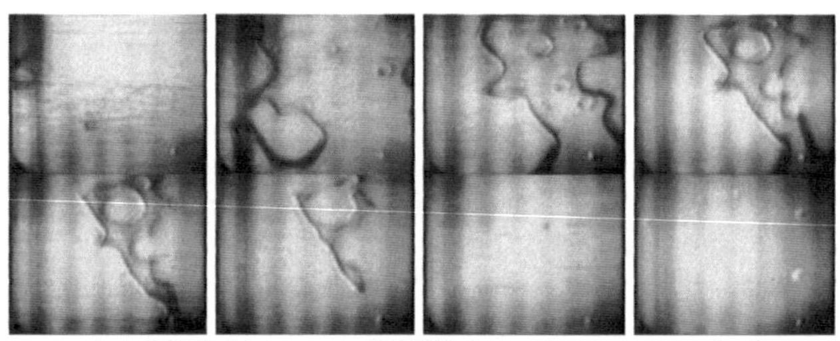

Figure 3. Fragments of SEM-images of TGS domain structure during complete cycle of polarization reorientation in potential electron contact mode [2] characterized by fractal dimension distribution over the range of (1.012, 1.214).

The fractal analysis of domain boundary geometry with use of slit island method (4) yields change in dimension in the interval $D\in(1.012,1.214)$ with a growth of both positive and negative domains. The minimum value of fractal dimension corresponds to the initial stage of domain nucleation with practically regular boundaries. The maximum value is attributed to critical state: the square of the switched area is equal to the square of nonswitched area. To analyse the fractal dimension of images computed by box-counting method (2) the temporal dependence of fractal dimension is represented in figure 4 a. These data can be acceptable approximated by quadratic function.

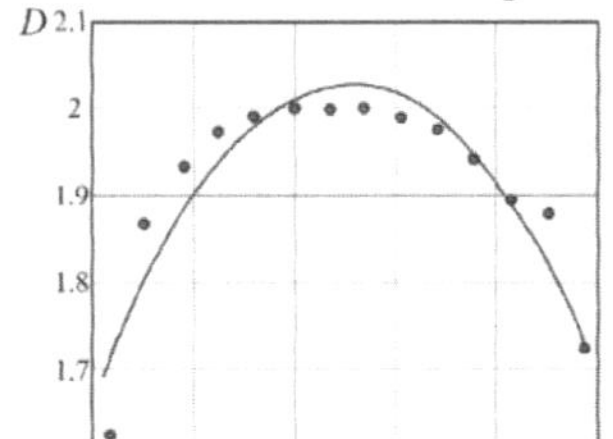

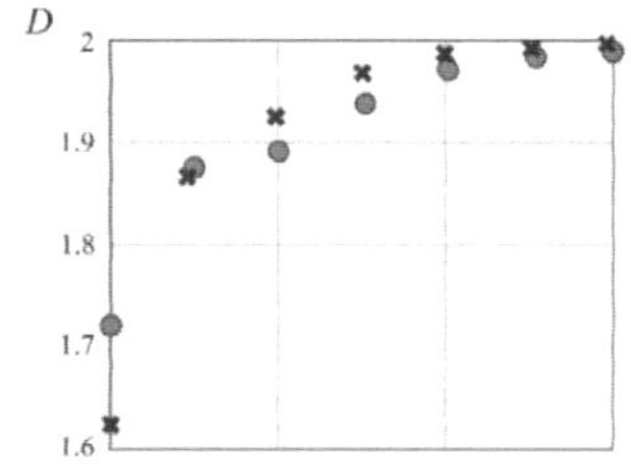

a *b*

Figure 4. The temporal dependence of fractal dimension calculated by box-counting method – a, the dynamics of changes in fractal dimension for positive (o) and negative (x) domains – b.

Providing we consider temporal dependence of fractal dimension for growth of positive and negative domains on the one graphic we can conclude that complication of domain structure for growth of both positive and negative domains occurs almost identically as shown in figure 4 *b*.
The evidence from this study suggests that visually observed features of domain structure dynamics can be attributed to peculiarities of change in fractal dimension. In general, therefore, it seems that not only static and quasi-static domain configurations are characterized by a fractal structure but also domain dynamics induced by an electron probe demonstrates self-organization process.

5. Concluding remarks

The most obvious finding to emerge from this study is that fractal and multifractal methods are sensitive techniques to study degree of domain structure and domain boundary imperfections and can be used for the quantitative characteristic of observed modifications of ferroelectrics in SEM, in particular, are caused by accumulation of the injected charge, defect distribution, and stochastic nucleation process (Barkhausen effect).

Our data suggest electron beam-induced erosion effect of domain boundaries in TGS can be characterized by increase in fractal dimension from D=1.245 до D=1.521. Also complication of domain structure geometry during its dynamics is characterized by increase in value of fractal dimension of the image and insignificant increase in fractal dimension of boundaries when analyzing ferroelectrics with dynamic contrast modes of SEM.

The evidence from fractal analysis for step-by-step images of a full cycle of polarization switching in TGS in the potential electron contact mode indicates that complication of domain structure for the lateral growth of positive, as well as negative domains occurs similarly: fractal dimension of the image changes within the interval $1.63<D<1.99$ and domain boundaries $1.012<D<1.214$ respectively.

Acknowledgments

The work has been supported by a scientific program of the Ministry of Education and Science of Russia (project No. 1158.2014/9).

References

[1] Robinson G Y and White R M 1967 *Appl. Phys. Lett.* **10** 320
[2] Sogr A A, Maslovskaya A G and Kopylova I B 2006 *Ferroelectrics* **341** 29
[3] Ozaki T, Fujii K and Ohgami J 1995 *J. Phys. Soc. Jpn.* **64** 2282
[4] Isupov VA 1998 *Physics of the Solid State* **40** 1188
[5] Jeng Y-R, Tsai P-C and Fang T-H 2003 *Microelectronic Engineering* **65** 406
[6] Shur V Ya, Kuznetsov D K, Lobov A I et al 2006 *Ferroelectrics* **341** 85
[7] Wu Z, Duan W, Wu J and Gu B-L 2007 *Nanotechnology* **18** 325703
[8] Roy M K, Paul J and Dattagupta S 2010 *IEEE Xplore* **109** 014108
[9] Galiyarova N M, Bey A B, Kuznetzov E A and Korchmariyuk Y I 2004 *Ferroelectrics* **30** 205
[10] Tadic B 2002 *Eur. Phys. J. B.* **28** 81
[11] Maslovskaya A G and Barabash T K 2012 *Physics Procedia* **23** 81
[12] Maslovskaya A G and Barabash T K 2013 *Ferroelectrics* **442** 18
[13] Pavelchuk A, Barabash T and Maslovskaya A G 2016 IOP Conf. Series: Materials Science and Engineering, **110** 012080
[14] Bojokin S V and Parshin D A 2001 *Fractals and multifractals* (Moscow: R&C Dynamics)
[15] Falconer K 1990 *Fractal Geometry: Mathematical Foundations and Applications* (Chichester: Wiley)
[16] Stauffer D and Aharony A 1992 *Introduction to Percolation Theory* (London: Taylor & Francis)
[17] Arneodo A, Decoster N and Roux S G 2000 *Eur. Phys. J. B.* **15** 567

Irradiation-Induced Thermal Effects in Alloyed Metal Fuel of Fast Reactors

F N Kryukov, O N Nikitin, S V Kuzmin, A V Belyaeva, I F Gilmutdinov, P I Grin and I Yu Zhemkov
Joint Stock Company «State Scientific Center – Research Institute of Atomic Reactors», Zapadnoye Shosse 9, 433510, Dimitrovgrad, Ulyanovsk region, Russian Federation

E-mail: kryukov@niiar.ru

Abstract. The paper presents the results of studying alloyed metal fuel after irradiation in a fast reactor. Determined is the mechanism of fuel irradiation swelling, mechanical interaction between fuel and cladding, and distribution of fission products. Experience gained in fuel properties and behavior under irradiation as well as in irradiation-induced thermal effects occurred in alloyed metal fuel provides for a fuel pin design to have a burnup not less than 20% h. a.

1. Introduction

Metal fuel remains the most attractive fuel to be used in fast reactors due to its simple manufacture process with the use of existing metallurgic techniques, high density of fission isotopes, high thermal conductivity, and compatibility with sodium coolant. Further electrochemical reprocessing of metal fuel and a metallurgical method of manufacturing new fuel slugs may be the decisive factors in minimizing the cost of the fuel cycle and generated electric power [1]. This fuel is considered to be one of the most promising fuels to be used in Generation IV Reactors. A large-scale program on studying metal fuel has been conducted in the U.S. [2-6]. The U.S. program focused on physical properties of uranium-based and uranium-plutonium-based alloys, their irradiation characteristics, optimal fuel composition and fuel design. Thorough work has been carried out in the feasibility of fuel performance including life tests and studying fuel behaviour under transient and abnormal conditions. Among the examined fuel compositions were U-5%Fs (uranium alloyed with metals – 5wt.% fission products), U-10%Zr, and U-Pu-Zr. The final choice of the fuel composition was made in favor of U-Pu-Zr. Zirconium was introduced into the alloy to increase the melting point and decrease chemical fuel-to-cladding interaction. The density of U-19%Pu-10%Zr in an extruded state is equal to 15.82 g/cm^3. The solidus temperature makes up approximately 1100°C, and the thermal conductivity ranges from 15 to 30 W/mK in the temperature interval from 400 to 1000°C. The alloy strength decreases rapidly as the temperature increases, e.g., the tensile strength of U-15%Pu-10%Zr ranges from 480 MPa at room temperature to 97 MPa at 400 °C. The yield strength at 400°C is equal to 69 MPa, and the relative elongation makes up 53 %. Low strength and high plasticity of U-Pu-Zr alloys favorably manifest themselves in swelling under irradiation exerting low pressure on the fuel cladding. In free swelling the U-Pu-Zr fuel density makes up approximately 75 % of the theoretical value at a burnup of 2-3 % h.a. By that time, a gap between fuel and cladding disappears, and fuel contacts the cladding.

Porous fuel appears to be of so high plasticity to provide weak mechanical fuel-to-cladding interaction. This effect became a decisive factor in maintaining the performance of fuel irradiated up to a high burnup. As the fuel swelling mechanisms were revealed, its quantitative characteristics were specified and the fuel pin design was improved, the burnup increased consistently from ~1 to ~18 % h.a. On single fuel pins the peak burnup of 19.9 % h.a. was achieved. At a fuel burnup of 16-18 % h.a. there was an 8% increase in the cladding diameter of fuel pins made of D9 austenitic steel that is likely to undergo irradiation swelling. As for fuel pins made of HT9 ferritic-martensitic steel that is not likely to undergo swelling under the same conditions, this value made up about 1%. Based on these data American scientists concluded that in fast reactors operated with metal fuel a burnup of 200 GW/t HM can be achieved, and the simple manufacture process and reprocessing technique of U-Pu-Zr fuel give benefits for its wide use.

In Russia the first experiments in the development of alloyed metal uranium- and plutonium-based fuel for fast reactors were launched in 1950-60 [7]. Later on, tenacious work related to these issues stopped. From 1988 to 2001 a range of activities has been completed including manufacture of four fuel assemblies (FAs) and their irradiation in the BOR-60 reactor. These FAs contained experimental fuel pins (U-Zr and U-Pu-Zr). Post-irradiation examination (PIE) data of three FAs are presented in papers [8-10]. This paper presents the results of studying FA VS-174e.

2. Fuel characteristics and irradiation conditions

Table 1 gives the main fuel parameters, design features and irradiation parameters.

Table 1. Parameters of experimental fuel pins and irradiation conditions.

Parameter	Value
FA shroud	EP-450
Fuel cladding	0Kh16N15M3B
Cladding wall diameter and thickness, mm	6.0×0.3 mm
Fuel column height, mm	450±3
Fuel composition	U-18%Pu-10%Zr
Fuel pin diameter, mm	4.7±0.1
Average relative effective fuel density, %	75±3
Heat-transfer fuel-cladding layer	Na
Peak burnup, % h.a.	9.7
Peak neutron fluence (E>0.1 MeV), cm^{-2}	1.28×10^{23}
Peak linear thermal power, W/cm	400
Peak temperature on the cladding inner surface,°C	600

3. PIE data

3.1. Irradiation swelling and fuel-cladding mechanical interaction

The cladding diameter was measured on ten fuel pins by profilometry. Figure 1 shows a profile diagram of one fuel pin.

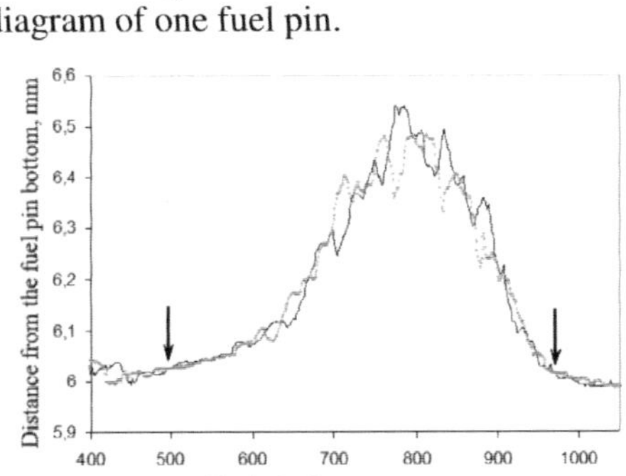

Figure 1. Typical profile diagram in two mutually perpendicular planes of a fuel pin with metal fuel after irradiation in BOR-60 to achieve the peak burnup of 9.7 % h.a. The arrows show the fuel slug boundaries.

IOP Conf. Series: Materials Science and Engineering **168** (2017) 012029 doi:10.1088/1757-899X/168/1/012029

This profile diagram was obtained by scanning along the fuel pin in two mutually perpendicular planes. The ovality of fuel cladding is seen in the area of the maximum diameter increase resulted from mechanical interaction with neighboring fuel pins through the spacer wire. According to the measurement data obtained on ten fuel pins in the deformation maximum areas, the average value of the relative diameter increase is equal to 7.1±0.1 %. The absence of the "steps" in the profile diagram on the boundaries of fuel slugs shows that there is insignificant pressure of fuel on the cladding. It agrees with the conclusions of American scientists about the results of metal fuel irradiation. The maximum swelling of steel 0X16H15M3Б in a solution treated state makes up 15% for the examined fuel pins under irradiation. Therefore, an increase in the cladding diameter due to irradiation swelling is 5%. About 2% of diameter growth results from irradiation-induced creep under stress due to fission gas pressure. Based on gamma scanning results the relative elongation of the fuel slugs makes up 4.2±0.7 %. The relative increase in the fuel volume under fuel-to-cladding interaction is equal to approximately 46 %, and the relative fission gas release from fuel into a free volume of the fuel pins makes up about 80 % according to the results of cladding puncture and gas phase analysis, which is close to the data obtained in the EBR-2 reactor [3].

3.2. Fuel microstructure

In examining cross-sections of the fuel pins of different levels relative to the fuel slug height with the use of a metallographic microscope, it was found that in all cross-sections the fuel slug diameter increased before fuel contacts the cladding. Sodium did not prevent swelling, and it was displaced from the gap. In most cross-sections fuel has a layered ring structure (Figure 2).

a) b) c)

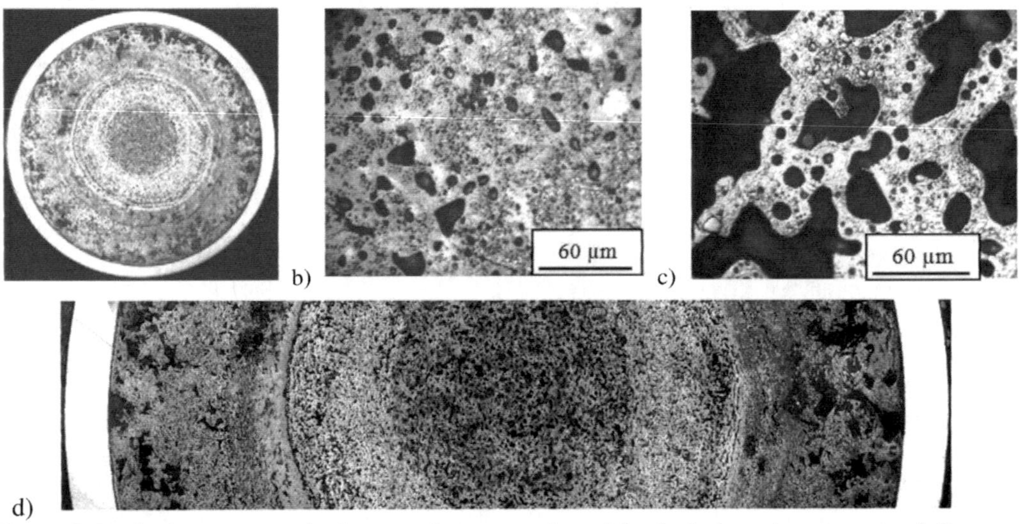

d)

Figure 2. Fuel microstructure in the central cross-section of the fuel pin active core: a) – full cross-section macrostr d) b) – bubble area, c) – cavitation swelling area, d) – panoramic image in a diametrical direction.

The porosity nature depends on the fuel slug operating temperature. It changes both radially (in the fuel pin cross-section) and axially (fuel pin heightwise). When the fuel temperature is low, i.e. in the outer ring area of the fuel slug and in the lower parts of the fuel slug, cavitation swelling is observed with the presence of large-size pores with zigzag boundaries. At a higher temperature bubble swelling area is seen with pores of mainly smooth and round shape. A change in the porosity nature is clearly seen as ring layers on macrographs and panoramic photographs of the fuel pin cross-sections. This change arises from phase formation in fuel verses temperature. At a temperature lower ≈670 °C in alloy U-Pu-Zr an orthorhombic alpha phase is generated showing a tendency to strong anisotropy that manifests itself in the anisotropic irradiation-induced growth. This effect is based on anisotropic distribution of irradiation defects in the crystal lattice [15]. Under irradiation crystals grow in size in direction [010] and decrease in size in direction [100]. As a result, along with generation and growth of gas bubbles in the low-temperature outer ring area of the fuel slug there is redistribution of the material in the grain volume due to irradiation growth in a certain crystallographic direction. This leads to a break in the grain boundaries and formation of microscopic discontinuity flaws. Fission gas makes stable the generated discontinuity flaws that could further grow and change their shape. These processes result in a typical microstructure that is always observed in a low-temperature area of the fuel slug.

3.3. Distribution of fuel components and fission products

Distribution of the fuel components along the radius of the fuel slug, accumulation and state of fission products may affect the melting point and fuel thermal conductivity. Therefore, in the course of PIEs attention was paid on these issues. Figure 3 shows a typical microstructure obtained by scanning electron microscopy and distribution of the fuel components along the fuel slug radius in the fuel pin cross-section with the peak local burnup of 9.7 % h.a. obtained by electron probe microanalysis. The given results show layering of alloy U-Pu-Zr in composition and peculiarities of its structure formation affected by the radial temperature gradient. At a distance from 900 to 1400 μm from the inside of the cladding a layer was formed in which the zirconium weight fraction makes up less than 1 %, plutonium content is decreased and uranium content is increased. It is evident that zirconium migrated in the fuel central part. In addition, in distribution of zirconium the local maximum is observed due to the presence of zirconium alpha phase inclusions. On the periphery of the fuel slug zirconium migrated towards the surface with the generated alpha phase layer that is of high importance in terms of preventing fuel-to-cladding interaction.

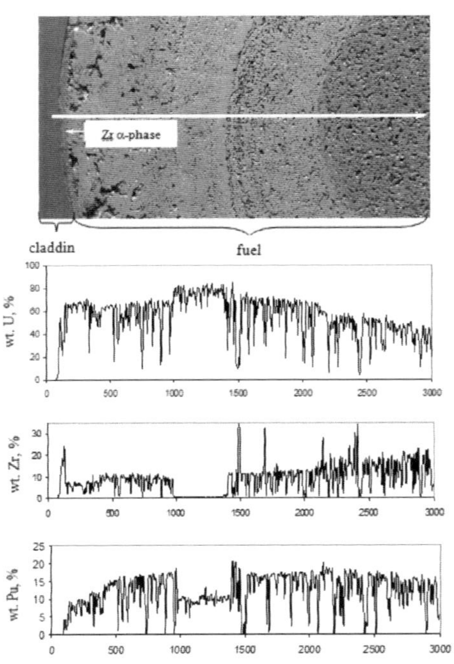

Figure 3. Electron microscope image of the fuel microstructure and distribution of the components along the fuel slug radius in the fuel pin cross-section with a burnup of 9.7 % h.a. The arrow shows scanning direction from the cladding to the fuel slug center.

The redistribution of components results from the diagram structure of the alloy U-Pu-Zr state. Despite the apparent complexity of structure formation the basic mechanisms are well-known, and their influence on fuel and fuel pin performance is determined. For instance, in studying chemical fuel-to-cladding interaction a favorable effect of zirconium alpha phase was observed on the fuel slug surface. There were some concerns about formation of a ring area of depleted zirconium in the fuel slug due to a possibility of a solidus temperature decrease in this area. However, it turned out that the main component in this area is uranium. Plutonium content was lower that the set value, and the correlation of components corresponds to the solidus temperature of about 1000 °C, which is significantly higher than the operating temperature in this fuel slug area.

Study of the fission product distribution shows that the content of the fission products in a solid solution is low except for molybdenum, and their main fraction presents in fuel as metal inclusions. Among all fission products molybdenum is described by the highest solubility in the alloy, and the nature of its distribution along the slug radius is similar to that of zirconium distribution.

4. Conclusion

At the early stage of irradiation under free swelling (without mechanical suppression) alloyed metal fuel increases rapidly in volume mainly due to formation and growth of gas pores. At a burnup of about 3 % h.a. the volumetric fuel swelling achieves 30-35 %, which corresponds to the effective density of about 75 %. The fuel state undergoes changes described by generated open porosity, high fission gas release, and as a result, a sharp slowdown in swelling. Further long-term irradiation does not change significantly the fuel state. Fuel swelling occurs due to solid fission products, and the swelling rate is low. Such fuel state that does not change much under irradiation was observed up to a burnup of about 20 % h.a. If the effective fuel density initially makes up no more than 75 % due to a slug-to-cladding gap, fuel contacts the cladding after the porous structure is formed. In the course of further irradiation this structure does not exert significant pressure on the cladding. No fuel cladding diameter increase was observed resulted from fuel-to-cladding mechanical interaction with the fuel density of 75 %. The major contribution to cladding loading was made by fission gas pressure that was compensated by an increase in the gas plenum volume. The PIE data fully confirm the described mechanism. Irradiation-induced thermal effects in alloyed metal fuel do not restrict fuel pin performance. Its operating life is determined by irradiation resistance of the cladding material.

References

[1] Tsykanov V A, Gadzhiev G I, Kirillov E V, Skiba O V, and Demidova L S 1989 *Atomic Energy* **67 (3)** 163-167

[2] Crawford D C, Porter D L and Hayes S L 2007 *J. of Nuclear Materials* **371** 202-231

[3] Hofman G L 1980 *J. Nucl. Technology* **47** 7-22

[4] Leibowitz L, Veleckis E, Blomquist R A and Pelton A D 1988 *J. of Nuclear Materials* **154** 145-153

[5] Pahl G, Porter D L, Lahm C E and Hofman G L 1990 *J. Metallurgical and Materials Transactions* **21A** 1863–70

[6] Cohen A B, Tsai H and Neimark L A 1993 *J. of Nuclear Materials* **204** 244–251

[7] Bochvar A A, Konobeevsky S T, Kutaitsev V I and al. 1958 *Proc. U.N. Intern. Conf. Peaceful Uses At. Energy (Geneva)* **6** 184–193

[8] Vaganov I V, Gadzhiev G I, Kosulin N S, and Syuzev V N 1998 *Proc. of the Fifth Inter-Industry Conference on Reactor Materials (Dimitrovgrad: RIAR)* **1 (2)** 20–28

[9] Vaganov I V, Gadzhiev G I, Kosulin N S, and Syuzev V N 2000 *Book of Reports of the Sixth Inter-Industry Conference on Reactor Materials (Dimitrovgrad: RIAR)* **2 (2)** 232–243

[10] Gadzhiev G I, Syuzev V N 2009 *Book of Reports of IX Russian Conference on Reactor Material Science (Dimitrovgrad: RIAR)* 335–344

[11] Sokurskii Yu N, Sterlin Ya M, and Fedorchenko V A 1971 Uranium and its Alloys

IOP Conf. Series: Materials Science and Engineering **168** (2017) 012030 doi:10.1088/1757-899X/168/1/012030

Research of the nonuniform layers by a reflectometry method near the Brewster's angle

V V Shagaev
Bauman Moscow State Technical University, Kaluga Branch, Russia

E-mail: shagaev_vv@rambler.ru

Abstract. The mathematical model of the electromagnetic wave reflection with *p*-polarization from the non-sharp boundary between two transparent media is built and studied. The feature of the model consisted in the fact that the thickness of the transition layer was much less than the electromagnetic wave length. The analytical ratio expressing the reflection coefficient through the coordinate dependence of permittivity is deduced. It is shown that the logarithmic derivative from the reflection coefficient with respect to angle can be used as the inhomogeneity indicator. The interval of the angles located near the Brewster's angle is the most informative in this case. The developed theoretical representations are illustrated by calculations executed for the terahertz range wave reflection from an implanted layer.

1. Introduction

It is known that the angular dependence of a reflection coefficient of the *p*-polarised electromagnetic wave from a homogeneous medium has the zero value minimum. The corresponding angle is named as Brewster's angle. The presence of thin transitive layer near a partition boundary results in the fact that the reflection coefficient, as before, has the minimum in angular dependence, which, however, doesn't reach zero. It is convenient in this case to use a logarithmic derivative instead of angular dependence of the reflection coefficient $R_p(\theta)$ [1]:

$$\eta(\theta) = \frac{1}{R_p(\theta)} \frac{dR_p(\theta)}{d\theta}. \tag{1}$$

The function $\eta(\theta)$ unlike the function $R_p(\theta)$ has not only the minimum but the maximum near the Brewster's angle. Besides that the experimental dependence $\eta(\theta)$ can be built in the values of the reflection coefficient expressed by the relative units. This circumstance simplifies execution of the necessary measurements because it is possible to ignore the change of the sounding radiation intensity as well as reflections from the input and the exit windows of the measuring system.

It is possible to deduce the analytical expression for electromagnetic wave reflection coefficient from a non-homogeneous medium only in the exclusive cases [2, 3]. The simplest example is the Drude model in which the permittivity changes by the jump at some depth. The reflection coefficient in this case can be expressed through the amplitude reflection coefficients from the semi-infinite homogeneous media with the sharp boundaries of partition (Fresnel's coefficients). The numerical methods for arbitrary profile of the permittivity go out on the first plan. The method in which the non-homogeneous part of a medium breaks into the thin homogeneous layers is the most often used [4].

The number of layers depends on the permittivity profile, and the reflection coefficient is calculated on the recurrent formula expressed through Fresnel's coefficients.

In this work the reflection coefficient was calculated by a perturbation theory method. It was necessary that the thickness of non-homogeneous layer was much smaller than the electromagnetic wave length. As an example the possibility of identification of the implanted layer by means of terahertz radiation is considered.

2. The development of a mathematical model

The electromagnetic wave with p-polarization is completely defined by strength of its magnetic field:

$$\boldsymbol{H} = \begin{bmatrix} 0 \\ h(z) \\ 0 \end{bmatrix} \cdot \exp i(\omega t - k_x x).$$

It was believed that the medium is homogeneous in the xy plane and the permittivity is a function of the z coordinate (Figure 1). Besides that the function $\varepsilon(z)$ approaches at $z \to \pm\infty$ to the permanent values $\varepsilon_1 = \varepsilon(-\infty)$ and $\varepsilon_2 = \varepsilon(+\infty)$. k_x is the x-component of a wave vector, and $k_x = \left(\omega\sqrt{\varepsilon_1}\sin\theta\right)/c$. ω is the wave circular frequency, θ is a wave angle of incidence on the media partition boundary, and c is the light velocity in vacuum. It was assumed that the magnetic permeability is equal to unit.

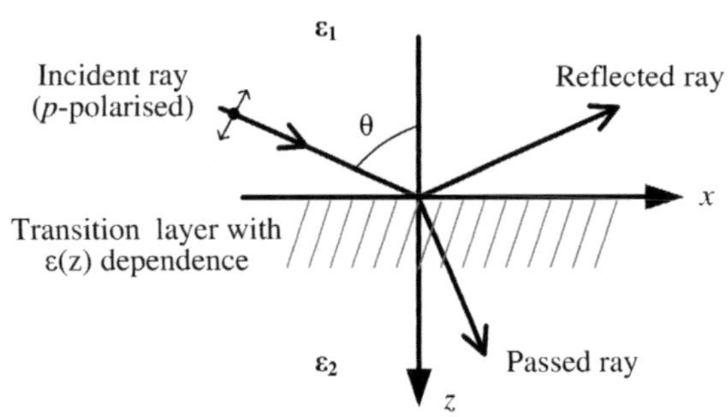

Figure 1. A structure of medium.

The differential equation follows from Maxwell's equations:

$$\frac{1}{\varepsilon(z)}\frac{\partial}{\partial z}\left[\frac{1}{\varepsilon(z)}\frac{\partial h(z)}{\partial z}\right] + p^2(z)h(z) = 0, \tag{2}$$

where

$$p^2(z) = \frac{1}{\varepsilon^2(z)}\left[\frac{\omega^2}{c^2}\varepsilon(z) - k_x^2\right]. \tag{3}$$

The function $h(z)$ was presented in the expression form:

$$h(z) = A(z)\exp[-i\varphi(z)] + B(z)\exp[i\varphi(z)], \tag{4}$$

where

$$\varphi(z) = \int_0^z p(z')\varepsilon(z')dz'. \tag{5}$$

The values $A(-\infty)$, $A(+\infty)$ and $B(-\infty)$ are the amplitudes of the incident, passed and reflected waves respectively.

Since two functions were injected instead of one required, it is possible to impose an additional condition on them. The condition was expressed by the equation:

$$\frac{dA(z)}{dz}\exp[-i\varphi(z)]+\frac{dB(z)}{dz}\exp[i\varphi(z)]=0. \tag{6}$$

For calculation of the reflection coefficient it is convenient to use a function

$$V(z)=\frac{B(z)}{A(z)}\exp[2i\varphi(z)].$$

Then a wave reflection coefficient is connected with $V(z)$ by equality

$$R_p=|V(-\infty)|^2. \tag{7}$$

From the Eqs. (2), (4), and (6) the equation follows:

$$\frac{d[V\exp(-2i\varphi)]}{dz}=\frac{1}{2p}\frac{dp}{dz}\left(1-V^2\right)\exp(-2i\varphi). \tag{8}$$

The boundary condition for $V(z)$ can be deduced from the physical restriction: $B(+\infty)=0$. Then

$$V(+\infty)=0. \tag{9}$$

Thus, the reflection coefficient from the media partition boundary with the non-uniform transition layer can be calculated from the differential Eq. (8) with the boundary condition (9) and from the formula (7).

The Eq. (8) generally hasn't an analytical solution. However approximate methods can be used [5]. The equation has the feature. Its right part approaches to zero at $z\to\pm\infty$, for $\varepsilon(z)$ approaches to the constant values, and according to Eq. (3) there is the limit $dp(z)/dz\to 0$. Dependences of $\varepsilon(z)$ with sharp change on one asymptotic value $\varepsilon_1=\varepsilon(-\infty)$ to another $\varepsilon_2=\varepsilon(+\infty)$ were considered. In this case the change interval of $\varepsilon(z)$ is defined by the d size, and the coordinate $z=0$ was situated inside this interval. Then the derivative $dp(z)/dz$ will have the greatest values nearby $z=0$. In this case for the reflection coefficient calculation it is possible to use the approximate solution of the Eq. (8), constructed in an interval $\Delta z\sim d$.

Let the value of d is so small that the inequality is carried out

$$d\max\{p(z)\varepsilon(z)\}<<1. \tag{10}$$

The inequality expresses in essence a condition of the phase thickness smallness of the transition layer. Then the function $\varphi(z)$ will be small for the values of z lying in the $\Delta z\sim d$ interval, and the function $\exp(-2i\varphi)$ can be approximated by a power series on $\varphi(z)$ in which the several first terms will be essential. The solution of the Eq. (8) can be also constructed in the form of a row which terms correspond to the $\varphi(z)$ powers taken into account. Besides at the large values of z the integral in expression of (5) can result in linear growth of $\varphi(z)$ dependence. In this case, for the profiles of $\varepsilon(z)$ aiming to asymptotic values not more slowly than exponentially, integration of the right side of Eq. (8), presented in the row form, leads to finite values of each term of the row even when the limits of integration will be infinite. Such feature of the equation arises due to the dp/dz derivative [6]. It allows to replace in final expressions the integration interval $\Delta z\sim d$ on integration with the infinite limits. Therefore the dependences $\varepsilon(z)$ aiming to asymptotic values at least exponentially were considered.

Decomposition of the function $V(z)$ into the row of the perturbations theory was used:

$$V(z)=V_0(z)+iV_1(z)+V_2(z). \tag{11}$$

The order of the used approximation is noted by the bottom index, and for convenience in corrections of an odd order the multiplier $i=\sqrt{-1}$ is entered.

It is assumed that $\exp[-2i\varphi(z)]\approx 1$ in zero approximation, and then from the Eq.(8) follows:

$$\frac{dV_0}{dz}=\frac{1}{2p}\frac{dp}{dz}\left(1-V_0^2\right),\ V_0(+\infty)=0,$$

$$V_0(z)=\frac{p(z)-p(+\infty)}{p(z)+p(+\infty)}. \tag{12}$$

Equality $\exp[-2i\varphi(z)]\approx 1-2i\varphi(z)$ was as a first approximation used, and the function $V_1(z)$ is defined by the equation:

$$\frac{dV_1}{dz}=-\frac{1}{p}\frac{dp}{dz}V_0V_1+2p\varepsilon V_0,\ V_1(+\infty)=0,$$

$$V_1(z)=-2\left[1-V_0^2(z)\right]\int_z^{+\infty}\frac{V_0(z')p(z')\varepsilon(z')}{1-V_0^2(z')}dz'. \tag{13}$$

In the second approximation it was assumed that $\exp[-2i\varphi(z)]\approx 1-2i\varphi(z)-2\varphi^2(z)$, and the function $V_2(z)$ is defined by the equation:

$$\frac{dV_2}{dz}=\frac{1}{2p}\frac{dp}{dz}\left(V_1^2-2V_0V_2\right)-2V_1p\varepsilon,\ V_2(+\infty)=0,$$

$$V_2(z)=4\left[1-V_0^2(z)\right]\left[2V_0(z)\int_z^{+\infty}\frac{V_0p\varepsilon I}{1-V_0^2}dz'-\int_z^{+\infty}\frac{1+V_0^2}{1-V_0^2}\cdot p\varepsilon I dz'\right], \tag{14}$$

where I designate the function

$$I(z')=\int_{z'}^{+\infty}\frac{V_0(z'')p(z'')\varepsilon(z'')}{1-V_0^2(z'')}dz''.$$

The purpose of calculation is the reflection coefficient $R_p=|V(-\infty)|^2$. Since the functions $\varepsilon(z)$, $V_0(z)$, $V_1(z)$, $V_2(z)$ has only the real values in a transparent medium, the ratio follows from the Eq. (11):

$$R_p=V_0^2(-\infty)+V_1^2(-\infty)+2V_0(-\infty)V_2(-\infty).$$

Substitutions of deduced expressions (12), (13), (14) and transformation allow to deduce an estimated formula [5]:

$$R_p=\left(\frac{p_1-p_2}{p_1+p_2}\right)^2+\frac{8p_1p_2}{(p_1+p_2)^4}\int_{-\infty}^{+\infty}\left[p^2(z)-p_1^2\right]\left\{\int_z^{+\infty}\left[p^2(z')-p_2^2\right]\varepsilon(z')dz'\right\}\varepsilon(z)dz, \tag{15}$$

where $p_1=p(-\infty)$ and $p_2=p(+\infty)$. The function $p(z)$ is connected to the function $\varepsilon(z)$ by formula (3).

Thus the ratio (15) expresses the reflection coefficient of the *p*-polarised electromagnetic wave by the analytical kind. The dielectric permittivity profile is given by the dependence of a general view $\varepsilon(z)$. In this case two restrictions were introduced. One assumed absence of the absorption and therefore the function $\varepsilon(z)$ can have only the real values. Another restriction is defined by the inequality (10).

3. Results of calculations

By means of the formula (15) the angular dependence of the reflection coefficient for the chosen functions $\varepsilon(z)$ was calculated. Further these dependences by means of the formula (1) were transformed to function $\eta(\theta)$.

In the non-uniform medium the dependence $\varepsilon(z)$ was simulated by Gauss' function:

$$\varepsilon(z)=\begin{cases}1, & z<0\\ \varepsilon_2+\delta\varepsilon\cdot\exp\left[-\dfrac{(z-d)^2}{2\Delta^2}\right], & z\geq 0\end{cases}.$$

The type of the function $\varepsilon(z)$ corresponds to the properties of the materials which underwent radiation by ions. Parameters $\delta\varepsilon$, d, and Δ are also defined by a radiation dose, energy, and dispersion of the ions run. Besides, the sign $\delta\varepsilon$ can be both the positive and the negative

Figures 1, 2 give an understanding about influence of each of the parameters on dependences $R_p(\theta)$ and $\eta(\theta)$. The domain of the angles located near the Brewster's angle is the most sensitive to the parameters. In this case the features of the dependence $\varepsilon(z)$ are shown in the position and in the extremums amplitudes of the $\eta(\theta)$ dependence. However the integral connection between $R_p(\theta)$ and $\varepsilon(z)$ doesn't allow to solve the reverse problem – it is impossible to construct dependence $\varepsilon(z)$ proceeding from the dependences $R_p(\theta)$ or $\eta(\theta)$. At the same time results of calculations show a possibility of the implanted layers identification with the help of the reflectometric characteristics.

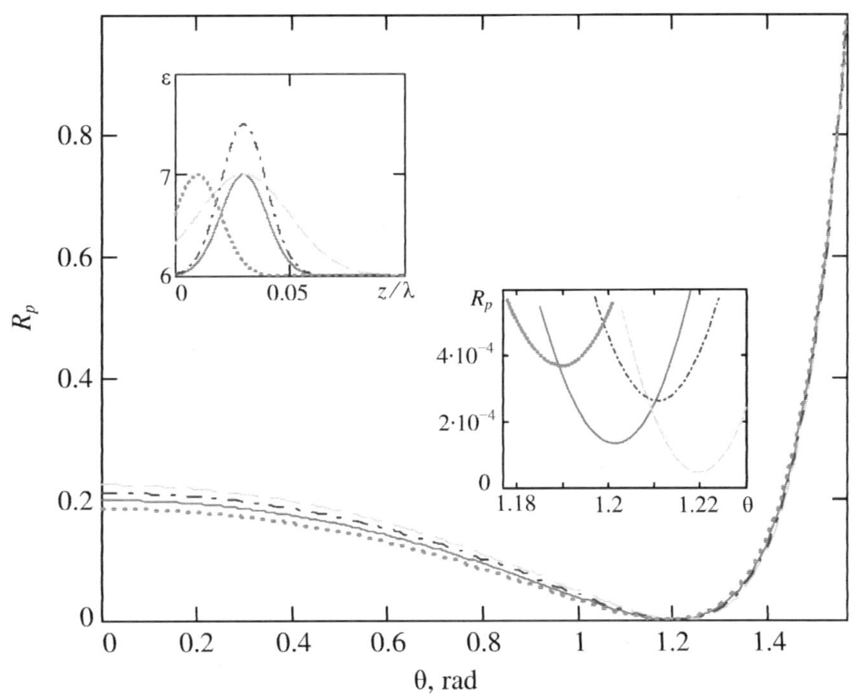

Figure 2. Angular dependences of the reflection coefficient $R_p(\theta)$. On the upper insertion the coordinate dependences of dielectric permittivity $\varepsilon(z/\lambda)$ which were used in calculations are figured (λ – the electromagnetic wave length), on lower – fragments of $R_p(\theta)$ dependences near the Brewster's angle.

It is necessary to note that the depth of penetration of the ions reaches several microns. Then the condition (10) can be satisfied if the electromagnetic wave has length into hundreds of microns. The frequencies of such waves have the values of $\sim 10^{12}$ Hz. Many dielectrics in this range are transparent. The transparency was supposed on deducing of formula (15).

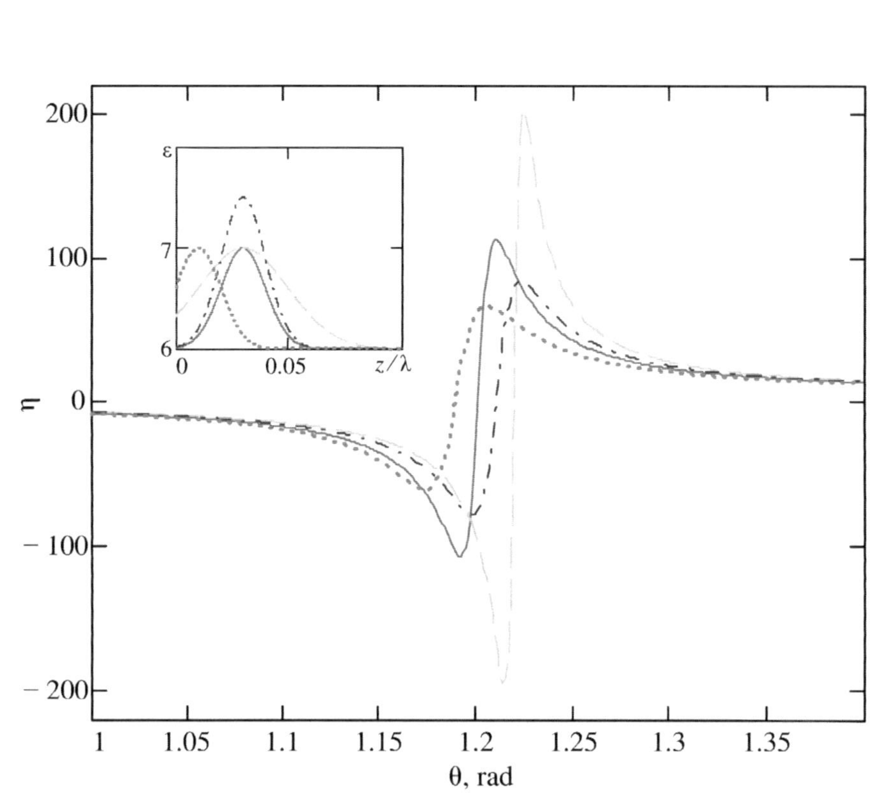

Figure 3. Angular dependences of the reflection coefficient logarithmic derivative near Brewster's angle $\eta(\theta)$. On an insert the dependences $\varepsilon(z/\lambda)$ which were used in calculations are represented.

4. Conclusions

1. The analysis of the electromagnetic wave reflection with *p*-polarization allows to detect thin layers with the non-uniform profile of dielectric permittivity. In this case the length of the wave can be into hundreds of times greater than the thickness of layer.

2. Estimated basis of such analysis is the deduced formula (15).

3. The reflectometric analysis allows to control the technological operations which are realized in case of the materials properties modification by an ion implantation technique.

The practical use of the expounded results is connected with a possibility of the reflection coefficient measurement in the terahertz range. The greatest difficulty arises in the interval of angles near the Brewster's angle where values of the coefficient become especially small. The problem can be solved both by increase in sensitivity of receivers and by increase in power of the used radiation sources.

5. References

[1] Zinchenko S P, Kovtun A P and Tolmachev G N 2009 *Technical Physics* **54** 1689
[2] Shvartsburg A B 2000 *Physics-Uspekhi (Advances in Physical Sciences)* **43** 1201
[3] Shvartsburg A B, Agranat M B, and Chefonov O V 2009 *Quantum electronics* **39** 948
[4] Bilenko D I, Polyanskaya V P, Getsman M A, Gorin D A, Neveshkin A A and Yaschenok A M 2005 *Technical Physics* **50** 742
[5] Shagaev V V 2015 *Technical Physics* **60** 1738
[6] Babikov V V 1976 *Method of Phase Functions in Quantum Mechanics* (Moscow: Nauka) 286

Model of convection mass transfer in titanium alloy at low energy high current electron beam action

V D Sarychev, A Yu Granovskii, S A Nevskii, S V Konovalov[1] and V E Gromov
Siberian State Industrial University, Novokuznetsk, 42 Kirov Street, 654007, Russia

E-mail: konovserg@gmail.com

Abstract. The convection mixing model is proposed for low-energy high-current electron beam treatment of titanium alloys, pre-processed by heterogeneous plasma flows generated via explosion of carbon tape and powder TiB_2. The model is based on the assumption vortices in the molten layer are formed due to the treatment by concentrated energy flows. These vortices evolve as the result of thermocapillary convection, arising because of the temperature gradient. The calculation of temperature gradient and penetration depth required solution of the heat problem with taking into account the surface evaporation. However, instead of the direct heat source the boundary conditions in phase transitions were changed in the thermal conductivity equation, assuming the evaporated material takes part in the heat exchange. The data on the penetration depth and temperature distribution are used for the thermocapillary model. The thermocapillary model embraces Navier-Stocks and convection heat transfer equations, as well as the boundary conditions with the outflow of evaporated material included. The solution of these equations by finite elements methods pointed at formation of a multi-vortices structure when electron-beam treatment and its expansion over new zones of material. As the result, strengthening particles are found at the depth exceeding manifold their penetration depth in terms of the diffusion mechanism.

1. Introduction

Titanium alloys are widely applied in various industries, e.g. in aircraft engines construction for manufacturing blades and discs. The thickness of damaged layers varies 10 to 20 μm as seen in the process of gas-turbine engines maintenance [1]. Therefore, it is necessary to develop technologies of surface modifying titanium alloys. At present a great variety of ways and methods of surface modification are developed, inter alia, the treatment by low energy high current electron beams [2]. This kind of treatment is advantageous as its pulse-periodic character allows of keeping surface layers in the molten state for a longer period of time, and it provides the effect of hardening, which furthers formation of submicro- and nanostructures [3]. The shortcoming of electron beam treatment is forming craters on the surface of materials [4], which are dangerous stress risers when cyclic stressing. Thus, treatment conditions with minimal influence of craters are to be selected; moreover, the thickness of the layer to be modified should not exceed it of the layer damaged in the process of maintenance. The information how electron beam treatment influences materials is necessary in order to search for the modes mentioned above. As shown in [4 – 10] electron beams have a multifactorial effect on

[1] Corresponding author

materials. It comprises heating, melting and evaporation of materials from the surface. When melting metals electron beams might cause formation of closed streams in the melt. The intensity and character of these streams are relevant for heat and substance transfer, which, in their turn, influences diverse technological processes. These streams might be formed and developed due to thermocapillary forces [8 –10]. The phenomenon of thermocapillary convection has been studied in various works. A double-vortices stream is formed for the definite, quite high Marangoni numbers [10]. The mathematical model of heat and mass transfer in the target irradiated by intense beam of charged particles was proposed in [8, 9], it involves equations of continuum mechanics, kinetic equation for fast particles and is enclosed by the wide-range equations of state. The basic mechanism of liquid phase heat and mass transfer is revealed to be thermocapillary convection and instability according to the results of calculations [9]. Moreover, it is possible when pulse duration is far shorter than the typical period of thermal diffusivity. Two oppositely directed vortices are formed in the sub-surface layer. The big vortex is the result of the surface tension force which depends on the temperature; whereas the small vortex arises as the inflow of liquid involved by big vortex exceeds the outflow, causing the formation of the dead zone and small vortex, as the consequence [8, 9]. However, evaporation from the material surface has not been taken into consideration in these works, although the high power density of the beam $q \sim 10^6 - 10^9$ W/cm^2 leads to heating up the metal surface, as the result, it starts melting and evaporating together with the forming plasma flame [5]. The thermal conductivity equations and the gas-dynamic problem of gas dispersion are to be solved to calculate the temperature distribution. The evaporation is irrelevant for power density $q \leq 10^6$ W/cm^2, so the power density of a heat flow is required for solution of the heat problem [5]. Otherwise, extraordinary high temperatures might be the result of the direct heat flow on the boundary [6]. Therefore, the authors [6] suggest the enthalpy statement of the problem with changing boundary conditions, which can register phase transformations automatically. The penetration depth increases linearly with the going up power density of electron beam according to the calculations, the pulse duration has principally no influence on it. In case electron beams influence the material exposed to heterogeneous conductor electric explosive-produced plasma flows, it is more difficult as heat, generated in the process of forming and dissolving phases, is to be considered. The effects of electron beam on titanium pre-processed by heterogeneous graphitize carbon fibers explosive-produced plasma flow is studied in [8]. The electron beam is found out to be the reason for dissolution of carbon particles, moreover, the period of dissolution depends on the radius of a particle. Therefore, we can conclude a mathematical model describing convection mixing, which involves phase transitions in conditions of electron beam treatment, is currently not available, and our work is aimed at the development of such a model.

2. Mathematical model

Let us consider the effect of electron beams on the plate (Figure 1). The input parameters of the problem are as follows: E_s – surficial power density, t_0 – pulse duration and radius of the impact – R_0 .

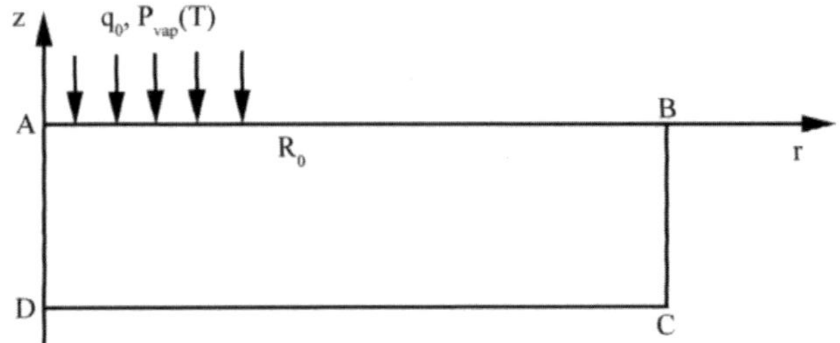

Figure 1. Computational domain scheme

The equation of thermal conductivity is written.

$$\rho \tilde{N}_p \frac{\partial T}{\partial t} = \nabla \cdot (k \nabla T) \tag{1}$$

As we focus on the distribution of temperature over the depth of the sample we consider the one-dimensional case only. We model phase transitions on the boundary via setting the temperature-dependent characteristics of material:

$$\tilde{N}_p(T)\rho(T) = \begin{cases} C_S \rho_S, T < T_L \\ \dfrac{L_L \rho_L}{\Delta T_L}, T_L \le T < T_L + \Delta T_L \\ C_L \rho_L, T_L + \Delta T_L \le T < T_V \\ \dfrac{L_V \rho_V}{\Delta T_V}, T_V \le T < T_V + \Delta T_V \\ C_V \rho_V, T_V + \Delta T_V \le T \end{cases}, \quad k(T) = \begin{cases} k_S, T < T_L \\ k_S + \dfrac{(k_L - k_S)(T - T_L)}{\Delta T_L}, T_L \le T < T_L + \Delta T_L \\ k_L, T_L + \Delta T_L \le T < T_V \\ k_S + \dfrac{(k_V - k_L)(T - T_V)}{\Delta T_V}, T_V \le T < T_V + \Delta T_V \\ k_V, T_V + \Delta T_V \le T \end{cases} \tag{2}$$

Heat losses caused by evaporating material were not taken into account in boundary conditions [7], therefore, the coefficient of thermal conductivity was selected in order to describe experimental data. We take these losses into consideration and use the model proposed in [12]. The conditions on the boundary AB with phase transitions taken into account are written in the form:

$$-\vec{n} \cdot \vec{q} = q_0(t) - q_{out}(T) = \frac{E_s}{t_0}\theta(t) - \dot{m}(T) L_V \tag{3}$$

Where $\dot{m}(T) = (1-\beta)\sqrt{\dfrac{M}{2\pi RT}}\, p_0 \exp\left(\dfrac{L_V M (T - T_V)}{R T T_V}\right)$ – mass flow through the boundary.

When the substrate evaporates the residual gas in the chamber takes part in the heat exchange. Some energy dissipates; the other part goes back to the substrate keeping it at the vaporization temperature longer than the period of impact. We assume heat capacity of the evaporated phase C_V =20 MJ kg^{-1} K^{-1} for numerical modelling of this gaseous layer. It allows of making a 1-2 μm thick heat "buffer", which will keep the putting in power and emit it gradually. Hence, setting the heat capacity of the gaseous phase we can model the boundary conditions [7], considering, however, heat losses caused by gaseous phase dissipation. It makes it possible to develop a more complete model of heat processes arising in the process of electron beam treatment, which describes experimental data on the penetration depth. The distribution of temperatures in titanium at various penetration depths and in diverse modes of irradiation is presented in Figure 2. As one can see the temperature on the surface does not reach the evaporation temperature at E_s = 45 J/cm^2, t_0 = 200 μs, and it drops in the point t_0 (Figure 2a). In conditions E_s = 65 J/cm^2, t_0 = 100 μs the temperature on the surface comes up to the evaporation temperature and is constant for the period 180 μs (Figure 2b), then it falls sharply.

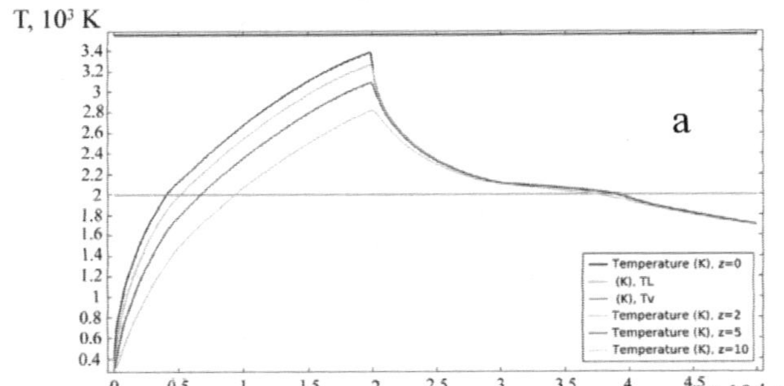

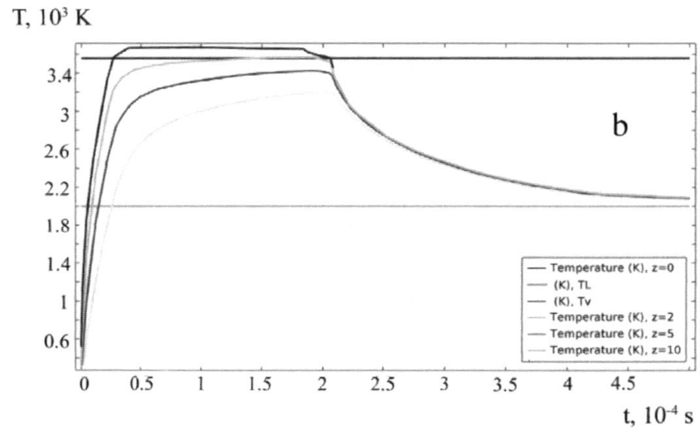

Figure 2. The distribution of temperature in titanium in the process of electron beam treatment a) $E_s = 45$ J/cm^2, $t_0 = 200$ µs; b) $E_s = 65$ J/cm^2, $t_0 = 100$ µs

Therefore, the heat "buffer" made by the evaporated material leads to stabilization of the temperature on the surface and over the depth of the sample for the period exceeding the pulse duration. As the result, the gradient of the undisturbed temperature throughout the depth is constant, developing the conditions for the thermocapillary instability. The initiation of this instability is analyzed in metals and binary alloys in the process of electron beam treatment [13], the dispersion equation is obtained too. On the base of this equation the critical wavelength with the maximum of the disturbance decrement was determined. Its value approximates to 100 nm. The crystallization cells are of the same range. Developing the numerical model of the thermocapillary convection we take into consideration the change in the surface curvature. We solve the axis-symmetrical problem for Navier Stocks equation and heat transfer to model the convection flows. The conditions on the boundary AB (Figure 1) are in the form:

$$-\vec{n}\cdot\vec{q} = q_0(r,t) - q_{out}(T) = \frac{E_s}{t_0}\exp\left(-\frac{r^2}{R_0^2}\right)\theta(t) - \dot{m}(T)L_V$$
$$p = p_0 + p_V, p_V = \frac{1+\beta}{2}(p_c - p_0) \tag{4}$$

where p_V – pressure of vapor output, p_c – Clapeyron pressure. On boundaries BC, DC: $\vec{n}\cdot\vec{q}=0, \vec{n}\cdot\vec{v}=0$. The problem was solved by the finite elements method in the software Comsol Multiphysics. To model the free surface we used the moving mesh method. The input parameters of the problem are given in Table 1.

Table 1. Material properties of the sample.

Symbol	Nomenclature	Value
ρ_S	Solid density	4500 kg m^{-3}
ρ_L	Liquid density	4100 kg m^{-3}
ρ_V	Vapor density	4100 kg m^{-3}
C_L	Liquid heat capacity	922 J kg^{-1} K^{-1}
C_S	Solid heat capacity	630 J kg^{-1} K^{-1}
C_V	Vapor heat capacity	$2\cdot10^7$ J kg^{-1} K^{-1}
k_L	Liquid thermal conductivity	30 W m^{-1} K^{-1}

k_S	Solid thermal conductivity	25 W m^{-1} K^{-1}
k_V	Vapor thermal conductivity	30 W m^{-1} K^{-1}
μ	Coefficient of dynamic viscosity	0.0021 kgm^{-1} s^{-1}
L_L	Melting latent heat	304 kJ kg^{-1}
L_V	Vaporization latent heat	8900 kJ kg^{-1}
σ_0	Surface tension coefficient	1.872 N m^{-1}
$d\sigma/dT$	Thermal gradient of surface tension	$-49\cdot10^{-5}$ N $m^{-1}\cdot K^{-1}$
E_s	Pulse energy	450 kJ m^{-2}
r_0	Radius of the impact	200 μm
M	Molar mass	0.047867 kg mol^{-1}
T_L	Melting temperature	1998 K
T_V	Vaporization temperature	3650 K

The results of calculations are given in Figure 3

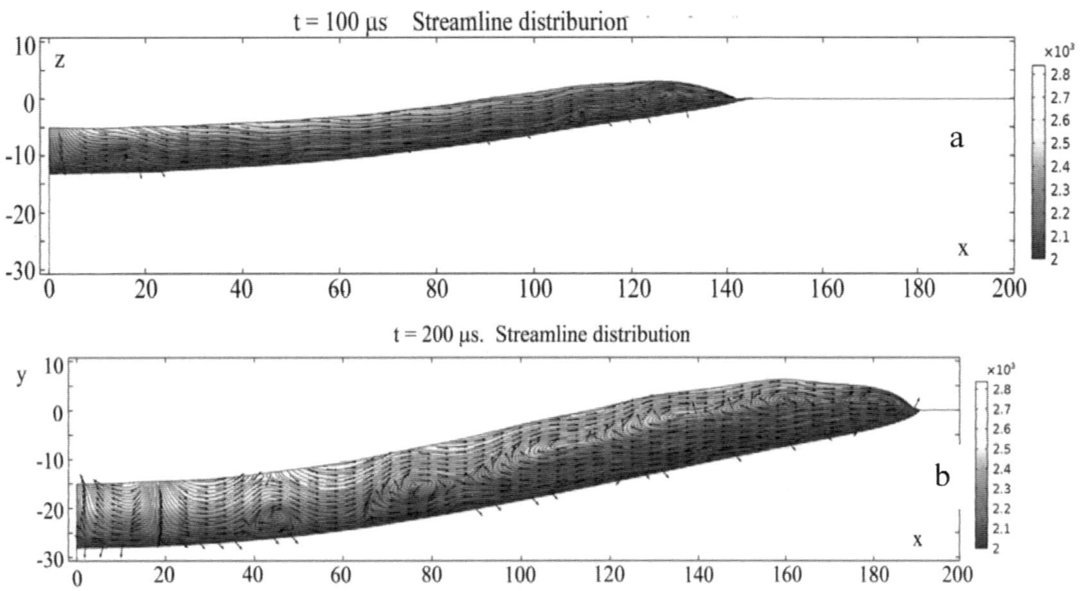

Figure 3. The distribution of streamlines at various time points
a) t =100 µs; b) t = 200 µs

The convection of material in the molten layer as well as its probable effect on heat transfer and geometry of the pool were not taken into consideration in all the works focused on electron beam melting, except [14]. The latest works on EBM modelling [15, 16] revealed it was not possible to obtain multi-vortices flow in the melt layer. In our calculations we have obtained parameters of the multi-vortices flow (Figure 4). This fact is principally important for mixing the melt and creating homogenous distribution of boron-containing particles, which is necessary for determining the mechanism of wear resistance improvement [11]. The distribution of microhardness over the depth and the thickness of the strengthening zone, calculated on its base, confirm, the proposed model of convection mass transfer can provide an adequate description of the processes when electron-beam treatment [11].

3. Acknowledgements
The research has been carried out under financial support of Russian Foundation for Basic Research within the projects № 15-08-03411a, and the State Orders of Ministry of Science and Education No 3.1496.2014/K submitted to conduct investigations.

References

[1] Fadeev A V, Bazhenov V E, Koltygin A V 2015 *Russian Journal of Non-Ferrous Metals* **56** 26
[2] Shulov V A, Gromov A N, Teryaev D A and Engel'ko V I 2016 *Russian Journal of Non-Ferrous Metals* **57** 256
[3] Novikov A S, Shulov V A and Engel'ko V I 2012 *Sil'notochnye impul'snye elektronnye puchki dlya aviatsionnogo dvigatelestroeniya (Pulsed High-Current Electron Beams for Aviation Propulsion Engineering)* (Moscow: DIPAK)
[4] Ozur G E, Proskurovsky D I and Rotshtein V P 2016 *Technical Physics Letters* **42** 328
[5] Bleicher G A, Krivobokov V P, Pashchenko O V 2014 *Erosiya Poverkhnosti tverdogo tela pod deistviem moshnykh puchkov zrayagennykh chastits (Erosion surface of solid state at power charged particles beams action)* (Novosibirsk: Nauka)
[6] Khaimzon B B, Sarychev V D, Soskova N A and Gromov V E 2013 *Steel in Translation* **43** 55
[7] Sarychev V D, Khaimzon B B, Gromov V E and Soskova N A 2012 *Titan* **1 (35)** 4
[8] Volkov N B, Leivi A Ya, Talala K A and Yalovets A P 2010 *Technical Physics* **55** 484
[9] Talala K A, Leivi A Ya and Yalovets A P 2012 Bulletin of the South Ural State University. Series Mathematics. Mechanics. Physics **34** 186
[10] Sanochkin Yu V 1984 *Fluid Dynamics* **19** 986
[11] Gromov V E, Kobzareva T Yu, Ivanov Yu F, Budovskikh E A, and Baschenko L P 2016 *AIP Conference Proceedings* **1698** 030006
[12] Knight C J 1979 *American Institute of Aeronautics and Astronautics* **17** 19–523
[13] Nevskii S , Sarychev V , Konovalov S , Kosinov D, Komissarova I 2016 *Mater. Sci. Forum* **870** 34
[14] Rai R, Burgardt P, Milewski J O, Lienert T J and DebRoy T 2009 *J. Phys. D: Appl. Phys.* **42**, 025503
[15] Jamshidinia M, Kong F, Kovacevic R 2013 *J. Manuf. Sci. Eng.* **135** 061010
[16] Markl M, Körner C 2016 *Annual Review of Materials Research* **46** 93

IOP Conf. Series: Materials Science and Engineering **168** (2017) 012032 doi:10.1088/1757-899X/168/1/012032

Peculiarities of structure and hardening of Ni-Ti alloy surface layers formed by $_{84}Kr^{15+}$ ions irradiation at 147 MeV energy at high temperatures

V Poltavtseva, A Larionov and G Zheltova
Institute of Nuclear Physics, Almaty, Republic of Kazakhstan

E-mail: poltavtseva@inp.rk

Abstract. The consistent patterns of changes in nanostructure and nanohardness of Ni-Ti alloy after irradiation with $_{84}Kr^{15+}$ ions with 147 MeV energy to the fluence of $1\cdot10^{19}\,m^{-2}$ at 250 and 300^{0}C temperatures depending on phase composition have been experimentally studied. It was shown that significant (44 – 94%) softening of surface layers for the single-phase and two-phase Ni-Ti alloys is connected with the formation of bubble nanostructured defects and complete sputtering of the process layers. The role of nanostructure in roughness of the irradiated Ni-Ti alloy surface of various phase composition has been established.

1. Introduction

During modification of NiTi alloys for medical application the great attention should be paid to the surface [1]. The papers [2-4] showed that as a result of irradiation with inert gases heavy ions of MeV-energy the structure and properties of the modified surface layers of Ni-Ti alloy vary considerably depending on exposure parameters and phase composition. So, the nanostructured defects in the form of bubbles ($_{40}Ar^{8+}$) or tracks ($_{84}Kr^{15+}$, $_{132}Xe^{22+}$) were revealed on the surface of Ni-Ti alloy of various phase composition after irradiation under comparable parameters of A/Z, E_{nucl}, Ft, J_{beam} and T_{irrad} [3, 4]. Transformation of the track structure into the granular structure occurs during irradiation with $_{84}Kr^{15+}$ ions to the fluences of $\leq 5\times10^{19}$ ion/m^2 [5]. In this case radiation hardening is observed near the Ni-Ti single-phase alloy surface layer modified with krypton ions, but in the case of argon ions irradiation it is observed at the depth of $\geq$ 100 nm only [2]. This is associated with the running processes of ion polishing and sputtering typical for krypton ions and lighter argon ions, respectively [2, 3]. The main reason for surface layers softening of Ni-Ti two-phase alloy is the radiation-induced phase transformation B19$'$→B2 [3, 5].

The represented results are obtained for Ni-Ti alloy of various phase composition, irradiated with heavy ions of MeV energy at the temperature of $\leq$ 100^{0}C. The exposure at high temperatures, as shown in [6, 7], contributes to stabilization or improving certain properties of the nickel-titanium based alloys retaining the shape memory effect.

The aim of this work is to study the effect of high temperature MeV-energy krypton ions exposure on the change in nanostructure and nanohardness of the surface layers of Ni-Ti alloy of various phase composition.

2. Experimental methods and material

The Ni-Ti alloy was studied in two structural-phase states: two-phase and single-phase. The experimental two-phase Ni-Ti alloy of the following composition: 53.46 wt.% Ni - 46.54 wt.% Ti, consisting primarily of NiTi with the B2 structure (austenite), NiTi with B19' structure (martensite), a small content of Ti, excessive Ni as a solid solution, and technological particles close in their composition to $Ti_2Ni(C)$ [2, 3]. The experimental samples of two-phase alloy Ni-Ti were cut from a massive forget plate across the rolling direction using spark cutting. The sample surfaces were mechanically ground and then polishing on a woolen cloth using GOI polishing paste. The surface quality was estimated by optical metallographic analysis. The samples size was $15 \times 3.9 \times 0.35$ mm^3. The experimental samples of single-phase Ni-Ti alloy with the B2 structure (austenite) were prepared by annealing the two-phase samples in the vacuum (~ 10^{-4} Pa) furnace chamber for 1 hour at 230 ± 30 °C [2].

Irradiation of $_{84}Kr^{15+}$ ions at 147 MeV (1.75 MeV/nucl.) energy up to the fluence 1×10^{19} ion/m^2 was performed with DC-60 accelerator (Astana, Kazakhstan) at 250 and 300^{0}C temperatures. Accuracy of irradiation temperature maintenance was ±5^0. The radiation treatment area was $\sim 1\times10^{-4}$ m^2.

The study of surface layers microstructure and elemental analysis was performed by scanning electron microscopy (SEM) using the microscope JSM-7500F (JEOL, Japan). The analyzed depth was about 1 µm. The degree of surface layers hardening was evaluated according to the results of nanohardness measurements by sclerometry [2] using the nanohardness gage "NanoScan-Compact" (Russia) at the applied load of 1 ± 200 mN. The accuracy was 1% for nanohardness measurement. The surface roughness was evaluated according to the scanned images.

3. Experimental results and discussion

SEM-images at low magnifications of surface layers structure of the single-phase Ni-Ti alloy after irradiation with $_{84}Kr^{15+}$ ions at 250 and 300^{0}C temperatures are shown in figure 1c,d. For comparison purpose also shows the microstructure of the single-phase Ni-Ti alloy before (figure 1a) and after irradiation with $_{84}Kr^{15+}$ ions at $T_{irrad} \leq 100^0C$ (figure 1b). Analysis of the obtained experimental data showed that interaction of $_{84}Kr^{15+}$ ions with the surface of the single-phase Ni-Ti alloy at high temperatures results in the sputtering processes of friable film consisting of Ni_4Ti_3 compound and ion surface polishing, according to X-ray diffraction studies.

It was found that the sputtering processes of the strain-hardened Ni_3Ti_3O [3] layer and the ion surface polishing also occur for the two-phase Ni-Ti alloy, however at all selected irradiation temperatures. Regardless of the structural-phase state the microstructure of Ni-Ti alloy surface is characterized by high quality after high temperatures exposure as the result of the $^{84}Kr^{15}$ ions polishing process (figure 1c,d).

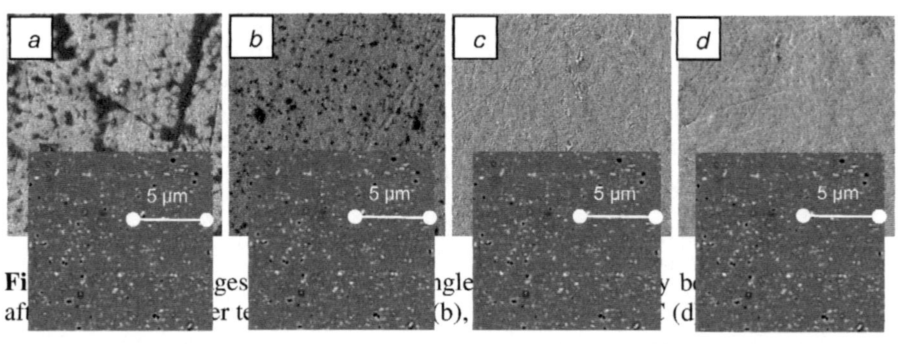

Fi ges ngle y b
af er t (b), (d

SEM-images at high magnifications of surface layers structure of the two-phase Ni-Ti alloy after irradiation with $_{84}Kr^{15+}$ ions at $T_{irrad} \leq 100^0C$ [3,4], 250 and 300^{0}C are shown in figure 2. We can see that as T_{irrad} increases to 250^{0}C formation of rare light bubbles up to 80 nm on the background of small bubbles (about 10 nm) occurs instead of tracks formation (figure 2b). At further increase of T_{irrad} to 300^{0}C the size of large bubbles is reduced to 50-60 nm, while the size of small bubbles grows to 25-30 and 15-20 nm (figure 2c). The number of larger bubbles decreases almost twice and the number of smaller bubbles remains small.

Similar patterns of exposure temperature effect on the surface layers structure were also revealed for the single-phase Ni-Ti alloy. However, compared to the two-phase Ni-Ti alloy, the size of large bubbles after irradiation at $T_{irrad} = 250^0C$ is less (to 70 nm), and the number of small bubbles (about 10 nm) grows. And on the contrary, the size of large bubbles grows to 100 nm, and the size of small bubbles grows to 50 and 20-30 nm during irradiation at $T_{irrad} = 300^0C$.

We shall note that the characteristic heterogeneity of elemental composition of the non-irradiated Ni-Ti alloy of various structural-phase states improves after irradiation with $_{84}Kr^{15+}$ ions at high temperatures. However, the almost uniform elemental composition was found only for the single-phase Ni-Ti alloy after irradiation at $T_{irrad} = 300^0C$.

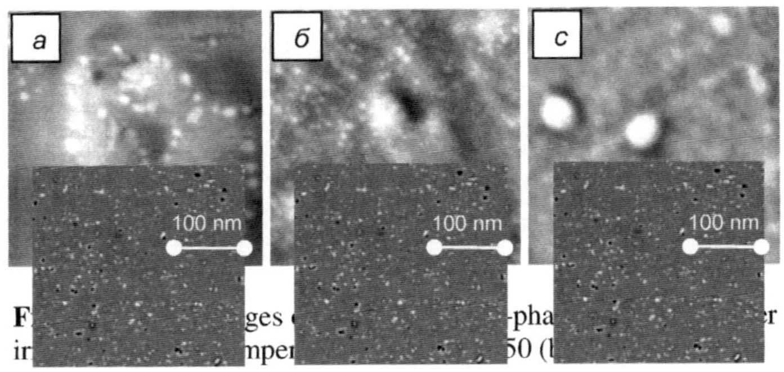

F ges -pha r
ir npe 50 (

The results of nanohardness measurements depending on surface layer thickness are shown in figure 3. According to these data, firstly, the sharp softening of Ni-Ti alloy occurs during exposure at high temperatures regardless of its structure-phase state. Near the surface the single-phase Ni-Ti alloy softens at 44 and 74% at 250 and 300^0C temperatures, respectively (figure 3a, curves 3 and 4) compared with non-irradiated alloy (figure 3a, curve 1). The degree of softening for the two-phase Ni-Ti alloy is much higher: 94 and 85% respectively after irradiation at 250 and 300^0C temperatures (figure 3b, curve 3 and 4). The hardening effect of the single-phase Ni-Ti alloy after irradiation at $T_{irrad} \leq 100^0C$ (figure 3a, curve 2) is noted in [2].

Secondly, the character of H (h) – curves modified at high temperatures of Ni-Ti alloys differs from similar curves for non-irradiated Ni-Ti alloys and after irradiation at $T \leq 100^0C$ by the following: there is no maximum near the surface (figure 3a,b). It points to complete sputtering during exposure process at high temperatures of the surface technological layers: Ni_4Ti_3 film and deformation-hardened layer of Ni_3Ti_3O 40 and 50 nm thick, respectively. Maxima shift on H (h) – curves to the modified surface is 20 and 30 nm, respectively, for the two-phase (figure 3a, curve 2) and the single-phase Ni-Ti alloy (figure 3b, curve 2) during irradiation at $T_{irrad} \leq 100^0C$, which indicates their partial sputtering

Thirdly, the nanohardness value near (up $\geq$130 nm) Ni-Ti alloy surface is practically independent from its structural-phase state after exposure at high temperatures (figure 3ab, curves 3 and 4). This fact is consistent with the formation of bubbles nanostructured defects at high temperatures, similar in size and concentration (figure 2), and sputtering of technological layers. It can be concluded that the surface layers of single-phase and two-phase Ni-Ti alloys, modified at high temperatures, are characterized by austenitic structure-phase state.

Let us consider the dependence of surface layers nanohardness value from the quality of exposed surface, i.e. its roughness. The results of the arithmetic mean values of roughness height R_a obtained by measuring of the roughness curves for the studied Ni-Ti alloys are shown in table 1. We shall mainly note a significant decrease in roughness for Ni-Ti alloys after irradiation at $T_{irrad} \leq 100^0C$ compared with the non-irradiated state, when the track nanostructured defects (figure 2 a) are observed on the surface [3, 4].

The roughness value for the single-phase Ni-Ti alloy is slightly lower than for the two-phase alloy, which is mainly due to the large size of track nanostructured defects at almost the same amount of them [4].

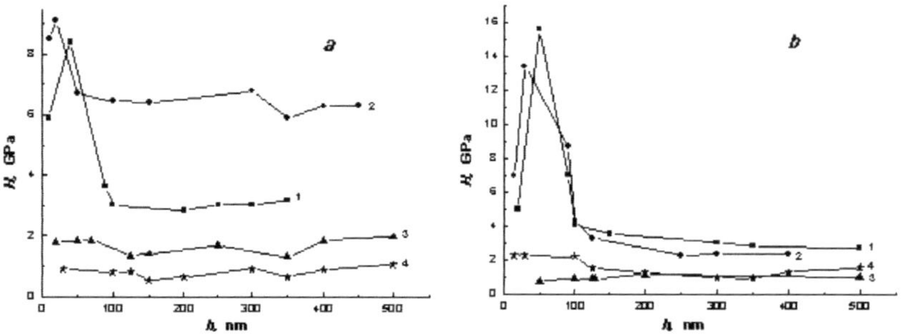

Figure 3. Dependence of nanohardness single-phase (a) and two-phase Ni-Ti alloy (b) before (curve 1) and after irradiation under temperature ≤ 100 (curve 2), 250 (curve 3) and 300^0C (curve 4) on thickness of surface layer.

For irradiation at high temperatures the roughness value, on the contrary, increases compared to non-irradiated Ni-Ti alloys, except for single-phase alloy irradiated at $T_{irrad} = 300^0$C (table 1). The main reason of roughness increase for the modified surface of both alloys is the formation of nanostructured bubble defects (figure 2b) with the size 1.6 - 2 times lower than for track nanostructured defects (figure 2a). According to analysis of the above-mentioned SEM - data, the roughness reduction is caused by contribution of small bubbles ($T_{irrad} = 250^0$C) or its increasing by reducing the number of large and small bubbles ($T_{irrad} = 300^0$C).

Thus, it was found that the roughness value of Ni-Ti alloy surface irradiated at various temperatures depends on the type of nanostructured defects, their size and number. The larger nanostructured defects, the lower the surface roughness. Decrease of nanostructured defects number leads to its increase.

Table 1. Roughness of Ni-Ti alloy surface, irradiated at various temperatures

Ni-Ti alloy	Roughness (nm)			
	Non-irradiated	≤ 100^0C	250^0C	300^0C
Single-phase	9.04	3.48	13.34	7.19
Two-phase	13.85	3.89	15.03	17.91

Changes of maximum nanohardness value for two-phase (curve 1) and single-phase (curve 2) Ni-Ti alloys, depending on the temperature of $_{84}Kr^{15+}$ ions irradiation, are presented in figure 4. Firstly we shall note that the magnitude of the surface layer nanohardness for the non-irradiated two-phase Ni-Ti alloy is determined by hardness of Ni_3Ti_3O compound, NiTi martensitic phase with the B19' structure and NiTi austenitic phase with the B2 structure, while the contribution in the non-irradiated single-phase Ni-Ti alloy nanohardness is stipulated by hardness of Ni_4Ti_3 compound and NiTi with the B2 structure.

As a result of irradiation at $T_{irrad} \leq 100^0$C the value of surface layer nanohardness for the two-phase Ni-Ti alloy is reduced due to competition of the contributions from the processes of track nanostructured defects accumulation (figure 2b), partial sputtering of the process layer (figure 3b), complete transfer of martensite to austenite B19'→B2 [3, 5] and contribution of NiTi with the B2 structure [3]. The predominance of the contribution from the process of track nanostructured defects accumulation [4] over the contribution from the partially sputtering of the process layer (figure 2a) is the cause of hardening of the single-phase Ni-Ti alloy surface layer after irradiation at $T_{irrad} \leq 100^0$C.

A significant (44 - 94%) softening of the surface layers of single-phase and two-phase Ni-Ti alloys after irradiation at high temperatures, as found in this study, is connected to the complete sputtering of the process layers and formation of bubble-type nanostructured defects, size and number of which is significantly less than for the track nanostructured defects.

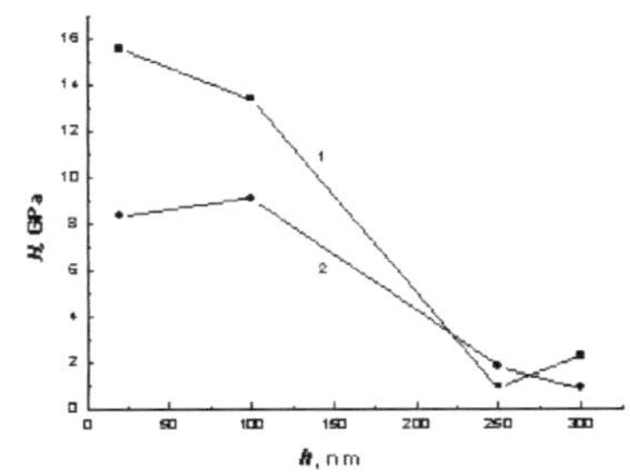

Figure 4. Dependence of H_{max} two-phase (curve 1) and single-phase Ni-Ti alloy (curve 2) on T_{irrd}.

4. Conclusion

Thus, as a result of the study it is shows that a significant (44 - 94%) softening near the surface of two-phase and single-phase Ni-Ti alloys after irradiation with ${}_{84}Kr^{15+}$ ions at 250 and 300^{0}C to a fluence of $1\cdot10^{19}$ m^{-2} is associated with a full sputtering of the technological layers and formation of nanostructured bubble defects. It is found that the value of roughness for the Ni-Ti alloy surface exposed at 100-300^{0}C temperatures depends on the type of nanostructured defects, their size and number. Thus, the larger nanostructured defects, the lower the surface roughness. Reduction of nanostructured defects number leads to roughness increase.

Acknowledgements
This work was partially supported by The Ministry of Energy, Republic of Kazakhstan within the framework of the budget program "Applied scientific research of technological character".

References

[1] Rather B and Hoffman A 2004 *Elsevier Academic Press*. 201
[2] Poltavtseva V, Larionov A, Satpaev D and Gyngazova M 2016 *J. Mater. Sci. and Engin.: Conf. Series* **110** 012011
[3] Poltavtseva V, Kislitsin S, Satpaev D, Mylnikova T and Chernyavskii A 2015 *J. Mater. Sci. and Engin.: Conf. Series* **81** 01234
[4] Poltavtseva V 2015 *Proc. Conf. on Interaction of Radiation with Solids* (Minsk: Belarus Stat University Press) 141
[5] Poltavtseva V, Kislitsin *S*, Koval N and Oskomov K 2012 *Izv. Vyssh. Uchebn. Zaved. Fiz.* **12/3** 41
[6] Karuta T. 1989 *Farid: First Atom. Pavel Ind. Group.* **121** p 19
[7] Kadyrzhanov K, Komarov F, Pogrebnyak A, Rusakov V and Turkebaev T. 2005 *Ion-beam and ion-plasma modification of materials* (Moscow: Moskow Stat University Press) p 126

Effect of irradiation and thermal annealing on quartz materials luminescence

M V Korovkin and L G Ananyeva
National Research Tomsk Polytechnic University, Tomsk, Russia

E-mail: mvk@tpu.ru, lga@tpu.ru

Abstract. X-ray and gamma-quanta irradiation of radiation-resistant quartz materials including natural and synthetic quartz crystals and high-purity quartzite causes the luminescence in the ultraviolet range (365 nm), thermally stimulated luminescence and radiofrequency electromagnetic emission. Preliminary radiation and thermal annealing improves luminescence properties of quartz materials.

1. Introduction
Quartz materials are widely used to manufacture various kinds of things including quartz radiodes and epireadies for compound semiconductors which must meet special requirements on defect population and on control over defect modifications under various radiation and/or thermal actions. Imperfection of natural quartz as a mineral is caused by its genetic formation conditions, whereas imperfection of synthetic quartz is represented by its biographic defects as a result of its synthesis. As a rule the purest and most transparent varieties of quartz materials of both the natural and synthetic origin contain few optically –inactive imperfections which can be detected with non-destructive luminescence methods [1, 2].

This paper covers the variation of luminescence properties of quartz materials when being exposed to a thermal and radiation action.

2. Research methodology
Quartz materials were exposed to X-ray radiation (X-ray luminescence) and to heat action at a temperature of up to 400 °C (thermally stimulated luminescence) and then their luminescence spectra in the range of 300...800 nm were studied. The research methodology was the same as common photometric procedures [3, 4]. In addition, a method of synchronous recording of radiofrequency (RF) electromagnetic emission was also used [4-6].

Samples of quartz materials including natural and synthetic quartz and highly-pure quartzite were in a form of thin plates, 10×10×1 mm in size.

3. Experimental results and discussion
When highly-pure synthetic and natural quartz crystals and natural quartzite are exposed to X-ray

radiation, a slight glow appears in the range of 620…650 nm (Fig.1, Curve 1). Irradiation of quartz crystals with X-rays or gamma-quanta results in ionization of silicon-oxygen tetrahedron and occurrence of recombination centers represented by oxygen ions. The luminescence recorded in the ultraviolet range and in the range of 630 nm is caused by recombination processes occurring in $[SiO_4]^{3-}$ -hole-centers and on nonbridging oxygen ions (Fig.1, Curve 2), respectively. Thermal pre-treatment of non-irradiated natural quartz samples even at rather low temperatures of 150°C results in drastic increase of the luminous intensity in the ultraviolet range (Fig.1, Curves 2 and 3).

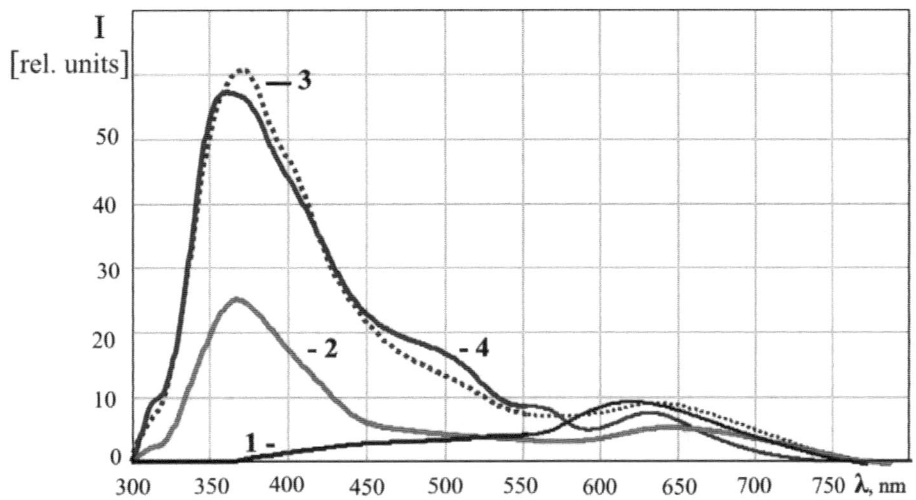

Figure 1. X-ray luminescence spectra of natural quartz crystals: 1 – virgin sample, 2 – irradiated by gamma-quantum, 3 – heated up to 150 °C, 4 – heated up to 400 °C.

Occurrence of thermally-stimulated luminescence (TSL) in crystals is associated with crystal imperfection, radioactive elements impurities and ionizing irradiation of them. The TSL mechanism is well explained by a band theory.

The TSL phenomenon is not observed in synthetic quartz samples. Highly-pure quartzites are characterized with a weak natural thermally-stimulated luminescence in the high-temperature range of 220-235 °C and 285-290 °C [4]. The TSL of natural quartz crystals is conditioned by the mineral formation environment and is widely used to restore the genetic information. In order to stimulate the mineral's "memory", crystals are irradiated with ionizing radiation that causes the induced TSL and significantly extends the TSL applications to solve mineralogy and material engineering practical tasks.

Irradiation of radiation-resistant quartz crystals with X-rays and gamma-quanta does not cause new defects in the crystalline lattice but results in ionization of silicon-oxygen tetrahedron and, probably, OH-groups in the quartz lattice. This also leads to developing various electrically active defects including dislocations, grain boundaries, blocks, point defects and color centers which "trap" electrons and "holes" [4, 5]. Further heating up of quartz crystals is accompanied both with the thermo luminescence and electromagnetic emission in the RF range (Fig.2). The electromagnetic emission is especially intensive in a low temperature range of 50…70 °C.

In nominally pure synthetic quartz crystals the electron-hole color centers are not stable due to a mobile hydrogen ion (proton) present in synthetic quartz crystals as an ion – compensator or formed in them due to ionizing processes; the mobile hydrogen ion, residing next to silicon-oxygen tetrahedra, does not allow creating a stable hole $[SiO_4]^{3-}$ center responsible for TSL in the temperature range of 50…70 °C.

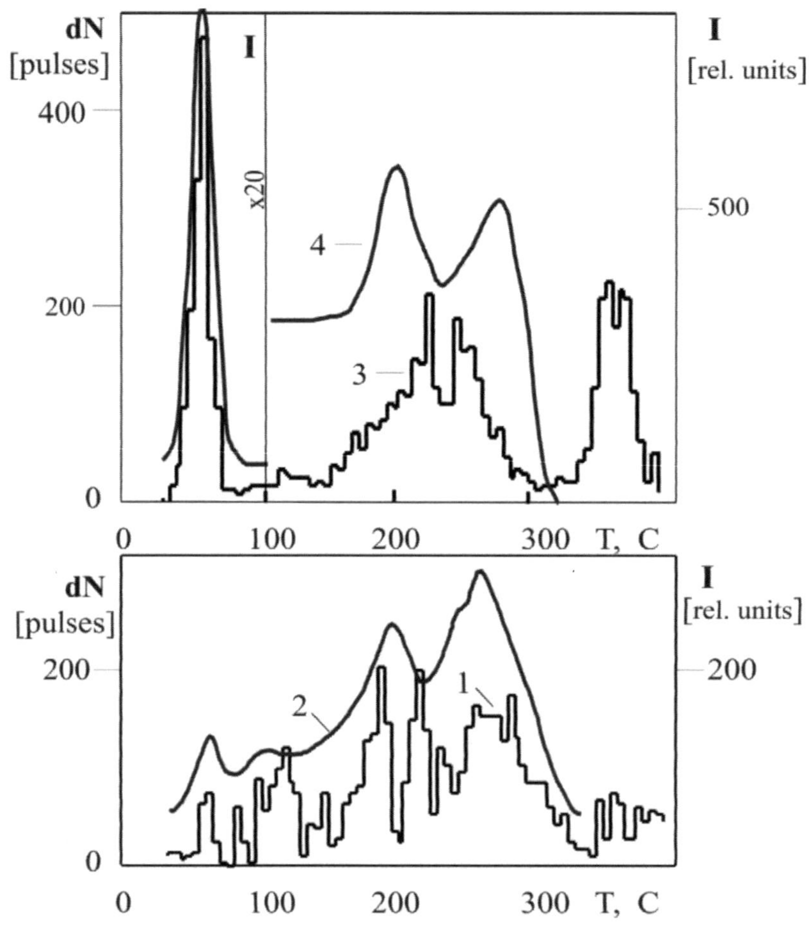

Figure 2. Radiofrequency electromagnetic emission (1, 3) and luminescence (2, 4) occurred in synthetic quartz crystals when being heated and the same dose (4.0×10^3 Gy) irradiated with gamma-quanta of the dose rate of 0,038 Gy/s (1, 2) and 3,6 Gy/s (3, 4).

When quartz crystals are irradiated with a high gamma dose rate of ~ 3,6 Gy/s, unlike the irradiation with gamma-quanta of low dose rate (~0,038 Gy/s), it causes more defects to occur due to high density of stimulation; such extra defects effectively stabilize fast hydrogen ions that results in creating a stable couple of centers, i.e. an electron $[SiO_4]^{5-}$ center and a hole center:

$[SiO_4]^{3-}$: $[SiO_4]^{4-} + e^-\} = [SiO_4]^{5-}$ and $\{[SiO_4]^{4-} + e^+\} = [SiO_4]^{3-}$.

The heat action destroys electron centers that causes the release of electron and, then, its recombination on oxygen ions according to the following scheme:

$$[SiO_4]^{5-} \rightarrow \{[SiO_4]^{4-} + e^-\} + [SiO_4]^{3-} \rightarrow [SiO_4]^{4-} + h\nu\ (365\ nm)\ \text{and}$$

$$[SiO_4]^{5-} \rightarrow \{[SiO_4]^{4-} + e^-\} + O^-\ (\text{nonbridging}) + h\nu\ (630\ nm),$$

This, in its turn, causes the intensive TSL and RF electromagnetic emission in the temperature range of 50…70 °C (Fig.2, Curves 3, 4).

It should be noted that the luminescence intensity in heat-treated samples increased in the high temperature range by 2-3 times with remaining the type of thermal luminescence curve the same, as mentioned in [7]. But a more significant difference of by 20-40 times is observed in the low-temperature luminescence peak at ~ 50…70 °C; moreover, when quartz crystals are thermally annealed, their luminescence intensity at this peak is higher for those crystals which were pre-irradiated than for samples subjected to heat treatment only.

4. Conclusion

The temperature dependences of luminescence and radiofrequency electromagnetic emission can characterize imperfection and electrical micro heterogeneity of quartz materials. Defects in quartz materials are developed during crystallization and various handling treatments and due to external irradiation or thermal actions. A crystal or a mineral, by assimilating external actions and accumulating genetic information, keeps, so to say, the memory about actions and events it has experienced [8].

Irradiation has a significantly effect on the imperfection status of quartz and the degree of such an effect depends on a genetic type of quartz and on pre-heat treatment of quartz materials. When modifying quartz materials with ionizing irradiation, one should take into consideration both the irradiation type and the dose rate. The heat treatment even at lower temperatures of ~ 150 °C results in changes in microdefects and the increase in luminescence sensibilization of quartz materials.

References

[1] Votyakov S, Krokhalev V, Purtov V and Krasnobaev A 1993 *Luminescence analysis of quartz structural imperfection* (Ekaterinburg: Ural Branch of the Russian Academy of Sciences)

[2] Gaft M, Reisfeld R and Panczer G 2005 Modern Luminescence Spectroscopy of Minerals and Materials. Springer, Berlin, New York

[3] Boroznovskaya N, Korneva A, Marfin A 2016 *Key Engineering Materials* **683** 168

[4] Ananyeva L and Korovkin M 2009 *Proc. Int. Conf. on 100-year anniversary of Professor Vorob'yov A.A. (Tomsk Polytechnic University)* **2** (Tomsk Russia) 457

[5] Sal'nikov V, Moninger G, Zavertkin S, Korovkin M and Dolgov I 1994 On some electrophysical properties of quartzites. *Fiziko-Tekhnicheskie Problemy Razrabotki Poleznykh Iskopaemykh* **3** 89

[6] Korovkin M and Galanov Yu 2015 *IOP Conf. Series: Materials Science and Engineering* **81** 012078

[7] Matrosov I and Pogorelov Yu 1978 *J. Appl. Spectroscopy* **29 (5)** 1327

[8] Yushkin N and Esterle O 1980 The nature of noises in the mineralogical information and information cleaning methods *Typomorphism and the genetic information content of minerals* (Syktyvkar, Russia) 15 - 35

IOP Conf. Series: Materials Science and Engineering **168** (2017) 012034 doi:10.1088/1757-899X/168/1/012034

Photo-induced currents in the sandwich metal-ferroelectric-metal structure

V Ivanov[1], Yu Karpets[2] and Yu Perkov[3]

[1] Professor, FESTU, Khabarovsk, Russia
[2] Professor, FESTU, Khabarovsk, Russia
[3] Graduate Student, FESTU, Khabarovsk, Russia

E-mail: ivanov@festu.khv.ru

Abstract. It was described the photoelectric effect in thin sandwich metal-ferroelectric-metal system. The effect was observed in doped lithium niobate crystals with two electrodes of different metals. The effect is observed only in doped lithium niobate crystals and has a maximum for concentrations of impurities of iron around 0.3 % weight. This paper proposed thermal model of the investigated phenomena resulting from field contact potential difference on the borders of section of metal-ferroelectric material. The results obtained can be used to develop radiation receivers, as well as in the interpretation of experimental results on studying the properties of sandwiched metal-ferroelectric-metal structure.

1. Introduction

It is well known the photovoltaic effect (PVE) in lithium niobate with iron and copper impurities. It has been established that the density of steady-state current J is proportional to the laser intensity $\Phi(t)$: $J_{\phi r} = \alpha K_G \Phi(t)$, where K_G is the Glass constant dependent on the nature of absorbing centers and wavelength of radiation and independent of the crystal geometry, electrode configuration and impurity concentration, α is the absorption factor [1]. In paper [2] the thermovoltaic effect consisting in initiation of the electric voltage between the opposite faces of the semiconductive sample of samaricsulphide (SmS) on heating to 400–500K is described. The effect has appeared in the presence of the donor impurity concentration gradient directed towards these faces. The electric voltage has initiated in the absence of temperature gradient along the direction of the electrodes' arrangement on the sample on heating and, therefore, cannot be explained by the classic Seebeck effect. There are a number of such nonequilibrium phenomena in the ferroelectrics which have no unambiguous theoretical description. In paper [3] the thermally-stimulated EMF arising only in the highly-alloyed with iron crystal of lithium niobate with the evaporated electrodes from a pair of different metals was described. A sign of the thermo-EMF is determined by location of electrodes applied by vacuum evaporation on the opposite faces of crystal and does not depend on orientation of the crystallographic axes of the sample in reference to electrodes.

In this paper, the results obtained in experimental studies of the light induced current in the thin-layer metal-lithium niobate-metal structure are presented. This effect is similar to photovoltaic effect, but it is characterized by thermal nature.

2. Thermovoltaic effect in the sandwich-like metal-lithium niobate-metal structure

In contrast with the another thermal classic pyroelectric effect (classic pyroelectric effect), the value of observed thermal response is proportional to the crystal temperature increase and does not depend on the rate of its change. Formula for the thermally-induced voltage across the load resistance R_n is, in general case, as follows:

$$U_n = \gamma S R_n \frac{dT}{dt} + U_n^0 + P_{el} S R_n \Delta T e^{-\frac{t}{\tau}}, \quad (1)$$

where S is the area of crystal surface covered with sputtered (evaporated) electrode; $P_{el} = (R_n S)^{-1} \Delta U_n / \Delta T$ is the coefficient characterizing the thermo-EMF value in the mode of voltage U_n measurement (A/K·cm^2); $\Delta U_н$ is the change of voltage across $R_н$ when the sample temperature changes by ΔT; ΔT is the difference between the initial and final temperatures of sample; U_n^0 is the initial voltage at T_0 (T_0 is the initial temperature of crystal); τ is the time of thermal response relaxation; γ is the pyroelectric coefficient; t is the time.

The first term in (1) corresponds to the classic pyroeffect while the second one describes the non-equilibrium thermovoltaic effect proper outstanding substantially of the general signal owing to large relaxation time ($\tau \geq 10^5$ s).

The study of the effect was performed using the slow modulation of the crystal temperature under the conditions of thermostat and the thermo-EMF was recorded, inter alios, under the stationary conditions (for observation time τ~10^5 *s*, relaxation was not observed).

The measurements showed that the thermo-voltaic effect has approximately equivalent values in the *Y*- and *Z*-cut crystals. Because the sufficiently high rates of heating (about 0.1 K/s) were used in the installation, the pyroelectric current for the polar *Z*-cut crystals reached the considerable value. Therefore, the principal measurements were performed for the *Y*-cut lithium niobate crystals where the thermal response could be observed in pure form.

The metal electrodes were applied by vacuum evaporation with thickness varying from 0.1 to 1 μm. The samples with different pairs of electrodes (aluminum (Al) – chromium (Cr), indium (In) – chromium (Cr), aluminum (Al) – copper (Cu), silver (Ag) – aluminum (Al)) were studied.

For studying the dependence of the thermo-EMF value on the concentration of impurity, the lithium niobate crystals with the same areas and thicknesses [S = 0.5 cm^2, d = 1 mm] were used. The iron concentration varied in the range of $1.3 \cdot 10^{-2}$ wt% to 0.6 wt%. The results are given in Fig. 1. For the impurity concentration of less than 0.25 wt%, the thermo-voltaic effect is not observed. Beginning with the impurity concentration of 0.25 wt%, the coefficient P_{el} rises sharply and reaches maximum ($P_{el} = 5 \cdot 10^{-12}$ A/cm^{2}K) at concentration of 0.3 ÷ 0.4 wt% of Fe.

The known contact phenomena in the sandwiched structures with ferroelectrics do not explain the observed quasi-stationary currents [4, 5]. The presence of initial voltage U_n^0 across the sample allows us to suggest the non-equilibrium nature of the thermally stimulated current.

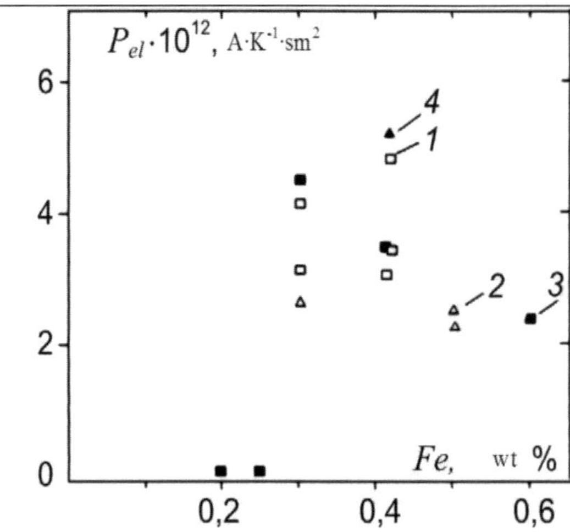

Figure 1. Dependence of coefficient P_{el} on concentration of iron impurity in the lithium niobate crystal for different cuts and contact materials: *1* – Al-Cr, *Z*-cut; *2* – In-Cr, *Z*-cut; *3* – Al-Cr, *Y*-cut; *4* – In-Cr, *Y*-cut).

When investigating the dependence of the coefficient P_{el} value on the sample geometry, it was identified that P_{el} value increases sharply with the crystal thickness decrease (Fig.2).

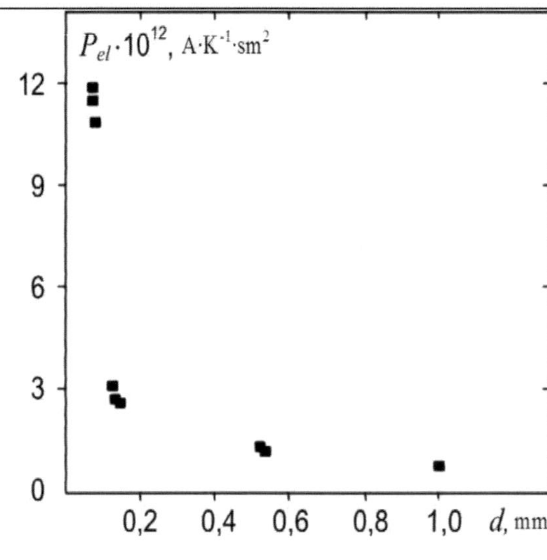

Figure 2. Dependence of coefficient P_{el} on the crystal thickness (S = 5 mm^2; 0.3 wt% of *Fe*; Al – Cr).

When increasing the heating range, it was found that P_{el} depends nonlinearly on the sample temperature (Fig. 3).

IOP Conf. Series: Materials Science and Engineering **168** (2017) 012034 doi:10.1088/1757-899X/168/1/012034

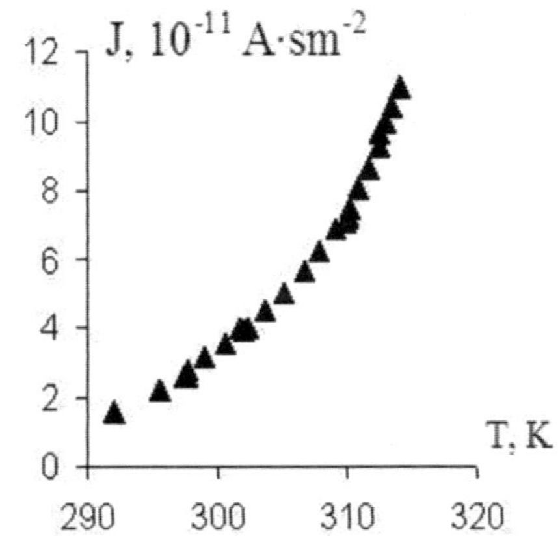

Figure 3. Experimental dependence of the thermally induced current (normalized to the electrode surface area) on crystal temperature ($LiNbO_3$:Fe – 0.43 wt%, 10x5x1mm^3, Y – cut; Al – Cr).

A sharp increase in the thermally induced current for small thicknesses of crystal (less than 100 μm) points to the crucial role of the near-contact zones with thicknesses of less than100 μm. In this paper, the thermo-chemical model of the studied phenomenon caused by the field of contact potential difference at the metal-ferroelectric interface was proposed [4]. In case of the same electrode materials, the contact potential differences are equal and oppositely directed. Therefore, the resulting EMF is equal to zero. The model assumes that the electrochemical oxidative reaction proceeds in the area of metal-ferroelectric contact which leads to the appearance of electrochemical potential gradient [5].

3. Conclusions

Therefore, it is evident from the research results that in the thin-layer metal-ferroelectric-metal system the thermo-voltaic response exists which increases essentially with decreasing the crystal thickness. This effect can be comparable to photovoltaic effect in light experiments[1].

The results obtained can be used to develop the radiation detectors as well as to interpret the experimental results of investigating the properties of the sandwiched pyroelectric structures [5, 6].

References

[1] Glass A, Von der Linde D, Nergran T 1974 *Appl. Phys.* **25** 233-236
[2] Kaminskii V 2013 *Technical Physics Letters* **39** 673-675
[3] Ivanov V, Karpets Yu, Kliment'ev S 2001 *Russian Physics Journal* **44** 119-121
[4] Itskovskii M, Shchedrina L, Kladkevich M 1980 Ferroelectrics **29** 167-174
[5] Ivanov V, Karpets Yu, Marchenkov N, Zdorovtsev G 2005 Proc. SPIE **5851** 415-419
[6] Zdorovtsev G, Ivanov V, Karpets Yu, Klimentyev S 2007 *Proceedings of Tomsk Polytechnic University* **311** 102-105

Modernization of electron accelerator with a large cross section beam for radiation effects on materials

M S Vorobyov[1,2], V V Denisov[2], N N Koval[2,3], S A Sulakshin[2], V V Shugurov[2] and V V Yakovlev[2]
[2]Institute of High Current Electronics SB RAS, Tomsk, Russia,
[3]National Research Tomsk Polytechnic University, Tomsk, Russia.

E-mail: vorobyovms@yandex.ru

Abstract. The results of the work on the creation of an automated wide-aperture electron accelerator with a grid plasma cathode based on the low-pressure arc discharge and outputting of a large section beam (750×150 mm) in the atmosphere through a outlet foil window. The distinctive feature of such electron accelerator is a weak correlation of beam parameters, as well as a high current beam extraction efficiency to air, reaching ≈ (80 ÷ 90)% of the current in the accelerating gap at an accelerating voltage of 200 kV, beam current amplitude in the atmosphere up to 30A, frequency and pulse duration up to 50 s^{-1} and 100 μs, respectively. The electron source provides a stable continuous operation for tens of hours in a repetitively pulsed modes at the maximum average beam power in the atmosphere is ≈5 kW. Examples of applications of such accelerator in the radiation-stimulated technology are given, showing the prospects of its using in scientific and technological purposes.

1. Introduction

Modern electron accelerators with beam in an atmosphere characterize by the principle of action, have a very wide range of the generated beam parameters, as well as its different configurations, etc. A wide variety of such accelerators due to the application prospect of such accelerators in the scientific and technological purposes [1–7]. Reliable and durable electron accelerators with stable parameters are required for the solution of existing problems still. Among the electron accelerators are accelerators based on a grid plasma cathode [1, 6–10]. This type of accelerators offers a series of advantages over those based on thermionic cathodes (longer service life, wider range of vacuum conditions, higher energy efficiency) and over accelerators based on explosive emission cathodes (longer service life, longer pulse duration, higher pulse repetition frequency, more uniform current density distribution). In view of the above advantages, we studied the possibilities of modification of an electron accelerator based on a grid plasma cathode for its automation and increasing its energy efficiency.

2. Experimental procedure

The electron accelerator scheme is shown in Fig.1. A plasma cathode *(2)* is located in a vacuum chamber *(1)*. The plasma cathode *(2)* of the accelerator is a hollow stainless-steel half-cylinder with two low-pressure-arc cathode units *(3,4)* fixed on the cylinder ends [11]. A stainless-steel mask *(6)* of

[1] To whom any correspondence should be addressed.

thickness 200 μm is put on the emission grid (*5*) of dimensions 750 mm×150 mm, which serves as the anode for the arc discharge. There are 344 round holes of diameter 8–12 mm in the mask, which are individual emission units of the plasma emitter, and the wide electron beam produced is a superposition of beamlets formed by individual emission units with the plasma boundary stabilized by the fine metal grid. The hollow anode (*7*) is electrically connected via a resistor to the emission grid (*5*), which is necessary to facilitate the discharge ignition, switching, and operation in the region of the emission grid. Electrons are extracted from the plasma surface of the emission units under the action of a dc accelerating voltage of up to 200 kV applied to the 120-mm gap between the plasma cathode (*2*) and the extraction foil window (*8,9*). The aligned holes in support grid (8), having a total geometric transparency of 56%, are of the same number as those in mask (*6*), but of larger diameter (15 mm). Alignment of the holes in mask *(6)* and in support grid (*8*), and thereby attainment of a coaxial plane-parallel geometry of the acceleration gap, minimizes the beam losses in the support grid of the extraction foil window. Support grid *(8)* is covered with 30-μm-thick foil (*9*) made of AlMg–2n alloy [10].

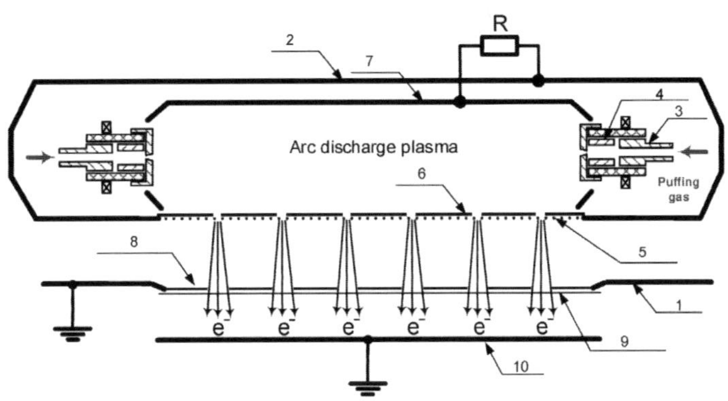

Figure 1. Scheme electron source with wide-aperture plasma emitter: 1 – vacuum chamber; 2 – plasma cathode; 3 – igniter; 4 – cathode; 5 – metallic mesh; 6 – metallic mask; 7 – hollow anode; 8 – support grid outlet foil windows; 9 – output foil; 10 – substrate.

As has been shown by theoretical and experimental studies [10, 12], the use of a multi-aperture two-electrode system in the source enhances the operation stability of the plasma cathode and the electric strength of the acceleration gap. Moreover, the system allows one to extend the range of electron beam parameters – namely, to enhance the beam extraction efficiency β, to increase the beam current density, and to narrow the beam energy spectrum within a pulse (through decreasing the capacitor bank voltage per pulse).

The electron current from the plasma is controlled by varying the plasma density for which the discharge current is varied, as clearly seen from the dependence in Fig. 2 *a*. From Fig. 2 *b* it follows that the accelerating voltage U_0 little affects the current in the acceleration gap I_0, and this, as noted, is one of the advantages of electron sources based a grid plasma cathode.

Reasoning from the previous study [10], we modified the wide-aperture electron accelerator with plasma cathode to increase its efficiency and its automation. Power and control block diagram of the modified electron accelerator is shown in Fig. 3. The modification has made possible an automated wide-aperture plasma cathode electron source capable of providing Main beam parameters, that the modified accelerator can provide, are presented in Table 1. It is important to note, that at a maximum average beam power in the atmosphere of ≈5 kW the maximum average power of the plasma cathode power supply is no greater than 100 W. This greatly simplifies the design of the accelerator.

IOP Conf. Series: Materials Science and Engineering **168** (2017) 012035 doi:10.1088/1757-899X/168/1/012035

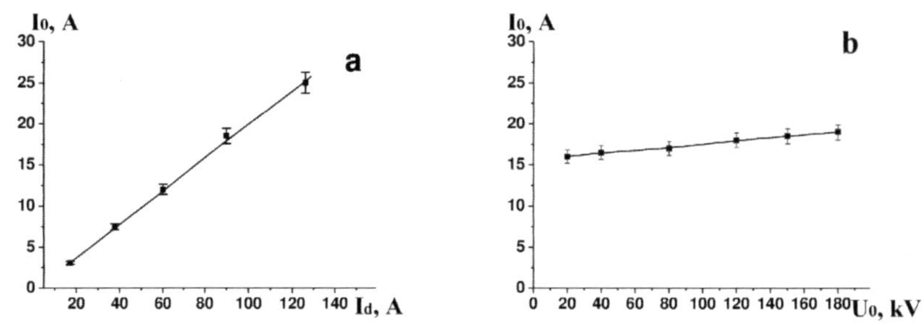

Figure 2. Current in the acceleration gap I_0 vs discharge current I_d at U_0=150 kV (a) and accelerating voltage U_0 at I_d=90 A (b)

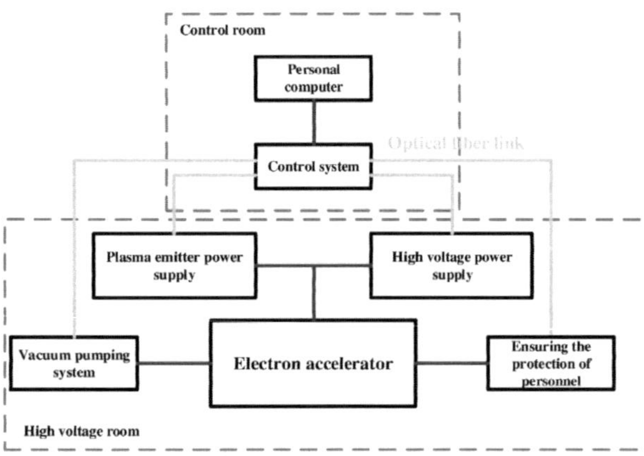

Table 1

Electron energy, keV	(100÷200)
Beam current, A	(5÷30)
Pulse duration, μs	(10÷100)
Pulse frequency, s^{-1}	(1÷50)
Beam dimensions, mm	750×150
Current density heterogeneity	±10
Average beam power	5

Figure 3. Power and control block diagram of the modified electron accelerator.

The main stages of the accelerator modernization are to automate the vacuum pumping system, to create of automated high-voltage and plasma cathode power supply systems, providing including additional safety system for staff operating with the electron source. Control of the vacuum and the high-voltage systems, as well as of the plasma cathode power supply system is carried out from the control room from a PC, using fiber optic communication channels.

The automated vacuum pumping system is based on a WAGO–750–312 industrial controller. This system allows to remotely manage and control the pressure in the vacuum chamber, and perform of emergency shutdown of the electron source in the event of a foil breakthrough.

Since a new high-voltage power supply was created by a new one built around modern high-frequency elements, this made it possible to decrease the overall dimensions and mass of the high-voltage source, to ensure a more stable retention of accelerating voltage, which is extremely important for operation at high average power, using which a smooth transition to the setting of the accelerating voltage was organized at 30 kV/s (the speed is chosen empirically). That is necessary to restore of the stable work of the electron accelerator at first accelerator starting or after electrical breakdown of accelerating gap.

Using this electron accelerator with a grid plasma cathode has been shown experimentally a principle possibility of silicon tetrafluoride molecules decomposition under the influence of low-energy (200 keV) electron beam to form a silicon film and fluorine free [5]. In addition, a possibility of efficient using of a wide-pulsed electron beam during the pre-radiation cross-linking of natural rubber latex without adding substances, accelerating the curing process, is demonstrated [7]. Also the paper [14] demonstrates the fundamental possibility of forming carbon structures in polyvinyl chloride films without additives and in polyvinyl chloride films added with 5 wt% of ferrocene by this electron accelerator, that produces the output response amorphous carbon material.

3. Conclusion

The automated plasma cathode electron source with its modern power supply system provides remote control of the beam parameters and their independent adjustment, stability of operation and personnel safety. The parameter range, reliability, and service life of the source meet the requirements that allow its successful use in electron beam technologies and small-scale industrial production.

Acknowledgement
This work was supported by the Russian Science Foundation (project № №14-29-00091).

References

[1] Bugaev S P, Kreindel Yu E and Schanin P M 1984 *Large section electron beams* (Moscow: EAI)
[2] Chmielewski A G, Zimek Z, Iller E, Tyminski B and Licki J 2000 *J. Tech. Phys. Spec. Iss.* **41(1)** 551
[3] Ershov B G 2013 *Herald of the Russian Academy of Sciences* **83(5)** 437
[4] Sokovnin S Yu 2007 *Nanosecond electron accelerators and radiation technologies based on them* (Yekaterinburg: UD RAS)
[5] Koval N N, Shugurov V V, Suslov A I, Denisov V V, Yakovlev V V, Vorobyov M S, Khandorin G P, Ledovskikh A K, Galata A A, Murlyshev A P and Volchkov V S 2010 *Proc. 16th Int. Symp. on High Current Electronics* 545.
[6] Bugaev A S, Koval N N, Lomaev M I, Mel'chenko S V, Ryzhov V V, Tarasenko V F, Turchanovsky I Yu, Fedenev A V and Shanin P M 1994 *Laser Part. Beams* **12(4)** 633
[7] Vorob'ev M S, Denisov V V, Koval' N N, Shugurov V V, Yakovlev V V, Uemura K and Raharjo P 2015 *High Energy Chem.* **49(3)** 143
[8] Bayless J R 1975 *Rev. Sci. Instrum.* **46(9)** 1158
[9] Gielkens S W A, Peters P J M, Witteman W J, Borovikov P V, Stepanov A V, Tskhai V N, Zavjalov M A, Gushenets V I and Koval N N 1996 *Rev. Sci. Instrum.* **67(7)** 2449
[10] Vorobyov M S, Koval N N and Sulakshin S A 2015 *Instrum. Exp. Tech.* **58(5)** 687
[11] Vorobyov M S, Devyatkov V N, Koval N N and Shugurov V V 2015 *J. Phys.: Conf. Ser.* **652** 012066 (1-6).
[12] Nguen Bao Hyng, Koval T V, Vorobyov M S and Koval N *N Izv.* 2015 *Vys. Ucheb. Zaved. Fizika.* **58(9/2)** 213
[13] Kryazhev Yu G, Koval N N, Vorobyov M S, Solodovnichenko V S, Trenihin M V, Sulakshin S A and Liholobov V A 2016 *Proc. 10 Intern. Conf.on Carbon in Moscow* 233

IOP Conf. Series: Materials Science and Engineering **168** (2017) 012036 doi:10.1088/1757-899X/168/1/012036

Crystallization kinetics of rapidly quenched $Cu_{50}Zr_{50}$ and $Cu_{46}Zr_{46}Al_8$ glass-forming alloys

T V Kulikova[1], **A A Ryltseva**[1], **V A Bykov**[1,2], **S Kh Estemirova**[1,2] **and K Yu Shuhyaev**[1,2]

[1]Institute of Metallurgy, Ural Branch of Russian Academy of Sciences, Ekaterinburg, Russia
[2]Ural Federal University named after First President of Russia B.N. Yeltsin, Ekaterinburg, Russia

E-mail: kuliko@gmail.com

Abstract. We studied the crystallization processes, the structure and thermal properties of amorphous alloys $Cu_{50}Zr_{50}$ and $Cu_{46}Zr_{46}Al_8$ in a wide temperature range. Comparative study of the crystallization kinetics of these amorphous alloys was carried out for the first time using multivariate non-linear regression. It was found that mechanisms of the crystallization of studied metallic glasses are substantially different. The binary alloy is crystallized by branched reaction complex in four steps. For the ternary system was proposed two-step kinetic model of the crystallization process with consecutive reactions. The values of the total energy of activation for each crystallization stage reach to $Cu_{50}Zr_{50}$: E_1 (345.2 kJ/mol); E_2 (307.9 kJ/mol), E_3 (281.1 kJ/mol), E_4 (259.51 kJ/mol) and $Cu_{46}Zr_{46}Al_8$: E_1 (350.7 kJ/mol); E_2 (150.4 kJ/mol).

1. Introduction

The Cu-Zr system has unique glass-forming ability (GFA) and so it is the basis for the development of separate class of bulk metallic glasses (BMG) [1-3]. The binary $Cu_{50}Zr_{50}$ alloy, besides the BMG formation, demonstrates shape memory effects due to reversible martensitic transformation. It is well known that small additions of Al essentially improve both the GFA and mechanical properties of binary Cu-Zr alloys [4-7]. In this connection, it is interesting to study the impact of aluminum on the mentioned properties of $Cu_{50}Zr_{50}$ alloy. It was previously found that $Cu_{46}Zr_{46}Al_8$ alloy has the highest GFA among the $(Cu_{0.5}Zr_{0.5})_{100-x}Al_x$ ones [6,8]. So, in this paper we perform comparative study of structure, GFA and crystallization kinetics of binary $Cu_{50}Zr_{50}$ alloy and ternary $Cu_{46}Zr_{46}Al_8$ one.

2. Experimental

The Zirconium, copper and aluminum with purity of 99.98 mass.% were used to pre-pare the Cu-Zr-(Al) alloys with nominal compositions of $Cu_{50}Zr_{50}$ and $Cu_{46}Zr_{46}Al_8$. The initial polycrystalline alloys have been obtained using standard arc-melting technique under helium atmosphere. The initial samples were re-melted at least five times to ensure their homogeneity. The two series of samples were prepared for measurements. The first ones (referred to as as-cast) have been obtained by quenching of the melts on furnace mold with a cooling rate of about 100 K/sec. The samples of the second series (referred to as suction-cast) were obtained by suction casting of the melts into cylindrical copper mould of 2 mm in diameter. The estimated cooling rate in this case was of about 1000 K/s.

The compositions of the alloys as well as the concentrations of impurities were examined by atomic-emission method using the SpectroFlame Modula S analyzer.

X-ray diffraction (XRD) phase analysis and XRD structural analysis were performed using Shimadzu XRD-7000 diffractometer. Phase identification was carried out using PDF-2 database of the ICDD International Centre (The International Centre for Diffraction Data).

Thermal properties of the samples were studied by differential scanning calorimetry (DSC) during heating at different rates using a Netzsch STA 449C calorimeter. Kinetic analysis was performed with Thermokinetics Netzsch-Gerätebau GmbH software.

3. Results and discussion

Firstly, we compare the structure and crystallization process of as-cast and suction-cast samples for both the $Cu_{50}Zr_{50}$ and $Cu_{46}Zr_{46}Al_8$ alloys. In Fig.1 the XRD patterns for all the samples mentioned are shown. It can be seen that as-cast $Cu_{50}Zr_{50}$ alloy is the mixture of two monoclinic martensitic phases $P2_1/m$ and Cm (see Fig.1a). The XRD peaks are rather broad due to the high density of the defects caused by the existence of martensitic phases. The suction-cast $Cu_{50}Zr_{50}$ sample is mostly amorphous but small amounts of both B2 and B19' modifications of CuZr intermetallic compounds are present (Fig.1b). So, the sample may be treated as composite of the amorphous matrix with CuZr phase embedded. The XRD pattern for as-cast $Cu_{46}Zr_{46}Al_8$ alloy is a broad amorphous-like hump with the single pronounced crystalline peak corresponding to B2 CuZr phase (Fig1c). The suction-cast sample of this alloy is X-ray amorphous (Fig1d).

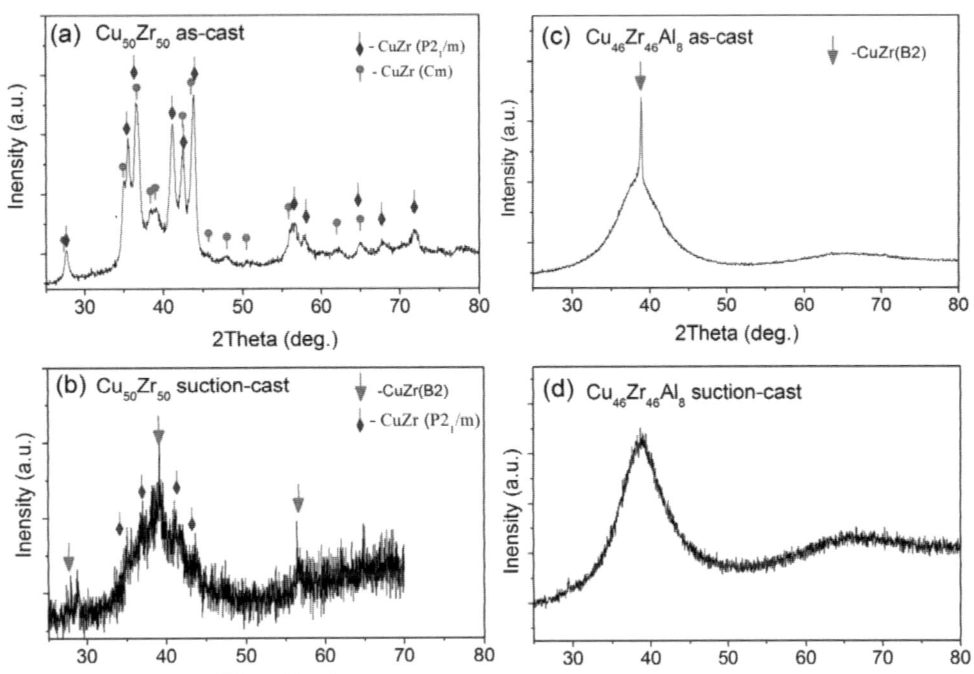

Figure 1. XRD patterns for binary $Cu_{50}Zr_{50}$ alloy (a,b) and triple $Cu_{46}Zr_{46}Al_8$ one (c,d) obtained for the as-cast (a,c) and suction-cast (b,d) samples.

The results of XRD analysis are confirmed by calorimetric measurements (Fig. 2). Indeed, the DSC curve of as-cast $Cu_{50}Zr_{50}$ sample shows the absent of amorphous phase. Instead we see austenitic (reverse martensitic) transformation (A_s) at T= 523 K and then the decomposition (T = 870 K) and formation (T = 1000 K) of CuZr B2 phase, respectively. The temperature of B2 phase formation is in accordance with phase diagram of Cu-Zr system [9]. The DSC curve of suction-cast $Cu_{50}Zr_{50}$ sample demonstrates a kink at the glass transition temperature T_g = 650 K and two peaks at T_1 = 710 K and T_2

= 750 K corresponding to crystallization process. The peculiarities of the latter will be described below.

DSC curves obtained for as-cast and suction-cast $Cu_{46}Zr_{46}Al_8$ samples are very similar (Fig.2b). A small difference between the transition temperatures is only observed. So we conclude that obtained XRD and DSC data are in close agreement with each other.

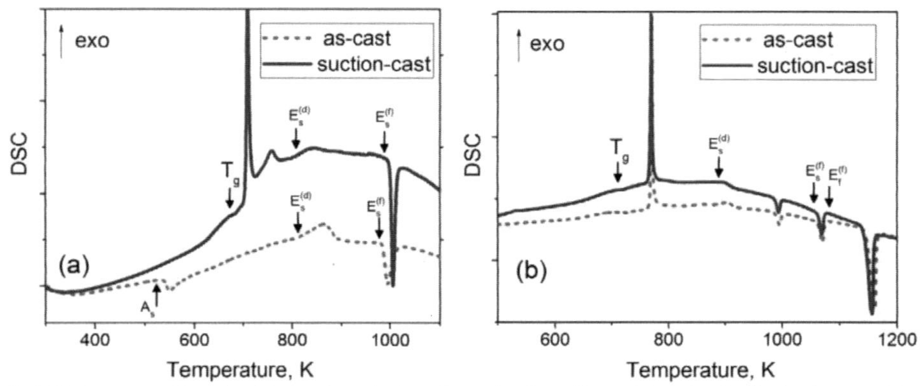

Figure 2. DSC curves for $Cu_{50}Zr_{50}$ (a) and $Cu_{46}Zr_{46}Al_8$ (b) alloys. The (red) dashed and (blue) solid lines correspond to as-cast and suction-cast samples, respectively. A_s - austenitic (reverse martensitic) transformation; $E_s^{(d)}$ – onset of eutectoid decomposition; $E_s^{(f)}$ – onset of eutectoid formation; $E_f^{(f)}$ - end of eutectoid formation; T_g - glass transition temperature.

In order to study the crystallization kinetics in the systems under investigation, we have performed DSC measurements of suction cast samples heating them from 300 K up to 823 K at different heating rates. The DSC data have been described by non-linear regression method with using Thermokinetics Netzsch-Gerätebau GmbH software. Verification of kinetic analysis data has been performed by both XRD and thermodynamic data obtained in [11,12].

The $Cu_{50}Zr_{50}$ sample has been heated at the rates of 1, 2, 4 and 8 K/min (see Fig. 3). During the kinetic analysis of a process, the optimal kinetic model describing the sequence of chemical reaction in the system under investigation is constructed [13]. For the $Cu_{50}Zr_{50}$ suction-cast alloy, the best model obtained is:

Here both the reaction $A \rightarrow B$ (1 and 2) correspond to the formation of CuZr B2 phase and obeys the extended Prout-Tompkins autocatalytic reaction equation:

$$\frac{d\alpha}{dt} = -k\alpha^n \beta^m . \quad (1)$$

Hereinafter $k = k(T) = A\exp(-E_a / RT)$ is the reaction rate constant which obeys the Arrhenius law; α is the concentration of an initial component; β the concentration of a reaction product; t is the time and T is the temperature. The catalyst of the reaction is the CuZr B2 phase which is present in amorphous matrix. The reaction $B \rightarrow C$ (3) corresponds to the transformation CuZr $\rightarrow$ $CuZr_2$ + $Cu_{10}Zr_7$ taking place in the form of dissusion-limited reaction obeying the Fik's law. In this case, the time dependence of reaction conversion is described by interpolation formula obtained from a solution of boundary-value problem for diffusion equation [14]. The transformation $C \rightarrow D$ (4) probably corresponds to re-crystallization of equilibrium phases $CuZr_2$ and $Cu_{10}Zr_7$ and proceeds as n-order autocatalytic reaction:

$$\frac{d\alpha}{dt} = -k\alpha^n(1 + k_{cat}\beta), \tag{2}$$

where k_{cat} is the coefficient of autocatalysis.

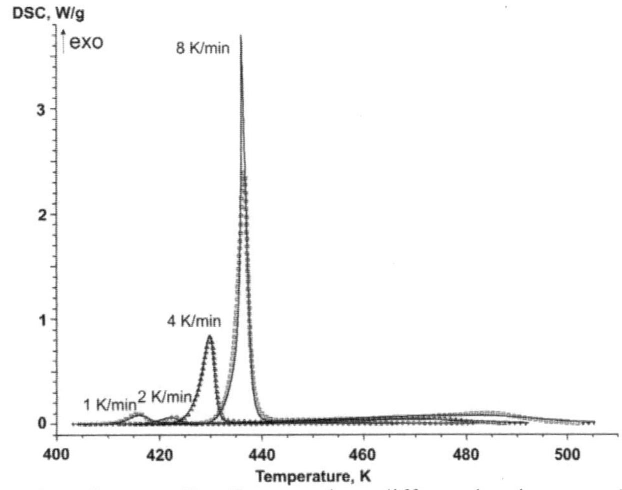

Figure 3. Thermograms of suction-cast $Cu_{50}Zr_{50}$ sample at different heating rates. Dotted and solid lines represent experimental data and the results of kinetic analysis, respectively.

The suction-cast $Cu_{46}Zr_{46}Al_8$ sample has been heated at the rates of 2, 4 and 8 K/min (see Fig. 4). Experimental DSC curves have been optimally described by the reaction sequence:

$$A \xrightarrow{1} B \xrightarrow{2} C \ .$$

The first stage corresponds to nucleation and growth which is described by classical Avrami model:

$$\frac{d\alpha}{dt} = -k\alpha(-\ln\alpha)^{\frac{n-1}{n}} \tag{3}$$

where, n is the parameter depending on both the nucleus geometry and their growth mechanism [15]. The second stage corresponds to n-order reaction:

$$\frac{d\alpha}{dt} = -k\alpha^n. \tag{4}$$

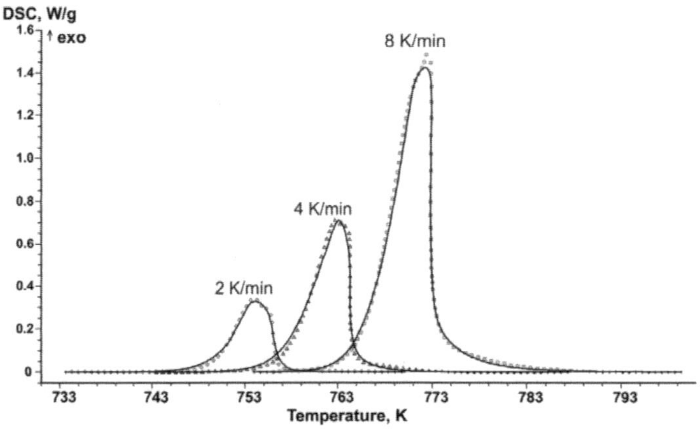

Figure 4. Thermograms of suction-cast $Cu_{46}Zr_{46}Al_8$ sample at different heating rates. Dotted and solid lines represent experimental data and the results of kinetic analysis, respectively.

The interpretation of the proposed reactions for $Cu_{46}Zr_{46}Al_8$ alloy is complicated due to the absence of both an accepted phase diagram of Cu-Zr-Al system and reliable high-temperature XRD data. In accordance with assessed phase diagram of Cu-Zr-Al system [12], the equilibrium state of the $Cu_{46}Zr_{46}Al_8$ alloy, at the temperatures in the vicinity of the crystallization DSC peak, is the mixture of three phases: $Cu_{10}Zr_7$, $CuZr_2$ и $AlCu_2Zr$. So we guess that the first reaction $A \rightarrow B$ corresponds to the formation of equilibrium phases and some metastable one; the second stage $B \rightarrow C$ is the formation of equilibrium phases $Cu_{10}Zr_7$, $CuZr_2$ и $AlCu_2Zr$. An additional study is needed to confirm this hypothesis.
The activation energies E_a and pre-exponent factors A, obtained from kinetic analysis are presented in Table 1.

Table 1. Kinetic parameters of crystallization of $Cu_{50}Zr_{50}$ and $Cu_{46}Zr_{46}Al_8$ amorphous alloys.

Sample	**E_A, kJ/mol**				**log A, s^{-1}**			
	1	2	3	4	1	2	3	4
$Cu_{50}Zr_{50}$	345.2	307.9	281.1	259.5	24.6	22.7	18.8	15.8
$Cu_{46}Zr_{46}Al_8$	350.7	150.4	-	-	19.4	9.7	-	-

4. Conclusions
The comparison of crystallization kinetics of $Cu_{50}Zr_{50}$ and $Cu_{46}Zr_{46}Al_8$ amorphous alloys has been carried out for the first time with using non-linear kinetic modeling. It has been found that mechanisms of crystallization of the systems under consideration are essentially different. The binary alloy crystallizes through the complex branched reaction sequence included autocatalytic reaction and diffusion-limited processes. Whereas, the crystallization of the triple alloy is the sequence of two consecutive reactions described by standard Avrami and n-order reaction equations.

Acknowledgments
The work was supported by Russian Science Foundation (Grant № 14-13-00676).

References
1. Inoue A 2000 *Acta Mater.* **48** 279
2. Inoue A and Zhang W 2004 *Mater.Trans.* **45** 584
3. Son C Y, Kim C K, Shin S Y, Lee S 2009 *Sci. Eng.* **508** 15

4. Lu F, Kong L T, Jiang Z *et al* 2014 *J. Mater. Sci.* **49** 496
5. Wang D, Tan H, Li Y 2005 *Acta Mater*. **53** 2969
6. Zhang Q, Zhang W, Xie G, Inoue A 2007 *Mater. Trans*. **48** (7) 1626
7. Das J, Pauly S, Duhamel C *et al* 2007 *J. Mater. Res.* **22** 326
8. Zhang Q, Zhang W, Inoue A 2007 *Mater. Sci. Forum*. **561-565** 1333
9. Kulikova T V, Bykov V A, Belozerova A A, Murzakaev A M, Ryltsev R E 2013 *J. Non-Cryst. Solids*. **378** 135
10. Zhang S, Ichitsubo T, Yokoyama Y *et al*, 2009 **50** (6) 1340
11. Kalay I, Kramer M J, Napolitano R E 2011 *Metallurg. Mater. Trans.* **42A** 1144-1153.
12. Raghavan V 2011 *J. Phase Equil. Diff.* **32** 452
13. Kulikova T V, Majorova A V, Shunyaev K Yu, Ryltsev R E 2015 *Physica B*. **466–467** 90
14. Serin B and Ellickson R T 1941 *J. Chem. Pys.* **9** 742
15. Liu C, Yu J, Xun S, Zhang J, He J 2003 *Polym. Degrad. Stabil.* **81** 197

Automated test bench for simulation of radiation electrification of spacecraft structural dielectrics

A M Vladimirov, A Yu Bezhayev, V M Zykov, V I Isaychenko, A A Lukashchuk and S E Lukonin

National Research Tomsk Polytechnic University, 30 Lenin av., Tomsk, 634050, Russia

E-mail: vadisay@gmail.com

Abstract. The paper describes the test bench "Prognoz-2" designed in Testing Center, Institute of Non-Destructive Testing, Tomsk Polytechnic University, which can be used: for ground testing of individual samples of spacecraft structural materials (e.g. thermal control coatings or cover glasses for solar batteries) or ceramics of the plasma thruster discharge channel), and whole spacecraft units or instruments (e.g. instruments of solar and stellar orientation or correcting plasma thrusters) exposed to radiation electrification factors; to verify the calculation mathematical models of radiation electrification of structural dielectrics under the impact of space factors in different orbits.

1. Introduction

At present, more and more information, communication and navigation systems make use of Earth-orbiting spacecraft. The development, manufacturing and orbiting of the spacecraft is extremely expensive and developers seek to maximize its active shelf life. However, it is difficult to implement, since the spacecraft is exposed to space factors (SF) such as high vacuum (up to $3 \cdot 10^{-10}$ Torr), temperature difference (from +120° C under sunlight to –130° C when the spacecraft enters the Earth's shadow) solar radiation (energy density of up to 1.38 $kW \cdot m^{-2}$) [1], and ionizing radiation of the Earth's radiation belts. In addition, when a spacecraft approaches the geomagnetic plasma, it provokes radiation charging of structural dielectric materials. Surface and volume electrical fields emerging in structural dielectrics can lead to radiation-induced electrostatic discharge (ESD) that causes electromagnetic interference and pulsed current noise in electrical circuits of the onboard equipment with an amplitude of up to 100 A and the rising edge duration of up to 10 ns. Insufficient protection measures against these electromagnetic interferences and current noise can lead to fatal failure of the onboard equipment. To evaluate the effectiveness of the means protecting the onboard spacecraft equipment from the effects of radiation electrification, ground tests are conducted in conditions simulating significant SF using special test benches. This test bench should provide:

- the study of radiation parameters of electrification of spacecraft structural dielectrics (surface potential distribution, frequency and electrical discharge characteristics) under conditions of space effects (vacuum, electron radiation simulating the energy spectrum in the near-earth orbit, low temperature and solar radiation);
- measurement of interference signals in the circuits providing control of the test spacecraft subsystems (such as the spacecraft correction system) caused by electrostatic discharges;

- development of methods and means of protection from the effects of radiation electrification factors.

2. Experimental techniques

Arrangement of the "Prognoz-2" test bench is shown in figure 1.

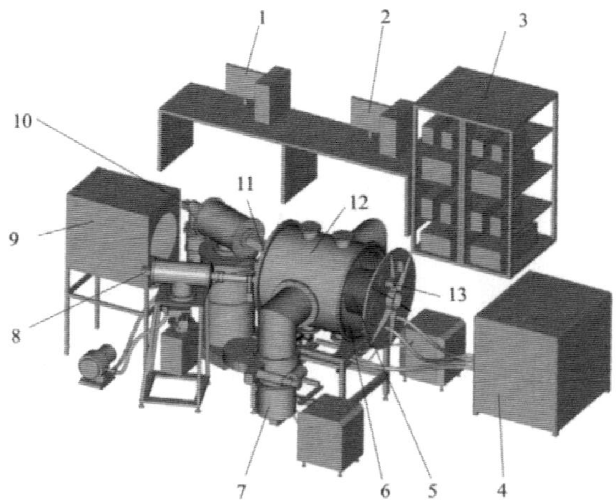

Figure 1. Arrangement of the stand "Prognoz -2".
1, 2 – Controlling and recording PC.
3 – Racks with measuring and control equipment.
4 – Refrigeration plant
5 – Cryo-plate with cryo-shield.
6 – Cryo-jacket.
7 – Cryogenic pump.
8 – Electron gun with energy of up to100 keV.
9 – Optical solar simulator in the range from 200 nm to 1.2 μm.
10 – Electron gun with energy of up to 300 keV.
11 – High-speed camera.
12 – Vacuum chamber.
13 – Vacuum gauge (3 pcs.).

The vacuum chamber of the test bench is a cylinder with a diameter of 1000 mm and a length of 1000 mm. The schematic of the vacuum system is shown in figure 2. Simulation of interaction of electronic radiation with structural dielectric materials of the spacecraft imposes restrictions on the type of the vacuum equipment, and vacuum pumps which do not use oil or air as working medium should be used. To ensure high suction rate during material testing, two high-vacuum cryogenic pumps CRYO-TORR 400 working with helium compressors CRYO-TORR 9600 are used. High suction rate enables: testing of materials having a porous structure; testing of materials with a high rate of gassing upon heating; testing of ion engines with direct inlet of the working material; testing of samples of large geometric dimensions.

A 100 KeV electron gun induces a monoenergetic electron beam with the given energy. A 300 keV electron gun induces a beam with the electron spectrum from 10 keV to 300 keV using a foil system [2]. The results of sample charging by monoenergetic electron beam and by beam with different electron spectrum differ from those obtained in exposure to monoenergetic beam only [3].

The solar simulator and temperature control system allow testing of solar batteries under conditions simulating space effects both in orbital eclipse period and under solar irradiation. The temperature control system has two independent loops for heating and cooling of the cryo-jacket and cryo-plate. The loop for cooling the sample is connected to the refrigeration unit with thermal capacity of 20 kW, which allows cooling of a sufficiently large sample to –80° C. For cooling the sample down to lower temperatures, liquid nitrogen supply in the cooling circuit is provided. An industrial 2 kW fluid heater is used to heat the sample up to +80° C in the heating loop.

The electrical discharges caused by charged dielectric surface are recorded by high-speed video camera.

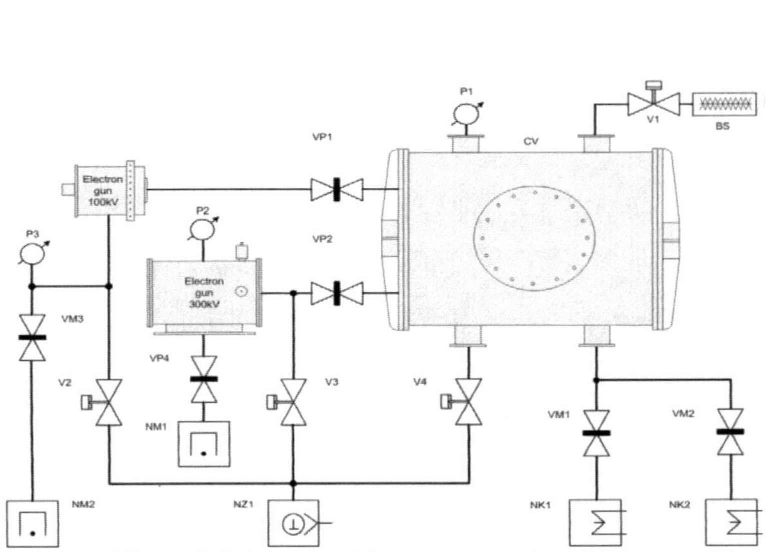

Figure 2. Schematic of the vacuum test bench "Prognoz-2".

VP1, VP2 – gate valves HTC 6VB-SS-KF50-P.;
VM1, VM2 – gate valves VAKMA23V7 – 400;
VP4 – gate valves HTC GVB-SS -ISO-160-P;
V1, V2, V3, V4 – vacuum valves HEC GVB-SS-KF40-P;
NK1, NK2 – high-vacuum cryogenic pumps CRYO-TORR 400;
NZ1 - forevacuum pump Kashiyama NeoDry 30E;
NM1, NM2 – ion water-cooled diode type vacuum pumps NMDO - 0.25 (NORD - 250);
BS – silica gel dehumidifier;
P1, P2, P3 – Granville-Phillips vacuum gauges 354 Micro-ION.

For ease of the control and protection against accidental errors of the operational personnel, the vacuum pumping system was automated. Implementation of the automatic control was performed using a PC, the National Instruments PCIe-6320 module and software created in LabView.

3. Results and discussion

The basic equations describing the dielectric electrification processes under external electron irradiation from the exposed surface follow from Maxwell's equations and include the laws of conservation of energy and electric charge, Poisson's equation and Ampere's law. Ampere's law implies that the total current density of the electric charge transfer (including the initial current of the accelerated electron beam) and the bias current intensity at any fixed time do not depend on the coordinate; it is a function of time. This total intensity of the total current (sum of the conduction and bias current) passes through the metal substrate of the test dielectric, and along with the surface potential, it is one of the main experimentally determined characteristics of dielectric electrification. The electrification parameters are calculated based on the data of the current density distribution and absorbed dose rate of the primary radiation electrons in the test dielectric determined based of the Monte Carlo method. The calculation must take into account the impact of the charged dielectric field on the energy of electrons which reach the dielectric surface.

Figure 3 presents the data acquisition systems required for testing of radiation electrification in dielectric materials. Depending on the aim of the experiment, a number of conventional schemes are used to measure the radiation electrification parameters in dielectrics. To determine the basic physical quantities required for establishment of initial conditions for a mathematical model, which considers the surface and bulk processes of radiation electrification in dielectric samples exposed to low-energy electron beam, the direct measurement is performed. The measured parameters are the current from metal substrate (6) placed on dielectric base (7) on which sample (4) is mounted, and the current from electrode (5) placed on the sample surface. Faraday cup (5) is used to measure electron beam Io (2) that irradiates sample (4). Faraday cup (5) is connected to KEITHLEY 6485 picoammeter (A1). The current from metal substrate (6) and the current from sample surface (3) are measured by KEITHLEY 6485 picoammeter A2, A3 [4]. The study of radiation-induced electrostatic discharges (ESD) implies measurement of the surface charge density by the electrostatic field sensor, which operates on the principle of dynamic capacitor [5] and the Tektronix DMM4040 voltmeter. To prevent failure of the picoammeter input stages during measurement of electrostatic discharges, the amplitude and time values of the discharge pulses are recorded by current shunts PA1, PA2 (Tektronix TCP0030) and high-voltage probes PU1, PU2 (Aktakom ACA-6039) connected to shunts R1 and R2. The current and high-voltage probes are connected to the Tektronix DPO4102B oscilloscope. The oscilloscope is placed in Faraday's cage (9); it has an independent power supply that is not connected to the common electric network to prevent measurement noise. [6] The sample is mounted on the thermoplate connected to thermal cover that allows changing the sample temperature from –80° C to +80° C. All the measuring devices are connected to personal computer (10) via the GPIB interface to measure physical quantities, and process and store the obtained experimental data in automatic mode. The program for measurement synchronization and data processing has been written in LabVIEW.

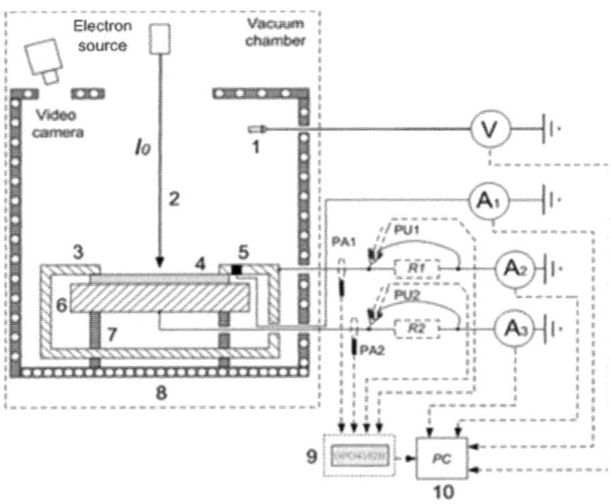

Figure 3. Scheme to measure basic parameters of dielectric charging.

The test bench "Prognoz-2" was used to test the electrification effect on solar battery sample (1) with GaAs-based photoconverters and borosilicate cover glass mounted on thermostat (2 and 3) (Figure 4).

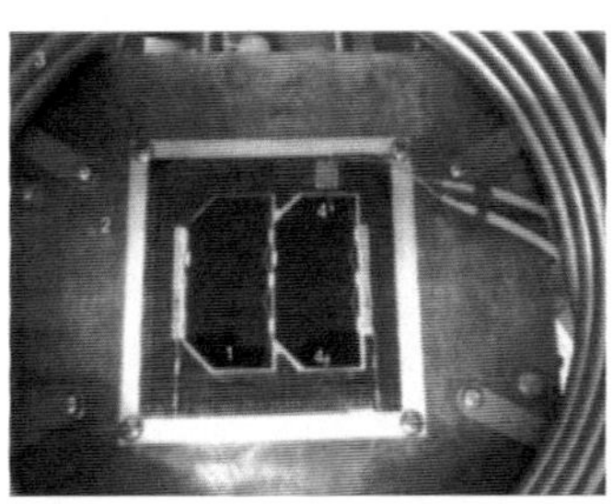

Figure 4. Solar cell sample with GaAs-based photoconverters.

The parameters of sample exposure to simulated magnetospheric plasma for standard tests were: electron irradiation Ee = 70 keV, Je = $1 \cdot 10^{-9}$ A cm^{-2}, beam stability ± 10%, beam non-uniformity across the area ≤10%. Irradiation was carried out on the front side of the solar battery. The angle between the electron beam axis and the sample plane was 90°. The temperature of the solar battery front side was of three values: –70°C, +20°C and +40°C. The temperature was recorded by thermocouples (4) mounted on the sample surface. The sample was in a vacuum P≤10^{-6} Torr.

The external field sensor [5] was used to measure the following parameters: the kinetics of the integral charge on the irradiated surface of the borosilicate glass sample; the average frequency of radiation-induced discharges.

Simultaneously, an oscilloscope was used to measure the following parameters with Rm=50Ω: the average and maximum amplitude of the discharge current pulses in the short-circuited electrodes of the borosilicate glass sample grounded through measuring resistance Rm=50Ω: average duration of the discharge current pulses in short-circuited electrodes of the borosilicate glass sample grounded through measuring resistance Rm=50Ω; average and minimum duration of the rising edge of the current pulses in short-circuited electrodes of the borosilicate glass sample grounded through measuring resistance Rm=50Ω. The sample irradiation was recorded by the video camera.

The results obtained in testing the electrification of the solar battery sample with GaAs-based photoconverters and borosilicate cover glass under SF effects are shown in Table 1. In the table, an asterisk * indicates accumulation of radiation charge and hence, potential which is limited to radiation-induced electrostatic discharges on the irradiated surface of the cover glass of the solar battery.

Table 1. Testing results of the sample electrification under various temperatures.

Sample temperature	Maximum integrated surface potential, kV	Maximum charge level of the BF sample, μC	Average time of yield of the discharge stationary mode, sec
- 60°C	12.2*	3.5*	400
+ 20°C	7.9*	2.0*	350
+ 50°C	1.2*	0.3*	200

The electrical discharges presented in figure 5 were recorded on the irradiated surface of the solar battery cover glass within the entire temperature range.

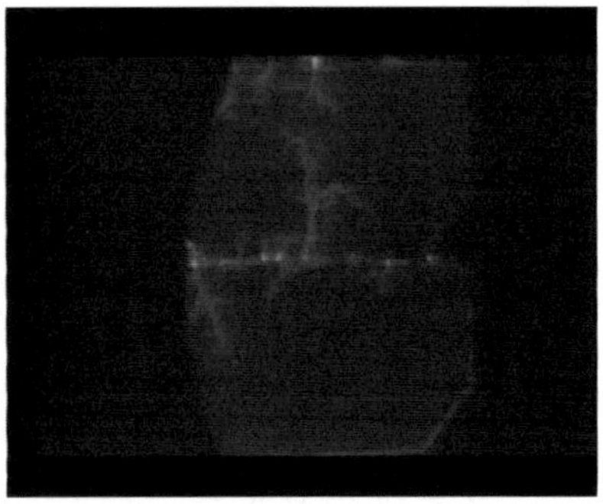

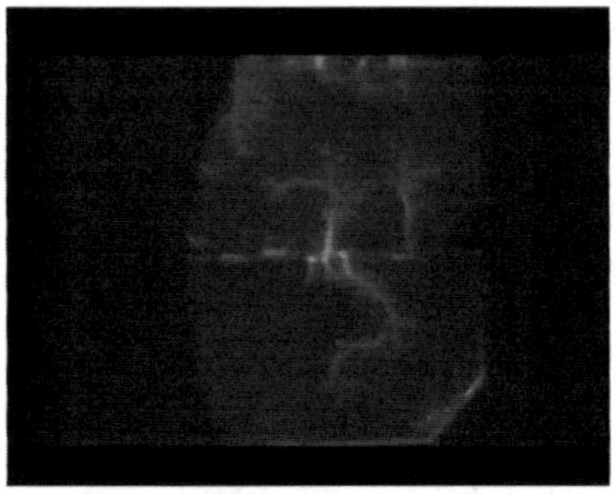

Figure 5. Electrical discharges on the surface of the solar battery cover glass.

The basic parameters of the recorded discharge pulses are presented in Table 2.

The conclusion drawn based on the results of testing the electrification effect on the solar cell sample is as follows. When exposed to electrons with energies up to 100 keV, the borosilicate cover glass accumulates significant radiation charge that causes electrical breakdowns on the cover glass surface within the entire temperature range. As a result, the current pulses with maximum amplitude of up to 12 A were recorded.

Table 2. The shape and parameters of the discharge pulses on the surface of the solar battery cover glass.

Parameter	Value			Form of a typical pulse
	– 60°C	+ 20°C	+ 50°C	
V_{max}, V	600	540	520	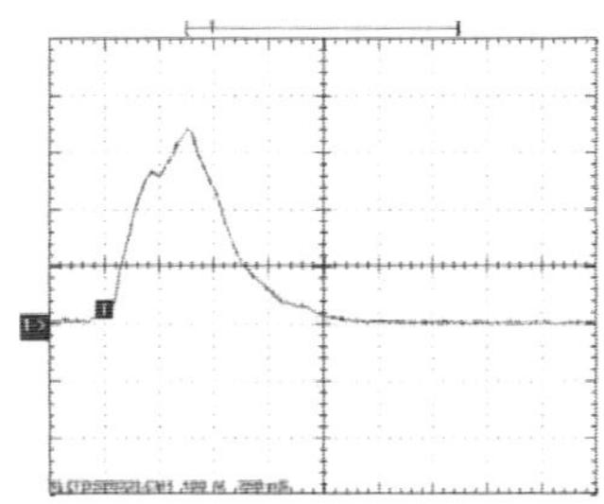
$T_{ed\ min}$, ns	25	25	50	
$T_{ed\ av.}$, ns	100	250	250	
$T_{dis.}$, ns	1 000	1 000	700	
$\nu_{dis.}$, MHz	37.5	31.0	0.5	

V_{max} — maximum amplitude of the discharge pulse voltage at $R_{vol.} = 50\ \Omega$;
$t_{ed.\ min}$ — minimum duration of the discharge pulse rising edge;
$t_{ed.\ av.}$ — average duration of the discharge pulse rising edge;
$t_{dis.}$ — average duration of the discharge pulse;
$\nu_{dis.}$ — discharge frequency.

4. Conclusions

The developed test bench "Prognoz-2" allowed testing dielectric structural materials of the spacecraft under simultaneous exposure to several space factors such as high vacuum $3 \div 6 \cdot 10^{-6}$ torr, variation of the sample temperature from –130° C to +80° C, spectrum of electrons with energy of 10÷300 keV, solar radiation with a wavelength of 0.2÷1.2 μm.

The dimensions of the vacuum chamber enable testing not only samples with large geometric dimensions and entire spacecraft modules, but micro and nanosatellites.

The test bench was used to test modified ceramics (based on boron nitride) of the discharge channel of the spacecraft correcting plasma thruster in cooperation with JSC "ISS" named after Academician M.F. Reshetnev in the framework of the development work "Katun". The spacecraft solar panels were tested in conjunction with OJSC "Saturn". The spacecraft sun sensor was tested together with JSC "Geophysics-Space".

References

[1] Novikov L S, Milev V N, Krupnilov K K, Maklentsov A A 2007 *Space model* (Moscow: KDU), 236–275
[2] Bespalov V I, Zykov V M, Nejman D A 2015 *Proc. Con. On Radioresistance of electronic systems (Lytkarino, Rus. Fed)* (Moscow) 87-89
[3] Paulmier Th, Dirassen B, Belhaj M, Payan D, Balcon N 2012 *IEEE Trans. on Dielec and Elect. Insul.* **3 19** 4 1215 – 1220
[4] NASA-HDBK-4002 Avoiding problems caused by spacecraft on-orniternal charging. http://everyspec.com/
[5] Shilov A M, Prokopiev Yu M, Prokopiev V Yu, Schepihin I V 2007 patent RU 2414717C1
[6] An American National Standard: Standard Test Methods for DC Resistance or Conductance of Insulating Materials 2005 (USA: ASTM Int'l)

Comparative physical-tribological properties of anti-friction ion-plasma Ti-C-Mo-S coating on VT6 alloy or 20X13 and 40X steels

A Y Shubin[1,3], A I Potekaev[2], V M Savostikov[2], S V Galsanov[2], V S Dmitriev[1], I B Stepanov[1], E B Kashkarov[1] and V H Dammer[3]

[1]National Research Tomsk Polytechnic University, 30 Prospect Lenina, Tomsk, 634050, Russian Federation
[2]National Research Tomsk State University, 36 Prospect Lenina, Tomsk, 634050, Russian Federation
[3]Joint Stock Company Scientific Production Centre «Polus», 56B Prospect Kirova, Tomsk, 634050, Russian Federation

E-mail: ayshubin@gmail.com

Abstract. Results of comparative tests mechanical and tribological properties of solid antifriction Ti-C-Mo-S coating, deposited by magnetron-plasma combined sputtering method on substrates of VT6 titanium alloy, 40X and 20X13 hardened steels are provided. Coating is sputtered using the same conditions and technological regimes on substrates of different materials. However, the friction tests results showed significant difference in tribological characteristics of coating depending on type of material used for substrate, first of all by wear-resistance ability. Authors suppose that this is due to difference between physical properties such as composition and structure of substrate materials that determines hardness and coating adhesion to surface.

1. Introduction

At the present time it is proved by the results of experimental tests that the optimal variant of antifriction coating producing is a gradient composition construction with relatively smooth change of composition, structure and basic physical properties (hardness, thermal expansion coefficient, shear resistance) from product's basis to outer coating. Developers use different technological methods to produce such compositions. For example, in article [1] pre-hardening of steel substrate by ion nitriding prior to magnetron deposition of hard Ti-N coating and final magnetron deposition of outer antifriction MoS_2 layer is performed. In article [2] leveling of both structure and properties between solid sublayer TiN and outer soft antifriction layer MoS_2 is performed by forming intermediate combined structure of TiN+MoS_2, and simultaneously by magnetron sputtering of Ti and MoS_2 cathodes. Combined structure layers technological method is provided in article [3]: solid anti-frictional CrN, TiN, TiCN layer is sputtered by means of magnetron sputtering outer MoST^{TM} layer that poses as a MoS_2 and titanium composition. Outer layer is characterized by nanocrystal or amorphous condition depending on titanium content in it and has high hardness, wear-resistance and moisture-resistance compare to fine MoS_2.

Over the last years tribological coating of combined structure with the use of different components heighten interest of developers and scientific researchers. For example, in article [4] magnetron a-C coating, doped Mo is subsided by magnetron. It is found out that depending on Mo content the physic-tribological properties of that coating may be materially changed. Similar researches using magnetron nanocomposited nc-TiC/a-C (nc – nanocrystal phase, a – amorphous phase) coating are conducted by the authors [5]. Instant connection between physical and tribological characteristics of that coating, namely that frictional coefficient and wear-resistance both depend on the ratio between hardness and Joung's efficient module is established.

The tribological Ti-C-Mo-S solid coating of combined structure is developed by Siberian Physical-Technical Institute of National Research Tomsk State University. Magnetron-plasma synthesis method of that coating for titanium alloys with gradient-compositional structure is patented as invention [6]. It is needed to estimate the possibility of coating for sliding friction pairs to practical using. In the present paper comparative studies results of physical-tribological properties of that coating on titanium alloy and hardened steels are presented.

2. Materials and methods of experiment

Research is carried out on titanium alloy VT6 (Ti – base, Al – 5.5-7.0%, V – 4.2-6.0%) and on industrially widely applied engineering 40X steel and corrosion-resistant 20X13 steel. These types of steels differ from each other compositionally by its chemical structure, primarily, by chrome content - approximately 1% in 40X and approximately 13% in 20X13. The hardness this of this steels are HRC40-42. VT6 alloy's hardness is about HRC36. Samples of material are produced in the shape of discs with 40-50 mm diameter and 4 mm thickness. Discs are grinded on abrasive disks with supply of quenching water and emery papers use. Then they are polished with use of diamond pastes and chrome oxide aqueous suspension. Roughness of polished samples (Ra) is controlled by profilograph model 296 and does not exceed 0.1 μm. Ti-C-Mo-S coating deposition is performed on KOMPOSIT-3 hybrid facility simultaneously for all three types of samples (in one load) during their continues rotation in regard to sputtering cathode targets. In forming of gradient-composite coating structure ion-plasma pre-doping technique of surface layer by Ti-C-Mo-S atoms set and PINK plasma generator assisting influence are used [7]. Technological sputtering process under of two multiple-component self-propagating high-temperature synthesis (SHS) targets Ti-C-Mo-S in argon atmosphere is performed. Magnetron discharge power for each target is approximately 1.0 kW. Bias potential is 200 V, coating sputtering time – 60 minutes. Hardness of doped layer and coating measurements are made on «CSEM Nano Hardness Tester» under 100÷250 mN load. Adhesion characteristics of coating (adhesion to surface) are conducted by «CSEM Micro Scratch Tester». Doped surface layer and coating component analysis is done on "Schuna-2" using auger spectral analysis. Friction tests are done on «CSEM High-Temperature Tribometer» using the pin-on-disk scheme. In the function of indenter HRC 62-63 steel balls are used. Indenter loading for all cases is 1 N. Counter bodies sliding velocity are within 0.5-0.6 m/s. Friction coefficient value is fixed depending on number of disk turns. Coating wear rate for surfaces of different materials comparison is realized immediately after friction coefficient increasing up to 0.2 and higher. Wear volume is defined by calculation of averaged in nine measures cross-sectional scans of wear track area using a surface profilometer «Micro Measure 3D Station». Friction tracks as well as areas around Rockwell prints on coated discs are examined using optical (on IMC and METAM RV-21) and electron scanning «Tescan Vega 3» microscopes.

3. Experimental investigations results and their consideration

Doped underlayer and coating hardness measurements showed that combined method of ion-plasma pre-doping and consequential Ti-C-Mo-S deposition allows relatively smooth hardness transition from surface to underlayer and then to coating for all cases. These values have insignificant difference for all three materials and are about 4-5 GPa, 6-7 GPa and 7-9 GPa. Doping elements depth dispatching comparative analysis using three types of materials showed that 40X steel has the highest one and approximately 500 nm [8] (figure 1). It is obvious that this is due to the difference between chemical structures of surfaces. Titanium in VT6 as well as chrome in 20X13 has high reactive capacity to form chemical compounds that limits diffusion depths during doping.

Ti-C-Mo-S coating friction tests on surfaces for two types of steels showed that 40X steel has friction coefficient is approximately 0.11 from the beginning, and right after 13000-14000 disc round μ decreased to 0.09. Friction coefficient is kept this value up to 38000 rounds (about 3000 meters of sliding path), then increased as consequence of critical coating wearout (figure 2a).

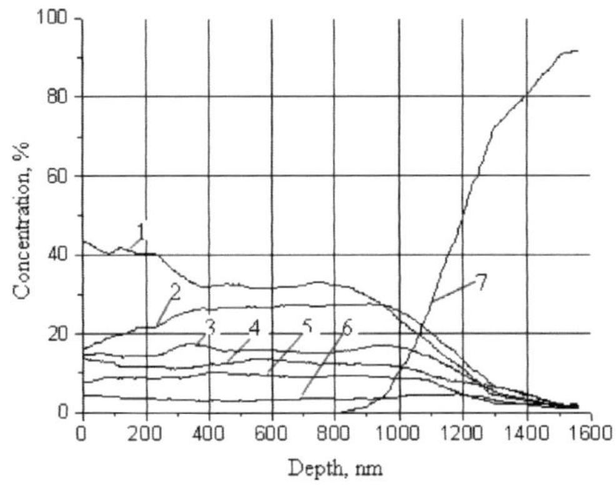

Figure 1. Elements allocation by depth of 40X sample after ion-plasma doping and magnetron Ti-C-Mo-S coating sputtering in argon atmosphere: 1-S; 2-Ti; 3-C; 4-Ni; 5-Mo; 6-O; 7-Fe. Accelerating voltage using doping is 1000 V; accelerating voltage using coating sputtering – 200 V.

During the same frictional tests for coating on 20X13 steel, it is established that μ is decreased from 0.09 to 0.07 and kept this value only to 18000 rounds of disc. Then μ highly increased due to coating wear (figure 2b). VT6 coating is withstood 31500 round with average value of μ equal to 0.115 (figure 3). Conducted optical tests of wear tracks revealed the significant difference in coating failure pattern on 40X (figure 4a) in comparison to 20X13 (figure 4b). From figure 4b it is seen on 20X13 coating damage as a result of friction contact to counter body appears as form of small flakes. Coating is scaled off by flakes prolonged to track's direction. On VT6 coating damage character is similar to the one that is also observed on 40X.

IOP Conf. Series: Materials Science and Engineering **168** (2017) 012038 doi:10.1088/1757-899X/168/1/012038

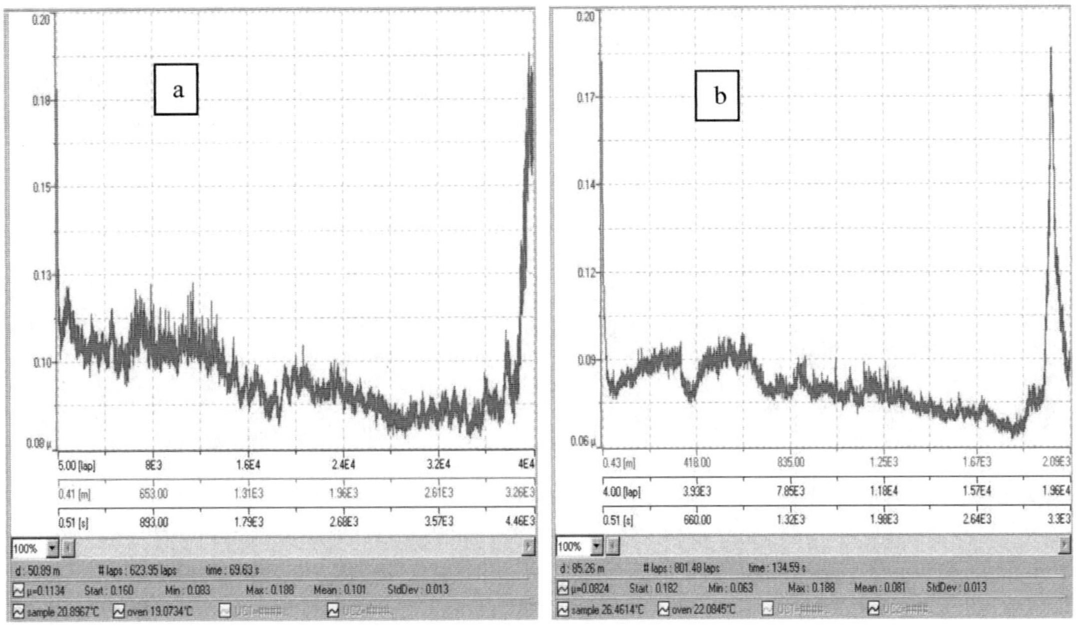

Figure 2. Friction coefficient value changings depending on number of sample-disc rounds with Ti-C-Mo-S coating, sputtered on 40X(a) and 20X13(b) steels.

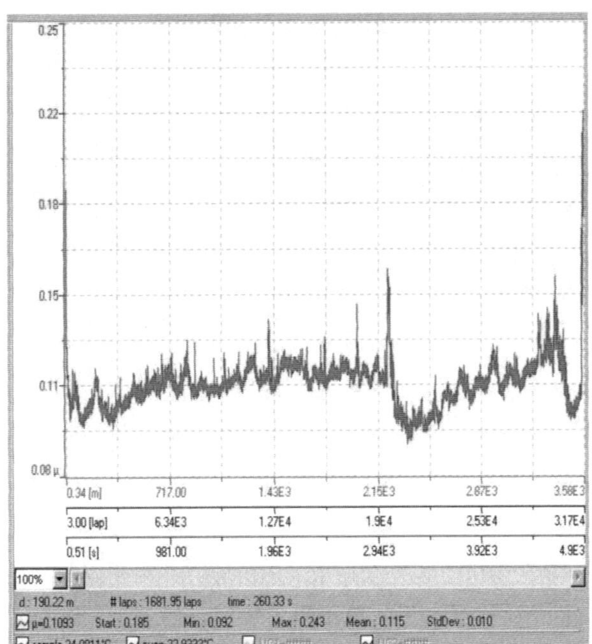

Figure 3. Friction coefficient value changings depending on number of sample-disc rounds with Ti-C-Mo-S coating, sputtered on titanium alloy VT6.

IOP Conf. Series: Materials Science and Engineering **168** (2017) 012038 doi:10.1088/1757-899X/168/1/012038

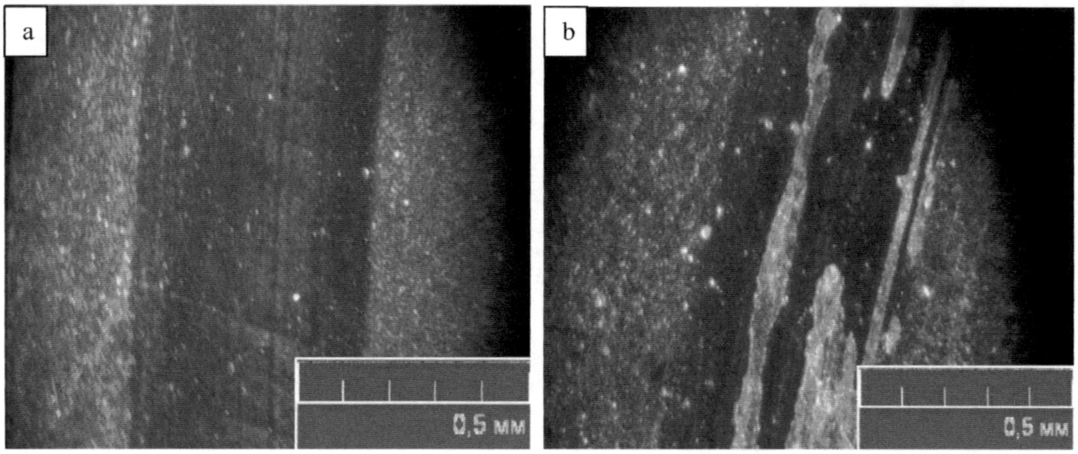

Figure 4. Ti-C-Mo-S coating damage character on wear track sputtered on layer of 40X(a) and 20X13(b) steels. Optical microscopy, IMC device.

Rockwell prints optical analysis after 4% nitric acid in ethanol etching showed significant difference in etching character of area around print for 20X13 in comparison to 40X (figure 5). According to almost identical coating scaling off character for both types of steel strongly marked difference of area livery bordered to prints (without coating), may evidence of chemical composition and their doped layer significant differences.

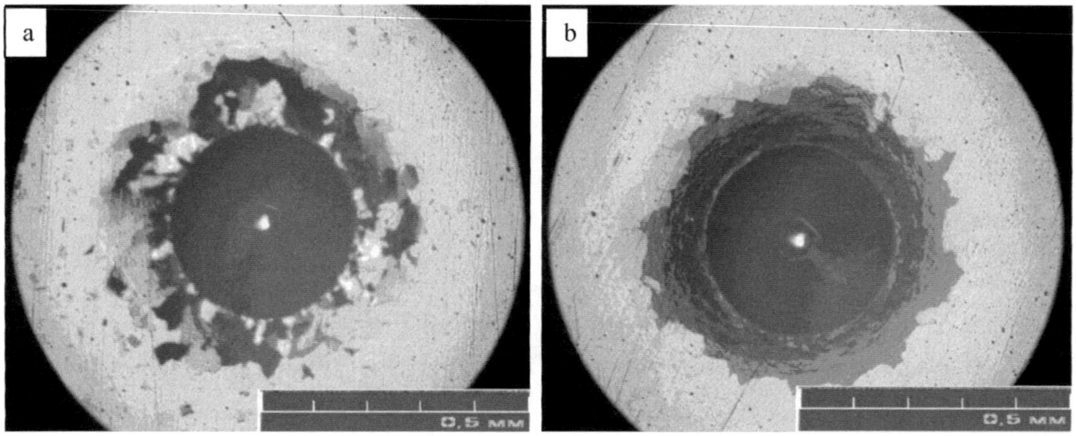

Figure 5. Etched area view around Rockwell's prints for 40X(a) and 20X13(b) steels with Ti-C-Mo-S coating. Optical microscopy by METAM RV-21.

It is revealed by electronic scanning microscopy the significant difference in observed steels initial structure (figure 6).

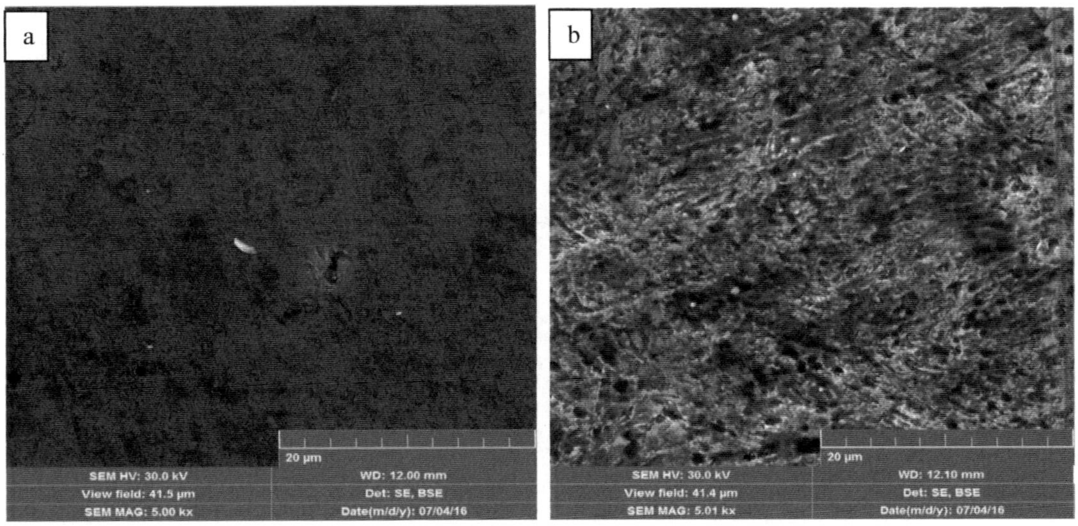

Figure 6. Microscopic structure appearance of 40X(a) and 20X13(b) steels by electron scanning microscope «Tescan Vega 3».

The wear density of Ti-C-Mo-S coating comparative evaluation by averaged crosscut area of wear tracks showed that 40X has approximately 3.3 μm^2 per 1000 rounds, for VT6 this value is equal to 7.65 μm^2 and for 20X13 – 12 μm^2 per 1000 rounds.

Conclusion

Ti-C-Mo-S coating frictional tests proved the significant difference of its wear-resistance depending on substrate material. The most signification difference among coating life value is fixed in comparison to low-alloyed 40X steel, where wear-resistance is the highest and high-alloyed 20X13 steel. Authors suggest that it is determined by difference of such physical properties as surface material chemical structure and coating adhesion to surface.

References

[1] Rahman M, Haider J, Dowling D P, Duggan P and Hashmi M S J 2005 *Surface and Coatings Technology* **200** 1451–57
[2] Gangopadhyay S, Acharya R, Chattopadhyay A K and Paul S 2009 *Surface and Coatings Technology* **203** 3297–305
[3] Renevier N M, Hamphire J, Fox V C, Witts J, Allen T and Teer D G 2011 *Surface and Coating Technology* **142-144** 67-77
[4] Novák P, Musil J, Čerstvý R and Jäger A 2010 *Surface and Coatings Technology* 2010 **205** 1486–90
[5] Musil J, Novák P, Čerstvý R and Soukup Z 2010 *J. Vac. Sci. Technol.* A **28** 244-9
[6] Savostikov V M, Tabachenko A N, Potekaev A I and Dudarev E F A Method for Deposition of Antifrictional Wear-Resistant Coatings on Titanium Alloys, RF Patent №. 2502828. Priority of 18.06.2012. Publ. 27.12.2013, Bull. № 36
[7] Savostikov V M, Potekaev A I, Tabachenko A N 2011 *Izv. VUZov. Fis.* **7** 26-34
[8] Savostikov V M, Potekaev A I, Tabachenko A N, Shulepov I A, Kuzmichenko V M and Didenko A 2012 *Russian Physics Journal* **11** 1232–40

Mathematical modelling of convective processes in a weld pool under electric arc surfacing

V D Sarychev, A Yu Granovskii, S A Nevskii and S V Konovalov[1],
Siberian State Industrial University, Novokuznetsk, 42 Kirov Street, 654007, Russia

E-mail: konovserg@gmail.com

Abstract. The authors develop the mathematical model of convective processes in a molten pool under electric arc surfacing with flux-cored wire. The model is based on the ideas of how convective flows appear due to temperature gradient and action of electromagnetic forces. Influence of alloying elements in the molten metal was modeled as a non-linear dependence of surface tension upon temperature. Surface tension and its temperature coefficient were calculated according to the electron density functional method with consideration to asymmetric electron distribution at the interface "molten metal / shielding gas". Simultaneous solution of Navier-Stokes and Maxwell equations according to finite elements method with consideration to the moving heat source at the interface showed that there is a multi-vortex structure in the molten metal. This structure gives rise to a downward heat flux which, at the stage of heating, moves from the centre of the pool and stirs it full width. At the cooling stage this flux moves towards the centre of the pool and a single vortex is formed near the symmetry centre. This flux penetration is ~ 10 mm. Formation of the downward heat flux is determined by sign reversal of the temperature coefficient of surface tension due to the presence of alloying elements.

1. Introduction

At present time the method of arc surfacing with flux-cored wire is gaining popularity for producing wear-resistance coating [1]. It allows producing bimetallic products characterized by high strength and low production cost combined with long service life. The main part of the product can be manufactured from low-alloy steels. Application of this method for repairing worn-out parts allows reducing the amount of spare parts produced for the operated equipment, reducing the breakdown time during the repairs, reducing maintenance costs. On the other hand, in the process of arc surfacing, in spite of the simple construction of the applied units and their high productivity, coarsely dispersed non-uniform structures with low impact strength are formed in the surface layer, quality of the surface requires its further machining [2]. This determines the necessity of searching for such surfacing conditions that ensure forming a structure with high physical-mechanic properties [3]. To achieve this goal we have to understand the hydrodynamic processes taking place in the molten pool. The specific features of these processes will determine the geometry of the surfaced layer, its structure, phase composition and mechanic properties [3]. The hydrodynamic processes in the molten pool under arc surfacing are studied in many works, [4 – 9] to be mentioned among them. In the given works the

[1] Corresponding author

authors complete a computational investigation of heat exchange in the liquid and liquid flowing under pulse-arc welding and direct current welding. In [4] the authors come to the conclusion that surface-tension gradient has the most significant impact upon the weld pool geometry while other factors, such as magnetic conductivity, thermal expansion coefficient of the material are less important. In [5] the authors obtain the results which demonstrate that application of pulse current results in a more qualitative weld joint. Thermocapillary convection in the weld pool was studied in [6 – 8]. In [6] the authors conducted the computational investigation of Marangoni effect influence upon the parameters of various metals under closed arc welding. As a result the current intensity ranges under which the thermocapillary convection plays the dominating role were established. Further increase of current intensity does not result in increase of penetration depth, thus, the linear dependence of surface tension coefficient of pure molten metals upon the temperature deteriorates the penetrating ability of the electric arc. In [7] it is shown that, for pure metals, thermocapillary convection plays an important role only at the initial stages of metal melting and then gravity-induced deformation of liquid becomes the source of convective flows. On the other hand study of the concentrated energy flows action upon metals shows that heat-gravitational convection is insignificant under the great values of power density [8]. It is also necessary to take the action of alloying elements into consideration [8]. In [9] the researchers study influence of these elements upon surface tension of liquid iron. It is shown that addition of sulfur results in surface tension increase as temperature grows and its further decrease, i/e/ there is a non-linear thermal dependence. If sulfur is introduced directly into the molten pool it leads to tension growth, so the pattern of vortex flows will change significantly. The substances involved into vortex flow will move to the centre, then downward, to the bottom of the pool. The downward flow of hot metal in the centre of the pool works like a thermal drill bit. This results in penetration depth increase and pool geometry change. In [10] the scientists suggest the mathematical model of pulse radiation impact with consideration to the dependence of surface tension upon the superficially active substance in the molten metal. The numerical modeling results showed that in the presence of superficially active substance multi-vortex structure of flows is formed ensuring more uniform distribution of alloying component.

Thus, while developing the model of convective flows in the molten pool it is important to consider action of alloying elements upon the hydrodynamic flow. It is relevant for studying the convective flows in the weld pool when surfacing with flux-cored wires. The action of the alloying elements is modeled by the dependence of surface tension coefficient upon the temperature. The alloying elements in the composition of flux change the character of this dependence which results in alteration of the convective flow pattern and this, in its turn, effects the structure-phase composition of the surfaced metal. That is why the aim of our work is developing the mathematical model of convective flows which take these elements into consideration.

2. Problem formulation

It is relevant to start modeling of convective flows from determining the volume forces effecting the liquid as well as from calculating the electromagnetic fields. The volume force effecting the liquid comprises the gravity and the electromagnetic forces.

$$\vec{F} = \vec{F}_g + \vec{j} \times \vec{B} = \rho_0 \vec{g} - \rho_0 \vec{g} \beta (T - T_{ref}) + \vec{j} \times \vec{B} \quad (1)$$

Electromagnetic fields can be calculated according to Maxwell-Lorentz equation

$$\nabla \cdot \left(\sigma \nabla V + \sigma \frac{\partial \vec{A}}{\partial t} \right) = 0, \ \sigma \frac{\partial \vec{A}}{\partial t} + \frac{1}{\mu_0} \nabla \times \left(\nabla \times \vec{A} \right) + \sigma \nabla V$$
$$\vec{E} = -\nabla V - \frac{\partial \vec{A}}{\partial t}, \quad \vec{j} = \sigma \vec{E}, \ \vec{B} = \nabla \times \vec{A} \quad (2)$$

With consideration to (1), (2) let us put down Navier-Stokes and thermal conductivity equations

$$\rho\left(\frac{\partial \vec{u}}{\partial t} + \vec{u}\nabla\vec{u}\right) = -\nabla p + \mu\Delta\vec{u} + \vec{F}_V, \nabla\cdot\vec{u} = 0,$$
$$\rho C_p\left(\frac{\partial T}{\partial t} + \vec{u}\nabla T\right) = \nabla\cdot(k\nabla T) + Q_V \qquad (3)$$

where $\vec{u}$ - velocity vector, p - pressure, ρ - density, μ – dynamic viscosity, $\vec{F}_V$ - volume forces which are given by the total of Lorentz force and gravity force, T- temperature, C_p – specific heat capacity, k – thermal conductivity coefficient, Q_V – volume heat sources. For cathode and anode only Joule effect is the volume heat source $Q_V = \vec{j}\cdot\vec{E}$.

Let us consider the action of electric arc plasma upon a steel plate (Figure 1).

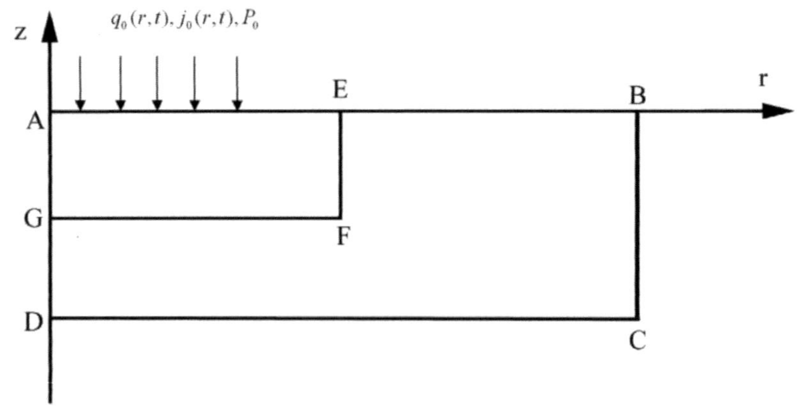

Figure 1. Computational region

On the top surface AB we set density of the current, pressure and heat flow which result from plasma action upon the metal surface like in [3, 5, 11].

$$-\vec{n}\cdot\vec{q} = q_0(r) = \frac{dUI\eta}{\pi r_0^2}G(r,t),\ -\vec{n}\cdot\vec{j} = j_0(r) = \frac{dI}{\pi r_0^2}G(r,t)$$
$$p = p_0(r) = \frac{\mu I}{4\pi}j_0(r) \qquad (3)$$

where q_0, j_0, p_0 - heat flow, density of current, pressure accordingly; U, I - напряжениеvoltage and intensity of current; η - process efficiency coefficient; G(r,t) – function of surface and time distribution of the action:

$$G(r,t) = \exp\left(-\frac{d(r^2 + (r_0 - V_e t)^2}{r_0^2}\right) \qquad (4)$$

V_e - electrode motion speed, d, r_0 - parameters of distribution. On the bottom surface CD we set the condition of convective heat transfer and electromagnetic continuity:

$$-\vec{q}\cdot\vec{n} = h_0(T - T_0), \vec{j}\cdot\vec{n} = 0 \qquad (5)$$

On the side surface BC we set the condition of convective heat transfer and scalar electric potential equaling zero:

$$-\vec{q}\cdot\vec{n}=h_0(T-T_0),\ V=0 \tag{6}$$

For the magnetic field we set the condition of continuity at all boundaries: $\vec{n}\times\vec{A}=0$.
At the boundaries EF and FG the slip condition was fulfilled for the speed.

The model of phase transition from the solid into the liquid state was based on the idea of dynamic viscosity coefficient growth in the solid region. It looked as follows:

$$\mu=\mu_L(T)(1-f_s)+\mu_S f_s \tag{7}$$

where f_s - the coefficient determining the solid phase, μ_L - viscosity of the liquid phase, μ_S - viscosity of the solid phase large enough to stop motion in the solid phase.

The calculations were completed according to finite elements method in the pack Comsol Multiphysics. In Table 1 we provide the characteristics of the material and treatment conditions.

Table 1. The input parameters of the problem

Symbol	Description	Value
U	Voltage	30V
I	Intensity of current	250 A
r_0	Effective action radius of the plasma jet	10mm
T_l	Temperature of the liquid phase	1723K
ρ_s	Density of the solid phase	7500kg m^{-3}
ρ_l	density of the liquid phase	6350kg m^{-3}
C_{pl}	Heat capacity ratio of the liquid phase	720 J kg^{-1} K^{-1}
C_{ps}	Heat capacity ratio of the solid phase	602 J kg^{-1} K^{-1}
k_l	Thermal conductivity coefficient	20 W m^{-1} K^{-1}
k_s	Thermal conductivity coefficient	26 W m^{-1} K^{-1}
σ	Electric conductivity of metal	$7.7\times10^5\ \Omega^{-1}\,\mathrm{m}^{-1}$
μ	Dynamic viscosity coefficient	0.05 kg m^{-1} s^{-1}
L_f	Melting heat	247×10^3 J kg^{-1}

Surface tension and its temperature coefficient were calculated according to the methods described in [12, 13]. The method of electron density functional was chosen as the basic method. The functions which take into consideration the asymmetrical distribution of electrons near the surface were taken as testing functions [13]. The possible combinations of iron and second element in the surfaced material were considered. They were determined according to the constitution diagrams. In Figure 2 we show the dependence of surface tension which looks like a parabola.

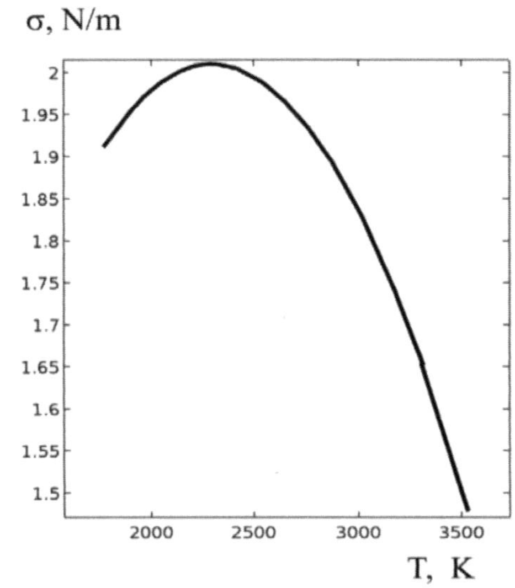

Figure 2. Dependence of surface tension upon temperature

3. Results and discussion

It is known that the result in the molten pool is determined by marangoni convection which is characterized by the temperature gradient of surface tension $d\sigma/dT$. In Figure 2 we can see that $d\sigma/dT>0$ under the temperatures below critical $T<T_c=2300\text{K}$ and $d\sigma/dT<0$ for $T>T_c$. In Figure 3 the evolution of molten metal flow is shown at various times. We can see that before the heat flow reaches its peak values the downward flux formed by two symmetrical vortexes moves from the centre of the pool stirring the molten metal full-width. When the heat action of the plasma jet decreases the cooling stage starts characterized by movement of the downward flux towards the centre of the pool and further formation of the single vortex near the centre of symmetry (Figure 3b). This flux penetrates the depth of 6mm. Formation of the given downward flux can be explained by the sign change of the temperature gradient of surface tension $\partial\sigma/\partial T$.

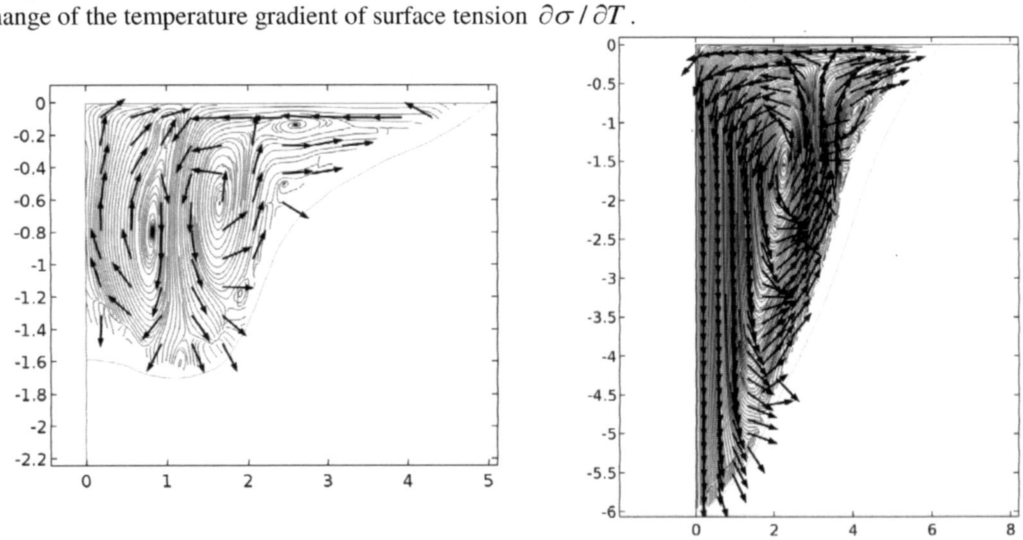

Figure 3. The lines of current of the molten metal and the field of velocities at various times: a) t=2.7 s; b) t = 5.3 s

4. Conclusion

We completed numerical modeling of the hydrodynamic flow in the molten pool under the action of electric arc discharge with consideration to Marangoni effect. Presence of alloying elements in the metal has a significant impact upon the dependence of surface tension on temperature which, in its turn, determines the pattern of convective flow of the molten metal.

5. Acknowledgements

The research was performed with a grant from the Russian Science Foundation (No. of project 15-19-00065)

References

[1] Zahiri R, Sundaramoorthy R, Lysz P, Subramanian C, 2014 *Surface & Coatings Technology* **260** 220
[2] Berger L-M 2015 *Int. Journal of Refractory Metals and Hard Materials* **49** 350
[3] Traidia A, Roger F, Guyot E, Schroeder J, Lubineau G, 2012 *International Journal of Heat and Mass Transfer* **55** 3946
[4] Tong L G, Gu J C, Wang L, Yin S W 2015 *International Journal of Heat Mass Transfer* **90** 968
[5] Traidia A, Roger F 2011 *International Journal of Heat Mass Transfer* **54** 2163
[6] Sultangazieva R, Amankulova N 2015 *Vestnik KGUSTA* **2** 78
[7] Kim Y, Hossain A and Nakamura Y 2013 Journal of Thermal Science and Technology **8** 136
[8] Volkov N B, Leivi A Ya, Talala K A and Yalovets A P 2010 *Technical Physics* **55** 484
[9] DebRoy T 2015 *Welding Journal* **94** 58
[10] Popov V N, Kovalev O B, Smirnova E M and Tsivinskaya Yu S 2012 *Vestnik NSU* **7** 114
[11] Wu C S, Zhao P C and Zhang Y M 2004 *Welding Journal* **83** 330
[12] Toshihiro T 2014 Surface tension models Treatise on process metallurgy (Process Phenomena vol 2) ed S Seetharaman (Amsterdam: Elsevier) chapter **1.4** 35–59
[13] Kanchukoev V Z, Kashezhev A Z, Mambetov A Kh and Sozaev V A 2002 *Technical Physics Letters* **28** 515

IOP Conf. Series: Materials Science and Engineering **168** (2017) 012040 doi:10.1088/1757-899X/168/1/012040

Research into processes of production of hydrides of materials containing rare-earth metals and their corrosion

V L Sofronov, E Y Kartashov, P B Molokov, A N Zhiganov and M E Kalaev

National Research Nuclear University MEPhI, 115409, Russia, Moscow, Kashirskoye Shosse, 31

E-mail: VLSofronov@mephi.ru

Abstract. Production of permanent magnets on basis of rare earth elements (REE) is implemented by means of powder metallurgy, therefore a technologically important operation is the multistage mechanical crushing of materials to the extent of domains. The promising technique of crushing of magnetic materials is their consistent hydrogenation-dehydrogenation that allows obtaining nano-dispersed powders which are stable enough in air. Hydrogenation apparatuses, as opposed to conventional grinding machines, do not comprise motion works and their producing capacity is much higher. Hydrogenation process does not require any additional preparation of materials and it excludes undermilling and overmilling as well as material oxidation. The paper presents the results of investigation on the temperature effect on the hydrogenation process of Nd-Fe alloys. The study results on the corrosion stability of ligature hydrides under various conditions are also given. Kinetic parameters of the hydrogenation process of ligatures are determined. The phase composition of corrosion products is detected. Guidelines on hydride powder storage are given.

1. Introduction

One of the methods of improving the quality of Nd-Fe-B alloys, used for the production of rare earth permanent magnets, is the solid-phase alloying technique based on the combined grinding of magnetic alloys with alloying additions. At the same time, there is the optimal allocation of constituents by volume and the particles surface activation which improves the growth of the master magnetic phase of $Nd_2Fe_{14}B$, with the subsequent compression and agglomeration of powders taking place.

The base material, used for the correction of the composition of magnetic alloys, is the Nd-Fe ligature. However, this material has an essential fault, i.e. high strength. It results in impossibility of its usage for the combined grinding with the fragile Nd-Fe-B alloy [1].

Separate ligature grinding with subsequent adding into the composition of magnetic alloys is technologically difficult, as the powders are pyrophoric and easy-to- oxidize in air. To solve the problems described, we suggest using the hydrogen destruction method which is used for the grinding of magnetic alloys [2].

The process of the hydride grinding of Nd-Fe-B alloys is carried out by means of crushing a magnetic alloy ingot into pieces of 10-20 mm and their processing with hydrogen. This process is known as the hydrogen decrepitation (HD) [3]. It results in the hydrogenation of intergranular high-neodymium phase and its destruction. The hydrogenation being done, the material is easily destructed into the powder having domains of 1-3 micrometers. During the process the oxidation of the hydrogenated product does not occur, the grinding taking place generally along the grain boundaries of the $Nd_2Fe_{14}B$ magnetic phase.

The paper presents the results of investigation of both the temperature effect on the hydrogenation process of Nd-Fe alloy and the corrosion stability of hydrogenation products of magnetic materials containing a rare-earth component (40 – 80 % wt) under various conditions.

2. Research into the temperature effect on the Nd-Fe alloy hydrogenation process

2.1. The scheme of the experimental device

The subject of inquiry was the Nd-Fe alloy produced by means of out-of-furnace calcium-thermal reduction of neodymium and iron fluorides.

The research into the hydrogenation process was conducted with the help of the apparatus presented in Fig. 1.

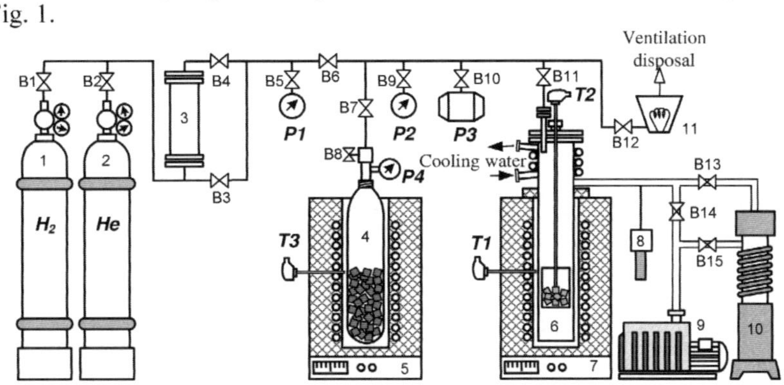

1, 2 – hydrogen and helium cylinders; 3 – column with dehumidifier; 4 - container with lanthanum nickelid (hydrogen accumulator); 5, 7 - shaft furnaces; 6 - hydrogenation reactor; 8 - thermocouple vacuum gauge ; 9 – fore pump BH-1; 10 - diffusion pump; 11 - hydrogen afterburner; T1, T3 - furnace temperature sensors; T2 - sample temperature sensor; P1, P2 - pressure-and-vacuum gages; P3 – pressure gauge "Sapfir"; P4 – manometer; B1 – B15 valves

Figure 1. The scheme of the experimental apparatus.

The apparatus consists of a vacuum post which provides the system degassing to the residual pressure of 10 Pa, a hydrogenation reactor and a reactor with a hydrogen accumulator (lanthanum nickelid). The usage of lanthanum nickelid allowed purifying technical hydrogen from traces of water and other impurityies.

The rate of the hydrogenation process was controlled via pressure changes in the reactor with the help of an automatic manometer "Sapfir" accurate within (±1)%, and the general pressure in the supply pipeline was controlled by means of precision pointer-type combined pressure-and-vacuum gages.

2.2. Procedure of the experiment

Hydrogen treatment. The system was vacuumized to the residual pressure of 10 Pa with the help of a fore pump. Hydrogen from the cylinder ran through the column with silica gel and filled the container with lanthanum nickelid. After that hydrogen pressure was increased up to 1 MPa. Supervision was being carried out to monitor hydrogen absorption by lanthanum nickelid. Pressure decrease having been stopped, the container with lanthanum nickelid was ready for operation.

Alloy treatment. Alloy ingots were chipped with the help of a hydraulic press. Then they were ground into pieces of 5-10 mm by means of a jaw breaker.

Experiments. The required quantity of the alloy was taken and put into the pan, made of stainless steel. Then it was weighed on the high-accuracy weighing machine. In a number of cases, metallographic samples of the alloy were used for the subsequent metallographic study. The pan was put into the reactor. Then the reactor was vacuumized to the residual pressure of 100-1000 Pa. In a number of experiments higher vacuum was used, but it had almost no impact on the results of the experiments.

Further, the retort was filled with pretreated hydrogen and the recording of pressure changes was started. The effect of the process temperature on the hydrogenation rate was studied.

Based on the results of the conducted experiments there have been designed the diagrams of dependence of the hydrogenation degree on the duration at various temperatures of the process. Then mathematical treatment of the diagrams was done.

To investigate structural changes, taking place in the material, the X-ray phase analysis was used (Dron 4.0 with a copper cathode).

The specific area of samples was estimated with the help of BET technology using thermal argon desorption as consistent with the internal standard. Before the beginning of measuring operations all samples were heated in Ar and He current at the temperature of 120 °C for 30 minutes.

2.3. Consideration of results

Based on the experimental data there have been designed the diagrams of dependence of the hydrogenation degree changes on temperature for samples having different masses. The diagrams are presented in Figure 2.

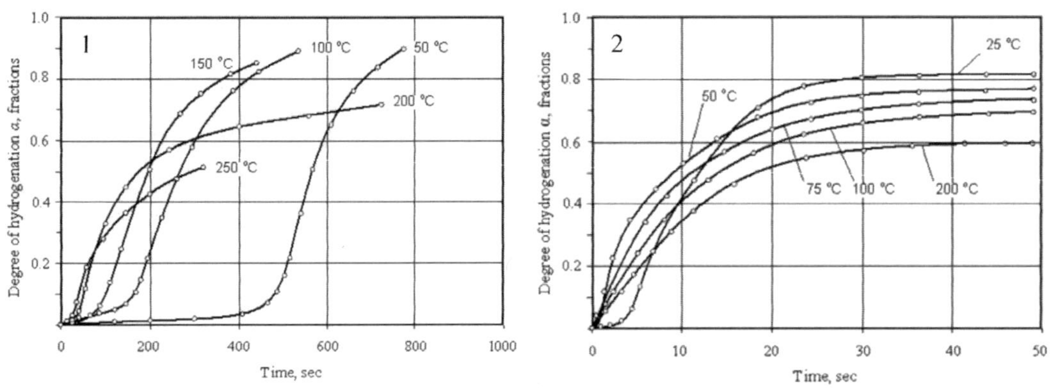

The charge is: 1 – 3-5 g; 2 – 150-200 g

Figure 2. Dependence of the alloy hydrogenation degree on temperature.

Dependence of the alloy hydrogenation degree on the duration of the process at different temperatures. The diagrams presented in fig. 2 show that temperature has a material effect on the duration of the induction period and does not on the rate of the process in whole. It is particularly detected when the hydrogenation of small alloy charges (3-5 g) is taking place (fig. 2.1).

The maximum induction period (longer than 6 min) is observed at the temperature of 50 °C. The charge being enlarged to 200 g (fig. 2.2), the induction period almost disappears.

From our point of view, such influence of temperature on the duration of the induction period is conditioned by the rate of hydrogen intrusion deep into the material along the grain boundaries.

At cold temperatures the rate of hydrogen intrusion into the material is low; therefore, the reaction surface is negligible. The hydrogenation rate races up only after forming the sufficient number of transporting fractures which provide unimpeded hydrogen penetration to the reaction surface. metallographic examinations confirm that.

Calculations, carried out with the help of formal kinetic equation for heterogeneous reactions of the type "solid + gas = solid", show that in the range of 25-250°C the temperature does not have any material effect on the hydrogenation rate and the process is controlled by the hydrogen diffusion to the reaction surface.

The apparent activation energy of the process is (7.0±0.5) kJ/mol.

The formal equation of contractive sphere $1-(1-\alpha)^{1/3} = k\tau$ describes the experimental data of the Nd-Fe alloy hydrogenation process particularly precise (correlation coefficient is 0.8-0.9).

3. Research into the corrosion stability of hydrides under different atmospheric humidity

The subjects of inquiry were nano-disperse hydrides of Nd-Fe alloys, having different compositions, as well as hydrides of Dy-Fe alloys obtained in work [1], which were used in the solid-phase alloyage process in the sequel.

The research technique was the following: the polydisperse hydrogenation product, having the weight up to 60 g, was transferred to the boat with the size 50×30×20 mm and put into the temperature-controlled exiccator filled with water. Over all the time of the experiment the temperature in the exiccator was maintained in the range of (20±1) °C. In this case the absolute humidity was 100%, the water vapor pressure was constant and equaled 3.2 kPa.

Periodically (once a day) the exiccator was opened, the changes which took place on the product surface during this time were registered, the boat was pulled out of the exiccator and weighed, the sample of the material was taken to carry out the X-ray phase analysis.

The conducted experiments showed the following: being put under such extreme conditions, the hydrogenation products have sufficiently high hygroscopicity. For three days the weight of the sample has increased by more than 10% (fig. 3, curve 1); then it almost stopped changing.

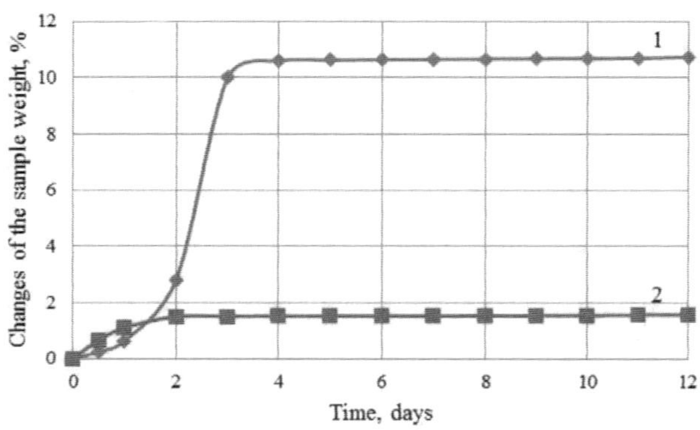

1 - damp atmosphere, 2 – flow air atmosphere

Figure 3. Dependence of changes of the sample weight on the duration of the corrosion process of Nd-Fe hydrides.

The diffraction pattern presented in figure 4 shows that a virtually complete surface amorphization of the sample takes place in twenty-four hours. In two days sufficiently well-defined $Nd(OH)_3$ phase reflexes appear in the diffraction pattern.

In the course of further damp atmosphere storage the materials do not undergo changes in the phase composition. It should be mentioned, that owing to high amorphization of these hydride powders, X-ray phase analysis does not allow complete identification of the forming crystalline phases. To consider the details of this process a complex physicochemical analysis is required.

Figure 5 presents the results of the derivatographic analysis of the alloy hydride.

Figure 5 shows that adsorbed water extraction takes place at initial time. The temperature being raised up to 120 °C, the dehydrogenation process starts. The dehydrogenation process has several stages and it ends at temperatures of 350-400 °C. It means that the NdH_3 dehydrogenation process proceeds according to the following probable scheme:

$$NdH_3 \rightarrow NdH_2 \rightarrow NdH_{2-x} \rightarrow Nd.$$

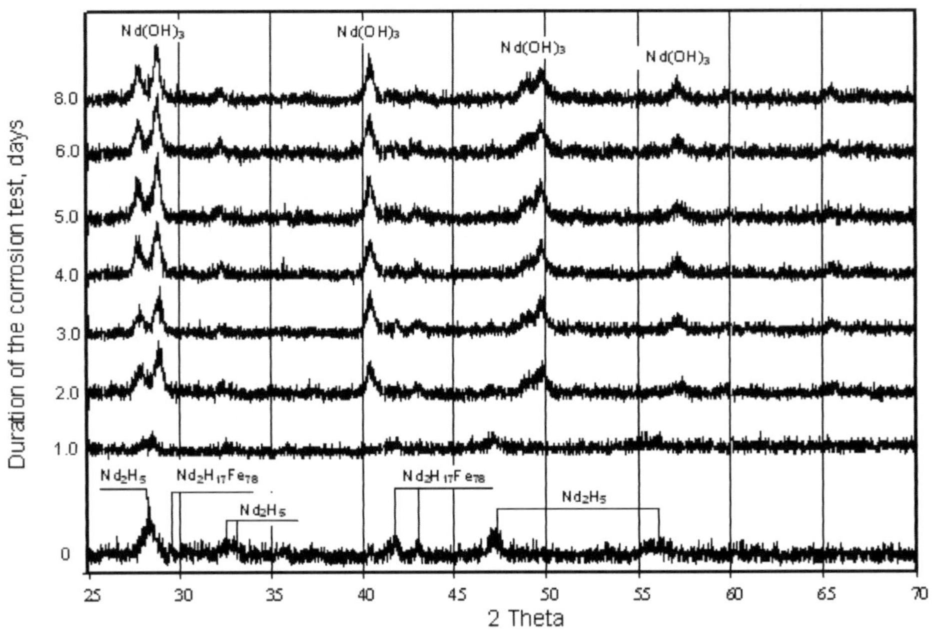

Figure 4. Phase composition of corrosion products of Nd-Fe hydrides (according to the X-ray fluorescence analysis).

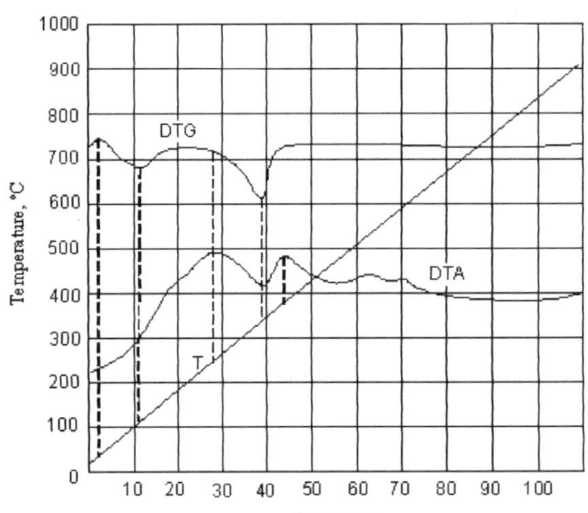

Figure 5. Derivatogram of decomposition of the hydrogenation product of the high-neodymium alloy Nd-Fe.

The investigated sample being put in inert atmosphere, oxidation of the dehydrogenated product did not take place.

In the course of the research the physical properties of the sample have also undergone changes. Thus, the hydride which initially looked like black plates having different shapes turned into a grey powder (fig. 6) having high specific surface (about 15 m^2/g).

The neodymium hydroxide production process is accompanied by increase in molar volume. Owing to this the volume of the charge increased visually one and a half or twice as much.

To investigate hydride storability and to detect optimal hydride storage conditions there have been conducted tests in flow air atmosphere, isolated air atmosphere, inert atmosphere and hydrogen atmosphere. The results of investigations on neodymium hydride in flow air atmosphere are presented in figure 3 (curve 2).

The diagram (2) presented in figure 3 shows that corrosion takes place in flow air atmosphere at room temperature and humidity fluctuations from 70% to 85% during the two-day initial period. In the sequel the corrosion rate goes down sharply, practically to zero. The mass increase of samples is 1.0-1.5% wt. In isolated air atmosphere (a covered plastic container) and inert helium and hydrogen atmospheres the corrosion process was not observed.

a - source alloy; b - after hydrogenation; c - after damp atmosphere storage

Figure 6. Exterior view of the products of hydrogenation and corrosion of Nd-Fe alloy.

4. Conclusions

The temperature effect on the alloy hydrogenation process has been studied. It has been determined that this process is limited by hydrogen diffusion to the reaction surface, apparent activation energy of the process in the temperature range 25-300 °C is (7.0±0.5)kJ/ mol.

The research into corrosion stability of powders of Nd-Fe and Dy-Fe hydrides under different conditions shows the following:

- being in dry atmosphere of air, helium and hydrogen the materials are totally stable;
- with the atmosphere humidity being approximate to 100%, the metal hydrides undergo complete amorphization in 24 hours and turn into hydroxides in 48 hours.

Thus, storage and usage of powders of nano-disperse alloys of rare earth metals should be done in dry air or inert atmosphere.

Acknowledgments

This work was supported by the Federal Target Program "Research and development on priority directions of scientific-technological complex of Russia for 2014-2020" (RFMEFI57814X0002).

References

[1] Buinovskiy A, Sofronov V, Kartashov E, Kalaev M 2016 Hydrogenation of Nd-Fe Alloys under Conditions of Different Pressure and Hydrogen Concentration Key Engineering Materials **683** 44-52

[2] Harris R, Williams A J 1999 Rare Earth Magnets (Materials World vol 7) 478-481

[3] Shatalov V V, Melnickov S A, Nickonov V I, Parshin A P 2007 Development of safe practice of hydrogen processing of different functional materials Functional metal materials (Mining Information Analysis bulletin issue 1) 232-244

[4] Karnaukhov A P 1999 Adsorption. Texture of disperse and porous materials (Novosibirsk: Science. Siberian Enterprise of Russian Academy of Sciences) 470

[5] Barre P Kinetics of heterogeneous processes (Moscow: World) 400

Production of Nd-Fe-B magnets by the hydride technique

Y N Makaseev, A S Buinovskiy, E Y Kartashov, V I Sachkov, V L Sofronov

National Research Nuclear University «MEPhI», 115409, Russia, Moscow, Kashirskoe sh. 31

VLSofronov@mephi.ru

Annotation. Rare-earth permanent Nd-Fe-B magnets are the most promising ones because they have the highest magnetic and operating characteristics. These magnets are produced from alloys having the respective chemical composition principally by the powder metallurgy technology. The paper presents the results of study of the manufacturing process of high-quality permanent magnets by the method of solid-phase alloying of magnetic alloys obtained by means of the out-of-furnace fluoride technique. In the present work, metallic foundry alloys and alloy hydrides of Nd-Fe, Dy-Fe and Tb-Fe systems are considered as alloying additions.

1. Introduction

One of the ways to improve the quality of magnetic materials is the technique of solid-phase alloying (SPA) of magnetic materials. This technique is based on intergrinding of base Nd-Fe-B magnetic alloys and alloying additions. SPA makes it possible to correct both the chemical composition and the structure of base magnetic alloys.

Rare-earth foundry alloys are considered to be one of the most promising materials for alloying additions. However, these materials have serious disadvantages – high strength and ductility, making it practically impossible for them to be used during mechanical intergrinding with the brittle Nd-Fe-B alloy.

The promising method of grinding such materials is the hydrogenation-dehydrogenation technique [1, 2].

After hydrogenation the material is easily destructed. The main advantages of this method, which define its wide application for magnetic materials production, are high efficiency and simplicity of the process, the possibility of obtaining powder materials of stable quality, ease of grinding and short reaction time.

The paper considers the process of solid-phase alloying (SPA) of Nd-Fe-B magnetic alloys by metallic foundry alloys and alloy hydrides having the following composition: 75%Nd-Fe, 42Dy-Fe and 70Tb-Fe. Alloy hydrides were produced by hydrogenation of corresponding foundry alloys by means of gaseous hydrogen, using the technique mentioned in [3].

Alloys and foundry alloys, used for the research, were obtained by the technique of out-of-furnace calcium-thermic reduction smelting of metal fluorides [4].

The schematic diagram of production of Nd-Fe-B permanent magnets by the solid-phase alloying technique is presented in Fig. 1. We consider these stages in detail.

2. Reduction

To obtain alloys and foundry alloys we used the calcium-thermic technique. This technique has no expensive stages of production and refinement of rare-earth metals (REM). REM fluorides, transition metal (Fe, Co, etc.) fluorides and alloying elements were reduced. It is an out-of-furnace process, i.e. the quantity of heat, released during the reduction reaction of REM compounds and transition metals, is enough to transform all reaction products into the molten state and to form compact metal ingots. Due to the high exothermicity of the reduction process of transition metal fluorides, it is possible to introduce a part of elements into the charge mixture in the form of metal powder, and boron -in the form of ferroboron. REM fluorides, used for this method, were obtained either by precipitation from solutions by means of the hydrofluoric acid and subsequent precipitate dehumidification and baking, or, like transition metal fluorides, by metal oxide processing with gaseous fluorine.

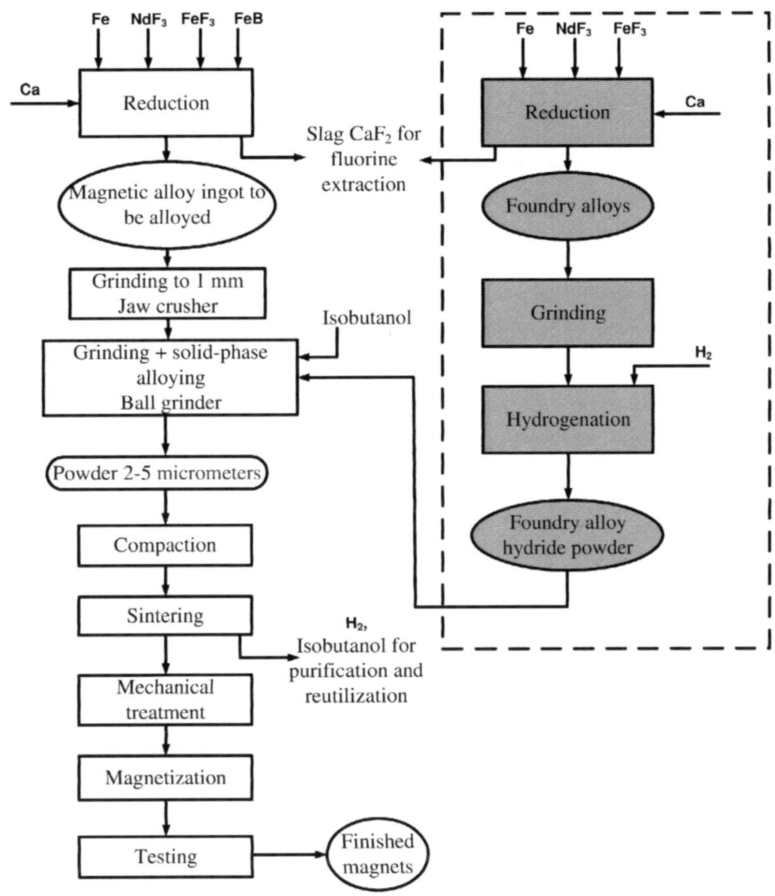

Figure 1. The schematic diagram of Nd-Fe-B magnets production using solid-phase alloying.

3. Grinding of Nd-Fe-B alloys

Ingots grinding was performed in two steps: crushing and milling. Preliminary crushing to pieces ranging in size from 20.0 to 30.0 mm was done by the press; small fraction fines were periodically separated; crushing to pieces ranging in size from 1.0 to 2.0 mm was continued in jaw crushers. To prevent powder oxidation, milling was performed by a vibrating mill UB in a liquid medium or inert atmosphere at the ratio Solid : Liquid = 1:1.7 (by weight). Ethyl and isopropyl alcohols served as inert liquid. Used inert liquids were rectified for reutilization.

The duration of grinding was chosen experimentally for each type of material being ground and the mill, so as not less than 90.0 % of weight of ground powder had particle size ranging from 2.0 to 5.0 micrometers.

4. Centrifugation and drying
The pulp from the vibrating mill was separated by a centrifuge. The liquid phase (alcohol) was sent to regeneration, wet powder was sent to vacuum drying and pressing.

5. Compaction in the magnetic field
The pressing operation was applied in the transverse magnetic field of 900 kA/m (11.2 kOe). The direction of the pressing force was orthogonally related with the direction of the texturing magnetic field.

Moulding blanks was performed on presses equipped with electromagnets. After the die, made of diamagnetic alloy, had been filled with alloy powder, it was put between the poles of the electromagnet. Imposition of vibration on the die during powder texturing in the constant magnetic field or substitution of the constant magnetic field by the pulsed magnetic field promoted more qualitative alignment of particles in the pressed powder and improved magnetic properties of future magnets.

6. Sintering blanks
This operation is necessary for the final texture consolidation, increase in density and mechanical strength of magnet blanks. To obtain sufficient density and strength of the sintered product, the following main parameters are of great importance: temperature and duration of sintering, gaseous medium, size of powder being moulded and degree of compaction during pressing and sintering.

Pressed blanks were packaged in boats, made of alloyed steel 12X18H10T, for loading into vacuum furnaces with tungsten heaters. They were sintered at temperature of 1060-1200 °C under residual pressure of about 0.1 Pa during 90-120 minutes.

7. Mechanical treatment
To obtain products of desired sizes, sintered and thermal treated (or annealed as appropriate) magnet blanks were cut by vulcanite wheels. Then they were ground by surface-grinding and circular grinding machines with corundum wheels having bakelite bond, with lubrication-cooling fluids being used. The cutting process was conducted very carefully to minimize overheating, oxidation and edge chipping.

8. Magnetizing and control
After the manufacturing was completed, the blanks required pulsed magnetization in the homogeneous magnetic field with intensity of 20-30 kOe. Finished magnets were 100% tested, with induction and magnetic linkage being measured.

Before hydrogenation, foundry alloys were treated the same way as alloys. Hydrogenation was performed by gaseous hydrogen under residual pressure of 100-1000 Pa at temperature of 50-100 °C. To obtain the desired structure of magnetic materials, foundry alloy hydride powder was sent to the stage of alloy grinding; then the mixture obtained was moulded and sintered at temperature of 1060-1200 °C. At this temperature, complete dehydrogenation takes place and hydrogen is eliminated from enclosed volume. Nd, Dy and Tb, obtained by dehydrogenation, complement the actual composition and they are involved in the recombination of $P3_2Fe_{14}B$ phase. It results in producing blanks of permanent magnets which have desired composition and structure.

9. Solid-phase alloying (SPA) of base magnetic alloys by metallic foundry alloys
The experiments were conducted using the technique described above. The only difference was the following: base magnetic alloys and foundry alloys, obtained by calcium-thermic reduction smelting, were subjected to two-step mechanical trituration. Then these powders were mixed for SPA; their

intergrinding continued during 30-40 minutes. It has been found experimentally that optimum amount of alloy additives for SPA is (5.0±1.0)%. Hysteresis characteristics for the second quarter (demagnetization curves) are presented in Fig. 2.

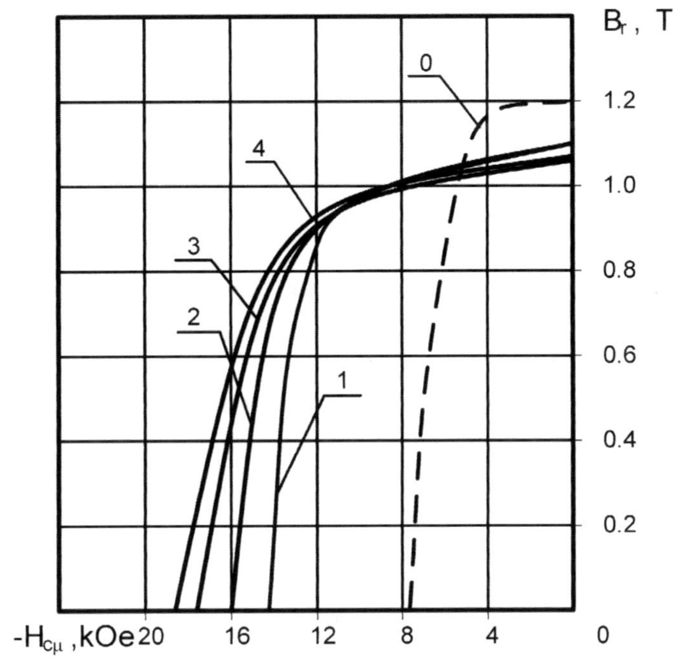

0 – B_r = 1.20 T, $H_{c\mu}$ = 7.5 kOe; 1 – B_r = 1.10 T, $H_{c\mu}$ = 14.0 kOe; 2 – B_r = 1.08 T, $H_{c\mu}$ = 15.9 kOe; 3 – B_r = 1.05 T, $H_{c\mu}$ = 17.6 kOe; 4 – B_r = 1.05 T, $H_{c\mu}$ = 18.5 kOe

Figure 2. Demagnetization curves of magnets obtained by SPA with metallic foundry alloys.

Magnet "0" was made of the base magnetic alloy 32Nd-Fe-1.1B; magnets "1-4" were made by the solid-phase alloying with foundry alloys 75%Nd-Fe, 42Dy-Fe, 40Tb-Fe and 70Tb-Fe, respectively.

The diagrams presented in Fig. 2 show that addition of metallic foundry alloy during the stage of fine grinding considerably increases the coercive force of magnets from 1.5 to 3 times. The value of residual magnetic induction either remains at the same level, or decreases insignificantly.

Most likely, it occurs owing to partial oxidation of the chemically active foundry alloy, its undermilling, causing it to fulfill a role of interphase ballast, rather than alloying material for the magnetic phase.

10. Solid-phase alloying (SPA) of base magnetic alloys by hydrides of foundry alloys

The research of the effect of solid-phase alloying on properties of magnets began with calibration of the crushing machine. Figure 3 presents the dependence of residual magnetic induction (B_r) and the magnet coercive force ($H_{c\mu}$) on the duration of intergrinding of the alloy and the alloying addition in the form of 75Nd-Fe and in the amount of 5% by weight.

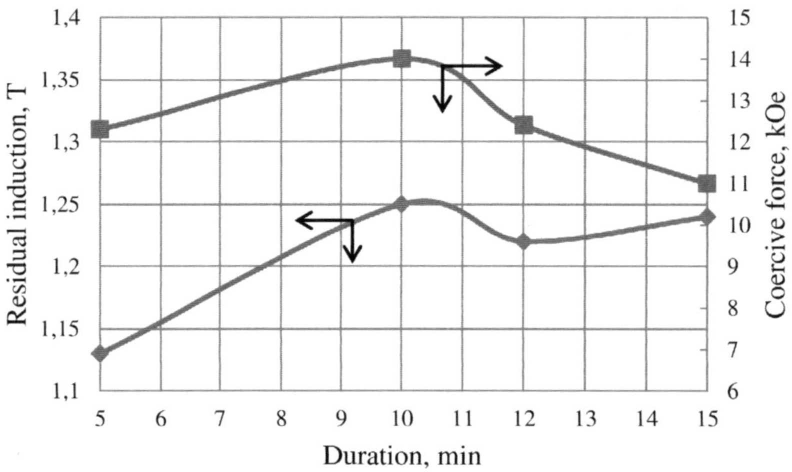

Figure 3. Dependence of coercive force and residual magnetic induction on duration of grinding.

The diagrams presented in figure 3 show that the optimal duration of intergrinding of the base magnetic alloy and the alloying addition in the form of hydrides by a vibrating mill UB is 10 minutes.

Hysteresis characteristics of the magnets, obtained by solid-phase alloying of base alloys by means of hydrides of foundry alloys with the composition 75%Nd-Fe, 42Dy-Fe, 40Tb-Fe and 70Tb-Fe for the second quarter (demagnetization curves) are presented in Fig. 4.

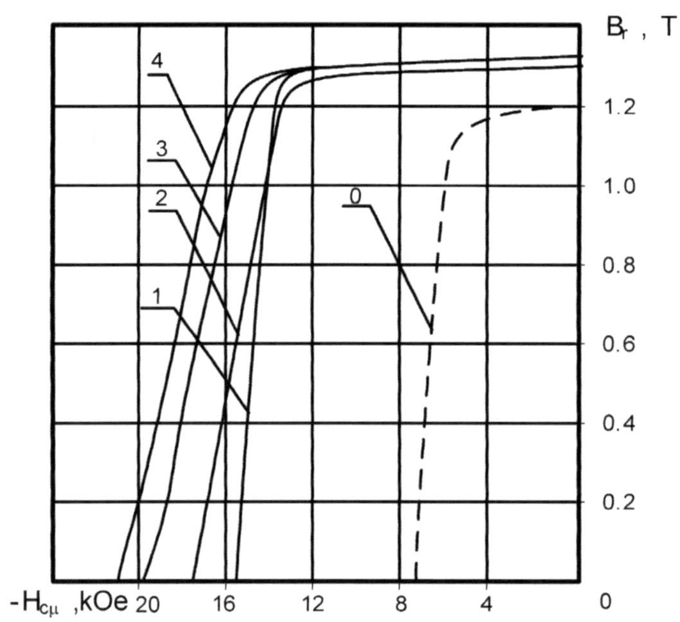

0 – B_r = 1.20 T, $H_{c\mu}$ = 7.5 kOe; 1 – B_r = 1.25 T, $H_{c\mu}$ = 15.5 kOe; 2 – B_r = 1.32 T, $H_{c\mu}$ = 17.5 kOe; 3 – B_r = 1.32 T, $H_{c\mu}$ = 19.2 kOe; 4 – B_r = 1.32 T, $H_{c\mu}$ = 21.0 kOe

Figure 4. Demagnetization curves of magnets obtained by SPA by mean of hydrides of foundry alloys.

The diagrams presented in Fig. 4 show that during solid-phase alloying by hydride powders, one can observe a considerable increase in both the coercive force of magnets (up to 3-5 times) and their residual induction (about 5−15 %). Dy and Tb are the most efficient materials which allow increasing the coercive force of Nd-Fe-B permanent magnets.

Thus, in the course of the experimental research of production of magnets using solid-phase alloying of base magnetic Nd-Fe-B alloys by powders of REM-containing addition alloys obtained with the help of mechanical and hydride comminution, it has been ascertained:

- the process of solid-phase alloying by hydride powders is a quite effective technique of improving properties of Nd-Fe-B magnets;
- the use of mechanically comminuted metallic powders of foundry alloys for solid-phase alloying mainly results in increase of the coercive force of alloyed magnets; at the same time, the residual magnetic induction either remains the same or decreases insignificantly.

It is important that the process of work with powders of foundry alloys hydrides is simpler than that with metallic powders of foundry alloys, because the latter are highly active, pyrophoric and must be stored and used in the inert atmosphere only. Hydride powders do not have these disadvantages, and they can be stored in dry hermetically sealed containers.

11. Conclusion

There has been studied the production process of Nd-Fe-B magnets using solid-phase alloying by REM-containing foundry alloys in the form of metallic powders and their hydrides. The process of solid-phase alloying by hydride powders is the most effective technique of improving properties of Nd-Fe-B magnets.

Acknowledgments

This work was supported by the Federal Target Program "Research and development on priority directions of scientific-technological complex of Russia for 2014-2020" (RFMEFI57814X0002).

References

[1] Dolukhanyan S K, Aleksanyan A G, Shekhtman V Sh 2010 Synthesis of Transition Metal Hydrides and a New Process for Production of Refractory Metal Alloys An Autoreview *International J. of Self-Propagating High-Temperature Synthesis* **19** 2 85–93

[2] Shatalov V V, Melnickov S A, Nickonov V I, Parshin A P 2007 Design of a safe technique of hydrogen processing of different functional materials *Mining information analysis newsletter* **12** 232-244

[3] Buinovskiy A, Sofronov V, Kartashov E, Kalaev M 2016 Hydrogenation of Nd-Fe Alloys under Conditions of Different Pressure and Hydrogen Concentration *Key Engineering Materials* **683** 44-52

[4] Bujnovskij A S, Sachkov V I, Sofronov V L, Anufrieva A V 2015 Basic stages of magnet production by fluoride technology *Advanced Materials Research* **1085** 209-213

IOP Conf. Series: Materials Science and Engineering **168** (2017) 012042 doi:10.1088/1757-899X/168/1/012042

The Mathematical Model of Image, Generated by Scanning Digital Radiography System

V A Udod[1,2], S P Osipov[1] and Wang Yanzhao[1]

[1]National research Tomsk polytechnic university, Lenina str., 30, Tomsk, 634050, Russia
[2]National research Tomsk state university, Lenina str., 36, Tomsk, 634050, Russia

E-mail: pr. udod@mail.ru

Abstract. The mathematical model of image, generated by scanning digital radiography system is present. This model takes into account the X-ray energy spectrum transformation of the test object and a noise due to the quantum nature of radiation. The calculation results confirm the importance of fluctuations of the absorbed energy of the registered photon for the small size of the scintillation detectors.

1. Introduction

Digital radiography in its various implementations continues to be one of the most popular and developing NDT methods [1–3]. The scanning systems of digital radiography (SSDR) are widely used in industry, medicine and customs control. Quality of primary digital images (PDI) generated by these systems is determined by a significant number of parameters [4–6], characterizing the X-ray source and multi-channel X-ray detector, the test object (TO), the scanning geometry, etc. Designing SSDR is impossible without the development of an adequate PDI mathematical model describing the main characteristics of the image with the system's parameters. The PDI mathematical models available in the articles [7–9] does not fully take into account all the features of SSDR, therefore, the model need to be further improved.

2. Assumptions and constraints

To construct a mathematical model of PDI formed by a multi-scanning system of digital radiography, in analogy with [7] we assume:

– carried out a uniform continuous movement along the Ox axis of TO, and the system of source – line of detectors is fixed;

– measuring channels (combinations: a radiation detector – temporal integrator – analog-to-digital converter) are identical;

– line of detectors is formed in the direction of the Oy axis (perpendicular to the direction of the TO movement) and, moreover, symmetrically about the beam axis;

– the radiation source and the line of detectors are collimated by slit collimators, so that scattered by the TO radiation contribution to the registration results is negligible;

– the beam axis coincides with the Oz axis;

– the radiation source has azimuthally symmetrical angular distribution;

– the transverse dimensions of the apertures and detectors of the source focal spot are much less than the focal distance (distance from the source to the detectors along the line of Oz axis);

– the radiation registration is carried out in an analog (mean current) mode;

– the quantization step of analog-digital converters is small compared to fluctuations in the radiation flux.

3. The Mathematical Model of Image, Generated by Scanning System of Digital Radiography

The primary digital image, generated by SSDR, is a matrix **B**, consists of M lines and K columns. The number of columns K is equal to the number of detectors (measurement channels) in the line of detectors. In view of the assumptions and limitations the matrix **B** is equal to the sum of two matrices $\overline{\mathbf{B}}$ and **N**

$$\mathbf{B} = \overline{\mathbf{B}} + \mathbf{N}. \tag{1}$$

The elements of matrix $\overline{\mathbf{B}}$ are equal to the corresponding mean values of the radiometric signals, and the elements of matrix **N** are characterized an image noise.

The expression for elements of matrix $\overline{\mathbf{B}}$ has a form

$$\overline{B}_{mk} = A_{mk}\lambda_{mk}\int_{-\infty}^{\infty}\varphi(t)dt, \tag{2}$$

where A_{mk} is a mean value of the electrical pulse amplitudes (electrical charges, coulomb) from output of k-th detector from the line of detectors for m-th image line; λ_{mk} is density of electrical pulses from output k-th detector at time t_m, 1/sec; $\varphi(t)$ is a temporal integrator impulse response.

For the system of ideal integrators with reset the impulse response $\varphi(t)$ is described by a step function

$$\varphi(t) = \begin{cases} 1, & 0 \le t \le T; \\ 0, & \text{otherwise}, \end{cases} \tag{3}$$

where T is the time constant of the integrator (time radiation measurement).

The values A_{mk} are connected with mean value of the absorbed energy by k-th detector of the registered quantum of radiation $\overline{E_{mk}^{ab}}$, MeV by expression

$$A_{mk} = \gamma_c \overline{E_{mk}^{ab}}, \tag{4}$$

here γ_c is the conversion factor of absorbed X-ray energy into an electric charge, coulomb/MeV.

The dependence $\overline{E_{mk}^{ab}}$ from the TO parameters, X-ray source with the maximal energy E_0 and k-th detector has form

$$\overline{E_{mk}^{ab}} = \frac{\int_0^{E_0} P(E)e^{-\mu(E)H_{mk}} E_{ab\,k}(E)\varepsilon(E)dE}{\int_0^{E_0} P(E)e^{-\mu(E)H_{mk}}\varepsilon(E)dE}, \tag{5}$$

where $P(E)$ is the energy spectrum of the radiation source by the number of photons, 1/(MeV·sec); $\int_0^{E_0} P(E)dE$ is the total output of photons from all sources per unit time in the solid angle 4π; $\mu(E)$ is the linear attenuation coefficient of photons with energy E for the TO material; H_{mk} is the TO beam size along the line connecting the transmitter center and the center of the k-th detector at time t_m (using ideal integrators with reset $t_m = mT$); $E_{ab\,k}$ is the mean value of the absorbed energy by k-th detector for one registered quantum with the energy E, MeV; $\varepsilon(E)$ is the detection efficiency of photons with energy E by a separate detector from the line within its aperture.

The intensity flow λ_{mk} is calculated by using the formula

$$\lambda_{\mathrm{mk}} = \frac{\psi(\theta_k)}{|\mathbf{r}_k|^2} S \int_0^{E_0} P(E) \mathrm{e}^{-\mu(E) H_{mk}} \varepsilon(E) dE, \tag{6}$$

where θ_k is an angle between Oz axis (axis of the radiation flow symmetry) and the vector $\mathbf{\Omega}_k$ is the direction of beam from X-ray source center to the center of k-th detector of line; θ is the angle between Oz axis and the direction of photon's flight; $\Psi(\theta)$ is the angular radiation distribution, which considered normalized on 1 quantum for solid angle 4π; $\mathbf{r}_k$ is the radius-vector of the aperture center of k-th detector; S is the square of aperture of a single detector from the line of detectors.

The elements $\mathrm{N_{mk}}$ of matrix $\mathbf{N}$ are characterized a noise due to the quantum nature of radiation. Its mean value, covariance and variance are equal correspondingly:

$$\overline{N_{mk}} = 0;\ \overline{N_{mk} N_{ij}} = \begin{cases} 0, & j \neq k, \\ \overline{A_{mk}^2} \lambda_{mk} \int_{-\infty}^{\infty} \varphi(t)\varphi(t + |m-i|\Delta t) dt, & j = k \end{cases};\ \sigma^2[N_{mk}] = \overline{A_{mk}^2} \lambda_{mk} \int_{-\infty}^{\infty} \varphi^2(t) dt. \tag{7}$$

The mean value of square of the electrical pulse amplitudes from output of k-th detector $\overline{A_{mk}^2}$ is connected with the mean value of square absorbed energy by k-th detector for the single registered quantum of radiation $\overline{E_{mk}^{ab\,2}}$, MeV^2 by expression

$$\overline{A_{mk}^2} = \gamma_c^2 \overline{E_{mk}^{ab\,2}}. \tag{8}$$

The value $\overline{E_{mk}^{ab\,2}}$ is calculated similar to the expression (5)

$$\overline{E_{mk}^{ab\,2}} = \frac{\int_0^{E_0} P(E) \mathrm{e}^{-\mu(E) H_{mk}} \overline{E_{ab}^2}(E) \varepsilon(E) dE}{\int_0^{E_0} P(E) \mathrm{e}^{-\mu(E) H_{mk}} \varepsilon(E) dE}, \tag{9}$$

where $\overline{E_{ab}^2}(E)$ is the mean square of the absorbed energy by separate detector for one registered quantum with energy E, MeV^2.

For completeness of the model description, we present a expressions to calcule the values $\overline{E_{ab}}(E)$ and $\overline{E_{ab}^2}(E)$. We consider the case use SSDR scintillation detectors containing scintillators (crystals) of cylindrical shape of radius r and thickness h (size in the direction of the incident X-rays) only. Then according to [10] we get:

$$\begin{aligned} \overline{E_{ab}}(E) &= \overline{E_{\min}(E)} + (\overline{E_{\max}(E,h)} - \overline{E_{\min}(E)})(1 - \mathrm{e}^{-g(E,h)r}), \\ \overline{E_{ab}^2}(E) &= \overline{E_{\min}^2(E)} + (\overline{E_{\max}^2(E,h)} - \overline{E_{\min}^2(E)})(1 - \mathrm{e}^{-g(E,h)r}). \end{aligned} \tag{10}$$

Here E is the quantum of energy, experienced the interaction with the scintillator material, MeV; $\overline{E_{\min}}$, $\overline{E_{\min}^2}$ and $\overline{E_{\max}}$, $\overline{E_{\max}^2}$ are minimal and maximal (corresponding to minimal and maximal radius of scintillator) mean values of absorbed and square of absorbed energy of registered quantum;

the coefficient g is characterized the rate of increase of the relevant function of the radius of the sensitive volume of the detector (scintillator).

The values $\overline{E_{\min}}$ и $\overline{E^2_{\min}}$ are depend from energy of registered quantum and the material of scintillator. The calculation expressions are [10]:

$$\begin{aligned}
\overline{E_{min}} &= E\,\mu_{foto}/\mu_{sc} + (E-1{,}02)\,\mu_{par}/\mu_{sc} + \pi r_0^2 N_e 0{,}511/\mu_{sc} \times \\
&\times\left[\frac{-20\alpha^4+102\alpha^3+186\alpha^2+102\alpha+18}{3\alpha(1+2\alpha)^3} - \frac{2\alpha+3-\alpha^2}{\alpha^2}\ln(1+2\alpha)\right]; \\
\overline{E^2_{min}} &= E^2\,\mu_{foto}/\mu_{sc} + (E-1{,}02)^2\,\mu_{par}/\mu_{sc} + \pi r_0^2 N_e 0{,}511^2/\mu_{sc} \times \\
&\times\left[\frac{-68\alpha^5+184\alpha^4+566\alpha^3+494\alpha^2+180\alpha+24}{3(1+2\alpha)^4} - \frac{2\alpha+4-\alpha^2}{\alpha}\ln(1+2\alpha)\right].
\end{aligned} \tag{11}$$

Here $r_0^2 = 7.94\cdot 10^{-26}\,\text{cm}^2$; N_e is number of electrons in 1 cm^3 of scintillator's material; μ_{sc}, μ_{foto}, μ_{par} are the full linear attenuation factor of radiation in the scintillator's material and the linear attenuation factors due to the photo-absorption effect and the effect of pair creation; $\alpha = E/0.511$, where energy E is presented in MeV.

The dependencies $\overline{E_{max}}$ and $\overline{E^2_{max}}$ from thickness h of scintillator are described with sufficient accuracy for practical purposes (up to 3%) by expressions [10]:

$$\begin{aligned}
\overline{E_{max}} &= \overline{E_{min}} + (1-\overline{E_{min}})\left(1-e^{-t\mu_{sc}(E)h}\right); \\
\overline{E^2_{max}} &= \overline{E^2_{min}} + (1-\overline{E^2_{min}})\left(1-e^{-t\mu_{sc}(E)h}\right).
\end{aligned} \tag{12}$$

Here t is a coefficient, depending from detector's material (see Table 1).

The formula to calculate the coefficient $g(E,h)$ has form [10]:

$$g(E,h) = b_1\mu_{sc}(E) + \frac{b_2}{\mu_{sc}(E)h}, \tag{13}$$

where b_1, b_2 are the approximation coefficients, depending from the material of the sensitive volume of detector (scintillator). The values of coefficients for some scintillators are given in Table 1.

Table 1. The approximation coefficients t, b_1, b_2 for different scintillators [10].

Coefficient	Plastic	CsI	$CdWO_4$
T	0.51	0.85	0.75
b_1	1.51	2.01	2.33
b_2	0.25	0.61	0.73

Note that the expressions (10) – (13) can be used to estimate $\overline{E^{ab}_{mk}}$, $\overline{E^{ab\,2}_{mk}}$ not only for the cylindrical scintillators, but also for scintillators with square cross-section. This conclusion is following from monotony increasing of $\overline{E^{ab}_{mk}}$, $\overline{E^{ab\,2}_{mk}}$ and η_{mk} from the radius. The values of analyzed variables take the intermediate position for the cylindrical scintillators with radiuses of inscribed and circumscribed in a square circles. The preliminary calculation results for scintillators with transversal sizes 1 mm or smaller, show the closeness of the top and bottom estimations of the analyzed variables. The deviations are less then 0.001 %.

The set of expressions (1) – (13) and Table 1 allow to simulate the primary digital radiographic image of TO with arbitrary cross-section.

4. Calculation Example

On base of formulas (10) – (13) and (5), (9) and with abovementioned note, as example, we calculated values $\overline{E^{ab}_{mk}}$, $\overline{E^{ab\,2}_{mk}}$ and coefficient η_{mk} of electrical pulse amplitude variation from output k-th detector of line, which defined by expression [10]

$$\eta_{mk} = \frac{\sqrt{\overline{A^2_{mk}}}}{\overline{A_{mk}}} = \frac{\sqrt{\overline{E^{ab\,2}_{mk}}}}{\overline{E^{ab}_{mk}}}.$$

With respect to the central detector of detector's line (k=K/2, θ_k=0) under the following assumptions:
– TO has shape of plate from carbon filled plastic with the density 1 g/cm^2 and with thickness 5 mm;
– scintillator from cadmium tungstate ($CdWO_4$) in shape of cuboid with cross-section 0.6 mm×0.6 mm and with thickness (size in direction of incident radiation) 3 mm;
– the energy spectrum of the radiation source by number of quanta is described by expression

$$P(E) = \frac{1}{E}\varphi(E) = \begin{cases} \frac{2}{EE_0}\left(1-\frac{E}{E_0}\right), \text{if } 0 \le E \le E_0; \\ 0, \quad \text{otherwise,} \end{cases},$$

where $\varphi(E)$ is the energy spectrum of radiation intensity, described by Kramers formula (in normalized form)

$$\varphi(E) = \begin{cases} \frac{2}{E_0}\left(1-\frac{E}{E_0}\right), \text{if } 0 \le E \le E_0; \\ 0, \quad \text{otherwise.} \end{cases}$$

The energy spectrum of radiation by number of quanta after TO is a continuous function in point E= 0, for any thickness of the attenuating barrier H≠0

$$\lim_{E \to 0+} P(E)e^{-\mu(E)H} = 0.$$

Under those assumptions and with using the data base [11] about the photon radiation attenuation, we get:

for E_0 = 200 KeV – $\overline{E_{ab}} = 0.333E_0 = 66.6\,\text{KeV}$, $\overline{E^2_{ab}} = 0.171E_0^2 = 6840\,\text{KeV}^2$, $\eta_0 = 1.241$;

for E_0 = 250 KeV – $\overline{E_{ab}} = 0.292E_0 = 73\,\text{KeV}$; $\overline{E^2_{ab}} = 0.138E_0^2 = 8625\,\text{KeV}^2$; $\eta_0 = 1.273$;

for E_0 = 300 KeV – $\overline{E_{ab}} = 0.258E_0 = 77.4\,\text{KeV}$; $\overline{E^2_{ab}} = 0.112E_0^2 = 10080\,\text{KeV}^2$; $\eta_0 = 1.298$.

From result analysis we can conclude about significance of the absorbed energy fluctuations of the registered photon for the small size of the scintillation detectors.

5. Conclusion

The paper shows the mathematical model of the image generated by the scanning system of the digital radiography. The model allows for the transformation of the X-ray energy spectrum by the test object, a noise due to the quantum nature of radiation, and the radiation registration with considering the leakage of secondary photons. The calculation results confirm the importance of fluctuations of the absorbed energy of the registered photon for the small size of the scintillation detectors. The model can be easily adapted to the panel X-ray detectors.

6. Acknowledgments

This work was financially supported by The Ministry of Education and Science of the Russian Federation in part of the science program (project 5.1485.2015) and by the National Reseach Tomsk Polytechnic University (project VIU INK 66 2014).

IOP Conf. Series: Materials Science and Engineering **168** (2017) 012042 doi:10.1088/1757-899X/168/1/012042

References

[1] Rusca S 2012 Digital radiography: Characterizing the CR systems for industrial applications *Rivista Italiana della Saldatura* **64** 641–9
[2] Harara W 2010 Low cost digital radiography system for weld joints and castings testing *Russian Journal of Nondestructive Testing* **46** 618–25
[3] Opolonin O D and Ryzhikov V D 2013 Increasing informativity of digital radiographic systems *Functional Materials* **20** 528–33
[4] Kalivas N, Costaridou L, Kandarakis I, Cavouras D, Nomicos C D and Panayiotakis G 1999 Effect of intrinsic-gain fluctuations on quantum noise of phosphor materials used in medical X-ray imaging *Applied Physics A* **69** 337–41
[5] Yaffe M J and Rowlands J A 1997 X-ray detectors for digital radiography *Physics in Medicine and Biology* **42** 1–39
[6] Wang Q, Zhu Y and Li H 2015 Imaging model for the scintillator and its application to digital radiography image enhancement *Optics Express* **23** 33753–76
[7] Lebedev M B, Sidulenko O A and Udod V A 2007 A mathematical model of multichannel continuously scanning systems for digital X-ray analysis *Russian Journal of Nondestructive Testing* **43** 401–6
[8] Ryzhikov V D, Opolonin O D, Galkin S M, Voronkin Y F, Lysetska O K and Kostioukevitch S A 2009 Development of Receiving-Detecting Circuit for Digital Radiographic Systems with Improved Spatial Resolution *Proceedings SPIE* **7450** 74500J/1–6.
[9] Lindgren E and Wirdelius H 2012 X-ray modeling of realistic synthetic radiographs of thin titanium welds *NDT & E International* **51** 111–9
[10] Zav'yalkin F M and Osipov S P 1985 Dependence of the mean value and fluctuations of the absorbed energy on the scintillator dimensions *At Energ.* **59** 281–3
[11] Gamma data for elements. http://www.ippe.ru/podr/abbn/libr/groupkon.php

Combined surface modification of commercial aluminum

Yu Ivanov [1,2,3,a)], I Lopatin[4,b)], Yu Akhmadeev[5,c)], E Petrikova[6,d)], A Teresov[7,e)], V Shugurov[8,f)], O Tolkachev[9,g)] and N Koval[10,h)]
[1]Leading Researcher, Institute of High Current Electronics SB RAS, Tomsk, Russia
[2]Leading Researcher, Laboratory of Low Temperature Plasmas, National Research Tomsk State University, Tomsk, Russia
[3]Professor, National Research Tomsk Polytechnic University, Tomsk, Russia
[4]Reserach Fellow, Institute of High Current Electronics SB RAS, Tomsk, Russia
[5]Reserach Fellow, Institute of High Current Electronics SB RAS, Tomsk, Russia
[6]Junior Researcher, Laboratory of Low Temperature Plasmas, National Research Tomsk State University, Tomsk, Russia
[7]Lead Electronic Engineer, Institute of High Current Electronics SB RAS, Tomsk, Russia
[8]Junior Researcher, Institute of High Current Electronics SB RAS, Tomsk, Russia
[9]Postgraduate, National Research Tomsk Polytechnic University, Tomsk, Russia
[10]Deputy Director, Institute of High Current Electronics SB RAS, Tomsk, Russia

E-mail: [a)] yufi55@mail.ru, [b)]lopatin@opee.hcei.tsc.ru, [c)]ahmadeev@opee.hcei.tsc.ru, [d)]petrikova@opee.hcei.tsc.ru, [e)]tad514@sibmail.com, [f)]shugurov@inbox.ru, [g)]ole.ts@mail.ru, [h)]koval@opee.hcei.tsc.ru

Abstract. The paper analyzes research data on the structure and properties of surface layers of commercially pure A7-grade aluminum subjected to treatment that combines deposition of a thin metal film, intense pulsed electron beam irradiation, and nitriding in low-pressure arc plasma. The analysis shows that the combined method of surface modification provides the formation of a multilayer structure with submicro- and nano-sized phases in the material through a depth of up to 40 μm, allowing a manifold increase in its surface microhardness and wear resistance (up to 4 and 9 times, respectively) compared to the material core. The main factors responsible for the high surface strength are the saturation of the aluminum lattice with nitrogen atoms and the formation of nano-sized particles of aluminum nitride and iron aluminides.

1. Introduction

The low hardness and the low wear resistance of aluminum limit the application of the material and its alloys in industry [1]. Generally, the service characteristics of materials are increased by hardening of their surface layers [2–4]. One of the widely used methods of surface hardening is thermochemical treatment [5]. Although the method has a series of advantages (possibilities to treat elements of any shape, large difference between surface and core properties, etc.), its use for treatment of aluminum suffers form serious shortcomings: the long process time; the presence of an oxide film on the surface of aluminum and its alloys, being a barrier to nitrogen supply to the surface; the formation of an

[1] To whom any correspondence should be addressed.

aluminum nitride film on the surface, which prevents the penetration of nitrogen deep into the material and complicates the supply of potential to it during ion nitriding.

Here we analyze the structure and properties of commercially pure aluminum subjected to treatment that combines deposition of a thin metal film, intense pulsed electron beam irradiation, and nitriding in low-pressure arc plasma.

2. Material and research technique

The material under study was commercially pure aluminum of grade A7 (aluminium1070: 0.16Fe, 0.15Si, 0.04Zn, 0.03Mn, 0.02Mg, 0.01Ti, 0.01Cu, and the rest is Al). The Al specimens were 4.5-mm thick plates sized to 15×15 mm. Their modification was as follows. First, a steel film 0.5 μm thick was deposited on the Al surface using a Kvinta setup [6]; the steel was 12Cr18Ni10Ti (AISI 321: 0.12C, 11Ni, 19Cr, 1Ti, 2Mn, 0.8Si, 0.3Cu, 0.02S, 0.035P, and the rest is Fe). Then, the film–substrate system was irradiated with an intense pulsed electron beam on a SOLO setup at 17 keV, 30 J/cm^2, 200 μs, and 0.3 Hz; the number of pulses was 10 and 20 [7]. Finally, the specimens were nitrided on a NNV-6.6-I1 setup equipped with a PINK plasma generator [8]. For nitriding, a negative bias with a frequency of 50 kHz was applied to the specimens; the bias amplitude and the pulse duty factor were varied. The nitriding temperature was measured with a chromel-alumel thermocouple fixed on the specimen holder and was 540 ^{0}C. The nitriding times was 8 h. The hardness and the elastic (Young's) modulus of the specimens were measured in mechanical tests using a PMT-3 device with a Vickers indenter at a load of 0.2 N and 0.5 N and a Shimadzu DUH-211S ultra micro hardness tester at an indenter load of 10–300 mN. The wear rate was determined in tribological tests on a CSEM S/N 07-142 high-temperature tribometer by estimating the wear track cross-sectional area with a Micro Measure 3D station. The defect structure was examined using a Philips SEM-515 scanning electron microscope (SEM) and a JEOL JEM-2100F transmission electron microscope (TEM). The phase composition and the lattice state of surface layers were analyzed using a Shimadzu XRD-7000s diffractometer.

3. Results and discussion

3.1. Nitriding of aluminium in low-pressure arc plasma

Our study shows that when nitrided, the Al surface layer assumes an island structure (Fig. 1, a). The lateral size of islands is up to 5 μm and the size of their substructure is 100–500 nm (Fig. 1, b). The average height of islands estimated by interference microscopy (MNP-1 device) is 1.75 μm.

According to X-ray diffraction analysis, the main phase (≈84 mass%) in the nitrided layer is AlN (space group $P6_3mc$ [9, 10]) with lattice parameters a = 0.31165 nm and c = 0.49879 nm; the size of coherent scattering regions is D = 21.9 nm; the microdistortion of the AlN lattice is □d/d = $3.5*10^{-3}$. The second phase is Al with a = 0.40487 nm, D = 73.6 nm, and □d/d = $0.37*10^{-3}$.

Electron diffraction analysis of thin foils shows that the AlN islands have a columnar structure with lateral sizes of columns of 0.4–0.5 μm (Fig. 2, a). The columns have a nanocrystalline substructure with a crystallite size of 15–25 nm (Fig. 2, b).

Indexing of the diffraction patterns (Fig. 2, c, d) reveals reflections that belong to the AlN lattice (space group $P6_3mc$, main phase) and Al lattice (space group Fm3m). In most cases, the reflections of the AlN and Al lattices are overlapped, and their confident identification by electron diffraction analysis is difficult. Comparing the data of X-ray and electron diffraction analyses, it can be noted that the sizes of coherent scattering regions of AlN agree well with the sizes of crystallites revealed by electron microscopy.

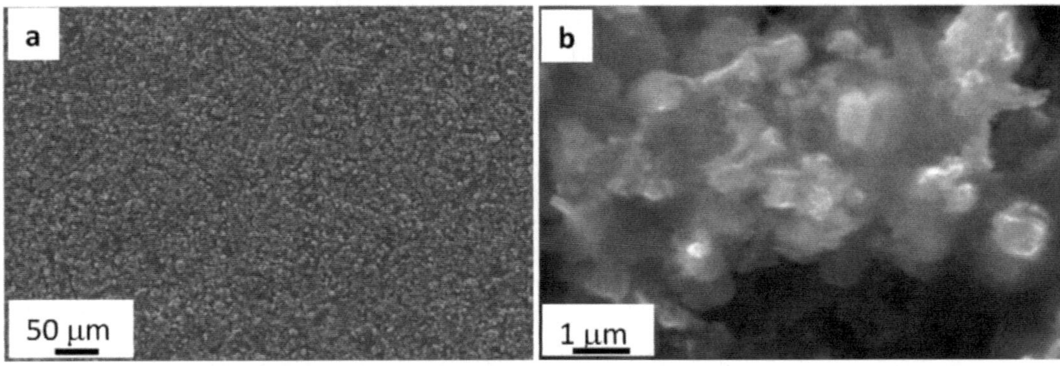

Figure 1. SEM images of the Al surface structure after nitriding in low-pressure arc plasma.

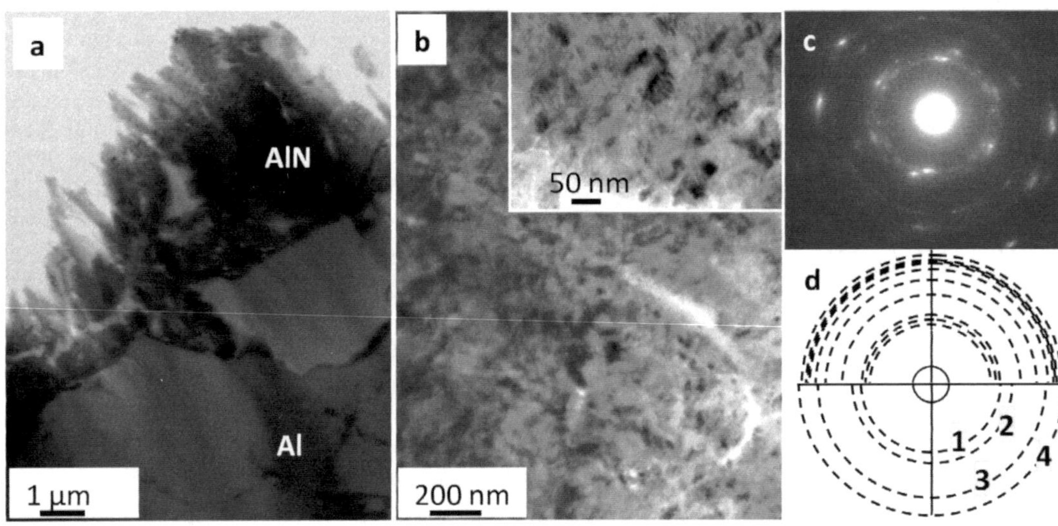

Figure 2. Structure of commercially pure aluminum nitrided in low-pressure arc plasma: a, b – bright fields; c – electron diffraction pattern; d – indexing (diffraction rings in the upper part correspond to AlN, and in the lower pat, to Al with reflections: *1* – [111], *2* – [200], *3* – [220], *4* – [113]).

The nitride layer is separated from the adjacent Al volume by a sharp boundary along which chains of pores with a size of 10–50 nm are observed (Fig. 3, white arrows). The Al volume adjacent to the nitride layer has a fragmented (subgrain) structure, suggesting the formation of thermal stresses in the nitrided material and their subsequent relaxation (Fig. 3, b). The fragment size varies from 1 μm to 2 μm. In the fragment volume, there is a substructure of chaotically distributed dislocations.

Thus, the study demonstrates that nitriding in low-pressure arc plasma provides the formation of a thin (2–4 μm) nanocrystalline surface layer with an island structure in commercially pure aluminum. The nitride layer is separated from the main material volume by an interlayer containing micropores.

3.2. Nitriding of aluminium in low-pressure arc plasma

To improve the nitriding efficiency, a 12Cr18Ni10Ti film was deposited on the surface of A7 aluminum, and then, the film–substrate system was nitrided in low-pressure arc plasma. Data of nanoindentation showed that when nitrided, the film–substrate system was hardened through a depth of no more than 1 μm. Therefore, before nitriding, the film–substrate system was irradiated with an intense electron beam, and the thus formed surface alloy was nitrided in low-pressure arc plasma.

Measurements of the hardness show that a hardened layer of thickness up 30 μm is formed in the system during nitriding; the microhardness of the nitrided surface layer is more than 4 times higher than that of the specimen volume. The modified specimens were subjected to tribological testing. It is shown that the combined method of surface modification increases the wear resistance of the material ≈9 times and decreases the friction coefficient ≈1.3 times.

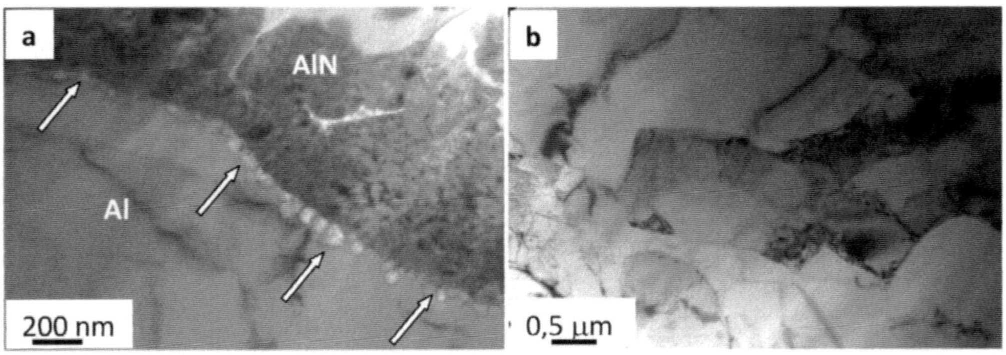

Figure 3. Structure of commercially pure aluminum nitrided in low-pressure arc plasma; the pores along the interface between the nitride layer and the main Al volume are shown by arrows.

After combined modification, the material reveals a multilayer structure consisting of a surface layer (Fig. 4, layer *1*), a transition layer (Fig. 4, layer *2*), and a heat-affected layer.

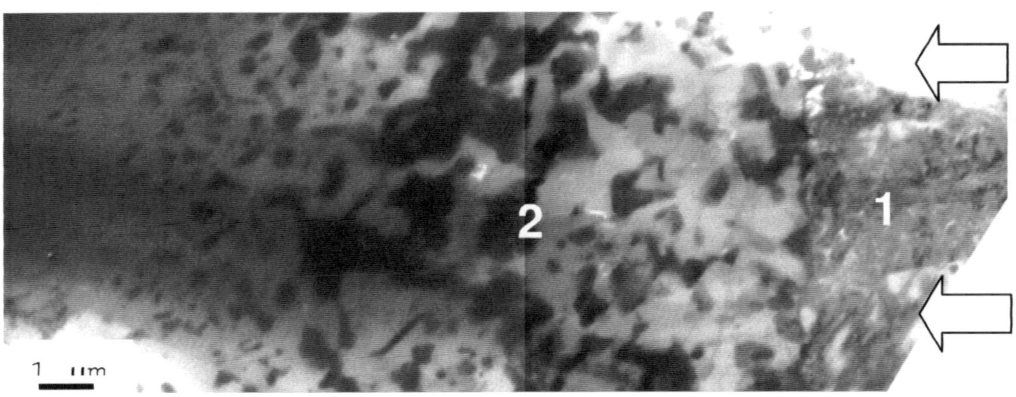

Figure 4. Structure of 12Cr18Ni10Ti film-coated A7 aluminum after electron beam treatment and subsequent nitriding in low-pressure arc plasma (the nitrided surface is marked by arrows): *1* – surface layer; *2* – transition layer.

The surface layer has an island structure similar to that presented in Fig. 1 and Fig. 2; the average height of islands is 3.45 μm. The islands represent aluminum nitride, as evidenced by transmission electron microscopy, scanning transmission electron microscopy, and X-ray diffraction analysis. There is also a small amount of particles of α-Fe-based solid solution.

The transition layer is up to 40 μm and represents aluminum hardened by second phase particles (Fig. 4, layer *2*). As shown by TEM and STEM, the inclusions are aluminum nitrides, aluminides, and α-Fe(Cr, Ni, Al). In the layer adjacent to the columnar structure, the size of second phase particles is 100–250 nm. With depth from the modified surface, the number of particles per unit area decreases and so does their size (Fig. 4).

The main volume of the material and its modified surface layer are separated by a heat-affected zone in which a band structure typical for deformed aluminum arises. Apparently, the strains in the material are induced by thermoelastic stresses that arise in its surface layer at the stage of intense electron beam irradiation [11, 12].

Reasoning from the mechanical properties, tribological characteristics, and structural phase state of the modified material, it can be concluded that the increase in its surface microhardness and wear resistance is due to several physical mechanisms of hardening: (1) saturation of the Al lattice with nitrogen atoms, (2) precipitation of submicro- and nano-sized second phase particles, including aluminum nitride particles, and (3) band substructure formation in its heat affected zone.

4. Conclusion

Thus, we analyzed the structure and properties of commercially pure A7 aluminum after combined surface modification that included electron beam irradiation of the material coated with a 12Cr18Ni10Ti steel film and its subsequent nitriding in low-pressure arc plasma. It is shown that the combined method of surface modification increases the wear resistance of the material more than 9 times and the microhardness more than 4 times while decreasing the friction coefficient □1.3 times. The increase in the strength and tribological properties of the material is due to several physical mechanisms of hardening: (1) saturation of the Al lattice with nitrogen atoms, (2) precipitation of submicro- and nano-sized second phase particles, including aluminum nitride particles, and (3) band substructure formation in the heat affected zone.

Acknowledgments
The work was supported by the Russian Foundation for Basic Research (project No. 16-58-00075 Bel_a).

5. References

[1] Belov N A 2010 *Phase composition of commercial and advanced aluminum alloys* (Moscow, MISiS)
[2] Berlin E B, Koval N N and Seidman L A 2012 *Plasma thermochemical surface treatment of steel parts* (Moscow: Tekhnosfera)
[3] Budilov V V, Koval N N, Kireev R M and Ramazanov K N 2013 *Integrated methods of treatment of structural and tool materials with glow and vacuum arc discharges* (Moscow: Mashinostroenie).
[4] Gribkov V A, Grigoriev F I, Kalin B A and Yakushin V L 2001 *Advanced radiation beam technologies for material treatment* (Moscow: Kruglyi God)
[5] Lakhtin Yu M and Arzamasov VN 1984 *Thermochemical treatment of metals* (Moscow: Metallurgia)
[6] Shugurov V V, Kalushevich A A, Koval N N, Denisov V V and Yakovlev V V 2012 *Izv. Vyssh. Uchebn. Zaved. Fiz.* **12/3** 118
[7] Ivanov Yu F and Koval N N 2007 (Structure and properties of advanced metal materials), A I Potekaev, Ed. (Tomsk: Izd. Nauch. Tekh. Lit.) 345
[8] Ivanov Yu F, Akhmadeev Yu H, Lopatin I V, Petrikova E A, Krysina O V and Koval N N 2015 *J. Phys.: Conf. Ser.* **652 0**12013
[9] Wriedt H A 1986 *Bulletin of Alloy Phase Diagrams* **7** 4 RIVERS 329
[10] Vol A E 1959 *Structure and properties of binary metal systems*. **1** (Moscow: Gos. Izd. Fiz. Mat. Lit.)
[11] Rotshtein V, Ivanov Yu and Markov A 2006 Materials surface processing by directed energy techniques Y. Pauleau, Ed. (London: Elsevier) chapter **6** 205.
[12] Laskovnev A P, Ivanov Yu F, Petrikova E A e.a. 2013 *Electron-ion plasma modification of the structure and properties of eutectic silumin* (Minsk: Belarus. Nauka)

IOP Conf. Series: Materials Science and Engineering **168** (2017) 012044 doi:10.1088/1757-899X/168/1/012044

Numerical simulation of discharge plasma generation and nitriding the metals and alloys

T V Koval[1], R A Manakov[2], Nguyen Bao Hung[3] and Tran My Kim An[4]

[1] Professor, National Research Tomsk Polytechnic University, 30, Lenin Ave., Tomsk, 634050, Russia.
[2] Student, National Research Tomsk Polytechnic University, 30, Lenin Ave., Tomsk, 634050, Russia.
[3,4] Post-graduate students, National Research Tomsk Polytechnic University, 30, Lenin Ave., Tomsk, 634050, Russia.

Email: tvkoval@mail.ru

Abstract. This research provides the numerical simulation of the plasma generation in a hollow cathode as well as the diffusion of nitrogen atoms into the metal in the low-pressure glow discharge plasma. The characteristics of the gas discharge were obtained and the relation of the basic technological parameters and the structural and phase state of the nitrided material were defined. Authors provided the comparison of calculations with the experimental results of titanium nitriding by low-pressure glow discharge plasma in a hollow cathode.

1. Introduction

The ion-plasma surface treatment in plasma gas discharge, particularly the nitriding treatment have been increasingly developed [1-9] in recent years. The vacuum plasma technology on the ground of low temperature glow discharge plasma is used for the surface modification of materials and bulk items [4-9], when the ion current density of ~ 1 mA/cm^2 over the treated surface as well as the discharge voltage of hundreds volt to be provided [8-11].

The target material for plasma treatment is immersed into the gas plasma where ions accelerated towards the target in the double electric layer between the target and the discharge column. A treating agent (gas atoms) is propagated in the target material due to the activated thermal diffusion which rate is dependent on the temperature of the treatment and the agent density distribution profile. The main parameters for the plasma treatment are gas mixture, temperature and time of the process, operating pressure, ion current density to the target surface and the ion energy. Most of the parameters are interdependent. Therefore, the proportion of components in gas mixture impacts the voltage, which triggers the discharge, and the ion energy as well. The voltage and the current density are bounded with the value of the allowed treatment temperature and the gas pressure. The nitriding in the range of 200-300 °C provides the solid nitride layers up to 500 μm thick [4, 5].

The multiple complex processes, which take place during the nitriding, make it difficult to reveal the common trends in structuring the modified layers and their characteristics [11-14]. Simulating the processes, which occur during the plasma generation and nitriding of materials, provides the efficient opportunity to identify the mechanism of the modified layer formation with predefined characteristics and their relation to the technological parameters such as gas type and pressure, discharge current and voltage, ion current density. It is fair to assume that the precise quantitative ratio of process parameters and the phase-structural condition of the material surface will help to improve the established nitriding techniques and develop new ways of modification based on the data available.

The current research provides the simulation of plasma generation with the non-self-sustained glow discharge in a hollow cathode at the low gas pressure as well as the diffusion penetration of the nitrogen atoms into the metal. The authors also made a comparison of the numerical simulation with the nitriding of titanium in the real experiment.

2. Simulation of gas discharge in a hollow cathode

As the experiments [8, 9] demonstrated, plasma was generated in a hollow cylindrical cathode (figure 1). The anode *1*, a set of two tubes with a total area of S_a, is located at the side face of the cathode, which hollow volume is $V_c = 2\times10^5$ cm^3, $S_a = 125$-500 cm^2. Plasma might be generated in the main self-sustained discharge mode and in the electron beam assisted discharge mode. In latter case, the electron beam is extracted from plasma of the auxiliary arc discharge *3*. Plasma is generated inside the hollow cathode, the cathode voltage drop U_c is about the discharge voltage U. The plasma ions accelerated in the cathode layer provide the ion-electron emission from the cathode surface. The electrons, oscillating in the hollow cathode, ionize the gas and provide the discharge self-sustainability.

The equation for the main discharge in non-self-sustained mode can be obtained from the equation of the high-energy electron balance

$$u = \frac{P}{P(\gamma+\delta)-1}, \tag{1}$$

which combines the dimensionless discharge voltage $u = e\,U_c/W$, gas pressure $P = (p/kT_e)\sigma_i L$ and the effective length of the hollow cathode $L = 4V/S_a$, where W – total energy loss of an electron during the gas ionizing; γ – the coefficient of the ion-electron emission; δ – the ratio of the external electron current involved in ionization to the discharge current; V – the cathode effective volume with targets; σ_i – the cross-section of the gas ionization with electrons.

To define the spatial distribution of the electron temperature and plasma concentration as well as the impact of geometry, the size and material of the parts immersed into the hollow cathode the numerical model was used that described the density of electrons (n_e) and their average energy (n_ε) as the temporal and spatial functions [9].

$$\frac{\partial n_{e,\varepsilon}}{\partial t} + \nabla\vec{\Gamma}_{e,\varepsilon} + \vec{E}\cdot\vec{\Gamma}_{e,\varepsilon} = R_{e,\varepsilon}, \quad \vec{\Gamma}_{e,\varepsilon} = -\left(\mu_{e,\varepsilon}\cdot\vec{E}\right)n_{e,\varepsilon} - D_{e,\varepsilon}\cdot\nabla n_{e,\varepsilon} \tag{2}$$

where $\vec{\Gamma}_{e,\varepsilon}$ – flows of electrons and energy, $\mu_{e,\varepsilon}$ – mobility, $\vec{E}$ – electrical field, $D_{e,\varepsilon}$ – diffusion coefficients, $R_{e,\varepsilon}$ – the ionization rate and energy loss/gain of electrons due to the non-elastic collisions. The plasma generation in the hollow cathode was simulated regardless of the collisionless layer of the positive charge, which separates the cathode sides and targets from the plasma column. On condition of the current continuity at the computational domain boundaries, the input parameters for the numerical model are the discharge characteristics obtained from the model (1).

The high voltage of the main discharge in the self-sustained mode is the disadvantage of the hollow cathode technique since it results in the intensive ion-beam etching of the treated surface. As is shown in the equation (1) and figure 2, the characteristic of the main discharge tends to the lower voltage and gas pressure, when the auxiliary discharge is present. Figure 2 presents the calculated (by 1) and the experimental [8, 9] relations between the discharge voltage and the gas (nitrogen) pressure at different effective lengths of the hollow cathode $L = 4V/S_a$ and a discharge current of 30 A. Reducing the effective length of the hollow cathode $L = 4V/S_a$ (by extending anode area S_a or reducing the cathode volume due to targets $V_c - V_d$) as well as the decline of the operation pressure of the main discharge results in the increase of discharge voltage. It occurs due to the deposition of the high-energy electrons to the anode. The predefined current of the self-sustained discharge is provided by the emission current (due to the coefficient $\gamma_{N_2} = 0.04\text{-}0.2$).

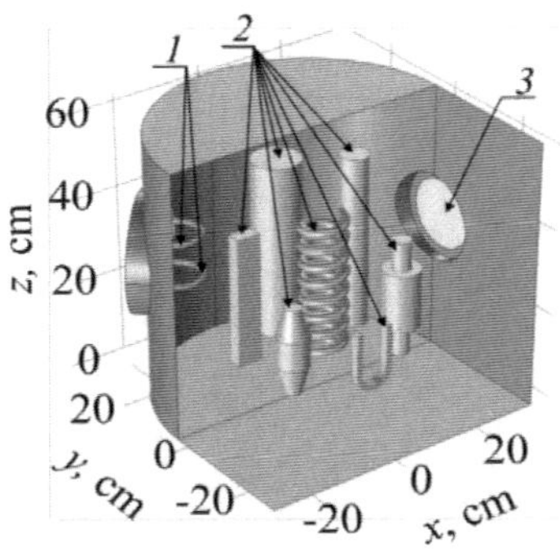

Figure 1. Hollow cathode model: *1* – anode; *2* – targets; *3* – auxiliary discharge window.

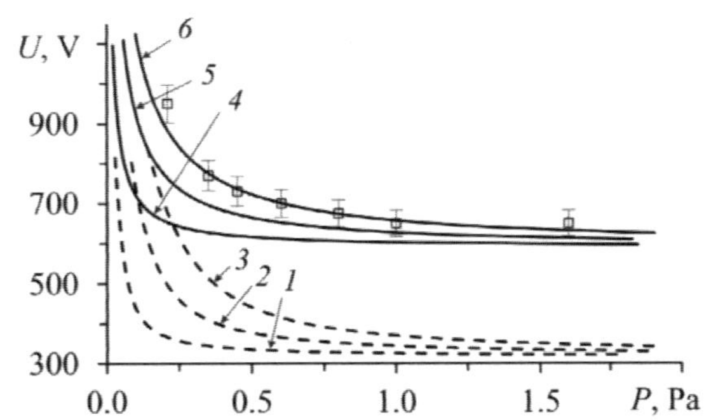

Figure 2. Dependence of discharge voltage on gas (nitrogen) pressure in non-self-sustained (*1,2,3*) ($\delta = 0.11$) and self-sustained discharge (*4,5,6*) modes; *1,4* – $L = 6.4\times10^3$cm; *2,5* – 4.0×10^3cm; *3,6* – 1.6×10^3cm; ⫱ – experiment; $I = 30$ A.

The numerical simulation revealed that the plasma concentration was about 10^{11} cm^{-3} (nitrogen) under the discharge voltage of 300-400 V and the gas pressure of 0.65 Pa. The plasma temperature was within the order of 1 eV.

The control of the ion current density and the main discharge voltage could be provided by the external injection current without regard to the surface area and target material. The dependence of the discharge voltage U on the volume of target specimen (figure 3) was altered inessentially in the self-sustained as well as in the non-self-sustained discharge under the nitrogen pressure of p_{N2}>0.65 Pa.

The dependence of the external injection current and the plasma potential on the number of treated targets N for $p_{N2} = 0.65$ Pa is shown in figure 4. The discharge voltage of 370 V and the ion current density was stabilized when increasing the number of targets (N = 1-8) by changing the external emission current from 9 to 14 A. The plasma potential slightly increased since $(V_c - V_d) << V_c$ (figure 4), the total target volume was $(0.25\text{-}2)\times10^4$ cm^3, the relative surface area of the target parts was $S_d/S_c = 0.071\text{-}0.57$.

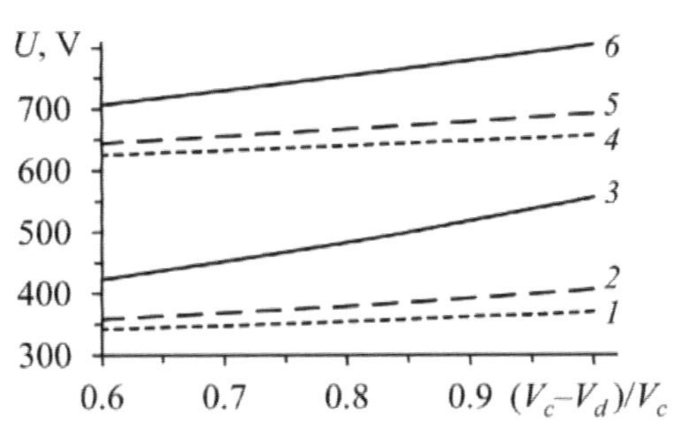

Figure 3. Discharge voltage in nitrogen vs. the relative volume in the non-self-sustained (*1,2,3*) and self-sustained (*4,5,6*) modes; *1,4* – p = 1; *2,5* – 0.65; *3,6* – 0.35 Pa; $\delta = 0.11$.

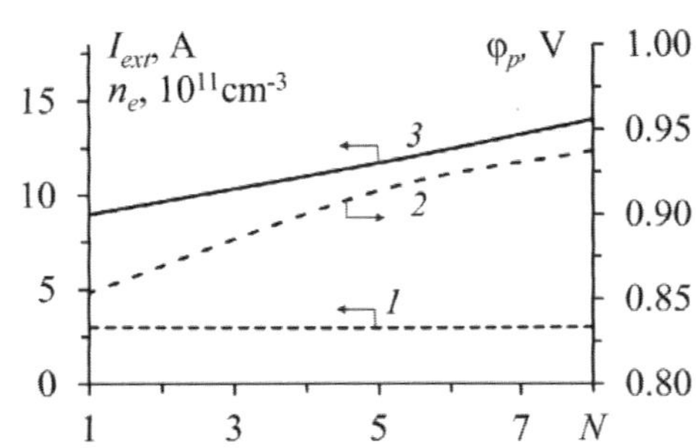

Figure 4. Plasma density (*1*), plasma potential (*2*) and external injection current (*3*), vs. the number of targets; working gas – nitrogen.

The shape of the targets and their positioning relatively to the cathode surface impact the flow density of the accelerated ions to the surface of the treated targets and the nitriding uniformity, respectively. The calculated lines of the plasma concentration level when I_{ext} = 14 A and U = 370 V in two cross sections of the hollow cathode are presented in figure 5. The small gradient in plasma concentration occurred around the emission window due to the external injection electrons. The calculations suggested that the change in the distance between the target frame and the cathode side from 5 mm to 2 mm resulted in the halving of the plasma concentration near the target's end face with respect to the face side.

During the treatment, the target was used as a cathode. A space charge layer formed between the plasma and the target, and the plasma ions accelerated in the electric field of the charge. The targets were heated due to the ion impact. The numerical solution of the heat conduction problem with regard to the water cooling of the cathode walls (figure 1) was performed for the Ti and Fe targets of ∅ = 5-13 cm and l = 20-40 cm in size.

The calculations demonstrated that using the active screen placed 2 cm apart from the cathode and around all targets led to the time reduction of the target heating up to 400 °C for 1 hour (figure 6). That provided the time difference in the nitriding as well as in the nitrogen penetration depth. The control of the target heating was performed by the external injection current and by auxiliary screens as well.

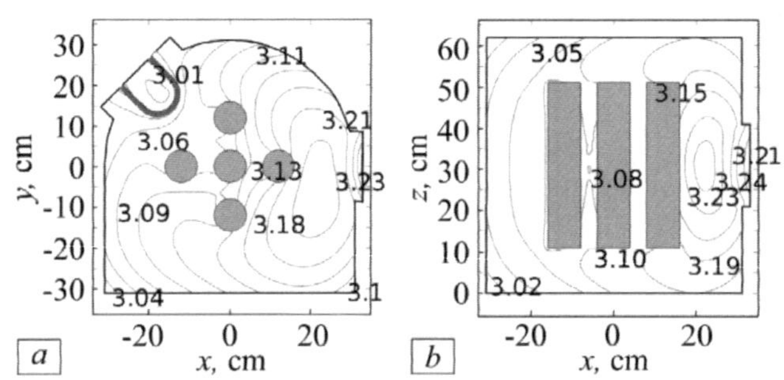

Figure 5. Plasma density distribution for nitrogen ($cm^{-3}\times10^{11}$) in the lateral (*a*) and longitudinal (*b*) sections of the hollow cathode with targets; non-self-sustained mode.

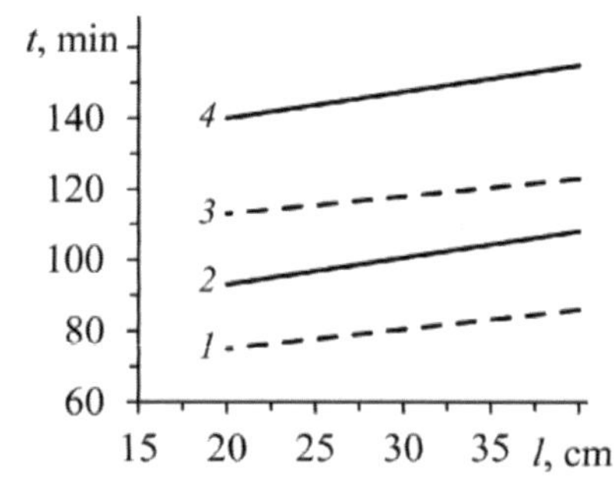

Figure 6. Heating time vs. detail length; *1,3* – ∅ = 10 cm; *2,4* – 13 cm; *1,2* – Ti; *3,4* – Fe.

The discharge condition (1) and the numerical simulation of plasma generation (2) make it possible to determine the technological parameters such as gas pressure, ion current density and discharge voltage required for the plasma treatment of the target parts.

3. Simulation of diffusion and kinetic processes

The experiment demonstrated that the increase of the nitrogen concentration correlates to the increase of the surface microhardness, and the thickening the nitrated layer provides the increase of durability [6, 7]. Currently, there is no universal model, which explains the process of the material nitriding by the discharge plasma treatment. One of the perspective approaches to clarify the nitriding process is as follows: the main part in the process belongs to atomic nitrogen, and the nitriding intensity in a glow discharge is defined by the amount of atomic nitrogen [11]. Considering that the increase of nitride layers is controlled by the diffusion of the nitrogen atoms, let us take a look at the following diffusion model:

$$\frac{\partial C}{\partial t} = \frac{\partial}{\partial x} D \frac{\partial C}{\partial x} . \tag{3}$$

where $C = C(x, t)$ is the nitrogen concentration when the ion nitriding of the metal, $D = D_0\exp(-E_a/RT)$ is the nitrogen diffusion coefficient dependent on the temperature T and concentration C, R is the gas constant, E_a is the energy of the diffusion activation. When simulating the diffusion of nitrogen to titanium, the initial condition $C(0, t) = 0$ and the third boundary condition were used:

$$-D\frac{\partial C(x=0,t)}{\partial x} = \alpha(C_S - C), \qquad -D\frac{\partial C(x=l,t)}{\partial x} = 0, \tag{4}$$

where α is the effective mass transfer coefficient; C_S is the nitrogen concentration at the target surface.

Experimentally [6], the microhardness profiles for the near-surface layers of titanium BT1-0 were obtained. The titanium BT1-0 was nitrided in the non-self-sustained glow discharge plasma with a hollow cathode under a low pressure and different temperatures T = 550, 650, 850 °C during five hours (figure 7). The backside of the specimen was positioned at the distance of 0.5 mm from the hollow cathode surface. The microhardness was tested from both sides of the specimen. The experimental results on the surface microhardness of the titanium test objects treated under different parameters of the glow discharge and the nitriding temperature demonstrated that the microhardness increased when increasing the density of cathode current, discharge voltage and the test object's temperature [6]. In experiment [7], the titanium nitriding under 480-500 °C did not change the microhardness of the near-surface layers.

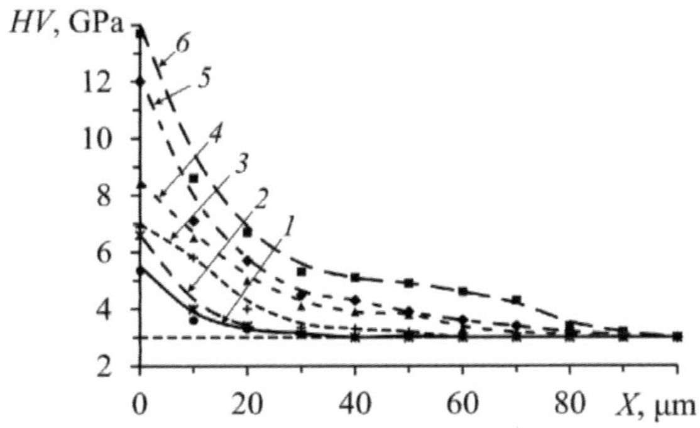

Figure 7. Microhardness profiles of the of the near-surface titanium layers BT1-0 under temperatures (°C) 550 (*1,2*), 650 (*3,4*) and 850 (*5,6*); *2,4,5* – backside of the test object; *1,3,6* – face side; dash line – object microhardness before nitriding [6].

The simulation results are presented in figure 8. The results of the research [14] at the temperature range of 540-610 °C were used to define the diffusion coefficient. Applying the extrapolation, the following values of the diffusion coefficient were obtained: $D(T = 550\ °C) = 5\times10^{-15}\ \mu m^2/s$, $D(T = 650\ °C) = 4\times10^{-14}\ \mu m^2/s$, $D(T = 850\ °C) = 7\times10^{-14}\ \mu m^2/s$. The calculated distribution of the nitrogen relative concentration C/C_S after five hours of processing is presented in figure 8a, where C_S is the concentration of nitrogen atoms at the object face side. The concentration dependence on time for the face side and backside of the test object are presented in figure 8b.

As figure 7 shows, the values of the microhardness at the face and side of the test object differ slightly which points the core role of the nitrogen atoms in the titanium nitriding and the nitride layer formation. The depth of the hardened layer depends on the nitriding time and the object temperature. The lower values of microhardeness at the object face side (figure 7, curves 1 and 3) compared to the backside might be caused by the ion etching of the surface. This effect was not considered in the numerical simulation.

The comparing of the experimental results (figure 7) and the numerical diffusion simulation (figure 8) shows that the time-and temperature-dependent microhardness profile of the titanium surface and the nitrogen concentration are in agreement with each other. This means that the growth of the nitride layers is defined by the nitrogen atomic diffusion. The nitrogen atomic diffusion into titanium was well demonstrated by the diffusion model (3)-(4).

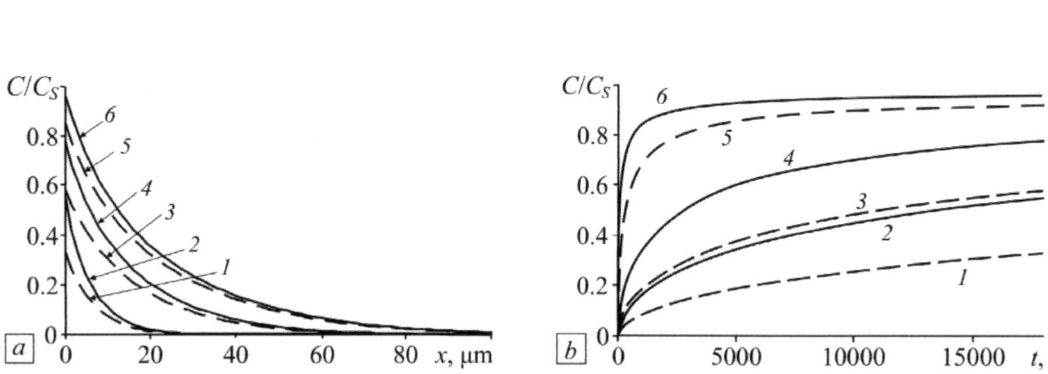

Figure 8. Calculated profiles of the nitrogen concentration in titanium (*a*) and its dependence on the processing time (*b*) under different temperatures (°C) 550 (*1,2*), 650 (*3,4*) and 850 (*5,6*) without considering the spraying surface by ions; *1,3,5* – test object backside; *2,4,6* – face side.

4. Conclusion

The numerical simulation of the plasma generation and the plasma-chemical nitriding covers the solution for the problems of the non-self-sustained discharge, gas ionization, thermal conductivity and diffusion. The model provides relations between main technological parameters such as gas pressure, electrical characteristics of a discharge, ion current density and the structure of the modified near-surface layer (the nitrogen concentration profile, the width of nitride layers and the diffusion zone) formed while nitriding metals or alloys by the non-self-sustained low-pressure glow discharge.

References

[1] Panajoti A 2003 *Fiz. Khim. Obr. Mater.* **4** 70–8
[2] Schanin P M, Koval N N, Goncharenko I M and Grigoriev S V 2001 *Fiz. Khim. Obr. Mater.* **3** 16–9
[3] Hosseini S R and Ashrafizadeh F 2009 *Vacuum* **83** (9) 1174–8
[4] Zagonel L F, Figueroa C A, Droppa J R and Alvarez F 2006 *Surf. Coat. Tech.* **201** (1-2) 452–7
[5] Berlin E V, Koval N N and Seidman L A 2012 *Plasma thermochemical surface treatment of steel parts* (Moscow: Tekhnosfera) p 462 (in Russian)
[6] Akhmadeev, Yu H 2007 *Non-self-sustained glow discharge with a hollow cathode for the nitriding of titanium* (PhD thesis, Institute of High Current Electronics SB RAS, Tomsk, Russia) Retrieved from http://www.dissercat.com/content/nesamostoyatelnyi-tleyushchii-razryad-s-polym-katodom-dlya-azotirovaniya-titana (in Russian)
[7] Akhmadeev Yu H, Lopatin I V, Koval N N, Schanin P M, Kolobov Yu R, Vershinin D S and Smolyakova M Yu 2010 *Modification of Materials with Particle Beams and Plasma Flows: Proc. 10th International Conference* (Tomsk, Russia, 19–24 September 2010) 228–31
[8] Lopatin I V, Schanin P M, Akhmadeev Y H, Kovalsky S S and Koval N N 2012 *Plasma Phys. Rep.* **38** (7) 583–9
[9] Koval T V, Lopatin I V, Nguyen B H and Ogorodnikov A S 2015 *Radiation and Nuclear Techniques in Material Science – Adv. Mater. Res.* **1084** 196–9
[10] Korolev Yu D, Frants O B, Landl N V, Shemyakin I A and Geyman V G 2013 *IEEE Trans. Plasma Sci.* **41** (8) 2087–96
[11] Pastuh I M 2006 *Theory and Practice of Nitriding in a Glow Discharge Without Hydrogen,* (Kharkiv: KIPT) p 364 (in Russian)
[12] Mufu Yan Qingchang Meng and Jihong Yan 2003 *J. Mater. Sci. Technol.* **19** (1) 164–6
[13] Bernal J, Medina A, Bejar L, Rangel S and Juanico A 2011 *International Journal of Mathematical Models and Methods in Applied Sciences* **5** (2) 395–403
[14] Sadliy T P, Milowa L G, Baranowa T A and Kurjatnikov S V 2004 *Physics of aerodisperse systems* **41** 124–9 (in Russian)

IOP Conf. Series: Materials Science and Engineering **168** (2017) 012045 doi:10.1088/1757-899X/168/1/012045

Light-induced thermodiffusion in two-component media

V Ivanov[1], **G Ivanova**[2], **K Okishev**[3] **and V Khe**[4]

[1] Professor, FESTU, Khabarovsk, Russia
[2] Senior Lecturer, FESTU, Khabarovsk, Russia
[3] Associate Professor, FESTU, Khabarovsk, Russia
[4] Engineer, FESTU, Khabarovsk, Russia

E-mail: khe@ngs.ru

Abstract. We have theoretically studied the optical transmittance response of thin cell with liquid containing absorbing nanoparticles in a Gaussian beam field. The transmittance spatial changing is caused by thermal diffusion phenomenon (Soret effect) which produces the variations of concentration of absorbing nanoparticles. The thickness of optical cell (including windows) is significantly less than the size of the beam. As a result, an exact analytical expression for the one dimensional thermal task is derived, taking into account the Soret feedback that leads to the temperature rising on the axis of a Gaussian beam. We have experimentally studied this phenomenon in carbon nanosuspension.

1. Introduction

The light-induced phenomena in nanosuspensions are widely used in nonlinear optics [1-2], dynamic holography [3-4], optical diagnostics of materials [5]. There are special mechanisms of optical nonlinearity in such media associated with the concentration flows due to light-induced phenomena [2-4]. The redistribution of the components concentration results in a corresponding change in the refractive index (and absorption) of the liquid. The thermal diffusion (Soret effect) is one of such phenomenon which was investigated theoretically and experimentally [1, 5]. We proposed a new thermodiffusion scheme for a thin-layer cell [6]. In this work we have theoretically studied the optical transmittance response of thin cell with liquid containing absorbing nanoparticles. The experimental results are represented for carbon nanosuspension.

2. Theoretical model

We consider one-beam scheme measurement of the thermal diffusion response in the two-component liquid layer (Fig. 1). Let's the absorption coefficient α is determined entirely by one component of liquid with mass concentration C ($\alpha = \beta C$, where $\beta = (\partial\beta / \partial C)$ – constant) and liquid is located in a thin cell with thickness d.

We have for a Gaussian beam distribution in the cell plane perpendicular to the optical axis z:

$$I = I_0 \exp(-r^2 / \omega^2)\,, \tag{1}$$

where ω is the radius of the beam, r is the distance from the axis of the beam, I_0 is intensity on the axis of a Gaussian beam.

We assume a small thicknesses of the liquid layer d and of the cell window $L: (d+2L) << \omega$. In this case the radial heat flux is negligible and we get one-dimensional heat propagation task:

$$c_m \rho_m \partial T_m / \partial t = \chi_m \partial^2 T_m / \partial z^2 + \alpha I_0 \exp(-r^2 / \omega^2), \tag{2}$$

where c_m, ρ_m are the specific heat capacity and density of the liquid medium, T_m is the liquid temperature, χ_m is the heat conductivity coefficient. The absorption is low ($\alpha d << 1$).
The similar thermal task was also considered for the temperature change in the window of the cell T_w :

$$c_w \rho_w \partial T_w / \partial t = \chi_w \partial^2 T_w / \partial z^2, \tag{3}$$

where c_w, ρ_w, χ_w are thermo-physical parameters of the window material.

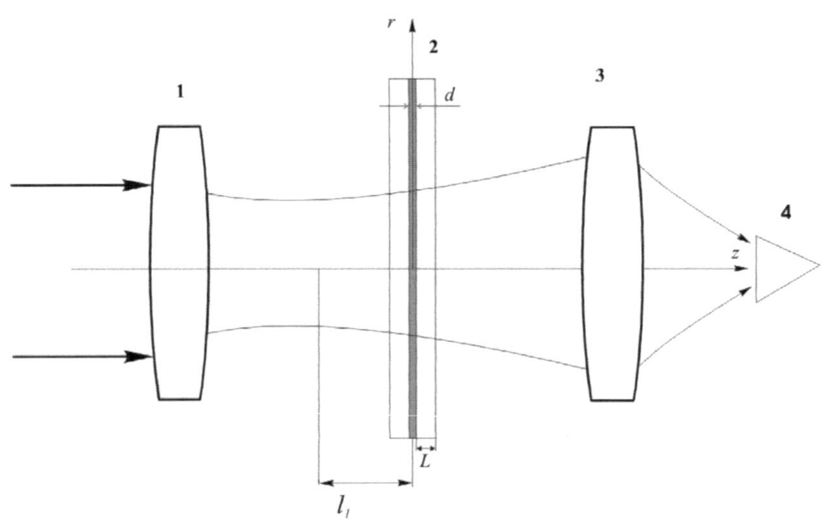

Figure 1. The scheme for the thermal task in “Soret cell”: 1, 3 – lens, 2 – cell with liquid, 4 – photo-detector, l_l- distance between beam waist and cell center.

Boundary conditions are as following:

$$\chi_w (\partial T_w / \partial z)_{z=L+d/2} = -\gamma (T_z - T_0). \tag{4}$$

$$T_m(d/2) = T_w(d/2). \tag{5}$$

$$\chi_m (\partial T_m / \partial z)_{z=d/2} = \chi_w (\partial T_w / \partial z)_{z=d/2}. \tag{6}$$

where γ, T_0 are the convective heat transfer coefficient and ambient temperature accordingly, $T_z = T_w(L + d/2)$.
In the stationary mode we have the following solution of the system (2-6) for temperature distribution in the liquid $T_m^s(z)$ and the cell window $T_w^s(z)$:

$$T_m^s = T_0 + T(d/2) + \alpha I(r) \chi_m^{-1} \{(d/2)^2 - z^2\}, \tag{7}$$

$$T_w^s = T_0 + T(d/2) + \alpha d I(r) \chi_w^{-1} \{(d/2) - z\}, \tag{8}$$

$$T(d/2) = \alpha I(r) d (\gamma^{-1} + L \chi_w^{-1}). \tag{9}$$

For the layer thicknesses $d << L$ we can neglect the temperature change along the thickness of the cell and take it equal $T_m(r)$:

$$T_m(r) = T_0 + \alpha d A I_0 \exp(-r^2/\omega^2), \tag{10}$$

where $A = (L\chi_w^{-1} + \gamma^{-1} + d\chi_m^{-1}/2)$.

Now let's consider the effect of thermal diffusion. Since the diffusion processes are much orders of magnitude slower than the heat transfer processes, we can assume that thermal diffusion of absorbing components occurs in stationary temperature field that is determined by formulas (7-9), but with a concentration-dependent absorption coefficient.

Balanced equation for the concentration of absorbing particles C is as following:

$$\partial C/\partial t = D\Delta_r C + D_r C(1-C)\Delta_r T, \tag{11}$$

here: D is the diffusion coefficient of absorbing particles, and D_T is the thermal diffusion coefficient, Δ_r is the Laplacian operator in cylindrical coordinates.

The change in the concentration of the absorbing components is small compared to the primary one: $C = C_0(1+C_1)$, where $C_1 \ll 1$. Let us initial concentration of the absorbing particles is small too ($C_0 \ll 1$), thus the task is linearised. We have the following equation with boundary conditions:

$$\partial C_1/\partial t = D\Delta_r C_1 + D_T \Delta_r T, \tag{12}$$

$$\nabla_r C_1(r)\big|_{r=0} = 0, \tag{13}$$

$$C_1(r=\infty) = 0. \tag{14}$$

The solution is taken by using the Green function method:

$$C_1(r,t) = -\frac{I(r)}{I_S}\left\{\exp\left[-\frac{r^2}{\omega^2}\right] - \left(1+\frac{t}{\tau}\right)^{-1}\exp\left[-\frac{r^2}{\omega^2}\left(1+\frac{t}{\tau}\right)^{-1}\right]\right\}. \tag{15}$$

where $I_S = (C_0\beta d A S_T)^{-1}$, $S_T = D_T D^{-1}$ is the Soret coefficient, $\tau = \omega^2(4D)^{-1}$ is the relaxation time.

Then we get for the intensity of the passed radiation $I_{pass}(r)$:

$$I_{pass}(r) = \{1 - \beta C_0 d[1 + C_1(r,t)]\} I_0 \exp\left[-\frac{r^2}{\omega^2}\right], \tag{16}$$

We use this term to calculate the total passed power:

$$P_{pass}(t) = (1-\alpha_0 d)P_0 + \left(\frac{I_0}{I_s}\right)\frac{\alpha_0 d P_0 t}{2(\tau + t)}, \tag{17}$$

where first term is the passed radiation without Soret effect. We neglect the Fresnel reflections for simplicity.

Using (10,15) we get for the temperature of the nanosuspension:

$$T_m(r) = T_0 + S_T^{-1}(1 + C_1(r))\left[\frac{I(r)}{I_S}\right]. \tag{18}$$

The last expression describes a so-called 'Soret feedback' that leads to the temperature rising near the axis of a Gaussian beam[7].

3. Experimental results and discussions

We used the suspension of carbon nanoparticles (with a diameter of 0.1-0.2 μm) in water. The initial mass fraction of nanoparticles was $C_0 = 6\cdot 10^{-3}$. The vertical laser beam (He-Ne laser, wavelength 0.63 μm, 60 mW power) was directed on the horizontal optical cell. The thikness of the cell window (130 μm) and liquid laer (70 μm) was smaller than the light beam radius (1200 μm). The temperature of the upper window was controlled by the infrared thermograph "IRTIS-2000".

Figure 2a shows the time evolution of the total signal (passed beam power) after beginning of the cell illumination. The disperse phase concentration and absorption coefficient decline both as a result of action of thermal diffusion. Figer 2b shows the time evolution of surface temperature in the center of the window.

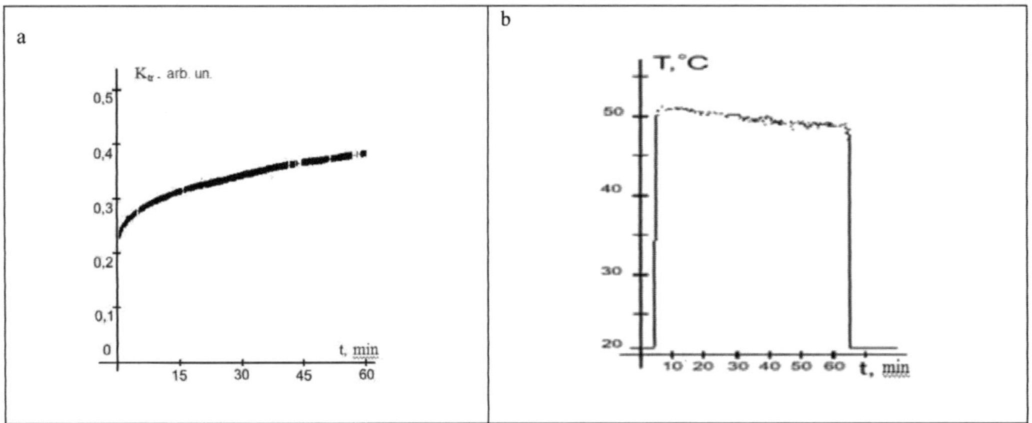

Figure 2. The time evolution of the total signal (passed beam power) after beginning of the cell illumination (2a); the time evolution of surface temperature in the center of the window (2b).

The enlightenment time corresponds to the diffusion processes. The real value of the absorbing factor was sufficiently high ($\alpha d \approx 2$). Thus the theoretical model corresponds to the experimental dates only approximately. We have the evaluation for the Soret coefficient $S_T \approx 2 \cdot 10^{-3} K^{-1}$ which corresponds to the characteristic values for nanosuspensions[8].

4. Conclusions

It is shown that thermal diffusion response in the two-component liquid contains a Soret feedback contribution due to thermodiffusion concentration change of absorbing components. The magnitude of this contribution can be quite large for the low radiation intensity. Thus, a modulation of the self-induced absorption coefficient should be taken into account in the thermal lens spectrometry analysis of multi-component liquids [4-6]. Received expressions can be used for experimental determination of the heat and mass transfer coefficients in such multi-component materials [7-9].

References

[1] Cabrera H, Sira E, Rahn K, García-Sucre M 2009 *Appl. Phys. Let.* **94** 051103
[2] Ivanov V, Okishev K 2006 *Technical Physics Letters* **32** 967-8
[3] Vicary L 2002 *Philos. Mag. B.* **82** 447-52
[4] Ivanov V, Lobov A, Livashvili A, Simakov S 2004 *J. of Opt. Tech.* **71** 586-7
[5] Okishev K, Ivanov V, Kliment'ev S, Kuzin A, Livashvili A 2010 *Opt. Atm. and Ocean* **23** 106-7
[6] Ivanov V, Ivanova G, Khe V 2015 *Proc. SPIE* **9680** 968042
[7] Tabiryan N, Luo W 1998 *Phys. Rev.* E. **57** 4431–8
[8] Leppla C, Wiegand S 2003 *Philosoph. Mag.* **83** 1989–99
[9] Hoffmann B, Kohler W, Krekhova M 2003 *J. Chem. Phys.* **118** 3237–42

The methods of receiving coal water suspension and its use as the modifying additive in concrete

S L Buyantuyev[1], L A Urkhanova[1], S A Lkhasaranov[1], Y Y Stebenkova[1], A B Khmelev[1] and A S Kondratenko[2]
[1]East Siberian state university of technologies and management, Ulan-Ude
[2]Buryat state university, Ulan-Ude

E-mail: buyantuevsl@mail.ru

Abstract. Results of research of the coal water suspension (CWS) from a cake received in the electrodigit ways in the fluid environment and gas are given in article and also the possibilities of its use as the modifying additive in concrete are considered. Use of a coal cake is perspective as it is a withdrawal of the coal and concentrating enterprises and has extremely low cost. Methods of receiving CWS and possibility of formation of carbon nanomaterials (CNM) are given in their structure. Research and the analysis of a microstructure of a surface of exemplars before electrodigit processing, their element structure, dependence of durability of a cement stone on a look and quantity of an additive of CWS is conducted. For modification of cement the carbon nanomaterials received from the following exemplars of water coal suspensions were used: foams from a cake from a scrubber of the plasma modular reactor, coal water suspension from a cake from electrodigit installation. The product which can find further application for a power engineering as fuel for combustion, and also in structural materials science, in particular, as the modifying additive in concrete allows to receive these methods.

1. Introduction

Use as a part of composition of carbon nanomaterials, as a result of their complex physical and chemical impact on modes of formation and concreting of cementitious matters, leads to increase of strength indexes of termination products. In literature works on research of introduction of carbon nanomaterials to composition of cement concretes [1-4] are rather widely lit. In researches it is noted that gel-forming participation and the modifying influence of carbon nanomaterials can be result of the following interdependent mechanisms:

1) the mechanism providing decrease of the common porosity, structural change of porosity of material. The nanodimensional particles which are present at system are capable due to increase in volume is adsorptive and hemosorbtsionno the water connected by them to reduce the volume of the capillary and bound and free water, to lead to change of rheological behavior of the cement test;

2) mechanism, the bound to a catalytic role of nanodimensional particles as crystallization centers and acceleration of process of hydration;

3) mechanism of zoning of structure of concreting;

4) the mechanism, the bound to a possibility of immediate chemical participation of nanodimensional particles in heterogeneous processes of phases formation of hydrated connections (such opportunity is defined as a substantional sign – by chemical and mineralogical structure of

particles, and the increased values of the specific area of their surface and a specific surface energy) [4].

2. The pilot unit and technique of researches

In the conducted researches for modification of cement the carbon nanomaterials received from the following exemplars of water coal suspensions were used: foams from a cake from a scrubber of the plasma modular reactor (fig. 1), coal water suspension (CWS) from a cake from electrodigit installation (fig. 2).

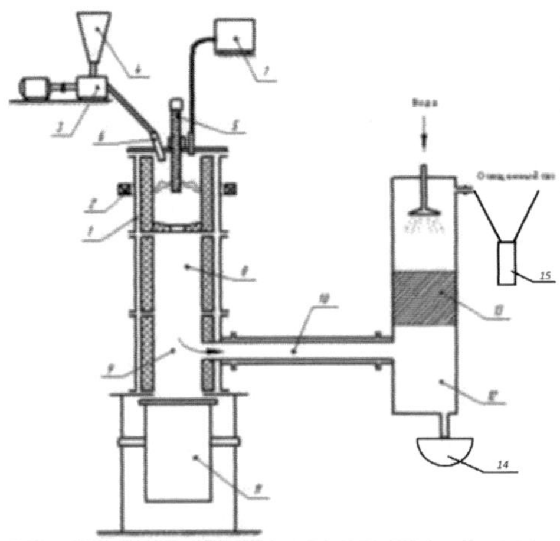

Figure 1. Plasma modular reactor.
1 – reactor; 2 – magnetic coil; 3 – batcher; 4 – raw materials bunker; 5 – cathode; 6 – ejector; 7 – steam generator; 8 – muffle chamber; 9 – division chamber; 10 – gas conclusion pipe; 11 – activation camera; 12 – scrubber; 13 – filter; 14 - collection of dust of a scrubber; 15 – cyclone.

In this plasmochemical installation, in the form of soot, having formed nanocarbon substances under the influence of electric arc plasma of the material of electrodes and coal given for heat treatment settle on the water-cooled top cover of the chamber having lower temperature that promotes preservation of the formed nanostructures of carbon. Also there is their ablation in a prerefining scrubber because the exhaust blower for the removal formed at plasma processing of coal of synthesis gas of an electric arc zone of the reactor is connected to installation.

Heavy particles of coal of a past plasma processing fall in an activation camera, but the mild ash unit together with synthesis gas "is sucked in" in a scrubber where there is its final condensation in the form of water coal suspension. There is a small negative pressure, gives the chance to separate large particles of coal from the sooty unit, owing to the fact that soot having low weight goes with gases to a scrubber.

At a temperature from 2000 to 4500 ^{0}C during time 0,2 – 1 sec. occur sublimation desublimation a microcomponent about coal surfaces, and also the processes of gasification and maximal opening of micropores in the center of a coal particle combined with optimum energy consumption.

Distinctive design signs of the plasma modular reactor are:

- receiving the uniform profile of temperature 2000-4500 ^{0}C in a transverse section of a camera of the combined plasma reactor by means of formation of the rotating electric arch that allows to reduce time of processing of coal up to seconds and to give to termination products special characteristics which are difficult for receiving in the existing technologies of receiving carbon nanomaterials;
- serial installation behind the plasma reactor of a chamber of a muffle, a chamber of division, a chamber of the activation combined with a pyrolysis chamber provides more flexible process control of receiving an absorbite;

- in the course of plasma processing carbon nanomaterials can be formed not only of material of electrodes (by the known methods), but also that is very important, of the coal passing processing by plasma. This fact gives particular advantages for receiving carbon nanomaterials.

When carrying out experiment on electrodigit installation in capacity coal dust, with fraction from 100 microns was filled up, mixed up with water and then the received mix was processed by electric current. After giving of tension process of electrochemical processing of mix in the course of which fuming was observed proceeded. Electric discharge proceeds between an internal electrode 3 and a housing of capacity. For the uniform electrochemical processing of coal water suspension the internal electrode is executed in spherical shape and evenly rotates the el ectric motor 2.

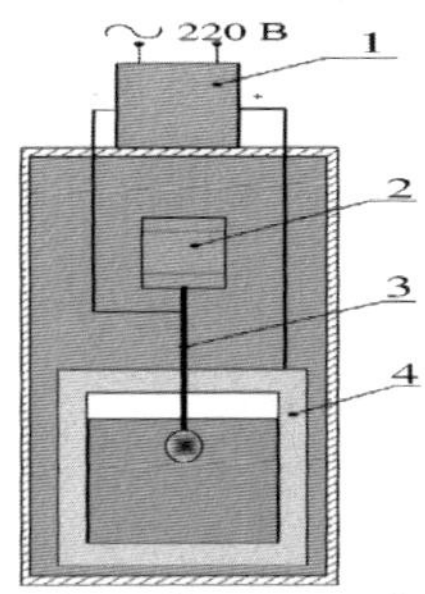

Figure 2. The pilot unit for preparation of water coal fuel by the electrolytic method:
1 - autotransformer; 2 - electric motor; 3-electrode-mixer; 4 - container with water coal suspension.

The following stage of selection of carbon nanomaterials from exemplars of coal water suspension is their processing by non-polar solvent (toluene, benzene, etc.) for the purpose of extraction and division of substances. Soluble carbon nanomaterials (fullerens, nanotubes) are extracted in solvent, and the insoluble part of soot settles on the bottom of a vessel. For the purpose of increase of extraction it is possible to apply a mechanical agitation, stirring or heating in Sakslet's device, it is also possible to apply ultrasonic extraction to acceleration of process and increase in concentration.
The further stage consists in

3. Results of researches and their discussion

Analyses of carbon nanomaterials were carried out by means of a microscopic and spectral ultimate analysis on the scanning raster microscope JEOL JSM-6510LV.
The element structure and structure of a surface of a cake before processing are presented in tab. 1 and fig. 3,4.

Table 1. Element structure of a cake.

Element	Weight %	Atomic %
C	43.98	55.26
O	38.69	36.50
Mg	0.16	0.10
Al	4.76	2.66
Si	7.77	4.17
S	0.33	0.16
K	0.41	0.16
Ca	0.62	0.23
Ti	0.25	0.08
Fe	1.66	0.45
Zr	1.37	0.23
Total	100.00	

Figure 3. Structure of the cake surface.

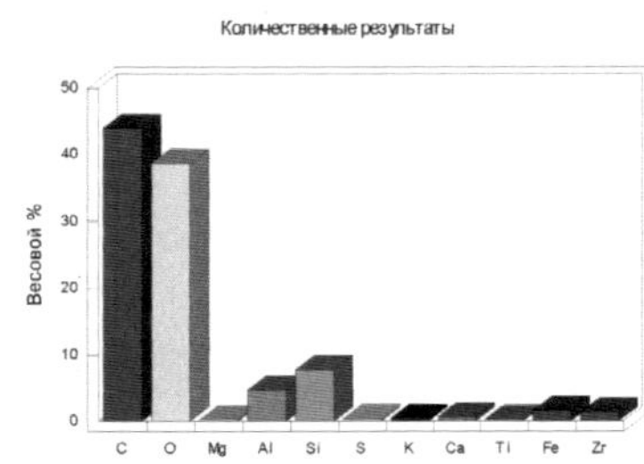

Figure 4. The diagramm of distribution of elements of a cake (with the indication of weight % of the found elements).

The element structure and structure of a surface of extraction of a cake after processing in elektrodigit installation are presented in tab. 2 and fig. 5,6.

Table 2. Element structure of a cake later electrodigit processing.

Element	Weight %	Atomic %
C	84.05	88.43
O	13.47	10.64
Al	0.13	0.06
Si	0.42	0.19
S	0.98	0.38
Cl	0.44	0.16
Ca	0.32	0.10
Fe	0.20	0.04
Total	100.00	

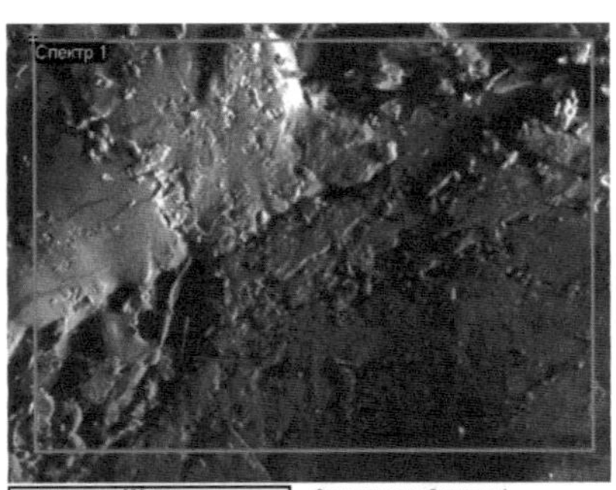

Figure 5. Extraction of CWS from a cake after electrodigit processing.

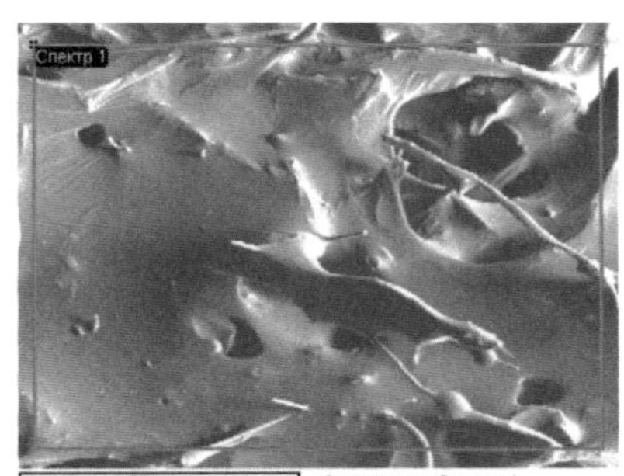

Figure 6. The chart of distribution of elements of extraction of a cake after electrodigit processing (with the indication of weight % of the found elements).

The element structure and structure of a surface of extraction of a cake after processing in the plasma modular reactor are presented in tab. 3 and fig. 7,8.

Table 3. Element structure of CWS extraction.

Element	Weight %	Atomic %
C	76.26	84.94
O	14.98	12.53
Al	0.64	0.32
Si	2.04	0.97
S	1.20	0.50
Cl	0.28	0.11
K	0.22	0.07
Ca	0.52	0.17
Fe	0.70	0.17
Pt	3.14	0.22
Total	100.00	

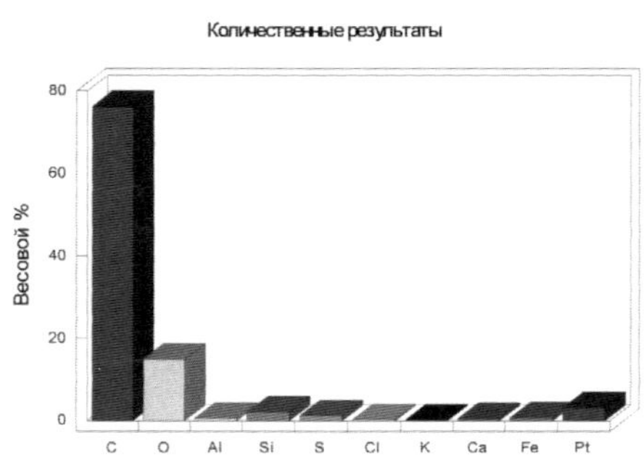

Figure 7. Extraction of CWS from a cake after processing in the plasma modular reactor.

Figure 8. The chart of distribution of elements of extraction of a cake after processing in the plasma modular reactor (with the indication of weight % of the found elements).

From micrographs it is visible that after electrodigit and plasma processing the surface of a cake has more dispersed and homogeneous structure, and results of an ultimate analysis show increase in a composition by weight of carbon by 32-40% and decrease in a composition by weight of oxygen by 23,71-25,22%.

Changes of element structure and microstructure of a cake as a result of electrodigit processing in liquid and gas have similar character though the mechanism of influence of gas discharge plasma and electric discharge in liquid significantly differ.

Positive influence of carbon nanomaterials on characteristics of a cement stone and concrete is in full shown on condition of their uniform distribution on all volume of a composite. Cavitational and pulse processing (ultrasonic processing, a hydrodynamic cavitation) of mixing water with nanodispersible additives is for this purpose used. For distribution of particles of coal water suspension in water volume of a zatvoreniye ultrasonic processing within 15 min. was carried out.

When adding in a cement matrix there is a change of physicomechanical properties of a cement stone. As a result of researches dependences of durability of the modified cement stone on a look and quantity of additives of exemplars of water coal suspensions (fig.9) were filled.

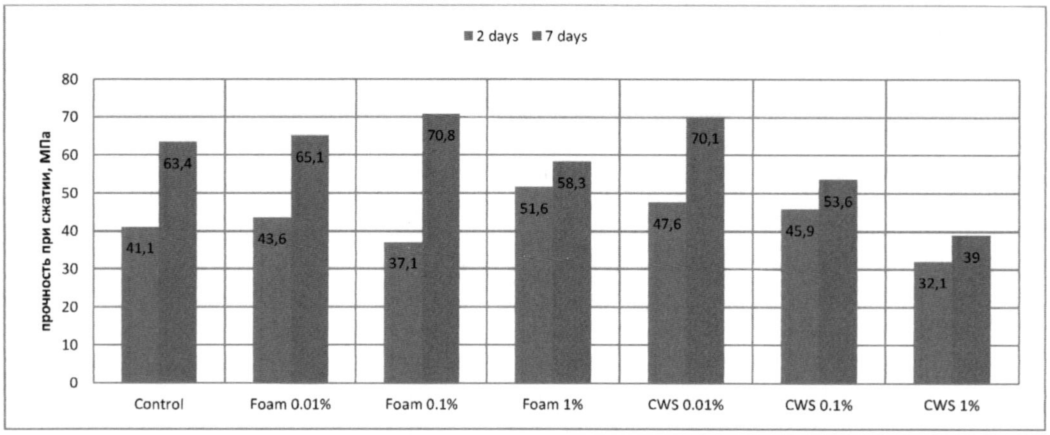

Figure 9. Dependence of durability of a cement stone on a look and quantity of an additive.

Research of physicomechanical properties of a cement stone showed that for different types of additives the effect of increase in durability varies depending on their quantity. So, for exemplars of a cement stone with the carbon nanomaterials received by cake foam from a gasifier scrubber, optimum results are received at the maintenance of an additive 0,1 Mas. %. In too time, for exemplars of a cement stone from coal water suspension from electrodigit installation more good results are received at the maintenance of an additive 0,01 Mas. %. At increase the maintenance of additives to 1% cement strength loss is observed, at the same time the received results are lower than control. In our opinion, it is bound to various range of distribution of particles in exemplars of additives. Particles of the carbon nanomodifier serve as crystallization centers of products of hydration of cement that accelerates processes of hydration and concreting of cement, especially in initial terms of concreting. The effect of increase in durability of cement is shown in small dosages of additives. At the increased content of carbon nanomaterials there is so-called "poisoning" of system when nanoparticles block cement grain, reducing extent of hydration of cement.

4. Conclusion

Thus, adding of water coal suspensions positively affects improvement of properties of cement. The greatest effect of improvement of properties is shown at dosages of additives of 0,01-0,1% depending on a type of water coal suspension. Improvement of properties of cement at introduction of trace amounts of carbon nanomaterials will promote creation of high-strength and long-lived composites.

References

[1] Poods I A, Yakovlev G I, Lushnikova A A, etc. 2011 A hydrodynamic way of a dispergation of multilayer carbon nanotubes at modification mineral the knitting *Intellectual systems in production* **2** 285-93

[2] Pukharenko Yu V, Nikitin V A, Letenko D G 2006 Nano-structuring mixing water as a way of increase of effectiveness of softeners of concrete mixes *Structural materials Science* **8** 11-13

[3] Ponomarev A N 2009 High-quality concrete. Analysis of opportunities and practician of use of methods of nanotechnology *Construction magazine* **6** 8 25-33

[4] Artamonova O V, Sergutkin O R 2013 Structural nano-materials: tendencies of developments and prospect *Scientific Bulletin the Voronezh state architectural and structural un-that* **1** 13-23

IOP Conf. Series: Materials Science and Engineering **168** (2017) 012047 doi:10.1088/1757-899X/168/1/012047

Investigation of radiation influence on LED

O I Rabinovich, S A Legotin, S I Didenko, A A Krasnov, A N Kovalev and S V Podgornaya
National university of science and technology "MISiS" 119049, Leninskiy pr., bl.4, Moscow, Russia

E-mail: rawork2008@mail.ru

Abstract. In this paper the investigation goal was to study the radiation effect (irradiation by fast electrons) on detectors, LEDs electro-physical parameters as they are widely used in space shuttles, satellites and airplanes on which space radiation influence. It was detected the critical value of irradiation for the irreversible changes in LED. The way for reducing characteristics degradation is suggested.

1. Introduction

High technical and economical performance of modern semiconductor sign indicators, detectors and light-emitting diodes (LEDs) make their usage one of the most promising areas in the visual display systems development. Using high efficiency InAlGaN Heterostructures (Het.) is an important milestone in the semiconductor devices development.

Investigation results that are discussed in current paper are relevant, due to necessity to improve the radiation hardness of devices designed to operate in extreme conditions, with increased lifetime.

The simulation methods, radiation process (RP) usage in the production of semiconductor devices (SD) and integrated circuits (IC) can effectively regulate the amplifier and pulse parameters of transistors (discrete and composed chip), LEDs quantum efficiency as well as eliminate the lightly controlled operation of the gold and platinum diffusion, to achieve new combinations of devices parameters (amplifier, efficiency, pulse, frequency), to study and improve their radiation resistance and stability [1-8].

SD parameters depend on many technological factors, it explains the large spread of finished products property values. RP is based on the effect of radiation centers formation in semiconductors irradiated by fast electrons or by other forms of ionizing radiation. Caused radiation centers, characterized by high stability in the operating temperatures ranges and electrical loads change electrical properties of semiconductor materials in various areas of the device structures and, accordingly, electric characteristics of semiconductor devices and integrated circuits. In the interaction, the electron energy is consumed mainly for the inelastic scattering on atoms, causing their ionization. The electron irradiation at low integrated flows creates mainly as primary defects the simple defects types with a uniform distribution in the volume of irradiated device.

Radiation process consists of the following operations:

- controlled irradiation of semiconductor structures or devices by fast electrons;
- annealing the irradiated semiconductor structures or devices.

The LED samples were irradiated by an electron flux with an energy from 4 to 5 MeV and doze upto 10^{15} e/cm^2 as the same average level in space. The minimum energy is determined by threshold energy of shift (bias). It is allowed the irradiation with a higher electron flux density if the structures or devices temperature at the subsequent irradiation does not exceed the stabilizing treatment temperature.

Required integral electron flux for specific types of SD and ICs may be approximately selected by calculation or, by nomograms which are obtained experimentally. The nomograms are created based on the requirement to obtain a certain value of the controlled parameter, or a device parameters combination after irradiation operations and stabilizing treatment.

The irradiated sample annealing is used to stabilize the surface and bulk properties of the device structure and the optimum combination of device parameters after the irradiation and annealing operations. Stabilizing annealing may be performed immediately after the irradiation at increased temperatures or during subsequent processing operations. However, experience has shown that most effectively it is the carrying out separate operations annealing irradiated SD structures and IC under strictly controlled conditions. Irradiated structures on wafers or devices located in a heating chamber in which the temperature is set not less than 150 °C. For a particular device type the annealing conditions (temperature, time, environment) are determined by experimentation, based on the need for stable parameters and their optimal combination at the maximum allowed temperature.

For the stability improvement of the device parameters, the last RP irradiation is advisably carried out by a flux $6 \cdot 10^{14}$ cm^2, which is caused by an increase in the most stable of radiation centers, the probability of their creation increases with the concentration of vacancies.

For RP it is advisably apply linear electron accelerators with energies up to 8 MeV, due to their main advantages: the possibility of obtaining high electron flux power and current control over a wide range, relatively small size and a high reliability. Linear accelerator that was used in the experiment is a controlled pulse source of monoenergetic electrons with the energy - 5 MeV. Electrons appear due to emission in the electron source, then they are drawn into the injector that ejects electrons in accelerating tube of accelerator. Electron acceleration in the accelerating tube is under the influence of high-frequency field, synchronized in phase with the input moment of the electrons bunch into the accelerating tube. From the theory of free electron movement in the high frequency (HF) field, it follows that for sustainable progressive motion of the electron in the HF field it is necessary that the initial phase of the electron motion is multiple divisible by π. In this case, the free electron will have the energy E from the HF field. Since the free electron energy, increased (flight time of the accelerating gap is decreased), it is necessary to increase the next accelerating gap to maintain the free electron energy increment from the HF field. Practically the implementation of electrons resonance acceleration by the HF field linear accelerator is a diaphragmatic waveguide. Since from the injector into the waveguide the electron bunch with a different phase angle enters at the initial section of the acceleration the automatic selection of free electrons occurs, the phase of which is different from a predetermined equilibrium phases. Since the accelerating HF field has a maximum value along the waveguide axis for focusing the electrons it is used a longitudinal magnetic field causing electrons to move close to the centerline of the waveguide line to obtain the maximum accelerating effect.

Electromagnetic oscillations of a microwave range are produced by generator, power is supplied to the diaphragmatic waveguide. The generator operates in a pulsed mode, since the generator power should be significant for the creation of the necessary strength of the accelerating field. As a generator of microwave oscillations it is used powerful pulsed magnetrons and klystrons, the voltage to the generator is supplied by short pulses of up to 5 μs with a predetermined sending frequency, which is produced by a modulator formed on pulsed hydrogen thyratrons. At the same time from the modulator voltage is supplied to the injector, injecting electrons into the accelerating disk loaded waveguide.

Thermal processing of irradiated LEDs in order to stabilize their parameters for a given dose was carried out by isothermal annealing in a laboratory tube furnace. Temperature constancy in the work area is 5 ^{0}C. The specified temperature value is maintained automatically. Operating furnace chamber heating is a quartz tube fitted with external heating elements and the ceramic end caps.

2. The experimental procedure

The main LED characteristic is a I-V and efficiency. The main reasons for theoretical and experimental deviations are:

- the devices and material surfaces influence;
- generation and recombination of carriers in the depleted layer;
- tunneling of carriers between the states in the band gap;
- a high level of injection observed even at relatively low forward bias;
- the effect of sequential contact resistance.

The surface has an effect on p-n–junction mainly due to ionic charges on it or near it, which induces charges in the semiconductor. This leads to the formation of so-called surface channels or depleted surface layer. The presence of the channel region affects on the depleted p-n – junction surface and causes an increase in leakage currents. In the planar p-n– junctions the surface leakage currents are typically much less than a generation current in surface leakage region.

Actual experimental results in the general case can be described by the following expression:

$$I = \exp\left\{\frac{e\cdot(U - I\cdot R_K)}{nkT}\right\}, \quad (1)$$

where I – current, mA; e – electron charge, C; U – voltage, V; R_K– consistent contact resistance; n – nonideality factor; k– Boltzmann constant, J/K; T – temperature, K.

The product $I \cdot R_k$ is the voltage drop on the LED contacts. Nonideality factor takes into account the deviation of the experimental current-voltage characteristics from ideal linear (logarithmic scale) characteristics. At the predominance of the recombination current n = 2; at the predominance of the diffusion current n = 1. If both currents are comparable in magnitude, then the value of n is between 1 and 2.

The investigation was made by two irradiations and annealings to detects critical irradiation doze.

3. Results and discussion

It has been investigated the radiation influence on the blue and green LEDs electrical parameters: I-V, watt-voltage and spectral characteristics after irradiation (average doze according space influence) and after subsequent annealing as shown in Figure 1-3.

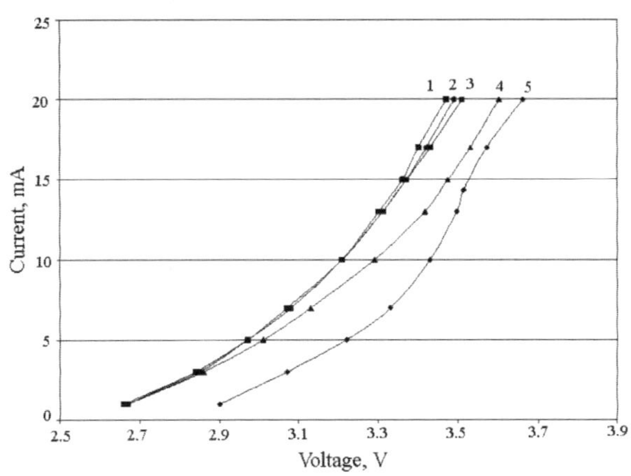

Figure 1. The typical I-V trend for blue and green LED: 5-initial, 4-after irradiation with doze 10^{14} cm^2, 3-after 1st annealing, 2-after irradiation with doze $6\cdot10^{14}$ cm^2, 1-after 2nd annealing.

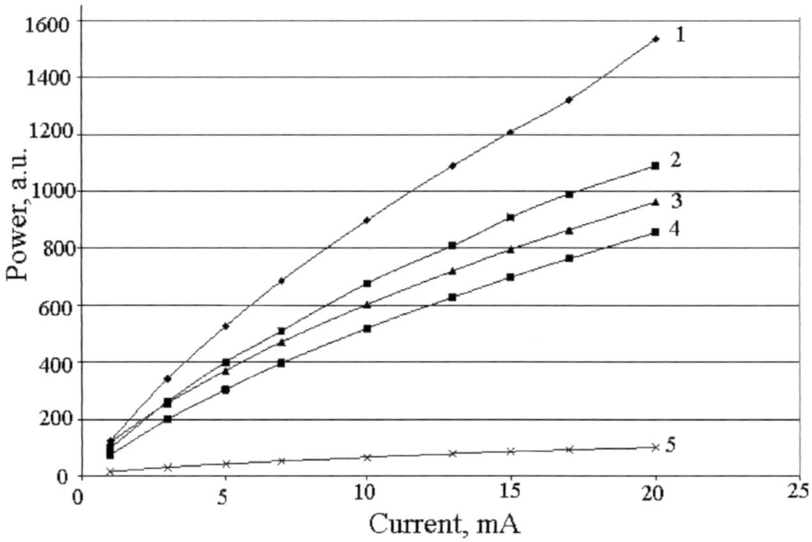

Figure 2. The typical trend for the watt-ampere blue and green LED characteristic: 1-initial, 2-after irradiation with doze 10 ^{14} cm^2, 3-after 1st annealing, 4-after irradiation with doze 6·10 ^{14} cm^2, 5-after 2nd annealing.

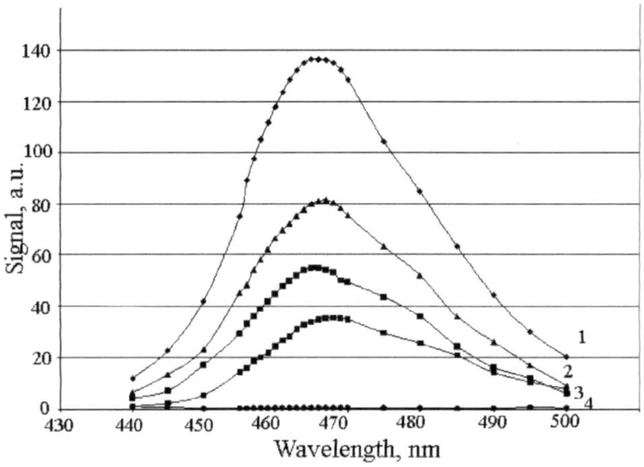

Figure 3. Spectral distribution vs wavelengths for the blue LED:

1-initial, 2-after irradiation with doze 10 ^{14} cm^2, 3-after 1st annealing, 3-after irradiation with doze 6·10 ^{14} cm^{2}g (no 5 curve as no effect was detected).

It was detected:

- after the first electron flux irradiation 10^{14} e / cm^2 it was occurred the characteristics degradation, the radiation power decreased;
- after annealing there was a slight recovery.

- after the second electron flux irradiation $6 \cdot 10^{14}$ e / cm^2 the power of radiation has decreased;
- after even the second annealing the characteristics recovery was unconvertible.

The performance characteristics degradation is a consequence of changes in the Het. composition and structure. Knocking out of the impurities from the interstitial sites or replacement of the gallium impurity atoms in the crystal lattice, as a result, leads to stress in lattice breaking impurity complexes.

"Red" shift to longer wavelengths is a consequence of the fact that the areas of low-atoms In comparison with the average value, creating a short-wavelength part of the radiation, are subject to destruction. The areas are like "burn" or In atoms are redistributed, since on the one hand through them high-density currents pass, on the other hand due to external irradiation influences (fields, irradiation influence (II)) and the induced by them effects. It appears more areas with high In atoms concentration and, consequently, the long-wavelength part of the electroluminescence spectrum increases. It may be also due to the piezoelectric effect (PE) and spontaneous polarization (SP) caused by the II, current or long running time. This effect may also be due to Stark and Stocks effects. The conclusions of some studies confirm this model. The significant decrease in the forward voltage at high current densities after II is well explained, because the elements with a reduced In atoms content have the greatest E_g, and consequently, the highest value U_f. However, an increase in the half-width of the spectral distribution is associated with the change in certain areas of the radiation parameters, it is possible to have the shorter waves than before the II, for example due to changes in the distribution of the intensity of the applied field and the higher density currents flow through this element. Figure 3 shows that the degree of change in the composition of the spectral distribution may depend on irradiation doze and it is associated with the occurrence of the resonance properties of SP effect.

The growth of long-wavelength components leads to the radiation efficiency of those parts of the structures that emit them, which however is not proportional to substantial degradation of sectors radiation with the central wavelengths and therefore the total integrated luminous flux decreases. After II the shunt effect appears as described above. However, different power characteristics changes may be due to a significant parameters mismatch of the crystal lattice of the Het. and the substrate.

Considered part of the model said that the SP is a major contributor to the "red" Stark energies shift in the GaN quantum well. Known effects associated with external electric fields (Bloch oscillations, Stark localization states, band-band tunneling), are significantly modified in the presence of internal polarization, and can occur even with a small number of ultra-thin layers of Het.

At high current densities, the difference in the behavior of the quantum efficiency with respect to the initial moment significantly increases. It is also due to the shunting effect of the described areas - the results of breakdown action of big potentials arising due to the PE. The current flow is already accompanied by a significantly greater proportion of non-radiative acts. However, the difference gradient decrease of the optical power between the Het. with a variety of substrates can be caused by a difference in the degree of mismatch periods of the substrate gratings and structures. This is clearly seen when considering the figures for the LED on sapphire and silicon substrates. Large drop values of optical power will meet the increasing mismatch lattices parameters. Mechanical II intensities this mismatch by defects formation caused by the difference in the mechanical parameters of materials with different lattice constants.

A marked reduction of optical power and voltage, respectively, at the same time II is seen in Figures 1-3. On variation in performance the changes of the applied external field and the high current density flowing through the Het. influence. It depends on irradiation doze, which in turn is connected to the resonance SP nature effect. The above part of the model pointed out that the PE field influence leads to a "red" shift of luminescence with respect to the E_g width of bulk material and the absorption edge, and a noticeable decrease in the oscillator strength. On the other hand it is likely that II activates the impurities in the AlGaInN Het.: donor oxygen which presents in the of $Al_xGa_{1-x}N$ layers and its activation amplifies the effect of DX-centers, as it can replace N_2 atoms in the lattice sites, which will create additional defects; it is probable the $Mg\text{-}H_2$ complexes disintegration, which will create additional parasitic resistance in the LED. Irradiation creates high-frequency piezoelectric field which is superimposed on the static piezoelectric field and the SP, e.g. there is a combined effect. These phenomena could extra added for above results explanation.

4. Conclusions
It is determined the maximum unconvertible influenced by radiation electron flux $6 \cdot 10^{14}$ cm^{-2} with an energy of 5 MeV on LEDs. It leads to the characteristics degradation. It is shown that the resistance was increased performance degradation of I-V, watt-ampere and spectral characteristics. This degradation relates to the formation of defects and atoms redistribution in the Heterostructure, which could be due to the lattice mismatch of substrate and active region. Thermal annealing at 150 ^{0}C did not lead to the full samples recovery.

This could be solved by GaN substrates usage.

References
[1] Tringe J W, Conway A M, Felter E T, et al. 2008 *IEEE Transactions on nuclear science* **55**(6) 3633
[2] Rabinovich O I 2014 *Journal of Alloys and Compounds* **586(1)** S258
[3] Rabinovich O I 2014 *Bulletin of the Russian Academy of Science* **78**(10) 979
[4] Rabinovich O, Didenko S, Legotin S 2015 *Advanced Materials Research* **1070-1072** 600
[5] Rabinovich O, Legotin S, Didenko S, Yakimov E et. al 2016 *Japanese Journal of Applied Physics* **55** 05FJ131
[6] Rabinovich O I, Didenko S I, Legotin S A 2015 *Journal of nano and electronic physics* 7(4) 040351
[7] Legotin S A, Murashev V N, Didenko S I, Rabinovich O I et al 2014 *Journal of nano and electronic physics* **6**(3) 030201
[8] Murashev V N, Legotin S A, Didenko S I, Rabinovich O I et. al 2015 *Journal of nano and electronic physics* 7(1) 010111

Processing of ash and slag waste of heating plants by arc plasma to produce construction materials and nanomodifiers

S L Buyantuev[1], L A Urkhanova [1], A S Kondratenko [2], S Yu Shishulkin[1], S A Lkhasaranov[1] and A B Khmelev[1]

[1] East Siberia State University of Technology and Management,
[2] Buryat State University

E-mail: buyantuevsl@mail.ru

Abstract. The resultsare presented of plasma processing slag and ash waste from coal combustion in heating plants. Melting mechanism of ashand slagraw material is considered by an electromagnetic technological reactor.The analysis was conducted of temperature and phase transformations of raw material when it is heated up to the melting point, and also determination of specific energy consumption by using a generalized model of the thermodynamic analysis of TERRA.The study of materials melting temperature conditions and plum of melt was carried with high-temperature thermal imaging method, followed by mapping and 3D-modeling of the temperature fields.The investigations to establish the principal possibilities of using slag waste of local coal as raw material for the production of mineral (ash and slag) fibers found that by chemical composition there are oxides in the following ranges: 45-65% SiO_2; 10-25% Al_2O_3; 10-45% CaO; 5-10% MgO; other minerals (less than 5%).Thus, these technological wastes are principally suitable for melts to produce mineral wool by the plasma method. An analysis of the results shows the melting point of ash and slag waste - 1800-2000 °C.In this case the specific energy consumption of these processes keeps within the limits of 1.1-1.3 kW*h/kg. For comparison it should be noted that the unit cost of electricity in the known high-melting industrial installations 5-6 kW*h/kg.Upon melting ash and slag waste, which contains up to 2-5% of unburned carbon, carbon nanomaterials were discovered.in the form of ultrafine soot accumulating as a plaque on the water-cooled surfaces in the gas cleaning chamber.The process of formation of soot consists in sublimation-desublimation of part of carbon which is in ash and slag, and graphite electrode.Thus, upon melting of ash and slag in the electromagnetic reactor it is possible to obtain melt, and in the subsequent mineral high quality fiber, which satisfies the requirements of normative documents, and simultaneously to receive a condensed product in the form of carbon sublimated nanoparticles, which can be found further use in construction materials, in particular in high-strength concrete and other materials.

1. Introduction

The demand of construction industry is constantly growing in mineral fiber materials and nanomodifiers. For the production of mineral fibers, it is appropriate to use technological wastes generated during the combustion of solid fuels, due to the fact that the formation of technogenic waste tends to increase significantly in the future, which contributes to their steady accumulation [1, 2]. Processing of solid technogenic waste in fiber materials is an important scientific and technical challenge, as these types of waste have a higher melting temperature, and therefore cannot be used for the production of mineral wool by known heat units (cupola furnaces, tank furnaces, etc.) [3].

One promising direction in this field is the use of a plasma arc to melt the raw material in order to obtain fibrous heat insulation material, as when used as a thermal energy source of the electric arc due to the high temperature sharply decreases the time of melt producing [4]. Also, in the processing of ash and slag, which usually contains up to 2-5% of unburned carbon, carbon nanomaterials were found in the form of ultrafine soot accumulating on water-cooled surfaces in gas cleaning chamber. Soot formation process consists in sublimation-desublimation of the carbon which is in ash and slag, and graphite electrode surface.

Thus, the article examined melting mechanism of ash and slag in a plasma reactor, in which is possible to obtain a mineral high quality fibers while simultaneously obtain the condensed product in the form of carbon sublimated nanoparticles that may find further application in construction materials, in particular, as additives in high-strength concrete and other materials.

2. The pilot unit and technique of researches

As noted above one of the promising methods is the use of electro-thermal melting method to melt the raw material in order to obtain mineral fiber materials.

The experimental setup was used to produce the melt and the fibers of the ash wastes, shown schematically in Fig. 1 [5, 6]. An advantage of this reactor is that the overlapping zones and release its heat absorption during the flow process, provides an intensification of heat and mass transfer from the plasma arc recyclable materials in the reaction chamber through the reactor chamber to create a large area - of the plasma volume, filling all its transverse section.

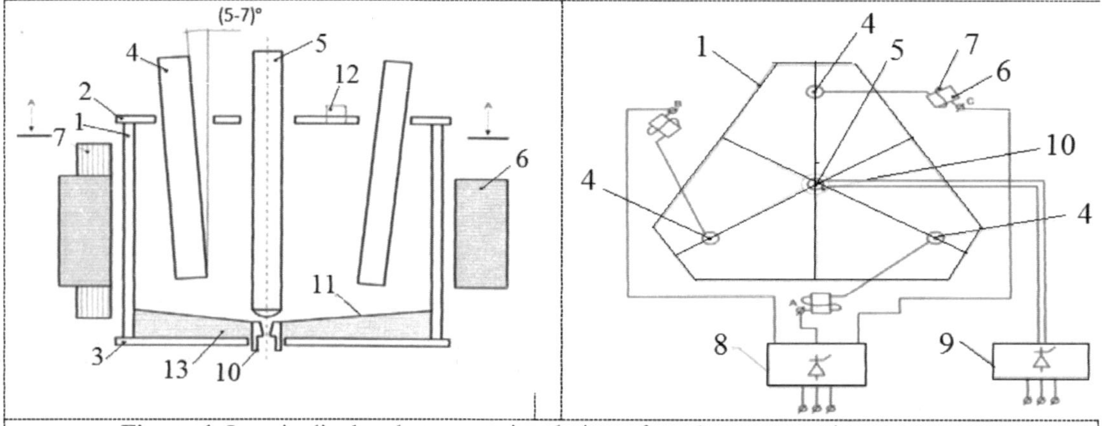

Figure 1. Longitudinal and cross-sectional view of an electromagnetic process reactor
1 - the reaction chamber; 2 - water-cooled cover; 3 - water-cooled bottom; 4 - rod electrodes (3 pcs.); 5 - electrode rod locking; 6 - pole piece; 7 - seriesnaya winding; 8 - power supply; 9 - an additional power source for heating the jet; 10 - to display the melt device (tap-hole); 11 - Lined bottom of the chamber; 12 - nozzle into the reaction chamber for supplying raw material; 13 – lining.

Thermal zone in the reactor chamber of the electromagnetic studied thermal imaging method for subsequent mapping with modeling of temperature fields (Figure 2) using the high-temperature thermal imager model MSC-640 by Luma Spec company.

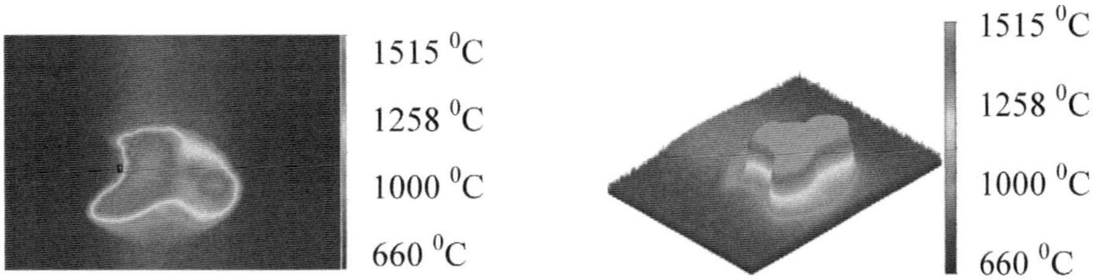

Figure 2. The temperature field of electromagnetic reactor in the range of 1000-2000 ^{0}C with the construction of the bulk of the temperature profile and the temperature distribution histogram.

Smelting mode is performed in a single step, consisting of the combined effects of, when starting occurs by electrical plasma heating and melting the raw material, and in the future, as the penetration and the formation of electrically conductive melt bowl comes bedding materials and the flow of current through the molten aluminum silicate mass, while electromagnetic agitation and homogenising with a series-connected series electromagnets, which significantly reduces time-to-operation and reduces the energy intensity of production. Residual carbon in the form mechanical underburningin melting ash and a part of graphite from the surface electrodes in electrothermal erosion under the influence of high temperature and uniform melt density actively evaporate and form ultrafine carbon black is removed from the reaction zone of the reactor in the gas cleaning chamber where their accumulation with further collection and treatment in non-polar solvents for use in the form of carbon nanomodifiers. Some practice expended power required to produce a melt with a weight up to 150 kg / h of 1.1-1.3 kWh / kg (for comparison, the power expended working induction furnaces - 6 kW / kg).

3. The analysis of temperature transformations of raw materials in the melting

Mineral fiber production process includes two major stages - preparation of a homogeneous melt and feed it into the fiber blowing. For computational and theoretical results on the melting of ash and slag waste from the definition of technological conditions in the preparation of the melts and the fiber necessary to carry out calculations of the process of high-temperature melting raw materials. In the study of the mechanism of melting raw materials used TERRA modeling environment for calculation of melting slag waste processes [7, 8].

The described software package allows you to simulate the maximum equilibrium states and markets created for his method and algorithm calculations. For each material property set should consist of its chemical formula, the limits of approximation of the thermodynamic functions Tmin, Tmax, the seven factors for the reduced thermodynamic potential Φi (T) (φ1 - φi), standard enthalpy of formation H_f^0 (298).

Specific energy consumption in the heat treatment process consists of heating energy costs for raw materials to a predetermined temperature and melting resulting in chemical reactions, thermodynamic to establish equilibrium in the system. The expression for Q_{spec} is:

$$Q_{spec} = I_{eq} - I_{origin}, \text{kW·h/kg}, \tag{1}$$

where I_{eq} and I_{origin} - total enthalpy, referred to 1 kg of the working fluid (batch + oxidizer) which is in the initial and equilibrium (after all transformations) states.

For specific thermodynamic systems I_{eq} value calculating methods of chemical thermodynamics of multicomponent heterogeneous systems using the TERRA program.

The enthalpy of the feedstock can be written expression:

$$I_{origin} = \Delta H^o_f(T_0)_{origin} + \int_{Torigin}^{To} Cp\,(T)dT, \tag{2}$$

where $\Delta H^o_f(T_0)_{origin}$ - the standard heat of formation of a working body, T_0 = 298,15 K - standard temperature; T_{origin}- initial process temperature at which the reactants enter the reaction zone.

At equality $T_{origin} = T_0$, the expression will be:

$$I_{origin} = \Delta H^o_f(298)_{origin} \quad (3)$$

Then, it is necessary for the reference data to determine system I_{origin} this summation enthalpy of formation of all components included in the raw material (ash):

$$\Delta H^o_f(298)_{origin} = \Delta H^o_f(298)_{SiO2} + \Delta H^o_f(298)_{Al2O3} + \Delta H^o_f(298)_{TiO2} + \Delta H^o_f(298)_{Fe2O3} + \Delta H^o_f(298)_{CaO} + \Delta H^o_f(298)_{MgO} + \Delta H^o_f(298)_{Na2O} + \Delta H^o_f(298)_{K2O} + \Delta H^o_f(298)_{MnO} + \Delta H^o_f(298)_{P2O5} + \Delta H^o_f(298)_{SO3}$$

The enthalpy of formation of the above, depending on the system components get value Iorigin = 112074.3 kJ / kg.

When calculating Q_{spec} necessary to consider mass fractions are part of the components [5], so partite composition of the ash on the basis of the chemical (in the capture of 100% per unit) distributed as follows: SiO_2 = 0,565; Al_2O_3 =0,2087; TiO_2 = 0,0071; Fe_2O_3 = 0,1438; CaO = 0,0380; MgO = 0,0186; Na_2O = 0,0057; K_2O = 0,0082; P_2O_5 = 0,0028; SO_3 = 0,0021.

Defining Q_{spec} (kWh / kg), calculated the minimum de- posited electrical power P_{el} in kW:

$$P_{el} = Q_{spec} * G_{charge}, \quad (4)$$

Where G_{charge} - the batch weight, kg.

As a result of the melting of the charge in electric plasma airdoes not participate. Weight average loaded batch of 150 kg, the melting time and tentative - 1 hour.

The results of thermodynamic calculation of the total enthalpy I_{eq}, condensed phase amount Z (%), and the equilibrium composition via TERRA program shown in Figures 3 a, b; energy dependence on the temperature is shown in Table 1.

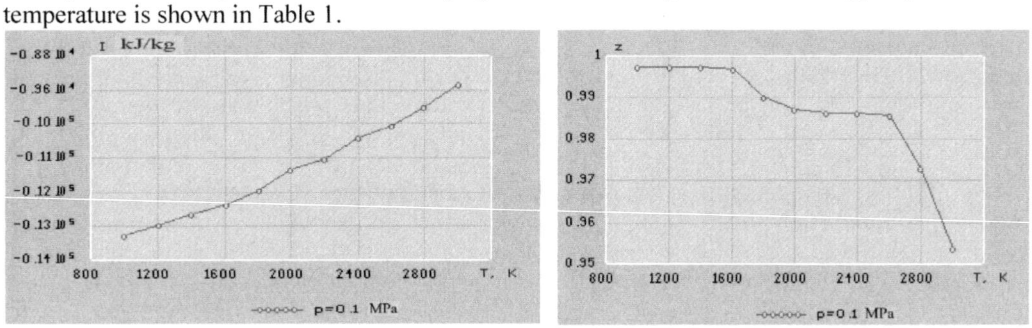

A B

Figure 3. Complete melting enthalpy (A) and the amount of the condensed phase (B) Okino-Kliuchevskoi ash.

Table 1. Dependence of specific energy consumption from ash melting point.

T, °C	1000	1200	1400	1600	1800	2000	2200	2400	2600	2800	3000
Q_{spec}, kW*h/kg	0,285	0,34	0,424	0,68	0,96	1,23	1,58	2,42	2,91	3,02	3,18
P, kW	37	48,1	64,8	75,9	92,5	120,3	137	165	181,4	204,8	235

4. The research results

Earlier studies were conducted by authors to obtain high-strength concrete with the introduction of carbon nanomaterials, simultaneously obtained upon melting carbon-containing wastes in the form of unburned carbon in electromagnetic reactor [9, 10]. According to the research the nature of the impact is identified of carbon nanomaterials on processes of hydration and hardening of cement and composite binders.Compositions are developed of modified concrete on cement and composite binders using carbon nanomaterials. Another promising direction of research is the use of mineral fiber which is obtained from the ash and slag waste to disperse reinforcement of concrete. It is known that concrete structures are not having sufficiently high tensile strength in flexure. This problem can be solved by disperse reinforcement of concrete by various fibers [11-13]. Disperse reinforcement by metal and non-metallic fibers allow qualitatively improves the indicators of composites. It is necessary to consider the possible negative impact of hardening cement environment when using non-metallic mineral fibers.

The most effective of non-metallic mineral fibers for concrete dispersed reinforcement is the basalt roving and finely staple fiber. However, the technologies of production of these types of mineral fibers are characterized by high energy costs. Use of these mineral fibers with less stable and homogeneous properties compared to the roving and staple fibers for dispersed reinforcement of cement composites is a promising direction in the technology of their production.

The introduction of the mineral fibers increases the compressive strength after 28 days by 16%, the flexural strength by 50% compared to control composition. Noticeable increase of flexural strength is associated with the reinforcing action of mineral fibers (fig.4).

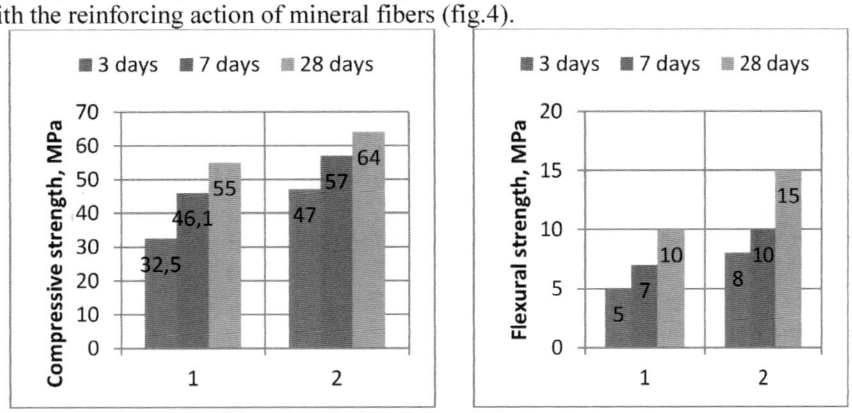

A B

Figure 4. Compressive (A) andflexural (B) strength of cement stone:
1 – without mineral fiber, 2 – with mineral fiber on the basis of ash and slag waste, 4 wt. %.

Studies have been conducted on the definition of physical, mechanical and performance properties of the fine fiber-reinforced concrete using ash and slag fibers (Table. 2, 3).

Table 2. The composition and technological parameters of fiber-reinforced concretes with mineral fiber.

Material consumption per 1 m3 of concrete, kg			W/C ratio	Segregation by water separation,%
cement	sand	fiber		
550	1375	-	0,4	0,6
550	1375	22	0,4	0,45

Table 3. Mechanical and performance and technological properties of fiber-reinforced concrete.

Properties	Units	Indicators	
		Reference composition of concrete	Fiber-reinforced concrete
Theaveragedensity	kg/m^3	2400	2460
Compressive strength	MPa	44	50
Flexural strength	MPa	9	13
Waterabsorption	% by weight	4,0	3,5
Frost resistance	cycles	150	200
Shrinkage	mm/m	2,7	1,6

3. Conclusions

Presented electric smelting reactor, which is a type of plasma reactors, allowed to regulate smoothly of melt temperature and maintain it at the exit of the jet from the tap hole, which made it possible to lower the viscosity and increase the melt flow. It may be recommended ash and slag waste heating plants to process plasma method in the melt for mineral casting and thin fibers as a component of composite materials, as well as for the production of carbon nanoparticles for modification [14].
Introduction of mineral fibers increases the strength characteristics for compression strength at 13%, flexural strength - 40%. The use of mineral fiber, obtained using plasma arc and characterized by inhomogeneity of on the main characteristics, it allows obtaining high-strength fine fiber-reinforced concrete with improved properties. In the Baikal region the use of fiber-reinforced concrete will improve the quality and reduce resource consumption in construction, such as road and bridge construction. Electric smelting reactor allows obtaining from ash and slag waste sublimated at a high temperature carbon nano sized additive, which make it possible to receive high strength properties of concrete and other construction materials.

References

[1] Buyantuev S L, Sultimova V D 2004 Production of thermal insulation materials made of ash and slag waste TPP using low-temperature plasma *Construction Materials* **10** 51-53
[2] Buyantuev S L, Sultimova V D, Dondokov A T, Volokitin G G, Zayahanov M E, Tsyrenov S A 2006 Manufacture of heat-insulating building materials using elektroplazmennoy processing *Building Complex of Russia: science , education, practice: Proceedings of the Intern. scientific. Conf. Ulan-Ude* 90-93
[3] Bobrov Y L, Ovcharenko E G, Shoikhet B M, Petukhova E J 2003 Thermal insulation materials and construction. M .: INFRA-M 268
[4] Sergeev P V 1978 An electric arc in electric reactors *Alma-Ata. The science* 140
[5] Buyantuev S L, Sultimova V D 2010 The use of low-temperature plasma to produce fibrous thermal insulation materials from ash and slag waste of thermal power plants *Ulan-Ude: VSGTU* 132 p
[6] Buyantuev S L, Malykhl A V, Pashinsky S G, Ivanov A A, Kitaev V V Russian patent for invention №2432719 Electromagnetic Technology reactor, published 27.10.2011 Bull. №21
[7] Vatolin N A, Trusov B G, Moiseev G K 1994 Thermodynamic modeling in high-inorganic systems. M.: Metallurgy
[8] Trusov B G 2002 TERRA program complex for calculation of plasma-chemical processes *Mater. 3 Intern. Symp. in Theoretical and Applied Plasma Chemistry. Ples* 217-218
[9] Buyantuev S L, Kondratenko A S, Khmelev A B 2013 The use of carbon nanomaterials produced plasma treatment of coal as modifying additives in concrete *Wschodnie Partnerstwo - 2013: materialy IX miedzynarodowej naukowi-praktycznej konferencji, 07-15 wrzesnia 2013 roky. - Przemysl: Nauka studia*. **34**: Techniczne nauki 49-55
[10] Urkhanova L A, Buyantuev S L, Lkhasaranov S A, Kondratenko A S 2012 Concrete on composite binders with nanostructured fullerene additive *Nanotechnology in construction* **1** 22-25
[11] Pukharenko Y V 2012 Restoration and construction of: the potential fiber reinforced materials and products *Modern problems of science and education* **4**
[12] Borovskikh I V, Khozin V G 2009 Changing the length of basalt fibers in its distribution in the composite binder of high bazaltofibrobetonov *Proceedings of Kazan State Architectural University* **2 (12)** 233-237
[13] Stepanova V F, Buchkin A V 2011 Corrosion behavior of basalt fibers in the cement matrix of concrete *Building materials, equipment, technologies of XXI century* **9** 22-26
[14] Buyantuev S L, Kondratenko A S 2014 Thermal insulation materials of the fibrous structure of the rock melts and technogenic waste of ash: monograph *Ulan-Ude: VSGUTU* 180

IOP Conf. Series: Materials Science and Engineering **168** (2017) 012049 doi:10.1088/1757-899X/168/1/012049

Determination of stress concentration coefficients upon cyclic loading by photometric analysis of structure images

N A Minina[1], V A Emishkin[1] and I N Ovchinnikov[2]

[1] Federal State Institution of Science Baykov Institute of Metallurgy and Materials Russian Academy of Sciences, Moscow, Russia
[2]Bauman Moscow State Technical University Moscow, Russia

E-mail: minina1951@rambler.ru

Abstract. The development of an experimental method for the determination of stress concentration coefficient by the photometric analysis of surface reflectivity of material near stress concentrator and at a distance from it is considered in the paper. The experiments were performed with the AD-19 aluminum alloy samples under fatigue test conditions realized at a vibrating table with cantilevered sample. Notches 5 mm long and 0.4 mm wide spaced at constant intervals were applied by electrical discharge along one edge of the samples. The reflections of light from the fragments in the vicinity of the stress concentrator and from the opposite edge of the sample were compared with the help of photometric image analyzer. It is shown that such analysis can be used to determine the stress concentration coefficients.

1. Introduction

In metal structures, the presence of local cuts, notches, and grooves as well as the abrupt thickness differentials in the joints of metal sheets differing in thickness results in a significant reduction in structural strength. All they cause local overstresses in the metal and act as stress concentrators, which can lead to premature failure upon operation. The extent of their danger to operational capability of structures is characterized by stress concentration coefficient (K), which is the ratio of stress in the vicinity of the stress concentrator (σ_s) to the stress in a close position, in which its effect is negligible (σ_n):

$$K = \frac{\sigma_c}{\sigma_n} \quad (1)$$

However, Eq. (1) can be used if the experimental data obtained by strain gauge measurements are available. The theoretical estimates of the stress concentration coefficients were fulfilled within the frameworks of the theory of elasticity and allow for the geometry and mechanical properties of the material [1-3]. As a rule, the real material in such estimates is represented by solid structure less isotropic medium whose properties do not change upon plastic deformation. For this reason, the development of an experimental method for determining the stress concentration coefficient (K) from the data obtained by optical measurements of structure elements before and after loading is actual and important.

2. Experimental

The experimental part of the work was performed by fatigue tests of D19 aluminum alloy samples. Its chemical composition is given in Table 1.

Table 1. Chemical composition of the D19 alloy, mass. %.

Al	Cu	Mg	Mn	Be
base	3.6-4.3	1.7-2.3	0.5-1	0.0002-0.005

The samples were cut from a sheet 1.80 mm thick. The samples were 20 mm in width and 223 mm in total length. A thin photometric coating was applied on the sample surface, which before and after the test was examined by photometric analysis of structure images (PHASI). The sample is shown in Figure 1.

Figure 1. General view of the sample for fatigue tests.

Four notches 5 mm long at intervals of 41 mm were applied by electroerosion method at one end of the sample. The first cut was applied at a distance of 53 mm from the sample edge at the head side. The samples were tested on a vibrating table with electrodynamic excitation. They were attached to the vibrating table rod in a cantilever mode. The sample was excited by polyharmonic vibrations at the frequencies of the first three self-oscillation harmonics. The data on the vibration mode excited in the samples are given in Table 2. Letter A denoted the oscillation amplitude measured at the free end of the sample, and v means oscillation frequency.

Table 2. Loading modes of the samples.

Sample No.	Mode of load fluctuations					
	Mode I		Mode II		Mode III	
	$2A_I$, mm	v_I, Hz	$2A_{II}$, mm	v_{II}, Hz	$2A_{III}$, mm	v_{III}, Hz
1-к	40	32-31	10	198-197	3.5	550
2-к	40	32-31	10	198-197	3.5	550
3-к	57	31.5-29	-	-	-	-

The effective stresses were calculated by the formulas of strength of materials [4] as a superposition of three bending stresses from three loading modes. Loading of the sample by alternating stresses according to the cantilever bending scheme leads to the formation of distributed loads with the intensity q determined by Eq. (2). Equation (3) was obtained from Eq. (2) of a flexible line describing the dependence of deflections, which are on the horizontal axis of the sample, on the coordinate z:

$$y(z) = \frac{q}{EJ_x}\left(\frac{lz^3}{6} - \frac{z^4}{24}\right) \quad (2)$$

where: q is the intensity of the distributed load, E is the normal module of the sample material, l is the sample length, J_x is the moment of inertia of the sample relative to the horizontal axis of symmetry of the sample cross section. Therefore, the expression for determining the intensity of a distributed inertial load q takes the form:

$$q = \frac{yE\,J_x}{\left(\frac{lz^3}{6} - \frac{z^4}{24}\right)} \tag{3}$$

In general, the effective stresses in each cross section of the sample can be expressed by the formula:

$$\sigma(z,t) = \sum_{i=1}^{i=3} \sigma_i(z) \cdot \sin(2\pi\nu_i t) \tag{4}$$

After the determination of $q(z)$ by Eq. (3) and using the data from Table 2, we have found the distribution of bending moments over the sample length for the loading scheme used in the experiment:

$$M(z) = \frac{q}{2} \cdot (l^2 - z^2) \tag{5}$$

Then the amplitude distribution along the sample length is expressed by the formula:

$$\sigma(z) = \frac{M(z)}{W} = \frac{q \cdot (l^2 - z^2)}{2W} \tag{6}$$

where: W is the moment resistance to be determined by the formula:

$$W = \frac{bh^2}{6} \tag{7}$$

According to the results of the calculations by the above method, the resultant stresses operating in a sample were determined as a function of time and location of cross section in the samples. Figure 2 shows the time dependence of the resulting stresses at the time interval corresponding to 5 oscillation periods of the i-th harmonic for self oscillations of the samples of selected shapes and sizes.

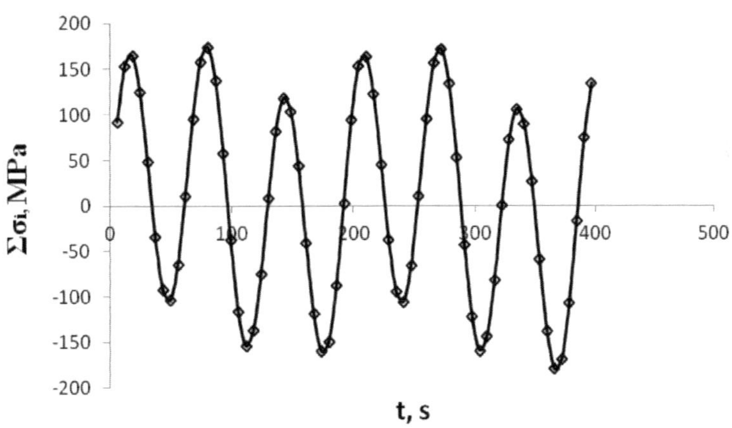

Figure 2. Time dependence of the resulting stresses in the sample upon fatigue test.

Polyharmonic loading at the selected oscillation parameters insignificantly changed the sinusoidal character of the resulting load. It is seen that the amplitude of the resulting load is variable. Its maximum value of 149.76 MPa was used for the calculations. Figure 3 shows the effective stress distribution of the over the sample length. Hereafter, we are interested in the stress in the cross sections weakened by the notch reducing its moment resistance by a factor of 1.33. The sample fragments containing notches were selected for the photometric analysis of structural images (PHASI) taken before and after the fatigue tests up to failure of the samples.

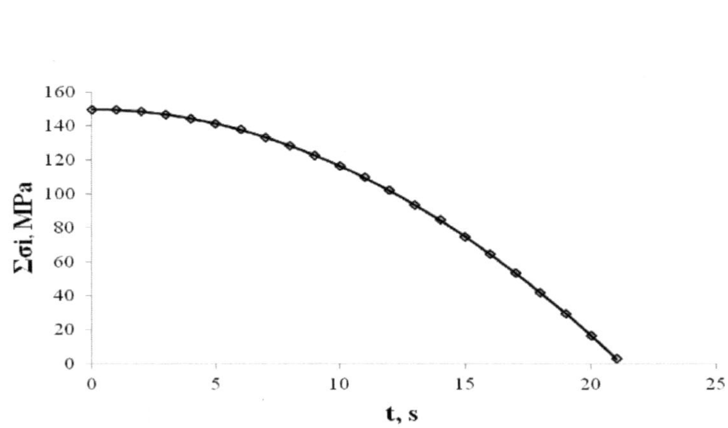

Figure 3. Distribution of the maximum effective stress over the sample length upon fatigue test.

In the selected fragments, the zones adjacent to the notch were cut off, and the remainder of the fragment was divided into two equal parts as shown in Figure 4. The PHASI is a software-analytical complex that analyzes the brightness spectra of the visible light reflection from the surface fragments examined before and after the actions of different physical nature. The areas under the spectral curves in physical sense represent the light energy reflected from the test fragment in arbitrary units. The difference in the areas under the spectral curves obtained from each fragment is determined by the intensity of external action (in our case, by the effective stress [4, 5]). According to [6], this energy (U) is related to the internal energy of the object (E) by the formula:

$$U = A \cdot E \tag{8}$$

where: A is the coefficient characterizing the probability of spontaneous emission of radiant energy by the solid body. The difference in the areas under the spectral curves recorded from each notched fragment before and after fatigue test in arbitrary units is related to the elastic energy of the corresponding micro fragments by Eq. (9):

$$A_i = \frac{U_i}{E_i} \tag{9}$$

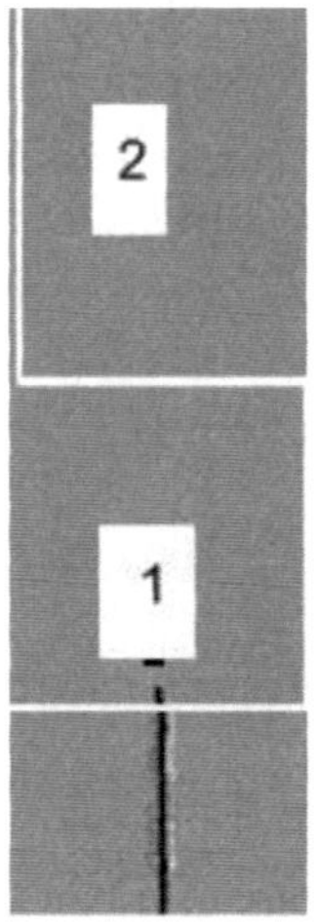

Figure 4. Scheme of the division of the sample fragment with a notch for PHASI: (1) micro fragment closest to the notch and (2) notch-free fragment near the sample edge.

3. Results and Discussion

The areas under the spectral curves have been identified by PHASI for all micro fragments of types 1 and 2. For each micro fragment, the specific surface energy E_i was calculated by Eq. (10):

$$E_i = \frac{\sigma_i^2}{2E} \quad (10)$$

Then the A_i values were calculated by Eq. (9) and the $A_i = f(\sigma)$ dependence was constructed (Figure 5).

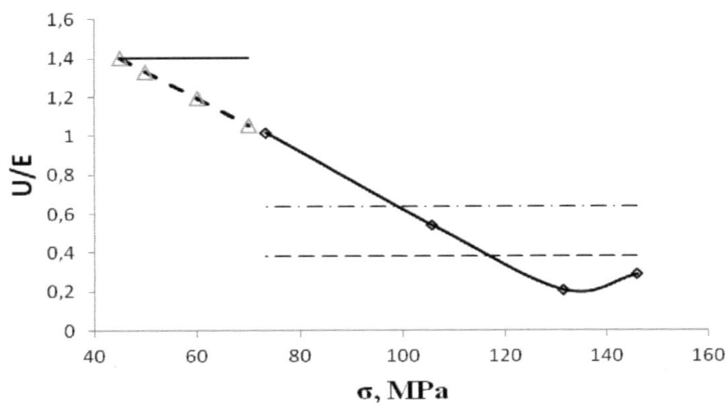

Figure 5. Dependence of the size coefficient A_i on stress σ_i for the fragments with notches.

The horizontal lines in Figure 5 are drawn through the A_i values corresponding to micro fragments of type 2. It is clear that the nominal stresses in the fragments of both types (1 and 2) are the same and they are calculated with allowance for the weakening of cross sections by the notches. For this reason, the difference between the A_i values of these micro fragments can be explained by the effect of stress concentrators in the immediate vicinity of the micro fragments of type 1. Then K can be defined as the relationship between the stresses corresponding to the A_i values for micro fragments 1 and 2 of the same cross section containing the notch, i.e.,

$$K = \frac{\sigma_{1i}}{\sigma_{2i}} \quad (11)$$

The obtained K values are given in Table 3.

Table 3. Stress concentration coefficient determined from the PHASI data on the sample tested for fatigue.

No. of cross section with a notch	σ, MPa	K without allowance for weakening the cross section	K with allowance for weakening the cross section
1	190.77	1.28	1.70
2	169.23	1.31	1.74
4	132.88	2.84	3.79

The last column of Table 3 shows the K values determined with allowance for weakening the cross section by the notch. The data show that, with increasing stress, K tends to decrease. This can be caused by the development of local plastic deformations at the top of the notch.

4. Conclusions

1. The developed method for evaluating stress concentration coefficient K is based on the measurements of the brightness of the visible light reflection from the objects under study.

2. The method is suitable for the study of K upon the development of plastic deformation at the notch tip.

3. The development of plastic deformation at the stress concentrator tip leads to the reduction of stress concentration coefficient K.

Acknowledgments
This work was supported by the Russian Foundation for Basic Research, project 14-08-00417a and 15-08-04272a.

References

[1] Popkovich PF 1939 *The theory of elasticity*. (M-L: Gos. Tech. Izdat. Oboronnoi Promyshlennost) 643
[2] Neuber H 1947 *Stress concentration* [Russian translation]. (M: Gostekhizdat) 204
[3] Kelly A 1975 *High-Strength Materials* [Russian translation] (M: Mir) 236
[4] Feodos'ev VI 1999 *Strength of Materials*. (M: Izd-vo MVTU im. N.E. Baumana) 589
[5] Ermishkin VA, Murat DP, Podbel'skii VV 2007 Information technologies of photometric analysis of materials fatigue damage. *Informatsionnye Tekhnologii*. **11** 65-70
[6] Ermishkin VA, Murat DP, Podbel'skii VV 2008 System of photometric analysis of structural imaging and its application to the study of materials upon fatigue. *Pribory i Sistemy. Upravlenie, Kontrol', Diagnostika*. **10** 38-44a
[7] Astapenko VA 2010 *Interaction of radiation with atoms and nanoparticles*. (Dolgoprudnyi: Izd-vo Intellekt) 492a

IOP Conf. Series: Materials Science and Engineering **168** (2017) 012050 doi:10.1088/1757-899X/168/1/012050

Research of the temperature fields influence on the stress state of internal cylindrical surfaces modified by magnetron sputtering

A I Blesman[1,3], D V Postnikov[2,3], E A Tkachenko[4] and D A Polonyankin[2,3]

[1]Head of the Physics Department, *Physics Department*,
Omsk State Technical University (OmSTU), 11 Mira Avenue, Omsk, 644050, Russia
[2]Associate Professor, *Physics Department*,
Omsk State Technical University (OmSTU), 11 Mira Avenue, Omsk, 644050, Russia
[3]Research scientist, *Scientific-educational resource center of nanotechnology*,
Omsk State Technical University (OmSTU), 11 Mira Avenue, Omsk, 644050, Russia
[4]Postgraduate, *Physics Department*,
Omsk State Technical University (OmSTU), 11 Mira Avenue, Omsk, 644050, Russia

E-mail: nano@omgtu.ru

Abstract. For material selection of protective metal coating, deposited on the products' surface with cylindrical symmetry by magnetron sputtering and exposed to intense heat loads during pulsed operation mode, it was executed numerical simulations of the temperature fields and the stress state caused by pressure and uneven temperature distribution. It was determined a coatings' material satisfying the minimum of internal stresses, which is one of the main criteria of their service life. Based on the calculation results titanium provides the maximum reduction in the internal stress while the usage of molybdenum as a coatings' material leads to the greatest decrease of pulsed temperature impact to the binary "steel–coating" systems with cylindrical symmetry.

1. Introduction

Numerical modeling of the temperature and stress fields in cable articles [1], heat pipes [2], thermal tubes [3] and shells [4] with cylindrical symmetry may be used at designing the thermal–stress modes for control systems. Formation of thermally induced residual stress due to plastic deformation during plasma etching step of the symmetric RF–MEMS fabrication process was explained and modeled using the Bauschinger effect [5]. There are several investigations reporting about residual stress evaluation on a cylindrical surface by use of the conventional X-ray stress measuring technique [6, 7].

Thus, the problem relevance of the temperature decrease and, as a result, thermal–stresses reduction in various fields of manufacturing industry caused by the wide usage of products with cylindrical symmetry, operating under dynamically changing temperature fields and aggressive environments. Surface modification of products for their service life extension by physical and chemical methods of thin coatings forming is one of the possible ways to solve this problem. Data on the temperature and internal stresses distribution require for determining the optimal coatings' composition. Simulations of the temperature fields, radial and tangential stresses in the binary "steel coating" systems with cylindrical symmetry was conducted. Titanium, molybdenum, tungsten and tantalum were used as coating materials for calculations.

2. Method of the temperature fields computation in systems with cylindrical symmetry

To determine the temperature fields we used the equation of thermal conductivity. As the temperature changes can reach 1000 K during heating, we cannot neglect the thermal parameters' dependence from the temperature; therefore the heat equation was solved numerically. All thermo–physical parameters of material were introduced as a piecewise linear function at intervals of 100 K, so it was used an implicit finite–difference scheme for the heat equation solution. To solve the heat equation the following initial and boundary conditions were used:

$$\left\{ \begin{array}{l} \varepsilon_1 \sigma (T_1^4 - T_{gas1}^4) + \alpha_1 (T_1 - T_{gas1}) = \lambda_1 \dfrac{\partial T}{\partial r_{(r=r_1)}} + Q(r, t) \\ T_1(r = R_{2-0}) = T_2(r = R_{2+0}); \; -\lambda_1 r_{R_{2-\frac{1}{2}}} \dfrac{\partial T_1}{\partial r} = -\lambda_2 r_{R_{2+\frac{1}{2}}} \dfrac{\partial T_2}{\partial r} \\ \varepsilon_2 \sigma (T_N^4 - T_{gas2}^4) + \alpha_2 (T_N - T_{gas2}) = \lambda_2 \dfrac{\partial T}{\partial r_{(r=r_2)}} \end{array} \right. \quad (1)$$

Here T_1 is the temperature of a sample surface, from influence of a thermal flow; T_{gas1} is the gas temperature, from a heated surface; T_N is the surface temperature, from the outside, not subjected to thermal influence; T_{gas2} is the gas temperature, from the outside, not subjected to thermal influence; ε is a surface blackness degree; σ is a Stephan-Boltzmann's constant; α is the coefficient of proportionality in Newton–Rikhman's equation; $Q(x, t)$ is the heat flow.

The first equation from system (1) reflects the radiation and convection heat transfer on the internal cylindrical surface with coating. The second equation relates to the interface between the coating and matrix. This boundary condition introduces the continuity of the temperature distribution and the heat flow equality on the coatings' border from matrix [8]. As initial conditions, the temperature is evenly distributed radially and is equal to 300 K. On the basis of the proposed calculation method was designed a computer program. Verification of its operation was carried out by comparison of analytical solution for the case of a fixed temperature from the outside and inside of cylinder with the constant thermal parameters. The verification was considered stationary when the temperature distribution did not change over time. As the computation results and the analytical solution show, a relative error has the differences of 0.089 %. To analyze the coatings behavior at the impulse impact of concentrated energy flows within the system with cylindrical symmetry, consider a pipe with internal diameter of 9 mm, the system pressure up to 100 MPa, the gas temperature of 3000 K. Determination of the heat transfer coefficient is performed by the empirical equation (2) at pressures in the range from 100 to 500 MPa:

$$\alpha_1 = 90 \times 4.18 \times 1000 \times (1000\,\tau)^{0.8384} \exp(-1.200\tau) \quad (2)$$

Here τ is the pulse duration [9].

Due to substitution of thermo–physical parameters' dependency from the temperature as the result of pulsed heating process we obtained the following temperature distribution (see figure 1). It was conducted a calculations of the pressure pulsed propagation with value of 100 MPa in a pipe with a diameter of 9 mm, at a gaseous medium temperature of 3000 K and 0.3 ms pulse duration. As calculations show titanium coating with a 10 μm thickness on a steel matrix has a maximum temperature while molybdenum coating has a minimum temperature. To analyze the protective coatings' application under different conditions, cooling water vapor was calculated in the system with

1000 K vapor temperature. In this case the maximum integral coefficient of heat transfer from the water vapor to the surface takes the value of 104 W/(m^{2}K). As the calculation results indicated the maximum temperature amounted to 618 K for 2 seconds heating duration. The temperature difference between the steel and the coated pipes was 1 K. The temperature distribution profile is completely determined by the matrix thermal conductivity. Analysis of different time intervals indicates that 1 ms impulse duration is the high value of the time range for protective coatings using to reduce the temperature maximum.

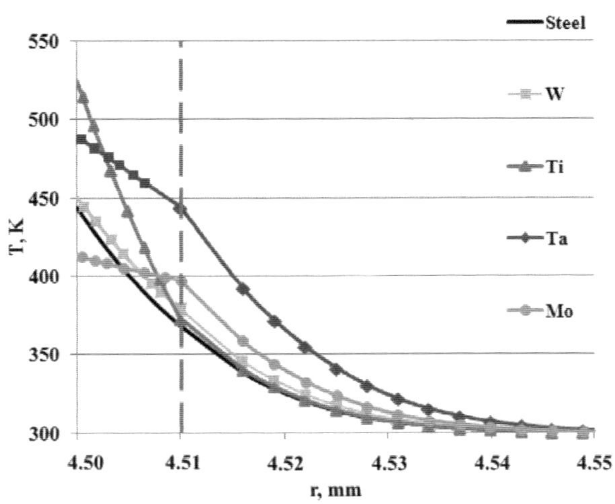

Figure 1. The temperature distribution through the radial direction of matrix with cylindrical symmetry and internal coating at 3000 K gas temperature, 0.3 ms pulse duration; the dashed line is the coatings' boundary.

To protect the surfaces of steel products with cylindrical symmetry from destruction during its operation at ultrafast pulsed heating and cooling impact from concentrated energy flows, the fields of temperature and internal stresses are the key determining factors. Internal stresses often lead to the mesh cracks formation with followed crumbling and product failure [10].

3. Method of the internal stress state computation in systems with cylindrical symmetry

The initial equations we use for internal stress calculation at uneven temperature distribution is the system borrowed from the reference [11]:

$$\left\{ \begin{aligned} \varepsilon_r &= \frac{1}{E_T}(\sigma - \upsilon_T \sigma_\tau) + \varepsilon_{T,r} \\ \varepsilon_\tau &= \frac{1}{E_T}(\sigma - \upsilon \sigma_r) + \varepsilon_{T,r} \\ \varepsilon_{T,r} &= \int_{T_0}^{T} \alpha_T dT \end{aligned} \right. \tag{3}$$

Here ε_r and ε_τ are the relative deformations in the radial and tangential directions correspondently; E_T is the Young's modulus (index indicating the parameter dependence from the temperature); σ_r and σ_τ are the radial and tangential stresses correspondently; v_T is the Poisson's coefficient; $\varepsilon_{T,r}$ is the deformation caused by thermal expansion; α_T is the linear expansion coefficient and T is the temperature.

To solve the system of equations (3) this paper proposes the following method. We use a combination of analytical and numerical scheme and divide the system by finite differences method on the radial sections so that it can be considered within these areas the physical and mechanical properties (E_T, v_T, α_T) do not depend on temperature. As a result of integration with using the boundary conditions we obtain:

$$\sigma_r = [P_0 - \frac{r_2}{r_2 - r_1}(P_0 - \frac{\alpha E}{(1-v)r_2^2}\int_{r_1}^{r_2} Trdr) - \frac{\alpha E}{(1-v)r^2}\int_{r_1}^{r} Trdr + \\ + \frac{1}{r}(\frac{r_1 r_2}{r_2 - r_1})(P_0 - \frac{\alpha E}{(1-v)r_2^2}\int_{r_1}^{r_2} Trdr)] \qquad (4)$$

$$\sigma_\tau = -[P_0 - \frac{r_2}{r_2 - r_1}(P_0 - \frac{\alpha E}{(1-v)r_2^2}\int_{r_1}^{r_2} Trdr) - \frac{\alpha E}{(1-v)r^2}\int_{r_1}^{r} Trdr + \\ + \frac{1}{r}(\frac{r_1 r_2}{r_2 - r_1})(P_0 - \frac{\alpha E}{(1-v)r_2^2}\int_{r_1}^{r_2} Trdr)] - \frac{\alpha E}{(1-v)} + \\ + P_0 - \frac{r_2}{r_2 - r_1}(P_0 - \frac{\alpha E}{(1-v)r_2^2}\int_{r_1}^{r_2} Trdr) \qquad (5)$$

Here P_0 is pressure inside the pipe, r_1 is the internal radius and r_2 is the external radius.

4. Results and discussion

The initial data we use for calculations are the following values: the pipes' diameter of 9 mm, at a temperature 3000 K of the gaseous medium, the pressure of 100 MPa, 0.3 ms pulse duration and 10 μm coating thickness at a different coatings composition. In accordance with calculation results (see figure 2), stresses are composed of two competitive components in the radial direction (4): the first type caused by the pressure inside the cylinder (compressive mechanical stresses) and the second induced by the temperature (stretching stresses). Thus, the resulting stress is decreased. With this type of mechanical loading stresses are determinative, in addition, they considerably less than a tensile strength of material and cannot lead to the cylinder destruction. Calculation of tangential (stretching) stresses show its reduction by products' with protective coatings compared with the unmodified steel parts (see figure 3). It has a great importance for increasing the products lifetime because the tangential stresses are responsible for the mesh cracks formation on the surface of parts with followed crumbling and product failure. Reducing the maximum of tangential stresses is determined by thermo–physical properties of the coatings' material. Based on the calculation results titanium provides the maximum reduction of the internal stress while the usage of molybdenum as a coating leads to the greatest decrease of the pulsed temperature impact in the binary "steel–coating" system with cylindrical symmetry.

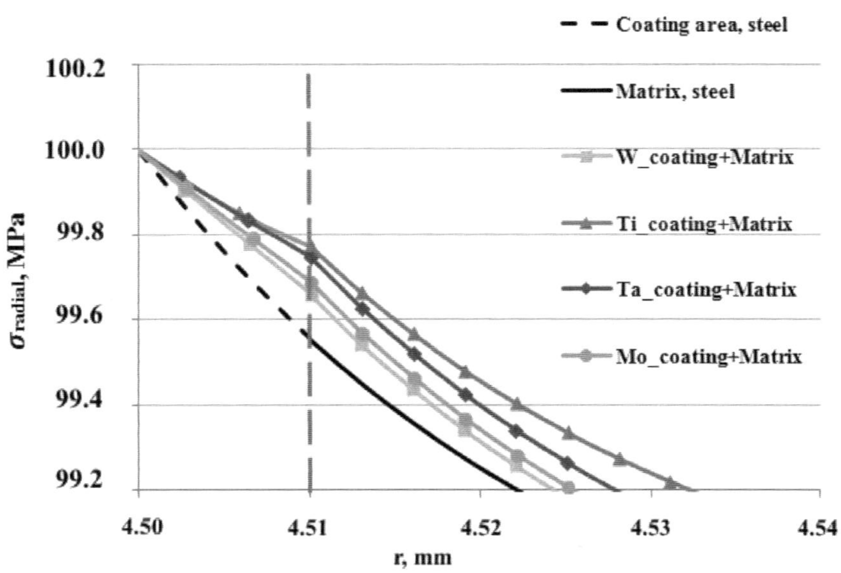

Figure 2. The radial stresses distribution through the matrix with cylindrical symmetry and internal coating at 3000 K gas temperature, pressure of 100 MPa, 0.3 ms pulse duration; the dashed line is the coatings' boundary.

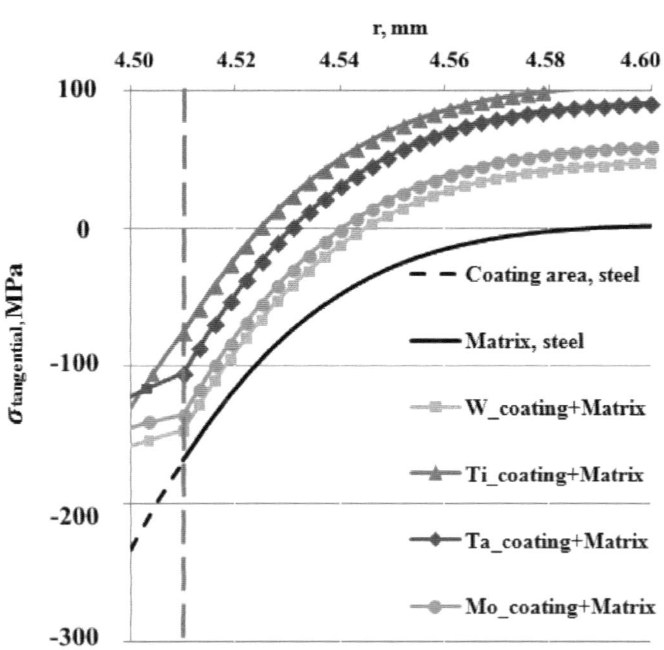

Figure 3. The tangential stresses distribution through the matrix with cylindrical symmetry and internal coating at 3000 K gas temperature, pressure of 100 MPa, 0.3 ms pulse duration; the dashed line is the coatings' boundary.

5. Conclusion

Analysis of different time intervals indicates that 1 ms impulse duration is the high value of the time range for protective coatings usage to reduce the temperature maximum. The greatest decrease of the temperature maximum provides by molybdenum coatings that reveals the optimal thermo–physical properties in the case when temperature is decisive for the products' service life (e.g. corrosion protection [12]). Tangential stresses take the maximum value and can reach 250 MPa under pulsed heating impact on the products with cylindrical symmetry. Application of protective coatings formed by physical and chemical methods (including magnetron sputtering) reduces tangential stresses up to 125 MPa that is quite significant for their service life.

References

[1] Kuznetsov G V and Ivanova E V 2011 *Chem. Petrol. Eng.* **47** 447
[2] Kuznetsov G V and Sitnikov A E 2002 *High Temp.–Engl. Tr.* **40** 898
[3] Kolousova A A and Kuznetsov G V 2006 *J. Eng. Phys. Thermophys.* **79** 377
[4] Chernopiskii D I 2012 *Strength Mater.–Engl. Tr.* **44** 40
[5] Somà A and Saleem M M 2015 *J. Micromech. Microeng.* **25** 055007
[6] Oguri T, Murata K, Uegami K and Sato Y 2002 *J. Strain Anal. Eng. Des.* **37** 427
[7] Oguri T, Murata K, and Sato Y 2003 *J. Strain Anal. Eng. Des.* **38** 459
[8] Blesman A I, Postnikov D V and Polonyankin D A 2015 *IOP Conf. Ser.: Mater. Sci. Eng.* **81** 012031
[9] Aralov A D, Nekrasova E I and Yudaev B N 1985 *Glass Ceram.* **42** 190
[10] Blesman A I, Postnikov D V and Polonyankin D A 2015 *Procedia Eng.* **113** 413
[11] Andreev V I, Yazyev B M and Chepurnenko A S 2014 *Advanced Materials Research* **900** 707
[12] Tkachenko E A, Postnikov D V, Blesman A I and Polonyankin D A 2016 *IOP Conf. Ser.: Mater. Sci. Eng.* **110** 012009

Modelling of chemical reactions in plasma

N E Aktaev[1], G E Remnev[1] and A P Yalovets[2]
[1]National Research Tomsk Polytechnic University, Tomsk, Russia
[2]National Research South Ural State University, Chelyabinsk, Russia

E-mail: nurkenaktaev@gmail.com

Abstract. The paper is devoted to theoretical investigation of interaction of pulsed high current electron beam with gas substance. As a result of the interaction the formation of chemical active plasma can be observed. One of the key parameter for theoretical analyze of the process is the electron distribution function. Within the framework of the Boltzmann approach we obtained the dynamical equation for electron distribution function depending on the electron energy, coordinate and time.

1. Introduction

The influence of concentrated of energy fluxes on the gas substance attracts the interest of many theoretical groups [1-9]. As a source of the energy fluxes it is often used the discharge and lasers. The action time of the energy fluxes is small ($\approx$ $10^{-3} - 10^{-6}$ sec) compared with typical time of physical and chemical processes in created plasma ($10^{-7} - 10^{-12}$ sec). In these cases the stationary equations for describing the electron distribution function (EDF) is applicable [1, 2, 4, 6, 8].

Another method of action on gas is the using of pulsed high current electron beams [10]. The pulsed duration of such accelerators is nearly $10^{-6} - 10^{-10}$ sec and the power flow of $10^6 - 10^9$ W/cm^2. Under the action of the beams on a chemical active gas the nonequlibrium plasma is formed. Note that the physical and chemical processes occur simultaneously with spreading of beam. In this case the stationary describing is not applicable. Moreover, the concentration of chemical active particles changes significantly for small time interval. In addition if one use the power flow of $10^6 - 10^9$ W/cm^2 the priority and rates of physical and chemical processes are changed.

In present paper we obtained the dynamical equation for time evolution of EDF based on the Boltzmann equation. The equation allows investigating the electron beam spreading in the gas substance.

2. Boltzmann equation

The time evolution of electron distribution function $f = f(\vec{r}, \vec{\mathrm{v}}, t,)$ in the space of the coordinates $\vec{r}$ and the velocities $\vec{\mathrm{v}}$ is described by means of Boltzmann equation

$$\frac{\partial f}{\partial t} + \vec{\mathrm{v}}\nabla_r f + \frac{\vec{F}}{m}\nabla_{\mathrm{v}} f = \left(\frac{\delta f}{\delta t}\right), \qquad (1)$$

where e, m are the elementary charge and mass of electron respectively, $\vec{F}$ is the electromagnetic force created by beam electrons, $(\delta f / \delta t)$ is the collision integral. Under the assumption of weak anisotropy it can be used a Legendre polynomial expansion P_n for EDF [1]

$$f = \sum_{n=0}^{\infty} f_n P_n\left(\frac{\vec{\mathrm{v}}}{\mathrm{v}}\right). \tag{2}$$

Inserting in Boltzmann equation (1) the expansion (2) to first order the second term is written as

$$\vec{\mathrm{v}}\nabla_r f = \vec{\mathrm{v}}\left(\frac{\vec{r}}{r}\frac{\partial f_0}{\partial r} + \frac{\vec{\mathrm{v}}}{\mathrm{v}}\frac{\vec{r}}{r}\frac{\partial \vec{f}_1}{\partial r}\right), \tag{3}$$

and the second term is

$$\frac{\vec{F}}{m}\nabla_{\mathrm{v}} f = \frac{\vec{F}}{m}\left[\frac{\vec{\mathrm{v}}}{\mathrm{v}}\frac{\partial f_0}{\partial \mathrm{v}} + \frac{1}{3\mathrm{v}^2}\frac{\partial}{\partial \mathrm{v}}\left(\mathrm{v}^2 \vec{f}_1\right)\right], \tag{4}$$

where f_0, f_1 are the isotropic and anisotropic parts of EDF respectively. Thereby, after EDF expansion the (1) can be represented as the two equations for zero and first order of velocity

$$\begin{cases} \dfrac{\partial f_0}{\partial t} + \dfrac{\mathrm{v}}{3}\dfrac{\vec{r}}{r}\dfrac{\partial \vec{f}_1}{\partial r} + \dfrac{\vec{F}}{3m\mathrm{v}^2}\dfrac{\partial}{\partial \mathrm{v}}\left(\mathrm{v}^2 \vec{f}_1\right) = \left(\dfrac{\delta f_0}{\delta t}\right) \\ \mathrm{v}^{-1}\dfrac{\partial \vec{f}_1}{\partial t} + \dfrac{\vec{r}}{r}\dfrac{\partial f_0}{\partial r} + \dfrac{\vec{F}}{m\mathrm{v}}\dfrac{\partial f_0}{\partial \mathrm{v}} = \mathrm{v}^{-1}\left(\dfrac{\delta \vec{f}_1}{\delta t}\right) \end{cases}. \tag{5 a, b}$$

3. Dynamical equation

The left part of (5b) after substitution (2) is written as

$$\left(\frac{\delta \vec{f}_1}{\delta t}\right) = -\mathrm{v}\vec{f}_1 \cdot \sum_k N_k(t)\left[Q_{el}^k(\mathrm{v}) + \sum_j Q_{inel}^{jk}(\mathrm{v})\right], \tag{6}$$

where $N_k(t)$ is the time function of molecular concentration of various sorts k, Q_{el}^k is the cross section of elastic interaction between electrons and k-molecules, Q_{inel}^{jk} is the cross section of j-type inelastic interaction. For obtaining (6) we supposed that the decreasing velocity of f_1 is proportional to the collision frequency between electrons and molecules. Substituting the (6) in (5 b) it is gets

$$\mathrm{v}^{-1}\frac{\partial \vec{f}_1}{\partial t} + \frac{\vec{r}}{r}\frac{\partial f_0}{\partial r} + \frac{\vec{F}}{m\mathrm{v}}\frac{\partial f_0}{\partial \mathrm{v}} = -\vec{f}_1 H(\mathrm{v},t), \tag{7}$$

$$H(\mathrm{v},t) = \sum_k N_k(t)\left[Q_{el}^k(\mathrm{v}) + \sum_j Q_{inel}^{jk}(\mathrm{v})\right], \tag{8}$$

The solution of the differential equation (8) is

$$\vec{f}_1=\left[\mathrm{v}\frac{\vec{r}}{r}\frac{\partial f_0}{\partial r}+\frac{\vec{F}}{m}\frac{\partial f_0}{\partial \mathrm{v}}\right](H\mathrm{v})^{-1}+\exp(-H\mathrm{v}t). \qquad (9)$$

The second term can be neglected because the $H\mathrm{v}$ has the order of magnitude is nearly 10^{-12} sec that ever for pulsed regime (10^{-7} sec) is very small time. Then the (9) is rewritten

$$\vec{f}_1=\left[\mathrm{v}\frac{\vec{r}}{r}\frac{\partial f_0}{\partial r}+\frac{\vec{F}}{m}\frac{\partial f_0}{\partial \mathrm{v}}\right](H\mathrm{v})^{-1}. \qquad (10)$$

Substituting the (10) in the (5 a) we get

$$\frac{\partial f_0}{\partial t}+\frac{1}{3H}\frac{\vec{r}}{r}\frac{\partial}{\partial r}\left[\mathrm{v}\frac{\vec{r}}{r}\frac{\partial f_0}{\partial r}+\frac{\vec{F}}{m}\frac{\partial f_0}{\partial \mathrm{v}}\right]+\frac{\vec{F}}{3m\mathrm{v}^2}\frac{\partial}{\partial \mathrm{v}}\left[\frac{\mathrm{v}}{H}\left(\mathrm{v}\frac{\vec{r}}{r}\frac{\partial f_0}{\partial r}+\frac{\vec{F}}{m}\frac{\partial f_0}{\partial \mathrm{v}}\right)\right]=\left(\frac{\delta f_0}{\delta t}\right). \qquad (11)$$

For convenient let us transform from velocity space to energy space $u=m\mathrm{v}^2/2e$, where $\partial/\partial\mathrm{v}=m\mathrm{v}e^{-1}\partial/\partial u$ [2]. Then dynamical equation (11) can be rewritten as

$$\frac{\partial f_0}{\partial t}+\frac{1}{3H}\sqrt{\frac{2u}{m}}\frac{\vec{r}}{r}\frac{\partial}{\partial r}\left[\sqrt{e}\frac{\vec{r}}{r}\frac{\partial f_0}{\partial r}+\frac{\vec{F}}{\sqrt{e}}\frac{\partial f_0}{\partial u}\right]+\frac{\vec{F}}{3e}\sqrt{\frac{2}{mu}}\frac{\partial}{\partial u}\left[\frac{u}{H}\left(\sqrt{e}\frac{\vec{r}}{r}\frac{\partial f_0}{\partial r}+\frac{\vec{F}}{\sqrt{e}}\frac{\partial f_0}{\partial u}\right)\right]=\left(\frac{\delta f_0}{\delta t}\right). \qquad (12)$$

The right path of the equation (12) contains the inelastic and the hyper elastic terms

$$\left(\frac{\delta f_0}{\delta t}\right)=\left(\frac{\delta f_0}{\delta t}\right)_{inel}+\left(\frac{\delta f_0}{\delta t}\right)_{hypel}. \qquad (13)$$

In work [2] the relation for the inelastic term was derived. We have generalized this relation for the time dependent case and have obtained the relation for hyper elastic term

$$\left(\frac{\delta f_0}{\delta t}\right)_{inel}=\sqrt{\frac{2e}{mu}}\sum_j\sum_k\left[(u+u_{jk})f_0(u+u_{jk})N_k(t)Q_{inel}^{jk}(u+u_{jk})-uf_0(u)N_k(t)Q_{inel}^{jk}(u)\right], \qquad (14)$$

$$\left(\frac{\delta f_0}{\delta t}\right)_{hypel}=\sqrt{\frac{2e}{mu}}\sum_j\sum_k\left[(u+u_{jk})f_0(u+u_{jk})N_{jk}(t)Q_{inel}^{(-j)k}(u+u_{jk})-uf_0(u)N_{jk}(t)Q_{inel}^{(-j)k}(u)\right].$$
(15)

where u_{jk} is the energy of j-type excitation state of k-molecule, $N_{jk}(t)$ is the time function of molecular concentration of various sorts k at the j-type excitation state. The cross section of the hyper elastic collisions is related with the Q_{inel}^{jk} by means of the ratio

$$Q_{inel}^{(-j)k}=\frac{u+u_{jk}}{u}Q_{inel}^{jk}(u+u_{jk}). \qquad (16)$$

4. Correspondence principle

The relations (12)-(15) are described the time evolution of electron distribution function. The main features are the possibilities to imply the equation (12) for the gas substance and the time dependence of the concentration of molecules at the various stages: ground and excited. If in (12) we eliminate the mixed derivatives, the coordinate and time dependences in (14) and (15) we obtain the "classical" formulae derived by W.L. Nighan [2].

5. Conclusion

The dynamical equation for describing of spreading process of pulsed high current electron beam in a gas substance is obtained. The relation was derived by using Boltzmann equation. The time dependence of chemical active particle concentration is a main feature of the dynamical equation. It is argued that the obtaining relation in limit case transform to the equation previously received in other works.

Acknowledgment

The reported study was funded by RFBR according to the research project No. 16-32-00071 mol_a. and within the framework of the Programme on National Research Tomsk Polytechnic University Competitiveness Enhancement in the Group of Top Level World Research and Academic Institutions.

References

[1] Petrov G, Winkler R 1997 *Jour. Phys. D: Appl. Phys.* **30** 53
[2] Nighan W L 1970 *Phys. Rev A* 2 1989
[3] Aktaev N E 2014 *Jour. of Phys.: Conf. Ser.* **552** 012033
[4] Winkler R, Arndt S, Loffhagen D, Sigeneger F, Uhrlandt D 2004 *Contrib. Plasma Phys.* **44** 437
[5] Loffhagen D, Arndt S, Sigeneger F, Uhrlandt D, Winkler R 2005 *Contrib. Plasma Phys.* **45** 309
[6] Sigeneger F, Winkler R 1996 *Contrib. Plasma Phys.* **36** 551
[7] Maulois M, Ribiere M, Eichwald O, Yousfi M, Azais B 2016 *Phys. of Plasmas* **23** 043501
[8] Starikovskiy A, Aleksandrov N 2013 *Prog. In Energy and Combust. Science* **39** 61
[9] Starikovskaia S M, Starikovskii A Yu, Zatsepin D V 2001 *Contrib. Plasma Phys.* **5** 97
[10] Pushkarev A I, Remnev G E 2005 *Combust. Explos. and Shock Waves* **41** 278

IOP Conf. Series: Materials Science and Engineering **168** (2017) 012052 doi:10.1088/1757-899X/168/1/012052

Influence of laser irradiation on change properties of bulk amorphous Zr-Pd metallic alloys

V A Fedorov[1], A V Yakovlev, T N Pluzhnikova, A A Shlikova and A D Berezner
Tambov State University n.a. G R Derzhavin, 392000, Tambov, Russia

E-mail: feodorov@tsu.tmb.ru

Abstract. We study the morphological features of laser irradiation zones formed on the surface of the bulk metallic glasses. We use the nanoindentation method for estimation alloys properties caused by impulse heating during irradiation.

1. Introduction

There are many investigations [1-6] of new class of amorphous materials so-called bulk amorphous metal alloys (AMS) or metallic glasses (MS) during last decade. Alloys based on ZrC are especially interesting among bulk MS [7, 8] because they have very wide supercooled liquid region which exceeds 100K for a number of alloys. These alloys are of interest in connection with the possibility to obtain nanocrystalline structure [9] by heating. Palladium-based alloys are highly resistant to damages. They combine hardness inherent in glasses with typical metal resistant for cracking. The high cost of palladium [10] is the root cause of narrow usage such glasses in technique.

A comprehensive study of the evolution structure regularities and mechanical properties of the heating bulk MS is an actual area of applied and fundamental research. MS evaluation of mechanical properties in the areas of local action, in particular, pulses of coherent radiation is practically significant challenge due to the fact that the laser radiation is one of ways to heating treatment.

Objective: To establish the mechanical properties of bulk MS based on Pd and Zr, and a study of morphological features of regions formed as a result of action of laser radiation.

2. Materials and methods

We investigate bulk MSs based on zirconium and palladium. We use in experiments samples with dimensions 2x5x4 mm for both systems: Zr - Ti -Cu - Ni - Al (52,5% Zr) and Pd - Cu - Ni - P (40% Pd). We use as a laser light source "LTA-4-1" with the active element based on yttrium aluminum garnet doped with neodymium (Nd: YAG), with a wavelength λ = 1064 nm, which allows to obtain different forms of impulses and energies. We use device IEM-4-1 for the energy and influence time of the laser impulse measuring. The pulse duration ~ 2-4 ms. We use pulses with an energy of 0.6 J / mm2. We affect on surfaces of amorphous alloys by laser radiation in a mixture of argon and air. We prepare working surfaces as a metallographic grinding. We measure material properties changing after laser radiation by the method of nanohardness measuring by Nano Indenter G200 with diamond Berkovich indenter with a radius of blunting on the top R = 20 nm and the loading mode with constant relative deformation rate of 0.1 s-1. Amplitude (2 nm) and frequency (45 Hz) of harmonic component of the load keep constant. We use for investigation of morphological features and elemental surfaces composition by scanning ion-electron microscope Quanta 200 3D.

The focused pulsed laser radiation forms local areas of irradiated material on bulk MS surfaces (Fig. 1). Each area has its own morphological features.

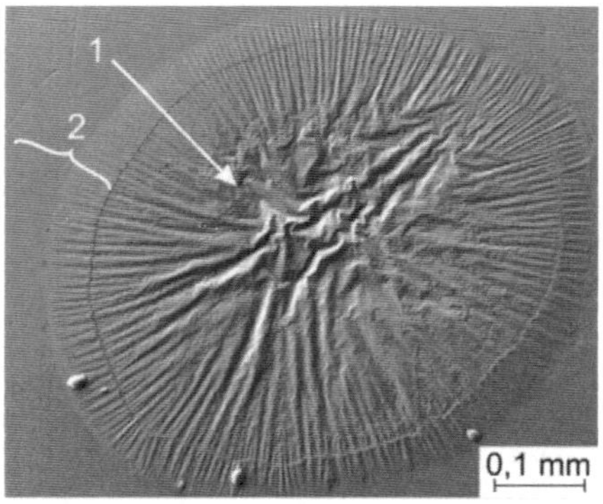

a)

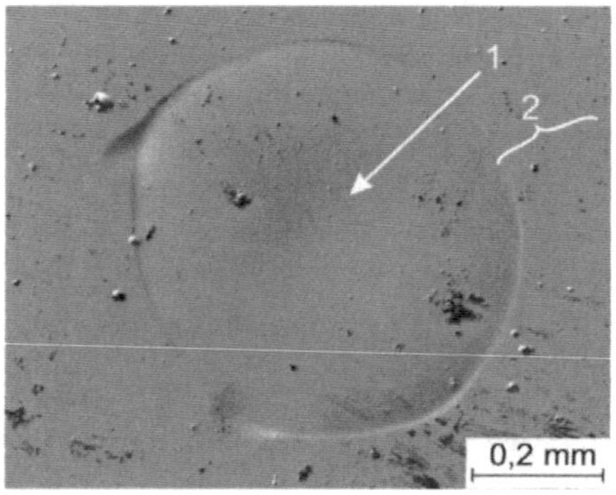

b)

Figure 1. The morphology of the laser irradiation areas: a) Zr -based alloy MS; b) Pd -based alloy MS. 1– region of the melting, 2– heat-affected area.

Affected areas in bulk AMS have a “rosette”-appearance consists of radially growing crystals formed in the central region of the melting. The affected areas dimensions are vary depending upon the impulse energy and it ranges from 0.4 mm to 0.6 mm. There are two areas: melting area and heat-affected area. We see crystallization out of melting area, which is not related to material melting. The surface temperature at the center of the radiation-affected area is not lower than the melting temperature (~ 18550 C). The temperature of melted abroad area is not less than crystallization temperature (450^0 C by DSC researches).

The relief formed on the surface relates with volume effect during crystallization. Dilatometric researches confirm it. Our dilatometric researches demonstrate that the crystallization of the sample based on alloy Zr occurs in two stages. Volume effect corresponding to the second conversion step is more than volume effect in the first stage. Full volume effect of crystallization is about 2%, which is a typical value for the crystallization of metallic glasses [11].

We see the presence of the crystallization in the laser affected area [12] (Fig. 2). Radiographic researches confirm this.

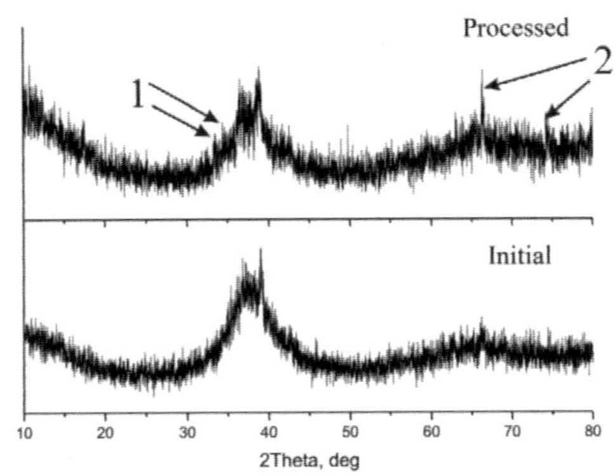

Figure 2. The diffractograms of Zr-based alloy surface: in laser irradiation area and fresh surface. Arrow 1 shows the peaks linked with crystallization, 2 - peaks linked with the oxides of zirconium formation.

We do not observe any structural changes in the center of the affected by radiation zone in Pd-based alloy (Fig. 2b). The impact zone has appearance of a "moon crater". Metallographic method cannot detect affected by heating zone. Crystals are growing after the rising of impulse quantity in Zr-based alloys. Changes in affected area in Pd-based alloys are not significant.

We determine element compound in different points of affected areas. It presents that oxygen contents in the middle of affected areas in Zr-based alloys is growing more than 4 times in accordance with fresh material. There are not any same changes in Pd-based alloys.

We use the nanoindentation-method for researching of mechanical properties in laser affected areas which are formed on bulk MS surfaces in comparison with fresh condition.

Figure 3 shows P-H diagram for Zr-based alloy before irradiation. We can select several typical sections on this diagram: monotonic increase h during increase of load and creep. Diagram P-h for fresh material has differences from diagram inside irradiation zones for both values – P and h. In addition, P-h diagrams occur into irradiation zones and are followed by irregular deformation during indenter indentation (fig.3b). Values of P and h are close to values of P-h of fresh material.

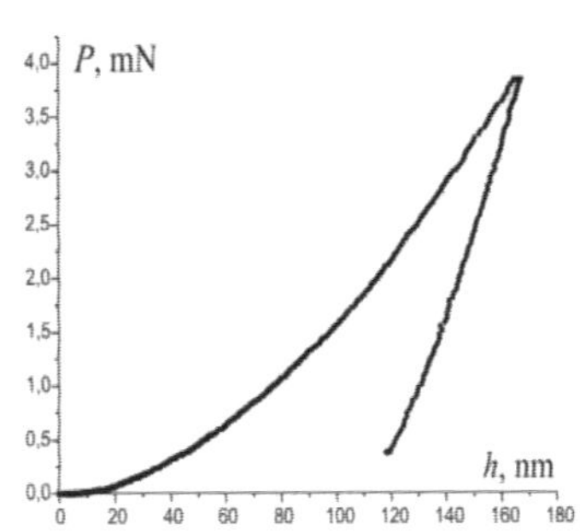

a)

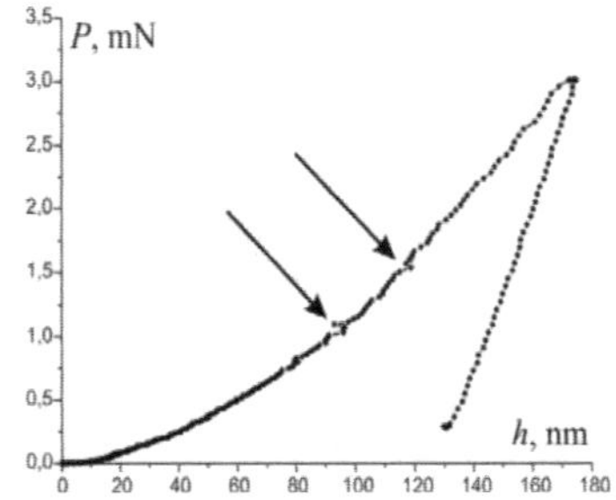

b)

Figure 3. $P-h$ diagram for Zr-based alloy a) fresh condition б) in affected area; P – load, h – the indentation depth, arrows mark the plot of changing deformation.

We can see P-h diagrams on fig.4 for Pd-based alloy inside and outside of affected area. They are similar. We can separate typical sections: monotonic increase h during increase of load and irregular increment of h.

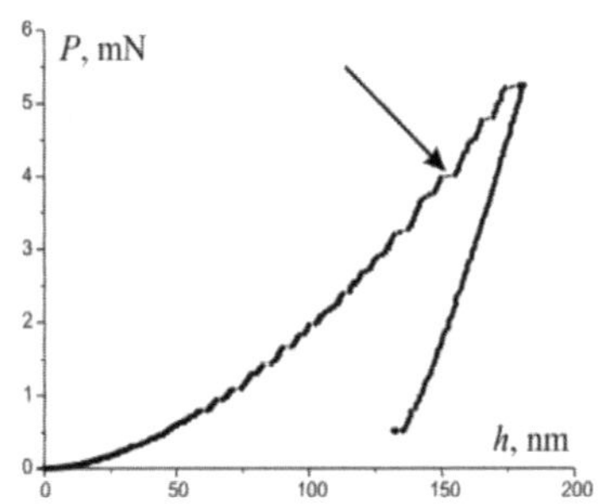

a)

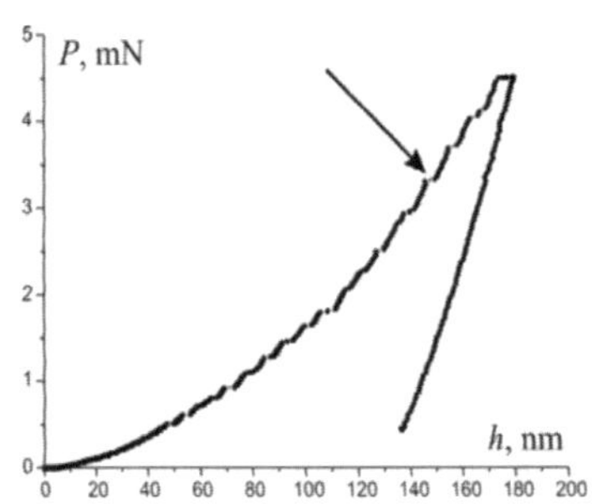

b)

Figure 4. $P-h$ diagram for Pd-based alloy: a) fresh condition, б) in affected area arrows mark the plot of changing increment h.

Based on our experiments results we calculate nanohardness (H) and Young's modulus (E) inside irradiation zone and in fresh condition for researched alloys (see table).

Table 1. The mechanical characteristics of alloys in irradiation area and in fresh condition.

	Alloy	H (GPa)	E (GPa)	H_{out}/H_{in}	E_{out}/E_{in}
Zr-based alloy	in irradiation area	4,3±0,5	90,2±5	1,8	1,08
MS	fresh material	7,8±0,5	98±5		
Pd-based alloy	in irradiation area	6	96,5	1,1	1,05
MS	fresh material	6,7	101		

3. Discussion of results

We interpretive this results next ways. Structure changed bulk glass→GPU crystal is the root cause of nanohardness (H) and Young's modulus (E) variation in Zr-based alloys. GPU- crystals growth is parallel to observe plane along the main c-axes. In connection with this the basal plane (0001) – easy glide plane – becomes the area of crystals deformation. We connect this reason with lower values of nanohardness and Young's modulus. Based on chemical composition of Zr-based alloys we can make a conclusion that oxygen, which is rising into irradiation zone, react with Zr. This reaction produces poorly soluble oxides which become central points of crystallization. Also such oxides are the root cause of crystals rising into the heat affected area. We research laser affected zones by a probe microscope. We can determine linear crystallization central points along the border of melting. Values are in the range from 900 to 2100 cm^{-1} (Fig.5).

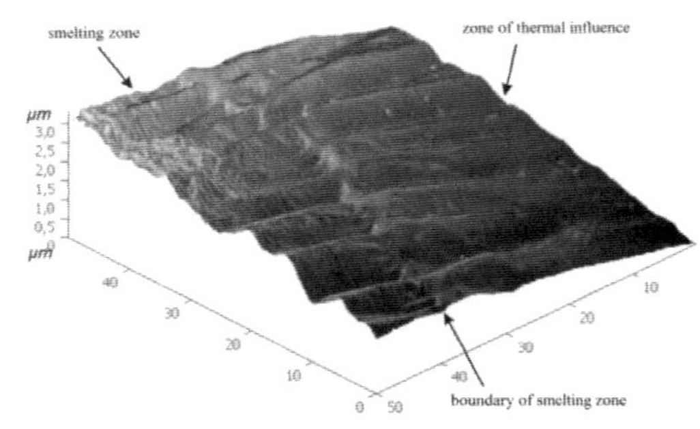

a)

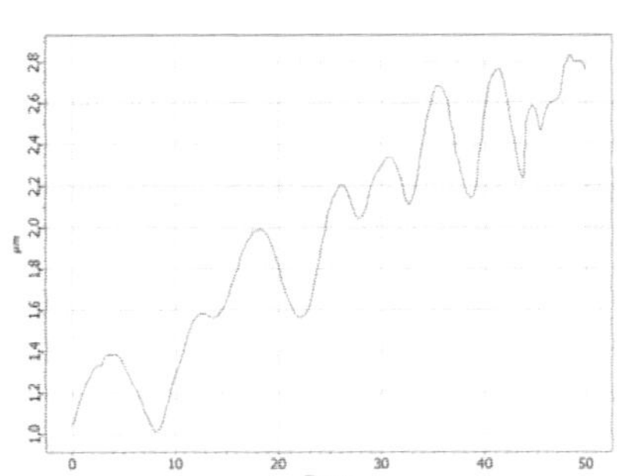

b)

Figure 5. The crystallization on the surface of the heat affected area (a) and its profilogram in the direction of transverse to the crystal growth direction (b).

We can confirm of ZrO_2 inclusion inside the irradiation affected area by $P - h$ diagram (Fig. 3b). Deformation jumps onto diagram (Fig. 3b) are connected with the formation of microcracks. The values of the Young's modulus and nanohardness, which we calculate for this chart, are close to the corresponding values of zirconium ceramics [13]. We can see processes of the secondary vitrification in irradiation affected areas of Pd-based alloy. This is due to high values of viscosity coefficient of Pd-based alloy and thermal conductivity [14]. First complicates crystal growth the second - ensures rapid dissipation of thermal energy in the sample. Derating E and H for 5% and 10%, respectively, may be associated with a decrease of initial hardening stresses inside irradiation affected zones. Thus we make a conclusion that laser pulse changes mechanical properties of researched material by the result of material local heating in affected area. Kind of irradiation affected zones in these cases is different, due to differences in thermal properties of materials. The directional solidification is coming in the center of the affected zone of Zr-based alloy. Crystals grow from the center to the periphery. The affected area on Pd-based alloy surface looks like "moon crater" without any heat affected area due to the high viscosity of the alloy, which prevents processes crystallization. The morphology of the affected zone on the surface of a Zr-based alloy depends on the shape and the number of impulses of the irradiation by laser. The laser irradiation forms on the Pd-based surface relief which form does not depend from shape and quantity of laser impulses. Nanoindentation method shows that affected zones values of nanohardness and elasticity module for volume Zr,Pd-based MSs are changed in comparing with the original material in the area of laser irradiation. In the Zr -based alloy the reduction of nanohardness and Young's modulus are related to the structural transformation of metallic glass=>GPU chip. In Pd-based alloy such reduction processes associated with secondary vitrification, which is going on the surface.

Acknowledgments
This work has been supported by PFBR (grant № 15-42-03206). Specimens have been prepared by Professor V.A. Khonik.

Refrences:

[1] Berlev A E, Ota M, Khonik V A 2003 *Tambov University Reports. Series: Natural and Technical Sciences* **8** 522
[2] Kobelev N P, Kovivanov E L, Khonik V A 2005 *Solid State Physics* **47** 395
[3] Kobelev N P, Kovivanov E L, Khonik V A 2005 *Solid State Physics* **47** 400
[4] Fedorov V A, Yakovlev A V, Pluzhnikova T N, Kapustin A N 2009 *Basic problems of materials science* **6** 87
[5] Yakovlev A V, Chernova I V, Fedorov V A, Barishev V A 2009 *Tambov University Reports. Series: Natural and Technical Sciences* **14** 211
[6] Yakovlev A V, Pluzhnikova T N, Cheremisina Yu V, Fedorov V A, Tarasova I D 2010 *Tambov University Reports. Series: Natural and Technical Sciences* **15** 1105
[7] Kornevistiy Yu K 1999 Space-amorphysed metallic glasses (Moscow: Science) p 80
[8] Gleser A M, Permyakova I E, Gromov V E, Kovalenko V V 2006 Mechanical behavior amorphous alloys (Novokuznetsk: SibSIU) 2006 p 416
[9] Abrosimova G E, Aronin A S, Matveev D V, Molokanov V V 2004 *Solid State Physics* **46** 2119
[10] Demetriou M, Launey M, Garrett G, Schramm J, Hofmann D, Johnson W, Ritchie R 2011 *Nature Materials* doi:10.1038/nmat2930
[11] Bakai A S, Bakai S A, Michailovskaya I M, Nekludov I M, Stoev P I, Macht M P 2002 JETP Letters. **76** 254
[12] Sun H, Flores K 2013 *Intermetallics* **43** 53
[13] Rutman D, Tropov Yu, Pliner S, Neymin A, Polegaev Yu 1985 High-refractory materials of zirconium dioxide (Moscow: Metallurgy) p 136
[14] Kikoin I 1976 Reference of physical values (Moscow: Atomizdat) p 1008

IOP Conf. Series: Materials Science and Engineering **168** (2017) 012053 doi:10.1088/1757-899X/168/1/012053

Study of the physical properties of materials by the photometric analysis of structure images

V A Ermishkin, S P Kulagin, A K Tomenko, N A Minina and J B Solov'eva

Federal State Institution of Science Baykov Institute of Metallurgy and Materials Russian Academy of Sciences, Moscow, Russia

E-mail: minina1951@rambler.ru

Abstract. Several results of using the photometric analysis of structure images (PHASI) for the determination of the physical properties of materials and for the estimation of some specific features of their structural state are considered. In particular, it is shown that, using PHASI, one can construct the fatigue curve of material from standard fatigue tests by cantilever bending of 3-4 flat samples. Application of PHASI for the determination of the thermal properties of materials is shown by the example of determining the coefficient of thermal diffusivity of a copper foil subjected to the action of single laser pulse. The structure features of molybdenum single crystals oriented according to the corners of the standard stereographic triangle have been identified by PHASI and serve as example of PHASI use for the structure examination.

1. Introduction

Man gets visually more than 80% of the information necessary for productive activities. Such information it not only gives a general idea about the observed object such as its shape, color, and design features, but also allows one to quantitatively estimate its size, weight, and mechanical characteristics. Therefore, it is natural to assume that the visible light flux, which is emitted by a source with known parameters, reflected from the object under study, and analyzed by objective methods, can give information about the structure and properties of material. In fact, upon the reflection from the surface of the object, the incident light is converted into the reflected light, which carries the information about the structure elements interacting with the photons of the incident light. As follows from modern physics [1, 2], the elementary act of light reflection is a photon-electron interaction. The fact that the phenomenon of light reflection is due to the photon-electron interaction is confirmed by the experimental results obtained by photometric analysis of structure images (PHASI) [3] (Figure 1). The reflection brightness spectrum is plotted in the «reflection brightness I - spectral density $p(I)$" coordinates. Both coordinates are measured in arbitrary dimensionless units. The zero value on the abscissa axis corresponds to the complete absorption of the incident light flux by the object surface, and the unit corresponds to the complete reflection of light from the surface. The spectral density is taken as the ratio of the number of pixels with the brightness corresponding to the predetermined brightness scale micro range to the total number of pixels of the image. The ordinate axis also ranges between 0 and 1. It is seen from Fig. 1 that, in the initial state, the reflectivity of ebonite is zero, but its electrification by friction with the help of wool fabric results in the appearance of reflection brightness spectrum from the surface areas identified by the color detection of non-zero spectral frequency range in the spectrum. The experiments conducted using PHASI have shown that

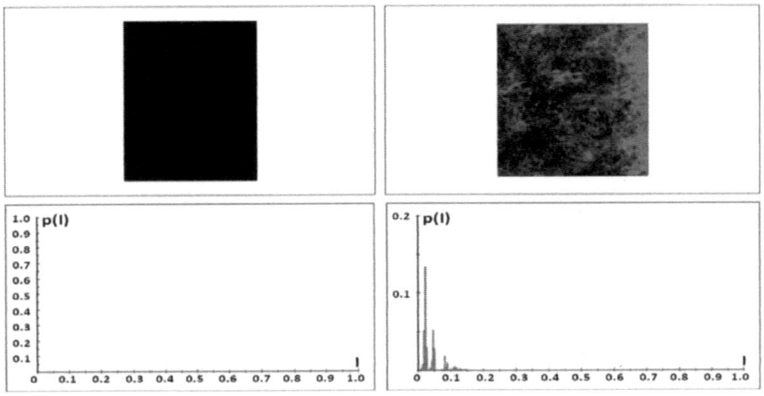

Figure 1. Results of the photometric analysis of ebonite: (left) image fragment and brightness spectrum from its surface and (right) the same fragment and its spectrum after ebonite electrification by friction.

external actions cause changes in the density of electron states at the surface of the object. Such actions can increase the surface roughness (mechanical actions), change the chemical composition (corrosive actions), intensify the lattice vibrations (temperature actions), and directly change the energy states of electrons (electromagnetic actions). Since the properties of materials quantitatively measure the changes in their structural state with changing intensity of the external action, the application of PHASI opens the possibilities for measuring various physical properties corresponding to the nature of the action. Some results of such studies are presented in our work.

2. PHASI - a rapid method of fatigue tests

In particular, PHASI was the basis for the development of a rapid method of fatigue tests on flat samples subjected to alternating bending by cantilever loading scheme. The gage surface image recorded prior to the fatigue testing is fed into computer and is broken into fragments of 10×10 mm^2 in size. After bringing the sample to failure, the sample surface is repeatedly scanned and the images in the digital code are fed into the personal computer, where they are analyzed by PHASI. Figure 2 shows the distribution of structural damage over the sample length according to the PHASI results. The structural damage (*Ds*) is understood as the expression:

$$D_s = \frac{p_i(t_r) - p_i(0)}{(p_{max}(t_r) - p_{max}(0))} \qquad (1)$$

where: $p_i(t_r)$ is the spectral density of the i-th fragment at the time of the sample failure, $p_i(0)$ is the same, but before the start of the test, $p_{max}(t_r)$ is the spectral density in the fragment, in which the sample failed, and $p_{max}(0)$ is the same, but before the start of the test. The amplitude of operating stresses in each fragment was taken as the maximum bending stress in the middle of the fragment. Such maximum stress was defined by the formula of strength of materials for the sample loaded by the distributed load q determined by the deflection at its free end. The final formula for calculating the effective stress is as follows: where: W is moment resistance of the cross section, l is the gage length of the sample, and z is the coordinate along its length from the sample head. The time to failure of the unbroken fragments, t_{ri}, was determined according to photometric measurements of damage by Eq. (1) and the time to failure of the sample, t_r.

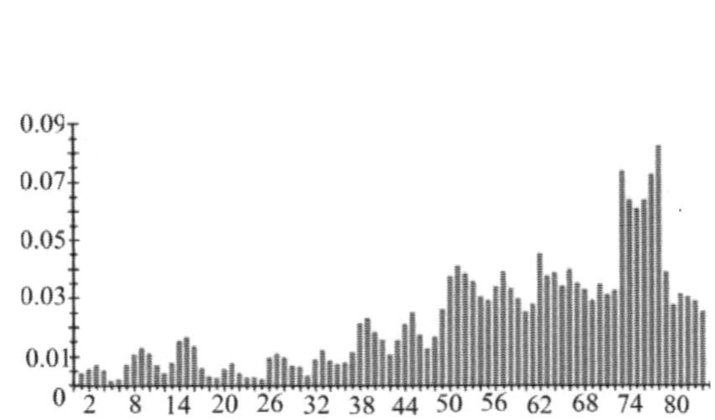

Figure 2. Damage distribution in the 08KP steel sample tested for fatigue.

$$\sigma(z) = \frac{M(z)}{W} = \frac{q \cdot (l^2 - z^2)}{2W} \quad (2)$$

The formula for estimating the time to failure t_{ri} was obtained from the equation of structural damage, Eq. (1), and from the relationship between the times to failure of fragments and the sample as a whole:

$$t_{ri} = t_r \cdot \frac{D_S}{D_{Si}} \quad (3)$$

The fatigue curve (Figure 3) was constructed for the data sets obtained from Eqs. (2) and (3). It was found that the experimental data shown in Fig. 3 are well approximated by the dependence:

$$\sigma = \sigma_R + \frac{B}{\sqrt{-t_r}} \quad (4)$$

Note that PHASI allows one to substantially reduce the number of samples due to the possibility to use the structural damage measurement at both gage surfaces of samples and their subdivision into fragments. At a gage length of 140 mm, one sample tested becomes equivalent to 28 standard samples.

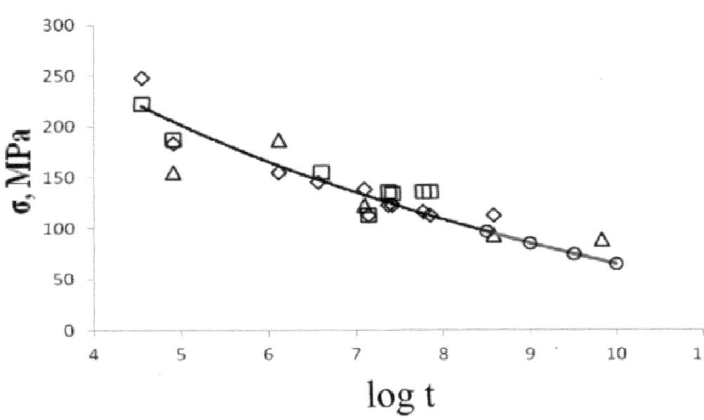

Figure 3. Fatigue curve constructed according to the PHASI data for the 08KP steel: (♦) results of PHASI measurements on the lower sample surface, (■) data obtained by standard fatigue tests, (▲) results of PHASI measurements on the upper sample surface.

3. PHASI in the study of thermo physical properties of materials

The capability of PHASI in the study of thermo physical properties of materials can be illustrated by the example of determining the thermal diffusivity coefficient of copper foil by the photometric analysis of the propagation of thermal field excited by a single laser pulse. The temperature field excited by laser is oscillating in character. This is shown in Figure 4 representing the time dependence of the radiation energy (in arbitrary units) measured by PHASI in the vicinity of the focus, which is not treated by the laser pulse.

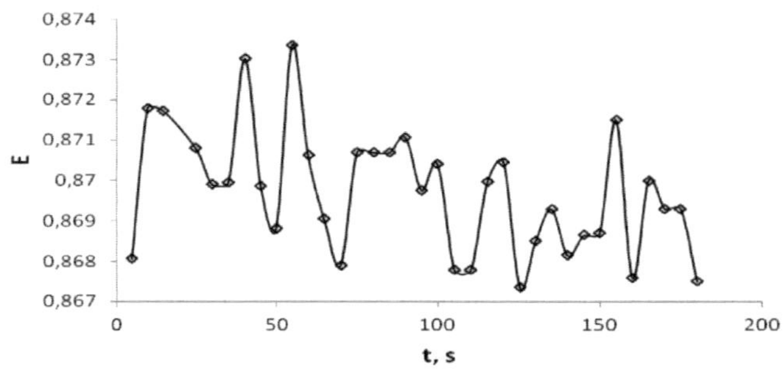

Figure 4. Time dependence of the radiation energy of the target after laser irradiation.

The local internal energy can be estimated from Eq. (5) [4] using the radiation energy (U) measured by PHASI:

$$U = A \cdot E \tag{5}$$

where: A is the coefficient, which characterizes the probability of spontaneous radiation upon oscillations of the charged particles in the body of E in internal energy. The coefficient A may be determined by measuring U at room temperature and internal energy of the target fragment with the availability of the data on the heat capacity of the target material. However, this is not necessary for the evaluation of the thermal diffusivity, since the substitution of thermal energy into the heat conduction equation, Eq. (5) cancels the constant A, and only the derivatives of the variable U, which is measured by PHASI, enters Eq. (5) except the thermal diffusivity. In this case, the equation is written for the two-dimensional case as:

$$\frac{\partial U}{\partial t} = a\left(\frac{\partial^2 U}{\partial x^2} + \frac{\partial^2 U}{\partial y^2}\right) \tag{6}$$

The derivatives of the variable were determined from the dependences $U = f(x)$, $U = f(y)$, and $U = f(t)$ constructed from the data measured at two successive micro fragments of 0.254×0.254 cm^2 in size, diverging from the center of the ellipse in two mutually perpendicular directions. Figure 5 shows the distribution of temperature field along the 0X and 0Y axes in the center of the target 10 sec after the laser pulse action.

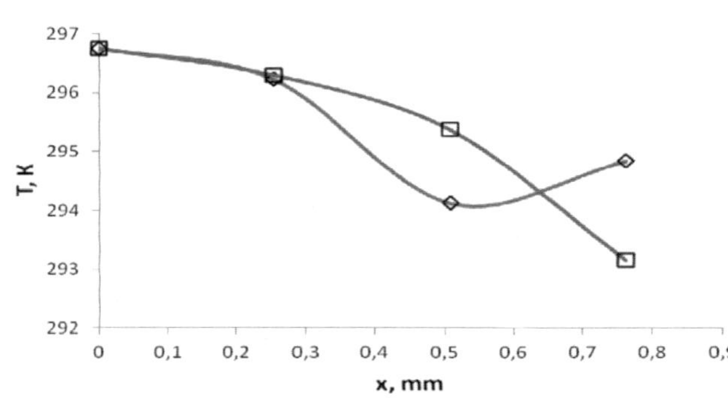

Figure 5. Temperature distribution of in the center of the ellipse 10 sec after the laser pulse action on the (□) 0X and (◊) 0Y axes.

Note that, evaluating the energies of micro fragments, one should allow for corrections for elastic energy, which results from target bending upon oscillations of the temperature field. The thermal diffusivity coefficients estimated in a series of experiments are given in Table 1.

Table 1. Evaluated thermal characteristics.

No.	Characteristics determined by PHASI	a, m^2/s	Λ, W/m K
1	on the compressed surface	$6.56 \cdot 10^{-5}$	212.91
2	on the tensioned surface	$2.243 \cdot 10^{-4}$	775.76
3	average value	$1.44 \cdot 10^{-4}$	498.03
4	according to literature data [6]	$1.15 \cdot 10^{-4}$	397.73

The thermal conductivity coefficients and the data given in the fourth row of Table 1 are calculated from the tabulated data on copper from the handbook [5]. Table 1 shows fairly good agreement between the results of our experiments and the literature data.

4. The study of the structural features of single crystals by PHASI

PHASI is useful also for the study of single crystals for the determination of their structure perfection and the crystallographic orientation of the plane of observation. Figure 6 shows the structure fragments of molybdenum single crystals after their treatment by PHASI at a magnification of 250. The crystals are oriented according to the corners of the standard stereographic triangle.

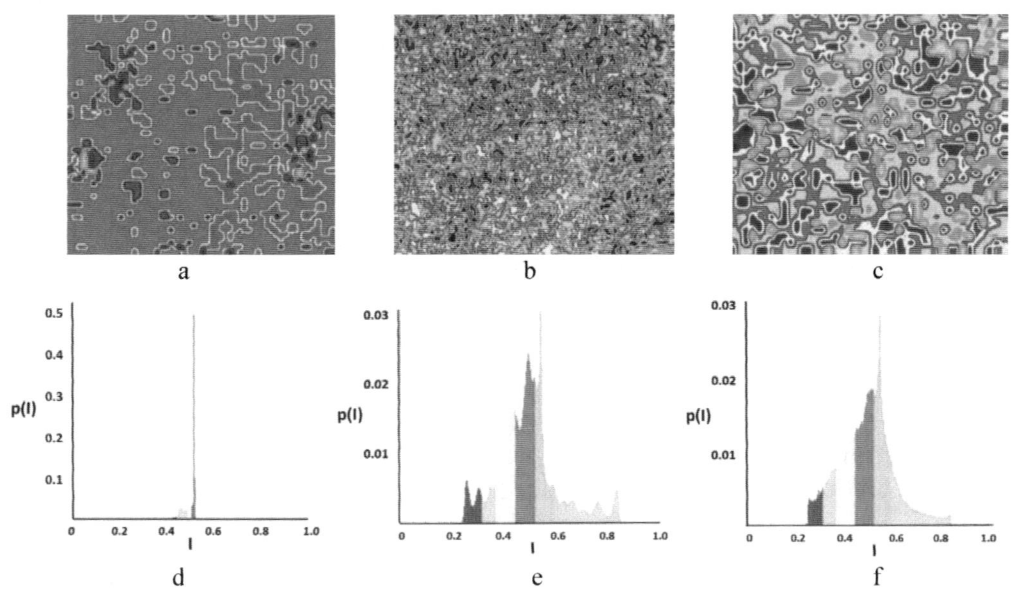

Figure 6. Structure fragments of molybdenum single crystals and the brightness spectra of the reflected light: (a) [100]; (b) [110]; (c) [111], (d) [100]; (e) [110]; (f) [111].

Figure 6 shows that the PHASI method allows one to distinguish the single crystal orientations not only by the reflection brightness spectra, but also by color of the fragments of structure images. It is seen that the width and shape of the spectra and the spectral density of their maxima depend on the order of symmetry axes coinciding with the normal to the observation surfaces. Such differences can be quantified by the fractions of the areas of certain color. This is clearly seen (Figure 7) from the histograms of the colors of fragments shown in Figures 6 a, 6 b, and 6 c.

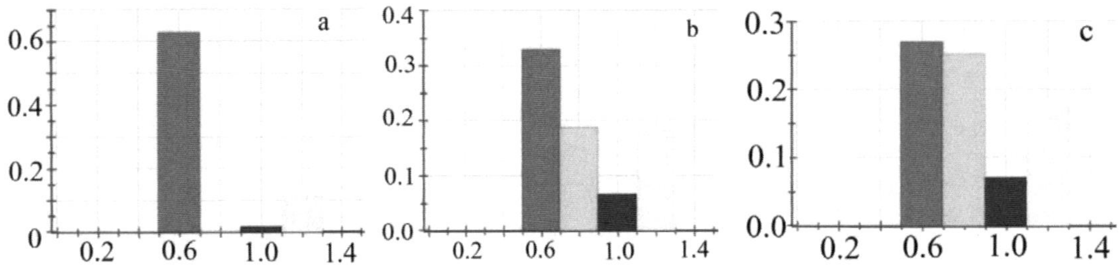

Figure 7. Histograms of the colors of fragments in Figures 6.

The parameters, which characterize the orientation of the molybdenum single crystals, were determined by PHASI method (Table 2). The reflection brightness spectra are sensible to external physical and chemical actions of any nature. This allows one to use them for the measurement of the physical properties and structural damage of construction materials.

Table 2. PHASI data on the parameters defining the orientation of the molybdenum single crystals.

[hkl]	Symmetry order	Selected ranges				Number of peaks	$p(I)_{max}$	Spectrum	
		I	II	III	IV			Start	End
[100]	4	0.9361	0.0033	0.0046	0.3291	1	0.466	0.442	0.562
[110]	2	0.3291	0.1874	0.0660	0.1803	1	0.229	0.251	0.851
[111]	3	0.2691	0.2518	0.0702	0.1733	12	0.0168	0.266	0.854

The PHASI software and analytical complex is of multifunctional use, which includes the estimation of the structural state and properties of materials as well as of the crystallographic features of single crystals. Such opportunities of PHASI have been demonstrated above by several examples, which do not exhaust all possibilities of the method.

5. Conclusions:

1. A method for the photometric analysis of the structural imaging (PHASI), based on an assessment of changes in the reflectivity of the surface of the object being studied.

2. To implement the method Fasi photometric analyzer structural image was created on a personal computer, allowing to carry out mathematical operations on digitized data arrays corresponding to the analyzed images at certain times and under specified conditions.

3. Procedures for allowing digital information according to the brightness of the reflection spectra obtained from the surface of the objects during exposure to the physical nature of certain fields, to determine their physical characteristics.

Acknowledgments
The work was supported by the Russian Foundation for Basic Research, projects nos. 14-08-00417a and 15-08-04272a.

References

[1] Orir J 1981 *Physics* [Russian translation] (Moscow: Mir) **1** 335 **2** 622
[2] Animalu A 1981 *Quantum theory of crystalline solids*. [Russian translation] (Moscow: Mir) 575
[3] Ermishkin V A, Murat D P, Podbel'skii V V 2007 Information technologies of photometric analysis of materials fatigue damage. *Informatsionnye Tekhnologi* **11** 65-70
[4] Novikov I I, Ermishkin V A, Kudryavtsev E M, Minina N A 2014 The study of thermal diffusivity by the photometric analysis of the target after laser treatment *Bulletin of the Kazan Technological University* **17** 22 152-154
[5] Smithells C J 1980 *Metals* Reference Book [Russian translation] (Moscow: Metallurgiya) 415

Influence of annealing on optical and photovoltaic properties of nanostructured TiO_2 films

T M Serikov[1], N Kh Ibrayev[1], Zh Kh Smagulov[1] and K.A. Kuterbekov[2]
[1] Karaganda, 100028, Republic of Kazakhstan, Institute of Molecular Nanophotonics, E.A. Buketov Karaganda State University
[2] Astana, 010000, Republic of Kazakhstan, L.N. Gumilyov Eurasian National University

E-mail: serikov-timur@mail.ru

Abstract. Spectral and kinetic characteristics of the photoluminescence of TiO_2 films obtained from TiO_2 nanoparticles and nanotubes were studied. Luminescence spectra typical for the TiO_2 with anatase structure were observed under UV excitation of the films. Heat treatment of the films at T=1273 K leads to a long-wavelength shift of the photoluminescence band with maximum at 850 nm, which corresponds to the rutile structure. The luminescence duration of rutile films is longer than the luminescence duration of the anatase films as for nanoparticles and for nanotubes. The photovoltaic properties of TiO_2 films with different structures were investigated. It was established that anatase structured films have a higher photocurrent than the rutile structured film. By impedance spectroscopy method it was found that the electron transport resistance in the nanotube films is higher but the recombination rate is lower than in the TiO_2 nanoparticle films.

1. Introduction

Titanium dioxide (TiO_2) - belongs to a class of wide-band semiconductors widely used in photocatalysis, medicine, ecology, and dye-sensitized solar cells (DSSC) [1-4]. The speed of electrons transport through the oxide semiconductor to the collecting electrode has a great importance for efficient operation of solar cells. Using the various methods we can get TiO_2 nanostructures in the form of nanoparticles (NPs) [6], nanotubes (NTs) [7], nanorods [8] and nanowires [9]. In DSSC TiO_2 NPs are the most commonly used. However, the relationship between the NPs formed by thermal annealing, affects the ability of electron transition from nanoparticle to nanoparticle. Unformed contacts lead to a decrease in the efficiency of electron transport. When using NTs one-dimensional electrons transport is possible along the sides. In addition, with proper design of the electrodes there will be fewer defects hindering electrons transport [10]. The presence of defects in nanostructures can create additional trapped levels of energy in the forbidden gap. Defects in the TiO_2 crystal lattice have a significant impact on its optical and electrical properties that can affect the efficiency of converting solar energy into electrical one. Therefore, the need to obtain information about the defect states in the bulk and on the surface of TiO_2 nanostructures is essential [11-13]. One of the sensitive methods in the study of structural defects is photoluminescence spectroscopy. It is characterized by high sensitivity to the presence of impurities and defects in the crystalline structure that alter the spectral composition and intensity of the luminescence. There is a number of papers devoted to the study of defects in nanostructures of titanium dioxide [14-17]. It should be noted that the majority of studies were performed for the anatase and rutile of titanium dioxide nanoparticles [18]. Luminescence centers of titanium dioxide nanotubes obtained by electrochemical anodization, remains little explored. This paper presents the results of studies of the effect of heat treatment of films formed by NTs titanium dioxide on their luminescent and electrophysical properties.

2. Experimental part

2.1. Obtaining films from NPs TiO_2

Colloidal TiO_2 powder (Sigma Aldrich) was triturated in a porcelain mortar with deionized water and acetone in a volume ratio of 10:1. After the formation of a homogeneous viscous paste NPs solution of titanium dioxide was deposited on the substrate surface by the doctor-blading method. For optical measurements polished plates from non-luminescent quartz were used. For measurements of current-voltage characteristics (CVC) glass plates with a conductive layer of the FTO (glass coated with tin oxide doped with fluorine) were used. The average size of NPs is ~ 21 nm. The resulting films were subjected to heat treatment at a temperature of 773 and 1273 K during 2 hours.

2.2. Synthesis of NTs TiO_2

The titanium foil with a thickness of 60 microns was subjected to chemical polishing. As the basis of electrolyte the $C_2H_6O_2$ was used (content of NH_4F is 0,5 wt%, H_2O is 3 wt%). Anodic oxidation of titanium was carried out in an electrochemical cell in potentiostatic mode at a temperature of 5-7 ^{0}C. The platinum foil was as a cathode. The distance between the anode and cathode was 3 cm.
The electrochemical oxidation process consists from three steps. First is anodizing for 2 hours. Separation the formed film at the first step is carried out in an ultrasonic bath of 1 M hydrochloric acid solution. Then the sample was washed with copious amounts of deionized water and dried at a temperature of 80 ^{0}C. The duration of the second step of anodizing was 24 hours. Anodizing voltage was 50 V. After the second stage of the anodizing the obtained TiO_2 nanotubes were sonicated to remove the byproducts and the surface oxide layer. Then they were washed with copious amounts of distilled water and dried in a stream of nitrogen. Crystallization of the amorphous phase of TiO_2 was performed by the thermal treatment of samples in a muffle furnace at 450 ° C for 2 hours with a temperature increase of 5 ^{0}C per minute. The third stage of the anodizing was used for films separation formed by the NTs titanium dioxide from titanium foil [20]. Anodizing was carried out at a temperature of electrolyte of 70^{0}C and the voltage of 60 V. The process came to an independent film separation from the titanium foil. Then, the resulting films were subjected to heat treatment at a temperature of 773 and 1273 K during 2 hours. The images of samples surface were obtained on the scanning electron microscope MIRA 3LMU (Tescan, Czech Republic). Measurements of spectral and kinetic characteristics of the films were carried out on an automated spectral and kinetic installation with registration in photon counting mode at the boiling point of liquid nitrogen. Excitation was carried out by a nitrogen laser AIL -3 (λ_{gen} = 337 nm, E = 30 mJ, τ_p= 10 ns). Before measurement the sample was placed in an optical cryostat, which had previously been evacuated to a residual pressure.

2.3. Assembling DSSC

Solar cells based on NPs and NTs titanium dioxide were obtained following the procedure described in the paper [21, 22]. To increase the spectral sensitivity range the ruthenium dye N719 was used (Di-tetrabutylammonium cis-bis (isothiocyanato) bis (2.2' - bipyridyl - 4.4' - dicarboxylato) ruthenium (II), Sigma Aldrich). Sorption occurred within 24 hours. Platinum electrodes were deposited from an ethanol solution H_2PtCl_6 by the electrochemical method on FTO. Gasket between the electrodes served the film Meltonix (Solaronix, Switzerland) with 30 microns thick. As electrolyte Iodolyte H30 (Solaronix, Switzerland) was used. IV-characteristics of solar cells were measured under illumination of cells with light of xenon lamp with a light output of 100 mW / cm^2 (Air Mass (AM) 1.5) at the measuring complex CT50AAA (PET PHOTO Emission TECH. INC., USA). Spectra of electrochemical cell impedance was measured under standard simulated solar radiation (Air Mass

(AM) 1.5) at Z-500PRO (Elins, Russia). The amplitude of the applied sinusoidal signal was 20 mV and the frequency was varied from 1 MHz to 100 MHz.

3. Results and discussion

It is known that before NTs heat treatment, TiO_2 structure is amorphous [23]. Crystallization of the structure occurs during thermal annealing of the sample, starting with the temperature of 553K. At a temperature of 773 K the amorphous phase proceeds anatase structure. At further annealing to 1273 K anatase structure is completely transformed into the rutile phase. The morphology of the films surface formed by the NPs and NTs titanium dioxide and heat-treated at a temperature of 773 and 1273 K is shown in Figure 1.

a) б) в) г)

Figure 1. SEM images of the titanium dioxide films.

From Figure 1, a and 1, b it is seen that the films from NPs titanium dioxide have distinct granular structure. For the same magnification it is noticeable that the film annealed at 1273 K consists of larger particles than the film that had been subjected to thermal annealing at a temperature of 773 K. This is probably due to the fact that under high temperature some of the particles sintered forming larger agglomerates. Figure 1 shows a SEM image of a film formed by TiO_2 NTs. The figure shows that NTs are tightly packed and their ends are open. Increase of the annealing temperature up to 1273 K also results in a significant change of the structure (Fig. 1, d). It is evident that although the film retains a tubular structure, the open channels are not observed on the surface. Figure 2 shows the luminescence spectra and kinetics of films luminescence decay of anatase structure formed by the NPs and NTs titanium dioxide.

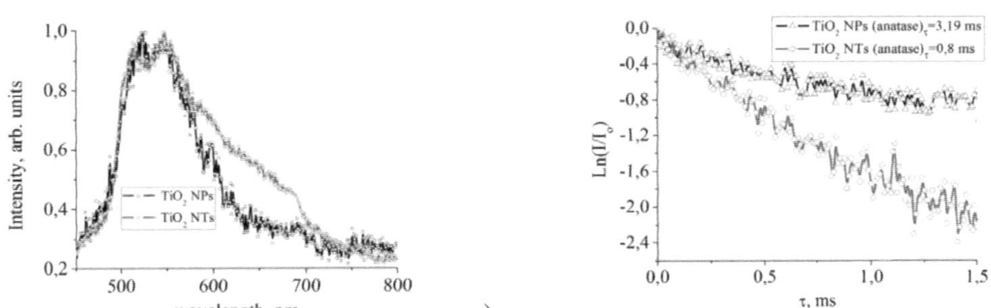

a) б)

Figure 2. Normalized luminescence spectra (a) and luminescence kinetics (b) of anatase films formed by NPs and NTs titanium dioxide.

At room temperature luminescence was not observed for both samples. When cooling film to T=160 K in the wavelength range of 400 - 800 nm a broad band of emission from maximum 510 and 540 nm is recorded. By lowering the substrate temperature to 90 K the intensity of the luminescence increases. The nature of this luminescence is associated with defects located in the surface area of the crystallites [23]. From a comparison of the luminescence spectra it is evident that in NTs luminescence intensity of 600-700 nm is higher than in the case of NPs. Using the approximation by the Gaussian functions luminescence spectra were divided into components, the form of which is shown in Figure 3. The results of approximation show that the observed luminescence spectra are formed by three bands with maximum at 510, 540 and 600 nm.

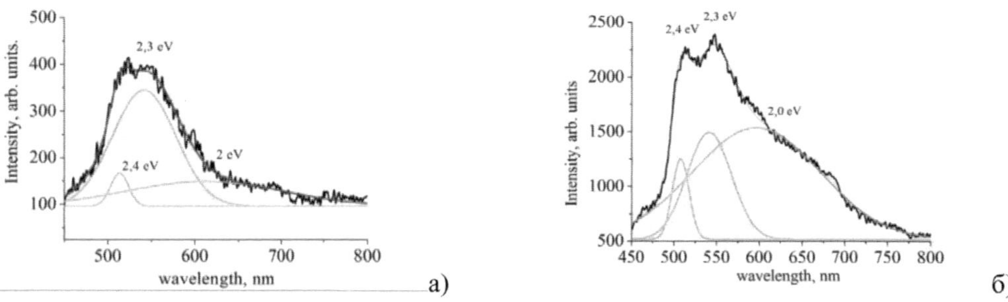

a) NPs TiO_2 (773 K); б) NTs TiO_2 (773 K)

Figure 3. Gaussian approximation of the photoluminescence spectrum of nanoparticles and nanotubes of titanium dioxide measured at 90 K.

For all of the observed defect luminescence centers the kinetics of luminescence decay were measured. Figure 4 shows the luminescence decay kinetics for NPs and NTs measured at a wavelength of 540 nm. Kinetic curves are generally non-exponential. Lifetimes of excited states calculated from the exponential part of the decay curves are shown in Table 1.

Table 1. Duration of luminescence of films from NPs and NTs on different wavelengths.

Sample	Excitation wavelength		
	510 nm	540 nm	600 nm
NPs (773 K)	τ=6.2 ms	τ=3.0 ms	τ= 6.7 ms
NTs (773 K)	τ=2.5 ms	τ=0.8 ms	τ=6.3 ms

The presented data shows that the lifetimes of the excited states of the defect centers are different as for NPs and for NTs. The obtained spectral results are correlated well with the published data [23], where the luminescence of the anatase structure of NT titanium dioxide is associated with three possible defect states: self-trapped excitons STE, Ti^{3+}/F и F^{+}. For rutile phase photoluminescence is associated with oxygen vacancies [14]. Oxygen vacancies are important defects in nanocrystalline TiO_2, because the efficiency of electron transport and recombination processes depends on their number. Figure 4 shows the spectra and kinetics of luminescence decay films from NPs and NTs, annealed at T = 1273 K.

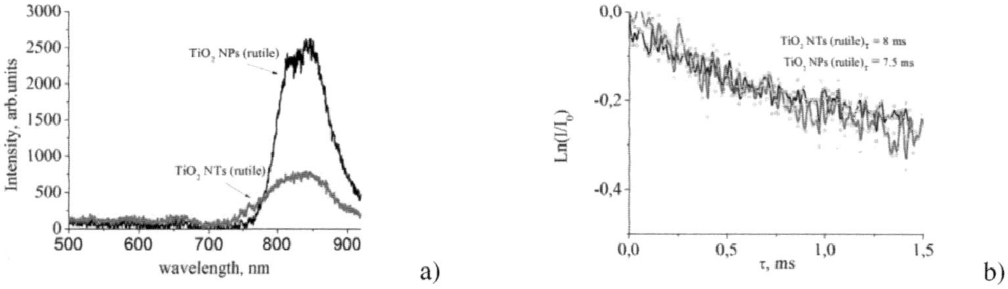

Figure 4. Spectral and kinetic properties of the film annealed at 1273 K.

For both samples luminescence of rutile modification with a maximum range of 850 nm is observed (Fig. 1, a). It should be noted that the light intensity for NPs is much higher than for NTs. Figure 4 (b) also shows the kinetics of luminescence decay of NPs and NTs. Excitation wavelength was 850 nm. Lifetimes of excited states make up 7.5 and 8.0 ms for NPs and NTs respectively. For the investigation of the possibility of using nano-structured films of titanium dioxide in DSSC sandwich structures on the standard method were designed [23] and measured their IV- characteristics which are presented in Figure 6.

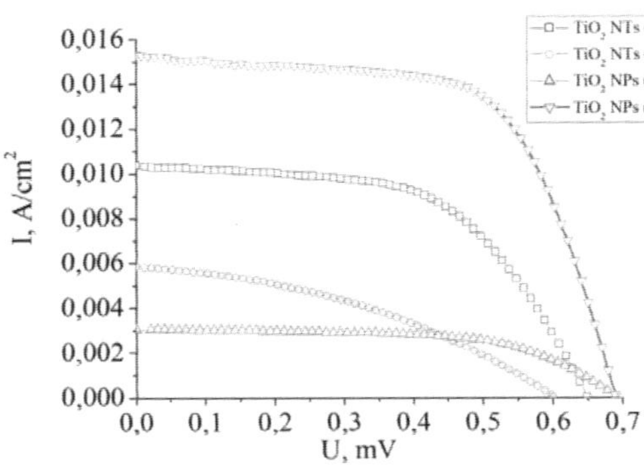

Figure 6. I-V characteristics of DSSC based on TiO_2 films and the dye N719.

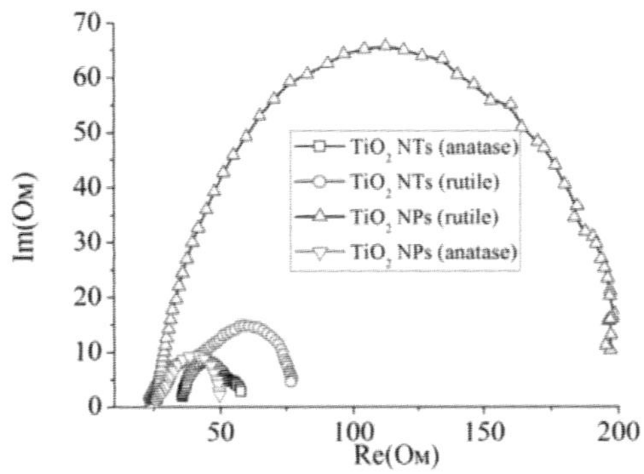

Figure 7. Impedance spectra of DSSC.

Based on the obtained curves of IV- characteristics main photovoltaic solar cell parameters were determined which are presented in Table 3.

Table 3. Photovoltaic properties of DSSC.

Sample	U_{oc}, B	J_{sc}, (A/cm^2)	FF	Efficiency, %
NPs (773 K)	0.69	0.014	0.62	4.10
NPs (1273 K)	0.68	0.003	0.60	1.32
NTs (773 K)	0.65	0.010	0.56	3.80
NTs (1273 K)	0.60	0.005	0.38	1.36

From this tabulated data it is evident that the photovoltaic characteristics of cells based on films from NPs and NTs anatase are higher than for the films of rutile structure. The difference of U_{oc} values for cells based on rutile and anatase is associated with the location of Fermi quasilevel. Figure 7 shows hodographs of impedance in Nyquist coordinates for solar cells based on TiO_2 films. Using the method described in [25, 26] from the impedance spectra were calculated effective diffusion coefficient of *Deff* electrons, the effective speed of k_{eff} recombination, effective lifetime of τ_{eff} electron, resistance of electron transport in the film of R_w titanium dioxide, R_k charge transfer resistance, associated with recombination of electron. The results are shown in Table 4.

Table 4. The electron transport properties of TiO_2 films in the DSSC.

Annealing temperature	D_{eff}	k_{eff} (c^{-1})	τ_{eff} (c)	R_k (O_M)	R_w (O_M)
NPs (anatase)	$9.9*10^{-5}$	14.0	0.07	22.0	26
NPs (rutile)	$4.7*10^{-5}$	57.0	0.02	175.0	26
NTs(anatase)	$5.5*10^{-5}$	3.6	0.27	17,5	36
NTs ((rutile)	$2.9*10^{-5}$	25.3	0.04	38.0	40

From the tabulated data it is evident that the electron transport resistance in TiO_2 (Rw) of NTs films is higher than in NPs. From the table 4 it is evident that the rate of recombination (*keff*) in the films based on NTs is less than *keff* cells based on NPs similar modification of TiO_2. This is probably due to SCS manufacturing technology based on NTs. In accordance with the formula $k_{eff}=2N_sk_r$ (*Ns* - density of electrons on the defect levels of the energy (*cm-3*), k_r - recombination rate constant (cm^3s^{-1}) to electrolyte of electrons from TiO_2 defect levels [26]) it can be concluded that in the rutile films of NPs and NTs the concentration of the defects through which recombination processes occur is higher than in anatase.

4. Conclusion
In the paper the spectral and kinetic properties of the photoluminescence films formed from NPs and NTs of titanium dioxide were investigated. For samples annealed at T=773 K, when excited by a nitrogen laser luminescence is observed with a spectrum characteristic of the anatase structure. Annealing of the samples at T=1273 K leads to a long-wavelength shift of the photoluminescence band with maximum at 850 nm wavelength which corresponds to the rutile structure. The researches of the photovoltaic properties sensitized by the dye of solar cells have shown that the anatase-based films have a higher cell efficiencies than the cells with rutile films. It is found that in the films based on NTs the recombination rate is less but the resistance to the electron transport is higher than in the films from NPs. In the rutile films of NPs and NTs the concentration of defects through which the recombination processes occur is higher than in anatase.

5. References
[1] Fujishima A, Honda K 1972 *Nature* **238** 37
[2] Hasan M et al. 2008 *Chemical and Biological Engineering* **1** 92
[3] Ito S, Murakami T N, Comte P 2008 *Thin Solid Films* **516** 4613
[4] Lee J, Sang K, Shin K 2012 *Journal of Industrial and Engineering Chemistry* **18** 19
[5] Yang C 2008 *Applied Surface Science* **23** 2685
[6] Yang J, Mei S, Ferreira J 2001 *Materials Science and Engineering: C* **15** 183
[7] Wu X, Jiang Q Z, Ma Z F 2005 *Solid State Communications* **136** 513
[8] Fujihara K, Kumar A, Jose R 2007 *Nanotechnology* **18** 37
[9] Arcadipane E, Sanz R, Miritello M 2016 *Materials Science in Semiconductor Processing* **42** 24
[10] Mukul D, Hongshan H 2012 *Scanning Electron Microscopy* **27** 537
[11] Mo S D, Ching W Y 1995 *Physical review* **51** 15
[12] Carpa O, Huisman C L, Reller A 2012 *Progress in Solid State Chemistry* **32** 33
[13] Landmann M, Rauls E, Schmidt W G 2012 *Journal of Physics: Condensed Matter* **24** 195503
[14] Ivlev V, Gusev S, Latyshev A 2012 *Condensed Matters and the interphase boundaries* **14** 141
[15] Wang X, Feng Z, Shi J 2010 *Physical Chemistry Chemical Physics* **12** 7083
[16] Nakajimaa H, Mori T 2004 *Journal of applied physics* **96** 1
[17] Serpone N, Lawless D, Khairutdinov R 1995 *Journal Physics Chemystry* **99** 16646
[18] Mercado C, Seeley Z, Bandyopadhyay A 2011 *ACS Applied Materials & Interfaces* **3** 2281
[19] Sanjinés R R, Tang H, Berger H 1994 *Applied Physics* **75** 2945
[20] Lin J, Chen J, Chen X 2010 *Electrochemistry Communications* **12** 1062
[21] Bach U, Lupo D, Comte P, Moser J E and M Gratzel 1998 *Nature* **395** 583
[22] Lin J, Chen J Chen X 2011 *Nanoscale Research Letters* **6** 475
[23] Zhang H, Zhou M, Fu Q 2014 *Nanotechnology* **25** 275603
[24] Brunauer S, Deming L S, Deming W S 1940 *Jour. of the American Chemical Society* **62** 1723
[25] F-Santiago F, Bisquert J, G-Belmonte G 2005 *Solar Energy Materials & Solar Cells.* **87** 117
[26] Adachi M, Sakamoto M, Jiu J 2006 *Journal Physics Chemistry* B **110** 13872

IOP Conf. Series: Materials Science and Engineering **168** (2017) 012055 doi:10.1088/1757-899X/168/1/012055

Luminescent Response to the Phase Composition of Nd^{3+}:Y_2O_3-Al_2O_3 System

P V Toropova[1,3], A V Spirina[1], V I Solomonov[1,2], V A Shitov[1], A V Chukin[2] and S O Cholakh[2]

[1] The Institute of Electrophysics of the Ural Branch of the Russian Academy of Sciences, 106 Amundsen Str., Yekaterinburg, Russia, 620016
[2] Ural Federal University named after the first President of Russia B.N.Yeltsin, 19 Mira Str., Yekaterinburg, Russia, 620002

E-mail: greenpaulin@gmail.com

Abstract. The work demonstrates the luminescent method of identifying the phase composition of Nd^{3+}:Y_2O_3-Al_2O_3 system by building calibration curves using standard samples.

1. Introduction

The most prominent problem in the synthesis of optical ceramics is the one of controlling the presence of impurity phases, as even in small amounts they impact the optical properties severely [1]. The presence of a large amount of structural modifications and polymorphic changes in the process of ceramics synthesis requires the use of rapid and sensitive analytic methods for synthesis technology correction and final ceramics quality inspection. The standard method of phase composition identification is X-ray phase analysis. Meanwhile, it is the luminescent method that is a rapid and qualitatively more sensitive one for phase composition identification [2,3]. In paper [2], the luminescent method was used for quantitative phase analysis of Nd^{3+}:YAG (yttrium-aluminium garnet Nd^{3+}:$Y_3Al_5O_{12}$) ceramics and monocrystals with Nd^{3+}:YAP (yttrium-aluminium perovskite Nd^{3+}:$YAlO_3$) impurity phase. Physically, the method is based on different locations of Stark emission bands of Nd^{3+} ion in crystal fields of various structures [4]. The luminescent parameter determining the concentration of the impurity phase is the ratio of light sums taken from two different spectrum ranges that include Nd^{3+} ion emission bands localized both in YAG and YAP structures.

The purpose of this paper is to identify the quantitative phase composition using the luminescent method for Nd^{3+}:Y_2O_3-Al_2O_3 system.

2. Objects of research and experimental equipment

Specially prepared multiphase Nd^{3+}:Y_2O_3-Al_2O_3 systems used as the objects of the research. The starting material used were nanopowders of monoclinic Al_2O_3 and Nd^{3+}:Y_2O_3 prepared with the

[3] To whom any correspondence should be addressed.

method of target laser evaporation with a repetitively pulsed CO_2 LAERT laser, the mean emission power of which is 500 W and the emission wavelength is 10.6 μm [5]. Used as targets were commercial Al_2O_3 micron powder and a mixture of Y_2O_3 microsized powder and 1 mol. % of Nd_2O_3 powders, respectively. The resulting Nd^{3+}:Y_2O_3 and Al_2O_3 nanopowders were annealed at 1000 °C during 30 min and at 1250 °C during 3 h, respectively. Then, Nd^{3+}:Y_2O_3 and Al_2O_3 nanopowders were blended, with the ratio (Nd + Y) : Al = 3 : 5, which corresponds to yttrium-aluminium garnet stoichiometry. To prepare a multiphase system, a certain excessive amount of yttrium oxide or aluminium oxide was added to the resulting mixtures. The blending was carried out in ethanol during 48 h. After drying, the nanopowder mixtures were pressed into compacts with the method of dry monoaxial static pressing at 200 mPa. The prepared compact was annealed at 800 °C during 3 h, with following sintering at 1780°C during 20 h. The ceramics obtained was additionally annealed at 1300 °C for 5 h to reach transparency.

The additionally used one-phase samples were a Nd^{3+}:YAP monocrystal, monoclinic Al_2O_3 and cubic Nd^{3+}:Y_2O_3 nanopowders. The latter was obtained by annealing the starting monoclinic nanopowder in the air at $T = 1500$ °C during 3 h.

Photoluminescence (PL) was excited by DMH 808-4000 laser diode, emission wavelength 808 nm, emission power 4W. A spectrograph photodetector was represented by Hamamatsu G9214-512S semiconductor InGaAs line that included 512 elements with the discretion of 0.5 nm at one unit of the array. The detection was carried out in the range 890-1150 nm in the air at room temperature. Pulsed cathodoluminescence (PCL) was excited by an electronic beam with the mean energy 180 keV and pulse duration of 2 ns which was emitted by RADAN-220 high-current electron accelerator [3]. The spectograph photodetector contains 2,048 elements with the discretion of 0.2 nm per one unit of the line. The detection was carried out in the range 350-890 nm in the air at room temperature. Precision for wavelength identification did not exceed 0.5 nm.

The X-ray phase analysis was performed at D8 Discover diffractometer in copper emission with a graphite monochromator at a diffracted beam and also at XPert PRO MRD Panalitical diffractometer.

3. The results and discussion

Table 1 presents the tested samples and their phase composition according to the X-ray phase analysis.

Table 1. X-ray results for tested samples

	YAG, %	YAP, %	$Y_4Al_2O_9$, %	cubic Y_2O_3, %	monoclinic γ-Al_2O_3, %
Ceramics no. 1	56.5	25.3	1	3.6	13.6
Ceramics no. 2	48.6	37	1	3.7	12.8
Ceramics no. 3	33.4	52.7	2.5	5.6	5.8
Ceramics no. 4	39	47	1.7	4.7	7.6
Monocryst. Nd^{3+}:YAP	-	100	-	-	-
Nd^{3+}:YAG ceramics	100	-	-	-	-
Al_2O_3 nanopowder	-	-	-	-	100
Nd^{3+}:Y_2O_3 nanopowder	-	-	-	100	-

In Figure 1, PCL and PL spectra of the tested samples are presented according to Table 1. The spectra of ceramics no. 1,2 and 3,4 are qualitatively identical, so for illustrative purposes only the spectra of the samples of ceramics no. 1 and no. 3 are given, with the concentration of YAG and YAP phases, respectively, exceeding 50 % (Table 1).

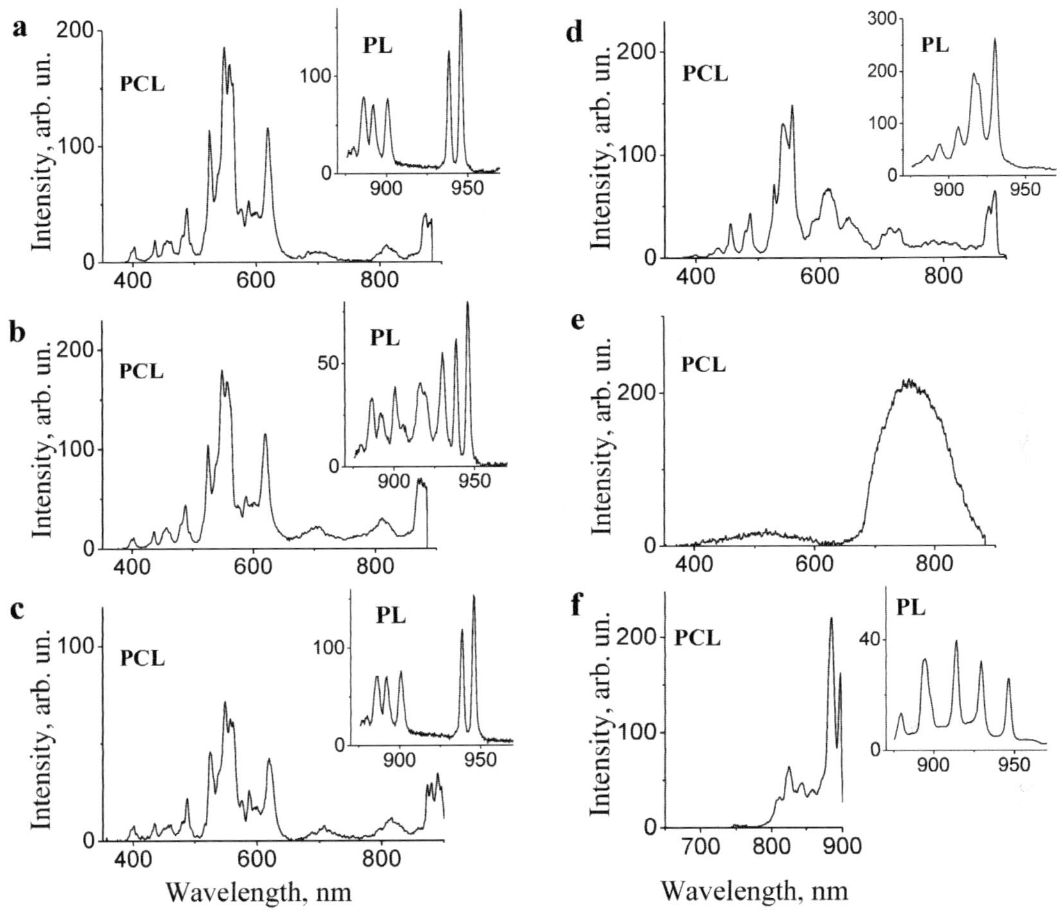

Figure 1. PCL and PL spectra: a – ceramics no. 1, b – ceramics no. 3, c – Nd^{3+}:YAG ceramics, d – Nd^{3+}:YAP monocrystal, e – monoclinic Al_2O_3 nanopowder, f - cubic $Nd^{3+}:Y_2O_3$ nanopowder

Assume the Nd^{3+}:YAG phase is the main one, whereas the rest of them are impurity phases. Comparative qualitative analysis of single-phase samples spectra (Fig. 1 c, d, e, f) discovered ranges where Nd^{3+}:YAG emission bands were pair-wise covered with the bands of only one of the impurity phases. That fact led to creation of a simple scheme for identifying $Nd^{3+}:Y_2O_3$-Al_2O_3 system phase composition with the luminescent method. It should be noted that the concentration of $Nd^{3+}:Y_4Al_2O_9$ phases was low relatively to the rest of them, which resulted in absence of any significant Nd^{3+} luminescence in $Y_4Al_2O_9$ structure in the standard samples, although the spectrum included $Nd^{3+}:Y_4Al_2O_9$ bands covered only with Nd^{3+}: YAG bands. The scheme makes it possible to determine the composition of all phases except $Nd^{3+}:Y_4Al_2O_9$.

The scheme for determination of phase composition is as follows. On the basis of the obtained pair-wise covered ranges, multiphase systems can be presented as two-phase systems:

Nd^{3+}:YAG/Nd^{3+}:YAP, Nd^{3+}:YAG/ Al_2O_3, Nd^{3+}:YAG/Nd^{3+}:Y_2O_3. In the first case the ceramics has only Nd^{3+}:YAG phases and Nd^{3+}:YAP impurity phase, in the second case it contains Nd^{3+}:YAG and Al_2O_3 impurity phase, in the third case - Nd^{3+}:YAG and Y_2O_3 impurity phase. Based on this, a recalculation of the initial percentages of phase concentrations was carried out with their ratio preserved. In Table 2 the example for the sample of ceramics no. 1 is given, where C_G, C_P, C_Y, C_A are the initial percentages of Nd^{3+}:YAG, Nd^{3+}:YAP, Nd^{3+}:Y_2O_3, Al_2O_3 concentrations according to Table 1, while $\tilde{C}_{G1}$, $\tilde{C}_{G2}$, $\tilde{C}_{G3}$, $\tilde{C}_P$, $\tilde{C}_Y$, $\tilde{C}_A$ are the recalculated concentration percentages for the three cases of two-phase systems without regard to other phases.

Table 2. Recalculation of phase concentrations for ceramics no. 1

	Ratio of initial phase concentrations	Recalculated phase concentrations ratio
Nd^{3+}:YAG/Nd^{3+}:YAP	$\frac{C_G}{C_P} = \frac{56.5}{25.3} = k_1$	$\frac{\tilde{C}_{G1}}{\tilde{C}_P} = \frac{69.071}{30.929} = k_1$
Nd^{3+}:YAG/ Al_2O_3	$\frac{C_G}{C_A} = \frac{56.5}{13.6} = k_2$	$\frac{\tilde{C}_{G2}}{\tilde{C}_A} = \frac{80.599}{19.401} = k_2$
Nd^{3+}:YAG/Nd^{3+}:Y_2O_3	$\frac{C_G}{C_Y} = \frac{56.5}{3.6} = k_3$	$\frac{\tilde{C}_{G3}}{\tilde{C}_A} = \frac{94.01}{5.99} = k_3$

For each ceramics, the following system of equations was obtained:

$$\begin{cases} \frac{C_G}{C_P} = \frac{\tilde{C}_{G1}}{\tilde{C}_P} = k_1, \\ \frac{C_G}{C_A} = \frac{\tilde{C}_{G2}}{\tilde{C}_A} = k_2, \\ \frac{C_G}{C_Y} = \frac{\tilde{C}_{G3}}{\tilde{C}_A} = k_3, \\ C_G + C_P + C_A + C_Y = 100\ \% \end{cases} \quad (1)$$

To determine concentrations of phases $\tilde{C}_{G1}$, $\tilde{C}_{G2}$, $\tilde{C}_{G3}$, $\tilde{C}_P$, $\tilde{C}_Y$, $\tilde{C}_A$ the luminescent method for two-phase systems was used [2], where the parameter determining the concentration of the impurity phase is the ratio of light sums taken from two different spectrum ranges that include emission bands only from two phases. For a system consisting only of Nd^{3+}:YAG and Nd^{3+}:YAP, the luminescent parameter was calculated the following way:

$$l_{G/P} = \frac{S_1}{S_2} = \frac{\int_{420}^{580} I(\lambda)\, d\lambda}{\int_{420}^{505} I(\lambda)\, d\lambda}, \quad (2)$$

where $I(\lambda)$ – is the luminescence intensity at wavelength λ measured in nm. After finding coefficient $l_{G/P}$ for the researched multiphase systems and for single-phase Nd^{3+}:YAG and Nd^{3+}:YAP, dependency $\tilde{C}_P(l_{G/P})$ was built (Figure 2a). With the help of approximation of the experimental points on the hyperbolic curve [2], a calibration curve was obtained, equation of which allows to find the ratio Nd^{3+}:YAG/Nd^{3+}:YAP using the luminescent method (Figure 2a).

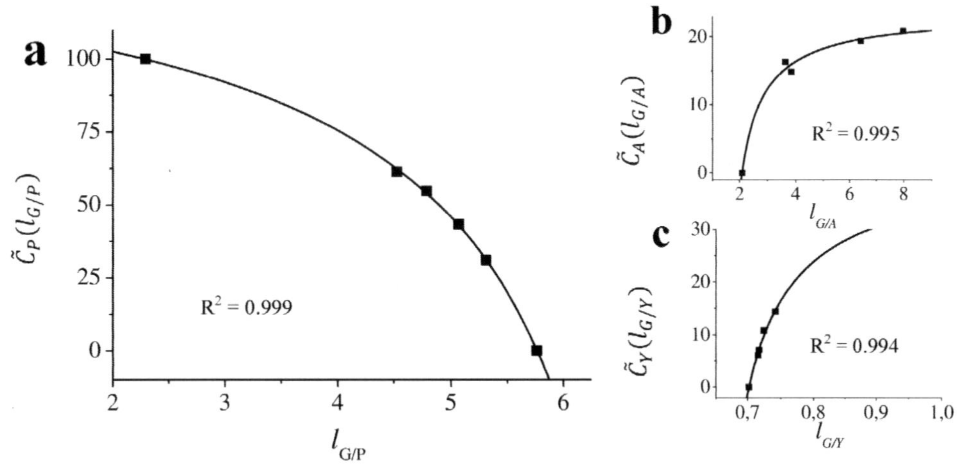

Figure 2. Calibration curves for determining the ratio of Nd^{3+}:YAG main phase to the impurity phases: a - Nd^{3+}:YAP, b - γ-Al_2O_3, c - Nd^{3+}:Y_2O_3

The same coefficients $l_{G/A}$ and $l_{G/Y}$ were calculated for the ratios Nd^{3+}:YAG/ Al_2O_3 and Nd^{3+}:YAG/Nd^{3+}:Y_2O_3 respectively. Unlike (2), in these cases the intensity ratio was used instead of the light sums ratio, as the spectrum ranges featuring the luminescence of only two phases appeared to be short. For parameter $l_{G/A}$ the ratio of intensities at wavelengths 577.0 nm and 688.0 nm was used. For $l_{G/Y}$ – the ratio of intensities at 938.7 nm and 945.8 nm. The resulting calibration curves $\tilde{C}_A(l_{G/A})$ and $\tilde{C}_Y(l_{G/Y})$ are resented in Fig. 2 b,c.

To find the initial concentration of all the phases C_G, C_P, C_Y and C_A the obtained values $\tilde{C}_P(l_{G/P})$, $\tilde{C}_A(l_{G/A})$, $\tilde{C}_Y(l_{G/Y})$ should be put into (1).

Table 3 shows the percentage of concentrations of all phases according to X-ray phase analysis and the luminescent method (LM) for the multiphase ceramics, which was described above. The table shows that the found values proved to be close, but the present difference is associated with not taking into consideration $Y_4Al_2O_9$ phase. When it is necessary to solve a problem of phase composition identification, including this structure, one has to create the corresponding standard samples where its concentration is enough to build a calibration curve.

Table 3. The composition of multiphase samples according to X-ray phase analysis and LM

	Ceramics no. 1		Ceramics no. 2		Ceramics no. 3		Ceramics no. 4	
	X-ray, %	LM, %	X-ray, %	LM, %	X-ray, %	LM, %	X-ray, %	LM, %
Nd^{3+}:YAG	56.5	56.6	48.6	47.9	33.4	33.2	39.0	40.8
Nd^{3+}:YAP	25.3	25.6	37.0	36.0	52.7	54.8	47.0	47.5
Al_2O_3	13.6	13.9	12.8	12.4	5.8	6.3	7.6	7.3
Nd^{3+}:Y_2O_3	3.6	4.0	3.7	3.7	5.6	5.7	4.7	4.4
Nd^{3+}:$Y_4Al_2O_9$	1	-	1	-	2.5	-	1.7	-

Thus, using the example of Nd^{3+}:Y_2O_3-Al_2O_3 multiphase system, the possibility of rapid determination of the qualitative phase composition with the luminescent method is demonstrated, involving the building of calibration curves and featuring the use of standard samples.

The research has been conducted with the support of the President Grant of the Russian Federation MK-3669.2015.2

References

[1] Ikesue A and Aung Y L 2008 *Nature photonics* **2** 721
[2] Osipov V V, Solomonov V I and Spirina A V 2011 *J. Opt. Technol* **78** 408
[3] Solomonov V I, Michailov S G, Lipchak A I, Osipov V V, Shpak V G, Shunailov S A, Yalandin M I and Ulmaskulov M R 2006 *Laser Physics* **16** 126
[4] Kaminskii A A 1990 *Laser crystals* (Berlin: Springer)
[5] Osipov V V, Platonov V V, Lisenkov V V, Podkin A V and Zakharova E E 2013 *Phys. Stat. Sol. C.* **10** 926

IOP Conf. Series: Materials Science and Engineering **168** (2017) 012056 doi:10.1088/1757-899X/168/1/012056

Ionizing Radiation Effects in Ni Nanotubes

D Shlimas[1,2], A Kozlovsky[1,2], A Shumskaya[3], E Kaniukov[*3], M Ibragimova[2], M Zdorovets[1,2] and K Kadyrzhanov[1]

[1] L.N. Gumilyov Eurasian National University, Mirzoyan street, 2, Astana 010008, Kazakhstan
[2] Institute of Nuclear Physics, Ibragimov street, 1, Almaty 050032, Kazakhstan
[3] Scientific-Practical Materials Research Centre NAS of Belarus, P. Brovki street, 19, Minsk 220072, Republic of Belarus

[*]E-mail: Ka.Egor@mail.ru

Abstract. Polycrystalline nickel nanotubes with diameter of 380 nm and wall thickness 95 nm were synthesized by electrochemical method using PET track-etched membranes with thickness of 12 μm. A comprehensive study of the structural, morphological and electrical characteristics of Ni nanotubes irradiated with C^{+13} ions with energy 1.75 MeV/nucleon and fluence ranging from 10^9 to 5×10^{11} cm^{-2} was carried out. The ability of modification of structural parameters such as lattice parameter and the average size of crystallites and conductivity of Ni nanotubes by irradiation was shown.

1. Introduction

Nowadays, a huge attention is paid to nanostructures and one of the most promising of them is magnetic nanotubes (NTs). NTs are prospective for a wide range of its practical application: from drug delivery to the magnetic storage [1-4]. For the devices functioning at extreme conditions it is necessary to understand the influence of ionizing irradiation on the structure and electrical properties of them. It is well known that defect formation mechanism in irradiated nanostructures is different from the same processes in bulk materials [5]. Accordingly that, it is not possible to apply standard approaches for nanodimension systems. Unfortunately, nowadays only limited number of works are devoted to study of influence of various types of radiation on the structure and physical properties of nanoobjects [eg, 6, 7]. To improve the understanding of irradiation effect in this field, we present the results of studies of the structural and electrical properties of nickel NTs, which were treated by ionizing radiation.

2. Experimental

Track-etched membranes based on polyethylene terephthalate (PET) were used as templates with following dimensions: thickness of 12 microns, nominal pore diameter of 380 nm and a density of 4×10^7 cm^{-2}. Electrochemical deposition in pours was carried out at voltage of 1.5 V in a potentiostatic mode. The electrolyte composition is $NiSO_4 \times 6H_2O$ (100,14 g/l), H_3BO_3 (45 g/l), $C_6H_8O_6$ (1,5 g/l). Control of the deposition was carried out by chronoamperometry using a multimeter Agilent 34410A.

Ionizing radiation effects in Ni NTs was studied on samples irradiated on cyclotron DC-60 with C^{+13} ions with energy 1.75 MeV/nucleon with fluence ranging from 10^9 to 5×10^{11} cm^{-2}. Estimation of the appropriate irradiation energy was calculated in SRIM Pro 2013 program.

Characterization of structural features was carried out by scanning electron microscopy (SEM, Hitachi TM3030), energy dispersive analysis (EDA, Bruker XFlash MIN SVE) and X-ray diffraction (XRD, D8 ADVANCE) using Cu K*α*-irradiation. Determination of internal diameters of Ni NTs was carried out by determination of gas permeability method (Sartocheck® 3 Plus 16290) at a pressure in the range from 8 to 20 kPa [8]. Evaluation of electrical conductivity of NTs was carried out studying of current-voltage characteristics of samples using a current source HP 66312A and amperemeter 34401A Agilent. To provide SEM and EDA investigations of individual NTs templates were etched in 5M solution of sodium hydroxide at a temperature of 90°C during 60 minutes.

3. Results and Discussion

There are a lot of promising NTs synthesis techniques, but the simplest and cheapest one is electrochemical deposition in pores of templates from silicon, aluminum oxide, polycarbonate or polyethylene terephthalate [9-11]. Features of the electrodeposition of metals into the pores of track-etched membranes, as well as the stages of formation of the NTs are described in detail in [12-15]. Morphology of irradiated and non-irradiated NTs arrays were determined by SEM and gas permeation methods. Analysis of the SEM images showed that the nanostructure has tubular shape, their length is coincided with the templates thickness (12 microns) and NTs diameters correspond to pore diameter (380 nm, Figure 1). The results of experiments of gas permeability allow to calculate the inner NTs diameters which are ~ 190 nm, and the wall thickness is ~ 95 nm.

Figure 1. SEM microphotographs of Ni NTs: a) initial; b) irradiated at fluence 5×10^{10} cm^{-2}; c) 10^{11} cm^{-2}; d) 5×10^{11} cm^{-2}.

Effect of irradiation of carbon ions with high energy leads to a modification of the NTs wall structure, herewith the size and number of defects would rise with increasing radiation fluence. This fact is associated with multiple processes occurring during the interaction of ions with solids. For example, the interaction of the particles of high energy with solids lead to restructuring NTs walls by knocking out nickel particles and embedding of carbon particles. However, analysis of the SEM image shows the formation of vacancies and roughness on NTs surface take place only by irradiation with the maximum fluence (Figure 1d).

Determination of carbon presence in NTs structure was held by EDA. According to obtained data, the atomic ratio of nickel in the initial NTs sample is 100% (Figure 2a) and remains constant up to fluence of 5×10^{10} cm^{-2}. Exceeding this fluence leads to registration in EDA spectra carbon peak (Kα - series - 0.277 keV, Figure 2b). Future expanding of the fluence provides not only the increasing of carbon peaks intensity, but also changing in the intensities of Ni Kα - series lines at 7,474 keV and Lα - Series 0.849 keV, which caused by changes in the structure of NTs as a result of the interaction of carbon ions with the crystal lattice (Figure 2c-d).

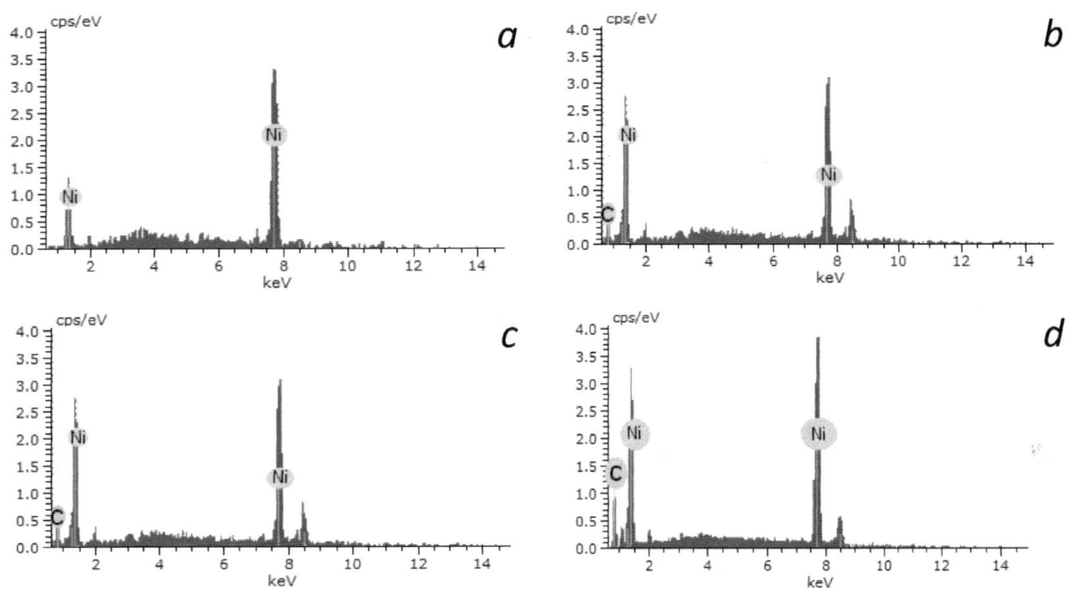

Figure 2. EDA spectra of Ni NTs: a) initial; b) irradiated at fluence 5×10^{10} cm^{-2}; c) 10^{11} cm^{-2}; d) 5×10^{11} cm^{-2}.

Determination of the influence of ionizing radiation on the crystal structure was carried out by XRD (Figure 3). Analysis of the observed XRD peaks (111) (200) and (220) showed that the obtained NTs have a polycrystalline structure with a predominant crystallographic direction [111], typical for electrodeposited nickel nanostructures. At the same time, future increasing radiation fluence more than 5×10^{10} cm^{-2}, the typical for nonstoichiometric carbides NiC_x peak with Miller indices (114) at 2θ = 36.2° appears on spectra. This fact confirms with the previously showed results of the EDA.

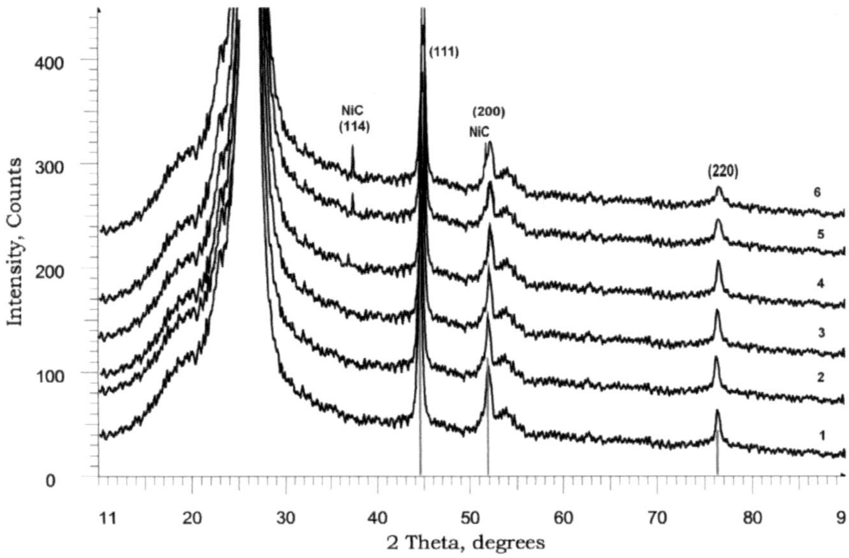

Figure 3. XRD spectra of Ni NTs in PET template at different irradiation regimes.
1) initial;
2) 10^9 cm^{-2};
3) 10^{10} cm^{-2};
4) 5×10^{10} cm^{-2};
5) 10^{11} cm^{-2};
6) 5×10^{11} cm^{-2}.

It is worth to notice that on the every diffraction diagram of the samples broadening peaks typical for nanodimension objects are observed. Calculations of Ni NTs structural characteristics showed that both the original and irradiated samples have FCC structure with lattice parameter *a*, which is typical of the nickel, but different from the standard value. The results of *a* calculation and the average crystallite sizes are shown in table 1.

Table 1. Parameters of Ni NTs obtained at different irradiation regimes.

Irradiation fluence, cm^{-2}	*a*, Å	The average size of crystallites, nm
0 (initial)	3,5223±0,0007	25,3±1,1
10^9	3,5190±0,0057	24,1±0,7
10^{10}	3,5173±0,0040	23,2±0,9
5×10^{10}	3,5160±0,0008	21,5±1,2
10^{11}	3,5157±0,0023	20,3±1,0
5×10^{11}	3,5244±0,0021	29,9±1,5

To determine the influence of the irradiation conditions on the electrical properties (conductivity σ) of Ni NTs array in PET template were studied. The electrical conductivity was calculated by the formula:

$$\sigma = \frac{dI}{dU}\frac{l}{A},$$

where l – length of NTs, A – conducting surface area, dI/dU – tangent of the angle depending I–U. Graph of dependence σ from deposition potential is shown in figure 4.

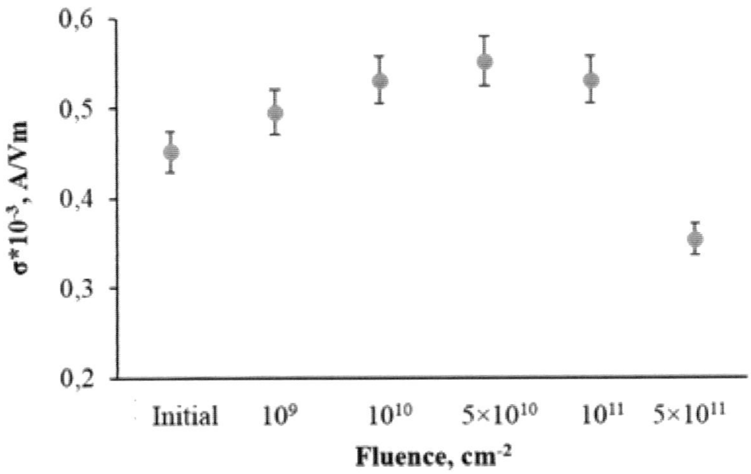

Figure 4. Dependence of Ni NTs conductivity on irradiation regimes.

The diagram of dependence of the conductivity on the radiation fluence has two parts. On the first part, from a zero to the fluence of 5×10^{10} cm^{-2} the conductivity slightly increases to a value on 20% more then the conductivity of the initial sample. On the second part of the diagram conductivity decreases sharply to value less on 23% than the non-irradiated sample. The growth of the conductivity on the first section, most likely related with the relaxation of defects of the NTs structure and the formation of less-defected wirings along the main NTs axis. At a fluence of 5×10^{10} cm^{-2} carbon in the form of carbides compounds accumulates appreciably in the structure (figures 2 and 3). The large quantity of carbon causes both the decreasing in the conductivity and the destruction of the wall structure, as is observed in figure 1d.

Conclusions

Nickel NTs with length of 12 microns, an external diameter of 380 nm and wall thickness about 95 nm were obtained by template synthesis method in the pores ion-track PET membranes. Modification of NTs structures was held on by irradiating by ions C^{+13} with energy of 1.75 MeV/nucleon and fluence ranging from 10^9 to 5×10^{11} cm^{-2}.

It was shown that irradiation with fluences up to 5×10^{10} cm^{-2} led to decrease of the lattice parameter and the average crystallites size, which form the NTs walls. Together with that increasing in conductivity of samples on 20% more relative to non-irradiated structures was observed. At fluences higher than 5×10^{10} cm^{-2} accumulation of carbon in the structure of NTs in a form of a new phase - carbon carbide took place. Growing the amount of carbon led to degradation of the NTs structure and, consequently, to reduction of their conductivity.

Acknowledgments

The work was done within the №561 Project of the Ministry of Education and Science of the Republic of Kazakhstan approved on April 07, 2015.

References

[1] Abedini A, Larki F, Saion E, Zakaria A and Hussein M Z 2012 *Radiat. Phys. Chem.* **81** 1653
[2] Ayres V M, Jacobs B W, Englund M E, Carey E H, Crimp M A, Ronningen R M, Zeller A F, Halpern J B, He M-Q, Harris G L, Liu D, Shaw H C, and Petkov M P 2006 *Diam. Relat. Materi.* **15** 1117
[3] Banhart F 1999 *Rep. Prog. Phys.* **62** 1181

[4] Bid A, Bora A and Raychaudhuri A K 2006 *Phys. Rev. B* **74** 1

[5] Ivanov L I and Platov Yu M 2004 *Radiation Physics of Metals and Its Application* (Cambridge, Cambridge International Science Publishing, UK)

[6] Singh A K, Das B, Sen P, Bandopadhyay S K and Mandal K 2014 *IEEE Transaction on Magn.* **11** 2302104.

[7] Kaurn A and Chauhan R P 2014 *Radiat. Phys. Chem.* **100** 59

[8] Kaniukov E, Kozlovsky A, Shlimas D, Yakimchuk D, Zdorovets M and Kadyrzhanov K 2016 *IOP Conf. Series: Materials Science and Engineering* **110** 012013

[9] Coa G and Wang Y 2004 Nanostructures and Nanomaterials—Synthesis, Properties, and Applications. (Imperial College Press, London: Toh Tuck Link Ltd., Singapore)

[10] Kaniukov E Yu, Ustarroz J, Yakimchuk D V, Petrova M, Terryn H, Sivakov V and Petrov A V 2016 *Nanotechnology* **27** 115305

[11] Dalchiele E A, Marotti R E, Cortes A, Riveros G, Gomez H, Martinez L, Romero R, Leinen D, Martin F and Ramos-Barrado J R 2007 *Physica E* **37** 184

[12] Kozlovsky A L, Meyrimova T Y, Mashentseva A A, Zdorovets M V, Kaniukov M V, Yakymchuk D V, Petrov A V and Kadyrzhanov K K 2016 *Chem. Bulletin of Kazakh National University* **3** 4

[13] Kozlovskiy A L, Shlimas D I, Shumskaya E E, Kaniukov E Yu, Zdorovets M V, Kadyrzhanov K K 2017 Influence of electrodeposition parameters on structural and morphological features of Ni nanotubes *The Physics of Metals and Metallography* **2** (in press)

[14] Shumskaya E E, Kaniukov E Yu, Kozlovskiy A L, Zdorovets M V, Rusakov V S, Kadyrzhanov K K 2017 Structural and physical properties of template-synthesized iron nanotubes *Physics of Solid State* **4** 766

[15] Kozlovskiy A, Zhanbotin A, Zdorovets M, Manakova I, Ozernoy A, Kiseleva T, Kadyrzhanov K, Rusakov V and Kanyukov E 2016 *Nucl. Instruments Methods Phys. Res. Sect. B Beam Interact. with Mater. Atoms* **381** 103

Method of stress and measurement current levels for MIS structures researching and modifying under high-field injection of electrons

D V Andreev[1], G G Bondarenko[2], V V Andreev[1] and A A Stolyarov[1]

[1] Bauman Moscow Technical University, Kaluga Branch, 2, Bazhenov st., Kaluga 248000, Russia
[2] National Research University Higher School of Economics, 20, Myasnitskaya Ulitsa, Moscow 101000, Russia

E-mail: vladimir_andreev@bmstu.ru

Abstract. We developed new method of stress and measurement current levels to research and modify thin dielectric of MIS structures under high-field injection of electrons. This method allows to take into account processes of MIS structure capacity charging and trapping of charge in gate dielectric of MIS structure when injection mode. It is shown that when a high density of the injection current, a characteristics control of charge, which accumulates in gate dielectric, when realizing the method of stress and measurement current, is necessary to implement by monitoring of voltage change on MIS structure when amplitude of measurement injection current is lower than amplitude of stress current. In order to rise performance of the method and implement a possibility of researching of fast relaxation charges, which is accumulated in gate dielectric during the process of high-field stress influence, we suggest to realize charging and discharging of MIS structure in accelerated regime when higher density of measurement current takes place.

1. Introduction

At present, in order to research of charge effects in MIS structures under high-field stress conditions methods of controlled current tests are widely used. The most frequently applicable methods are Constant Current method, Current-Ramp method (J-Ramp), Bounded J-Ramp method, and the method of the multi-level current test [1–6]. In order to investigate the charge effects in MIS structures, one most often uses Constant Current method, which concludes in monitoring of change of charge state, determined by dependence of voltage, applied to the sample, on time [1–6]. When using the method of constant current, short interrupting of injection process and measuring of voltage shift at opposite polarity of current pulse allow to determine a density of charge accumulated in gate dielectric and its centroid [6–8]. However, such measurements do not always have enough accuracy, and in a number of cases can lead to appearance of significant mistakes. J-Ramp and Bounded J-Ramp methods are mostly used to realize tests aimed to quality control of gate dielectric and determining of its defectiveness [1, 2]. To research thin gate dielectric films of MIS structures, in paper [6] we suggest a new method of controlled current stress. A distinctive feature of method proposed is using both stress and measuring levels of current in order to have a possibility to take into account accumulating of charge in gate dielectric not only during process of stress influence, but as well at time of stress mode reaching when injection currents less than stress values flow through gate dielectric. Usage of this method is more preferable when researching of charge effects in MIS structures at high values of stress current. Therefore, further improvement of this method, aimed to improving of its metrological characteristics, is a state-of-the-art problem and it will allow to receive a new information about charge effects in thin dielectric films of MIS structures when they are in conditions of high-field stress influences.

2. Description of methods

When using the method of constant current for the study of charge effects in MIS structures under high-field conditions, most researchers [5–7] commences the analysis of the results from the time when the injection current becomes equal to the amplitude of the applied current influence. This approach can be applied for low electric fields and densities of the injection current at which the rate of change of the charge state of the MIS structures is low.

However, at high densities, injection current increases the rate of generation and accumulation of charge in the gate dielectric, including case when MIS structure reaches the injection mode, which is corresponding to defined stress level of current I_s. Fig. 1 (b) shows for the method of constant current how the currents of MIS structure charging (I_c) and injection currents (I_{inj}) change, when an impulse of constant current with amplitude I_a is applied to MIS structure. Fig. 1 (b) shows that before the moment of reaching of stable mode, when I_{inj} is equal to I_a, charge Q_{inj} is injected into gate dielectric (shown by shaded region). This charge could lead to accumulation of a significant amount of charge in gate dielectric. Therefore, in to improve an accuracy and reliability of the constant current method, one has to begin an analysis of experimental results from the process of MIS structure charging.

The method of constant current supposes that the linear rising part of $V_I(t)$ dependence corresponds to charging of MIS structure. In this part all the current flowing through the gate dielectric is capacitive (fig. 1 (a, b). When a rectangular impulse is applied to sample researched, the equation of charges neutrality for MIS structure can be written as follows [6]:

$$Q_a = Q_c + Q_{inj} + Q_t\,, \tag{1}$$

where Q_a – charge supplied to the sample ($Q_a = I_a \cdot t$); Q_c – capacitive component of the charge of MIS structure; Q_t – the charge, which accumulates in gate dielectric during the process of high-field tunnel injection of electrons (can be calculated by means of preliminary studies).

Solving the equation (1) for Q_{inj} and assuming that the capacitance of MIS structure in the mode of high-field injection is equal to capacitance of gate dielectric (usually in frames of the method the accumulation mode is used; at the same time the mode of inversion can be used if the construction or the factor like heat or illumination provide a source of minority carriers in inversion layer), we can get an expression to determine an amount of charge injected into gate dielectric:

$$Q_{inj}(t) = \int i_{inj}(t)\mathrm{d}t = I_a \cdot t - C_i \cdot V_I(t) - Q_t(t)\,, \tag{2}$$

where i_{inj} – the injection current, I_a – the amplitude of the current pulse applied. Differentiation the expression (2) for time allows to obtain an equation to find the injection current [6, 7]:

$$i_{inj}(t) = I_a - C_i \cdot \frac{\mathrm{d}V_I(t)}{\mathrm{d}t} - Q_t(t)\,. \tag{3}$$

With the expression (3), taking into account time dependence time the dependence $V_I(t)$, it is possible to find time dependence of injection current when MIS structure reaches the stable mode of injection ($I_{inj} = I_a$). At the same time the minimal level of injected current is limited by accuracy of $V_I(t)$ measuring and the maximum level is limited by I_a [6]. Therefore, in order to accelerate the measuring process, the proposed method assumes to realize the charging of MIS structure by using of higher amplitude of current with respect to amplitude of measurement current.

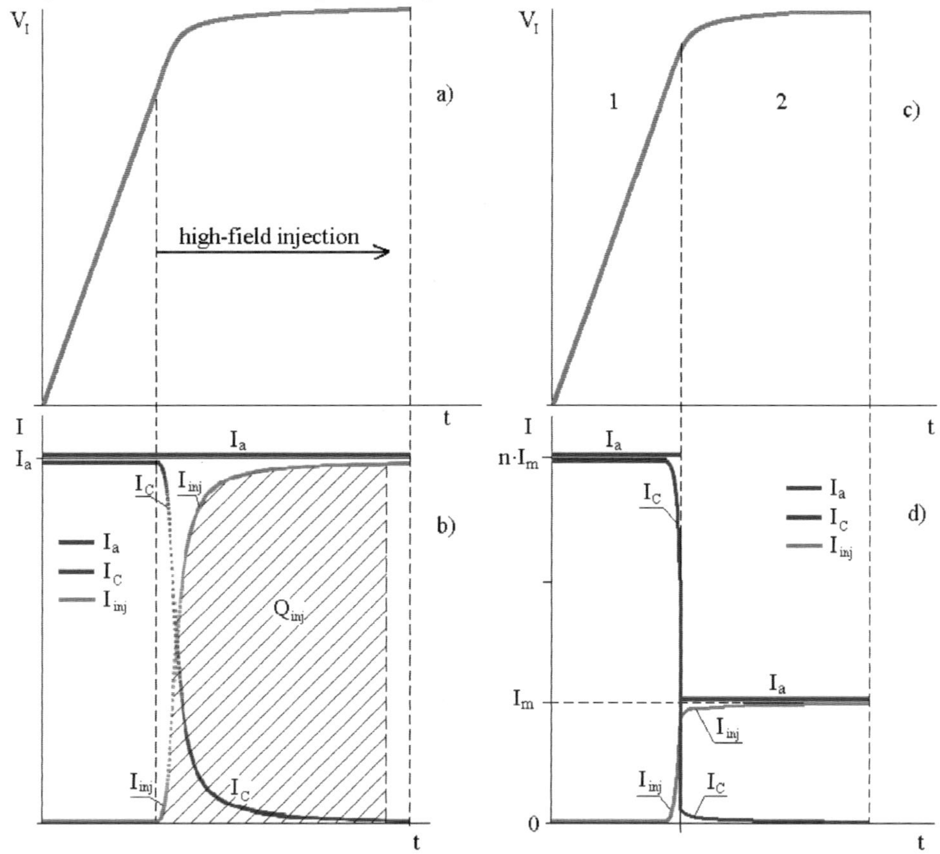

Figure 1. Time dependencies of voltage dropping on MIS structure (a, c) and currents flowing through MIS structure (b, d) in case of applying to MIS structure the impulse of constant current (a, b) and two-level current influence (c, d): I_a – amplitude of current impulse applied to MIS structure; I_{inj} – high-field tunnel injection current flowing through gate dielectric.

Fig. 1 (c, d) shows the time dependencies of voltage dropped on the MIS structure and dependencies of the currents, flowing through it, when applying the two-level current stress to the sample. The charging of MIS structure (Fig. 1, section 1) is carried out at a current density which is in n times than density of measuring current. At the same time, by measuring of $V_I(t)$ dependence and using expression (3), we gather a time dependence of the injection current. When injection current reaches the value of 0.9 of measurement current density, we carry out the switching of current and then to MIS structure a current impulse with amplitude of $I_a = I_m$ is applied (section 2 of figure 1). The proposed algorithm of current applying allows to reduce time of MIS structure charging and discharging in n times and, as a result, increase the accuracy of measurements and give a possibility to research fast relaxing charges.

The proposed method suggests that the dominating mechanism is Fowler-Nordheim tunnel injection of electrons in both the stress influence and the measurement influence [1]. Spatial distribution of charge trapped is supposed to be described as lateral equal, i.e. trapped charge locates enough distantly from injecting interface. Moreover, we neglect discrete character of accumulating charge, and we assume that possible presence of charge with opposite sign before has only insignificant influence to currents being measured.

Method of MIS structure charging by current with amplitude which is in n times greater than I_m, considered by using of example (Fig. 1 (c, d)), can be applied for discharging of MIS structure. In that case, the algorithm of applying of exhaustive current stress, used in the proposed method of stress and measurement currents, can be realized as Fig. 2 (a) shows. Fig. 2 presents time dependencies of current stress (a) and voltage (b) measured on MIS structure. This algorithm proposes to use three levels of current influence. First of them is with amplitude of I_s, which is according to the stress mode; this level is used when researching a change of MIS structure charge state. Second of them is with amplitude of I_m, which is according to measurement mode; at this level of current significant changes of charge state gate dielectric is not observed (typically $I_m \ll I_s$). Third of them is with amplitude of $n \cdot I_m$; at this level processes of MIS structure charging and discharging take place. Using of currents with opposite polarity in measurement mode ($-I_m$) allows to determine density, cross section of charge traps and location of charge centroid in volume of dielectric [3, 8].

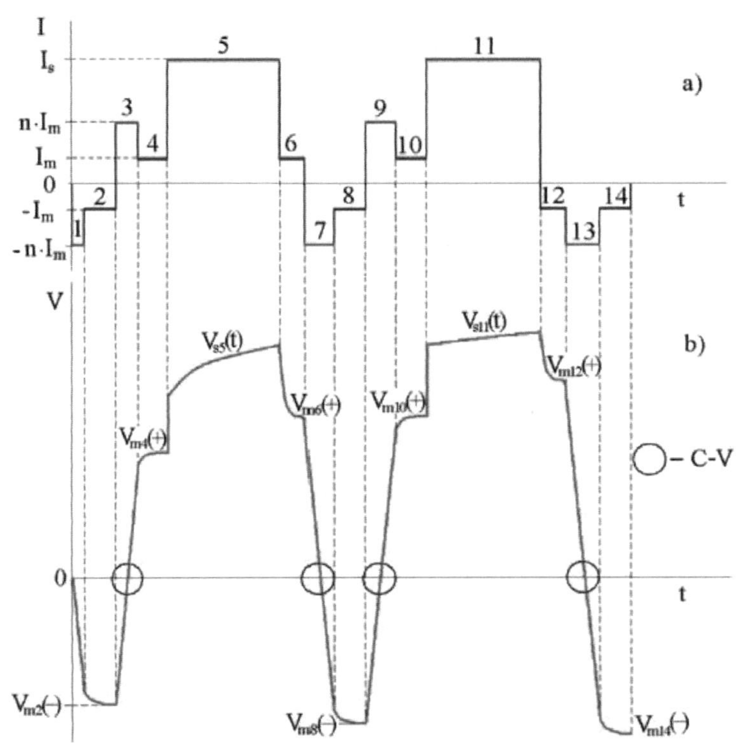

Figure 2. Time dependencies of multilevel current stress (a) and voltage (b), which is measured at MIS structure, for the stress and measurement current levels method.

The algorithm suggests to realize injection of charge in stress mode into gate dielectric by portions as fig. 2 shows (sections 5 and 11). Fig. 2 demonstrates only two sections corresponding to stress mode. In case of usage proposed method in practice, the method advises to choose amount and time of stress sections on the basis of research tasks. After the charge injection at each portion of stress influence, charging or discharging of MIS structure is implemented and injection by currents with amplitudes of I_m and $-I_m$ is taken place. Such an algorithm of current influence applying allows to measure voltage change on MIS structure at both positive and negative polarities of gate at measurement amplitude of injection current. In order to realize the algorithm proposed, the following condition should be satisfied. Switching of voltage polarity at gate, when injection, which is implemented to measure $V_m(+)$ and $V_m(+)$, should not influence to processes of changing of charge state of gate dielectric, which take place when injection by stress current level. To this purpose, injection at negative polarity of gate has minimum time and injection is implemented at measurement current amplitude, which is much lower than amplitude of stress current. In order to receive proper experimental results, it is needed to gage few MIS structures at different durations of sections of stress and measurement currents. Fig. 2 shows the algorithm, which can be implemented without realizing of voltage changing measurements at negative polarity of measurement current $-I_m$. However, in this case, informativity of proposed method is significantly decreases. Influence of measurements at negative polarity of measurement current $-I_m$ can be estimated by form of dependencies of voltage shift ΔV_s (at I_s amplitude of current), which become presented when realization of the algorithm shown in fig. 2 in both cases: first, at realization of the exhaustive algorithm, and, second, without realization of measurements of voltage changing at negative polarity of measurement current $-I_m$.

At high densities of injection current an increasing of rate of charge state changing for gate dielectric, when MIS structure reaches the injection mode which in accordance to stress current determined (I_s), is observed. As a result, when standing of stationary injection mode, gate dielectric can accumulate some portion of charge. As a consequence, time dependence $V_s(t)$ does not consider this accumulated charge. This can make for significant inaccuracies when rating of change of MIS structure charge state.

Measuring of voltage changing dropped on MIS structure at current with amplitude of I_m allows to take into account the charge accumulated in gate dielectric when MIS structure reaches injection mode, which is corresponding to stress current, and, thus, significantly reduce inaccuracy, which is inherent in method of constant current. As an example, voltage changing $\Delta V_m = V_{m6} - V_{m4}$ consider both charge accumulated in gate dielectric at the time of reaching of injection mode by MIS structure when stress current level I_s and charge accumulated in gate dielectric due to injection of electrons in section 5 (fig. 2). At the same time, voltage changing on MIS structure ΔV_{s5} at the end of section 5 considers only charge accumulated in gate dielectric when injection of electrons by stress level of current I_s at section 5 takes place (fig. 2).

The method proposed, when it realizes the algorithm of measurements shown in fig. 2, allows to monitor low-frequency C-V characteristics of MIS structure by simultaneous analysis $C(t)$ and $V_I(t)$ at sections of charging and discharging of sample (these sections shown in fig. 2 (b) by circles) [5–7].

The method proposed in the implementation of the measurement algorithm is shown in Fig. 2, also allows to control the low frequency C-V characteristics MIS structure analyzing jointly $C(t)$ and $V_I(t)$ of the charge portions and the discharge capacity of the sample (areas shown by circles in Fig. 2, b) [5–7]. The proposed method of producing C-V and I-V characteristics in a single method allows you to control the parameters of the degradation of the charge immediately after the tunnel injection, reducing the influence of relaxation processes. The proposed way of acquiring of C-V and I-V characteristics within the one method allows to monitor parameters of charge degradation immediately after tunnel injection, thus reducing influence of relaxation processes.

3. Conclusions

This paper proposes the new method of stress and measurement currents level to research charge effects under high-field stress influence. The method takes into account processes of charging and discharging of MIS structures capacitance and also charge trapping in gate dielectric of MIS structure in process of stress injection reaching.

The paper shows that when high density of injection current and high rate of charge state changing of MIS structure, the control of characteristics of charge, which accumulates in gate dielectric, when realizing the method of stress and measurement currents level, is better to implement by monitoring of change of voltage on MIS structure at amplitude of measurement injection current much less than amplitude of stress current.

In order to heighten the performance of the method of stress and measurement current level and rise possibility of observation of the fast relaxation charges, which accumulate in thin gate dielectric under high-field stress influence, charging and discharging of MIS structure capacitance, which is essential to monitor of changing of sample charge state in measurement mode, is important to implement in accelerated mode, besides density of stress current should be more than density of measurement current.

Acknowledgements

This study was supported by the Ministry of Education and Science of the Russian Federation, state task of the Moscow State Technical University, and the Kaluga Region Administration (grant № 16-42-400791). Support from the Basic Research Program of the National Research University Higher School of Economics is gratefully acknowledged.

References

[1] JEDEC Standard, JESD35–A: Procedure for the Wafer–Level Testing of Thin Dielectrics. 2001
[2] Strong A W, Wu E Y, Vollertsen R, Sune J, Rosa G L, Rauch S E and Sullivan T D 2009 *Reliability wearout mechanisms in advanced CMOS technologies* (Wiley-IEEE press) p 624
[3] Chen C, Chang I Y, Lee J Y, Chiu F, Chiouand Y, Wu T 2007 *Applied Physics Letters* **91** 123507-3
[4] Andreev V V, Bondarenko G G, Maslovsky V M, Stolyarov A A 2012 *IOP Conf. Series: Materials Science and Engineering* **41** 012017-6
[5] Bondarenko G G, Andreev V V, Loskutov S A, Stolyarov A A 1999 *Surface and Interface Analysis* **28** 142-5
[6] Andreev V V, Bondarenko G G, Maslovsky V M, Stolyarov A A, Andreev D V 2015 *Phys. Status Solidi C* **12** 299–3
[7] Fischetti M V 1985 *J.Appl.Phys.* **57** 2860-79
[8] Nissan-Cohen Y, Shappir J, Frohman-Bentchkowsky D 1985 *J. Appl. Phys.* **57** 2830-9
[9] Andreev D V, Bondarenko G G, Andreev V V, Stolyarov A A 2016 *IOP Conference Series: Materials Science and Engineering* **110** 012041-6
[10] Nasyrov K A, Gritsenko V A 2013 *Physics-Uspekhi (Advances in Physical Sciences)* **56** 999-13
[11] Vexler M I, Grekhov I V 2016 *Semiconductors* **50** 671-7

Thermodynamic modelling of phase equilibrium in system Ti-B-Si-C, synthesis and phases composition of borides and carbides layers on titanic alloyVT-1 at electron beam treatment in vacuum

N N Smirnyagina[1,2], V M Khaltanova[1,2], A E Lapina[1] andD E Dasheev[1]
[1]Institute of Physical Materials ScienceSB RAS, 670047, Ulan-Ude, Russia.
[2]Buryat State Universities, 670000,Ulan-Ude Russia

E-mail smirnyagina09@mail.ru

Abstract. Composite layers on the basis of carbides and borides the titan and silicon on titanic alloy VT-1 are generated at diffused saturation in vacuum. Formation in a composite of MAX phase Ti_3SiC_2 is shown. Thermodynamic research of phase equilibrium in systems Ti-Si-C and Ti-B-C in the conditions of high vacuum is executed. The thermodynamics, formation mechanisms of superfirm layers borides and carbides of the titan and silicon are investigated.

1. Introduction

Last year's interests to the titan-silicon-boron-carbon alloys increases, as they, possess ease, high durability, excellent electro conducting and stability at the raised temperatures. The development of new compositions according to phase diagrams of the Ti-Si-B-C system is one of the main ways to obtain and investigate new titanium alloys with improved properties. By present time of the MAX phases (double carbides of silicon and titan Ti_3SiC_2,$Ti_5Si_3C_x$) are synthesized by various methods in a nano structural status in the form of layers, films or ceramic.

The crystal structure of thin layers and physical and chemical, mechanical properties etc. are investigated [1-4].$Ti_3SiC_2$is a rare example of so called "plastic ceramics". It combines propertiestypical of ceramic compounds, such as high melting temperature, thermal and chemicalstability, with mechanical properties of metallic systems.The main difficulty while synthesizing$Ti_3SiC_2$is the loss of silicon during reaction. The reaction features depend on the Ti_3SiC_2crystal structure that consists of titanium carbide blocks divided by silicon layers. The inter-atomic dimension analysis shows that they are larger for silicon in Ti_3SiC_2than for other compounds Ti-Si-C. At the same time,Ti-Cdimensions are close to those of titanium carbide. The weak bond of silicon in the Ti_3SiC_2lattice can be the cause of its loss in thermal reactions.

In [5] the conclusion is drawn on synthesis possibility complex carbide phases, analogues of MAX phases in system Ti-Si-C as a result of high-intensity influences, for example electron beams.

To improve properties of the material, informationabout the phase diagram and thermodynamic properties in theTi–Si–C system is essential [6].The authors strongly recommend the simultaneous use of thermodynamiccalculations and experiments in order to establish technologicallyimportant phase diagrams highly efficiently.

Certainly, modeling phase equilibrium and revealing a field of borides and carbides in ternary systems Ti-Si-S and Ti-B-C represents to crystallization certain scientific interest. It is physicochemical basis for working out of technologyof functional coating formation at electron beam processing in vacuum.

In the present study revealed optimum conditions of formation the MAX phases at electron beam processing of titanic alloys.

2. Experimental procedures

Modeling phase equilibrium and revealing of fields of crystallization borides and carbides in ternary systems Ti-Si-S and Ti-B-C.

In the present study used interface of a program complexTERRA[7].Calculations are spent in a temperature interval 300-4500 K at a variation of the general pressure in system in a range from 10^5 to 10^{-3} Pa.

In calculations considered following phases:silisidesTi_3Si, Ti_5Si_3, Ti_5Si_4, TiSiи $TiSi_2$; carbides B_4C, SiC, TiC; borides TiB, TiB_2 and Ti_3SiC_2,$Ti_5Si_3C_x$.

Layers of composites formed on preliminary prepared surface of the samples made of titanic alloy VT-1 in the form of square plates in the size 15x15 mm and height of 7 mm. It was used sating daubs. Sating daubs contained reactionary a component(Si:2C) and the organic binding. As the organic binding applied a solution 1:10 glue BF-6 in acetone. Electron beam processing was carried out in a vacuum at most $2x10^{-3}$ Pa at power electron beam W = 250-450 W for 1-3 min.

X-ray phase analysis was carried out on the Phaser 2D Bruker (Cu $K\alpha_1$ – radiation). The microstructure of layers was investigated by METAM PB-22 microscope with the program NEXSYS Image Expert. The Vikers (HV) microhardness measurements are made by pressing a diamond indenter, of a specified shape, into the surface with a known force.

3. Results and discussion

Phase equilibrium and revealing of of crystallization fields of borides and carbides of the titan and silicon in ternary systems Ti-Si-C and Ti-B-C and their analysis representations about mechanisms and laws of education боридов and carbides in the conditions of vacuum allow to systematize and expand modeling. On fig. 1 the concentration tetrahedron of system Ti-Si-B-C is presented.

Phase equilibrium and revealing of of crystallization fields of borides and carbides of the titan and silicon in ternary systems Ti-Si-C and Ti-B-C and their analysis representations about mechanisms and laws of education боридов and carbides in the conditions of vacuum allow to systematize and expand modeling. On fig. 1 the concentration tetrahedron of system Ti-Si-B-C is presented.

3.1. Phase equilibrium in system Ti-Si-C.

On fig. 2 are presented phase balance in system Ti-Si-C at pressure $P=10^{-3}$ Pa. Areas of co-existing phases are allocated. It is shown, that the titan silisides titans of various compositions form квазибинарные cuts with titan carbide, instead of with silicon carbide.

3.2. PhasebalanceinsystemTi-B-C.

In this system cut Ti-B_4C is not binary. At him there can be various phases from ternary system Ti-B-C. Phase balance (isothermal sections at 1000 K) in system Ti-B-C are presented on fig. 3 atpressure 10^{-3} Pa, besides on fig. 4 are presented isotherms of this system at discussed pressure. Formation of ternary compounds is not revealed.

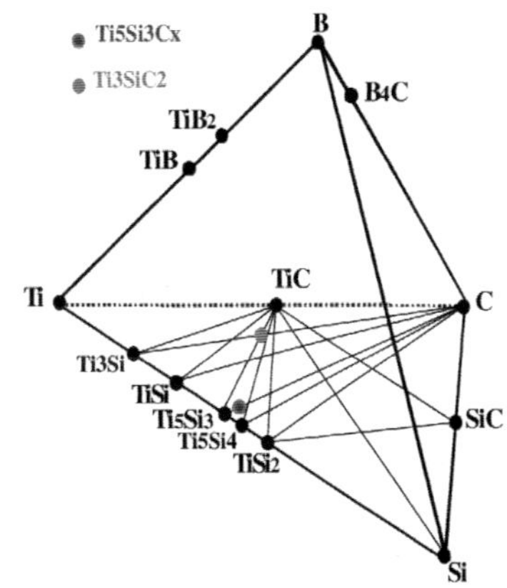

Figure 1. Concentration tetrahedron of system Ti-Si-B-C.

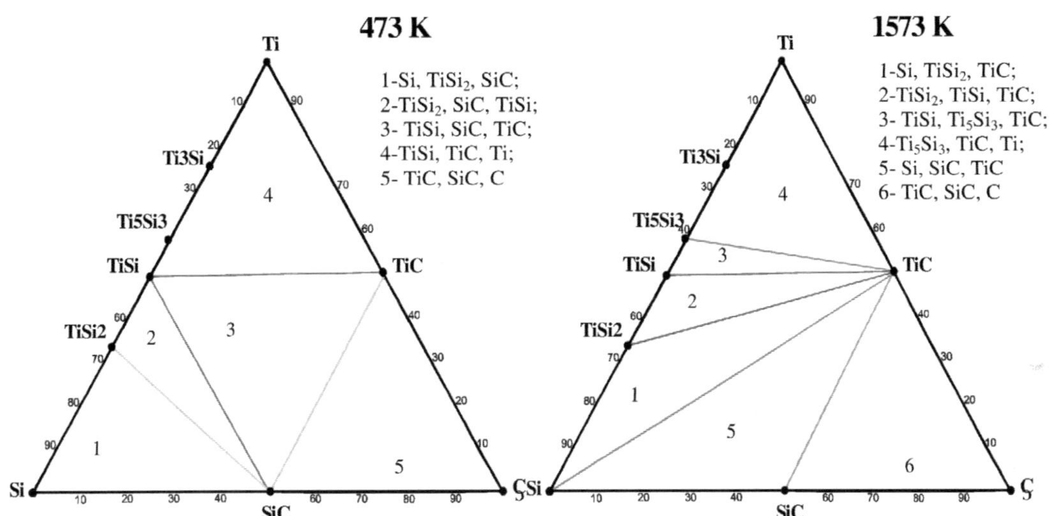

Figure 2. Phase equilibrium in system Ti-Si-C at 10^{-3}Pa.

According to thermodynamic calculations the coexistence of various areas of crystallization is possible: B_4C; C; C+TiB_2; B_4C+C+TiB_2; C+TiB_2+TiC; TiC, TiB_2+TiC, TiB+TiB_2+TiC; TiB+TiC and Ti +TiB (at pressure below 1 Pa). At atmospheric pressure at heats (above 3000K) areas single-phase boride TiB_2, boron, their joint presence, and three-phase area with B, C and TiB_2 are found out.

Formation processes of the carbides and borides titan proceed with allocation of a significant amount of energy, thereby raising temperature in system to 2000÷2150 K ($P=10^5$ Pa) and 1600÷1725 K ($P=10^{-3}$ Pa).

3.3. Titanic alloy VT-1.

Initial alloy VT-1 was represented a mix of two forms of the titan (α - and β-Ti). Titanic alloy VT-1 after electron beam treatment was also many phases. Lattice parameters of cells α-Ti: a=0,29504 *nm*and c=0,46833 *nm*(P63/mmc, PDF 03-065-3362), and parameters of a cubic cell β-Ti: a=0,3306 (Im-3m, PDF 00-044-1288).

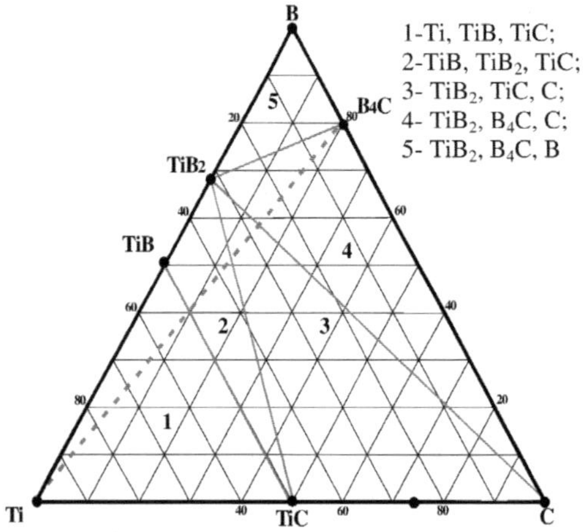

Figure 3. Phase equilibrium in system Ti-B-C at T=1000 K ($P=10^{-3}$ Pa).

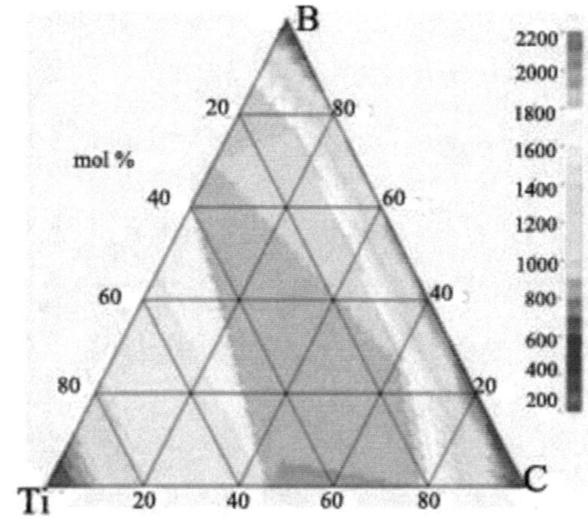

Figure 4. Isotherms in system Ti-B-C at $P=10^{-3}$ Pa.

Composite layers on the basis of titan and silicon carbides.
Layers of composites 80-100 microns are formed. A layers structure iseutectic (fig. 5).

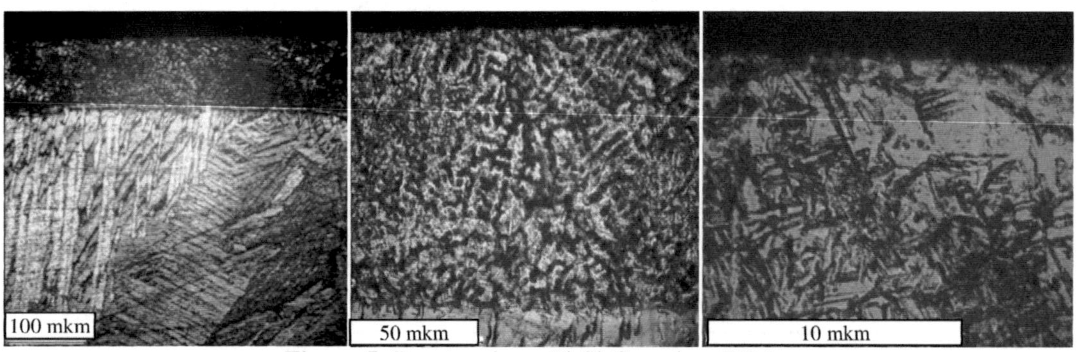

Figure 5. Structure layer Ti-Si-C on titan VT-1.

According to XRD results the composite layer are contain 76,5 mas. % crystalline and 23,5 mas. % amorphous phases. Thus the metal basis consists from 84,06 mas. % β- titan with and 4,06 mas. % α-titan. The layer contains double carbide Ti_3SiC_2 and silicon carbide β-SiC (PDF01-073-1665, F43m), with a cubic cell *a*=0,4358 *nm*.

3.4. Solid state boriding from sating daub of amorphous boron.
X-ray presence disordered phases of titan α-Ti and boride titan TiB_2 (PDF 01-073-2148) is revealed. There is no reflexes β-Ti that testifies to heats on a surface of a titan alloy at electron beam boriding. The considerable thickness of the modified layer (more than 700-900 microns) is observed.

Attempt of synthesis Ti_6Si_2B from sating daub of composition Si-B-2C is made. Layersofthickness 30-80 micronsareformed.

Measurement of microhardness of composite layers has shown achievement of values to 10 GPa.

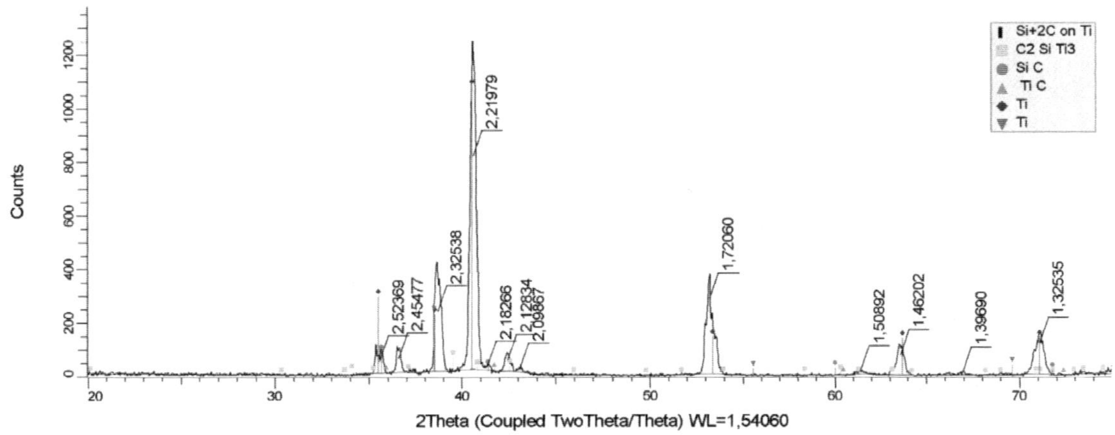

Figure 6. X-ray phase analysis of composite layersTi-Si-C on titan alloy VT-1.

4. Conclusions

In summary, preliminary results of electron beam boriding of titan alloy VT-1 are presented. The thermodynamic investigation of equilibrium in systems Ti-B-C-Si is carried out with the purpose of formation conditions optimization of functional layers on a surface of titan alloys as a result of electron beam boriding in vacuum. The thickness of MAX phase Ti_3SiC_2 coating is received100 μm. Composite layers will consist from Ti_3SiC_2. Research by means of a metallografic microscope has shown that carbide borides layers formation is difficult physical and chemical process. Their formation is difficult physical and chemical process. The structure, phase structure and strength characteristics are investigated.

Acknowledgements

This work was supported by the RFBR, project 15-48-04217-r-sibir-a.

References

[1] KorostelevaEN, Pribytkov GAandGurskikhAV2009*Powder Metallurgy and Metal Ceramics***48**8-12
[2] VahlasC, ChevalierPY, and BlanquetE1989*CALPHAD***13**(3) 273-92
[3] DebashisBandyopadhyay2004*Journal of Phase and Diffusion***25** (5) 415-20
[4] Yong Du, Julius C, Schuster, Hans J Seifert and Fritz 2000*Journal of the American Ceramic Society***83** (1) 197-203
[4] Ivanov IuF, TeresovAD, SoskovaNA 2012*IzvestiiavuzovFizika***55** (5/2) 130-39
[5]SmetkinAA, MaiorovaIuK 2015*Bulletin PNRPU Mechanical engineering, materials science* **17** (4) 120-37
[6] WakelkampWJJ, van LooFJJ, and MetselaarR 1991 *JEur. Ceram.Soc.***8**(1) 35–39
[7] Trusov BG 2002TERRA software package for calculation of plasma chemical processes *Matererials of III Intern. Symp.on Theoretical and Applied Plasma Chemistry*. Ples 217-18

Reception carbon nanomodifiers in arc discharge plasma and their application for modifying of building materials

A P Semenov[1], N N Smirnyagina[1], L A Urkhanova[2], S V Kanakin[3], S A Lkhasaranov[2], I A Semenova[1], B O Tsyrenov[1], D E Dasheev[1] and Z M Khaltarov[1]

[1]Institute of Physical Materials Science SB RAS, 670042, Ulan-Ude, Russia,
[2]East – Siberian state university of Technologies and Management, 670013,Ulan-Ude, Russia
[3]Geological Institute SB RAS,670042, Ulan-Ude, Russia

E-mail smirnyagina09@mail.ru

Abstract. Carbon nanomaterials are perspective additives for modifying cement composites. In this work the influence of carbon nanomodifier (CNM) formed in plasma chemical reactor on phase composition, structure and properties of cement stone was investigated. Method of dispersion of nanoparticles has been used, which consists in sonication mixing water with carbon nanomodifier and super plasticizers (SP). Change in phase composition, structure and properties of modifying cement stone were investigated.

1. Introduction

Modification of cement by various nanomodifiers is promising because their introduction allows regulatingthe structure of the material directionally through the various effects [1].

Use of carbon nanostructures in the modification of the cement matrix is promising and has been investigated in many studies [2-5]. It is noted that they can be used not only as centers of crystallization, but also as objects, changing the direction and rate-controlling physical and mechanical processes in hardening systems.Due to their beneficial mechanical, electrical and chemical properties carbon nanotubes (CNTs) are an attractive candidate for reinforcement of composite materials.

The possibility of using fullerenes for modifying of cement systems studied to a lesser extent. In general researches are devoted to study the influence of fullerenes on the properties of mixing water [6].

Mechanism of additives containing fullerenes not disclosed enough full, so the aim of the research is to study the phase composition, structure and properties of ordinary Portland cement with the introduction of the carbon nanomodifier, which obtained in the [7].

2.Materials and methods

In current research were used Portland cement (OPC), LLC "Timlyuisky Cement Plant" (Republic of Buryatia, Russian Federation) CEM I 32.5N, superplasticizerSikaViscoCrete 5 neu®, superplasticizerC-3®. SikaViscoCrete 5 neu ® - superplasticizing admixture on the basis of polycarboxylates(Switzerland). SuperplasticizerPolyplast C-3® (LLC "Poliplast-Uralsib", Russian Federation) is a synthetic compound based on polymeric naphthalene sulfonic acid.

The mechanical performance of modified cement stone was evaluated by compressive strength test. Samples of OPC 20*20*20 mm prepared with standard consistence, hardened under normal conditions at t = 20 ± 2 ° C and a humidity of 100%, were tested by at the age of 7 and 28 days.

Synthesis of initial carbon nanomodifier carried out on the apparatus - a plasma-chemical reactor [7] (Figure 1).The basis of the system is based on the erosion of graphite electrodes in the arc discharge plasma [8]. The discharge is initiated at a pressure of 10^5 Pa, by passing through the electrodes and current frequency 44 or 66 kHz. Erosion of rods occurs in closed sealed volume filled with helium.

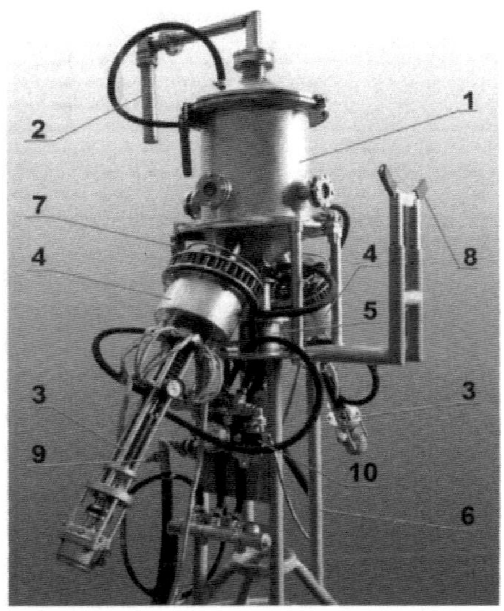

Figure 1. General view (plasma chemical reactor): 1 - camera for the synthesis 2-nitrogen trap, 3 - stocks, 4-block to meet the load, 5-carbon condensate storage device, 6-rack, 7-clips, 8- detent, 9-tap, 10 - the valve controlling the supply of cold water.

Apparatus contains a water-cooled chamber 1, in which there is synthesis of plasma. Towards the bottom and top of the chamber through the flanges mounted nitrogen traps 2. Within the camera body provided with an inspection windows for visual observation. Camera cover is removable. Towards the bottom of the camera connected removable stocks 3 with the system providing their movement as the combustion system of rods and to meet the load 4. Synthesized carbon condensate is collected from the chamber walls in the store 5. Camera is located on the rack 6. Electric power is carried out from voltage generator, with a rated capacity of 16 kW. Maximum current of erosion of graphite rods is 160 A. The average time of synthesis is 10 min.Carbon condensate containing 10-12 mass %fullerenes (80 % - C_{60}; 15 % - C_{70} and oxides $C_{60}O$).

For uniform distribution of CNM in water is used ultrasonic machine having the following characteristics: operating frequency of 43-45 kHz, power 30 W.

Indicators of conductivity and pH of modified water were determined on the instruments - conduct meter «HM Digital Aquapro», pH-meter «Pen type pH-009."

X-ray diffraction analysis was carried out on a diffractometerBrukerPhaser D2 using Cu $K_{\alpha 1}$ radiation.

Microstructure of cement stone was studied by scanning electron microscope LEO-1430VP (Carl Zeiss, Germany) with energy dispersive microanalysis system INCA Energy 350 (Oxford Instruments, UK).

3. Results and discussion

Introduction of CNM changes rheological properties and setting time of cement paste (table 1).

Table 1. Properties of cement pastes and mortars with CNM.

Indicators*	unit	without additives	CNM 0.01%	CNM 0.001%
Standard consistence	%	25	26	25,5
Initial setting	min	145	160	150
Final setting	min	245	250	255
Slump flow	mm	135	140	135
Slump flow through hours:	mm			
0,5		135	140	135
1		135	140	135
2		130	140	135
3		115	135	130
4		110	135	130

It is found that in cement with CNM at various concentrations is lengthened setting time (initial set - up to 5%, final set - up to 22%). Simultaneously, slump flow of cement mortar is increased and retained through some hours, which is important, for example in long transportation of the concrete mix based on the modified cement. Obviously, the change of setting time and slump flow of cement pastes and mortars is associated with the change in the properties of mixing water.

Introduction of CNM contributes to creation around particles the hydration shells, which are oriented and contributed to changing the rheological properties of cement pastes and mortars.

The greatest effect of the introduction of carbon nanomaterials is achieved while providing their uniform distribution in the composite structure.

Table 2. Strength parameters of cement stone with CNM.

Composition	Compressivestrength, Mpaafter	
	7 daysofhardening	28daysofhardening
Control	40	61
CNM 0.01%	38	67
CNM 0.001%	44	82
C-3® 0.7%	47	50
C-3® 0.7% + CNM 0.01%	42	67
C-3® 0.7% + CNM 0.001%	58	74
SikaViscoCrete 5 neu®0.3%	43	63
SikaViscoCrete 5 neu® 0.3% + CNM 0.01%	42	69
SikaViscoCrete 5 neu® 0.3% + CNM 0.001%	40	68

In work for the uniform distribution of the agglomerates in the mixing water CNM was used an ultrasonic method which is based on the use of cavitation. To increase the effect of distribution CNM in mixing water were introduced surfactants with different chemical compositions, which are traditionally used to modify Portland cement. We are determined the optimal time of water treatment with CNM.

IOP Conf. Series: Materials Science and Engineering **168** (2017) 012059 doi:10.1088/1757-899X/168/1/012059

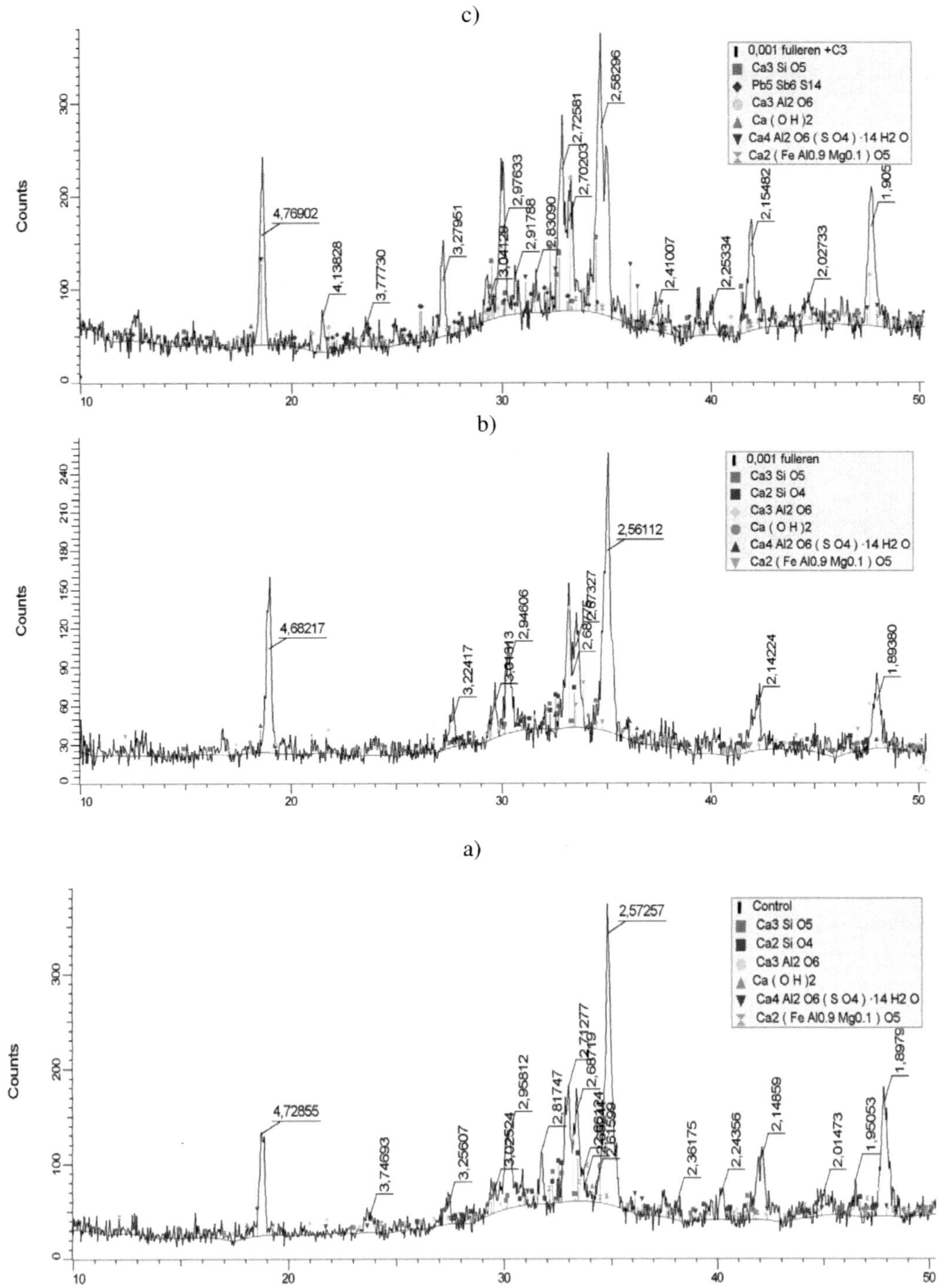

Figure 2.X-ray diffraction patterns of cement samples after hydration: a) control; b) 0.001 mass % CNM; c) 0.001 mass %CNM with C-3®;

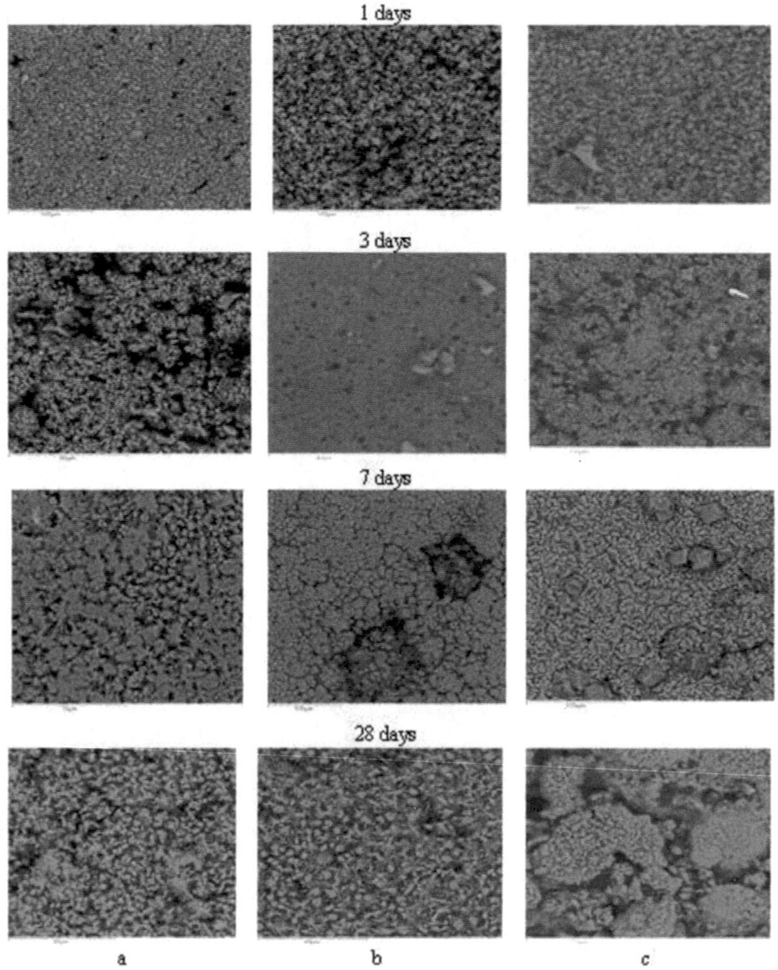

Figure 3. SEM patterns: a – hydrated OPC, 7 days; b – OPC+0,01% CNM, 3 days; c – OPC+0,01% CNM, 7 days.

Introduction of CNM in the cement leads to a change in the physical and mechanical characteristics. Table 2 shows the strength characteristics of the cement to CNM at age 7 days and 28 days of hardening.

Analysis of the results showed that the introduction of carbon nanomodifier in an amount of 0.01 mass % increases strength by 25-35%. Introduction in an amount of 0.001 mass % jointly with C-3® increases strength by 20-25%.

Changing the properties with introduction of CNM associated with changes in the phase composition and structure of modified cement stone.

According to X-ray diffraction analysis, the original Portland cement includes traditional phase characteristic of OPC: calcium silicates $Ca_3SiO_5$and Ca_2SiO_4, calcium aluminates $Ca_2(FeAl_{0,9}Mg_{0,1})O_5$ and $Ca_3Al_2O_6$ and calcium sulfate $CaSO_42H_2O$ (Figure 2). In hydrated cement decreased intensity of peaks of the original cement phases, appear peaks attributable to $Ca(OH)_2$ and calcium hydrosulfoaluminate in low- sulfate form $Ca_3Al_2O_6(SO_4)14H_2O$ (d=0,4729 nm) with expressive intensity of peak (Figure 2 a).Introduction of CNM into the cement causes a change in the hydrated samples of distance from d = 0.4682 nm to d = 0.4833 nm (Figure 2 b).

Contents of aluminates $Ca_2(FeAl_{0.9}Mg_{0.1})O_5$ and $Ca_3Al_2O_6$ is reduced, which is testified to intensive hydration of these phases, especially in the presence of CNM. For example, for phase $Ca_3Al_2O_6$ (PDF 00-006-0495 ICCD) a decrease in intensity of the reflection is d_{100} = 0.2700 nm. Furthermore, the introduction of CNM into hydrated cement leads to change in samples of distance from d = 0.2561 nm to d = 0.2604 nm for this plane. This variation in the d_{hkl} distances can be explained in terms of the degree of crystalline of hydrated cement samples. So in the control sample cement a content of the amorphous phase reaches 46% and the addition of fullerenes leads to an increase in crystalline up to 63%, which indicates an increase the degree of hydration of PCs in the initial period of hardening.

Introduction of CNM changes not only phase composition of hydrated cement but also the microstructure of cement stone. Figure 3 shows the surface of hydrated cement structure with CNM hardened for 3, 7 and 28 days.

Introduction of CNM reduces porosity of cement stone due to formation of gelatinous hydration products filling interporous space. It should be noted that an increase in hardening time from 3 to 7 days leads to a substantial reduction in porosity of the hydrated stone and reducing crystallite size (Figure 3 b, c). All of this favorable effect on the changes in physical and mechanical properties of modified cement stone.

4. Conclusions

Comparison of superplasticizers of various natures suggests that both superplasticizers ensure a uniform distribution of CNM in the volume of mixing water. However, evaluating the physic-mechanical properties of the modified cement stone it can be say that the superplasticizer C-3® is the most effective. Increasing of strength when it is used up to 25-35% compared with the control. Use of SikaViscoCreteneu ® is less effective (5-15%). This is due to the mechanism of action of this admixture, and enveloping and preventing interaction of the particles of CNMwith cement grains.With the introduction of CNM in a cement matrix, there is a change of the phase composition, structure and physic-mechanical properties of the cement stone. This is due to the complex action of carbon nanomodifier. It changes structure of mixing water, creating around their directionally oriented particles of hydrated shells that lead to changes in the rheological characteristics of the cement paste.Furthermore, particles of carbon nanomodifier act as nucleation sites of hydration products ofcement, which accelerates the hydration and hardening process of the cement, especially in the initial period of hardening. With introduction of CNM is reduced porosity of cement paste, which leads to high strength characteristics of modified cement.

Acknowledgements

This work was supported by the Russian Academy of Sciences (Artica, project №84).

References

[1] Artamonova O V, Sergutkina O R 2013 *Bulletin the Voronezh State University of Architecture and Construction* 13-23
[2] Lukutcova N P, Pykin A A, Karpikov E G 2011 *Building Materials* **9** 66-67
[3] Konsta-Gdoutos M S, Metaxa Z S, Shah S P 2010 *Cem. Concr. Res.* **40** 10 52-59
[4] Cwirzen A, Habermehl-Cwirzen K, PenttalaV 2008 *Adv. Cem. Res.* **20** (2) 65 -73
[5] Li G Y, Wang P M, Zhao X. 2005 *Carbon* **43** (6) 12 39–45
[6] Nasibulina L I, Anoshkin I V, Nasibulin A G, Cwirzen A, Penttala V, and KauppinenE I 2012 *Journal of Nanomaterials* 1-6
[7] ChurilovGN.*Patent RU2320536*2007 BI number 20
[8] Churilov G N, Bulina N V, Fedorov A S 2007Fullerenes: synthesis and theory of formation(Novosibirsk: Publishing House of Russian Academy of Sciences) 227

IOP Conf. Series: Materials Science and Engineering **168** (2017) 012060 doi:10.1088/1757-899X/168/1/012060

Role of spin states in the process of electron energy transformation in P3HT films, doped KI salt

A Nurmakhanova, D Afanasyev, N Ibrayev[1]
Institute of Molecular Nanophotonics, Academician E.A. Buketov Karaganda State University, Karaganda, Kazakhstan

E-mail: niazibraev@mail.ru.

Abstract. Influence of KI salt on the spectral-luminescent properties of polymer films of P3HT has been investigated in this work. Addition of KI doesn't lead to significant changes in optical density of the polymer films, but significantly changes the intensity and lifetime of the luminescence of P3HT. Adding KI reduces the lifetime of P3HT luminescence. Growth of the luminescence intensity occurs at a low concentration of KI and its quenching occurs at the high concentrations. A significant increase in the electrical resistance of the films takes place by adding of the KI. Type of magnetic effects on the photoluminescence of P3HT depends on the concentration of KI. Changes in the magnetic effect in P3HT-KI films are connected with increase in the concentration of triplet states in the polymer.

1. Introduction

Research of electronic processes in organic semiconductors actively conducting these days [1]. This interest is due to the great potential of practical application of such materials to create the elements of molecular electronics [2], elements for optoelectronics [3], in particular organic light-emitting diodes [4], organic photovoltaic elements [1], etc.

One of the possible ways to improve the efficiency of the polymer solar cells could be an increase in the singlet-triplet conversion in polymers. This can be achieved by changing the spin-orbit interaction, using external heavy atom effect [5]. It was shown earlier [6], that by adding KI-salt in epy polymer film of poly(3-hexylthiophene-2,5-diyl) (P3HT) occurs an increase of the intersystem crossing between electronic S_1 and T_1 states of the polymer. This represents a significant contribution to the excited triplet states of P3HT polymer of the organic solar cell in the process of transformation of light energy into electrical energy.

However, growth concentration of T_1 states in the polymer complicates the nature of the transformation of energy in the cells. Therefore, the mechanism of the effect of external magnetic field on the processes of the electron energy transformation in P3HT polymer films is not yet clear. The way to clarify the electron energy transformation in the organic semiconductor polymer could be the effect of the external magnetic field on the photoluminescence, electroluminescence and an electric current in the polymers.

The formation of excitons, polarons and bipolarons take place in the polymers with optical excitation [7,8]. Magnetic field (MF) can influence a number of processes involving these electronic

[1] To whom any correspondence should be addressed.

IOP Conf. Series: Materials Science and Engineering **168** (2017) 012060 doi:10.1088/1757-899X/168/1/012060

states. The MF influences on the interaction of triplet-polaron [9], on the conversion process of electron-hole precursor pairs [10] and on the spin conversion bipolaron precursor pairs [11]. It is necessary to take into account that the main change in the luminescence intensity and/or current can occur at small values of the induction of magnetic field (*B*), so-called low field effect, and at higher *B* - high field effect [8]. If the low field effect is usually associated with the hyperfine interaction [12], the high field effect can be caused by a number of different mechanisms, in particular triplet-triplet annihilation [13]. Therefore it is necessary to conduct comprehensive research for a deeper understanding of the electron energy transformation processes in semiconductor polymers. For example, in [6], we investigated the effect of potassium iodide salts on the photoluminescence and photovoltaic properties in the P3HT-PCBM organic solar cells. This article presents the results of studies of an effect of external magnetic field on the photoluminescence films of P3HT, doped with the KI salt. The results of the studies suggested a possible mechanism for the effect of the MF on the photoluminescence films of P3HT-KI.

2. Experiment

The P3HT polymer films based with the inorganic impurities additives of KI were prepared for study of spectral-luminescent properties of the polymer composites. The impurity concentration in the film of P3HT is in the range from 0.05 to 1 weight percent from mass of the polymer. Methods of preparing the polymer films P3HT–KI, is given in [6].

Registration the absorption spectra of the films was carried out using a spectrophotometer Cary 300. The excitation and luminescence spectra of the films were measured on a fluorescence spectrophotometer Cary Eclipce.

The kinetics of fluorescence of polymer films was measured using a pulsed spectrofluorometer with picosecond resolution and recording mode with time-correlated photon counting (Becker & Hickl). Fluorescence excitation of samples was carried out using a pulsed semiconductor laser with a wavelength of λgen = 488 nm and pulse width at half maximum of τ = 80 ps.

The effect of the external MF on the kinetics of a fast luminescence in the polymer films was measured by the pulsed spectrofluorometer Becker & Hickl. A neodymium magnet was used for the formation of the MF. Method of measuring of the MF effect on the luminescence kinetics of P3HT is described in detail in [6].

3. Results and discussion

Study of the spectral-luminescent properties of P3HT-KI films was carried out. The absorption spectrum shows a broad absorption band with the maximum on 520 nm and a shoulder at 600 nm (Figure 1, a). The adding impurities in the film leads to a slight change in shape of the absorption spectra on the short-wavelength part. The long-wavelength part of the spectrum is practically unchanged.

From the absorption spectra of the polymer films an energy value of bandgap (*Eg*) P3HT was determined by the addition of the inorganic impurities. To determine the width of the bandgap is plotted in the coordinates *eV* and $(A*h\nu)^2$. The bandgap is determined by the intersection of tangent to the curve of the absorption film and the ordinate axis corresponding to the energy of the absorbed light [14]. Results of determining the value of *Eg* are listed in Table 1. These results show that the inorganic impurity practically doesn't effect on value of the polymer bandgap. No change in the bandgap indicates a small effect KI impurity on structural and energy characteristics of P3HT film.

Influence of KI impurities on the photoluminescence spectra of the composite films is shown in Figure 1, b. The addition of KI results decreases an intensity of emission at 700 nm, compared with the intensity of emission at 650 nm. Low concentration adding of KI leads to an increase in the intensity of luminescence of the polymer (Figure 2). When the impurity concentration KI in the polymer was 0,5 %, the luminescence intensity is 3.5 times higher than from pure P3HT film. Further growth of KI concentration in the polymer films leads to a significant drop in light intensity [6].

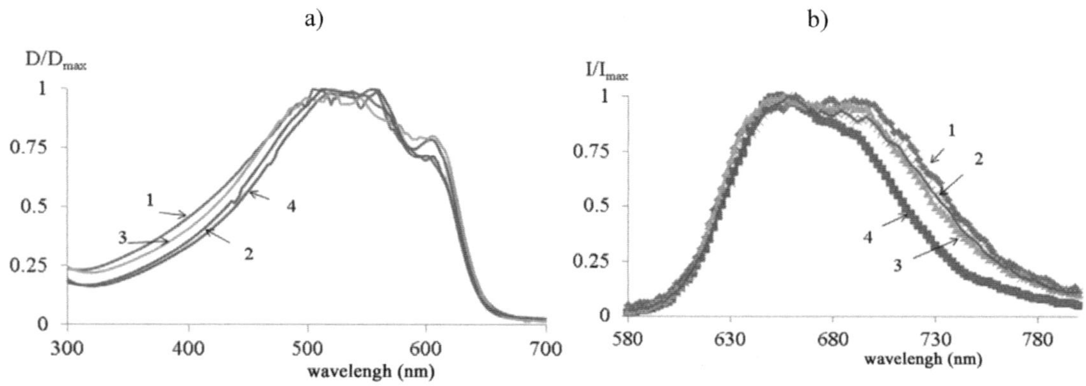

Figure 1. Relative absorption (a) and luminescence spectra (b) of P3HT-KI films. The concentration of KI impurity in the film: 1- 0; 2- 0,05 %; 3- 0,1 %; 4- 0,5 %.

Table 1. Influence of KI impurities on energy, fluorescent, electrical characteristics, and the magnetic effect in the composite films P3HT- KI

Samples	Value of bandgap (eV)	Lifetime of luminescence, τ, ps	Specific resistance ρ, Ω*cm	g_{max}(B) (%)
P3HT	1,875	460	$2{,}2*10^6$	+ 24 %
P3HT–KI 0,05 %	1,89	439	$4{,}5*10^8$	- 70 %
P3HT–KI 0,1 %	1,88	443	$4{,}5*10^8$	- 30 %
P3HT–KI 0,5 %	1,90	382	$1{,}1*10^9$	+ 30 %

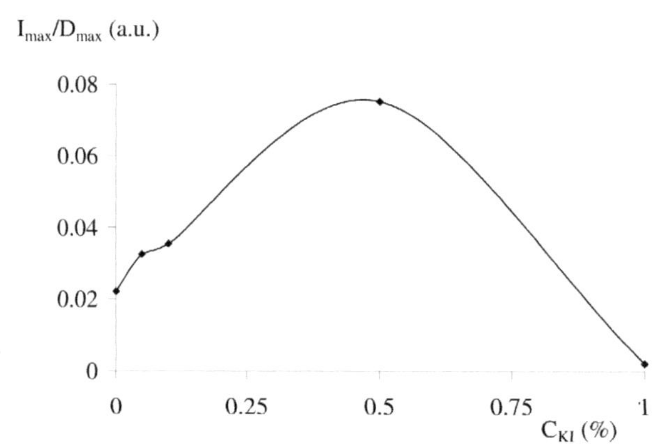

Figure 2. Dependence of the luminescence intensity of P3HT films from the KI concentration.

Kinetics of luminescence measured for P3HT-KI films in the nanosecond range. Signal registration is made at the maximum of the luminescence spectra (~ 650 nm). Lifetimes of luminescence (τ) was calculated from the linear portion of the logarithmic curve for various samples (Table 1). Maximum reduction in the lifetime observed for P3HT–KI film with the highest concentration of KI impurity. This behavior lifetime of luminescence indicates the dynamic nature of quenching of the excited singlet state polymer S_1. The most likely cause is the increase of the rate of intersystem crossing conversion from S_1 to T_1 state of the polymer [15].

Increasing of intensity luminescence at the low concentrations of KI and it reduction at the high additions of the impurity in the films shows a difficult influense of the heavy atom effect on the electron energy transformation in P3HT film.

Also measured influence of KI on an electrical resistance of P3HT films (Table 1). Measurement of the resistance on samples with ITO/P3HT/Al layers. Thickness of the polymer layer was constant. Control of the thickness implemented witch electron microscope. A contact area of the cells was 0,1 см^2. As it can be seen from the data shown in Table 1, KI addition leads to increase of electrical resistance of P3HT films. Qualitative assessment of the effect of the bandgap on a concentration of charge carriers was cited in formula [16]:

$$n_i^2 \sim \exp\left(-\varepsilon_g / kT\right) \tag{1}$$

Assessment of impact of the bandgap on the concentration of the charge carriers showed, that changing the band gap from 1.875 to 1,9 eV may lead to changing of concentration own of a free charge carriers in 2 times. While the resistance changes by 3 times of order of magnitude. Therefore, change of the resistance of films with KI addition is not connected only with change of the bandgap of the films.

More precisely, the mechanism of the effect of heavy atoms on photophysical processes in composite films P3HT- KI can be determined from measurements of influence of the external magnetic field on the kinetics of luminescence.

Measurements showed that the magnetic field affects on the intensity of the luminescence both pure P3HT film and P3HT-KI film. There is a time-dependent magnetic effect (Figure 3). As shown in [17], an external MF doesn't affect on the luminescence films of ITO/P3HT/Al. At the same time, in work [18] was observed effects on the magnetic photoluminescence of P3HT films with presence of oxygen. The appearance of the magnetic sensitivity explained by the appearance of oxygen-induced charge transfer complexes in the films of P3HT. The magnetic effect on P3HT photoluminescence observed in our work associated with formation of charge transfer complex in the presence of oxygen molecules. Type time dependence *g(B)* for P3HT film indicates on high rate of occurrence of the magneto-sensitive stages of transformation of electronic energy. Having high speed means that triplet state is not a major in formation of the magnetic effect in film of pure P3HT.

In the case of adding KI impurity occurs change of sign of magnetic effect and sharp increase of intensity of *g(B)* (Figure 3, table 1). With increasing concentration of KI decreases the value of *g(B)* and then sign of magnetic effect of the *g(B)* returns to plus.

Similar change of the sign of the magnetic depending on the current observed in [19]. In this work current density through polymer of film P3HT $i \approx 10^{-2}$ A*см^2 was critical. Above this current density the magnetic effect on the conductivity becomes positive and with lower value of i – negative. Change in magnetic effect explained by increasing of concentration of triplet excitons (N_T). The change occurred when the same concentration of triplet excitons and holes (N_p) in the polymer ($N_T \approx N_p$).

It should be noted that external magnetic field produces the opposite effect on the electrical conductivity and the photoluminescence in the polymer films. If there is a positive effect on the conductivity in the films, then there is a negative effect of the magnetic field on the luminescence and conversely.

These data in our work shows that addition of KI impurity does not lead to changes in the structure and energetics of polymer films of P3HT. Wherein increase in the KI concentration leads to an increase resistivity resistance of films. Also adding KI impurity leads to change of sign and value of magnetic effect *g(B)*.

As shown in [8, 9], the triplet state of the polymer can reduce the intensity of luminescence in the case of triplet-polaron dispersion in semiconducting polymers. So, it can be assumed that the observed magnetic effects on the luminescence in the P3HT-KI connected with triplet-polaron dispersion.

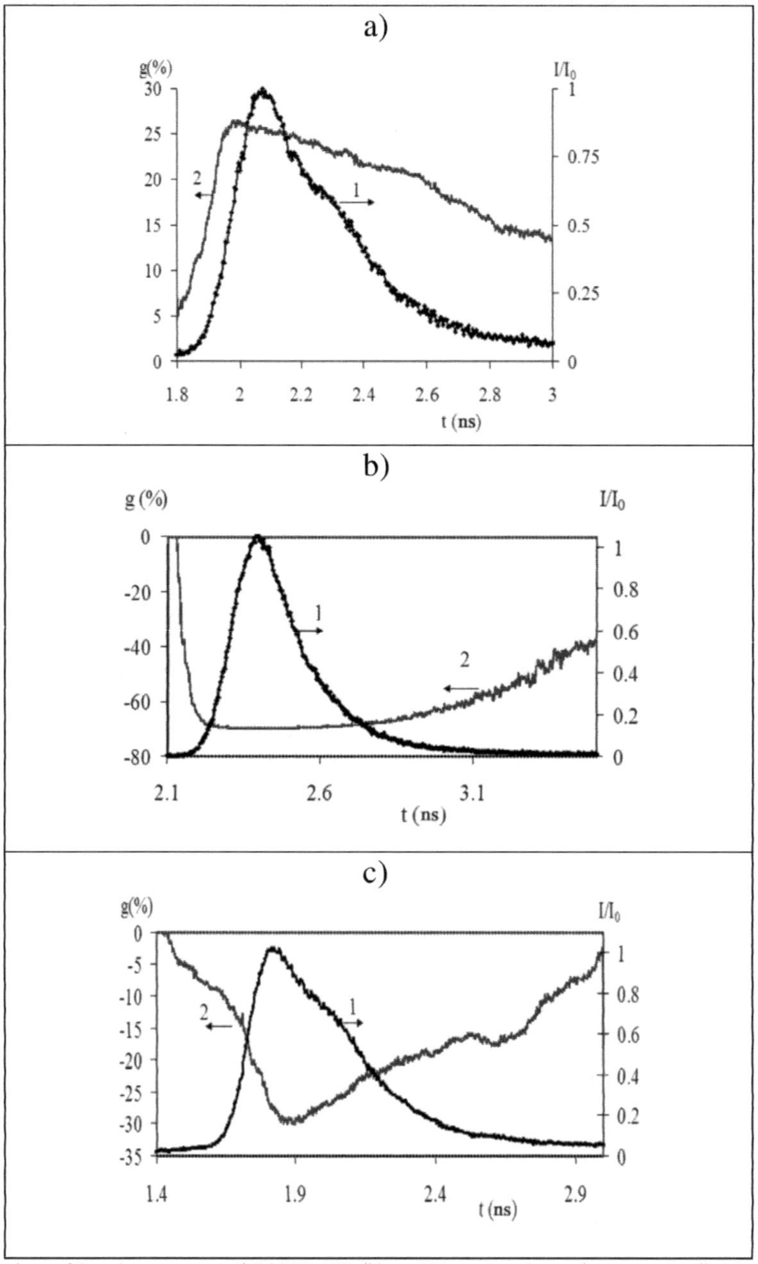

Figure 3. Kinetics of luminescence of P3HT–KI films (1) and value of magnetic field effect g(B) (2): a) P3HT; b) P3HT–KI 0,05 %; c) P3HT–KI 0,1 %

IOP Conf. Series: Materials Science and Engineering **168** (2017) 012060 doi:10.1088/1757-899X/168/1/012060

4. Conclusion

Research has shown different influense of KI impurities on the spectral-luminescent properties and the magnetic effect on the photoluminescence of P3HT semiconducting polymer. Adding impurity of KI almost does not affect on absorption spectra of P3HT films. Also there is little changes in the bandgap of the polymer. It points to the addition of KI impurity does not lead to changes in the structure and energy of P3HT. In contrast with absorption spectra, the luminescent properties of P3HT significantly changes with presence of KI. Addition of KI decreases the the lifetime of P3HT luminescence. The intensity of luminescence of the polymer depends on the impurity concentration. At a low concentration of KI occurs growth to the photoluminescence intensity and at the high concentrations (more than 1%) - decrease. Addition of KI in the film leads to increase of the electrical resistance in it. It is connected with decrease of the mobility of charge carriers in P3HT. Also character of magnetic dependence on the photoluminescence $g(B)$ depends on KI concentration. With first adding of KI (0,05 %) the magnetic effect of the changes sign from plus to minus. Value of $g(B)$ sharply increases to 70 %. Then with increasing of concertation of KI impurity a decrease of $g(B)$ value. Then the sign of the changes from minus to plus.

The observed changes in the magnetic effect on the photoluminescence P3HT-KI films are connected with increase of the concentration of the triplet states in the polymer. When observing the negative magnetic effect $g(B)$, the main process in the film magnetically P3HT-KI is the dispersion of the polarons on the triplet states T_1 of the polymer. In this way, KI impurity can be used to control electronic processes in semiconducting polymers.

References

[1] Deibel C, Dyakonov V 2010 *Rep. Prog. Phys.* **73** 096401

[2] Von Hauff E 2011 *Quantum Efficiency in Complex Systems, V.85 Chapter 7 – The Role of Molecular Structure and Conformation in Polymer Electronics Book Series: Semiconductors and Semimetals* (Amsterdam: Elsevier)

[3] O'Brien J A, Rallabandi S, Tripathy U, Paige M F, Steer R P 2009 *Chemical Physics Letters* **475(4-6)** 220

[4] Iwasaki Y, Osasa T, Asahi M, Matsumura M, Sakaguchi Y, Suzuki T 2006 *Physical review B* **74** 195209

[5] Berberan-Santos M N 2000 *Phys. Chem. Comm.* **5**(5)

[6] Ibrayev N Kh, Afanasyev D A, Zhapabaev K A 2016 *IOP Conf. Series: Materials Science and Engineering* **110** 012067

[7] Odzhaev V B, Popok V N, Azarko I I 2000 *The physics of conducting polymers* (Minsk: Belarusian State University) [in Belarus]

[8] Janssen P, Cox M, et. al. 2013 *Nature communications* **4** 2286

[9] Desai P, et al. 2007 *Phys. Rev. B* **75** 094423.

[10] Prigodin V N, Bergesona J D, Lincoln D M, Epstein A J 2006 *Synth. Met.* 156 757

[11] Bobbert P A, Nguyen T D, van Oost F W A, Koopmans B, Wohlgenannt M 2007 *Phys. Rev. Lett.* **99** 216801.

[12] Salikhov K M 2000 *10 Lectures on Spin Chemistry* (Kazan: UNIPRESS) [in Russian]

[13] Merrifield R E 1968 *J. Chem. Phys.* **48(9)** 4318

[14] Rasmussen S C, Ogawa K, Rothstein S D 2008 *Synthetic approaches to band gap control in conjugated polymeric materials. In: Nalwa H S (ed) Handbook of organic electronics and photonics* (Stevenson Ranch: American Scientific Publishers)

[15] Parker C 1968 *Photoluminescence of solutions* (Amsterdam: Elsevier).

[16] Shalimova K 1976 *Physics of semiconductors* (Moscow: Energy)

[17] Xu Zh, Hu B 2008 *Adv. Funct. Mater.* **18** 2611

[18] Sperlich A, Kraus H, Deibel C et. al. 2011 *J. Phys. Chem. B* **115** 13513

[19] Gu H, Kreouzis T, Gillin W P 2014 *Organic Electronics* **15(8)** 1711

Thermal effects of carbonated hydroxyapatite modified by glycine and albumin

S A Gerk[1], **O A Golovanova**[1] **and M V Kuimova**[2]

[1]F.M. Dostoevsky State University, Omsk, Russia
[2]National Research Tomsk Polytechnic University, Tomsk, Russia

E-mail: gerksa_11@mail.ru

Abstract. In this work calcium phosphate powders were obtained by precipitation method from simulated solutions of synovial fluid containing glycine and albumin. X-ray diffraction and IR spectroscopy determined that all samples are single-phase and are presented by carbonate containing hydroxyapatite (CHA). The thermograms of solid phases of CHA were obtained and analyzed; five stages of transformation in the temperature range of 25-1000°C were marked. It is shown that in this temperature range dehydration, decarboxylation and thermal degradation of amino acid and protein connected to the surface of solid phase occur. The tendency of temperature lowering of the decomposition of powders synthesized from a medium containing organic substances was determined. Results demonstrate a direct dependence between the concentration of the amino acid in a model solution and its content in the solid phase.

1. Introduction

Among a wide range of modern biomaterials used for the restoration of bone defects, composites, based on carbonated hydroxyapatite (CHA) and organic components such as collagen and non-collagenous proteins, polysaccharides, polymers, etc., are in wide demand [1-5]. Crystal-chemical similarity of CHA with inorganic basis of bone mineral provides a high biocompatibility of such materials used in implantation; namely, it induces biochemical reactions similar to those of bone remodeling [6, 7]. The use of organic compounds in the quality of dopants approximates the conditions of the synthesis of materials to bone mineralization *in vivo*, taking place on a collagen matrix and with biological fluids [1-3, 8]. Therefore, it is possible to obtain CHA similar in crystallinity and morphological characteristics to the native bone that also improves the osteointegration of the implant with the surrounding living tissues. Despite a large number of works on the preparation of biocomposites [1-8], the influence of organic materials on the structure and properties of CHA, the mineral component of bone tissue, are not adequately covered. Thus, the role of amino acids of the polypeptide chain of collagen, composing the crystalline segments -Gly-X-Y-, (Gly – glycine, proline, etc.) and amorphous regions (lysine, histidine and others) are not fully described [9, 10]. The impact of organic components of biofluids (albumins, globulins, lipids, carbohydrates, etc.) on the characteristics of the solid phase is not fully considered [11]. Therefore, the present work is devoted to the preparation of CHA powders from model solutions of human synovial fluid containing glycine and albumin, study of their composition and thermal effects.

2. Material and methods

CHA was produced by "wet synthesis" from the model solution approximated by electrolyte composition to the synovial fluid of an average healthy person at values pH = 7.40±0.05 and ionic strength - 0.174±0.001 [8]. This CHA was used as a control sample. Two series of model experiments were carried out. In the first series of syntheses, in the initial solution containing anions (HPO_4^{2-}, HCO_3^-, Cl^-, SO_4^{2-}) glycine was added (CH_2NH_2COOH, α- amino acetic acid of CP brand, the range of concentrations from 0.04 to 0.24 mol/l), the second – albumin (bovine serum albumin (BSA) of A brand; the range of concentration from 5 to 10 g/l). When selecting the concentrations of glycine and albumin, we were guided by their content in the collagen of the bone tissue and biological fluids (synovia, blood plasma), respectively [12, 13]. The solution containing the anions (HPO_4^{2-}, HCO_3^-, Cl^-, SO_4^{2-}) and the organic substance was spilled to the solution with calcium and magnesium cations at a rate of 5 ml/min. Further, the acidity of the model system was adjusted by adding a 10% solution of HCl until the value of pH = 7.40±0.05 was reached. Crystallization of the solid phase was carried out at room temperature (20-22 °C) for 7 days. At the end of this period, the precipitate was separated from the solution by filtration, washed three times with water, dried at 80 °C to completely remove the chemically not of the connected water, until a constant weight was achieved.

A study of the phase composition was conducted using the X-ray powder diffractometer D8 Advance (Bruker) under the following conditions: Cu-kα radiation (wavelength 0.15406 nm) scanning step – 0.05°, time of accumulation of a signal – 4 seconds/point, div. slit=0.5, voltage and filament current 40 kV and 40 mA, respectively; scan region 2θ – 5-80°. The transcripts of the diffractograms were obtained from the databases of powder diffraction ICDD PDF-2 in the program EVA (Bruker). IR spectra of the precipitates were recorded on a spectrophotometer "FSM-2202". The scanning range was 400-5000 cm^{-1}, the resolution was 8 cm^{-1}. The samples were prepared by compression into tablets with KBr. The thermal analysis was carried out on a STA-449 C synchronous thermal analyzer "NETZSCH". The samples were calcined in platinum crucibles in the atmosphere of the air from 25 to 1000 °C at the rate of 10 °C/min. The mass of the sample varied depending on the values of the recorded signals of the mass loss effect and the thermal effect. The mass was 10-20 mg. The quantitative data of mass loss during annealing were obtained using software Proteus 7.10. By obtained thermogravimetric (TG) and differential thermal (DTG) curves, the mass losses of the components of the samples were determined by heating. A differential gravimetric curve (DTA) was used for the characteristics of the observed thermal effects (changes in energy, °C/min). The specific surface areas of the samples were studied by the method of single-point nitrogen adsorption at 77.4 K on the adsorption apparatus "Sorbtometr", produced by Katakon LLC, Russia. The calculation of the obtained values of S_{BET} (m^2/g) was done using the BET method.

3. Results and Discussion

Using XRD and IR-spectroscopy, it was found that all precipitates are single-phase and are presented by CHA (figure 1a, b), regardless of the presence of organic substance in the initial model solution. On IR spectrums there are absorption of the bonds of inorganic groups and molecules: PO_4^{3-}, CO_3^{2-}, -OH и H_2O. The presence of modes of vibrations of the organic groups $-NH_2$, $-CH_n$, C=O, -N-H indicates that the glycine and albumin are in the composition of the precipitates.

However, according to data of the thermal analysis, the presence of organic substance in the initial medium reflects on the form of the curves TG, DTG and DTA (figure 2), which indicates different mass losses and thermal effects of these powders. Comparing the obtained thermal curves of the control sample (figure 1a) with the published data [14-20], we can distinguish five stages of thermo-transformation in the temperature range 25-1000 °C.

In the first stage (*I, 25-280 °C*, endothermic effect there is the removal of chemically free water and decomposition of volatile impurities (such as adsorbed CO_2).

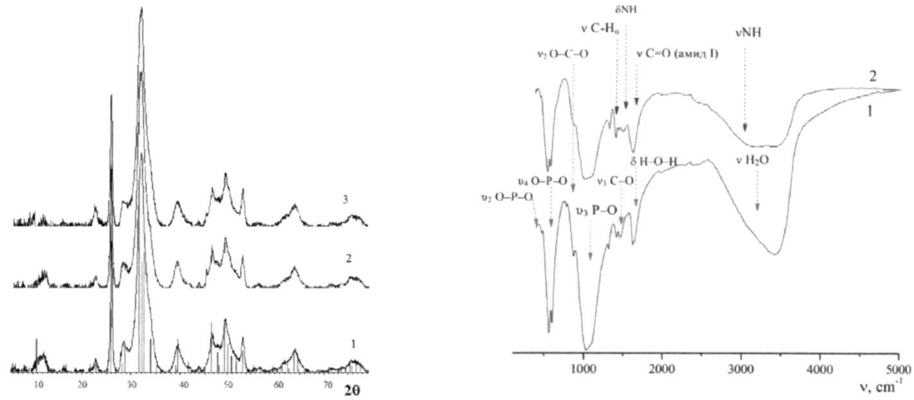

a *b*

Figure 1. Diffractogram (a) and IR spectra (b) CHA: 1) pure; 2) 0.16 mol/l glycine; 3) 5 g/l albumin.

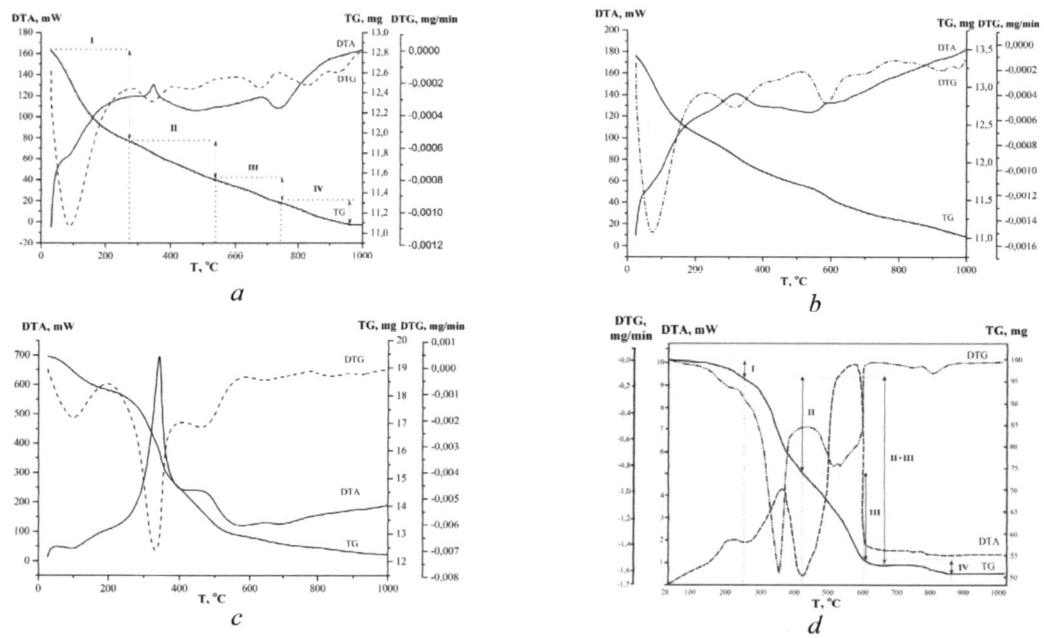

Figure 2. Thermograms CHA: a) pure; b) 0.04 mol/l glycine; c) 5 g/l albumin; d) human bone tissue «in the norm» [20].

The second and third stages (exothermic effects) are followed by the loss of water associated in various ways with the solid phase: physically adsorbed (*II, 280-470 °C*), chemisorbed and crystallization (*III, 470-750 °C*). Schematically the processes of water removal on the data can be represented as:

$$[\text{КГА} \cdot nH_2O] \cdot mH_2O_{(sol.)} \rightarrow \text{КГА} \cdot nH_2O_{(sol.)} + mH_2O_{(g.)} + Q; \ \text{КГА} \cdot nH_2O_{(sol.)} \rightarrow \text{КГА}_{(sol.)} + H_2O_{(g.)} + Q.$$

At higher temperatures, at the fourth and fifth stages (*IV and V, 750-900 °C and 900-1000 °C*) there are two endothermic peaks at 830 and 940 °C on DTA curves. In these temperature ranges, the transformation of CHA occurs, namely, the removal of carbonate ions from its structure in the form of CO_2 and transition of non-stoichiometric carbonated hydroxyapatite to the stoichiometric phase or β-$Ca_3(PO_4)_2$:

$$Ca_{10}(PO_4,CO_3)_6(OH)_{2(sol.)} \rightarrow Ca_{10}(PO_4)_6(OH)_{2(sol.)} + 6\ CO_{2(g.)} - Q;$$

$$Ca_9(PO_4)_{6-x-y}(HPO_4)_y(CO_3)_x(OH)_{2-y(sol.)} \xrightarrow{800\ °C} 3\ \beta\text{-}Ca_3(PO_4)_{2\ (sol.)} + x\ CO_{2(g.)} + 2\ H_2O_{(g.)} - Q\ [22].$$

In works [1, 23] it is marked that at a temperature of 850 °C, the transformation of GA hexagonal syngony into monoclinic form (hexagonal P63/m → monoclinic P21/b) is possible due to the reorientation of the dipolar groups OH^- in the triangular channels of Ca^{2+} ions of its crystal lattice. At 900 ^{0}C the formation of oxyhydroxyapatite from stoichiometric HA is possible according to the reaction [24]:

$$Ca_{10}(PO_4)_6(OH)_{2(sol.)} \rightarrow Ca_{10}(PO_4)_6(OH)_{2-2x}O_xn_{x(sol.)} + xH_2O_{(g.)},$$ где n_x – vacancy, $x<1$.

It was found that the powders, synthesized in the presence of an organic component, are characterized by the highest total losses of mass, and their transformation takes place at lower temperatures (I-IV, Table 1, figure 2 and 3a). Perhaps this is due to the presence of organic substance in the solid phase, which is consistent with the data of IR-spectroscopy.

Table 1. Temperatures of transformation and mass losses (Δm) of the components of powders.

Stages	T beginning of transformation (°C)			Δm (mg)							
					In the presence of glycine (mol/l)				In the presence of albumin (g/l)		
	Control sample	In the presence of glycine	In the presence of albumin	Control sample	0.04	0.08	0.16	0.24	5	7	10
I	25	25	25	7.098	7.367	7.300	8.352	10.647	6.282	5.381	5.430
II	280	240	200	2.340	3.302	3.408	3.728	4.095	18.280	18.130	20.440
III	470	450	400	2.496	2.964	3.015	3.057	3.744	9.166	9.195	12.449
IV	750	670	600	1.248	1.524	1.700	1.417	1.872	1.905	2.090	1.741
V	900	850	800	0.390	1.354	1.268	1.268	1.395	1.287	1.097	1.076
I+V		25-1000		13,572	16,511	16.733	17.822	21.753	35.920	35.893	41.136

It is known that thermal decomposition of aliphatic amino acids, to which belongs glycine, as well as peptides occurs at temperatures not exceeding 400 °C (T_m glycine = 232 - 262 °C; T_m albumin <90 °C.) [15, 16, 25]. Thus, the thermal degradation of glycine occurs in two steps. At the initial stage (fast stage, ΔH > 0) at 200-270 °C, oligopeptides and diketopiperazine (cyclic dipeptide) are formed. Methylamine, formamide, acetamide, propionamide, N-atsetilamid and acetic acid, water, carbon dioxide are by-products. Further, at 270-370 °C slow stage (ΔH < 0) of complete oxidation of degradation products proceeds. From data of Table 2 it follows that the thermal transformation of glycine, which is present in the obtained powders, occurs in stage III and corresponds at 2000 seconds to the exothermic peak on the DTA curve (321 °C). Exactly on the third stage with the increasing content of amino acids in the initial medium, the mass losses of glycine increase linearly. However, less energy is consumed for the destruction of the samples containing larger amounts of amino acids (figure 3*b*). Possibly at a lower concentration of glycine in an aqueous solution, components of zwitterions (NH_3^+- $C^{\alpha}H_2$ - COO^-) bond with the ions of calcium and PO_4^{3-} CHA more tightly and have a high adsorption activity in relation to their surface. This correlates well with the data of the specific surface area (Table 2). It is seen that the samples obtained from the model medium with low content of amino acid have the lowest specific surface.

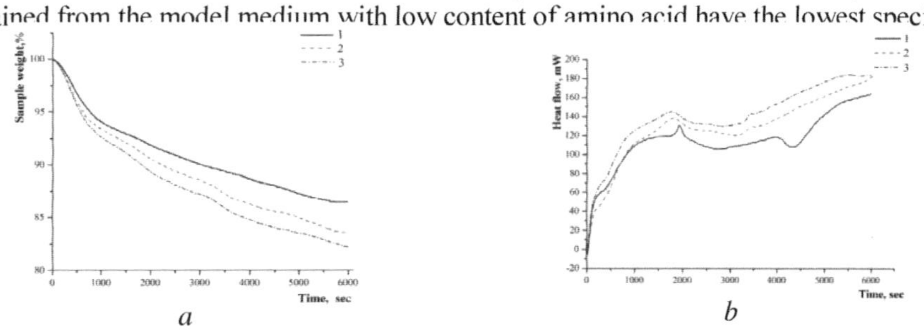

Figure 3. TG curves (a) and DTA (b) CHA: 1) pure; 2) glycine 0.04 mol/l; 3) 0.12 mol/l.

IOP Conf. Series: Materials Science and Engineering **168** (2017) 012061 doi:10.1088/1757-899X/168/1/012061

Table 2. The specific surface area of the powders.

CHA	Control sample	In the presence of glycine (mol/l)			In the presence of albumin (g/l)		
		0.04	0.08	0.016	5	7	10
S_{BET}, m^2/g	130±7	70±1	52±2	75±3	44±2	48±2	25±2

Interesting results were obtained in the analysis of the thermograms of the samples synthesized in the presence of albumin (figure 4, Table 1). It was shown that the presence of protein leads to higher mass loss at temperatures above 200 – 240 °C and does not depend on its mass in the model solution.

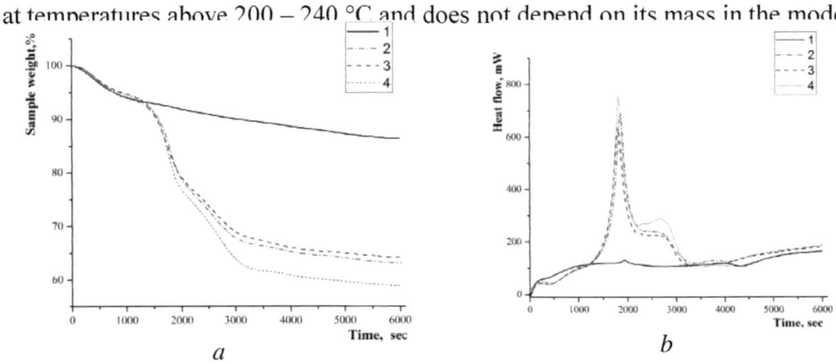

Figure 4. TG curves (a) and DTA (b) CHA: 1) pure; albumin – 2) 5 г/л; 3) 7 г/л; 4) 10 г/л.

We believe that, despite the low temperature of denaturation of pure albumin, the destruction of the polypeptide chains bounded with the surface of CHA occurs in the temperature interval 200-400 °C ... 450 (stage II, DTA maximum at 340 ... 250 °C and 2000 seconds figure 3a). We obtained similar results previously while studying the thermal curves of bone tissue (step II, figure 2d). In [21] it is shown that at this stage (DTA maximum/peak at 360 °C) there is a thermal oxidation of organic substances with less molecular mass than a collagen; for example, non-collagen proteins (albumin). In addition, high temperatures of thermal deterioration of albumin in the solid phase can be explained by a complex set of amino acids. Human and serum albumin are globular proteins, whose primary structure includes 585 and 582 residues respectively. These residues can be different amino acids: neutral, basic and acidic; their side radicals can be various functional groups (e.g., tryptophan, phenylalanine, tyrosine, etc.) [16]. It is known that the destruction of such amino acids may consist of four or five stages, and proceed at temperatures of 300 ... 400 °C and 600 ... 700 °C [26]. The presence of charged functional groups and zwitterions of amino acids provides a stronger interaction of albumin with the surface of CHA, which is in agreement with data from works about the chemisorption mechanism of their interaction [27]. The powders synthesized from the medium containing the protein, unlike the samples prepared in the presence of glycine, are characterized by the lowest specific surface area of powders. At the maximum content of albumin (10 g/l), the sample is formed, for which S_{micro} value is 5 times lower than the specific surface area of the control sample (Table 2).

In all samples, for which the precipitate was held in the presence of organic substance, a significant loss of mass at temperatures of more than 600 ... 670 °C (stage IV and V, Table 2) was recorded. At these temperatures, the removal of carbonate ions from CHA structure occurs. This fact indirectly indicates that in the crystalline grid of CHA of these precipitates, there are numerous carbonate ions; therefore, the samples with glycine characterized by low crystallinity and regularity of structure. At IR spectra of these powders, the reduction of intensity of mode υ_3(P–O) и υ_4(O–P–O) of phosphate groups (figure 1b) is indicated.

4. Conclusion

Thus, CHA powders were synthesized in the presence of albumin and glycine from prototypes of synovial fluid. It was shown that organic substance is present in the solid phase of the samples.

Investigated thermal effects and mass loss of the samples CHA obtained in the presence of glycine and albumin. It was found that the thermal decomposition of organic substance associated with surface area of CHA mainly takes place at temperatures of 200-400 ... 450°C. It was shown that with increasing content of amino acid in the initial medium, mass losses of glycine increased linearly. The mass loss for the powders obtained in albumin environment does not depend on its concentration in a model solution. For all samples obtained in the medium containing organic substance, unlike the control sample, the smallest specific surface area and crystallinity are characteristic.

Acknowledgments
The research was partially supported by the Russian Foundation for Basic Research (grant mol_a No. 16-33-00535 and No. SP-933.2015.4).

6. Reference

[1] Starikova V V and Rudchenko S O 2010 *Vest. Kharkov Un-ty* **14** 35
[2] Assumption W A, Zakharov V N and Villager M A 2014 *Patent.* 22534789
[3] Moysenovich M M, Arhipova A Yu, Orlova A A, Drutskaya M S, Volkova S V, Zaharov S E, Agapov I I and Kirpichnikov M P 2014 *Acta Naturae* **1(20)** 103
[4] Yanovska A A, Kuznetsov V N, Stanislaus A S, Gander E V, Pogorelov M V and Danilchenko S N 2015 *Chem., phys. and tech. of surface* **4** 535
[5] Samchenko Y M, Boldeskul I E, Sukhodub L B, Danilchenko S N, Birch L I and Ullberg Z R 2009 *Nanostruct. Materials* **1** 81
[6] Lukin Yu C 2010 *Dis. ... Cand. Techn. Sciences* (Moscow: Russian University of chemical technology) p 161
[7] Dorozhkin S V 2009 *Mater.* **2** 399
[8] Golovanova O A, Lemesheva (Gerk) S A and Ismailov R R 2013 *Patent.* 2496150(379).
[9] Green D W, Goto T K, Kim K S and Jung H S 2015 *J. Roy. Soc.* **3** 1
[10] Torbenko V P and Karsavina B S 1977 *Functional biochemistry of bone* (Moscow: Medicine) p 272
[11] Martin R B 1999 *Mater. Sci. Forum.* **1** 5
[12] Matveeva E L 2007 *Abstract. Dis. Doctor. Bio. Sciences* (Tyumen: Tyumen State University) 24
[13] Krylova N N 1954 *Meat Biochemistry* (Moscow: Pieprasit) p 320
[14] Goloshchapov Y N, Kashkarov V M, Rumyantsev N A, Seredin P B, Lenshin A C, Agapov B L and Domashevskaya E P 2011 *Condens. matter and interphase boundaries* **4** 427
[15] Smeltsova I L 2013 *Dis. ...Can. Chem. Sciences* (Nizhniy-Novgorod: Nizhny Novgorod State University of architecture and construction) 104
[16] Badelin V G, Tyunina E Y and Mezhevoy I N 2014 *Liq. Cryst. and their Appl.* **3** 43
[17] Nedoseko V B, Gorbunova I L and Drozdov V A 2004 *Dent.* **4** 13
[18] Sudakova T V, Litvinov S D, Baev E E and Gusev V V 2009 *Act. Prob. of new science* **10** 7
[19] Safronova T V, Putlyaev V I, Kurbatov S A, Shatalov T B, Larionov D S, Kozlov D A and Evdokimov P V 2015 *J. Inorg. Materials* **1** 1269
[20] Guzeeva T I, Guzeev V V, Leonova L A, Lelyuk O A, Krikunenko A S and Shatokhina Y M 2009 *Vest. Tomsk Polytech. Un-ty* **3** 47
[21] Lemesheva S A and Golovanova O A 2009 *Chem. for sustainable development* **3** 327
[22] Bogdanova E A, Sabirzyanov N A 2014 *Mater.* **10**
[23] Zakharov N A, Klyuev V A, Sentsov M Y and Toporov Y P 2012 *J. Techn. Physics.* **4** 153
[24] Barinov S M 2005 *Bioceramics of calcium phosphate* (Moscow: Science) p 204
[25] Sarnatskaya V V, Yushko LA, Korneeva L N, Sahno LA, Maslenny V N, Snezhkova E A, Mihalovskiy S V and Nikolaev V G 2005 *Effer. therapy* **11** 10
[26] Pshonkina N N 2011 *Fundament. Research* **12** 1067
[27] Morgues P A, Serro A P and Saramago B J 2003 *Biomaterials* **24** 451

Patterns of calcium oxalate monohydrate crystallization in complex biological systems

O A Golovanova[1], V V Korol'kov[1] and M V Kuimova[2]
[1]Department of Inorganic Chemistry Omsk F. M. Dostoevsky State University, Russia
[2]National Research Tomsk Polytechnic University, Tomsk, Russia

E-mail: Golovanoa2000@mail.ru

Abstract. The paper presents the features of calcium oxalate crystallization in the presence of additives revealed through experimental modeling. The patterns of phase formation are shown for the $Ca^{2+} – C_2O_4^{2-} – H_2O$ and $Ca^{2+} – C_2O_4^{2-} – PO_4^{3-} – H_2O$ systems with the components and pH of the saline varying over a wide concentrations range. The effect of additives on crystallization of calcium oxalate monohydrate was investigated. It was found that the ionic strength and magnesium ions are inhibitors, and calcium oxalate and hydroxyapatite crystals are catalysts of calcium oxalate monohydrate crystallization. The basic calcium phosphate (apatite) was found to be most thermodynamically stable, which indicates its special role in kidney stone formation since it is found in virtually all stones.

1. Introduction

The data gathered through the analysis of human body fluids is the basis for creation and construction of physical-chemical models to reveal the regularities of physiological processes. The experimental modeling techniques are considered to be promising to predict the behavior of biological systems upon changing its parameters, which is especially important when studying the processes caused by functional impairment, such as pathogenic biomineral formation in human tissues and organs. These pathogenic formations are kidney stones (uroliths). Urinary stone disease (USD) is defined as a disease caused by endogenous (including hereditary) and/or exogenous factors due to formation of stones in the urinary tract. USD is one of the most common diseases that are prone to relapse, and in many cases it is a severe acute disease [1–4]. The reported prevalence and incidence of USD in adults in the world is 1–5% [5–7]. The annual incidence rate of urinary stone formation is 1200–1400 per 100 000 population [8], and the average risk for stone formation throughout one's life ranges from 5 to 10% [9]. USD mostly affects the working-age population [3, 10]. According to many researchers, an increased incidence of USD in all population groups is currently observed in most of the industrialized countries [11–13].

The most common mineral phases [14–16] of human kidney stones are spherulite aggregates of calcium oxalate monohydrate ($CaC_2O_4·H_2O$). These are regular products of crystallization caused by high urine supersaturation. The $CaC_2O_4·H_2O$ crystals nucleate mainly heterogeneously and their growth occurs in a kinetic mode. Oxalate is crystallized from aqueous saline in the form of various hydrates, thermodynamically stable monohydrate ($CaC_2O_4·H_2O$, whewellite; COM) and metastable tetragonal hydrate ($CaC_2O_4·(2+x)H_2O \leq 0.5$, weddellite), occurring in the tissues of urinary stones and deposits of biological origin. Triclinic hydrate ($CaC_2O_4·xH_2O$, $3 \geq x > 2.5$;

COT) was prepared as well, and its crystal structure was studied. It is known that the formation of poorly soluble compounds is affected by the ions present in the saline [17] including analogous ions with deposits. Therefore, an increased concentration of oxalate ions in urine promotes the formation of calcium oxalate. When the concentration of oxalate > 1 mmol/l, calcium oxalate monohydrate is formed in urine.

The aim of this paper is to study the crystallization of calcium oxalate and to establish the effect of a series of additives on the size, rate and the number of the crystals formed.

2. Materials and methods

The studied samples were prepared by the technique developed in [16] in accordance with the chemical equation $Ca^{2+}+C_2O_4^{2-} = CaC_2O_4\downarrow$. The value of the final concentration of calcium and oxalate ions was 20 times above the concentration of the saturated saline. All the experimental series were carried out under continuous stirring, and the samples were taken and placed on a preparation glass at appropriate intervals. The crystallization time in each series varied from 2 to 580 hours. The concentration of additives was prepared with respect to the biological fluid composition [17].

Calcium oxalate crystallization was studied by practical metallography techniques using a Neophot-2 optical microscope. We used a linear method to measure the size of the microparticle cross sections (to measure the particle diameter) and an intercept method to estimate the number and sizes of the microparticle cross sections [18]. The length of the chords was determined with allowance for a linear magnification of the image of the microparticle structure.

The phase composition of the deposits was investigated by XRD (DRON-3) and optical microscopy. To perform a qualitative analysis of the sample phase composition, the experimental values of the interplanar spaces and relative intensities of the diffraction peaks were compared to a set of corresponding table values for each of the assumed phases. The sensitivity of the XRD method to carry out the measurements was 3%.

3. Results and Discussion

In the first stage, the study was performed under conditions close to the biological medium, namely, the samples were prepared at supersaturation γ=7, and the ionic strength was varied from 0.003 to 0.4 (corresponds to a biological fluid (human urine)). It was found that under these conditions, only calcium oxalate monohydrate is crystallized (figure 1 and 2); at ionic strength greater than 0.15, calcium oxalate crystallization does not occur. At ionic strength equal to 0.003–0.1, the microparticle size varies slightly, the average size being 10.83 μm. However, at ionic strength I=0.15, the microparticles grow up to 13.5 μm. It is known that the greater the rate of nucleation and growth, the faster the crystallization, and this depends on the degree of the saline supersaturation. Therefore, further investigation of the calcium oxalate crystallization was conducted for three series. For the first series, supersaturation γ=15, for the second series γ=20 and for the third series γ= 25. The crystal size growth was not found within one series irrespective of the crystallization time, however, during transition from the second to the third series, the crystals were observed to increase in size (figure. 3).

The obtained data was used to evaluate the average rate of the $CaC_2O_4 \cdot H_2O$ particle growth. It was found that when γ=25, the predominant process was the growth (ϖ=0.46 μm/h), but not the nucleation. The initial supersaturation γ=20 was chosen to reveal its effect on the additive growth rate.

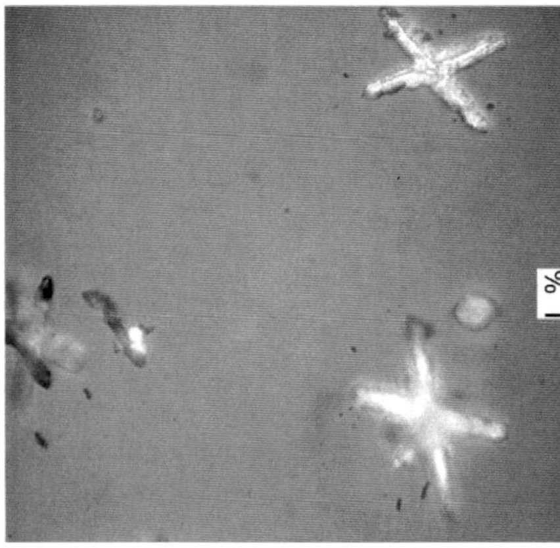

Figure 1. Optical image of calcium oxalate monohydrate crystals at γ=20 (x 450).

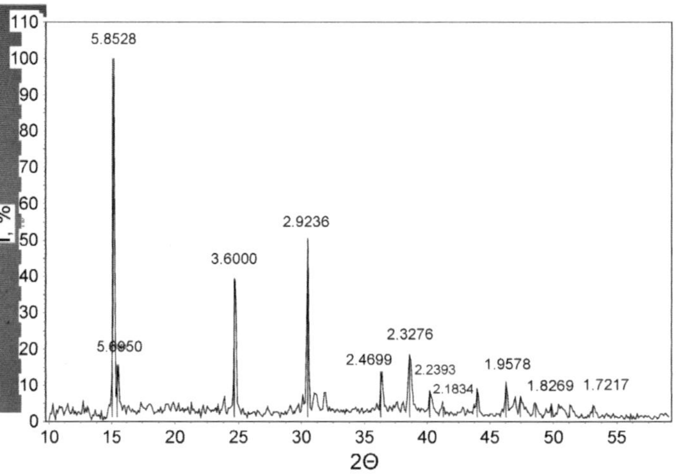

Figure 2. Diffraction pattern for $CaC_2O_4\ H_2O$.

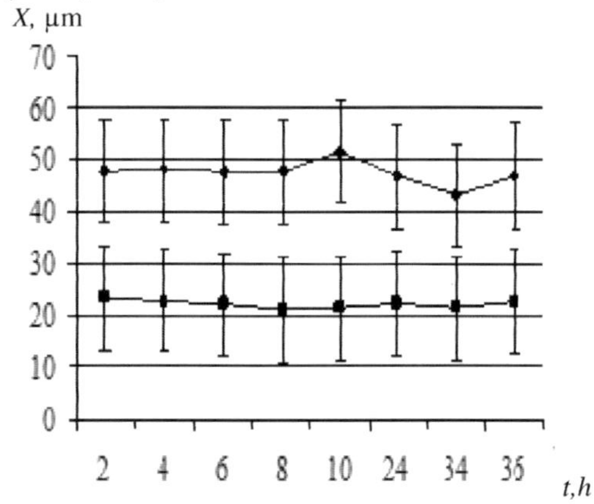

Figure 3. Dependence of the calcium oxalate monohydrate particle size x on t: γ=20 (1), γ=25 (2).

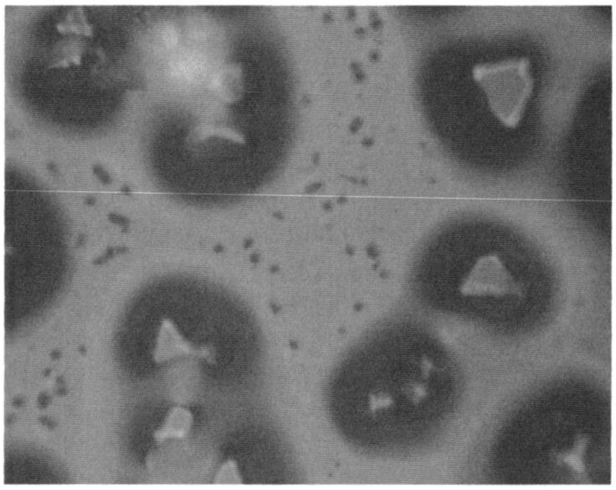

Figure 4. Microstructure of the particles (x450) crystallized from the saline solution after addition of magnesium chloride (=8.15 mmol/l, the crystallization time equal to 250 hours).

Biological fluids, in which kidney stones are formed, is a complex multi-component system [17] and a high ionic strength saline. To evaluate the effect of the ionic strength on the crystallization of calcium oxalate monohydrate, a series of experiments were conducted to study the patterns of the compound formation at constant ionic strength characteristic of the natural crystal-forming medium (urine) I=0.3. The formation of calcium oxalate monohydrate crystals at this ionic strength value was not observed. The obtained results can be attributed to a strong effect of repulsive electrostatic forces between similar ions in a high ionic strength saline, which inhibits chemical interaction between calcium and oxalate ions and hence reduces the nucleation probability [19].

Natural crystal-forming medium contains organic and inorganic compounds such as magnesium ions. Magnesium is one of the most important activators of many enzymatic processes. Although the substances that regulate the magnesium absorption are similar to those regulating the calcium absorption, in particular, proteins and vitamin D, there is definite antagonism between calcium and magnesium ions in the human body [17]. Addition of magnesium chloride to the feed saline (figure 4) inhibits the formation of calcium oxalate monohydrate crystals irrespective of the concentration (4.07 mmol/l ÷ 0.041 mmol/l) of the additive and crystallization time. The magnesium ions were found to inhibit crystallization, and as a result, calcium oxalate monohydrate crystals are not formed.

It is known that during kidney stone formation calcium oxalate monohydrate microcrystallites (crystalluria products) and phosphate micro-individuals are found in the physiological solution that can affect pathogenic mineral formation. The crystallization of calcium oxalate in the saline with the addition of calcium oxalate and hydroxyapatite seeds was studied. The crystallization time was varied from 10 minutes to 580 hours. It was found that seeding causes a slight growth of the formed calcium oxalate monohydrate crystals at the initial time moment. When the seed was calcium oxalate monohydrate, the crystals grew by 1.5 times within 50 hours, and in case of hydroxyapatite, the crystals grew by 2.5 times within 150 hours as compared to the initially formed crystals (figure 5). Thus, it was found that the presence of calcium oxalate hydroxyapatite crystals in the saline catalyzes crystallization of calcium oxalate monohydrate. The experimental data were used to estimate the average rate (Table. 1) of calcium oxalate monohydrate crystal formation, which was 4 times higher in the presence of hydroxyapatite crystallites than that obtained through the addition of $CaC_2O_4 \cdot H_2O$. Thus, the presence of hydroxyapatite crystals in the urine initiates the crystallization of calcium oxalate monohydrate at a higher speed.

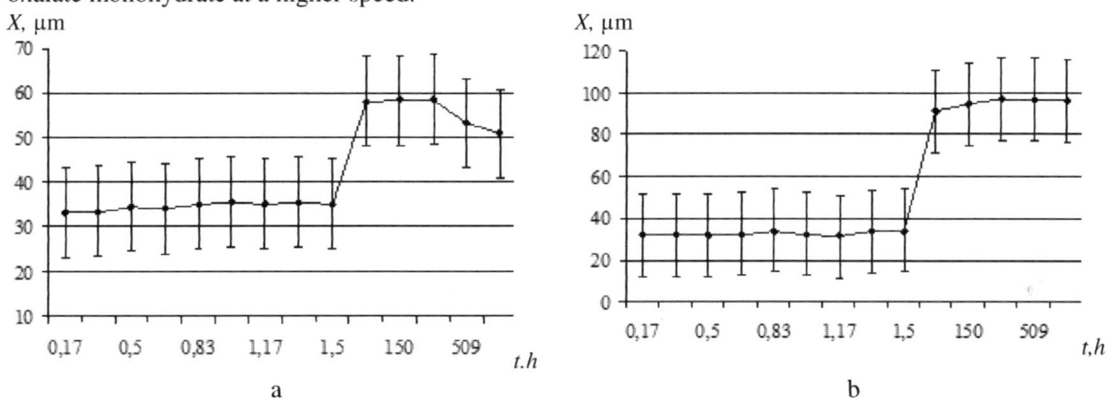

Figure 5. Dependence of calcium oxalate monohydrate particle sizes on the crystallization time in the presence of seeds: a) calcium oxalate; b) hydroxyapatite.

Table 1. The average rate of calcium oxalate monohydrate crystallization in the presence of seeds.

Average rate ϖ, μm/h	
Calcium oxalate monohydrate	Hydroxyapatite
0.03	0.12

The stability fields constructed for CaC_2O_4 calcium oxalate using the thermodynamic data from [20] showed that the surface of the thermodynamic equilibrium in the concentration range characteristic of the physiological solution is substantially invariant with respect to pH variations (figure 6a). The range of the saline concentrations and pH values lies in the range of a stable compound. The stability range of the basic calcium phosphate $Ca_5(PO_4)_3OH$ (Figure 6b) is significantly extended if the pH increases up to 12. The concentration range characteristic of saline also corresponds to that of the compound stability. During oxalate phase formation, phosphates can be formed even if the concentrations of oxalate ion are significantly high. This is mainly hydroxyapatite, especially if the pH of the saline s shifts to the alkaline range. The concentration range of oxalate and phosphate of the physiological saline lies almost entirely in the range of the $Ca_5(PO_4)_3OH$ stability.

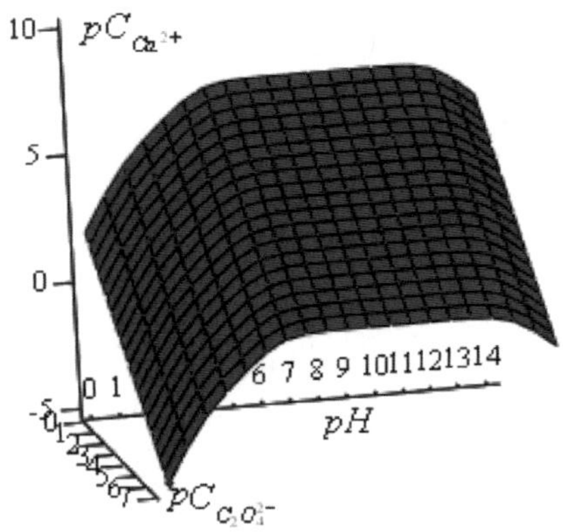

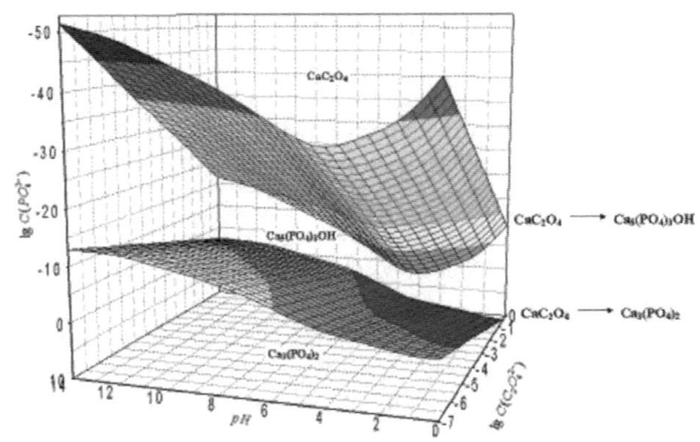

a b

Figure 6. Stability fields observed during formation of poorly soluble compounds from the saline solution at 37°C and ionic strength of 0.3 M for the systems: a) Ca^{2+}–$C_2O_4^{2-}$–H_2O; b) Ca^{2+}–$C_2O_4^{2-}$–PO_4^{3-}–H_2O.

The electron microscopy and microanalysis revealed separate apatite inclusions (figure 7), grains of hydroxyapatite, in oxalate type kidney stones (monomineral, according to XRD). The data yielded by the study showed that the Ca/P ratio in urolith apatites was less than that of stoichiometric apatite (1.67). Thus, the basic calcium phosphate apatite is the most stable compound, and it can be found in almost all kidney stones that indicates its special role in their formation. In real kidney stone samples, $CaC_2O_4{\cdot}H_2O$ plates form druse depositions on the kidney stone surfaces that traumatize the surrounding tissue upon the stone motion (figure 8).

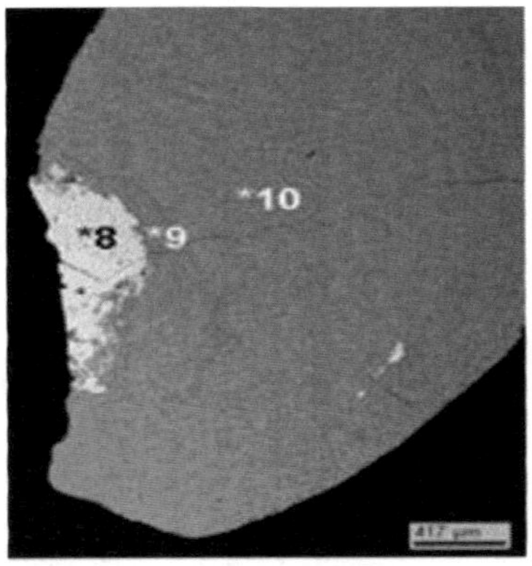

Figure 7. Apatite inclusions in oxalate type urolithes.

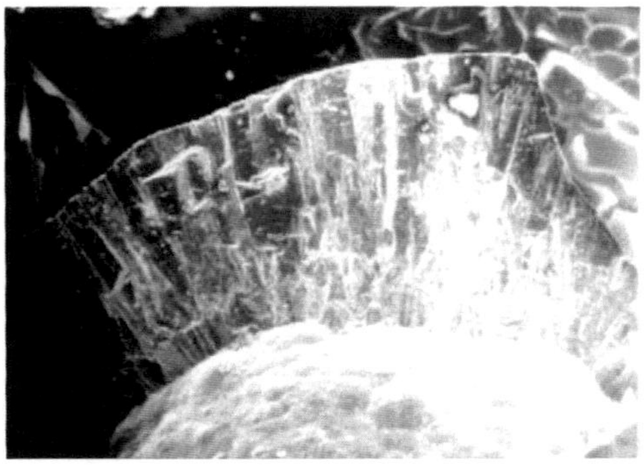

Figure 8. Whewellite crystals: on the apatite globule

4. Conclusion

The results of the investigation show that at supersaturation γ=20, 25 calcium oxalate monohydrate crystallizes only. The average size of its microparticles does not depend on the crystallization time. Ionic strength and magnesium ions are the inhibitors, and calcium oxalate and hydroxyapatite crystals are the catalysts of the calcium oxalate monohydrate crystallization. Basic calcium phosphate (apatite) is most thermodynamically stable and it can be found in almost all kidney stones that indicates its special role in their formation.

Acknowledgements
The research was supported in part by the Russian Foundation for Basic Research (grant No. 15-29-04839 ofi_m and No. 16-33-00406 mol_a).

References

[1] Borisov VV 2011 *Urolithiasis* (Moscow) p 96
[2] Kolpakov I S 2006 *Urolithiasis* (Moscow: Akademiya) p 224
[3] Aliev U G 2010 *Urolithiasis* (Moscow: GEOTAR Media) p 224
[4] Tiktinskiy O L 2000 *Urolithiasis* (St. Petersburg, Russia: Piter) p 384
[5] Dzyurak V S 2006 *Men's Health* **3** 98
[6] Golovanov S A, Sivkov A V, Dzeranov NK, et al. 2011. *Bulletin of Medical Internet conf.* **4** 265
[7] Vinarov A Z 2008 *Moscow surgical j.* **3** 65
[8] Edvardsson V 2005 *Pediatr Nephrol.* **20** 940
[9] Recommendations on the European Association of Urology (EAU) for the treatment of urolithiasis. 2008 (European Association of Urology) p 106
[10] Lopatkin N A 2000 *Urolithiasis* **8** 117
[11] Brikowski T H 2008 *Proc Natl Acad Sci USA* **28** 9841
[12] Chang I H 2011 *Korean J Urol.* **8** 548
[13] Kajander E O 1997 *Science* **11** 420
[14] Golovanova O A, Pyatanova P A, Pal'chik N A 2003 *Chem for sustainable development* **4** 593
[15] Pal'chik N A, Stolpovskaya V N, Leonova I V, et al. 2001 *Mineralogy of techno genesis,* (Miass: URAN) p 99
[16] Golovanova O A , Frank-Kamenetskaya O V, Punin Y O 2011 *Rus. J. of Gen. Chem.* **6** 1392
[17] Moskalev Yu I 1985 *Mineral metabolism* (Moscow: Medicine) p 288
[18] Saltykov S A 1976 *Stereometric metallography* (Moscow: Metallurgy) p 256
[19] Harin V S 1975 *TOXT* **9** 31
[20] Elnikov V Y, Rosseeva E V, Golovanova O A , et al. 2007 *Rus. J. of Inor. Chem.* **2** 190

IOP Conf. Series: Materials Science and Engineering **168** (2017) 012063 doi:10.1088/1757-899X/168/1/012063

A new approach to produce calcium-phosphate coatings on titanium

O A Golovanova[1], A V Zaits[1] and M V Kuimova[2]
[1]Department of Inorganic Chemistry Omsk F. M. Dostoevsky State University, Russia
[2]National Research Tomsk Polytechnic University, Tomsk, Russia

E-mail: Golovanoa2000@mail.ru

Abstract. In the study, hydroxyapatite-gelatin composite powders were synthesized from simulated body fluid (SBF) with gelatin content ranging from 1 to 3 wt. %. It was established that all the samples were single-phase and represented hydroxyapatite. The surface and morphological characteristics of the produced hydroxyapatite-gelatin (HAG) based coatings were studied. Uniform deposition of the composite on the titanium substrate surface (VT1-0) was found to occur on etched titanium samples. It is shown that exposure of titanium substrates with hydroxyapatite-gelatin (HAG) based coating to powerful ion beam can stimulate further growth of crystals and regeneration of the surface.

1. Introduction

Titanium and titanium alloys are currently used to replace bone defects; however, in some cases, these materials are rejected by the human body, which leads to repeat operations, extended rehabilitation of patients and increased cost of operation. This problem may be solved through the formation of a bioactive calcium phosphate layer on a metal substrate. A number of methods are currently used to form these coatings, including calcium-phosphate deposition on smooth and structured surfaces of metals and alloys (current-free deposition, isostatic compression, electrophoresis, chemical vapor deposition, plasma spraying and etc.) [1–8].

The disadvantage of all the above methods is insufficient adhesion of the coatings to the metal substrate. Strong chemical bonding between the coating and the substrate can be formed through fusion temperatures 1073–1273 K, which results in a hard diffusion layer. However, mismatches between the synthesized surface of the material and the implant result in significant stresses that cause damage of the coating during cooling.

An alternative method is formation of biomimetic coatings on metals and their alloys [9–11]. In this case, the implant–bone bonding develops through the biomimetic formation of an active carbonate-hydroxyapatite (HA) layer on the material surface. This layer is formed as a result of the transition of calcium ions from the implant material into the fluid which composition is similar, as an example, to that of the simulated body fluid (SBF). Biomimetic apatite coatings may be formed on an inert material stable to dissolution, polymer as an example. This method has been successfully used for coating various polymeric materials, including the surface of fibers or fabrics. These can be used to make implantable structures, such as matrices for bone regeneration through cellular techniques. The imparted properties of these structures can be similar to those of the natural bone tissue, including high fracture resistance and low modulus of elasticity.

Initially, the biomimetic method was used to form apatite layers on bioglass and bioglass-ceramics, which in themselves are the source of calcium ions. This method was extended to polymeric and metallic materials. Hydroxyapatite based coatings are highly effective for osteointegration of metal implants in the bone. Titanium implants with these coatings are used in dentistry and orthopedics.

Thus, this formation of hydroxyapatite based coatings by biomimetic method stands out from the existing variety of methods to form a bioactive calcium phosphate coating on titanium and titanium alloys as a controversial one, and it requires further investigation. In this research, we aimed to produce a biomimetic gelatin-calcium-phosphate coating on the VT1-0 titanium alloy and to determine its composition and physicochemical properties.

2. Materials and methods

Hydroxyapatite was synthesized from the solution with a model medium close in its electrolyte composition to the human extracellular fluid [12]. During the synthesis, 500 ml of the K_2HPO_4, $NaHCO_3$, Na_2SO_4 and NaCl solution with the addition of gelatin was introduced into 500 ml of the $CaCl_2$ and $MgCl_2$ solution. The total volume of the mixture was 1 liter and the pH of the solution was 7.40 (with acidity correction error $\pm$ 0.05) with HCl or NaOH solution (20%). The crystallization time was 48 hours. After settling, the solution was filtered using a folded filter (blue ribbon). A portion of the supernatant was taken for chemical analyzes and the pH of the final solution was measured. After filtration, the filter cake was washed with water (V=50 ml) and dried in a drying box.

A VT1-0 grade titanium alloy was used for the study. This material has high tensile strength; it is highly biocompatible, non-toxic and corrosion resistant. Its characteristics are similar to the mechanical properties of the bone tissue. The surface of the samples was polished and etched; the etchant composition was HNO_3, NaF (1:1).

For deposition of hydroxyapatite on the VT1-0 titanium surface, 15 mm*15 mm*1.2 mm titanium plates were made. A part of the titanium plates was further exposed to powerful ion beam (PIB) and subjected to laser ablation.

Synthesis of the coatings on the plates for the HA system was performed in the presence of 1%, 2% and 3% gelatin. A hydroxyapatite suspension was prepared with the addition of gelatin, and then the titanium substrate samples were immerged in the suspension. The pH was 7.4, which corresponds to the physiological pH value.

The dependencies of the HA crystal growth and the coated surface areas for the VT1-0 titanium alloy with etched and untreated surfaces was studied by optical microscopy. The first observation was carried out after 3 day soaking in the solution, then observations were performed with an interval of 6 days. After applying the GA-gelatin layer onto the titanium plates, they were exposed to powerful ion beam using the “Temp” setup with the ion current density j=100 A/cm^2 and the number of pulses n=1.

To measure the limiting wetting angle, we used a technique based on measuring the geometric parameters of the droplets (diameter and height) of the liquid wetting the surface. The titanium sample surface was pre-degreased with ethyl alcohol solution, and a droplet of the suspension was released from a glass capillary with a diameter of about 0.2 mm on the surface. The geometrical parameters of the droplet were measured with a Neophot 2 optical microscope using a micrometer screw.

The phase composition of the prepared suspend was investigated by XRD (DRON-3) and IR spectroscopy (“FT-02” spectrophotometer). The peaks in the diffraction patterns were identified using the JCPDS card files and the software DifWin4.0 and Crystallographica Search-Match. The surface morphology was studied using optical microscopes Neophot 2 and MBS-9, and a scanning electron microscope JEOL JSM-6610LV.

3. Results and Discussion

HA were synthesized in the presence of 1 to 3 wt.% gelatin. It was found that when the gelatin content was 1 wt.%, HA crystallization was insufficient, whereas when the gelatin content was 3 wt.%, modified HA crystalline was formed that was identified by optical microscopy (figure 1).

Figure 1. Crystallized HA powder in the presence of 3% gelatin solution (100x magnification).

As can be seen from the figure, the HA crystals formed in the presence of gelatin are of large sizes. The XRD results showed that the samples synthesized in the simulated body fluid under varying concentration of gelatin are single-phase and represent hydroxyapatite.

During formation of the coatings based on the synthesized composites (figure 2), dendritic crystals are seen to start growing on the plate edge.

a. b.

Figure 2. Surface morphology of the hydroxyapatite crystals grown on the VT1 titanium alloy surface in the presence of gelatin after 3 day soaking in the solution: etched surface (a), polished surface (b) (100x magnification).

The crystallization rate was found to depend on the technique used for treating the implant surface. More rapid growth of crystals was observed in the microsections of the polished samples, whereas on the etched samples, dendritic growth occurred in bulk defects caused by etching.

The next observation of the titanium sample surface was carried out after 9 day soaking in the model solution. Figure 3 shows the surface morphology of the samples coated with hydroxyapatite during crystallization in the 3 wt.% gelatin solution.

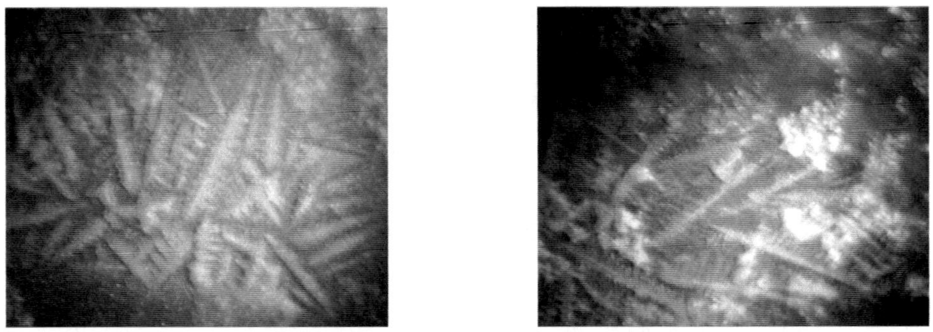

a. b.

Figure 3. Surface morphology of the hydroxyapatite crystals grown on the VT1-0 titanium alloy surface in the presence of gelatin after 9 day soaking in the solution: polished surface (a), etched surface (b) (100x magnification).

As can be seen from the figure, dendrites continue to grow on both the etched and polished surface of the titanium alloy; the dendrites grown on the polished surface are thicker and shorter, whereas those grown on the etched surface are longer and thinner. In our opinion, this is related to different treatment of the sample surface. The crystallization is found to start in the direction towards the sample center, and the HAG coated area on the metal substrate is ~20%. As the crystallization period was increased to 18 days (figure 4), the coating layer thickness increased. On the etched surface, it was ~70 μm, and on the polished one, it was ~50 μm. It is significant that when the HA-gelatin suspension was changed after three days, the increment in the thickness value was 20 μm for each sample, which is characteristic of the mechanism when crystals grow on the surface of the biopolymer that had already been formed.

On the polished sample, the coating was loose, and it could be easily removed from the surface. On the etched surface, cracks were formed due to increased thickness of the resulting layer.

According to the results obtained by scanning electron microscopy (figure 4), in all the cases, the HAG -gelatin coating was formed with characteristic hexagonal crystal structure. A greater layer thickness resulted in cracking. After applying the HAG-gelatin layer, the titanium plates were exposed to PIB using the "Temp" setup. Figure 5 shows the morphology of the HA-gelatin layer on the sample surfaces after exposure to PIB. Under PIB action, high temperature gradients caused melting of the titanium implant surface layer and partial mixing of the HAG -gelatin layer and the substrate. In some places, the HAG layer disintegrated. In our opinion, it was due to non-uniform coating thickness. It was found that PIB irradiation of the coatings synthesized from the model solution in the presence of gelatin lead to fusion of the coating layer and its reliable adhesion.

This increases the biocompatibility of titanium implants, and crystallization of the HAG -gelatin composite on the sample surface may continue.

The results obtained by scanning electron microscopy (figure 6) show that some part of the HAG gelatin coating particles are in the form of rods. The sample structure is porous and it has microcracks that also cause further crystal growth and in vivo biodegradation.

The XRD of the coating (figure 7) indicated the reflexes of the original model – the VT1-0 titanium alloy, HA-gelatin composite and the sample after irradiation. After irradiation, HA peaks could be observed, and the structure of the titanium substrate surface was not modified. Thus, it can be noted that the exposure of the coating to PIB results in firm adhesion of the HAG –gelatin layer crystallized from the model solution on the titanium substrate. The coating produced by this technique can be used for implant manufacturing.

Figure 4. Surface morphology of the hydroxyapatite crystals grown on the polished surface of the VT1-0 titanium alloy.

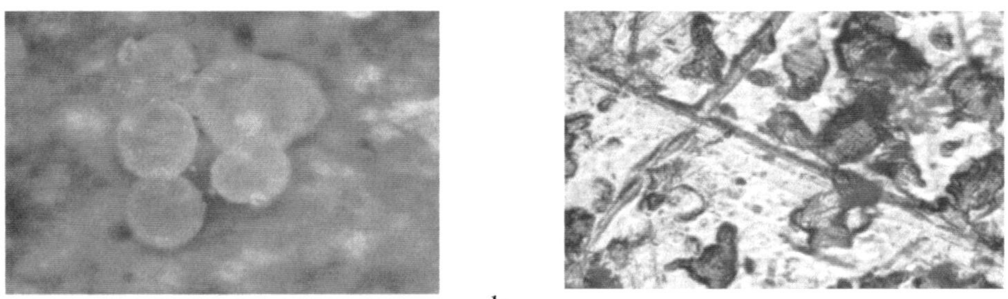

a. b.

Figure 5. Surface morphology of the hydroxyapatite crystals grown on the surface of the VT1 titanium alloy obtained by optical microscopy after exposure to PIB: polished surface (a), etched surface (b).

Figure 6. Surface morphology of the hydroxyapatite crystals grown on the polished surface of the VT1-0 titanium alloy after exposure to PIB.

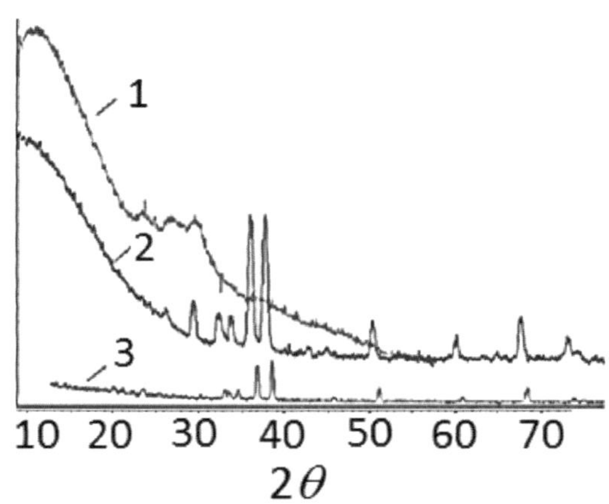

Figure 7. Diffraction pattern: GA-gelatin crystal sample (1); sample with HA crystals grown on the VT1-0 titanium alloy surface after irradiation (2); sample of the VT1-0 titanium alloy (3).

4. Conclusion

HAG-gelatin powders are synthesized from the SBF solution with gelatin content varying from 1 to 3 wt. %. It is established that all the samples synthesized in the medium of the model solution of the extracellular fluid at varying concentrations of gelatin are single-phase and represent hydroxyapatite.

Enhanced HAG-gelatin deposition on the titanium substrate surface is found to occur on etched samples. It is revealed that exposure of titanium substrates to PIB with j=100 A/cm^2 makes possible further growth of HA crystals and regeneration of the metal implant surface.

Acknowledgements

This research was supported in part by the Russian Foundation for Basic Research (grant No. 15-29-04839 ofi_m).

References

[1] Surmeneva M A, Surmenev R A, Chaikina M V et al. 2012 *Phys. and chem. of mater. Process.* **3** 51
[2] Khlusov I A, Pichugin V F, Gostishchev E A, Sharkeev Yu P et al. 2011 *Bulletin of Siberian Med.* **3** 72
[3] Hench L, Johns D 2007 *Technosphere* **1** 301
[4] Turov V V, Gun`ko V M, Bogatyrev V M, et al. 2005 *J. Coll. Interf. Sci.* **2** 329
[5] Murugan R, Ramakrishna S 2005 *Cryst Growth Des.* **5** 111
[6] Sundaram J, Durance T D, Wang R. 2008 *Acta Biomater.* **4** 932
[7] Li J, Chen Y, Yin Y, Yao F, et al. 2007 *Biomater.* **28** 781
[8] Chen M, Tan J, Lian Y, et al 2008 *Appl. Surf. Sci.* **254** 2730
[9] Shu C, Xianzhu Y, Zhangyin X, et al. 2007 *Ceram. international.* **33** 193
[10] Lyasnikova A V, Sakalla A M, et al. 2007 *Bulletin of Saratov State Technical Univ.* **2** 54
[11] Petrovskaya T S, Shakhov V P, Vereshchagin V I et al. 2011 (Tomsk: Publishing house TPU) p. 307
[12] Berdinskaya M V , Golovanova O A, Zaits A V, et al. 2014 *J. of Structural Chemistry* **5** 954

IOP Conf. Series: Materials Science and Engineering **168** (2017) 012064 doi:10.1088/1757-899X/168/1/012064

Study of the effects of heat-treatment of hydroxyapatite synthesized in gelatin matrix

A V Zaits[1], O A Golovanova[1] and M V Kuimova[2]
[1]Department of Inorganic Chemistry Omsk F. M. Dostoevsky State University, Russia
[2]National Research Tomsk Polytechnic University, Tomsk, Russia

E-mail: Golovanoa2000@mail.ru

Abstract. In the study, the isothermal thermogravimetric analysis (TGA) of hydroxyapatite synthesized in gelatin matrix (HAG) has been performed. 3 wt.% HAG samples were synthesized from the solution simulating the human extracellular fluid (SBF). X-ray diffraction and IR spectroscopy were used to determine the composition. During the experiment, increase in the calcination temperature up to 200°C–800°C was found to cause weight loss. The study of phase composition revealed that heat treatment does not affect the phase composition of the solid phase, which is composed of hydroxylapatite (HA). The prepared HAG (3 wt.% gelatin) samples are shown to have low thermal stability; the degradation of the samples occurs at 400° C.

1. Introduction

The global bioceramics demand on the market currently reaches ~40 billion euros per year, the estimated annual growth rate is of 7–12% and the required amount of the material to meet the demand is estimated at tens of tons [1–3]. The number of patients in need of surgery to restore the integrity of the bone is increasing every year: in the US, the annual number of patients exceeds more than 1 million people (of which over 300 thousand people need hip and knee prosthetics, the same holds true for dental implants).

Materials based on hydroxyapatite $Ca_{10}(PO_4)_6(OH)_2$ are analogs of the mineral component of bone, and they are considered the most promising biocomposites to be used as substitutes for bone defect repair due to their high bioactivity [4–9].

It is known that the main components of bone are type I collagen (~ 20%), a mineral phase (~ 60%), water (~ 9%) and non-collagenous proteins (~ 3%), the rest being polysaccharides and lipids [10, 11]. The organic part of bone is composed of collagen fibers and proteins, such as osteocalcin, osteonectin and fibronectin [12]. Gelatin is a natural polymer formed through hydrolysis of collagen. In contrast to the latter, it is more stable and exhibits less antigenicity, thereby can be effectively used as a calcium phosphate based organic matrix biomaterial [13]. Gelatin contains biologically active functional amino acid groups, and it is an advanced material for bone regeneration, including its combination with hydroxyapatite [14]. Since gelatin is a denatured form of collagen and it contains significant amounts of biologically functional amino acid groups, the investigation of properties of hydroxyapatite synthesized in the gelatin matrix (HAG) under near-physiological conditions is a challenging and relevant physicochemical problem.

2. Materials and Methods

The SBF was used as a model solution of extracellular fluid. A solid phase was prepared by precipitation from aqueous solutions; the starting components of the system were $CaCl_2$, $MgCl_2$, K_2HPO_4, $NaHCO_3$, Na_2SO_4, NaCl [15] solutions and 3 wt.% gelatin (figure 1).

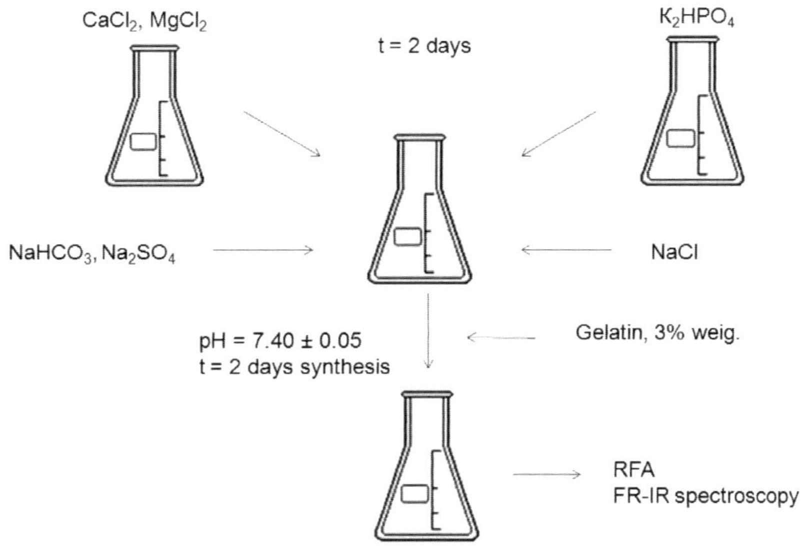

Figure 1. Scheme for HAG synthesis from SBF.

Thermal effects were studied by isothermal thermogravimetric analysis. Hydroxyapatite samples synthesized in the presence of gelatin were placed in dry ceramic crucibles. Then the samples were placed in a muffle furnace and kept for 2 h. The calcination temperature was varied from 200 to 800°C by increment of 200°C. After that, the crucibles were allowed to cool to room temperature, and then weighed on an analytical balance.

The solid precipitates were analyzed by FT-IR spectroscopy and X-ray diffraction (XRD) – qualitative and quantitative phase analysis, measurement of the crystallite sizes (coherent scattering regions – CSR). The IR spectra were recorded using a FT-801 spectrometer (the samples were prepared in the form of KBr pallets). The XRD of the powder samples was performed with the Bruker D8 Advance X-ray diffractometer, and the diffraction patterns were identified using the software package EVA (developed by Bruker AXS).

3. Results and Discussion

The XRD results showed that the samples synthesized in SBF under varying concentration of gelatin were single-phase and represented hydroxyapatite. The diffraction patterns indicated HA reflexes at 25.8° (002), 31.7° (211), 32.2° (112) and 32.9° (300) (figure 2).

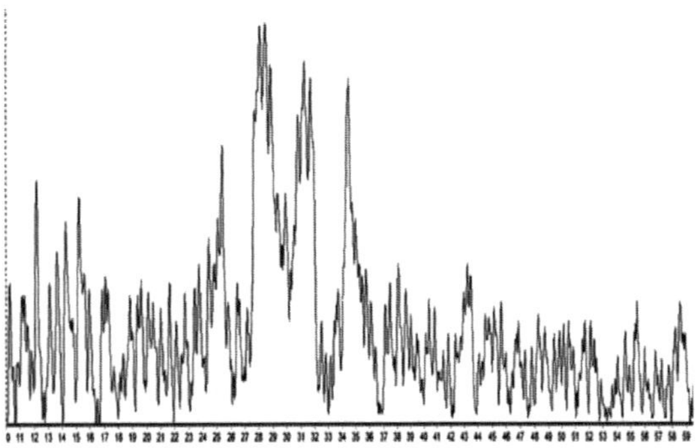

Figure 2. Diffraction pattern of the hydroxyapatite powder crystallized in the presence of 3% gelatin solution.

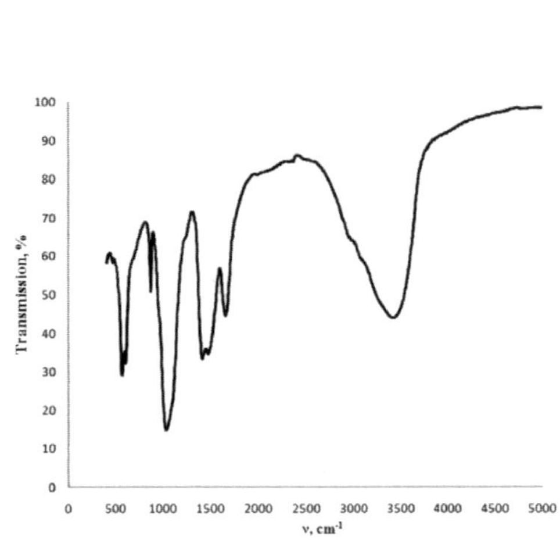

Figure 3. Results of FT-IR spectroscopy for the hydroxyapatite sample synthesized in a gelatin matrix.

The lattice parameters of the prepared solids are increased (Table. 1) as compared to those of the unmodified hydroxyapatite that can be attributed to possible HA sorption in the gelatin matrix.

Table 1. Lattice parameters.

Sample	Initial gelatin concentration, %	Lattice constants, Å	
		a	*c*
HA	-	9.414	6.865
Gelatin	3	9.432	6.881

The analysis of the IR-spectra (figure 3) identified the bands that determine the HA structure: 1040–1080 (ν_3), 960, 840 (ν_2), 602, 574 (ν4) and 473 (ν_2) cm^{-1} corresponding to the phosphate group vibrations, a broad band at 3440–3570 cm^{-1} related to deformation vibrations of the OH^--groups and valent vibrations of the adsorbed water. Carbonate ions are observed to localize in the phosphate ion lattice position with the corresponding absorption bands of the C–O deformation vibrations of the CO_3^{2-} ion bonds at 875 cm^{-1} and valent vibrations at 1415–1481 cm^{-1} in the IR-spectra [16]. A corresponding band of the deformation vibrations of the C=C bond in the region of 1658 cm^{-1}.

The study of the prepared hydroxyapatite samples synthesized in the presence of gelatin showed continuous weight loss (figure 4).

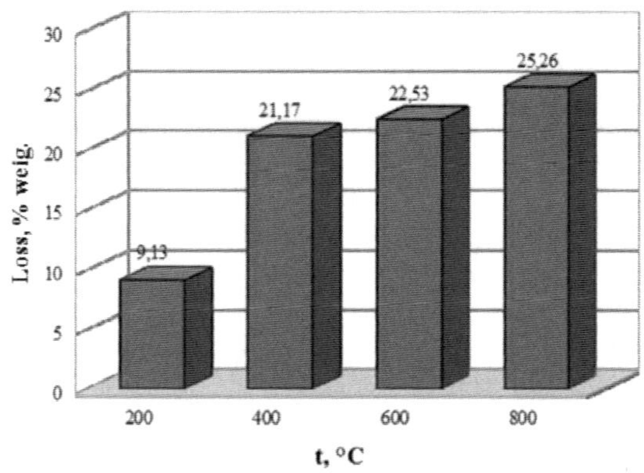

Figure 4. Diagram of TGA for 3 wt.% HAG samples.

The TGA diagram demonstrates the highest weight loss at 200°C and 400°C that can be attributed to adsorption and removal of the chemically bound water from the HAG samples. An increased weight loss at 400°C can be caused by burn out of the organic phase (gelatin) in the HA samples. At 600°C, the weight decreases insignificantly, while at 800°C, the weight loss increases by 2 times (compared to that at 600°C), which indicates the removal of the carbonate groups from the HA structure. These data is consistent with the results obtained by FT-IR spectroscopy (figure 5).

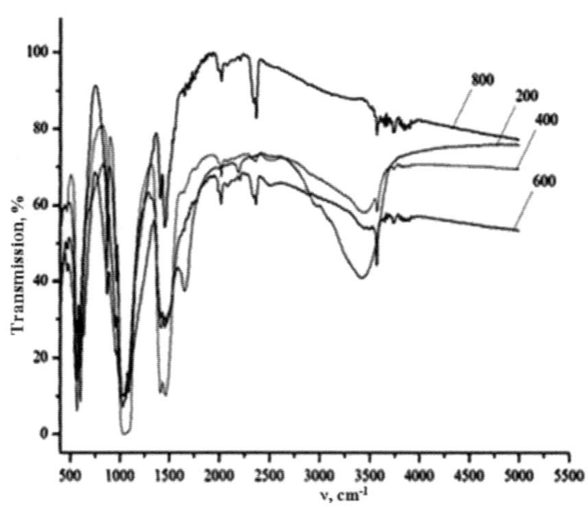

Figure 5. Results of FT-IR spectroscopy for 3% HAG samples subjected to heat treatment at 200, 400, 600 and 800°C.

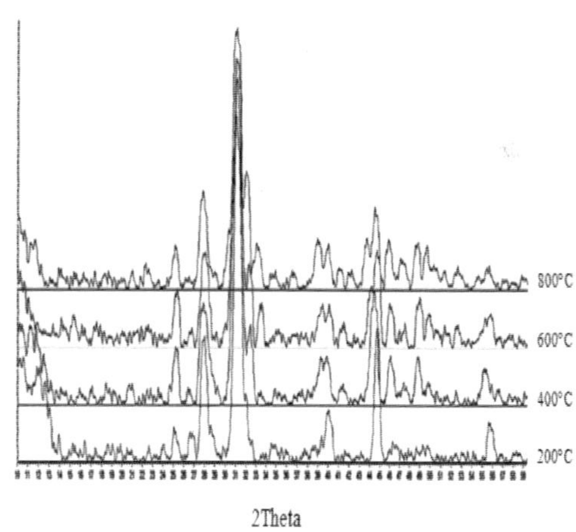

Figure 6. X-ray diffraction patterns for 3% HAG after heat treatment.

The X-ray diffraction patterns (figure 6) of the HAG samples subjected to heat treatment show peak splitting at 400°C. In our opinion, this is caused by degradation of the gelatin components (amino acids), and thereby release of the carbonate component that has been partially released in the form of carbon dioxide, and partially entered the calcium phosphate structure. At 600°C and 800°C, the intensity of peak splitting did not increase and remained at the same level. This is most likely due to the fact that no orbitals are left vacant to bind carbonate ions in calcium phosphate as compared to that in case of unsubstituted hydroxyapatite.

The analysis of the IR-spectra (figure 3) identified the bands that determine the HA structure: 1040–1080 (ν_3), 960, 840 (ν_2), 602, 574 (v4) and 473 (ν_2) cm^{-1} corresponding to the phosphate group vibrations, a broad band at 3440–3570 cm^{-1} related to deformation vibrations of the OH^--groups and valent vibrations of the adsorbed water. Carbonate ions are observed to localize in the phosphate ion lattice position with the corresponding absorption bands of the C–O deformation vibrations of the CO_3^{2-} ion bonds at 875 cm^{-1} and valent vibrations at 1415–1481 cm^{-1} in the IR-spectra [16]. A corresponding band of the deformation vibrations of the C=C bond in the region of 1658 cm^{-1}.

The results of FT-IR spectroscopy showed that the spectrogram of the samples not subjected to heat treatment (figure 3) is identical to that of the samples heated at 200°C (figure 5). This suggests that no significant change in the HAG structure occurs at 200°C. During the analysis, several overlapping spectra were observed: in the region from 2700 cm^{-1} to 3800 cm^{-1}, which, apart from the -OH water vibrations, include vibrations of the amino groups, and methyl and methylene groups. In addition, $-H_2$ groups overlap the carboxyl groups in the region of 1400 cm^{-1}.

As can be seen in Fig 4, at 400°C, the peak significantly decreases in the region of 3500 cm^{-1} that indicates removal of water from the structure and degradation of amino acids, and methyl and methylene groups. The shape of the carboxyl group doublet in the region of 1415–1466 cm^{-1} changed as well that can be confirmed by the removal of the $-CH_2$ group from the prepared sample.

When the temperature rises to 600°C, the intensity of the C=C band (1658 cm^{-1}) is not observed, and the intensity of the C-O band (1420–1466 cm^{-1}) decreases, which indicates the start of the carbonate component destruction. In addition, the O-H band (3569 cm^{-1}) reduces, which is also indicative of the start of the hydroxyapatite destruction.

At 800°C (figure 6), the spectrogram shows further reduction of the O-H band (3569 cm^{-1}) and C-O band (1420–1466 cm^{-1}).

At 200°C, a corresponding deformatio band of the C=C bond is observed in the region of 1658 cm^{-1}. At temperature increase up to 400°C, the peak intensity falla, and at temperature up to 600°C, it completely disappears, which is consistent with the data in [17].

4. Conclusion

The isothermal thermogravimetric analysis revealed four stages of thermal decomposition of composite materials: 1) removal of water and easily volatile impurities (t=25°C–280°C); 2) gelatin pyrolysis (t=280°C–400°C); 3) calcium pyrophosphate phase formation (t=400°C–600°C); 4) removal of carbonate ions from the HA structure and its decomposition (t > 600°C).

Acknowledgements
This research was supported in part by the Russian Foundation for Basic Research (grant No. 15-29-04839 ofi_m).

References

[1] Hench L, Johns D 2007 *Technosphere* **1** 301
[2] Suchanek W, Yashimura M 1998 *J. Mater. Res.* **13** 94
[3] Zethraeus N, Borgstorm F, Strom O et al. 2007 *Osteoporosis International* **18** 9
[4] Astakhov I N 2008 *Surgery.* **12** 34.
[5] Putlyaev V I 2004 *Soros Educational J.* **1** 44
[6] Fomin A S, Komlev V S, Barinov S M, et al. 2006 *Advanced Mater.* **2** 51
[7] Kosyakov A N 2010 *Orthopedics, Traumatology and Prosthetics* **4** 105
[8] Gnedenkov S V 2011 *Advanced Mater.* **2** 49
[9] Sidelnikov A I 2000 *Info-Dent.* **5** 10
[10] Orlovskii V P, Komlev V S, Barinov S M 2002 *Neorg. mater.* **38** 973

[11] Murugan R, Ramakrishna S 2005 *Cryst Growth Des.* **5** 111
[12] Li J, Chen Y, Yin Y, Yao F, Yao K 2007 *Biomaterials* .**28** 781
[13] Zhou Y, Zhao Y, Wang L, et al. 2012 *Radiat. Phys. Chem.* **81** 553
[14] Golovan A P , Rugal A A , Gun'ko V M, et al. 2010 *Surface* **2** 244
[15] Golovanova O A, Berdinskaya M V, Zaits A V, et al. 2014 *J. of Structural Chem.* **5** 992
[16] Khlusov I A, Surmeneva M A, Surmenev R A, et al. 2012 *Bulletin of Siberian medicine* **4** 78
[17] Solonenko A P, Golovanova O A 2013 *Rus. J. of Inor. Chem.* **58** 1420

Healing of discontinuities in the ionic crystals under complex thermo-electric influence

L Karyev, V Fedorov and A Zanina
Tambov State University n.a. G.R. Derzhavin, 33, Internatsionalnaya st., Tambov, Russia, 392622

E-mail: feodorov@tsu.tmb.ru

Abstract. It has been shown that the ionic current in the crystal with a defect in the form of a cleavage crack leads to the healing of the crack. The mechanism of healing is stipulated by the difference in mobility of the ions with different polarity as well as by the electrolysis and recombination crystallization.

1. Introduction

The strength of a real solid material is determined by the presence of in-structural defects, such as cracks and pores. Elimination of these makes it possible to improve the strength considerably. In the modern Materials' Science special attention is focused on the problem of elimination of such in-structural imperfections. In this context, identification of new methods for cracks' healing is important in principle, whereas the research on reconstructive kinetics results subsequently in the extension of the data on the properties of the crystals being investigated.

2. Electric field lines are perpendicular to the surface of the artificially induced cleavage cracks

2.1. Materials and experimental technique

We experimentally investigated a number of LiF single crystals with impurity concentration of Ca^{+2}, Mg^{+2}, Ba^{+2} totally not exceeding 10^{-3} wt. % and NaCl single crystals with impurity concentration of Cr^{+2} corresponding to 10^{-2} wt.%. Samples of 15×10×5 mm size were processed with riving of major blocks along their cleavage planes. Micro-cracks on the clipped surface were induced by indenter PMT-3 [1]. Macro-cracks were induced with blade pressure on the face of the crystal in planes {100} and {110}. The experiments were carried out in the air and in the vacuum (at the pressure ~ 0.01 Pa).

Healing of the micro-cracks' area was controlled by means of comparing the initial state of the defect and its state after the thermo-electrical treatment; in all the other cases – by investigating the structure of dislocations, detected with the chemical etching on the crystal cleavage, crossing the bed of the initial crack.

At the influence of heating and electrical field, the crystal (1, Figure 1) was placed into a closed-type furnace between Ni-Chrome electrodes. The current was supplied to the electrodes with nichrome wires (4). Reliable contacts between the crystal and electrodes were provided by using the fine-dispersed powder (∅ ~ 13 μm) of the examined crystal. This powder was sintered with the crystal and electrode in the process of heating. The speed of crystals' heating in the furnace up to the temperature T0 (673 K ≤ T_0 ≤ 983 K) at the passage of ionic current through specimen did not exceed 200 K/h. The healing was performed at the density of ionic current $J \approx 20 - 1{,}3 \cdot 10^3$ A/m^2. The time was varied within the range of 1 to 6 hours depending on the processing conditions - T_0 and U_0 (U_0 – the direct voltage on the sample). The processing modes were assigned on the basis of experimentally-obtained relationships of crystals' specific conductivity in relation to the temperature of the crystal γ *(T)*. As a criterion for defining the temperature range, we used the values of activation energy in the processes of the impurity and intrinsic conduction, respectively, for LiF – 0.45 and 4.2 eV; for NaCl – 0.49 and 1.7 eV.

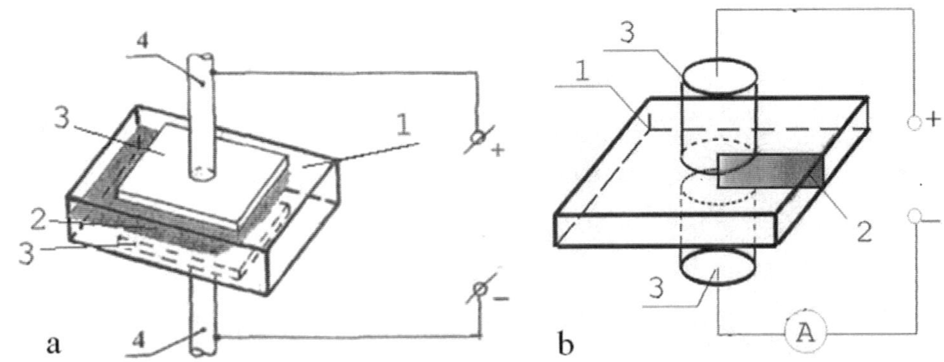

Figure 1. Schemes of experimental setups: electric field lines oriented perpendicular to the plane of the crack (a) and parallel to the plane of the crack (b) 1 - crystal; 2 - crack; 3 - electrodes; 4 – conductors.

2.2. Experimental results

In all the experiments the segment of the healed crack (2) (Figure 1) was generated directly under the electrodes (3). The increase of the treatment time did not change the effect nature, resulting just in an insignificant increase of the healed zone size.

While healing the micro-cracks induced by the indenter in planes {110}, we observed practically complete disappearance not only of the cracks but also of the dents. The heating of similar defects without the current – under otherwise equal conditions – did not result in the healing of cracks and disappearance of dents (refer to Figure 2). There was just a blunt of the crack tip due to the diffusion that was caused by the gradient of mechanical stressing [2].

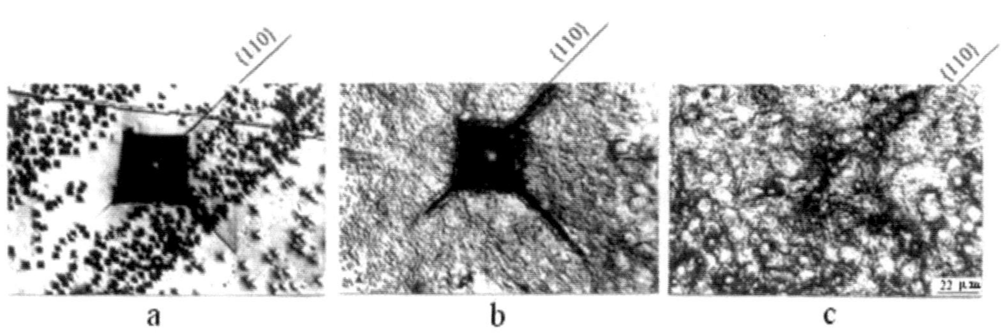

Figure 2. Dents with the cracks induced by indenter PMT-3: a -original sample; b - after heating T = 723 K, t = 4 h; c - after heating and passing the ionic current T = 723 K, t = 4 hours, j = 23 A /m^2.

Micro-structural investigations of the crystals with the healed macro-crack on {110} showed (refer to Figure 3), that in the place of the initial discontinuity there is formed a broken chain of dislocations, the linear density of which is maximal (~ $8 \cdot 10^3$ см$^{-1}$) near the mouth the healed crack; with a gradual decrease towards the crack tip (~ $2 \cdot 10^3$ см$^{-1}$); and in the area of the tip there is formed a zone of a practically defect-free crystal. The density of dislocations along the whole clipped surface was in compliance to the grown-in one (10^4–10^5 см$^{-2}$).

Investigations of the crystals with the healed macro-crack in plane {110} revealed chains of dislocations with the density of ~ 10^4см$^{-1}$ along the whole bed of the healed discontinuity (refer to Figure 3). The density of dislocations on the clipped surface was ~ 10^5см$^{-2}$.

Evaluation of the extent of the recovered material area [3] permits to conclude that it reaches up to ~ 97% of the total healed segment.

After the samples were treated in the range of temperatures of intrinsic conductivity, there were observed formations of micro-pores (in the bed of the initial crack and in its vicinity, mostly – from the side of the negative pole) extending towards <100> in planes {001}. Characteristic dimensions of the micro-pores were 2.5 – 11 microns (refer to Figure 4). When the crystals were heated in the vacuum, formation of the micro-pores was not observed

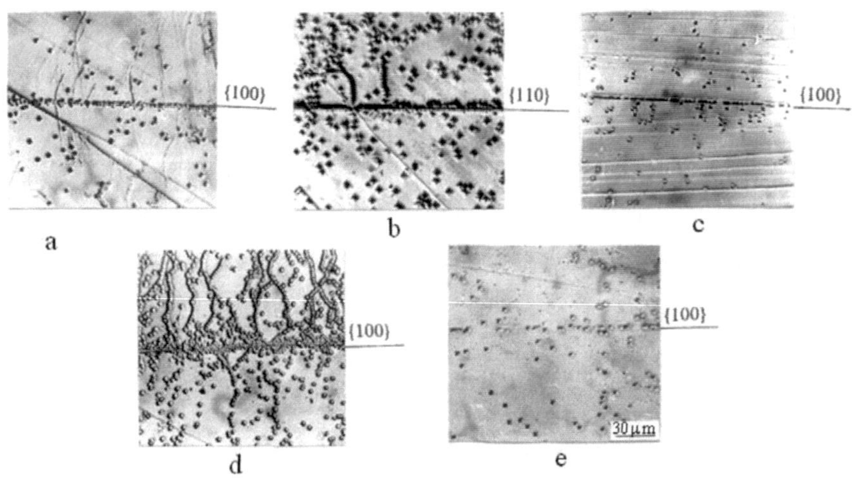

Figure 3. Dislocation structure of segments of the crack healed in the air: a - NaCl; b, c - LiF; and in the vacuum (0,01 Pa): d - NaCl; e – LiF.

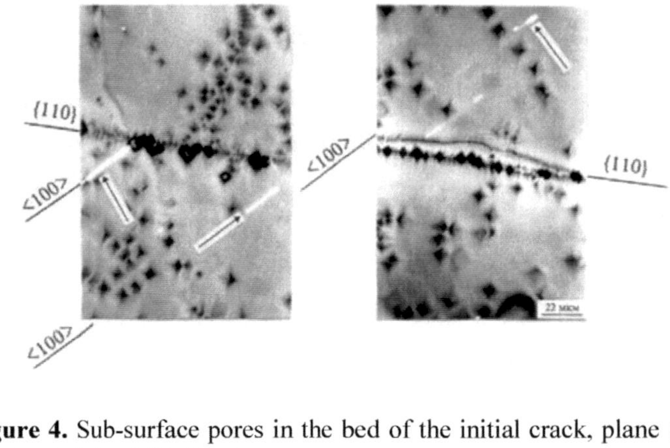

Figure 4. Sub-surface pores in the bed of the initial crack, plane {110} and its vicinity, LiF, the air.

2.3. Discussion on the results obtained

The suggested influence of heating and electrical field on the crystal makes it possible to restore the continuity of the alkali-halide crystals (AHCs) deformed on {100} and {110}. Healing is possible not only by shifting the substance mass, effected by the ionic current, but also by the difference in the mobility of ions of different polarity [4], resulting in unequal increase of the mass on the opposite sides of the crack being healed. The opposite inner surfaces of the crack react differently to the electrical field treatment.

In the crystal portion, located on the positive electrode, mostly higher-mobile ions of Li^+ and impurities ions [5, 4] move over towards the external electrical field. The ions of F^- are practically fixed. At that, the crack surface is charged with a positive charge without destructing the crystal lattice. Being the energy barrier for the ions of Li^+ and impurities' ions, the surface prevents from their emission. As contrasted to this, the negatively charged crack surface is depleted by the positive charge, diffusing to the negative electrode. Inter-ionic Coulomb forces and external electrical field help to the release of the ions of F^- [6], providing for the emission anionic current. It is to be pointed out that the structural changes of the clipped surface under consideration promote the decrease of F^- ions' sublimation energy, due to depletion of the surface by the positively-charged ions.

Bombarding the positively-charged surface, the originated anionic emission current results in deformation shifts. The F^- ions' energy, under the observed electrical field densities, can reach up to ~ 70 eV. The breakthrough takes place, primarily, in the points of electrical field concentration (i.e. on the cleavage steps). During the bombarding of the positively-charged crack surface by the F^- ions, there occurs mechano-emission of the ions of Li^+ [7] in the zones of plastic slips. As a result of the recombination crystallization, on the positively-charged surface there occurs the increase of the crystal mass along the whole surface of the crack. The crack cavity is filled with the crystal material till its complete disappearance.

The observed chains of dislocations in the bed of the initial crack can be identified with sub-boundaries of grains, because the procedure of crack introduction or joining of two crystals (in case of healing the cavity, limited by two halves of a crystal, placed on each other) makes the turning and shifting of the joined halves inevitable [8 – 11]. Evaluation of disorientation provides for the values of the angles, not exceeding 10′ for different samples.

The healing is not followed by the voltage gradients, as confirmed by the absence of clearings in the crystals while the crystals were examined in the polarized light. Besides, the high level of materials' structure restoration is also specified by the evolution of dislocation beams of the indenter hit. The passing of dislocations from the indenter hit through the healed segment and through the pure crystal is practically identical.

Formation of the pores is explained by a discharge which (within the specified range of temperatures) originates in the gas filling the space of discontinuity. The air ions penetrate into the crystal and diffuse to the electrodes under the influence of the electrical field.

3. Electrical field lines are parallel to the surface of the artificially induced crack

3.1. Materials and experimental technique

We experimentally investigated a number of LiF single crystals with impurity concentration of Ca^{+2}, Mg^{+2}, Ba^{+2} in the range 10^{-3}–10^{-5} wt. % and NaCl single crystals with impurity concentration of Fe^{+2} in the range of 10^{-2} – 10^{-5} wt.% . The samples of 20×10×5 mm size were processed with riving of major blocks along their cleavage planes {100}.

The experiment was performed on the set-up, the scheme of which is shown in Figure 1. The crystal was placed between cylindrical electrodes, made of nichrome. Reliable electrical contact between the crystal and electrodes was provided by the method similar to the one described in Item 2.1 (refer to the above). A macro-crack was artificially introduced into a crystal in the plane of primary cleavage. The sample was positioned in such a way that the crack tip was between the electrodes. The gap between the sides of the crack varied from 1 to 5 microns. Direct voltage, U_0 = 400 V, was supplied to the electrodes. The density of the external electrical field, *E0,* was equal to $8 \cdot 10^4$ В/м.

The set-up was placed into the furnace. The experiments were carried out in the range of temperatures 293 – 893 K. The heating speed was 200 K/hour. After the heating, the electrical field was shut off, and the sample subsequently cooled down with the furnace up to the ambient temperature.

Dislocation structure of the processed LiF and NaCl crystals was identified by means of chemical etching in the saturated solution of iron acid and in the ice-cold acetic acid respectively [12].

3.2. Experimental results

Within the temperature range of impurity conductivity (T < 823 K) there were no visible changes observed on the surfaces limiting the crack cavity along plane (001). Within the temperature range of intrinsic conductivity (T > 823 K) the crack, positioned between the electrodes, was healed (refer to Figure 5). The area of the segment of recovered continuity grew up with the increase of time of heating and electrical field influence on the crystal. At the duration of influence $t \approx 3.5$ hours, that is, corresponding to the heating up to the temperature T = 893 K, the area of the restored portion of the crack, located between the electrodes, practically reached the value of 100%.

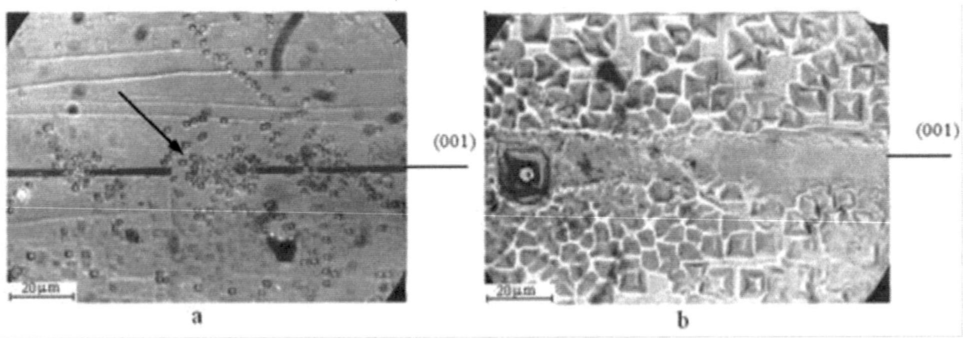

Figure 5. The healed segments of the crack: a – LiF (marked by the arrow), b – NaCl.

3.3. Discussion on the results obtained.

As such, while supplying the ionic current along the crack surfaces, these surface were influenced by the surface currents. The ionic current in the near-surface areas, limiting the crack cavity, is of higher value than the value of the current in the crystal volume, with all the conditions otherwise equal. This stipulates the unequal heating of the given surfaces and volume of the sample and, subsequently, is the reason of different thermal expansion in the surface layers, comparing to the parts of the crystal, distant from them. The process is of a growing nature: in course of time the power of the surface current increases, subsequently resulting in more intensive diffusion of the positively-charged ions to the surfaces limiting the crack cavity. The increased concentration of the positively-charged ions in the given areas is the reason of bringing the crack sides together till their complete joining. That is, the restoration of continuity takes place.

Acknowledgments
This work was supported by grant of Russian Foundation for Basic Research No.15-41-03166.

IOP Conf. Series: Materials Science and Engineering **168** (2017) 012065 doi:10.1088/1757-899X/168/1/012065

References

[1] Karyev L, Fedorov V, Glushkov A, Meksichev O and Tyalin Y 2000 Impact orientation Vickers indenter at the destruction of alkali halide crystals at microindentation *Vestnik TSU* **5** (2-3) 378-381

[2] Kosevich A, Feldman E and Streltsov V 1984The initial stage of the diffusion swimming thin end of the equilibrium cracks *Solid State Physics* **26** (10) 3006-13

[3] Fedorov V, Finkel V, Plotnikov V and Kuranova V 1991The mechanism and kinetics of healing "elastic" Rose channels in the calcite *Crystallography* **36** 5 1322-24

[4] Joffe A 1974 *Selected works* vol 1 *Mechanical Properties of Crystals* (Leningrad: Nauka) p 326

[5] Lydiard A 1962 *The ionic conductivity of the crystal* (Moskow: Foreign Literature) p 222

[6] Frisch S *1958The course of general physics*. T. II (Moscow: State Publishing House of Physical and Mathematical Literature) p 509

[7] Molotsky M 1978 The recombination of the electron emission mechanism Deryagina-Krotova-Karasev after cleavage *Doklady physics* **243** (6) 1438-41

[8] Finkel V, IvanovV, Zaitsev O and Tyalin Y 1985The healing of cracks in the bent crystals *Solid State Physics* **27** (10) 3119-21

[9] Finkel V, Kurgan L and Safronov V 1985 Some defects healed cracks *Solid State Physics* **27** (1) 189-191

[10] Finkel V and Konkin B 1983Types of dislocations on the healed fracture *Solid State Physics* **25** (5) 1553-55

[11] Dorokhov N 1994Formation processes of fractographic relief and its effect on the healing of cracks in a number of single crystals (Rostov-on-Don) p 154

[12] Pshenichnov Y 1974 *Identification solid structure of crystals* (Moscow: Metallurgy) p 528

Thermal stability of brushite with chitosan samples

E S Chikanova[1], O A Golovanova[1], T V Malikova[1] and M V Kuimova[2]

[1]F.M. Dostoevsky State University, 55A Prospekt Mira, Omsk, 644077, Russia
[2]National Research Tomsk Polytechnic University, 30 Lenin Avenue, Tomsk, 634050, Russia

E-mail: chikanova_es@mail.ru

Abstract. In this paper, the powders of brushite from an aqueous solution of $Ca(NO_3)_2$-$(NH_4)_2HPO_4$ with different content of chitosan were synthesized. XRD data revealed that all samples are single-phase and are brushite ($CaHPO_4 \cdot 2H_2O$). By FT-IR spectroscopy and BET methods, it was found that chitosan adsorbs onto the surface of powders. With increase of the content of the additive, the average size of crystallites increases 4.0 – 4.8 – 11.8 μm, respectively, and the dissolution rate in isotonic solution also decreases. The thermal stability of the composite powders was studied. It was established that the highest destruction of samples occurs in the range 473-673 K by removing of adsorption and crystallization water and partial change of the structure of the mineral and chitosan. At a temperature of 873 K, carbonization of the organic additive occurs.

1. Introduction

In the world of modern osteoplastic materials, much attention focuses on the creation of composite aggregates [1-3]. For this purposes, pure or mixed variants of calcium phosphates with various organic polymers are often used. For example, calcium phosphate cements with polyamides are used in tissue engineering to restore bone defects [4]. Phosphate based composite with polymethyl methacrylate is used to affix an antibiotic to reduce the risk of bacterial infection development during transplantation [5].

Chitosan, the deacetylated chitin derivative, is of particular interest as a biopolymer thanks to its multifunctionality, molecular structure (figure 1), presence of free amino groups, antibacterial properties and biocompatibility [6]. Chitosan is also a suitable object for immobilization of various compounds, particularly medical agents [7]. In this regard, it is important to synthesize and study the composite connections on the basis of chitosan and calcium phosphate of different nature. Therefore, the aim of this work is to obtain brushite samples with chitosan and study their properties and thermal stability.

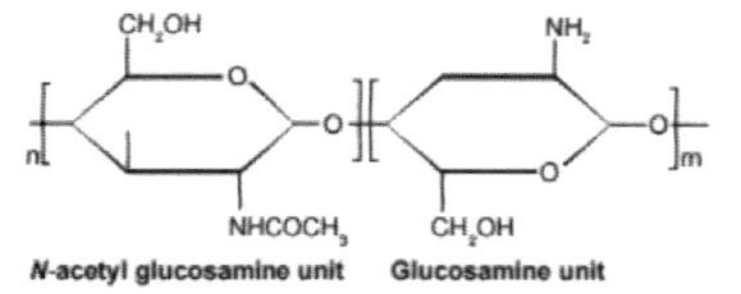

Figure 1. Chemical structure of chitosan [6].

2. Material and methods

2.1. Synthesis of powders

Brushite crystals in a chitosan matrix were synthesized by precipitation from aqueous solution at pH = 5.50±0.05, t = 25 °C by spontaneous crystallization. The powders were prepared by mixing of dilute solutions of calcium nitrate and ammonium hydrophosphate with equimolar concentrations (50 mmol/l) in accordance with equation (1).

$$Ca(NO_3)_2 + (NH_4)_2HPO_4 + 2H_2O \rightarrow CaHPO_4{\cdot}2H_2O\downarrow + 2NH_4NO_3 \quad (1)$$

The synthesis was carried out on three samples. The mass of chitosan varied in each sample by 0.01g, 0.04g and 0.08g. After each solution was left standing for 72 hours, the precipitate was filtered off and dried in the drying chamber at T = 80 °C until the complete removal of water.

2.2. XRD-analysis

The study by X-ray diffraction (XRD) method was performed on the powder X-ray diffractometer D8 Advance, Bruker in Cu-kα radiation (wavelength 0.15406 nm) using a position-sensitive linear detector LynxEye. The following measurement modes were used: scanning step – 0.05°, integration time – 5 seconds/point, the voltage and heater current are 40 kV and 40 mA, respectively. Scanning area 2θ: 4 – 90°. Interpretation of the obtained diffractograms was conducted using the database of powder diffraction ICDD PDF-2, 2006.

2.3. FT-IR spectroscopy

The study of powders by FT-IR spectroscopy method was carried out on the device FSM-2202. The scanning range was 400-5000 cm^{-1}, resolution was 8 cm^{-1}. Tablets of the sampling material with potassium bromide were made to conduct the study.

2.4. Specific surface area

The determination of the specific surface area of the samples was performed by the classical adsorption method. The specific surface area of the samples was measured by the method of one-point nitrogen adsorption at 77.4 K on the adsorption apparatus "Sorbtometer", production by Katakon LLC, Russia. Calculation of the obtained values of S_{BET} (m^2/g) was made by the BET method.

2.5. Optical microscopy

The synthesized powders were analyzed on the optical binocular microscope XSP-104, magnification 100. The images with the resolution of 2048*1536 were taken on the video ocular ToupCam UCMOS03100KPA.

2.6. Dissolution

To study the kinetic parameters of dissolution, 0.2 g of powder was poured with 100 ml of 0.9% NaCl at T=298 K and a constant stirring speed. After that, the pCa values (unit I-160-MI, LLC "Measuring equipment", Russia) and pH (pH unit 150 MI, LLC "Measuring equipment", Russia) were fixed at regular intervals before output of the value on the plateau.

2.7. TG analysis

Powders of m=0.2000±0.0005 were selected into dry ceramic crucibles. Then these samples were placed in a muffle furnace for 2 hours. The calcination temperature was 473-873 K. After a prespecified heat time, the crucibles were cooled to room temperature then weighed on the analytical balance. The samples were transferred to a specially marked container and dispatched for physical analysis.

3. Results and Discussion

Composites of brushite-chitosan composition in the form of powder were obtained. X-ray analysis revealed that the samples are represented by the presence of dicalcium phosphate dihydrate (DCPD) monoclinic system (figure 2). The phase transition of this calcium phosphate does not occur by adding chitosan to the initial solution.

In the IR spectra of the samples (figure 3) there are bands characteristic for valence and deformation vibrations of the groups and relationships HPO_4^{2-}, PO_4^{3}-, -OH, H_2O, $-NH_2$, $-CH_2$, -C=O, -C-N, -N-H. Thus, chitosan adsorbs on the surface of the DCPD. Reduction in the intensity of the 3500-3600 cm^{-1} (ν_{H-O-H}, ν_{O-H}) and 1133 cm^{-1}, 654 cm^{-1} (ν_{PO43-}, δ_{PO43-}) helps to determine the interaction in these positions. Adsorption facts were also confirmed by the specific surface area. In the absence of chitosan, brushite has S=12 m^2/g, with the addition of chitosan S=5 m^2/g (regardless of the initial mass). Since the initial samples are low-dispersive and semi-porous, it can be assumed that surface adsorption predominates. It is known that the surface of brushite at pH = 5.5 is positively charged, thus interaction occurs due to electrostatic attraction with electron-donating groups of chitosan.

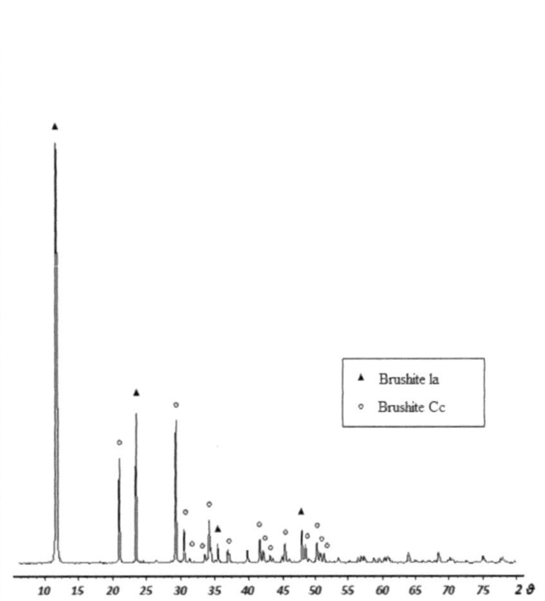

Figure 2. The results of XRD brushite obtained in the presence of chitosan.

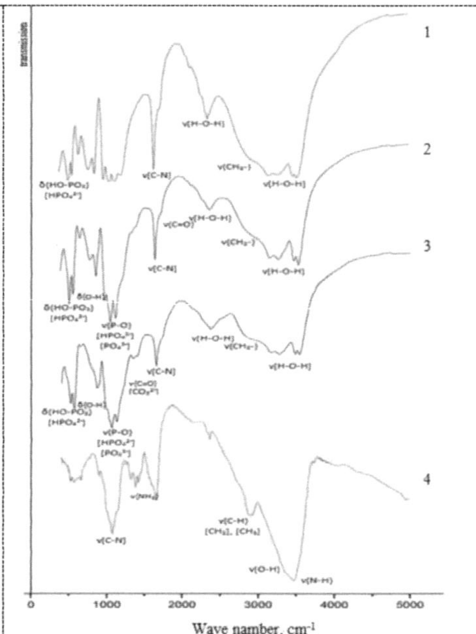

Figure 3. IR spectra of the samples:
1 – initial weight of chitosan is 0.01 g;
2 – initial weight of chitosan is 0.04 g;
3 – initial weight of chitosan is 0.08 g;
4 – pure chitosan.

The following plates show the morphology of the synthesized powders (figure 4). The average size of crystallites increases with the increasing mass of chitosan in the initial solution (figure 5) and has a value of 4–12 μm, but hereby their structure deteriorates. It should be noted that in the absence of the additive, the size of the DCPD crystals is ~ 60 μm [8].

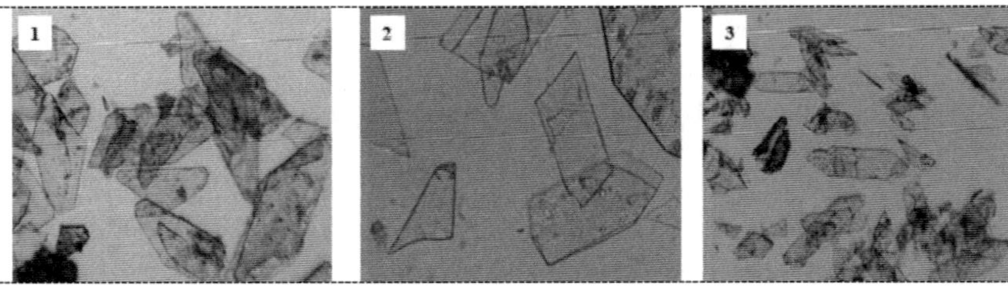

Figure 4. Optical microscopy of precipitation: weight of chitosan 1 – 0.01 g, 2 – 0.04 g, 3 – 0.08 g.

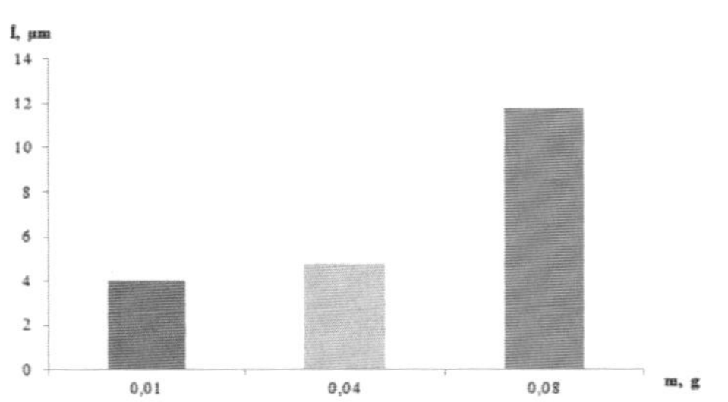

Figure 5. Dependence of the average crystal length of chitosan contained in the initial solution.

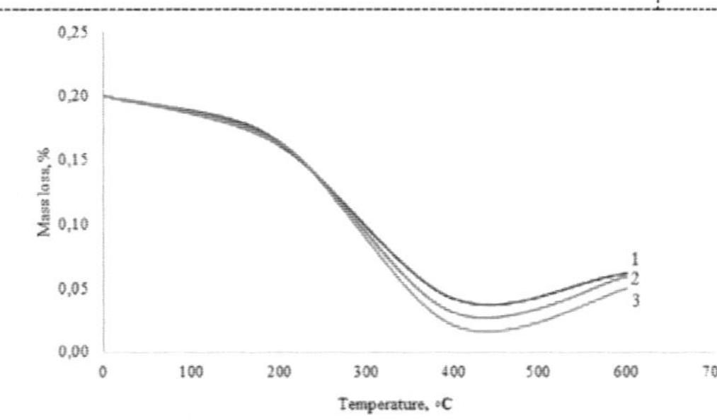

Figure 6. TG-graph of the solid phases: initial weight of chitosan is 1 – 0.01 g; 2 – 0.04 g; 3 – 0.08 g.

The results of the thermogravimetric analysis of the powders obtained at different weight of chitosan in the initial solution are shown in figure 6. As can be seen, the dependent variables are similar, but there are some differences. For example, a slight increase in the loss of mass at 473 K and a reduction in the range of 673–873 K with the increase of the initial mass of chitosan in the initial solution, the greatest mass loss is observed in the range of 473–673 K. The degradation of chitosan, forming part of the powders, starts from a temperature of 473 K. At that temperature, the color of the samples changes (figure 7.): beige (0.01g), pink (0.04g) and dirty pink (0.08g). At a temperature of 673 K, the discoloration is more intense. At a temperature of 873 K, chitosan burns out and carbonates.

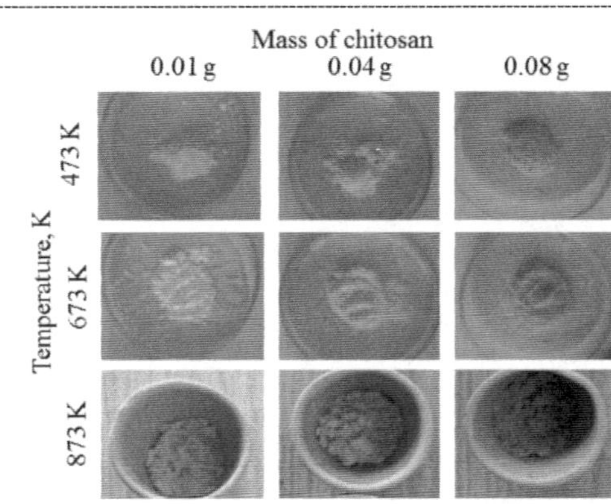

Figure 7. Change of the samples by heating.

The main processes that occur during the calcination of composites are presented in table 1 and are confirmed by IR spectroscopy (figure 8). Thus, for the further use of the composites as an additive to biomaterials or creation of ceramics based on them, the processing temperature must not exceed more than 673 K.

Table 1. Change of powders at calcination.

T, K	DCPD [8]	T, K	Chitosan [9]
298 – 373	Removal of free water $[CaHPO_4 \cdot 2H_2O]H_2O_{(ТВ)} \rightarrow CaHPO_4 \cdot 2H_2O_{(ТВ)} +$ $+ H_2O_{(ш)} + Q$	298 – 373	Removal of free water
353 – 503	Removal of adsorption and crystallization water $CaHPO_4 \cdot 2H_2O_{(ТВ)} \rightarrow CaHPO_{4(ТВ)} + 2H_2O_{(г)} + Q$	373 – 473	Removal of adsorption and chemically bounded water
473 – 633	Disintegration of ammonium nitrate $NH_4NO_{3(ТВ)} \rightarrow N_2O_{(г)} + 2H_2O_{(г)} - Q$ $NH_4NO_{3(ТВ)} \rightarrow N_{2(г)} + 1/2O_{2(г)} + 2H_2O_{(г)} - Q$	473 – 673	Thermal decomposition with the weight loss up to 40%
633 – 733	Melting of monetite and formation of calcium pyrophosphate $2CaHPO_{4(ТВ)} \rightarrow Ca_2P_2O_{7(ТВ)} + H_2O_{(г)} - Q$	> 673	Partial carbonization and burnout

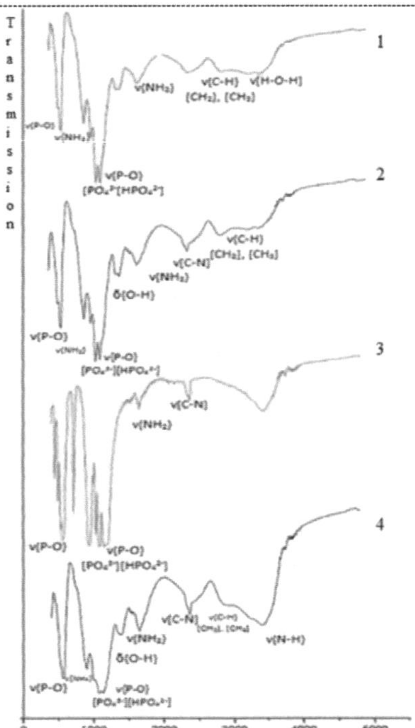

Figure 8. IR-spectra of the samples after heating: 1 – initial weight of chitosan is 0.04 g, T=473 K; 2 – initial weight of chitosan is 0.04g T=673 K; 3 – initial weight of chitosan is 0.04g, T=873 K; 4 – initial weight of chitosan is 0.08g, T=473 K.

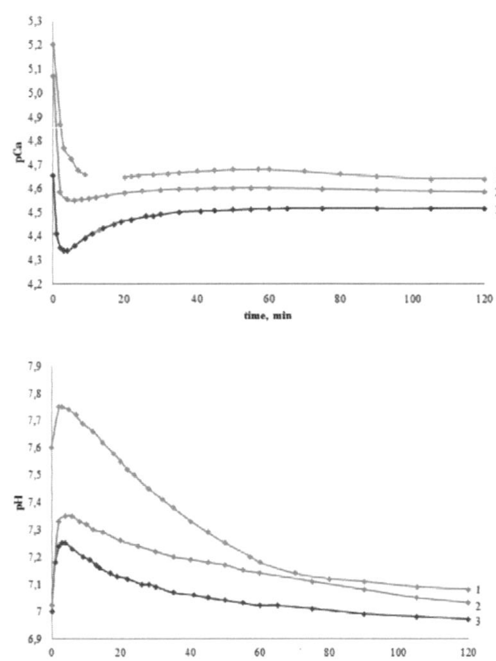

Figure 9. Kinetic dissolution curves. Initial weight of chitosan is 1 – 0.01 g; 2 – 0.04 g; 3 – 0.08 g.

In the study of the dissolution of powders in an isotonic solution (figure 9), it was revealed that the kinetic curves have the same dependences; the functions pCa = f (t) and pH = f (t) can be used to describe the process. As seen in Fig. 9, the maximum dissolution rate is observed in the first phase (up to 10 min.) Therefore, this curved segment is of interest. The equations calculated by the graphical method and the rate are presented in table 2. The values indicate that with the increase of the mass of chitosan in the initial solution, the dissolution rate decreases. The additive with a long acting (but not too slow) 0.01g and 0.04g is desirable for the use.

Table 2. Kinetic parameters of dissolution of powders in 0.9% NaCl.

Mass of chitosan, g	Kinetic equation	Rate
0.01	pCa = -0.0524t + 5.043	0.0524
0.04	pCa = -0.0382t + 4.834	0.0382
0.08	pCa = -0.009t + 4.447	0.009

4. Conclusion

Thus, the micro-sized composites of chitosan-brushite from the system $Ca(NO_3)_2$–$(NH_4)_2HPO_4$–H_2O were obtained in the work. Their morphology and composition were established. It was found that with the increase of the weight of chitosan in the initial solution, the size of crystallites increases, but the values are smaller compared to pure dicalcium phosphate dihydrate. The thermal stability of the powders was studied. The highest mass loss occurs in the range of 473–673 K. This is due to the removal of adsorption and crystallization water from the structure of samples and also due to the thermal decomposition of chitosan. The dissolution rate of the samples in an isotonic solution was determined. It is found that it decreases with the increasing mass of the organic additive.

Acknowledgments
This research was supported in part by the Russian Foundation for Basic Research (grant mol_a No. 16-33-00684 and No. 15-29-04839 ofi_m).

References

[1] Shastri V P 2003 *Curr. Pharm. Biotechnol.* **4** 331
[2] Chen H, Yuan L, Song W, Wu Z, Li D 2008 *Prog. Polym. Sci.* **33** 1059
[3] Agrawal C M, Ray R B 2001 *J. Biomed. Mater. Res.* **55** 141
[4] Canal C, Ginebra M P 2011 *J. Mech. Behav. Biomed. Mater.* **4** 1658
[5] Klemm K 1993 *Clin. Orthop.* **295** 63
[6] Tarek A A, Bader M A 2016 *Drug. Des. Devel. Ther.* **10** 483
[7] Sonia T A, Sharma C P 2011 *Adv. Polym. Sci.* **243** 23
[8] Solonenko A P, Golovanova O A, Phil'chenko M V, Ishutina V S, Leont'eva N N, Antonicheva N V, Buyal'skaya K S, Savel'eva G G 2012 *Vest. Om. Univ.* **2** 135
[9] Kuchina Yu A, Dolgopyatova N V, Novikov V Yu, Konovalova I N, Prinsteva M Yu, Sagaidachny V A 2015 *Vest. MSTU* **18** 94

IOP Conf. Series: Materials Science and Engineering **168** (2017) 012067 doi:10.1088/1757-899X/168/1/012067

Low-temperature volume radiation annealing of cold-worked bands of Al-Li-Cu-Mg alloy by 20-40 keV Ar^+ ion

V V Ovchinnikov[1, 2], N V Gushchina[1], S M Mozharovsky[1] and L I Kaigorodova[3]

[1] Institute of Electrophysics, Ural Branch of Russian Academy of Sciences, Amundsena Street 106, Yekaterinburg, 620016, Russia
[2] Ural Federal Technical University named after the First President of Russia B.E. Yeltsin, Mira Street 19, Yekaterinburg, 620002, Russia
[3] Institute of Metal Physics, Ural Branch of Russian Academy of Sciences, St. S. Kovalevskoy 18, Yekaterinburg, 620990, Russia

E-mail: viae05@rambler.ru

Abstract. The processes of radiation-dynamic nature (in contrast to the thermally-activated processes) in the course of short-term irradiation of 1 mm thick bands of cold-worked aluminum alloy 1441 (of system Al-Li-Cu-Mg) with Ar^+ 20-40 keV were studied. An effect of in-the-bulk (throughout the whole of metal bands thickness) low-temperature radiation annealing of the named alloy, multiply accelerated as compared with common thermal annealing processes was registered (with projected ranges of ions of considered energies definitely not exceeding 0.1 μm). The processes of recrystallization and intermetallic structure changes (occurring within a few seconds of Ar^+ irradiation) have the common features as well as the differences in comparison with the results of two hour standard thermal annealing.

1. Introduction

Utilization of accelerated ion beams with energies ranging from units to hundreds of keV has given rise to development of a number of innovative technologies currently finding broad application in treatment of structural materials with a view to obtain properties unattainable through application of traditional methods of treatment.

Unfortunately, the projected ranges of the above energy ions in material measure from a few tens to hundreds of nanometers only. The modified zone of such depth is definitely inadequate for the majority of technical applications.

Use of ions with energies of several tens to hundreds of MeV, whilst increasing the ion affected zone to several tens of micrometers [1, 2], however leads to significant complication and rise in the cost of the process. The same relates to attempts to combine different methods of coatings deposition and ion irradiation.

Use of powerful continuous low-energy [3] and nanosecond pulse power ion beams [4, 5, 6], also high-dose implantation (with generation of dislocations migrating deep into the material) [7, 8], allows the modified zone depth to be extended to several tens of nanometers.

However, for the purposes of many technical applications, it is more desirable to increase the depth of accelerated ions modifying action on the structure and properties of materials by at least 1-2 orders of magnitude, which would enable modification of *submillimeter* and *millimeter* layers of materials at their surface irradiation.

The record depth of action on materials may be reached at the expense of *"radiation-dynamic effects"* [9, 10]. Nanodomains of dense cascades of atomic displacements present the zones of explosive energy release with formation of thermal spikes. The temperature of such domains, with the time of thermalization of the order of 10^{-12} s, reaches 5000-6000 K and higher [11, 12]. Their fast expansion causes emission of post-cascade shock waves [13-15]. In thermodynamically *stable* media such waves are damped out quickly.

Phenomenological and hydrodynamic description of such processes in *metastable* media with high stored energy [9] indicates the possibility of the undamped mode of post-cascade waves propagation initiating structural-phase transformations at their front. At that, the registered depth of impact is by far (10^3-10^5 times) [9, 16, 17] greater than the projected ranges of heavy accelerated ions. This enables modification of *submillimeter* and *millimeter* subsurface layers of materials.

In the present work, the existence of a radiation-dynamic component in accelerated ions action on material is taken advantage of in recovery of ductility of sheets (1 mm thick) of cold-worked aluminum alloy 1441 of Al – 2.0 Li – 1.7 Cu – 1.0 Mg – 0.11 Zr system. As a result of the cold plastic deformation up to ~70% the alloy completely loses its plasticity, rendering its further cold rolling impossible without intermediate annealing.

2. Experimental

Samples were irradiated in ion implanter ILM-1 outfitted with technological low-pressure hollow-cathode glow discharge *ion source* PULSAR-1M [18] generating round cross section gas-ion beams (S ~ 100 cm^2, E = 10-50 keV, j = 50-500 $\mu A/cm^2$).

Mechanical tests and structural study were carried out on uniaxial tensile test samples (GOST 1497-84). There were four tensile samples irradiated simultaneously, to one of the samples a thin chromel-alumel thermocouple was welded (with laser welder "Quantum 15") for temperature monitoring. The initial and irradiated samples microstructure was studied under optical microscope Neophot-21 and electronic microscope JEM-200 CX. The irradiated samples structure was studied by these methods across two sections, parallel to the irradiated surface and square to it (cross-section). This allowed us to establish that structural changes and phase transformations initiated by ion irradiation occur throughout the bulk of alloy strips *1 mm thick*: this is more than 10^4 times greater than the implanted ions projected ranges which certainly not exceeding 0.1 μm [19].

3. Details of exposure, measurements and results

Cold-worked alloy 1441 features higher strength properties and lower ductility: $\sigma_в$ = 316 MPa, $\sigma_{0.2}$ = 296 MPa, δ = 3 %. In the initial state, this alloy microstructure features a characteristic structure, which points to the presence of elongated ~ 2 μm thin grains. A non-uniform dislocation cell structure is observed in the bulk of the grain (Figure 1 a). There are viewed intermetallides of crystallization origin Al_8Fe_2Si, on the average, up to 2 μm in size, equiaxial-shape β'-phase (Al_3Zr) particles 40-60 nm in diameter, and fine-disperse Al_3Li-phase particles less than 1 nm in diameter precipitated as a result of natural ageing process at room temperature.

Annealing (T = 370°C during 2 h) leads to an abrupt drop of strength characteristics and increase in ductility of alloy 1441: $\sigma_в$ = 245 MPa, $\sigma_{0.2}$ = 134 MPa, δ = 20 %, taking place at the expense of formation of recrystallized structure with a grain size of 3-8 μm. Besides, phase Al_8Fe_2Si and Al_3Zr particles considerably decrease in number with decrease in their size to 50 and 20-30 nm, respectively.

Alloy 1441 *irradiation* to relatively low Ar^+ ion fluences (E = 20-40 keV, j = 100-400 $\mu A/cm^2$, D = 10^{15}-10^{16} cm^{-2}, with the respective irradiation times τ = 1-10 s and temperature of sample heating with a beam T < 40-130°C) causes sample bulk structural transformation from the initial cell structure to a subgrain, similar to polygonal, structure (Figure 1 b). This leads to up to 6 % increase in relative elongation at constant strength characteristics.

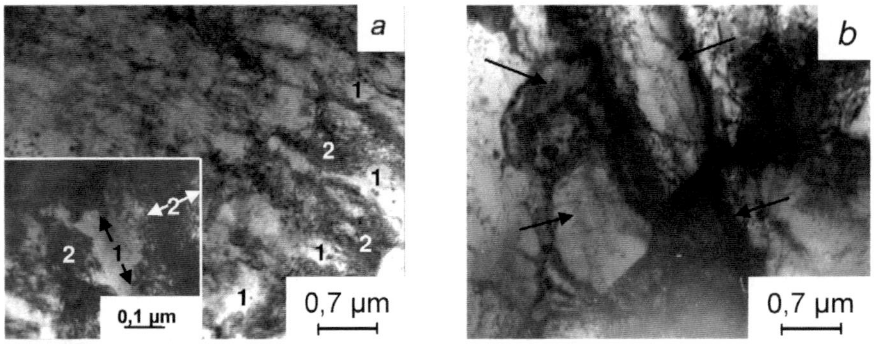

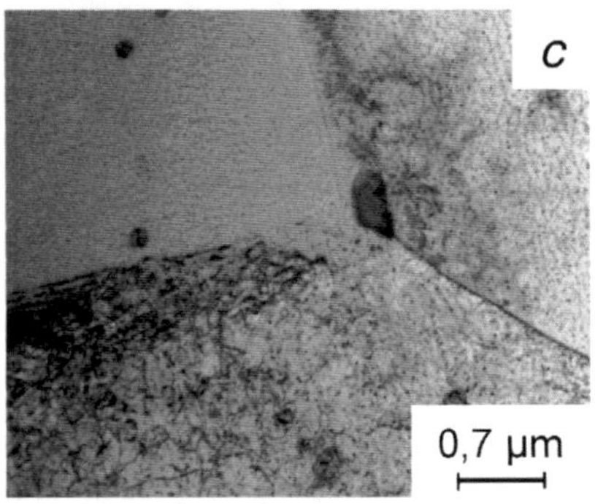

Figure 1. Light-field images of dislocation cell structure (cells central part free from dislocations is viewed as light regions marked 1; cell boundaries presenting dense dislocation tangles are marked 2) of cold-worked alloy 1441 (a); of subgrains (shown with arrows) after irradiation with Ar^+ ions to fluence 10^{15} cm^{-2} (b), and images of recrystallized grains after irradiation with Ar^+ ions to fluence $5.6 \cdot 10^{16}$ cm^{-2} (c); b, c – longitudinal section in the vicinity of *non-irradiated* surface at distance of ~ 150 μm from the latter.

At fluences of ~ (2-5)·10^{16} cm^{-2} (irradiation time τ ~ 7 s, T = 330°C at E = 40 keV, j=400 $\mu A/cm^2$, D = $1.75 \cdot 10^{16}$ cm^{-2} and τ ~ 60 s, T = 260°C at E = 20 keV, j=150 $\mu A/cm^2$, D = $5.6 \cdot 10^{16}$ cm^{-2}), formation of recrystallized grains (Figure 1 c) 12-20 μm in size and fine-disperse equiaxial grains ~ 2-5 μm in diameter takes place. This leads to significant softening of the alloy and, simultaneously, to increase in its ductility: $\sigma_{\text{в}}$ = 218-235 MPa, $\sigma_{0.2}$ = 130-138 MPa, δ = 17-18% (minor differences may occur due to conditions of irradiation). At that, changes in the mechanical properties, also in the dislocation and grain structure, are generally similar to those resulting from alloy 1441 furnace annealing at T = 370-400°C during 2-3 h.

With further fluence increase from $5 \cdot 10^{16}$ cm^{-2} to ~ 10^{17} cm^{-2}, *dissolution* of phase β' (Al_3Zr) as well as *decrease in size* and *partial dissolution* of Al_8Fe_2Si intermetallides take place throughout the bulk of the alloy. It should be noted that particles of β'-phase (Al_3Zr) of crystallization origin are preserved even after a two-hour furnace *annealing* at T~ 370°C, while *irradiation* during only units (or several tens) of seconds (in all cases, at T < 370°C) leads to dissolution of the above particles in the bulk of 1 mm thick samples.

Besides, irradiation to fluences of ~ $5 \cdot 10^{16}$ cm^{-2} and higher leads to formation of disperse particles of a new plate-like phase S_1 (Al_2LiMg) and particles of phase θ' (Al_2Cu).

Increase in fluence up to 10^{17} cm^{-2} following the mentioned *stage of softening*, again, *leads to increase* in strength characteristics: $\sigma_{\text{в}}$ = 315 MPa, $\sigma_{0.2}$ = 204 MPa (E = 20 keV, j = 150 $\mu A/cm^2$), with some drop of δ to ~13% (Figure 2), in spite of the observed noticeable growth of grain average size up to 30 μm (Figure 3) in the bulk of the irradiated samples. A similar change in properties is also observed at other ion beam parameters: E = 40 keV, j = 200 and 400 $\mu A/cm^2$.

The nonmonotonic nature of mechanical properties change in alloy 1441 under increased fluences (see Figure 2) is explained by simultaneous occurrence under irradiation of *two competing processes: recrystallization with formation of new grains*, on the one hand, and *solid solution decomposition* with formation of fine-disperse particles of strengthening phases θ' (Al_2Cu) and S_1 (Al_2LiMg), on the other.

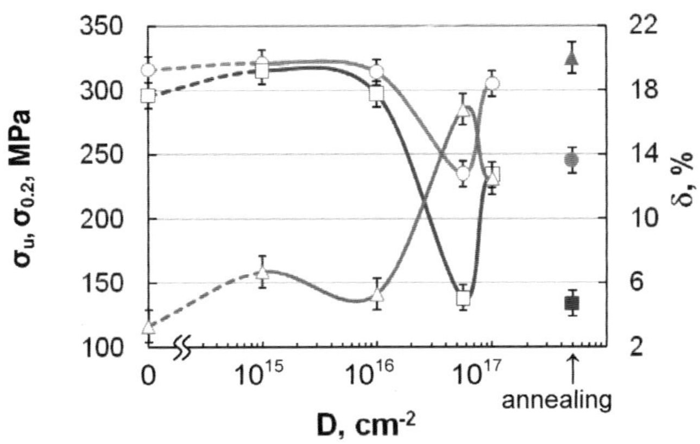

Figure 2. Dependence of alloy 1441 mechanical properties on Ar^+ ions fluence, E = 20 keV, j = 150 μA/cm^2.

The said processes take place throughout the bulk of 1 mm thick samples (Figure 3) in the course of their short-term heating with an ion beam to temperatures close to the temperatures of common furnace annealing (with no long-term exposure to such temperatures), or at lower temperatures, e.g., not higher than 260°C, in case of irradiation to fluences of $5.6 \cdot 10^{16}$ cm^{-2} at j = 150 μA/cm^2, which is 110°C lower than at intermediate furnace annealing.

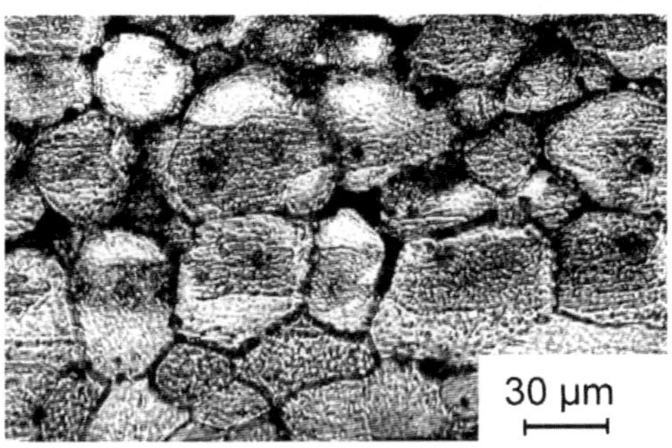

Figure 3. Microstructure of alloy 1441 irradiated with Ar^+ ions, E = 20 keV, j = 150 μA/cm^2, D = 10^{17} cm^{-2}, T < 370°C, in sample cross-section (near the *non-irradiated* side of the sample; the structure of regions in the vicinity of the irradiated side and in the sample central part is similar to that shown).

Short-term (8 s) heating of alloy 1441 samples to 130°C in furnace at a rate similar to that observed under ion irradiation to a fluence of 10^{16} cm^{-2} (E = 30 keV, j = 200 μA/cm^2, τ ~ 8 s, T = 130°C) exerts no influence on the structural state of cold-worked alloys. Samples preserve their cell structure. This positively points to a non-thermal nature of structural transformation in the bulk of alloy 1441 under low fluences (10^{15}-10^{16} cm^{-2}. It should be noted that at higher fluences (and, accordingly, at a substantial heating of the samples with ion beam exposure), thermally-activated processes get superimposed on the radiation-induced one.

As noted in the Introduction, in *metastable media* with higher stored energy, as is the case with cold-worked alloys, the conditions of undamped propagation of post-cascade waves may be realized, with such waves initiating structural and phase transformations at their front. This relates to the processes of single atoms rearrangement, dislocations unlocking, cooperative atomic restructuring in solids etc. It was shown [9], that practically in all metastable media, even in those with minimum stored energy ΔF of the order of hundredth, or even thousandth, fractions of eV (on the average, per atom of a medium), solitary waves of sufficiently high amplitude ε (ε ≥ Δf; where Δf is of any process energy barrier; normally, Δf < 0.2-0.3 eV) may propagate without damping, causing such media rearrangement.

Energy density in a *self-propagating* (undamped) solitary wave in steady-state conditions is such that the rate of this energy dissipation in a medium is exactly equal to the rate of energy release at the expense of structural rearrangements at its front.

4. Conclusions

Radiation annealing presents a new approach to recovery of ductility of cold-worked alloys. It takes place at lower temperatures and with multiple acceleration of the rate of the process. The energy requirement in treatment is 2-3 times lower than that in furnace annealing [20]. There exist acceptable engineering solutions to the problems related to necessity of implementing the process in vacuum. It was shown [21] that radiation annealing may run in the course of the sheets and sections moving process (relative to the ion source). The difference in structural changes, mostly at final stages of *thermal* and *radiation* annealing, provides additional opportunities for control of structure and properties of the metal.

Acknowledgments

The research was carried out within the state assignment of FASO of Russia (No. 0389-2014-0002), supported in part by RFBR (project No. 14-08-01049-A) and it was supported by the Act 211 of the Government of the Russian Federation (Agreement No. 02.A03.21.0006).

References

[1] Ohta J, Ohmura T, Kako K et al 1995 *J. Nucl. Mater.* **225** 187
[2] Amirkhanov I V, Didyk A Yu, Zemlyanaya E V et al 2006 *Letters to magazine "Fiziks elementarnykh chastits i atomnogo yadra"* **3** 63
[3] Anishchin V M and Uglov V V *Modifikatsiya instrumentalnykh materialov ionnymi i plazmennymi puchkami* (Minsk: BGU, 2003)
[4] Didenko A N, Ligachev A E and Kurakin I B *Vozdeistvie puchkov zaryazgennykh chastits td poverkhnost metallov i splavov* (M.: Energoatomizdat, 1987)
[5] Boiko V I, Valyaev A N and Pogrebnyak A D 1999 *Physics-Uspekhi* **42** 1139
[6] Ovchinnikov V V, Remnev G E, Gushchina N V et al 2011 *Russian Journal of Non-Ferrous Metals* **52** 304
[7] Sharkeev Yu P and Kozlov E V 2002 *Surface and Coatings Technology* **158-159** 219
[8] Didenko A N, Sharkeev Yu P, Kozlov E V and Ryabchikov A.I. *Effekty dalnodeistviya v ionno-implantirovannykh metallicheskikh materialakh* (Tomsk: Izd-vo NTL, 2004)
[9] Ovchinnikov V V 2008 *Physics-Uspekhi* **51** 955
[10] Goloborodsky B Yu, Ovchinnikov V V and Semenkin V A 2001 *Fusion Technology* **39** 1217
[11] Dremov V V, Sapozhnikov A F, Samarin S I et al 2007 *Journal of Alloys and Compounds* **444-445** 197
[12] Ovchinnikov V V, Makhinko F F, Solomonov V I, Gushchina N V and Kaigorodova O A 2012 *Technical Physics Letters* **38** 42
[13] Zhukov V P and Boldin A A 1987 *Atomnaya energiya* **63** 375
[14] Zhukov V P and Ryabenko A 1984 *Radiation Effects* **82** 85
[15] Zhukov V P and Demidov A V 1985 *Atomnaya energiya* **59** 29
[16] Ovchinnikov V V, Chernoborodov V I and Ignatenko Yu G 1995 *Nucl.Instrum. and Meth. in Phys.Res. B* **103** 313
[17] Ovchinnikov V V., Gushchina N V., Makhinko F F et al 2008 *The Physics of Metals and Metallography* **105** 375
[18] Gavrilov N V, Mesyats G A, Nikulin S P et al 1996 *J. Vac. Sci. Technol. A* **14** 1050
[19] Biersack J P and Haggmark L.G. 1980 *Nucl. Instrum. Methods Phys. Res. B* **174** 257
[20] Sposob polucheniya listovogo prokata is aluminiyevykh splavov (A method to produce flat rolled stock from aluminum alloys): Patent of RF No. 2363755 / Ovchinnikov V V, Gavrilov N V, Gushchia N V, Shkolnikov A R, Mozharovsky S M and Filippov A V; publ. 10.08.2009. Bull. No. 22; priority date 08.12.2006.
[21] Ovchinnikov V V, Gavrilov N V, Gushchina N V et al 2010 *Russian Metallurgy (Metally)* **3** 207

Nanowires of iron group metals: fabrication by matrix synthesis technique and investigation of structure and magnetic properties

D L Zagorskiy[1,5], S A Bedin[1,2], K V Frolov[1], V V Korotkov[3], V V Artemov[1], S N Sulyanov[1], S S Kruglikov[3], M A Chuev[4] and I N Mishenko[4]
[1]Centre of crystallography and photonics RAS, Moscow, Leninskii pr. 59.
[2]Moscow pedagogical state university, Moscow, Malaya Pirogovskaya st 1/1
[3]Mendeleev University of Chemical Technology, Moscow, Miusskaya sq. 9
[4]Institute of Physics and Technology RAS, Moscow, Nahimovskiy prosp., 36/1
[5]Gubkin Russian State University of Oil and Gas, Moscow, Leninsky prosp.,65/1

E-mail: dzagorskiy@gmail.com

Abstract. Nanowires (NW) of iron group metal alloys (Fe-Ni, Fe-Co) were obtained using matrix synthesis technique based on polymer track matrixes. The galvanic process was investigated and it was found that it consists of different stages. Deposition of metal inside the pores has non-linear character due to diffusion limitation. The specific features of the next part (formation and growing of the "caps') was also studied. Electron microscopy, X-rays analysis, Mössbauer spectroscopy and magnetic hysteresis were applied to investigate the dependence of structure and magnetic properties of the NW on electrodeposition conditions. It was found that the composition of two-component NWs differs from the composition of electrolyte and different at different parts of NW. Mössbauer spectroscopy gave possibility to estimate hyperfine parameters for Fe-Co NWs. For Fe-Ni NWs it was supposed that the spectra could be presented as superposition of at least three magnetic sextets with hyperfine parameters B_{hf} 27-33 T. It was shown that Fe-Co samples have "hard magnetic" properties, while Fe-Ni samples have "soft magnetic" parameters. The dependence of these parameters on the synthesis was demonstrated.

1. Introduction

Matrix synthesis. The obtaining and application of different types of nanomaterials are the main trends at the last time. The method of matrix synthesis are one of the most perspective among other techniques of nanomaterial fabrication [1]. The main idea of method is to fill the pores of specially prepared matrix by any desired material. The obtained structures are replicas of the pores [2]. Different mediums could be used as the matrixes-porous alumina, different zeolites, porous silica and polymer track membranes. Different materials could be used as a "filler"- polymers, dielectrics, some sorts of crystals and metals.

In this work polymer matrixes were used as the templates and their pores were loaded (electrodeposited) by metals in order to obtain so-called nanowires (NWs).

Track membranes. The process of fabrication of track membranes (TM) is well known- irradiation of thin polymer film with swift heavy ions followed by chemical etching of obtained latent tracks. The main parameters of TM are porous density (per sq. cm of the surface) and the diameter of the pores. These parameters could be specified at the stages of irradiation and etching. The main application of such membranes is fine filtration in medicine, biology, food production and so on [3]. Another application is matrix synthesis, mentioned above.

Filling of the pores. In this work electrochemical (galvanic) deposition of metal into the pores of TMs were used. For this electrolytes- the salt(s) of corresponding metal(s) with some additives – were used. By varying of the deposition parameters different types of the NWs could be obtained.

Additional possibility of changing the properties could be obtained by using multicomponent NWs- due to variation of composition. In this case the correlation between composition of electrolyte and deposition regimes with the properties of obtained materials should be investigated. Such processes and relationships were investigated for bulk materials and for thin films- for these cases some approaches for compositions of electrolytes and galvanic regimes were developed [4]. At the same time, the features of electrodeposition into the narrow pores were not studied for multicomponent compositions [5,6].

So, this work is devoted to features of obtaining of two-component NWs and to investigation of some properties of obtained materials.

2. Experiment

Matrixes and obtaining of NWs. Commercial TM (made in JINR, Dubna: PET (Polyetheleneterephthalate) film with thickness 10-15 μm and pores with surface density 10^8-10^9) were used in first experiments. For other experiments PET matrixes (parallel pores and lower density- 10^6-10^8) were specially prepared using U-400 (JINR, Dubna).

For deposition of bi-component NWs the following electrolytes were used:

1) iron-nickel NWs: $NiSO_4{\cdot}7H_2O$ - 16 g/l, $NiCl_2{\cdot}6H_2O$ - 40 g/l, $FeSO_4{\cdot}7H_2O$ - 16 g/l, ascorbic acid – 1-2 g/l, sodium lauryl sulfate – 1 g/l.

2) iron-cobalt NWs: $FeSO_4{\cdot}7H_2O$ – 150 g/l, $CoCl_2{\cdot}6H_2O$ – 43 g/l, $AlCl_3{\cdot}6H_2O$ – 40 g/l, ascorbic acid – 2 g/l, pH = 1.2.

Note, that the main problem was to found and maintain the optimal pH. Ascorbic acid was used mainly to prevent oxidation of Fe^{2+} to Fe^{3+}, while sodium lauryl sulfate stimulate penetration of electrolyte into narrow pores.

Electrodeposition was carried out in home-made galvanic cell. Potentiostat –galvanostat Elins P-30S was used as a current source for two-electrode connection. Electrodeposition was carried out using potentiostatic regime (constant voltage). The dependence of current of time was measured during the process, giving information about the stages of electrodeposition.

After formation of metal NWs they were removed from the host matrix by dissolving of polymer in concentrated alkali. If MWs remain in matrix so-called "metal-polymer composite" is formed.

Electron microscopy investigation of obtained NWs was carried out using standard techniques: SEM TESLA BS-340, FEI QUANTA 200 and FEI Scios working in secondary electron regimes were used. The elemental composition of some samples was estimated using EDAX analyzer (FEI QUANTA 200), while transmitted microscopy was done at TEM FEI Osiris.

Structure investigations. X-rays measurements of the samples were done at X'PERT PRO – difractometer (PANalytical, CuKα (λ = 1.5418 Å, with Ni filter) using standard methods.

Mössbauer spectroscopy was done at MS-1104Em spectrometer (with 57Co (Rh) source) at room temperature using standard technique. Isomer shifts were measured relative to metallic α-Fe foil.

Magnetic measurements were carried out for iron-nickel and iron-cobalt alloys at Vibrating Sample Magnetometr LakeShore 7407 at room temperature.

3. Results and discussions

Galvanic deposition of Fe-Co NWs was done at potentiostatic regime: the obtained dependences of current on time (potentiostatic curves) at different voltages are shown in Fig.1 for different pores (100 nm and 300 nm).

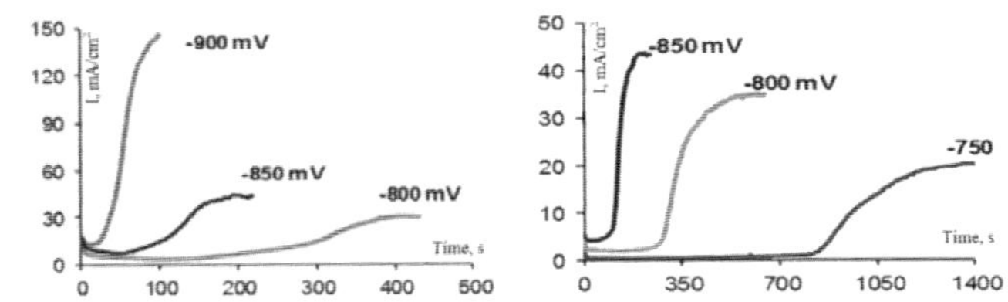

Figure 1. The dependence of current on time for deposition of Fe-Co alloy at different voltages. Temperature 200 C, pores diameter 100 nm (left) and 300 nm (right).

The grooving process could be estimated using these results: the speed of this process is rather fast and varied from 2 μm/min to 25 μm/min.

Galvanic deposition of Fe-Ni NWs was also done at potentiostatic regime and obtained results are shown in Fig.2.

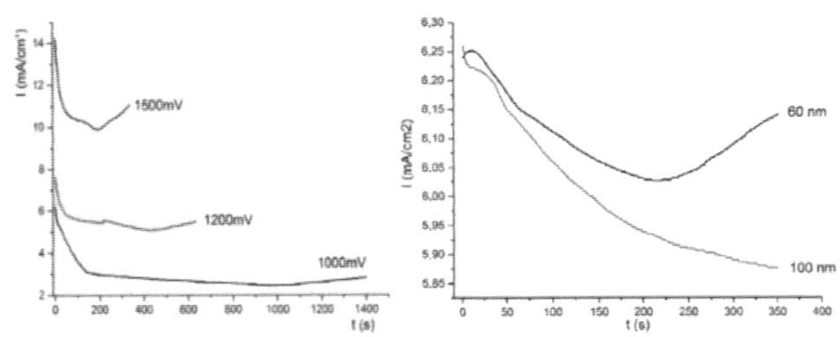

Figure 2. The dependence of current on time for deposition of Fe-Ni alloy at different conditiones. Pores diameter 100 nm, voltage 1 V, 1.2 V, 1.5 V (left) ; pores diameter 60 nm and 100 nm, voltage 1.5V (right). (Temperature 200 C).

It is found that the pores of small diameter are filled faster and increase of the voltage accelerate all the processes. It is easy to see that in all cases the process consists of different parts.

First part – initial reduction of current is caused by changing the concentration of electrolyte inside the pores. Second part (main part) – gradual growth of metal wires inside the pores. This process has non-linear character, with a minimum: the decreasing of current is found to be turned to increase; this effect is most pronounced for nano-size pores. It could be explained by changing of concentration due to diffusion limitation inside the narrow pore channels. Another explanation is inconstancy of the pores diameter: it was found that the channels are narrower in the middle part, and this is most apparent for thin pores. The third part of the curves – fast rise of current is related with the formation of metal deposit on the external surface outside the pores (formation of so-called "caps"). An evaluation of cathode current density has shown that in spite of a steep rise in the current in the third part of the chronoamperogram, the magnitude of the cathode current density is several times lower than that in the first and second of parts of the curves. The reduction of current density is caused by a sharp increase in the voltage drop across the cathode layer (≈ 1 mm thick).

We found that current increase (at the last stage) is not proportional to the increasing cathode surface area: while the surface area becomes about 20 times larger after the filling of pores with metal, the current value increases only by 3-5 times. This effect depends on electrode potential and doesn't depend on pores diameter. Possible explanation is the shift of the cathode potential in the positive direction due to increased value of ohmic voltage drop in the solution in the vicinity of the cathode. Electrical resistance of the solution should be taken into consideration to explain this phenomenon. So, metal deposition inside the pores proceeds at more negative potentials and, therefore, at higher rates rather than one on a plane surface which takes place at more positive potentials, i.e., at lower rates.

It is easy to see that growing of "the caps" outside the pores then leads to coalescence of these caps with formation of continuous surface.

It could be concluded that grooving of two-component NWs have the same character as grooving of pure metals. At the same time, the composition of NWs in alloy is different from the composition in electrolyte. It could be supposed that changing of growing condition during the deposition could lead to different composition of metals in different parts of NWs.

Electron microscopy study of the ensemble of free-standing Fe-Ni NWs is presented in Fig.3.

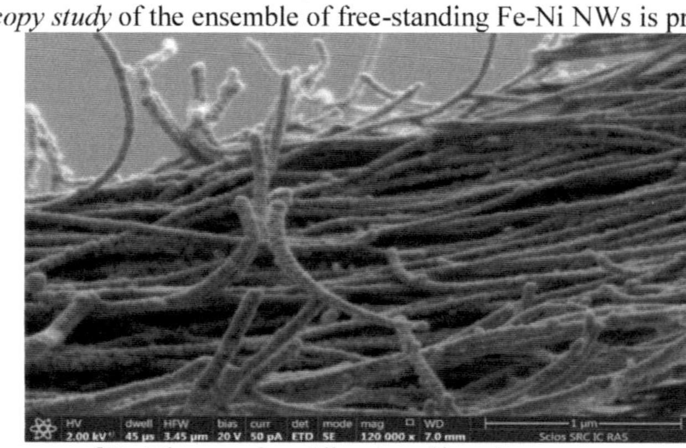

Figure 3. SEM images of Fe-Ni nanowires (diameter 60 nm).

Diameter (60 nm) and parallel orientation of NWs could be estimated. Using TEM it was found that NWs have crystal structure with small grains. Elemental analysis demonstrated that the main elements (Fe –approximately 55% and Ni- approximately 43%) distributed inside the bulk ofNW, while oxygen was found at the surfaces of NWs (obviously due to formation of oxides). It was also found that metals concentration slightly changed along the NW`s length. Fig 4. demonstrated SEM images of Fe-Co alloys NWs together with the results of elemental analysis.

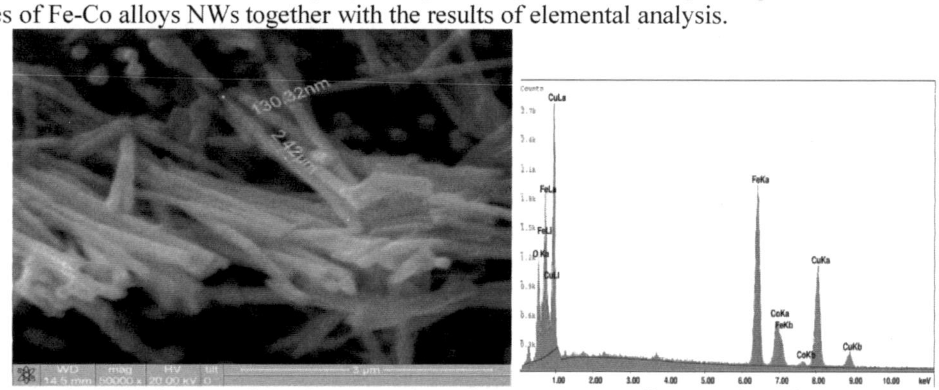

Figure 4. SEM image of Fe-Co nanowires (left), corresponding results of elemental analysis (right).

The obtained results (for both types of samples) demonstrated that NW`s diameters is higher than the diameter of matrix pores – it could be explained by compression of polymer during the electrical deposition of metal. It was also found that the composition of two-component NWs differs from the composition of electrolyte and different at different parts of NW. This effect is more pronounced for higher grooving voltage and for pores with small diameter. It could be explained by difference in deposition potentials of two metals and by diffusion limitation.

Structure investigations. X-rays measurements were done for Fe-Co samples. It was found that NWs have polycrystalline structure and cell parameters close to the parameters of pure Iron. The lines of Cobalt were not found. It could be explained by formation of solid solution on the base of Iron, where cobalt is interstitial impurity and slightly effects on the cell parameter. In a first approximation the dependence on growing voltage was not found. It was also found that the cell parameter slightly depends on pore diameter, but its average value (d = 2,854) is lower than the parameter of pure bcc Iron (d = 2,8664). It could be explained both by formation of solid solution and by compression during the alloy growth inside the pores of matrix.

Mössbauer spectra. Our first results obtained for NWs from pure Fe were published in [7]. In this work two groups of Fe-Co NWs samples were investigated: - the first one- three samples grown at pH=1.2 (constant pores diameter -100 nm and different deposition potentials -0.7 V, -0.75 V and -0.8 V); the second - four samples on Fe-Co NWs (constant deposition potential -0.75 V and different pores diameter 30 nm, 50 nm, 100 nm and 200 nm). Mössbauer spectra of all the samples have typical 6-peaks shape with different intensity ratio- the last one was used for estimation of hyperfine parameters B_{hf} at Fe^{57} nuclei.

For the first group samples the dependence of spectral parameters on deposition voltage was studied. It was shown that hyperfine parameters actually doesn`t change up to potential -0.75 V, but for higher potential (-0.8 V) parameters B_{hf} for all lines drop down at 0.5 T.

It should be noted that relative intensity of second and fifth lines are different for different samples, indicating the spontaneous magnetization of Fe. It is known that this relative intensity determined by equation $I_{2,5} = 4\sin^2\vartheta/(1 + \cos^2\vartheta)$, where ϑ is the angle between the magnetic moment M and the direction of gamma-quantum [8]. So, this ratio gave us possibility to determine the space orientation of M in NWs – the highest angle was found to be approx. 40°.

For second group of samples the dependence of spectral parameters of pores diameter was investigated –it was shown in Fig.5 (left).

For NWs with diameters 30 and 50 nm hyperfine parameters B_{hf} slightly differs from the same for other samples. Calculation shows that the angle ϑ (and, therefore, the sample magnetization) increased with the increase of NWs diameter. Fig.5 (right) demonstrated the Mössbauer absorption spectra for Fe-Ni samples with diameter 30 nm.

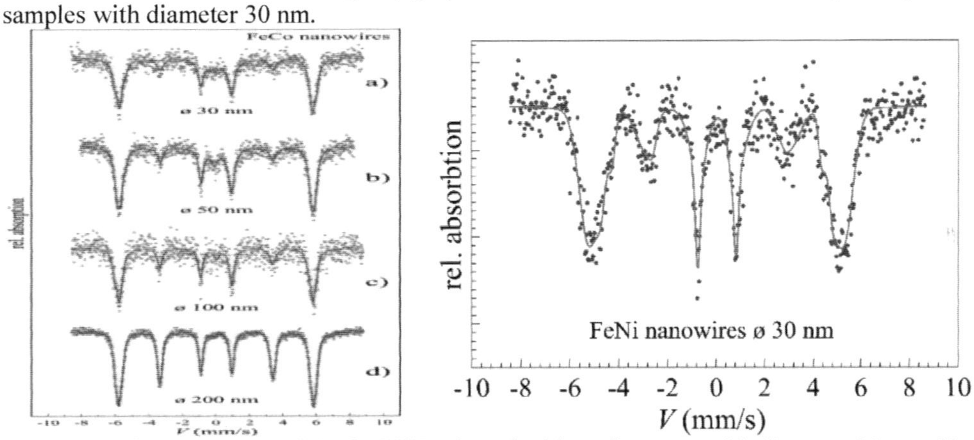

Figure 5. Mössbauer spectra of Fe-Co NWs, deposited into the pores with diameter 30 nm, 50 nm, 100 nm and 200 nm (a,b,c and d, resp.) at the deposition voltage -0.75 V (at the left side). Mössbauer spectra of Fe-Ni NWs (at the right side).

The typical magnetic splitting and changing of relative intensity of the lines, which indicated spontaneous magnetization of iron in Fe-Ni samples. It was found that the spectra could be presented as superposition of at least three magnetic sextets with hyperfine parameters B_{hf} 27-33 T. Supposedly they correspond to bcc and f_{cc} phases of alloy Fe-Ni with different local positions of Ni near Fe.

Finally, it should be mentioned that Mössbauer experiments results are in good agreement with the X-rays diffraction data and results of microanalysis.

Magnetic measurements were carried out for both Fe-Co and Fe-Ni NWs using vibrating magnetometer at room temperature. The obtained data are presented in Fig.6.

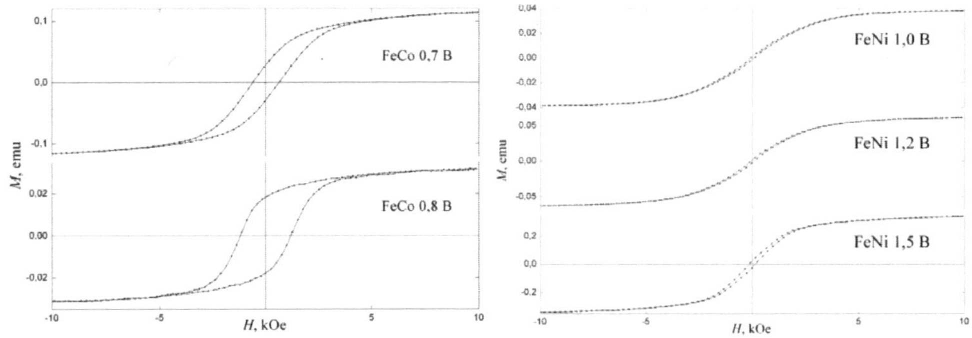

Figure 6. The hysteresis loops. At the left side - Fe-Co (growing voltage 0,7V and 0,8V); At the right side - Fe-Ni- (growing voltage 1, 1,2 and 1,5 V).

It could be concluded that all samples are ferromagnetic. Fe-Co samples have "hard magnetic" properties: coercetive force 630 Oe (and r. magnetization 27 memu) and 1100 Oe (r. magnetization 18 memu) - for grooving voltages 0.7 and 0.8 V (respectively). (For sample with grooving voltage 0.85V (not shown) the coercetive force is even higher -1200 Oe (r. magn. - 12 memu)).

At the same time Fe-Ni samples have "soft magnetic" parameters. Coercetive force was found to be 75 Oe (r.magn. 1 memu), 80 Oe (r.magn. 1.3 memu), and 140 Oe (r.magn. 2.2 memu), for growing voltage 1, 1.2 and 1.5V (respectively).

In all cases these parameters depend on growing conditions- it could be explained that increasing of voltage leads to acceleration of the process and therefore – to structure with small-grains.

4. Conclusions

It was found that the process of electrodeposition of two-component NWs has the same character as deposition of pure metals. For example, it has non-linear character, due to diffusion limitation inside the narrow pores. The composition of two-component NWs differs from the composition of electrolyte and different at different parts of NW.

Mössbauer spectroscopy demonstrated that hyperfine parameters (for Fe-Co NWs) actually doesn`t change up to growing potential -0.75 V, but for higher potential (-0.8 V) parameters B_{hf} for all lines drop down at 0.5 T. For Fe-Ni NWs the spectra could be presented as superposition of at least three magnetic sextets with hyperfine parameters B_{hf} 27-33 T.

It was shown that Fe-Co samples have "hard magnetic" properties, while Fe-Ni samples have "soft magnetic" parameters. Moreover, for all samples magnetic parameters depend on growing voltage because of increasing of voltage leads to acceleration of the process and to the formation of small-grains structure.

Acknowledgments

This work was supported by Grant RFBR 15-08-04949. We are also thankful to Prof. P. Apel (JINR, Dubna) for Polymer samples. The equipment of the Collective Centre of the Centre of crystallography and photonics of RAS was used.

References

[1] Martin C R 1994 *Science* **266** 1961
[2] Brock T Membrane filtration. Science Tech., Inc. Madison, 1983 462 p.
[3] Lupu N 2010 Electrodeposited Nanowires and Their Applications, InTech.
[4] Vinokurov E Phys. 2010 *Chem.Surf.&Mat. Protection* **46** 5 543 (in Russian)
[5] Korotkov V V et al. 2011 *Electroplating and Surface treatment* **XIX** 4 23-28
[6] Korotkov V V et al. 2015 *Electroplating and surface treatment* **XXIII** 1 24-33
[7] Frolov K et al. 2014 *JETP Letters* **99** 10 570–576
[8] R S Preston et al. 1962 *Phys. Rev.* **128** 2207

Research of niobium thin films with a predetermined thickness produced by RF magnetron sputtering

D A Polonyankin[1], A I Blesman[2], D V Postnikov[1], A I Logacheva[3], I A Logachev[4], A A Teplouhov[1] and A A Fedorov[5]
[1]Associate Professor, *Physics Department*,
Omsk State Technical University (OmSTU), 11 Mira Avenue, Omsk, 644050, Russia
[2]Head of the Physics Department, *Physics Department*,
Omsk State Technical University (OmSTU), 11 Mira Avenue, Omsk, 644050, Russia
[3]Deputy Director, *New Metallurgical technologies institute,*
Joint Stock Company Kompozit, 4 Pionerskaya St., Korolev, 141070, Russia
[4]Head of sector, *New Metallurgical technologies institute,*
Joint Stock Company Kompozit, 4 Pionerskaya St., Korolev, 141070, Russia
[5]Associate Professor, *Department of Mechanical engineering technology*,
Omsk State Technical University (OmSTU), 11 Mira Avenue, Omsk, 644050, Russia

E–mail: nano@omgtu.ru

Abstract. Niobium and niobium thin films are widely used in various fields of modern science and technology: in the electronics industry, in a nuclear medical imaging technique, in the information technology, in superconducting cavities technology etc. The grain size of thin niobium films depends on its thickness and the film's stoichiometry can be varied as a function of thickness. Thus the problem of thickness control has a great practical importance in all fields of niobium films application. The focus of this study was to perform an experimental calibration of STC–2000A deposition controller for niobium target on ADVAVAC VSM–200 setup and to conduct a grain size, roughness and stoichiometry research by scanning electron microscopy, X–ray diffraction and laser interference microscopy of niobium films produced by RF magnetron sputtering with the thickness range from 200 nm to 400 nm and 50 nm step.

1. Introduction

Niobium thin films produced by RF magnetron sputtering are widely used in various fields of modern science and technology. Niobium films have important potential applications in the electronics industry due to the threshold current reduction in resistive switching devices (memristor) which are promising emerging nanodevices for future information technology [1]. Nb films and Nb–Nb_2O_5 multilayer coatings may be used in a nuclear medical imaging technique (positron emission tomography, PET) because of the following features: it may prevent the aqueous corrosion of the chamber's wall as diffusion barrier at routine production of [^{18}F] (radiotracer in scanning technique), and it can withstand overpressure with value of 4 MPa and resist the proton beam energy absorption in the entrance foils [2]. Nb also is very available material for a bipolar plate in proton exchange membrane fuel cell (PEMFC) because it has the best corrosion resistance among transition metals in sulfuric solution [3].

Nb is a highly interesting transition metal that exhibits superconducting properties [4] and as a refractory superconductor is still considered as a perspective material for fabrication of robust and reliable Josephson devices for a number of key applications, such as high–speed superconducting digital circuits (RSFQ), mm–wave receivers and sub–millimeter wave mixers, and as voltage standard [5]. Niobium films are also applied for superconducting radiofrequency (SCRF) devices, such as superconducting cavity–stabilized oscillators (SCSO) [6], superconducting resonant cavities for particle accelerators [7], step–edge superconducting junctions with films thickness ranges from 25 to 200 nm [8] and superconducting quantum interference devices (SQUIDs) with nanobridge junctions [9]. Superconducting properties of niobium are also used in processors of quantum computers (superconducting qubit) [10]. Photocathodes with high quantum efficiency and electrical compatibility with superconducting cavities technology usually made of niobium [11].

Critical temperature of superconducting Nb transition is usually lower for thin films than in bulk sample and depends on various parameters such as thickness [12], stoichiometry [1] and grain size [13]. At the same time the grain size of thin niobium films depends on its thickness [2] and film's stoichiometry can be varied as a function of thickness [1].

Thus the problem of thickness control has a great practical importance in all fields of niobium films application. Magnetron sputtering seems to be the most suitable method to produce Nb films (i.e. a nanostructured material with optimal physical properties) [14] that cause the RF magnetron sputtering selection as an experimental method for thin niobium films deposition.

The focus of this study was to perform an experimental calibration of STC–2000A deposition controller for niobium target on ADVAVAC VSM–200 setup and to conduct a research of thin niobium films by scanning electron microscopy, X–ray diffraction and laser interference microscopy.

2. Calibration of STC–2000A deposition controller for niobium target

There are two experimental methods of thickness measurement: the first is online thickness control during the sputtering process by using deposition controller of magnetron system, and the second is the film's thickness evaluation by approved methods such as ellipsometry, profilometry, X–ray diffraction analysis etc. If necessary, these methods can be combined for data verification.

Calibration of STC–2000A deposition controller for niobium target was performed using three different material parameters: density, Z–Factor and tooling. Density and Z–Factor are material factors of niobium that take the values of 8.576 gm/cc and 0.493 correspondently. Tooling is a deposition system geometry correction (location of sensor to substrates, see figure 1). Initially the tooling was set to 100%.

Niobium deposition was done in RF mode at source power of 100 W under $3.4 \cdot 10^{-3}$ mbar argon pressure and pre ionic chamber cleaning for 10 minutes. Then tolling (*T*) was calculated by expression (1) after niobium film sputtering with displayed thickness of 5000 angstroms for relative position of sensor and soda–lime silicate glass ($Na_2O \cdot CaO \cdot 6SiO_2$) substrate which was sonicated in isopropyl alcohol prior to film deposition.

$$T = \frac{Measured\ thickness}{Displayed\ thikness} \times 100\ \% \qquad (1)$$

Niobium thin film thickness of 216 nm was obtained on the base of "film–substrate" line scanning profile by energy dispersive spectroscopy (EDS) performed with JED–2300 analyzer of JEOL JCM–5700 (see figure 2), thus calculated tooling had a value of 0.43%. As can be seen from figure 2, the substrate includes oxygen (shown in red), sodium (light green), silicon (turquoise) and calcium (black) that corresponds to composition of the soda–lime silicate glass. Niobium, which is film's element, is marked in dark blue.

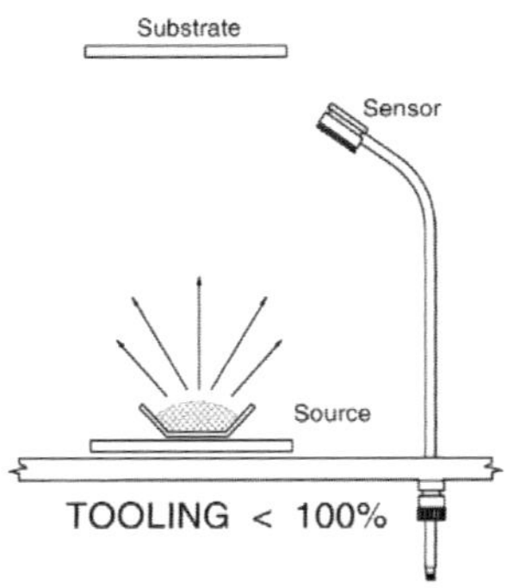

Figure 1. The relative position of the sensor and the substrate.

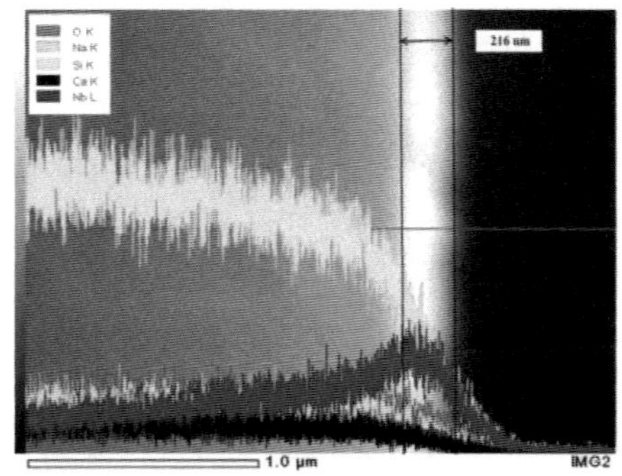

Figure 2. Line scanning profile of niobium thin film on soda–lime silicate glass substrate.

3. Experimental details

Nb films with the thickness range from 200 nm to 400 nm and 50 nm step were prepared by using the ADVAVAC VSM–200 setup at high vacuum RF mode with source power of 100 W under $3.4 \cdot 10^{-3}$ mbar argon pressure and pre ionic chamber cleaning for 10 minutes. Thickness range selection is caused by the reason that such thicknesses are far from quantum size effect region [15] and bulk properties are expected. Thin films were sputtered with pure argon (Ar) gas by using Nb target (99.95% in purity, 50 mm in diameter).

The XRD investigation of niobium films with thickness in the range from 200 nm to 400 nm and 50 nm step was carried out by a sliding X–ray beam method (fixed tube angle θ=5°) in Bragg–Brentano configuration with 2θ from 30° to 80° and Cu Kα radiation (λ=1.54 Å) using XRD–7000 diffractometer (Shimadzu). Shimadzu software allowed obtaining the peaks position and the according full–width at half–maximum (FWHM) for all 2θ angles.

Qualitative and quantitative analysis of thin niobium films was performed by scanning electron microscopy (SEM) with energy dispersive spectroscopy (EDS) and allowed to determine the composition of the surface layers.

The analysis of surface roughness has been provided by laser interference microscopy (LIM) performing root mean square (RMS) values in dependence of the film's thickness. Laser interference microscope MIM–340 (Shvabe) is designed for non–contact non–destructive investigations of submicron structures with nanometer level resolution in the field of materials science.

4. Results and discussion

The results of XRD and SEM with EDS investigations of thin niobium films are shown on the figures 3 and 4. The surface composition of 200 nm thickness niobium film is represented in table 1. LIM phase image of 200 nm niobium film's surface with roughness line profile is provided on the figure 5.

As seen from mixed XRD pattern it contains the peaks reflected by niobium crystal lattice and only lines related to planes with [(110), (211)] Miller indices was detected. The intensity corresponding to the grains oriented in the (110) plane surpasses in times similar intensity rates for the (211) plane. Moreover, as the film growth, signals reflected by the (211) plane does not practically change; while the intensity appropriate to the (110) orientation substantially increases. The shape of all peaks indicates a high crystallinity degree of sputtered films. Thus we can conclude that niobium films with the thickness range from 200 nm to 400 nm are polycrystalline with body–centred cubic structure of bulk Nb and crystallographic plane orientation of their grains preferential have (110) texture.

Applying a sliding X–ray beam allowed us to register scattering radiation only from niobium crystallite grains, because due to this method film's thickness is greater than the x–ray penetration depth so that a signal coming from the substrate is not observed.

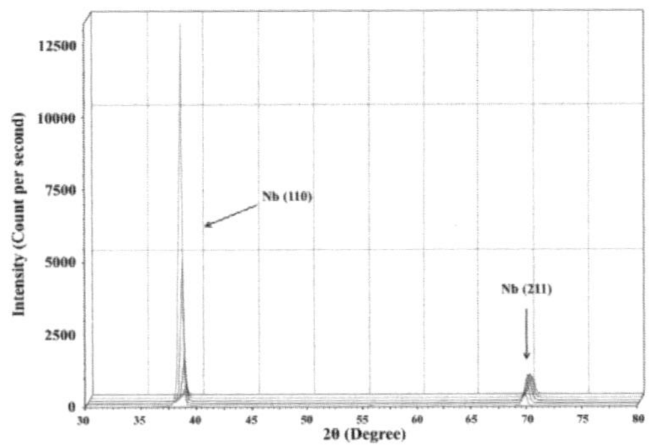

Figure 3. Mixed XRD pattern for niobium films with thickness in the range from 200 nm to 400 nm and 50 nm step.

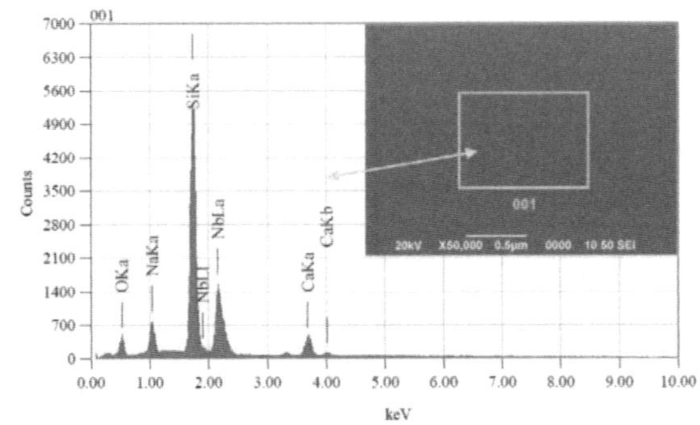

Figure 4. SEM image and EDS results of niobium film with 200 nm thickness on soda–lime silicate glass substrate.

In proof of niobium film's content but not completely coinciding to XRD method SEM results besides information of niobium concentration (19.03 % of atoms) include data about substrate's composition (proportions of Na, Ca, Si and O are presented in table 1). It can be explained by the penetration depth of the electron beam which exceeds even the highest thickness of film (400 nm).

Table 1. The surface layer composition of 200 nm niobium film.

Element	Energy (keV)	Mass %	Error %	Atom %
O	0.525	3.55	0.04	8.76
Na	1.041	4.78	0.04	8.22
Si	1.739	41.14	0.01	58.27
Ca	3.690	5.80	0.07	5.72
Nb	2.166	44.73	0.02	19.03

The roughness RMS values were calculated by using MIM–Visualizer program on the base length of 80 nm after bonding together four surface profiles by MIM–Series software for each film thickness.

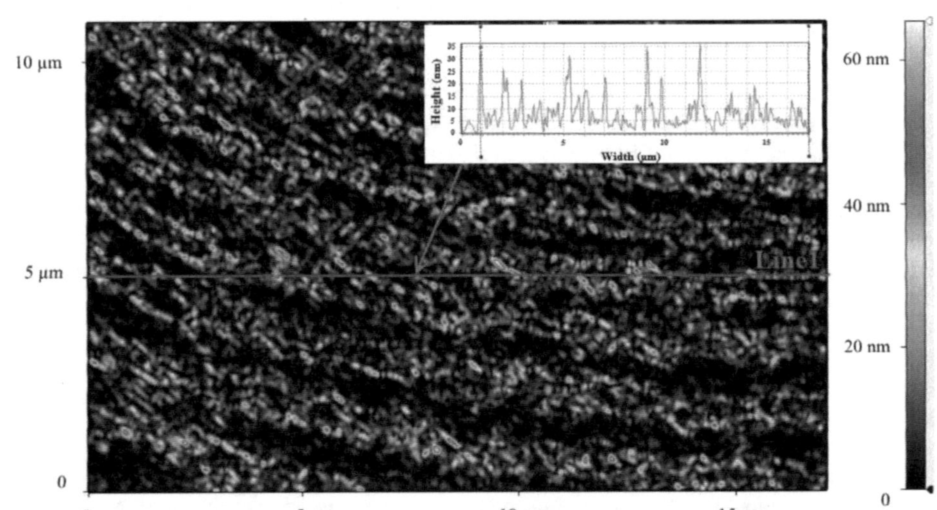

Figure 5. LIM phase image of 200 nm niobium film's surface with roughness profile along Line 1.

The average roughness of niobium films ranges from 4.49 nm (for 200 nm film thickness) to 6.86 nm (for 350 nm film thickness) in dependence of the film's thickness (see figure 6). All films produced by magnetron sputtering refer to highest (14) roughness level.

On the basis of XRD data according to the Debye–Scherrer formula the average grain size was calculated in dependence of the film's thickness (see figure 6). The average grain size increases with the growth of film thickness and ranges from the minimum 16 nm size (for 200 nm film thickness) to maximum value of 20 nm (for 350 nm film thickness). The roughness and the average grain size dependences from the film thickness can be explained by island mechanism [16] when adhesion force and energy to the substrate less than the mutual binding energy of the deposited atoms [17] and films properties with the thickness growth become close to the bulk material.

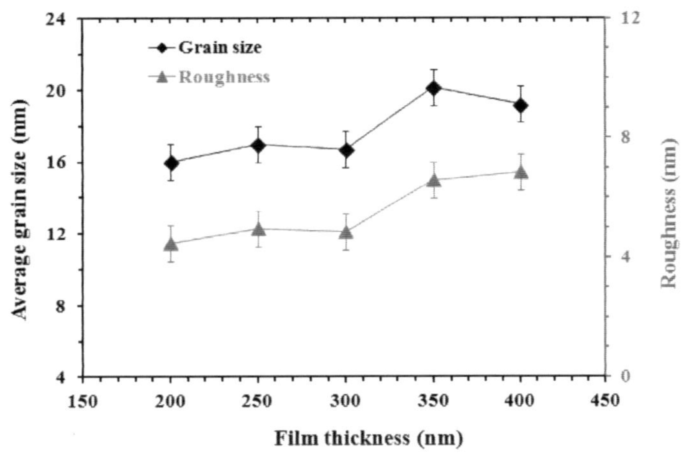

Figure 6. The average grain size and the roughness dependences from the thickness of niobium films.

5. Conclusion

Calibration of STC–2000A deposition controller for niobium target on ADVAVAC VSM–200 setup was performed using density (8.576 gm/cc), Z–Factor (0.493) and experimentally calculated tooling value (0.43%). Niobium films produced by RF magnetron sputtering with the thickness range from 200 nm to 400 nm are polycrystalline with body–centred cubic structure of bulk Nb and crystallographic plane orientation of their grains preferential have (110) texture. The average roughness of niobium films ranges from 4.49 nm (for 200 nm film thickness) to 6.86 nm (for 350 nm film thickness). The average grain size increases with the growth of film thickness and ranges from the minimum 16 nm size (for 200 nm film thickness) to maximum value of 20 nm (for 350 nm film thickness) which corresponds to the island mechanism of the films growth.

References

[1] Ganesan R, Treverrow B, Murdoch B, Xie D, Ross A E, Partridge J G, Falconer I S, McCulloch D G, McKenzie D R and Bilek M M M 2016 *J. Phys. D: Appl. Phys.* **49** 245201

[2] Skliarova H, Azzolini O, Cherenkova–Dousset O, Johnson R R and Palmieri V 2013 *J. Phys. D: Appl. Phys.* **47** 045306

[3] Kim J–H, Jung D–W, Kim S, Hong S, You Y and Kim D 2012 *Vacuum* **86** 1789

[4] Hála M, Čapek J, Zabeida O, Klemberg–Sapieha J E and Martinu L. 2012 *Surf. Coat. Technol.* **206** 4186

[5] A.Bruno, Mengucci P, Mercaldo L V and Lisitskiy M P 2012 *Physics Procedia* **36** 239

[6] Russo R 2007 *Meas. Sci. Technol.* **18** 2299
[7] Bemporad E, Carassiti F, Sebastiani M, Lanza G, Palmieri V and Padamsee H 2008 *Supercond. Sci. Technol.* **21** 125026
[8] Savvides N 2008 *Supercond. Sci. Technol.* **21** 045013
[9] Mitchell E E and Lam S K H 2012 *Physics Procedia* **36** 382
[10] Pristáš G, Gabáni S, Gažo E, Komanický V, Orendáč M, You H 2014 *Thin Solid Films* **556** 470
[11] Gontad F, Lorusso A, Panareo M, Monteduro A G, Maruccio G, Broitman E and Perrone A 2015 *Nucl. Instrum. Methods Phys. Res., Sect. A* **804** 132
[12] Pristáš G, Gabáni S, Orendáč M, Komanický V and Gažo E 2014 *Acta Phys. Pol., A* **126** 346
[13] Bose S, Raychaudhuri P, Banerjee R, Vasa P and Ayyub P 2005 *Phys. Rev. Lett.* **95** 147003
[14] De Freitas T C, Gonzalez J L, Nascimento V P and Passamani E C 2016 *Thin Solid Films* **611** 33
[15] Bose S, Galande C, Chockalingam S P, Banerjee R, Raychaudhuri P and Ayyub P 2009 *J. Phys.: Condens. Matter*, **21** 205702
[16] Wu Z and Dickey J M 2000 *Thin Solid Films* **371** 161
[17] Tkachenko E A, Postnikov D V, Blesman A I and Polonyankin D A 2016 *IOP Conf. Ser.: Mater. Sci. Eng.* **110** 012009

Electron-ion-plasma modification of the structure and properties of commercial steels

Yu Ivanov[1, 2], A Klopotov[2,3], A Potekaev[2], N Koval[1], V Vlasov[3,4]

[1] Institute of High Current Electronics of the SB RAS, 634055, Tomsk, 2/3 Akademicheskiy Ave., Tomsk, Russia
[2] National Research Tomsk State University, 634050, Tomsk, 36 Lenina str, Russia
[3] Tomsk State University of Architecture and Building, 634003, Tomsk, 2 Solyanaya Sq., Russia
[4] National Research Tomsk Polytechnic University, Tomsk, 30 Lenin str., Russia

E-mail: yufi55@mail.ru E-mail: klopotovaa@tsuab.ru

Annotation. The work is devoted to the structural-phase analysis of steels of the austenitic and martensitic grade, irradiated with a high-intensity pulsed electron beam of the submillisecond duration of exposure in the mode of the surface layer melting. A thermodynamic analysis of phase transformations occurring during heat treatment in alloys of the composition Fe-Cr-C and Fe-Cr-Ni-C, which are the basis of steels 20X13 and 12X18H10T, is carried out. It is shown that formation of solid solutions on the basis of α-iron (BCC crystalline lattice) and γ-iron (FCC crystalline lattice) as well as the entire range of carbide phases of a complex elemental composition ($M_{23}C_6$, M_7C_3 и M_3C, where symbol M refers to atoms of metallic elements Fe, Cr, and Ni) is possible in equilibrium conditions in given materials. The irradiation of steels 12X18H10T and 20X13 with a high-intensity pulsed electron beam of the submillisecond duration of exposure is carried out. It is shown that the electron-beam processing of steel in the melting mode and the subsequent rapid crystallization is accompanied by a significant transformation of the surface layer structure, consisting in complete dissolution of original carbide phase particles; in formation of dendritic crystallization cells of submicron sizes; in occurrence of martensitic $\gamma \rightarrow \alpha$ and $\gamma \rightarrow \varepsilon$ transformation; in re-allocation of nanosized particles of carbide and intermetallic phases.

1. Introduction

The development of new volume-doped materials, which are still the main way to improve the reliability and the longevity of details of mechanisms and machines, is becoming increasingly problematic due to the scarcity and the high cost of doping elements. Under the circumstances, it is economically and technically feasible to develop a fundamentally different approach to the development of materials when the mechanical strength of the part is achieved by using cost-effective, easily workable low-alloy steels, and special surface properties by a solid or a local formation of relatively thin doped layers, or by depositing coating with properties that meet performance requirements. Impulse fusion (high-frequency currents, laser emission, electron beams, plasma flows and jets) and saturation of surface layers of the material with doping elements with the subsequent crystallization and formation of strengthening phases (the so-called coatings of the intermediate type) is currently one of the most promising methods for modification of the structure and properties of metals and alloys [1, 2]. Generated layers have a metallurgical bond to the substrate with adhesion at a level of cohesion, and insignificantly change the dimensions of the item.

The aim of the study is the detection of evolution patterns of the structure and properties of the surface layer of industrial steels subjected to irradiation with a high-intensity pulsed electron beam of the submillisecond duration of exposure in the melting mode with the subsequent rapid crystallization of the modified layer.

2. Material and methods of research

Steels 12X18H10T (AISI 321) and 20X13 (AISI 420) in the hardened condition were chosen as the modified material. The samples had the shape of a plate with a thickness of 10 mm and a diameter of 15 mm. Formation of nanostructured, nanophased surface-doped layers in steel was carried out by the irradiation of samples with a high-intensity pulsed electron beam using the installation "SOLO". Exposure parameters: the electron energy – 18 keV, the density of the electron beam energy – (15-40) J/cm^2, the duration and the number of exposure pulses – (50-200) microseconds and (3-5) pulses, the pulse repetition rate – of 0.3 s^{-1}; the irradiation was carried out in argon at a residual pressure of 0.02 Pa. The given mode provides heating of the surface layer with a thickness of up to 30 micron to melting temperature and quenching from the melt at a rate of up to (10^6 – 10^8) K/s. The study of the phase and elemental composition of the defect substructure of the surface layer was carried out using methods of scanning and electron diffraction microscopy, X-ray diffraction analysis (Bragg-Brentano geometry, Cok$_\alpha$-radiation); physical-and-mechanical properties of the surface layer were studied by measuring the microhardness and the wear resistance.

3. Results and discussion

We shall briefly consider the features of structural and phase states in binary, ternary, and quaternary systems on the basis of elements Cr, Fe, Ni, and C, which are the main alloy-forming elements in steels of the austenitic class 12X18H10T and the martensitic class 20X13. The feature of structural-phase states of steels under the consideration is the formation of carbide phase particles, the composition, the distribution by sizes and the shape, the location and the morphology of which strongly depend on the mode of the material heat treatment. Carbides forming in steel are divided into two groups, depending on the complexity of the crystalline lattice. The first group is carbides with complex structures: M_3C, $M_{23}C_6$, M_7C_3, M_6C. Here, M in the chemical formula represents the sum of metallic elements forming carbide. For example, $(Cr,Fe,Ni,Mn)_{23}C_6$, $(Fe,Ni,Mn)_3C$, etc. The second group is carbides of the MC type (for example, VC) and M_2C (for example, Mo_2C), which are the interstitial phases. These carbides have simple crystalline lattices and have high thermal and mechanical stability. Under real conditions, they hardly dissolve in austenite. As a rule, carbides of the second group have higher melting point and greater hardness than carbides of the first group.

The investigated in this paper steels are complex-alloy materials, the basic elements of which are Fe, C, Cr, Ti, Ni. We shall briefly consider the options of phase transformations occurring during heat treatment of the given steels under equilibrium conditions.

3.1. System Cr-Fe-Ni [3]

Wide ranges of solid solutions based on the α-phase (disordered phase on the basis of the structure A2) and the γ-phase (disordered phase based on the structure A1) are present in the ternary system Cr-Fe-Ni at high temperatures (T = 1100 ^{0}C). A decrease in the temperature in the region of the equiatomic composition on the side Cr-Fe of the isothermal triangle leads to formation of the σ-phase. This leads to formation of two three-phase regions (α + γ + σ), two two-phase regions (α + σ), and one two-phase region (γ+σ) (Fig. 1).

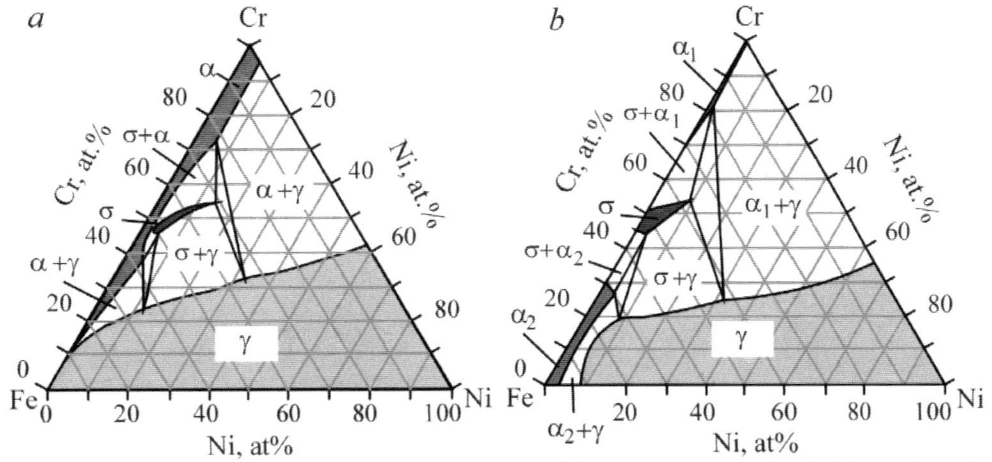

Figure 1. Isothermal sections at different temperatures of the ternary system Fe-Cr-Ni; *a* – T = 900 ^{0}C; *b* – T = 700 ^{0}C [3].

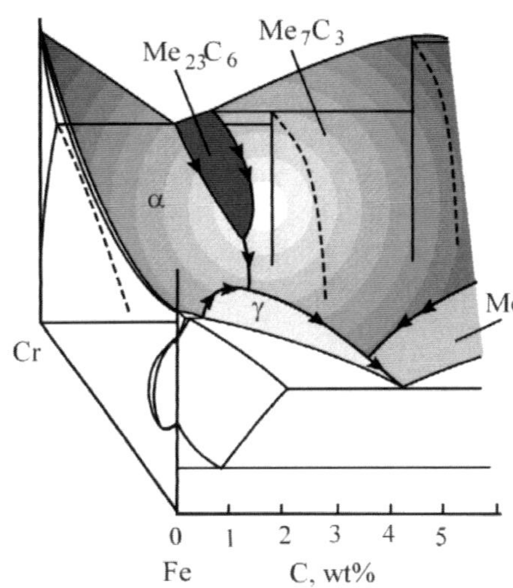

Figure 2. The metastable Fe-Cr-C phase diagram in the region of carbon concentration less than 5 wt. %. Thick black lines on the diagram separate regions of solid solutions. One arrow on thick lines corresponds to the eutectic type of the reaction. Two arrows on thick lines correspond to the peritectic type of the reaction. Dashed lines correspond to solidus lines [4].

3.2. System Cr-Fe-C

Fig. 2 schematically presents three-dimensional liquidus surfaces on the diagram Cr-Fe-C bounded by binary systems Fe-C and Cr-C [4].

The diagram in Fig. 2 shows five liquidus surfaces corresponding to formation of five phases in binary systems: Fe_3C, Cr_7C_3, $Cr_{23}C_6$, α–ferrite and γ–austenite. Here, ferrite means α–(Fe,Cr) and δ-(Fe,Cr) in the binary system Fe-Cr, as these two elements (Fe and Cr) in the given temperature range form solid solutions on the basis of a BCC lattice; γ–Fe in the binary system Fe-C. Three four-phase regions can be identified in Fig. 2: 1 – L + α + $M_{23}C_6$ + M_7C_3; 2 – L + α + γ + M_7C_3; 3 – L + α + M_7C_3 + M_3C. Here, M represents, as mentioned above, the sum of metal elements Cr and Fe that form carbide. L represents the alloy in the liquid state.

The state diagram in the system Fe-Cr-C (Fig. 3) along the section at constant carbon concentration C=0.1 wt.% (corresponds to carbon concentration in steel 12X18H10T) has been theoretically obtained in [4]. The data obtained in [4] allow determining regions of existence of two-phase and three-phase alloys. This diagram also shows that carbide $M_{23}S_6$ is more stable than carbides M_7S_3 and M_3S and has more extensive regions of existence.

Thus, it is possible to state that doping of alloys Fe-Cr with carbon leads to a significant change in the structure-phase state and has a significant impact on existence regions of carbides $M_{23}S_6$, M_7S_3, and M_3S with α-ferrite and γ-austenite.

System Fe-Ni-C is characterized by formation of a single carbide with the composition M_3C. The analysis of the equilibrium diagram Fe-Ni-C, shown in Fig. 4 , shows that in the Fe-corner on the

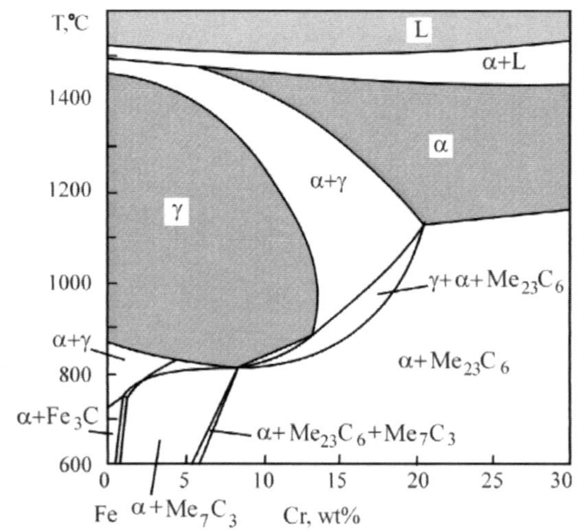

Figure 3. The calculated state diagram in the system Fe-Cr-C along the section at a constant value C=0.1 wt. % [4]

isothermal section there are three two-phase regions (α + M_3C, γ + M_3C and α + γ) and one three-phase region (α + γ + M_3C) [5]. Here, M denotes the sum of metal elements Ni and Fe that form carbide of the composition M_3C. It should also be noted that carbides $M_{23}C_6$ and M_7C_3, which are metastable, have been synthesized in alloys Fe-Ni-C at high pressures [5].

3.3. System Fe-Cr-Ni-C [6]

Carbides $M_{23}C_6$ and M_7C_3 with a complex structure, since the symbol M represents the sum of three metal elements Ni, Cr, and Fe that form the carbide, have been found in the system Fe-Cr-Ni-C [6].

Thus, the performed brief thermodynamic analysis of phase transformations that occur during heat treatment in alloys of the composition Fe-Cr-C and Fe-Cr-Ni-C, which are the basis of steels 20X13 and 12X18H10T, shows that under equilibrium conditions in these materials it is possible to form not only solid solutions based on α-iron (BCC crystalline lattice) and γ-iron (FCC crystalline lattice) but also the entire range of carbide phases of a complex elemental composition ($M_{23}C_6$, M_7C_3 and M_3C where the symbol M refers to atoms of metallic elements Fe, Cr, and Ni).

3.4. Results of experiment

The steels under the investigation in the initial state (before irradiation with an electron beam) are multiphase materials. Particles of carbide $M_{23}C_6$ act as the second phase. The particles have a globular shape, are distributed randomly in the volume of the grain; the average particle size is 167 nm (the size of actual particles ranges from 30 nm to 450 nm).

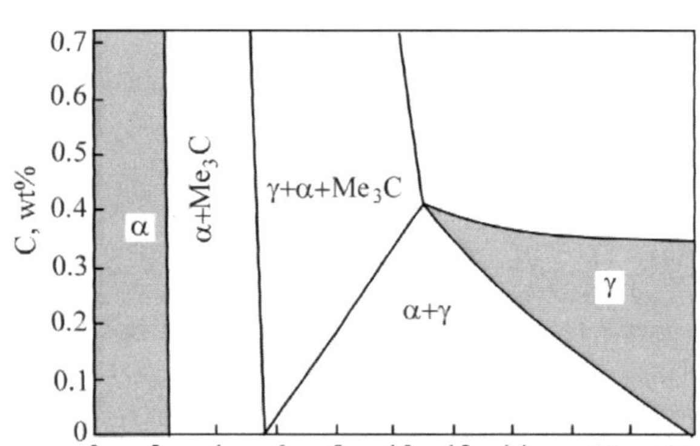

Figure 4. The isothermal section of the ternary system Fe-Ni-C at 600 C [5]

High-intensity electron beam treatment in the melting mode of the surface layer of the pre-hardened steel (regardless of the elemental composition of the material) leads to formation of a gradient multiphase structure. The surface layer, whose thickness ranges from units to tens of micrometers depending on parameters of the electron beam, is characterized by complete dissolution of submicron particles of carbide $M_{23}C_6$, which are present in steel of the initial state. The subsequent rapid crystallization is accompanied by formation of a multiphase structure of the cellular type with a cell size of (100-400) nm (Fig. 5).

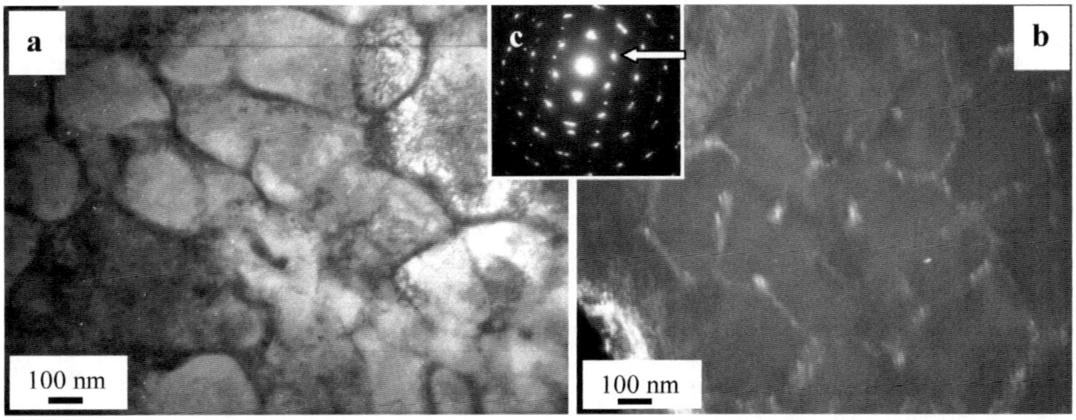

Figure 5. The structure of the cellular crystallization forming in the surface layer of steel 20X13 subjected to electron-beam treatment with an energy density of electron beams of 30 J/cm^2; a – light field; b – dark field, obtained in neighboring reflexes [110]α-Fe and [331]σ-FeCr; c – micro-electron-diffraction pattern, reflexes in which the dark field is obtained are indicated by the arrow.

Particles of the second phase are located along cell boundaries and in their volume. The study of the phase composition of the surface layer of steel, carried out using methods of electron-microscopy microdiffraction analysis with the use of the dark-field technique, has revealed the formation of particles of the intermetallic phase (σ-FeCr and Fe_9Cr_{15}) (Fig. 5) and the carbide phase of the composition M3C (cementite doped with chromium atoms), chromium carbides of the composition Cr_7C_3 and Cr_3C_2, and carbide of the type $M_{23}C_6$. An example of formation of carbide particles Cr_3C_2 in steel is shown in Fig. 6, b.

Rapid crystallization of steel is accompanied by martensitic transformation of the γ-phase. Formation of α-martensite, peculiar to steel 20X13 hardened with low cooling rates, after the irradiation with an electron beam is accompanied with formation of ε-martensite, i.e. the occurrence of martensitic transformation under the scheme $\gamma \rightarrow \varepsilon$. The typical image of ε-martensite crystals is shown in Fig. 6 a, b.

4. Conclusion

A brief thermodynamic analysis of phase transformations occurring during heat treatment in alloys of the composition Fe-Cr-C and Fe-Cr-Ni-C, which are the basis of steels 20X13 and 12X18H10T, has been carried out. It has been shown that formation of both solid solutions on the basis of α-iron (BCC crystalline lattice) and γ-iron (FCC crystalline lattice) and the whole range of carbide phases of the complex elemental composition ($M_{23}C_6$, M_7C_3 and M_3C, where the symbol M refers to atoms of metallic elements Fe, Cr, and Ni) is possible in the given materials under equilibrium conditions. The irradiation of steels 12X18H10T and 20X13 with a high-intensity pulsed electron beam of the submillisecond duration of exposure has been carried out. It has been shown that electron-beam treatment of steel in the melting mode and the subsequent rapid crystallization is accompanied by a significant transformation of the surface layer structure, consisting in:

- complete dissolution of particles of the initial carbide phase;
- formation of dendritic crystallization cells of submicron sizes;
- occurrence of martensitic $\gamma \rightarrow \alpha$ and $\gamma \rightarrow \varepsilon$ transformation;
- re-allocation of nanosized particles of carbide and intermetallic phases.

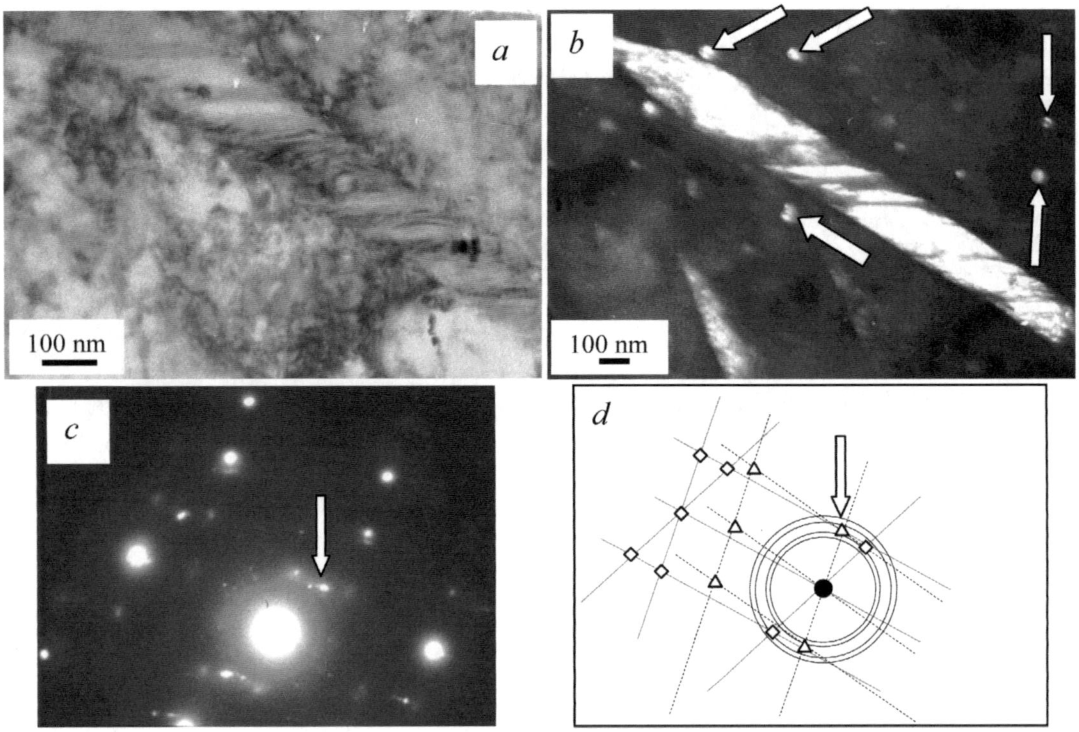

Fig. 6. The structure of the layer located at a depth of ≈5 micron in steel 20X13 treated with a high-intensity electron beam; a – light-field image, b – dark field obtained in closely spaced reflexes [101]ε-Fe+[320]Cr_3C_2; c – micro-electron-diffraction pattern (the arrow indicates the reflex in which the dark field is obtained), d – indexing scheme of the micro-electron-diffraction pattern (rings indicate reflexes that belong to carbide Cr_3C_2). In (b) the arrows indicate particles of carbide Cr_3C_2.

Acknowledgement
The work is carried out with partial financial support by grant RFBR (project No. 16-43-7006 59, in the course of the implementation of the project (No. 8.1.42.2015) and within the framework of the Program "Scientific Foundation named after D.I. Mendeleev of Tomsk State University" in 2015-2016.

References
[1] Gribkov V A, Grigoriev F I, Kalin B A, et al 2001 *Perspective radiation-beam metal treatment technologies* (Moscow: Krugliy stol) (in Russian)
[2] Laskovnev A P, Ivanov Yu F, Petrikova E A, et al 2013 *Modification of the structure and properties of eutectic silumin using electron-ion-plasma treatment* (Minsk: Belarus Navuka) (in Russian)
[3] Durand-Charre M 2004 *Microstructure of steels and cast irons* (Published by Springer Berlin Heidelberg)
[4] Lee B J 1993 *J. Korean Inst. Met.* **31** 480
[5] Romig A D, Goldstein J I 1978 *Metallurgical Transactons* A. **9** 1599
[6] Liu Y, Xiao L, Lin L 2002 *J. Mater. Sci. Technol.* **18** 171

IOP Conf. Series: Materials Science and Engineering **168** (2017) 012071 doi:10.1088/1757-899X/168/1/012071

Binary phase diagrams based on elements VIIIA and IB periods of the D.I. Mendeleev's table and features of crystallographic parameters

A I Potekaev[1], A A Klopotov[1,2], S A Porobova[3], V D Klopotov[3], T N Markova[2], M I Imanaliev[2]

[1]Tomsk State University, Tomsk, 634050, 36 Lenina str, Russia
[2]Tomsk State University of Architecture and Building, 634003, Tomsk, 2 Solyanaya Sq., Russia
[3]Tomsk Polytechnic University, Tomsk, 634050, 30 Lenina str, Russia

E-mail: klopotovaa@tsuab.ru

Abstract. The data obtained as a result of the analysis of crystallogeometric parameters and the structure of state diagrams of binary systems from elements VIIIA and IB periods of the D.I. Mendeleev's table are presented. It is shown that the classification of the evolution of phase diagrams of binary systems by types, proposed by T.A. Lebedev, correlates with features of concentration dependences of the deviation of atomic volumes in solid solutions from Zen law.

1. Introduction

The data on phase equilibria and structural-phase states in multicomponent systems is of greatest interest from the perspective of the development of functional materials. To identify common patterns in multicomponent systems it is necessary to obtain them in binary systems. Thus, functional materials based on binary alloys from elements VIIIA and IB periods of the D.I. Mendeleev's table are of keen interest. These alloys exhibit weak lattice resistance in temperature ranges preceding structural phase transitions (SPT) [1, 2].

Increasingly stringent requirements to materials for modern engineering structures require targeted search for systems with new functional properties. The conventional method, based on simple search of alloy-forming elements and alloying additives in multicomponent systems and standard technological methods, is no longer effective. Semi-empirical approach, widely developed in the classic materials science, also has limitations in forecasting new materials with a complex chemical composition. Thus, classical conceptual ideas on methods of designing the properties of new metallic and composite materials have exhausted all possibilities.

On the other hand, the search for general laws can be carried out using conventional classifications of binary and ternary state diagrams. This approach is productive since it reveals typical and special state diagrams. Namely, state diagrams that exhibit of a weak stability of the crystalline lattice in pre-transitional regions prior to phase transitions

It should be noted that the existence of a weak stability of the crystalline lattice in pre-transitional regions prior to phase transitions is manifested in features on temperature dependences of physical and mechanical properties. These weak-stable states affect features of the change in structural and phase states [3].

The aim of the work is to carry out a search for general laws by analyzing the structure of state diagrams of binary systems from elements VIIIA and IB periods of the D.I. Mendeleev's table and deviations of atomic volumes in solid solutions and ordered phases from Zen law

2. Atomic volume and metallic bond

The metallic bond is usually represented as a bond that arises due to interaction forces of positive ions surrounded by gas of "free" electrons. The presence of free electrons, due to their mobility in the interatomic space, results in high thermal and electrical conductivity. The energy of the crystal with a metallic nature of the bond can be written as [4]:

$$U = -Ae^2 \times \Omega^{-\frac{1}{3}} + B \times \Omega^{-\frac{2}{3}} + Ce^2 \times \Omega^{-1}, \quad (1)$$

where Ω is the atomic volume (volume per atom).

$-Ae^2 \times \Omega^{-\frac{1}{3}}$ this term in the equation (1) is the potential energy of free electrons, $B \times \Omega^{-\frac{2}{3}}$ this term is their kinetic energy, and $Ce^2 \times \Omega^{-1}$ characterizes the kinetic energy of electrons occupying lower energy states. The minimum value of the resulting energy can be related to the linear dimension r_0 that can be considered as the atomic radius $\Omega = \frac{4}{3}\pi r_0^3$.

This expression works well for alkali metals of the subgroup IA. However, for metals of the subgroup IB of the Periodic Table (ie, Cu, Ag, Au) an agreement between the theory and the experiment is not observed. The difference between structures and properties of metals of subgroups IA and IB reflects a phenomenon in which the metallic bond does not limit the number of nearest neighbors in nodes of the crystalline lattice, and that the values of the interaction between electrons and ions can be significantly different [4].

The comparison with the covalent bond shows that the metallic nature of the bond is characterized by the collectivization of electrons that become free, but the number of valence electrons is small (one to three), and CN $\geq$ 8,12 (CN - coordination number). This leads to the fact that all valence electrons can become free [4].

These types of bonds do not normally occur in the pure form. In fact, there is a situation where one type of interatomic bond contributes more to the interatomic interaction. A large number of compounds with a mixed type of bond have been revealed.

Thus, it can be stated that the concept of the representation of atoms in the form of non-compressible solid balls during the formation of crystalline structures that can exhibit different ratio between metal and covalent components in the interatomic interaction can be used in the search for general laws on the qualitative level [5]. On the other hand, the atomic volume affects the conditions of formation of solid solutions and intermetallic compounds.

3. Atomic volume. Zen rule

In binary systems with a continuous series of solid solutions it has been established that the atomic volume of the concentration $\Omega = f(C)$ has a functional dependence close to linear. This phenomenon well manifests itself on the example of solid solutions in systems Au-Ni, Au-Cu, and Cu-Ni [2]. This phenomenon has been formulated in the form of Zen rule [1, 2]:

$$\Omega = C_A \Omega_A + C_B \Omega_B \quad (2)$$

where C_A, C_B and Ω_A, Ω_B are concentrations and atomic volumes of pure components, respectively. The term "atomic volume" refers to the share Ω of the volume of the unit cell V, per atom ($\Omega = V/n$, n is the number of atoms in the unit cell).

In [6, 7] it has been established that atomic volumes of the same element in the FCC and BCC modification have similar values at the same temperature. The atomic volume of pure metals Ω is a more universal characteristic than the parameters of unit cells of pure metals. This allows the use of atomic volumes in the search for general laws of structural properties of alloys with solid solutions and with intermetallic compounds formed by elements with different crystalline structures. This approach has been successfully applied in the analysis of binary compounds based on Ti-Ni [1].

It has been established that the strict Zen rule is practically not executed, or executed in very rare cases. Despite these shortcomings, this rule is widely used in the analysis of features of the change in structural parameters from the concentration in various binary systems [1, 2, 6]. The experimentally determined values of average volumes per atom in different systems may be greater or less than those calculated by Zen rule. For the majority of ordered phases a negative deviation from Zen rule is observed. The value

$$\Delta\Omega = \Omega_i^{\exp} - \Omega_i^{\text{theor}} < 0 \quad (3)$$

characterizes the deviation from Zen rule. The value of the deviation from Zen rule is a nonconfigurational effect and plays a significant role in the stabilization of both ordered and disordered solid solutions, as well as intermediate phases due to the gain in the volume energy. The correlation between the value of the deviation of concentration dependences of atomic volumes from Zen law and Hume-Rothery factors is presented in the literature [7]. In the analysis of changes in the interatomic interaction in solid solutions and intermetallic compounds it is convenient to use the relative value of the deviation of the atomic volume from Zen law:

$$\frac{\Delta\Omega}{\Omega} = \frac{\Omega^{\exp} - \Omega^{\text{theor}}}{\Omega^{\exp}} = 1 - \frac{\Omega^{\text{theor}}}{\Omega^{\exp}} = \frac{\Delta\Omega}{\Omega^{\exp}}.$$

4. State diagrams and features of the interaction of Ag with neighboring elements in the Periodic Table

Silver belongs to metals of the group IB, and by the number of valence electrons in the s-shell of the atom to analogous metals of the group IA. An important point here is the fact that metals of the group IB have completely filled *3d, 4d, 5d* - electronic shells with 10 electrons, and atoms of alkali metals do not have d-peripheral shells. This affects features of the structure of atoms of metals and their physical and chemical properties, and features during the formation of state diagrams (Fig. 1). By the type of the interaction with other elements Ag is similar to its neighbor in the Periodic Table – Cu. However, silver forms continuous series of solid solutions with Pd and Au with a crystalline structure A1 (Fig. 1). Limited solid solutions based on silver usually have higher concentrations than solid solutions based on Ag in

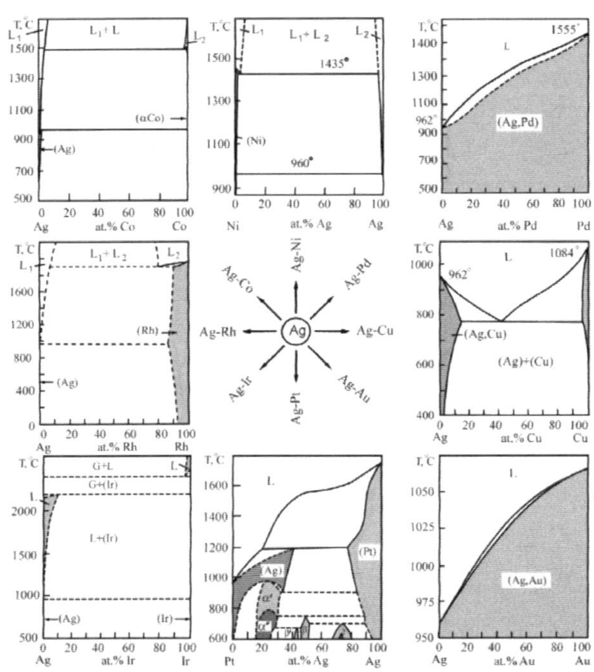

Figure 1. Phase diagrams of binary compounds based on Ag [4]

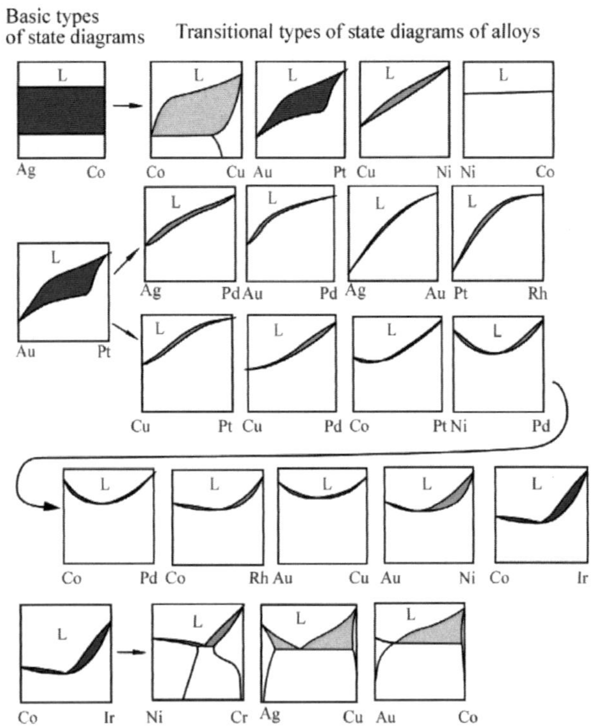

Figure 2. The schematic of the evolution of phase diagrams formed by elements VIIIA and IB of the D.I. Mendeleev's periodic table based on the T.A. Lebedev's classification [12]

electronegative metals and nonmetals [10]. Silver practically does not interact with some elements (V, Ta, W, Fe, Ir, C, N).

It is important to note that among all of the systems under consideration only two systems (Ag-Pd and Ag-Au) exhibit the formation of a continuous series of solid solutions during the crystallization from the liquid phase. In these systems, no structural phase transitions are observed at the decrease in the temperature. That is, a disordered solid solution is stable in systems Ag-Pd and Ag-Au in the entire temperature range after the crystallization from the liquid phase.

5. Deviation from Zen law in alloys of binary Ag-based systems

Concentration dependences of the atomic volume in systems Ag-Pd, Ag-Pt, and Ag-Au, obtained from data on lattice parameters of solid solutions and ordered phases, in these systems are shown in Fig. 2. From these diagrams it is evident that in disordered solid solutions in systems Ag-Pd, Ag-Pt, and Ag-Au a slight negative deviation from Zen law is observed. A significant negative deviation of atomic volumes from Zen law is observed in the system Ag-Pt for ordered phases. In Ag-based systems, in which the formation of broad regions of solid solutions is not observed, it is impossible to obtain the dependencies of the atomic volume from the concentration $\Omega = f(C)$ (Fig. 1).

6. Binary state diagrams from elements VIIIA and IB of the Periodic Table and deviations of atomic volumes of concentration dependencies from Zen rule

State diagrams from elements VIIIA and IB of the Periodic Table and deviations of atomic volumes of concentration dependencies from Zen law have been systematized. The analysis of phase diagrams from elements VIIIA and IB periods of the D.I. Mendeleyev's table shall be carried out based on the classification of diagrams proposed by T.A. Lebedev [12]. The classification is based on the change in ratios between the potential energy of interaction U_{AB}, U_{AA}, and U_{BB} between atoms of different types *A* and *B* and the atoms of the same type *AB* and *AB*. Moreover, metallic elements, their solid solutions, and intermetallic compounds typically have two types of bonds: covalent and metallic. It is known that in systems with unlimited solid solutions there are mainly three types of melting curves: 1) with a maximum; 2) monotonically increasing, the "cigar" type; 3) with a minimum (Fig. 1) [1]. A scheme of the evolution of binary state diagrams in solidus and liquidus curves (Fig. 2) has been built in this study (Fig. 1) according to the classification [12].in systems with unlimited solid solutions there are mainly three types of melting curves: 1) with a maximum; 2) monotonically increasing, the "cigar" type; 3) with a minimum (Fig. 1) [1]. A scheme of the evolution of binary state diagrams in solidus and liquidus curves (Fig. 2) has been built in this study (Fig. 1) according to the classification [12].

Moreover, the evolution of binary state diagrams must exhibit a phenomenon caused by a gradual transition from the metallic bond to the covalent bond, depending on the concentration of alloy-forming elements. This reflects the situation that, in most cases, two types of bonds are simultaneously present in the interaction of metal components [4-7]. As consequence, it should appear on concentration dependencies of atomic volumes in binary systems. This is the case in reality (Fig. 3).

The presented in Fig. 2 evolution of state diagrams of the type I under T.A. Lebedev's classification suggests that an increase in the energy of the interatomic interaction between atoms of different kinds *AB* U_{AB} relative to the energy of the interatomic interaction between atoms of the same kind U_{AA} and U_{BB} takes place. At the same time, there is a negative deviation in concentration dependences of atomic volumes from Zen law, which indicates a decrease in the atomic volume in the formation of solid solutions in alloys of the systems under consideration.

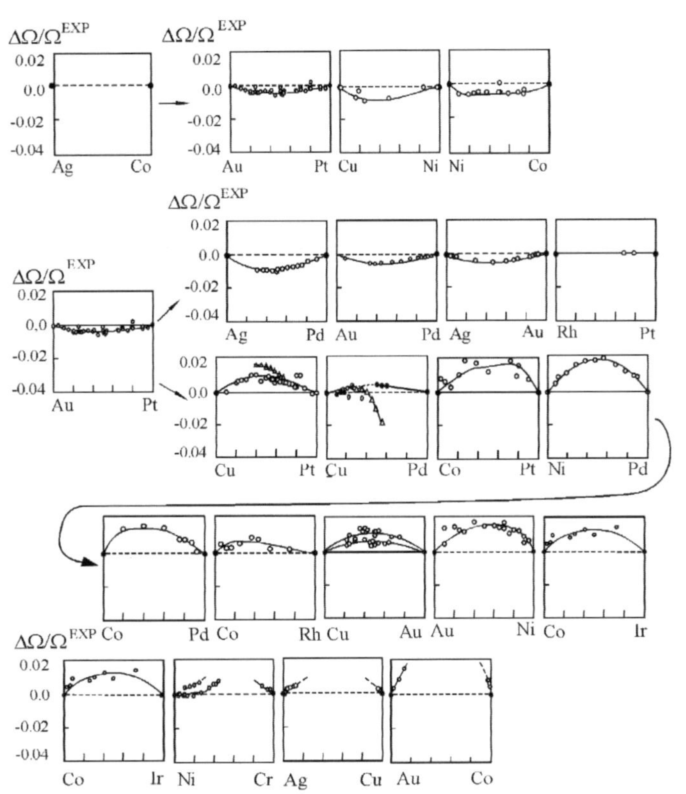

Figure 3. The schematic of the evolution of concentration dependences of atomic volume deviations in solid solutions and in ordered phases from Zen law in systems formed by elements VIIIA and IB of the periodic table

For type II of state diagrams (Fig. 2), the modification of phase diagrams takes place under two scenarios. The first scenario involves the modification of state diagrams due to the fact that the liquidus line goes on rapprochement with the solidus line. In second scenario, the solidus line goes on rapprochement with the liquidus line. Such evolution of state diagrams on solidus and liquidus lines leads to appearance maximum and minimum points.

Type III of state diagrams is represented by a series of state diagrams in which the degree of disintegration of liquid solutions increases to produce, in the end result, a mechanical mixture of pure components.

It is important that the minimum on the diagram of continuous solid solutions (Co-Ir, Fig. 2) cannot be mixed with the eutectic point on phase diagrams. The fact that at the beginning of the given diagram with the eutectic point there is a diagram with the minimum point on solidus and liquidus curves does not give grounds to identify the minimum point on this diagram with the eutectic point. There is a trend that reflects the process of a possible eutectic formation. This phenomenon is associated with the appropriate prerequisites that a eutectic transformation in the following systems may occur at the place of the unstable.

The distribution of concentration dependences of deviations of atomic volumes in solid solutions and in ordered phases from Zen law in the systems under consideration (Fig. 3) has been built according to the presented location of state diagrams (Fig. 2).

For diagrams of the type I under T.A. Lebedev's classification, only negative deviation of concentration dependences of atomic volumes from Zen law is observed (Fig. 2, 3).

Type II of state diagrams under T.A. Lebedev's classification, is characterized by both positive and negative deviation of atomic volumes in solid solutions and in ordered phases from Zen law, depending on the scenario of the evolution of state diagrams (Fig. 3).

Type III of state diagrams, for which the diagram of the system Co-Ir is the basis, is characterized by a positive deviation of atomic volumes in solid solutions from Zen law in places of their existence on state diagrams. At the same time, there is a significant increase in the deviation of atomic volumes from Zen law in regions presiding ruptures of the solubility on phase diagrams (Fig. 3).

7. Conclusion

The analysis of the structure of state diagrams and crystallogeometric parameters has allowed revealing a clear correlation between the type of the evolution of phase diagrams under T.A. Lebedev's classification, depending on the nature of deviations of atomic volumes in solid solutions and in ordered phases from Zen law in systems formed from elements VIIIA and IB periods of the D.I. Mendeleev's table.

Acknowledgement

The paper included the results obtained in the course of the implementation within the framework of the Program "Scientific Foundation named after D.I. Mendeleev of Tomsk State University" in 2015-2016.

References

[1] Potekaev A I, Klopotov A A, Kozlov E V, et al. 2004 *Pretransitional weak-stable structure in NiTi* (Tomsk: NTL) (in Russian).
[2] Potekaev A I, Klopotov A A, Morozov M M, et al. 2014 *Structural features of binary systems with weak-stable* (Tomsk: NTL) (in Russian).
[3] Olemskoy A I and Katznelson A A 2002 *Advances of phys. met. / Usp. Fiz. Met.* **3** 1-86.
[4] Madelung O 1980 *Solid State Theory*. (Moscow: Nauka) (in Russian).
[5] Vozdvizhinsky V M 1972 *Forecast of binary state diagrams. According to statistical criteria* (Moscow: Metallurgiya) (in Russian)
[6] Pearson W 1977 *Crystal chemistry and physics of metals and alloys* (Moscow: Mir) **1** 420 (in Russian)
[7] Hume-Rothery V, Raynor G 1959 *The structure of metals and their alloys* (Moscow: Metallurgiya) (in Russian)
[8] Savitskiy E M 1984 *Noble metals. Directory* (Moscow: Metallurgiya) (in Russian)
[9] Diagrams of binary metallic systems / Ed. by Lyakishev N P 1996-2000 (Moscow: Mashinostroeniye) **1-3** (in Russian)
[10] Kornilov I I, Matveeva N M, Pryakhina L I, Polyakova R S 1964 *Metallochemical properties of elements of the periodic system* (Moscow: Nauka) (in Russian)
[11] ViUars P and Calvert L D 1985 Pearson's *Handbook of Crystallography Data for Intermetallic Phases* (Ohio: Metals Park) **1**
[12] Lebedev T A 1951 *Some questions in the general theory of alloys* (Leningrad: Newspaper and magazine book) (in Russian)

Analysis and reoperation of the magnetic permeability spectra of textured composite based on Z-type hexaferrite by using Cramers-Kronig relations

O A Dotsenko[1], V I Suslyaev[2], K O Frolov[3] and E V Zhuravlyova[4]
[1] Assistant Professors, National Research Tomsk State University, Tomsk, Russia
[2] Assistant Professors, National Research Tomsk State University, Tomsk, Russia
[3] Undergraduate Student, National Research Tomsk State University, Tomsk, Russia
[4] Master Student, National Research Tomsk State University, Tomsk, Russia

E-mail: apr@mail.tsu.ru

Abstract. The frequency dependences of the magnetic permeability of textured composite based on $Ba_3Co_{2,4}Fe_{23,2}O_{41}$ ferrite are given. The magnetic permeability spectra were obtained by coaxial method in the frequency range 0.01–18 GHz. The expert judgement on the spectra was made by using Cramers – Kronig relations. It was shown, that the Cramers – Kronig relations can be used to correct magnetic permeability measurements by reciprocal recalculation of the frequency dependences of the real and imaginary parts.

1. Introduction

Field of materials science, additive technology and polymer engineering actively develop at the present time. New technologies are based on the use of composite materials. Composite materials are materials made from two or more ingredient materials with significantly different physical or chemical properties, that when combined, produce a material with characteristics different from the individual components. The separate components remain separate and clear within the finished structure. The new material may be preferred for many reasons: common examples include materials which are stronger, lighter or less expensive when compared to traditional materials [1]. Pure powdered materials are not suitable for use in electronics equipment so they should be used as filler in the composite materials. Composite microwave materials are widely used due to their actively interaction with electromagnetic radiation.

Electromagnetic (EM) characterization of materials is a pre-requisite step before their use in microwave applications: anechoic rooms, mobile phones, wireless communications, radar, etc. EM characteristics of the materials should know for the microwave electronics applications [2-4].

In this article composite, based on hexagonal ferrite, was studied. The fillers of composites were powders of hexaferrite $Ba_3Co_{2,4}Ti_{0,4}Fe_{23,2}O_{41}$. Epoxy resin was used for production of experimental samples. EM characteristics of materials are measured by a coaxial method and LCR method. Agilent's E4980A precision LCR meter was used to measure of resistance and inductance of experimental toroid samples. The imaginary permeability spectra of the samples were investigated at frequency range from 0.01 till 18.0 GHz using a universal spectrometer based on a P4M-18 MICRAN scalar network analyzer.

[1] To whom any correspondence should be addressed.

2. Materials and Methods

2.1. Nomenclature

Standard ceramic techniques were used for the fillers creation. The initial materials for the synthesis were barium oxide BaO powders, cobalt (II) oxide CoO, titan oxide TiO_2 and iron (III) oxide Fe_2O_3. Prior to synthesis, the oxides powders were dried for 3 h at temperature of 200 °C. After that powders were weighed according to a ratio.

$$3BaO + 2.4CoO + 0.4TiO_2 + 11.6Fe_2O_3 = Ba_3Co_{2.4}Ti_{0.4}Fe_{23.2}O_{41}.$$

Powders used for sintering experiments were placed for 4 h in a vibration ball mill. The ball-to-powder mass ratio was 5:1. Powders were die-pressed at about 1000 atm. Solid samples were heated up in 10–15 min to the sintering temperature (1150 °C), fired for 4 h and were cooled to the room temperature. After that samples were crushed. Powders were placed for 35 min in a vibration mill. Powders were die-pressed repeatedly at about 1000 atm. Solid samples fired in oxigen for 6 h 1200 °C finally. After that samples were placed once again. The powders were sifted by separator. The sizes of powders don't exceed 80 microns.

For production of experimental samples the polymer was used. It was epoxy resin EDP-20 by OKACHIM (Russia). In the liquid state its viscosity is known to be small. This allows a filler to be moved easily. Polyethylene-polyamine was used as the epoxy resin curing agent.

Samples were prepared as follows: hexagonal ferrite powder and epoxy resin in mass proportions of 66:34, respectively, are weighed in the desired proportions on scales Shimadzu AUX – 320 (accuracy 0.5 mg). Components of the composite were placed in the container and mixed for 15 minutes. Obtained mixture was placed in the form as washer. The mixture with ferrite is polymerized in constant magnetic field by special device for texture magnetic polymer materials [4]. It consists of a permanent magnet with the size of a field 0.5 kOe and the electric motor connected to a source of constant tension of 3V. To manufacture texture test samples, the process of polymerization was carried out for 5 h at the room temperature with the electric motor running. The final polymerization was carried out within 24 hours at the room temperature. The samples had the following sizes: thickness of 1.05 mm, an outer diameter of 7 mm and an inner diameter of 3,04 mm. The sample density was determined by weighing in a Shimadzu AUX-320 electronic balance. It was found to be around 2.45 g/cm^3.

2.2. Measuring equipment

Agilent's E4980A precision LCR meter was used to measure of resistance and inductance of experimental samples. The initial magnetic permeability was calculated by formulas [5]:

$$\mu' = \frac{2\pi r_{mean}}{w^2 S} L \quad , \qquad (1)$$

$$\mu'' = \frac{r_{mean}}{w^2 S} \frac{R}{f} \qquad (2)$$

where L is inductance, R is resistance, r_{mean} is the mean radius of the coaxial sample, S is the sectional area of the sample, f is the resonance frequency, w – is the number winding turns.

For microwave characterization of samples we have used a measurement procedure based on the coaxial techniques. P4M-48 MICRAN was used to measure of the module of the reflectivity and the transmission. Coaxial cell is selected due to the electromagnetic field interacts with sample as well as in free space.

The electromagnetic (EM) parameters (ε′, ε″, μ′, μ″) were calculated by the materials measurement software. So, the imaginary permeability was calculated by formulas [5]:

$$\mu''=\frac{15(1-K_r)}{\pi f d(1+K_r)} \tag{3}$$

where d is the thikness of sample, K_r is the reflectivity.

2.3. Reoperation of the magnetic permeability spectra
It is well known that the wider the frequency range of measurements, the more enough information is provided by Cramers – Kronig relations. The Cramers – Kronig relations are actually for the response functions of arbitrary linear systems.

The authors [7] have obtained formulas (4) for reoperation of the permeabilities:

$$\mu'(f)=1+\frac{1}{\pi}\sum_{i=1}^{N}\left\{2(\mu''_{i+1}-\mu''_i)+k_i f\,ln\left|\frac{(f_{i+1}-f)(f_i+f)}{(f_{i+1}+f)(f_i-f)}\right|+d_i\,ln\left|\frac{f_{i+1}^2-f^2}{f_i^2-f^2}\right|\right\},$$

$$\mu''(f)=-\frac{1}{\pi}\sum_{i=1}^{N}\left\{k_i f\ln\left|\frac{f_{i+1}^2-f^2}{f_i^2-f^2}\right|+d_i\ln\left|\frac{(f_{i+1}-f)(f_i+f)}{(f_{i+1}+f)(f_i-f)}\right|\right\}. \tag{4}$$

The results of calculations by the relations (4) involve an error related to the boundedness of the frequency interval in which approximation is performed. It is necessary to take advantage of the analytic properties of the function $\mu^*(f)$: (a) $\mu''(f)$ tends to zero as $f\to 0$ and $f\to\infty$ and the area bounded by the curve $\mu''(f)$ (it is proportional to the saturation magnetization) is finite and (b) $\mu'(f)$ is bounded: as $f\to 0$, $\mu'(f)$ tends to μ_0, the static initial permeability, which is determined by independent experiments (it is also proportional to the saturation magnetization), and $\mu'(f)\to 1$ as $f\to\infty$.

3. Results and discussion

At first, resistance R and inductance L were measured. Initial magnetic permeability was calculated by formulas (1) and (2). The equations linking the experimentally measured coefficients of transmission and reflection of electromagnetic waves with the values of the magnetic permeability are described in paper [6]. Initial magnetic permeability and imaginary part of magnetic permeability allow reoperation the spectrum of the real part of magnetic permeability with the help of the Cramers–Kronig relations [7]. The spectrum of the magnetic permeability of composite material is shown in figure 1.

As shown in figure 1, the real parts of the permeability decrease from 4.9 to 1.0 rel. units with increasing frequency from 0.1 to 18 GHz. The frequency dependences of the imaginary part of the permeability have two maximum. First maximum is region around 1 GHz. It is treated by the authors of [8] to be caused by the processes of displacement of domain boundaries. Second maximum is region around 4 GHz. It is considered to be related to natural ferrimagnetic resonance [8].

Note, that reflection coefficient was calculated according to the obtained values of permeability then compared with the experimentally measured and found that the permittivity of the composite material is equal to ε'= 6,7 rel. units, ε" = 0,6 rel. units.

4. Conclusions

The Z-type hexaferrites were synthesized using a ceramic technique. The composite has been prepared using epoxy as a matrix with 65 wt% the Z-type hexaferrite fillers concentration. Electromagnetic characteristics are measured in range of 0.01 – 18.0 GHz. It is show, that there are two maximum on permeability spectra. They are the processes of displacement of domain boundaries and the natural ferrimagnetic resonance.

Analysis of the spectra by using the Cramers – Kronig relations enables one to select the most reliable experimental results.

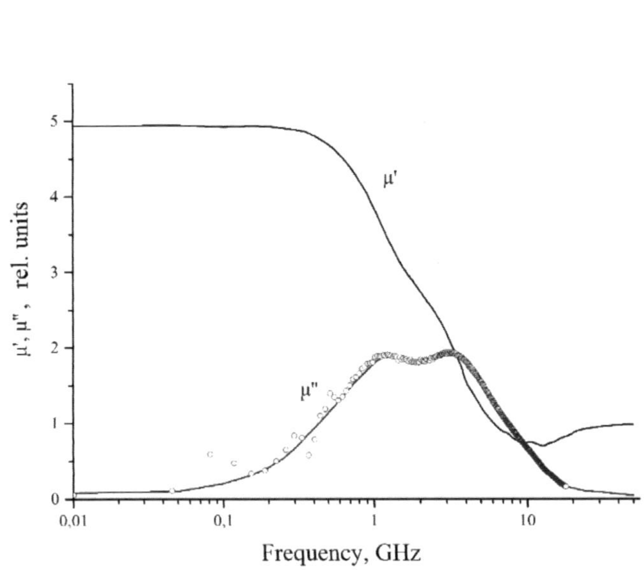

Fig.1 – The spectrum of the magnetic permeability of composite

Acknowledgment

The authors are thankful to PhD V.A. Zhuravlev and PhD E.Yu. Korovin of National Research Tomsk State University for help and support in our research.

This Research is supported by Tomsk State University Competitiveness Improvement Program.

References

[1] Birken J A, Duff W G, Pflug D R and Wallenberg R A 1981 *Composite material aircraft. Electromagnetic properties and design guidelines* (Washington: US Department of the navy)
[2] Imayev M F, Daminov R R, Popov V A and Kaibyshev O A 2005 *Physica C.* **422/1–2** 27
[3] Naiden E P, Zhuravlev V A, Itin V I, Minin R V, Suslyaev V I and Dotsenko O A 2013 *Russian Physics Journal* **55** 869
[4] Wagner D V, Ulyanova O A and Dotsenko O A 2014 *Adv. Mat. Res.* **1040** 29
[5] Chechernikov V I 1969 *Magnetic measurements* (Moscow: Moscow State University publishing house) (In Russian)
[6] Zhuravlev V A and Hackevich Yu A 2013 *Izv. visshih uchebnih zavedenii. Fizika* **56** 312 (In Russian)
[7] Zhuravlev V A and Suslyaev V I 2006 *Russ. Phys. J.* **49** 840
[8] Smit J and Wijn H P J 1959 *Ferrites* (NY: Philips Technical Library)

Effect of electron beam irradiation on thermal and mechanical properties of epoxy polymer

A T Nguyen[1], P M Visakh[1], O B Nazarenko[1], C S Chandran[2] and T V Melnikova[1]

[1]Tomsk Polytechnic University, Lenin av. 30, 634050 Tomsk, Russia
[2]School of Chemical Science, Mahathma Gandhi University, Kottayam, Kerala, India

E-mail: obnaz@mail.ru

Abstract. This study investigates the thermal and mechanical properties of epoxy polymer after exposure to different doses of electron beam irradiation. The epoxy polymer was prepared using epoxy-diane resin ED-20 cured by polyethylenepolyamine. The irradiation of the samples was carried out with doses of 30, 100 and 300 kGy. The effects of doses on thermal and mechanical properties of the epoxy polymer were investigated by the methods of thermal gravimetric analysis, tensile test, and dynamic mechanical analysis. The thermal properties of the epoxy polymer slightly increased after irradiation at the heating in air. The tensile strength and Young's modulus of the epoxy polymer increased by the action of electron beam up to dose of 100 kGy and then decreased. The elongation at break decreased with increasing the irradiation dose.

1. Introduction

The operation of nuclear power plants, mining and refining of uranium ore, industrial and medical use of isotopes generate radioactive waste. The nature of this kind of waste needs proper management and treatment methods in order to protect people and environment at present time and in the future [1]. The methods of nuclear waste management include the handling, pretreatment, treatment, conditioning, transport, storage and disposal of radioactive waste [2–4].

For immobilization of radioactive waste the following methods can be applied depending on the matrix materials: cementation, bituminization, vitrification, polymerization. Immobilization of radioactive waste using cement has such advantages as low cost and the use of relatively simple process plant. Cement is usually used for conditioning of large amount of low level radioactive waste. However, cementation is poorly incorporated with organic waste. Bitumenization is a process of mixing the waste with liquid bitumen in which free water evaporates and waste volume reduced [3, 4]. Bitumen could be used to immobilize organic waste. As a high molecular hydrocarbon, bitumen is expected to withstand well against environmental conditions. However, bitumen is flammable and requires special precautions to prevent its accidental ignition. Vitrification is a method for immobilization of radioactive waste for a long time in compact solid, insoluble form by combining solid waste with glass-forming material like borosilicates, and then heating the mixture under high temperature to form glass-like solid form [5, 6]. However, this method is expensive and can be applied for small scale facilities. For immobilization of radioactive waste, which cannot be treated by previously mentioned methods, polymers can be used. Polymers are chemically inert and have considerable resistance against corrosive elements and leaching the radionuclides [7, 8].

Epoxy resins are thermosetting polymers, which have cross-link polymer chain to form the solid form when adding curing elements. During the process of curing, the epoxy resins can take waste molecules into their molecular network and have tightly fixed them there, thus effectively immobilize radionuclides when treating radioactive wastes. However, the epoxy matrix can accumulate considerable radiation doses due to self-irradiation of the waste, which can deteriorate the properties of polymer. That is why it is necessary to study the possible effect of radiation on the changes in the properties of candidate polymeric materials for immobilization of radioactive waste. In this work, the effect of electron beam irradiation on the thermal and mechanical properties of epoxy polymer was studied.

2. Experimental

The epoxy polymer samples were prepared using epoxy-diane resin ED-20 cured by polyethylenepolyamine. The irradiation of the samples was carried out with doses of 30, 100 and 300 kGy using electron beam generated by the linear electron accelerator ELU-4 (Tomsk Polytechnic University). The sample coding and dose of radiation are given in table 1.

Table 1. The sample coding.

Samples	Radiation dose (kGy)	Code
Epoxy	0	E0-0
Epoxy	30	E0-1
Epoxy	100	E0-2
Epoxy	300	E0-3

The non-irradiated and irradiated epoxy samples were examined using the method of thermal gravimetric analysis (TG). TG was performed with a SDT Q600 thermal analyzer (TA instrument). The thermal decomposition in argon and air atmosphere was investigated at a heating rate of 10 °C/min from room temperature to 1000 °C, with a gas flow rate of 100 mL min^{-1}. The sample weight was about 10 mg.

The tensile properties of the epoxy polymer such as stress-strain behavior, tensile modulus and elongation at break were measured using a H50KT (Tinius Olsen) Universal Testing Machine (UTM) with a load cell of 1000 N. A crosshead speed was 50 mm min^{-1}. The sample size had a length 10 cm, a thickness 2 mm and a width 1 cm.

Dynamic mechanical analysis (DMA) of epoxy samples was carried out with a dynamic mechanical analyzer, DMA D800 (TA instrument), in the tensile mode. The rectangular strips with dimensions around 40×5×1 mm^3 from all nanocomposite films. The tests were performed at a heating rate of 2 °C under isochronal conditions at 1 Hz. The strain amplitude was limited to 0.05 % to be in the linear-viscoelastic range of the material. The storage tensile modulus (E') and loss angle tangent (tan δ) of the materials were evaluated as a function of temperature.

3. Results and discussion

During electron irradiation the color change of the epoxy polymer occurs from light brown to dark brown color, as it is shown in figure 1.

Figure 2(a) and (b) shows the TG results of the epoxy polymer before and after exposure to different doses of radiation, in the presence of argon and air, respectively. We can see from the TG results that there is no significant increment for the thermal stability after radiation treatment. The parameters of the thermal behavior of the studied samples are presented in table 2 (at the heating in argon) and in table 3 (at the heating in air).

The initial temperature of degradation ($T_{5\%}$) is considered as the temperature at which the mass loss is 5 wt. %. At the heating of the epoxy sample in argon the temperature $T_{5\%}$ decreased by 4 °C after exposure to dose of 100 kGy and then increased by 8 °C compared to that of non-irradiated sample. The mid-point temperature $T_{50\%}$ which is corresponded to a 50 % mass loss did not undergo any change and it was 377 °C for all samples. The thermal degradation of the epoxy polymer finished at ~600 °C. The residue at 600 °C was maximal for non-irradiated sample and decreased with increasing the irradiation dose.

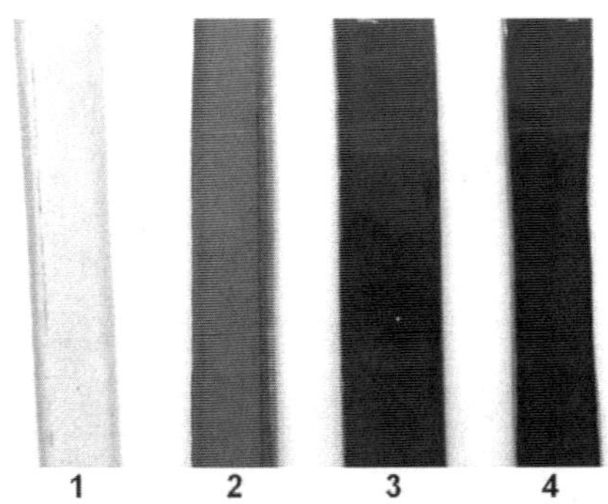

Figure 1. Epoxy polymer samples before (1) and after exposure to dose 30 kGy (2), 100 kGy (3) and 400 kGy (4)

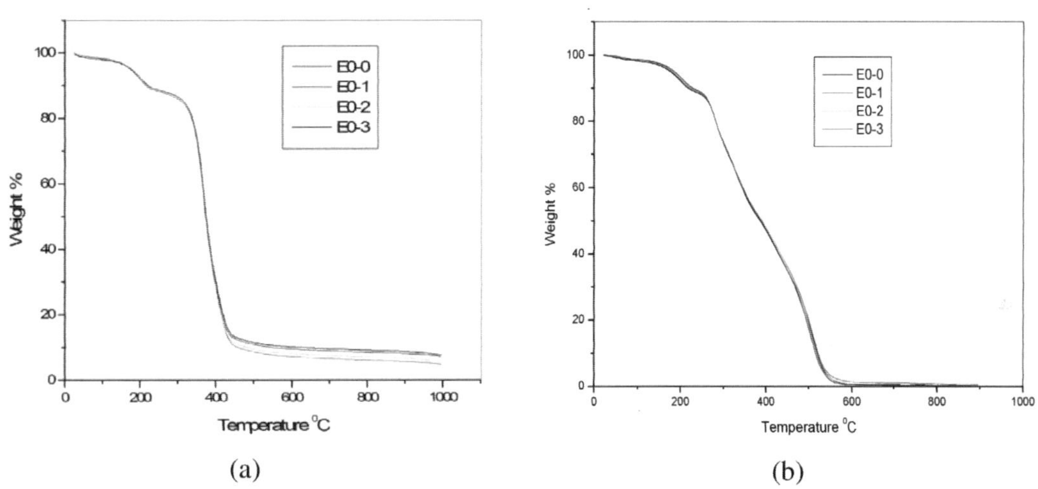

Figure 2. TG curves of the epoxy polymer after irradiation at the heating: (a) in argon; (b) in air.

Table 2. Thermal properties of the epoxy polymer after electron beam irradiation at the heating in argon.

Samples	Dose (kGy)	Degradation temperature (°C)		Residue at 600 °C (%)
		$T_{5\%}$	$T_{50\%}$	
E0-0	0	174	377	9.3
E0-1	30	174	377	7.5
E0-2	100	170	377	6.6
E0-3	300	178	377	6.9

The thermal stability of the epoxy polymer at the heating in air slightly increased after radiation treatment. The temperature $T_{5\%}$ increased by 4 °C after exposure to dose of 30 kGy and by 10 °C after increasing the dose up to 100 and 300 kGy compared to that of non-irradiated sample. The temperature $T_{50\%}$ increased by 4 °C at the dose of 30 kGy in comparison with the sample before irradiation, then it increased by 1 °C after increasing the dose up to 100 kGy. After further increasing the dose the temperature $T_{50\%}$ of the epoxy polymer decreased by 3 °C. The similar changes are typical for the residue at 600 °C.

Table 3. Thermal properties of the epoxy polymer after electron beam irradiation at the heating in air.

Samples	Dose (kGy)	Degradation temperature (°C)		Residue at 600 °C (%)
		$T_{5\%}$	$T_{50\%}$	
E0-0	0	176	386	0.4
E0-1	30	180	390	0.6
E0-2	100	186	391	0.6
E0-3	300	186	388	0.4

The typical processes that occur during irradiation of the epoxy polymer are cross-linking and chain scission reactions [9]. The cross-linking processes prevail in the dose range up to 100 kGy and leads to formation of chemical bonds between two adjacent polymer molecules. In the dose range higher than 100 kGy chain scission reactions dominate and the thermal stability decreases because of polymer chain break.

Typical stress vs strain curves for the epoxy polymer after exposure to different doses of radiation are shown in figure 3. The tensile properties, including tensile strength, elongation at break, and Young's modulus of the epoxy samples after irradiation are summarized in table 4.

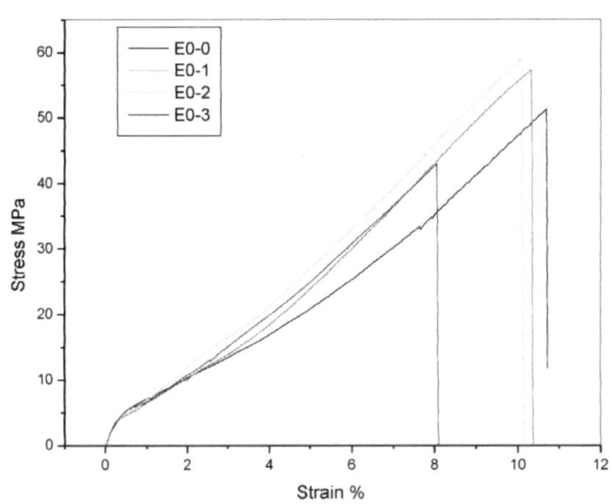

Figure 3. Mechanical properties of neat epoxy film and epoxy film with different dose of radiation treatment

Table 4. Tensile properties of epoxy polymer.

Dose (kGy)	Tensile strength (MPa)	Elongation at break (%)	Young's modulus (MPa)
0	51.1±3	10.55± 0.8	4.84
30	56.8±4	10.24±0.6	5.55
100	58.8±5	10.04±0.8	5.86
300	42.6±6	7.98±0.3	5.34

IOP Conf. Series: Materials Science and Engineering **168** (2017) 012073 doi:10.1088/1757-899X/168/1/012073

As followed from table 4, the tensile strength for non-irradiated sample is 51.1 MPa and increased after irradiation with doses 30 and 100 kGy to 56.8 and 58.8 MPa, correspondingly. However, after further increasing the dose to 300 kGy the tensile strength decreased to 42.6 MPa. A similar dependence was obtained for the Young's modulus of the epoxy polymer. The elongation at break decreased for the epoxy polymer after exposure to electron beam irradiation.

The improvement of the tensile properties after exposure to the doses of 30 and 100 kGy can be explained by the action of cross-linking reactions. At the higher doses up to 300 kGy, it was found that the irradiation of epoxy resins decreases the mechanical properties as a result of chain scissions reactions.

The representative curves of the storage modulus (E') of the epoxy samples before and after exposure to different doses of radiation are shown in figure 4(a) and the tanδ is shown in figure 4(b). The values of the glass transition temperature and the position and intensity of the tanδ peaks for the studied epoxy samples are shown in table 5.

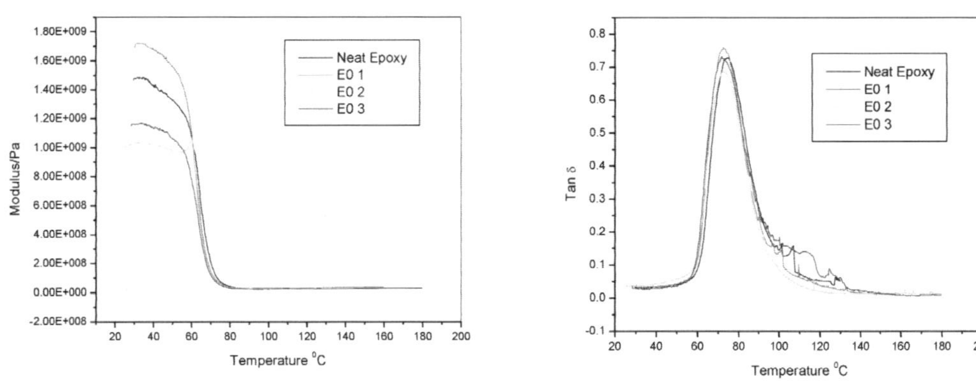

(a) (b)

Figure 4. Tensile modulus E' vs temperature (a) and tanδ vs temperature (b) of epoxy polymer before and after radiation treatment

Table 5. The values for the temperature and height of the tanδ peaks for neat epoxy and epoxy composites with different dose of radiation treatment.

Samples	Temperature tanδ peak (°C) (Tg)	Height tanδ peak
E0-0 (Neat epoxy)	74.6	0.72
E0-1	72.4	0.75
E0-2	72.9	0.68
E0-3	71.7	0.73

The samples of both non-irradiated epoxy polymer and irradiated epoxy samples are in the glass state at temperatures below 60 °C and in the rubber state at temperatures above 80 °C; thus, the glass transition region for these samples is from about 60 to 80 °C. In the glass transition region of all samples, the storage modulus drops sharply from its original value in the glass state to the lower value in the rubber state. At low temperatures (below 40 °C), modules are very different from one another (figure 4(a)). The value of modulus is higher for the sample irradiated with dose of 30 kGy than that of other samples. This is due to the fact that in the glassy state, molecular motions are largely restricted to vibration and short-range rotational motions. As the temperature increases, the studied samples show only a decrease in the storage modulus. The decrease in storage modulus can be assigned to the relaxation of the polymer chains at glass transition region. All samples showed a single transition in the temperature range of 60 to 80 °C.

4. Conclusions

The epoxy polymers were prepared using epoxy resin ED-20 cured by polyethylenepolyamine. These epoxy samples were exposed to electron beam irradiation with doses of 30, 100 and 300 kGy. The effect of electron beam irradiation on thermal and mechanical properties of the epoxy polymers was investigated. The results of thermal analysis showed that the thermal stability of the studied samples at the heating in air slightly increased. The tensile strength and Young's modulus increase up to dose of 100 kGy. All samples showed a single transition in the temperature range of 60 to 80 °C.

Thus, the obtained results showed that the mechanical characteristics of the epoxy resin are improved after the action of electron beam with dose up to 100 kGy. Epoxy resin ED-20 is considered as a polymer with high resistance to radiation and can be recommended to immobilize radioactive wastes after further additional investigation.

Acknowledgments

The authors are thankful to the Scientific and Analyzing Centre of Tomsk Polytechnic University for the providing the TG measurements.

References

[1] Rao K R 2001 *Curr. Sci.* **81** 1534
[2] Baisden P A and Atkins-Duffin C E 2011 Radioactive Waste Management *Handbook of Nuclear Chemistry* Eds A Vertes, S Nagy, Z Klencsar, R G Lovas and F Rösch (New York: Springer US) 2797–2835
[3] Tang Y S and Saling J H 1990 *Radioactive Waste Management* (Washington: Hemisphere Publishing Corporation)
[4] Nagasaki S and Nakayama S 2015 *Radioactive Waste Engineering and Management* (Tokyo: Springer Japan)
[5] Donald I W, Metcalfe B L and Taylor R N J 1997 *J. Mater. Sci.* **32** 5851
[6] Ojovan M I and Lee W E 2005 *An Introduction to Nuclear Waste Immobilisation* (Amsterdam: Elsevier)
[7] Dawson J, Smith V, Clifford J and Williams S J 2012 *Mineral. Mag.* **76** 2985.
[8] Efremenkov V M 1989 *IAEA Bull.* **4** 37
[9] Kircher J F and Bowman R E 1964 *Effects of Radiation on Materials and Components* (New York: Reinhold Publishing Corporation)

Generation of ionizing radiation from lithium niobate crystals

L N Orlikov, N L Orlikov, S I Arestov, K M Mambetova and S M Shandarov

Electronic devices Department, Tomsk State University of Control Systems and Radioelectronics, Russia.

E-mail: oln4@yandex.ru

Abstract. The work done experimentally explores generation of electron and x-ray radiation in the process of heating and cooling monolithic and iron-doped crystals of lithium niobate. Iron doping to the concentrations in the range of 10^{23} m^3 was carried out by adding ferric oxide into the melt during the process of crystal growth. The research into radiation generation was performed at 1-10 Pa. The speed of heating from -10 to 107^0 C was 10-20 degrees a minute. Current pulses appeared at 17, 38, 56, 94, 98, 100, 105, 106, 107^0 C with the interval of 1-3 minutes. The obtained electron current increased in direct proportion to the crystal surface area. The maximum current was 3mA at the design voltage 11 kV on the crystal with 14,5x10,5x10 mm^3 surface area. The article describes the possibility to control the start of generation by introducing priming pulse. The results achieved are explained by the domain repolarization while heating the crystal and the appearance of electric field local strength. Bias and overcharge currents contribute to the appearance of electric strength, which stimulates breakdown and plasma formation. X-ray radiation appears both at the stage of discharge formation and during electron deceleration on gas and target material.

1. Present state of the problem

Lithium niobate is a promising material for creating compact solid-state x-ray emitters and electron accelerators based on pyroelectric effect [1-4]. However, insufficient radiation power restrains solid-state emitters from their practical application. Irradiating crystal with laser, increasing the number of elements and engraving allow to intensify the electric field formation on the crystal surface. In [2] we show the impact of plasma in the formation of emission from piezoelectric crystals. In this respect, the range of pressure 1-10 Pa is quite attractive, as it allows to realize high-voltage discharge and charge current gas amplification.

2. Problem solution technique

The solution of the radiation intensification problem is based on changing crystal parameters via doping and analyzing radiation generation processes from the point of view of plasma emission electronics. Doping allows to change crystal capacity and its resistance, and to activate junctions which stimulate excited atom migration. Heating the crystal leads to formation of electric field local strength areas as a result of domain vector polarization change. Breakdown between these areas becomes the source of plasma. Short-wave radiation (ultraviolet, x-ray) appear both from plasma and as a result of electron deceleration on gas or target material. This paper considers generation of electron stream in vacuum chamber at 1-10 Pa.

3. Process modeling

Figure 1 shows the model structure of the radiation generation. Crystal with radiation surface S is placed at some distance d from target *1*. To make it simple, let's assume that crystal *2* is cylindrical and there is no electric field at the end faces. It's supposed that changing the crystal temperature T leads to the formation of local area *3* of electric field strength E because of crystal domain repolarization. This causes gas ionization, breakdown and electron emission from discharge plasma. Figure 2 shows the assumed diagram of current and voltage time change.

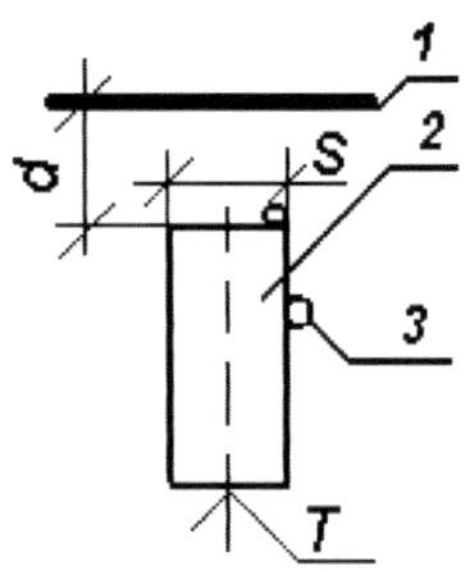

Figure 1. Emitter model structure: 1– anode; 2 – crystal; 3 – field strength area.

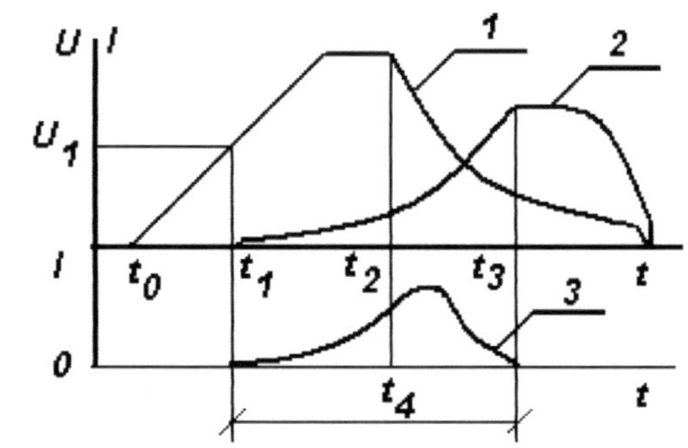

Figure 2. X-ray generation diagram: 1– voltage characteristic; 2 – discharge current characteristic; 3 – X-ray pulse.

The process of x-ray radiation generation during crystal heating proceeds in several stages. At time t_0 the crystal is excited and the areas of electric field strength appear on it. At a specific distance, the field strength gets sufficient enough to ionize residual gas. This enables the formation of electron avalanche with the voltage U_1 (characteristic *1*) and current (characteristic *2*). The ions produce plasma cloud, which results in the increase of current and gap breakdown. Beginning with time t_2, when the current gets reasonably large, the plasma finite conductivity, restricted by charged particles mobility and deionization processes, starts to show. During the times t_4 from t_1 to t_3 in the area where there is both current and voltage, we can observe shock processes with generation of x-ray, ultraviolet and other radiations (characteristic *3*). To implement electron emitter based on pyroelectric effect it's essential to provide the current-voltage correspondence. The absence of current (idle stroke) or the absence of voltage (short circuit mode) indicates the non-operability of an electron emitter. The correspondence of current $i(t)$ and accelerating voltage U at particle emission is set by the three-halves power law [5]:

$$i(t) = 2{,}33 \cdot 10^{-6} U^{3/2}(t)\frac{S}{d} \tag{1}$$

where S is current take-off area; d is the distance from emission zone to anode (target).

The charge presence is based on the Paschen Law, which correlates the ignition voltage U_3 and the product of pressure P and distance between electrodes d:

$$((U_3=f(Pd)).$$

The situation when the charge doesn't ignite can be described by the inequality [6]:

$$\gamma(e^{\alpha d} - 1) < 1, \tag{2}$$

where γ is the secondary ion-electron emission factor; α is Townsend ionization coefficient.

Electric field strength E appearing as a result of pyroelectric effect determines the breakdown voltage U:

$$U = E \cdot d \tag{3}$$

In the process of formation of electron avalanche, the shock ionization factor increases ($\alpha > 1$) and the conditions of self-maintained discharge are realized [7]:

$$\alpha d = \ln(1 + 1/\gamma) \tag{4}$$

In this case electron current component I_e can be found by [6]:

$$I_e = n_e V_e e S \,, \tag{5}$$

where n_e is plasma concentration; V_e, e are electron velocity and elementary charge respectively.
In the simplest case, electron velocity excluding gas encounter can be calculated from kinetic reasoning [8]:

$$I_e = n_e V_e e S \tag{6}$$

Parameters modeling was conducted with MATHCAD software.

4. Sample preparation technology

The samples of lithium niobate were obtained by Czochralski method. While doping some crystals, Fe_2O_3 was added to the initial mix. The concentration of doping material was determined with interference [9] and SIMS [10, 11] techniques. To conduct research we used the crystals of different size with Z-oriented plane of polarization. The samples were subjected to microscopic study to analyze polishing accuracy, chip and scratch defects, and surface contaminants. Then, the samples were dried at 500K for 24 hours.

5. Experimental studies

The studies into electron emission from lithium niobate were carried out with the standard general-purpose exhaust cart VUP-5. Figure 3 shows the experiment schematic course, figure 4 demonstrates sample parameters.

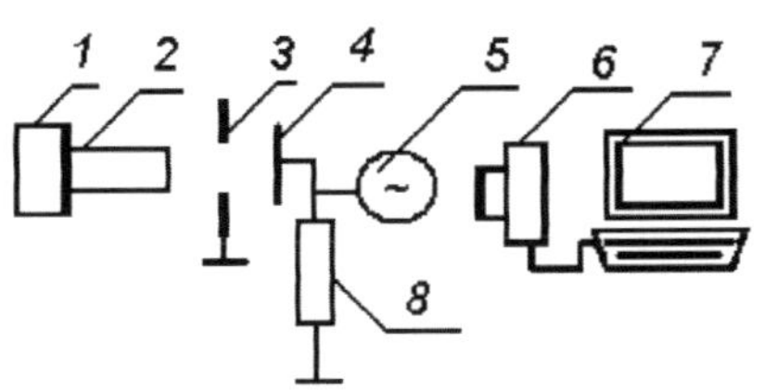

Figure 3. Experiment schematic course: 1- Peltier element; 2 – crystal; 3 – anode; 4 – collector; 5 – oscilloscope; 6 – camcorder; 7 – PC; 8 – resistor; 9 – power supply.

Sample parameters: length, width, thickness, concentration
1 – 15 x 6 x2 mm ; $C_{Fe\ 2+}$= $3{,}88\cdot10^{23}$ m^{-3};
2 – 18 x 21 x 3,5 mm; $C_{Fe\ 2+}$= $3{,}01\cdot10^{23}$ m^{-3};
3 – 14,5 x 10 x 10,5 mm; $C_{Fe\ 2+}$= $1{,}73\cdot10^{23}$ m^{-3};
4 – 10 x 4 x 4 mm undoped

Figure 4. Samples parameters.

Peltier element *1* of the type TEC1-12706 was connected to power supply unit HY3005-3. Crystal *2* was heated or cooled through copper plate. The temperature was measured with thermal converter attached to multimeter Protek 505. Anode *3* and electron collector *4* were arranged in the form of a Faraday pyramid at the distance of 3, 5, 10 mm from the crystal. Providing the pressure is ~1,33 Pa and heating the crystal proceeds at 10 - 20° /min, there appear surface charge areas on the crystal surface, which cause breakdown and plasma formation. Electron current flows towards anode *3*. Current signal from resistor 8 falls on the oscilloscope C8-17 (*5*). The fragments from oscilloscope pictures are recorded onto the camcorder *6* and processed with PC *7*. The resistor resistance (*R*=10 Ом) was chosen in accordance with the conditions of minimal impact of thermal noise on the current

IOP Conf. Series: Materials Science and Engineering **168** (2017) 012074 doi:10.1088/1757-899X/168/1/012074

detected at heating the crystal. The realizable breakdown voltage was estimated with the equation (1). The x-ray intensity was estimated by the blanket exposure of x-ray photographic film and compared with the negatives found in medicine. The pulse energy was estimated by the oscilloscope picture with the equation:

$$W = \int_0^t i(t) \cdot U(t)dt \qquad (7)$$

6. Results and outputs

6.1. Undoped crystal (No 4).

Changing crystal temperature is followed by the appearance of voltage in it. This results in self-induced, spontaneous generation of electrons, as well as bias and overcharge currents, occurring at the frequency ~ 1 MHz. The largest intensity of spontaneous electron stream generation is observed at the pressures ~ 1 Pa and the temperatures 425-375 K. Increasing the pressure to 10 Pa results in a slight increase of current, which can be explained by the "gas amplification" at which ionization factor increases with the pressure rise. The interval between radiating emissions is 1-3 minutes. The crystal cooled down through its own heat capacity at 10^0/min.

Figure 5 shows oscilloscope pictures of collector current at 1 Pa. Figure 6 shows the x-ray recorder.

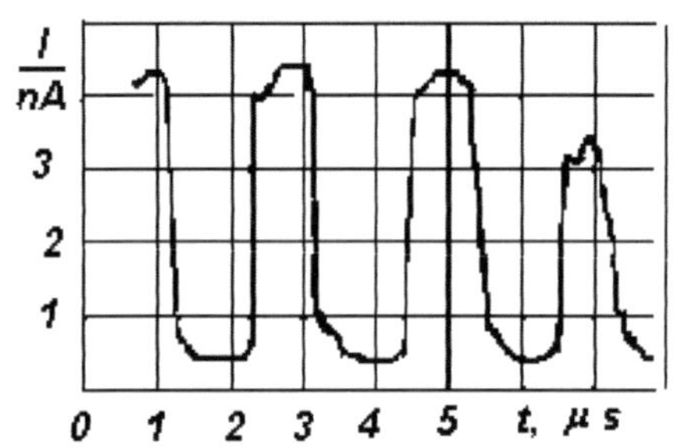

Figure 5. Oscilloscope pictures of collector current at 1 Pa for undoped crystal.

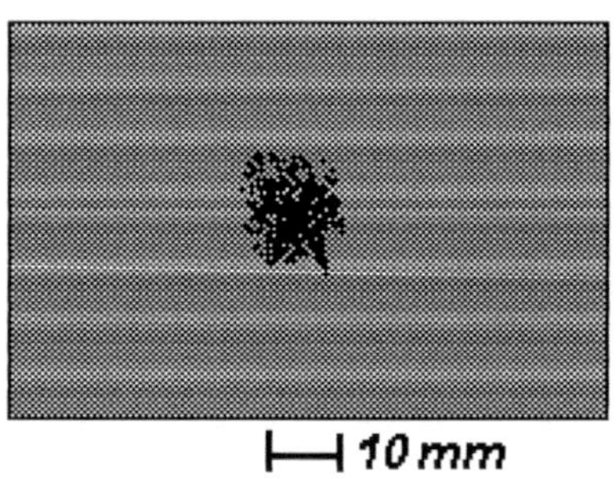

Figure 6. The x-ray recorder on a photographic film.

The series of oscilloscope pictures prove the gas gap breakdown along the crystal surface. Oscillogram "downfall" near the maximum tells about current take-off. Design stress was 20-45 kV with the current up to 10 nA.

The intensity of film blanket exposure after tenfold radiation treatment in vacuum is 0,04 mz according to the results of photometry and referencing. Based on [8] the deceleration radiation intensity is directly proportional to the current, target material atomic number and the square of accelerating voltage. No more than 1 per cent of input voltage turns into x-ray radiation.

Managing the moment of generating radiation.

The migration of electric field strength area due to spontaneous local repolarization under temperature causes the instability of radiation generation moment. Distance d from the emission bound to accelerating electrode grows with the rise of accelerating voltage and reduces with the densification of extractable particles.

Introducing additional electrode into the crystal region allows for the preliminary pulse electric spark from capacitive discharge (1000 V, 5 mA, 50 ms). Plasma generation in the region of crystal local electric field enables to initiate the main discharge along the surface of piezoelectric followed by the current take-off.

Preliminary pulse allows to reduce the discharge time and to weaken electron stream intensity.

6.2. Iron doped crystals

Figure 7 shows oscilloscope pictures of signals obtained while heating doped crystal №1 to 90^{0}C and cooling it to -2^{0}C. The distance from the crystal to anode was 10 mm.

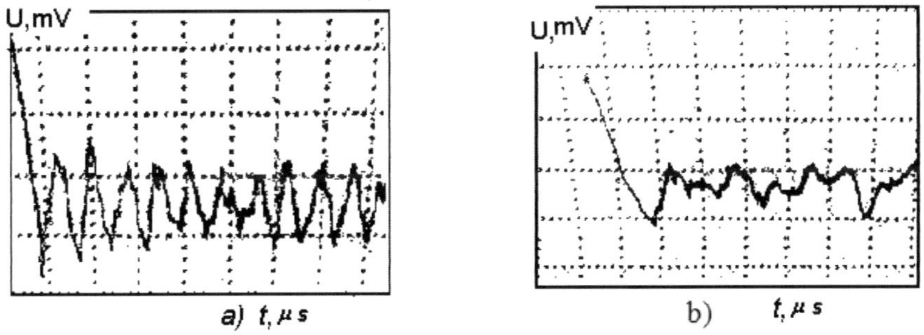

Figure 7. *a)* Oscilloscope picture of the current at heating the crystal; 1mV/div=0,5 mA, t – 2μ s/ div *b)* oscilloscope picture of the current at cooling the crystal 1 mV/ div =0,5 mA, t – 2 μs/ div

Increasing the temperature leads to increasing electric field strength and producing breakdown between crystal surface and anode. In pressures ~ 1Pa there is a possibility of a short non self-maintained glow discharge striking. Relatively permanent peak current intervals while heating can be explained by the gas gap breakdown between the areas of local strength. Current oscillations after the first breakdown result from the processes of domain recharge, deionization and recombination on the crystal surface.

Reducing the distance between the crystal and anode up to 3-5 mm results in transition of discharge into high voltage structure with higher breakdown voltages and more expressed current fluctuations. Current pulses at Pd = 0,5 – 1 correspond to voltages up to 5 kV. For copper anode (γ=7) impact-ionization rate is α= (5 – 23), U = 11 kV, I = 3 mA. The obtained values of current and voltage correlate to the equation (1). Higher values of voltage pulses correspond to Pd = 5 – 8, however, under these circumstances the conditions for maximum values of current and voltage are instable. This agrees with [7] and is probably caused by the slope of the left branch of Paschen curve with respect to reference axis when one Pd can have two appropriate ignition voltages.

Peak currents while heating and cooling the crystal are commeasurable. However, after several cycles of heating and cooling, discharge current at breakdown decreases fivefold or six fold. Changes in current and voltage result from sorption and desorption processes happening in the process of changing crystal temperature in the areas of fore vacuum pressure values.

Increasing the size of samples leads to increasing the current proportionally to crystal area. As a result, peak current value (3mA) was achieved with sample № 3 having the size 14,5x10,5x10 mm^3 while heating it at 20 ^{0}C/min from -10 to 107 ^{0}C. Current pulses appeared with the interval 1-3 minutes at temperatures 17, 38, 56, 94, 98, 100, 105, 106, 107^{0}C. Thus, iron-doped samples with $C_{Fe^{2+}}$ equal to 3,88$\cdot$10^{23} m^{-3} allow to increase emission current to several milliamps. Crystal cascading results in increasing breakdown voltage, but it restrains crystal heating and cooling possibilities.

7. Conclusions

During the research, the new data on possibilities to generate electron stream in the process of heating and cooling the crystal at pressures 1-10 Pa were obtained. Iterative intervals on oscilloscope pictures show the gas gap breakdown rather than crystal surface breakdown. After the breakdown, bias and overcharge currents are realized. These currents maintain remaining residues of electron avalanche and prevent the discharge from instantaneous extinction. The studies [8-9, 12] prove the feasibility of the suggested model for generating electron and x-ray radiation. The well-defined regularity of changes in radiation characteristics depending on the distance, crystal size, heating and cooling rates, and arranging system elements was established. Choosing doping material and its concentration allows for new prospects in developing solid-state optical electronic devices and x-ray equipment based on lithium niobate.

References

[1] Kukhterev N, Kukhtereva T, Stargell G, Wang J 2009. *Appl. Phys.* **106** 014111
[2] Mesyats G A 2008 *Advances in Physics* **178** 1 85-108
[3] Nagaychenko V, et.all. *Problems of atomic science and technology* 2010 2 Series: Nuclear Physics Investigations (53) 34-38
[4] Kisa T, Murakami K, Abo S, Wakaya F and Takai M 2010 *Vac. Scl. Technol.* **28** C2B27-C2B29
[5] Proskurovsky D Emission electronics 2010. Tomsk: Tomsk State University 288
[6] Burdovitsyn V, Klimov A, Medovnik A, Oks E, Yushkov Y 2014 Forvacuum plasma electron emitters. Tomsk: Tomsk State University 288
[7] Tkachev A, Fedenyev A, Yakovlenko S 2005 *ZhTF* **75** 4 60-65
[8] Vavilov S 1981 Pulse x-ray equipment S P Vavilov.– M: Energiya 120
[9] Orlikov L N et.all. 2011. TUSUR Proceedings, part 2(24) 135-138
[10] Surzhikov A P, Chernyavskii A V, Ghyngazov S A, Frangulyan T S 2002 *Russian Physics Journal* **45** 12 1190-94
[11] Gyngazov S A, Surzhikov A P, Frangul'yan T S, Chernyavskii A V 2002 *Russian Physics Journal* **45** 8 753-58
[12] Mambetova K et.all 2015 *Radiophysics and Quantum Electronics* **57** 8-9 603-609

IOP Conf. Series: Materials Science and Engineering **168** (2017) 012075 doi:10.1088/1757-899X/168/1/012075

Structure parameters and magnetic properties of nanosized strontium hexaferrite prepared by the sol-gel combustion method

V A Zhuravlev[1], R M Minin[2], V I Itin[2], M V Politov[1], I Yu Lilenko[1] and M R Ufimtsev[1]

[1]Tomsk State University, Lenin av., 36, Tomsk, 634050 Russia
[2] Tomsk Scientific Center SB RAS, pr. Academicheskii 10/4, Tomsk, 634021 Russia

E-mail: ptica@mail.tsu.ru

Abstract. The phase composition, structure parameters, and main magnetic characteristics of $SrFe_{12}O_{19}$ hexaferrite prepared by the sol-gel combustion method in combination with subsequent annealing at the temperature of 850°C for 6 hours are investigated. From the study of the magnetization curves in the pulsed magnetic fields the value of saturation magnetization of synthesized material was determined. The value of the magnetocrystalline anisotropy field was determined from the survey of the ferromagnetic resonance.

1. Introduction

Hexagonal ferrite of M-type $SrFe_{12}O_{19}$ is characterized by a large coercive force, comparatively large value of the saturation magnetization, high Curie temperature, large magnetocrystalline anisotropy, high electric resistance, high chemical stability and corrosion resistance. Due to the qualities listed and its low cost the ferrite is used in different spheres of science and technology as permanent magnets and bonded magnets, in various microwave devices (circulators, isolator, phase shifters, filters) and in magnetic and magneto-optic recorders of information with high density [1]. For the reasons listed above the materials which possess the controllable size and form of particles, narrow distribution of particles according to the size and with given magnetic characteristics are needed.

The traditional ceramic method of strontium hexaferrite synthesis is burning the mixture of $SrCO_3$ and Fe_2O_3 powders at a high temperature (> 1200 °C). The disadvantages of this method are high power consumption, uncontrolled morphology of particles, chemical heterogeneity and big size of particles and their agglomerates. Grinding of synthesized strontium hexaferrite powder leads to admixtures, elastic stresses and worsening of magnetic properties [2, 3].

To control the size and form of the particles as well as the magnetic properties of materials new energy-saving methods have been suggested. Such as sol-gel method and its modifications [4, 5], pyrolysis of aerosol, coprecipitation, hydrothermal method, SHS etc. The full description of these methods can be found in the review [1]. As a rule, using these methods the nanosized and nanostructured particles are obtained which possess the properties different from the properties of massive polycrystalline materials due to the significant part of atoms present in the defective surface layer. By means of additional annealing one can change the size of particles, their morphology and the magnetic properties.

In the present work, the sol-gel combustion method is used to prepare nanosized strontium hexaferrite powders of M-type with the $SrFe_{12}O_{19}$ composition. The method is based on the preparation of initial colloid nanosystem that can react in the combustion mode. This method is the most suitable for the preparation of powder materials with good stoichiometric control, narrow particle size distribution and it requires relatively short processing time at lower temperatures [6]. Results of investigations the phase composition, structural parameters, and magnetic properties of the synthesized materials are presented.

Research of the magnetocrystalline anisotropy of the samples was carried out by the method of ferromagnetic resonance (FMR). FMR is the only method for the study of the magnetocrystalline anisotropy of the multi-phase powder samples consisting of a phases with significantly different values of the anisotropy fields [7].

2. Preparation of samples

As the reagents the aqueous solutions of analytical grade strontium nitrate $Ba(NO_3)_2$, analytical grade 9-aqueous iron nitrate $Fe(NO_3)_3 \cdot 9H_2O$, and lemon acid $C_6H_8O_7$ with concentration of 1 M were mixed in proportions

$$\{[Sr(NO_3)_2]:[Fe(NO_3)_3 \cdot 9H_2O]\}:C_6H_8O_7 = \{1:11.5\}:2.$$

The concentrated solution of NH_4OH ammonium hydroxide was added dropwise to the prepared mixture with constant stirring until pH of the solution became equal to 7. An HI98103 checker portable digital tester (HANNA Instruments) was used to measure pH of the solution mixture. The prepared sol was heated in an ES-6120 magnetic stirrer (EKROS) at temperatures from 80 to 90°C for 3–5 h. As a result of heating, the sol was transformed into a viscous gel of brown color which foamed. Further heating of the gel foam to temperatures 150–160°C caused its ignition and combustion that lasted for several minutes. After combustion, a friable powder was obtained which was easily grinded in a mortar converting into a fine powder. The $SrFe_{12}O_{19}$ hexaferrite nanoparticles were finally formed by annealing of the powder at the temperature of 850°C for 6 h.

3. The research techniques

The X-ray diffraction analysis was performed with a SHIMADZU XRD-6000 polycrystalline diffractometer in the Bragg–Brentano geometry with a focusing pyrographite crystal-monochromator inserted into the secondary gamma-quantum beam. For qualitative analysis of the phase composition, the PDF4+ computer database of x-ray powder diffractometry of the International Center of Diffraction Data (ICDD, Denver, USA) was used. The quantitative analysis of the phase composition and the refinement of the structural parameters of the detected phases were performed using the Powder Cell 2.4 program of full profile analysis. The infrared spectrum was registered with a Nicolet 5700 infrared Fourier spectrometer (France) equipped with an attachment of diffuse reflection.

Magnetic measurements involved investigations of magnetization curves in pulsed magnetic fields up to 30 kOe using a setup described in [8] and measurements of ferromagnetic resonance (FMR) spectra by the standard waveguide passing through technique at frequencies in the range 37–53 GHz with an automated radio spectroscope. To investigate the FMR, powders of the examined samples were put in thin-walled quartz tubes with inner diameter of 0.7 mm and length of 10 mm. The tubes were arranged into a rectangular waveguide parallel to the wide wall of the waveguide in order that the alternating magnetic field was directed along the sample axis. A constant magnetizing field was directed perpendicularly to the wide wall of the waveguide.

4. The results of x-ray researches

The X-ray diffraction patterns of the synthesis product after ferritization at 850 °C for 6 hours are shown in Fig. 1. It can be seen that it contains basically diffraction peaks corresponding to $SrFe_{12}O_{19}$ hexaferrite. The traces of accompanying hematite phase (α-Fe_2O_3) are also present.

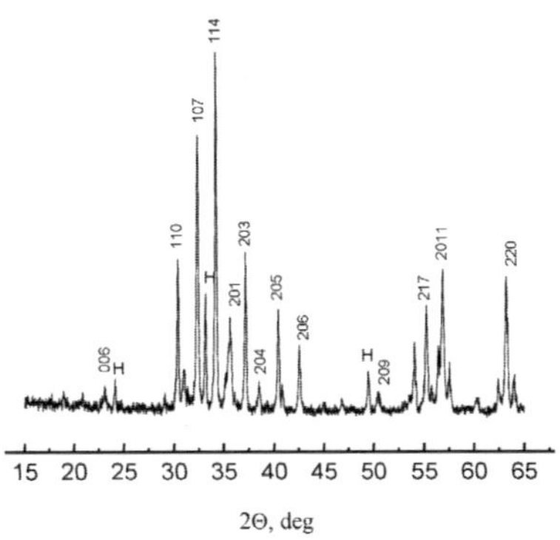

Figure 1. X-ray diffraction pattern of $SrFe_{12}O_{19}$ hexaferrite powder synthesized by sol-gel combustion method. The letters H show the traces of the hematite phase.

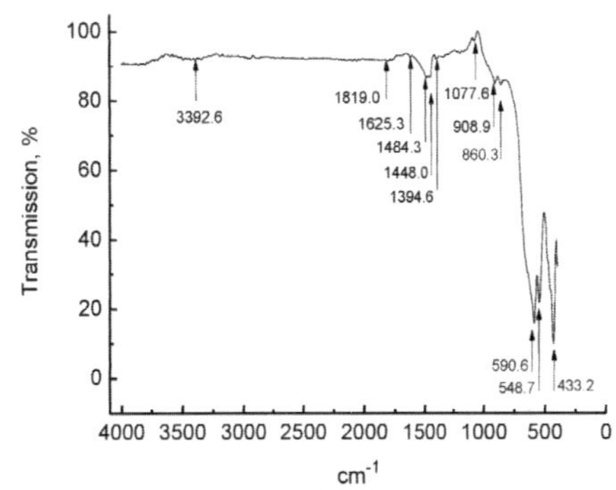

Figure 2. IR-spectrum of product after sol-gel combustion and heat treatment at 850 °C for 6 hours.

Results of the x-ray diffraction analysis of the sample are given in Table 1.

Table 1. Phase composition and structural parameters of synthesized material after calcinations.

Phase constitution, %		Lattice constant, Å		CSR, nm	$\Delta d/d$*10^3
$SrFe_{12}O_{19}$	α-Fe_2O_3	*a*	*c*		
99.4	0.6	5.9175	23.3130	98	2.7

The content of the Sr–M main phase exceeded 99% and hematite α-Fe_2O_3 was the additional phase. The average crystallite sizes (~ 100 nm) were estimated based on an analysis of the physical broadening of diffraction lines from sizes of coherent scattering region (CSR). In the last column of the Table 1 shown the relative changes in the interplanar distances $\Delta d/d$. The estimates of internal elastic microstrain $\sigma = E(\Delta d/d)$ can be carried out by that. Here *E* – Young's modulus.

5. FTIR spectrum of the sample

The Figure 2 shows the infrared spectrum of transmission coefficient of the obtained hexaferrite powder after annealing at T = 850 °C for 6 hours. At frequencies in the range 3500–3300 cm^{-1}, a very weak absorption band caused by strain-induced vibrations of O–H groups is observed, wide in comparison with other bands. It corresponds with a weak absorption band at 1625 cm^{-1} caused by H–O–H vibrations (bending vibrations of water molecules). The bands at 1484 - 1448 cm^{-1} and 860 cm^{-1} are due to the C–O vibrations due to the presence of lemon acid residues in the water.

The band localized at 909 cm^{-1} is probably attributed to the impurity phase α-Fe_2O_3 which corresponds with the data of X-ray diffraction analysis (Table 1). The bands observed at 433, 549 and 591cm^{-1} can be attributed to valence vibrations metal-oxygen of strontium hexaferrite. Similar results were obtained in [4] for strontium hexaferrite, synthesized by sol-gel combustion method with the usage of egg white.

Thus, the X-ray diffraction and spectrometric data given above prove that the synthesized material is strontium hexaferrite $SrFe_{12}O_{19}$.

6. Research of the magnetization curves

A study of the magnetization curves of $SrFe_{12}O_{19}$ hexaferrite powder was conducted at the room temperature in the pulsed magnetic field by the method described in [8]. The results are presented in the Figure 3. The arrows show the direction of the magnetizing field alteration. The linear extrapolation of high-field part of the magnetization curves on the zero magnetizing fields gives the estimation for the specific saturation magnetization (σ_s). The specific residual magnetization (σ_r) can be determined from the point of intersection of the drop-down portion of the magnetization curve with the Y-axis.

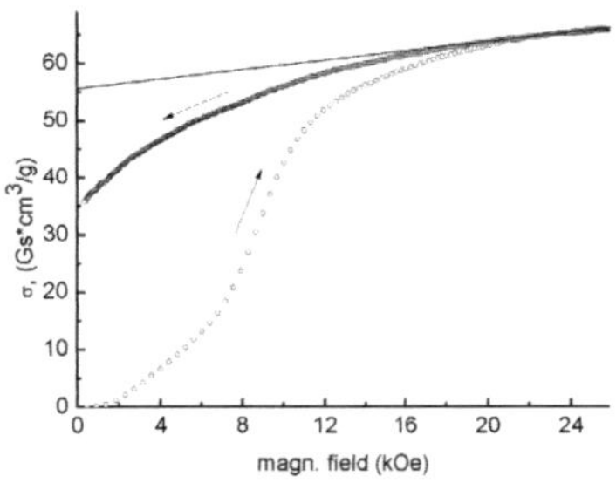

Figure 3. The magnetization curve of the synthesized strontium hexaferrite powder. The arrows show the direction of the magnetizing field alteration.

The values of the specific residual magnetization (σ_r) and the specific saturation magnetization (σ_s) are presented in the first row of the Table 2. In the last column of the table the calculations according to the formula $M_S = \sigma_S \cdot \rho$ are shown. Here $\rho = 5.11$ g/cm^3 is the density of strontium hexaferrite [1]. In the subsequent rows of the table for comparison are presented the results obtained in other studies.

Table 2. The research results of the magnetization curves

Sample	σ_r, Gs·cm^3/g	σ_s, Gs·cm^3/g	σ_r / σ_s	Size, nm	M_S, Gs
Sol-gel combustion, 850 °C	35.9	55.8 (0 kOe)	0.64	~ 100	285
Sol-gel [4], 1200 °C	-	56.2 (10 kOe)	-	≥ 1000	-
Sol-gel [5], 825 °C	-	62.6 (15 kOe) 70.8 (50 kOe)	0.49	135x50	-
Sol-gel [1], 1000-1200 °C	-	56 ÷ 70	-	-	-

In the paper [4] where the egg white was used as a fuel the value of σ_s measured in the field ~10 kOe is close to the one obtained in this study though the sizes of the particles differ significantly due to the different temperature of annealing. In the paper [5] the modified sol-gel synthesis is used and the hexaferrite particles were formed and annealed in the NaCl matrix. So a big value for the specific saturation magnetization was obtained. The data from the review [1] are given in the line 4 of the table. It is clear that the above values of σ_s are typical for strontium hexaferrites that synthesized by the sol-gel methods.

7. The study of the magnetic anisotropy by the FMR method

The study of ferromagnetic resonance spectrum for powder and polycrystalline hexaferrites gives an opportunity to experimentally determine the values and the sign of the fields of magnetocrystalline anisotropy (H_{ai}) as well as the value of the gyromagnetic ratio $\gamma = ge/2mc$, where g is the effective g-factor of the examined material, e is the electron charge, m is the electron mass, and c is the velocity of light. It has to be noted that in hexaferrites the γ parameter can be anisotropic [9]. The magnitudes and signs of the anisotropy fields H_{ai} define the type of magnetic ordering – the presence of either an easy magnetization plane (EMP) or an easy magnetization axis (EMA).

The method described in [9, 10] was used to calculate the resonance curves of FMR or components of permeability tensor of the single-domain polycrystalline and powder ferrimagnets with hexagonal crystal structure in the approach of non-interacting grains. The calculations showed:

- The shape of FMR curves makes it possible to define what type of anisotropy the material has: an EMA or an EMP.
- The resonance curves or field dependences of the imaginary parts of components the permeability tensor have two features: maximums and (or) steps. The low-field feature is caused by the resonance of crystallites in which the direction of the magnetic field ($\vec{H}$) is close to the direction of easy magnetization. The high-field feature corresponds to the resonance of the crystallites in which the field is oriented near the hard magnetization directions.

The features on the FMR curves are observed close to the values of the magnetizing fields which correspond to the stationary directions on the angular dependence of resonance field: $H_\parallel$ and $H_\perp$. The stationary directions are the maxima and minima on an angular dependence of the resonance field of monocrystalline grains. The values of these resonance fields (or frequencies) are determined by the formulas [10]:

$$\omega_\parallel = \gamma_\parallel\left[H_\parallel + (\gamma_\perp/\gamma_\parallel)H'_{a1}\right],\quad \omega_\perp = \gamma_\perp\left[H_\perp(H_\perp - H_\theta)\right]^{1/2}. \tag{1}$$

Here $\omega_\parallel$, $\gamma_\parallel$ and $\omega_\perp$, $\gamma_\perp$ are resonance frequencies and gyromagnetic ratios for directions along the hexagonal axis c of the crystalline lattice and in the base plane, H'_{a1}, H_θ are the magnetocrystalline anisotropy fields for these directions. These fields include contributions from magnetocrystalline anisotropy and anisotropy of the shape of crystallite:

$$H_\theta = H'_{a1} + H_{a2} + H_{a3},\quad H'_{a1} = H_{a1} + 4\pi M_S(N_\perp - (\gamma_\perp/\gamma_\parallel)^2 N_\parallel). \tag{2}$$

In the formulas (2) $H_{ai} = 2ik_i/M_S$ are the fields of magnetocrystalline anisotropy (i = 1, 2, 3); $N_\perp$, $N_\parallel$ are demagnetizing factors of the particle that has a form of ellipsoid of revolution, $2N_\perp + N_\parallel = 1$.

Figure 4 shows the experimental resonance curves (points) of the sample Sr-M, measured in three frequencies. In the range of magnetizing fields to 23 kOe one maximum is observed in the field close to $H_\parallel$, which corresponds with the direction of EMA. With the decrease of frequency the maxima shift to the smaller fields. The increase of losses in zero fields with the decrease of frequency is defined by the approach of the high-frequency magnetic field to the frequency of natural ferromagnetic resonance which is defined by the formula:

$$\omega_{NFMR} = \gamma_\perp H'_{a1}. \tag{3}$$

According to the Fig. 4, for the Sr-M hexaferrite this frequency is $\omega_{NFMR} \approx 2\pi$*48 GHz.

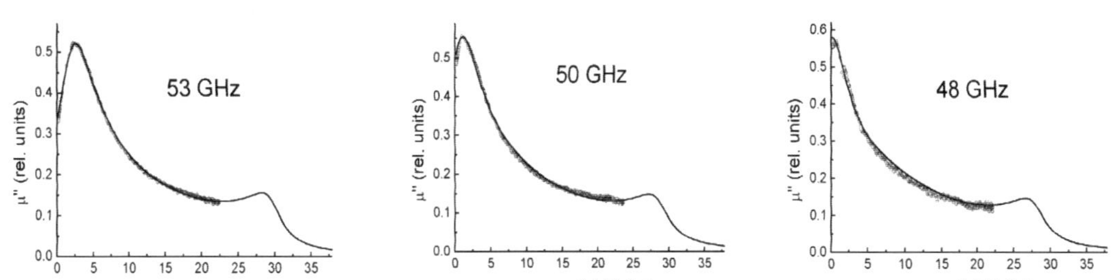

Figure 4. Curves of FMR Sr-M sample measured at different frequencies. The calculated curves are represented by lines, experimental data are represented by points.

The processing of experimental FMR spectra was carried out in two stages. In the first stage, the frequency dependences of the resonance fields were drawn. Then the estimations for magnetomechanical ratios γ and anisotropy fields (H'_{a1}) were performed by the least squares method for $\omega_\parallel$ according to the formula 1. At the second stage, more accurate values of these parameters were determined by a detailed comparison of the shapes of experimental and calculated resonance curves at different frequencies. In the resonance curves calculations the gyromagnetic ratio was considered to be isotropic $\gamma_{\parallel} = \gamma_{\perp} = \gamma$, and the value of anisotropy field $H_\theta = H'_{a1}$.

The calculated imaginary parts of the diagonal components of the permeability tensor are represented by lines in Figure 4. The calculated curves in the figure are shown in a wider range of magnetizing field than the experimental points. This is done in order to show a presence another peak near the field corresponding to the direction of hard magnetization ($H_\perp$). The resonance curves were calculated for values of γ, H'_{a1}, M_S and α, shown in the first row of Table 3. Here, α is a decay constant of monocrystalline grains in the Landau-Lifshitz-Gilbert equation of motion.

Table 3. Parameters of the calculated resonance curves

Sample	CSR, nm	$\gamma/2\pi$, GHz/kOe	H'_{a1}, kOe	M_S, Gs	M_r/M_S	α
$SrFe_{12}O_{19}$	98	2.78	17.25	285	0.64	0.12
$BaFe_{12}O_{19}$ [11]	87	2.80	15.40	366	0.58	-

The second row of Table 3 shows the characteristics of barium hexaferrite $BaFe_{12}O_{19}$. This sample was synthesized in [11] with the same procedure as in our work. According to the table, the particle size and the magnitude of the gyromagnetic ratio of strontium and barium hexaferrites of M-type differ slightly. The anisotropy field of hexaferrite Sr-M almost two kOe greater than anisotropy field of the BaM. This is consistent with the data for bulk materials [1]. Therefore, the ratio of M_r/M_S in SrM also higher, indicating the higher value of the coercive force.

8. Conclusion

Thus, the technology of synthesis of Sr-M hexaferrite by sol-gel combustion produces materials with magnetic properties close to the properties of the materials synthesized by conventional ceramic technology. The advantage of the proposed technology is a significant reduction in synthesis time and energy consuming process.

Acknowledgments
This Research is supported by Tomsk State University Competitiveness Improvement Program reg. N 8.1.23.2015

References

[1] Pillar R. C. 2012 *Progress in Material Science* **57** 1191
[2] Ding J, Maurice D, Miao W F, McCormick P G and Street R 1995 *J Magn. Magn. Mater.* **150** 417
[3] Mendoza-Suárez G, Matutes-Aquino J A, Escalante-García J I, Mancha-Molinar H, Ríos-Jara D and Johal K K 2001 *J. Magn. Magn. Mater.* **223** 55
[4] Li T, Li Y, Wu R, Zhou H, Fang X, Su S, Xia A, Jin C and Liu X. 2015 *J. Magn. Magn. Mater.* **393** 325
[5] Sapoletova N A, Kushnir S E, Li Y H, An S Y, Seo J W and Hur K H 2015 *J. Magn. Magn. Mater.* **389** 101
[6] Padampalle A S, Suryawanshi A D, Navarkhele V M and Birajdar D S 2013 *Int. J. of Recent Technology and Engineering* **2** 19
[7] Zhuravlev V A and Naiden E P 2009 *Physics of the Solid State* **51** 327
[8] Kreslin V Yu and Naiden E P 2002 *Prib. Tekh. Eksp.* **1** 63
[9] Zhuravlev V A 1999 *Physics of the Solid State* **41** 956
[10] Zhuravlev V A and Meshcheryakov V A 2014 *Russian Physics Journal* **56** 1387
[13] Naiden E P, Zhuravlev V A, Minin R V, Itin V I, Korovin E Yu 2015 *Russian Physics Journal* **58** 125

Welding of a corrosion-resistant composite material based on VT14 titanium alloy obtained using an electron beam emitted into the atmosphere

M G Golkovski[1], V V Samoylenko[2], I A Polyakov[2], O G Lenivtseva[2], I K Chakin[1], P N Komarov[2] and A A Ruktuev[2]

[1] Budker Institute of Nuclear Physics, Siberian Branch of Russian Academy of Sciences, 11, Academician Lavrentiev Avenue, 630090, Novosibirsk, Russia
[2] Novosibirsk State Technical University, 20, K. Marx Prospekt, 630073, Novosibirsk, Russia

e-mail: golkovski@mail.ru

Abstract. The study investigates the possibility of inert gas arc welding of a double layer composite material on a titanium base with an anti-corrosive layer obtained by fused deposition of a powder mix containing tantalum and niobium over a titanium base using an electron beam emitted into the atmosphere. Butt welding and fillet welding options were tested with two types of edge preparation. Welds were subjected to a metallographic examination including a structural study and an analysis of the chemical and phase composition of the welds. A conclusion was made regarding the possibility of using welding for manufacturing of items from the investigated composite material.

1. Introduction

The surface layer of a titanium base may be alloyed with refractory and corrosion resistant elements, such as tantalum and niobium using an electron beam with high penetrability emitted into the atmosphere [1-4]. These alloying elements have high solubility in the alpha and beta- phases of titanium, thus forming substitutional solutions [5, 6]. Surface layers formed using the powder metallurgy method with an electron beam have corrosion resistance in numerous aggressive media exceeding the resistance of titanium and stainless steel by many times [1-3, 7, 8]. The alloyed layers, and the VT14 titanium alloy base on which they are fused deposited have a set of mechanical properties enabling industrial use of the obtained composite material. The alloyed layer with a thickness of about 2 mm has high adhesion to base and tensile strength. Due to the peculiarities of fused deposition, however, this method may be used for manufacturing of plate materials only. The essential question regarding use, therefore, is whether normal operations with plate materials, such as welding, expansion, rolling, can be performed with the composite material. In [9] the authors performed the welding of low alloyed uniform material. This work investigates the possibilities of arc welding of the obtained composite material with inert gases: argon and helium.

2. Materials and Methods

A layer of powder containing a mixture of alloying components and flux was placed on the surface of the titanium base of Russian grade VT14 alloy to prepare the source samples. The following powder composition was used: 30%Ta+9%Nb+23%Ti+28.5%CaF_2+9.5%LiF. The alloying components of the source samples were fused deposited on an ELV-6 industrial electron accelerator, equipped with a device for electron beam emission into the atmosphere. The equipment was manufactured by the Budker Institute of Nuclear Physics, Novosibirsk [10, 11]. The energy of electrons in the beam was equal to 1.4 MeV. Powder mass density equal to 0.45 g/cm^{2}was selected so as to make sure that a beam completely penetrated the layer of powder and ensured its uniform heating. The source samples had dimensions of 12.7x50x100 mm after fused deposition with an electron beam. Welding was performed along the sample's longer edge with a 10 mm strip cut off in advance in order to prevent possible decrease of coating thickness at the sample edge. Thus, the dimensions of samples before welding were equal to 12.7x40x100 mm.

Both double-sided and single-sided welding with a non-consumable electrode and an argon or helium-shielded arc was used. The side opposite to the arc was protected with a special hood with argon fed into it. Source workpieces were butt or fillet welded. Rods with a section of 2x2 mm cut on an EDM machine from coatings of samples made especially for this purpose were used as filler for welding the coating. The composition of rods was close to the composition of the coatings being welded (16 % Ta and 5 % Nb). Tantalum wire was used as an alternative filler. Filler made of VT1sv of Russian grade welding wire was used for welding on the base. Weld edge preparation options are shown in figure 1. Welding modes used during the experiment are listed in table 1. The weld shown in figure 1a may be used when there is no access from the coating side, e.g. for pipe welding. In such weld the coating is welded first followed by weld reinforcement by welding on the base using VT1sv wire as filler.

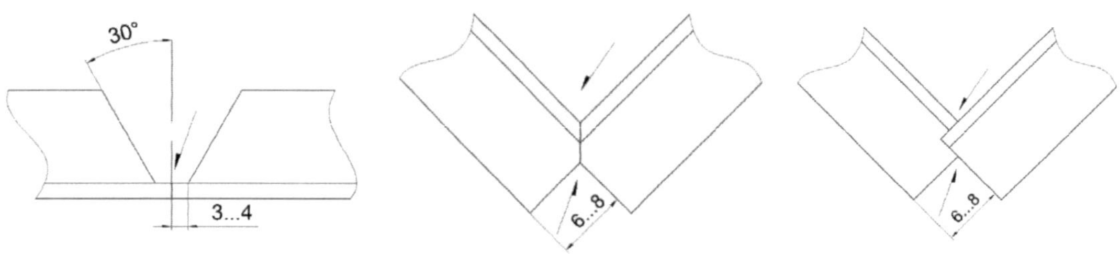

Figure 1. Weld edge preparation: a – single-sided butt weld, b – fillet weld with beveling, c – fillet tongue and groove weld.

Table 1. Welding parameters for VT14 titanium samples with a Ti-Ta-Nb coating.

Sample number	Weld configuration per figure 1	Filler wire in coating weld	Protective atmosphere for coating welding	Coating welding current, A
1	a	16Ta-5Nb-Ti	He	100
2	b	16Ta-5Nb-Ti	Ar	110
3	c	Ta	Ar	130

Metallographic studies were performed using a Carl Ziess Axio Observer A1m optical microscope and a Carl Zeiss EVO 50 XVP scanning electron microscope. The chemical composition of the welded composites was determined using an X-ACT energy-dispersive micro X-ray analyser (Oxford Instruments).

The phase analysis of the welds on the coating was performed using an ARL X'TRA X-ray diffractometer. Diffraction patterns were obtained using CuKα_1/α_2 radiation.

3. Results and Discussion

The external appearance of the samples after welding is shown in figure 2. A visual examination proved that there were no macro defects in the welds and HAZ both in the coating and in the titanium base.

a b c

Figure 2. The external appearance of samples after welding: a – single-sided butt weld on the coating side; b – single-sided butt weld on the base metal side; c – fillet tongue and groove weld on the coating side.

The structure of the welds obtained using butt and fillet welding is presented in figures 3 and 4. The cross-section of the created samples does not contain such defects as fractures, but has a small number of pores (samples 1, 2) and insoluble particles of the alloying components (sample 3). Pores in sample 1 are mostly confined to the areas of titanium layer fusion (figure 4a). The maximum cross-section size of pores is ~450 μm. The pores in sample 2 are observed at the boundary of the weld and surface-welded coating (figure 4b). The pore size in this case does not exceed 200 μm. It should be noted that pores found in the sample structure are individual and closed.

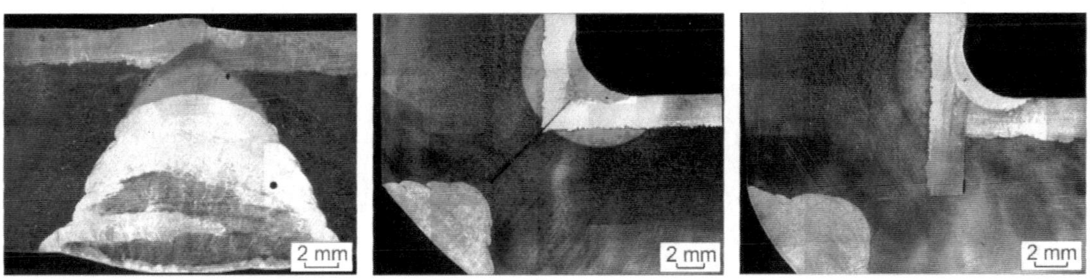

a b c

Figure 3. A panoramic view of samples: a – single-sided butt weld, b – fillet weld with beveling, c – fillet tongue and groove weld.

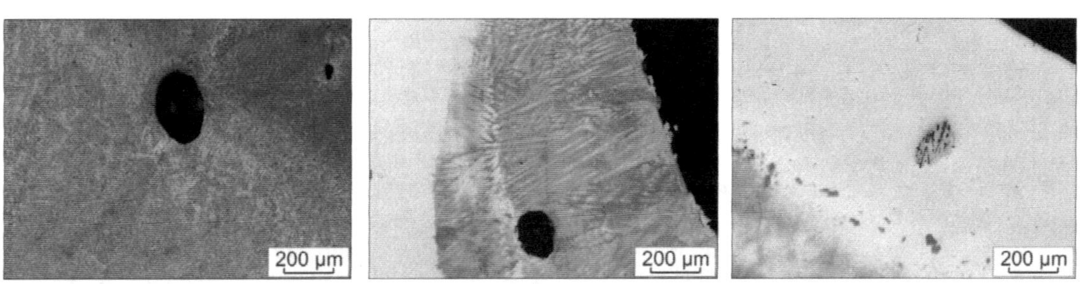

a b c

Figure 4. Weld defects: a – pores in sample 1; b - pores in sample 2; c – non-dissolved particles in sample 3.

The coating weld is in all cases exhibits a clear dendritic structure (figure 5a). The dendritic crystals mostly grow in the direction away from the molten pool towards the weld surface. With high magnifications, a martensite-structure typical for a majority of fused deposited titanium alloys is observed [2-4] (figure 5 b). The welds on the titanium base side are also characterized by a martensite-type structure, but the dispersion of plates is visually significantly lower than in the coating welds (figure 5c).

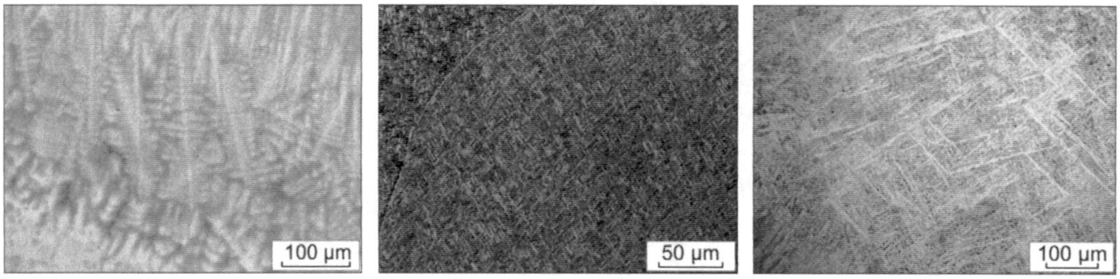

Figure 5. The structure of welds: a – dendritic structure in sample 3; b – martensite-type structure of the coating weld in sample 1; c – martensite-type structure of the titanium base of sample 1.

A chemical analysis of the weld was performed using the EDX method. The concentration of alloying elements for sample 1 was measured in the areas shown in figure 6. The analysis results are presented in table 2. The weld made with a filler wire from coating (spectrum 3) has a composition similar to the coating (table 2). Further filling with titanium results in a partial mixing of the layers and a reduction in the concentration of the alloying elements. The layers welded during the last pass consist mostly of titanium (spectrum 6).

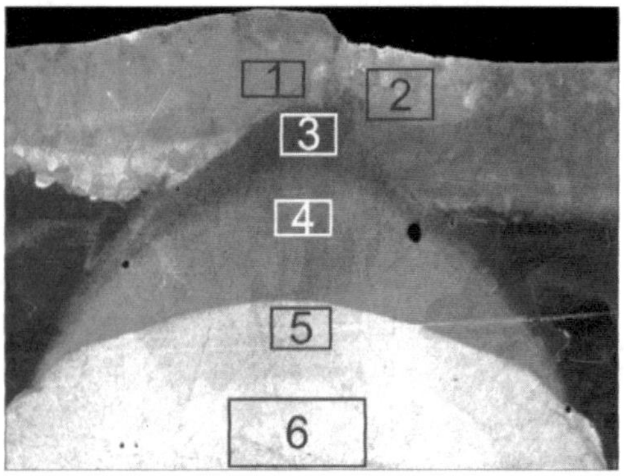

Figure 6. Weld chemical analysis: areas of spectrum accumulation of sample 1.

Table 2. Weld chemical compositions at areas of spectrum accumulation of sample 1.

Number of spectrum	Content, wt. %		
	Ta	Nb	Al
1	17,8	4,5	3,5
2	14,8	4,7	3,9
3	14,7	3,7	3,5
4	14,9		2
5	5,1		0,6
6			0,7

The EDX analysis results for the fillet welds are presented in table 3. The highest content of the alloying elements was recorded in the weld of sample 3. The tantalum and niobium concentration in it equals 47 wt.% and 2.9wt% respectively. The high tantalum content is explained by the use of tantalum wire for this sample as filler wire. The tantalum concentration in the weld of sample 2 is significantly lower than in sample 3 and equals 21.8 wt.%. The niobium content in the sample was 2.8 wt.%. A small amount of aluminum transferred from the titanium base to the coatings during electron beam fused deposition was also detected in all welds by EDX analysis.

Table 3. Chemical composition of the fillet welds.

Mode number	Chemical composition (Ti – the rest), wt %		
	Ta	Nb	Al
2	21,8	2,8	1,8
3	47	2,9	2,1

A structural X-ray analysis was performed in order to determine the phase composition of the coating welds. Sample 1 after butt welding was selected for the examination. The X-ray phase analysis identified the presence of titanium α- and β-phase peaks (figure 7). The diffraction pattern has the following features: a significant widening of the peaks, deviation of the relative peak intensity from the theoretical values, displacement of the peaks relative to the data in ICDD PDF-4+ database. Widening of the peaks may be explained by formation of a fine structure with a high level of microstresses and small size of coherent dispersion areas. On the other hand, wide peaks may be evidence of an emerging chemical composition gradient resulting in a variation of the primitive cell parameters of the formed solid solutions depending on the quantity of alloying elements in a local volume of the material. The deviation of the relative peak intensity from the theoretical values may occur as a result of formation of a texture [12-14] typical for the process of primary crystallization. The displacement of peaks relative to the database values serves as evidence of a change in primitive cell parameters. The parameter change is caused by formation of substitutional solid solutions and non-uniform cooling conditions.

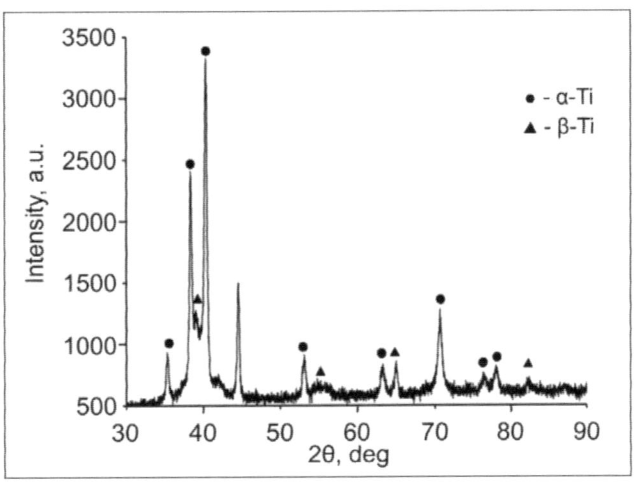

Figure 7. The X-ray pattern of the coating weld.

4. Conclusions

This study demonstrates the possibility of shielded-arc welding of titanium workpieces surface-alloyed by tantalum and niobium. The created welds have a uniform structure. There are no fractures, discontinuities or laminations observed in the weld structure. The detected defects such as pores and undissolved particles are individual and closed. It was established that welds mainly have a dendritic formation and martensite-type structure. The analysis of the X-ray patterns confirmed the presence of two phases: titanium α- and β-phases. The detected mismatch of integral intensities and the widening of peaks on the X-ray pattern prove that the weld contains texture and a fine structure with a high level of microstresses. The chemical analysis of the welds matches the analysis of coatings if the workpieces are welded with filler rod cut from the coating. The concentration of the alloying elements is significantly higher than at coating when samples are welded with tantalum wire.

Acknowledgments
This work was completed on a unique experimental facility based on an ELV-6 industrial electron accelerator for processing of materials with a concentrated electron beam emitted into the atmosphere with financial support from the Russian Ministry of Education and Science. Unique ID of the applied research (project): RFMEFI60414X0135.

References

[1] Golkovski M G, Bataev I A, Bataev A A, Ruktuev A A, Zhuravina T V, Kuksanov N K, Salimov R A, Bataev V A 2013 Atmospheric electron beam surface alloying of titanium with tantalum. *Materials Science and Engineering* A **578** 310–17

[2] Polyakov I A, Samoylenko V V, Lenivtseva O G, Golkovski M G 2015 The structure and corrosion resistance of the coatings obtained by nonvacuum electron beam cladding of the Ti-Nb powder mixture on a titanium substrates *Applied Mechanics and Materials* **788** 267–73 Trans Tech Publications, Switzerland doi:10.4028/www.scientific.net/AMM.788.267

[3] Ruktuev A A, Samoylenko V V, Golkovski M G 2014 Structure and corrosion resistance of Ti-Ta-Nb coatings obtained by electron beam cladding in the air-atmosphere *Applied Mechanics and Materials* **682** 100-3

[4] Samoylenko V V, Lenivtsiva O G, Polyakov I A, Laptev I S, Martyushev N V 2016 The influence of non-vacuum electron-beam facing on the structure of Ti-Ta layers formed on the surface of VT1-0 alloy *IOP Conf. Series: Materials Science and Engineering* **124** art. 012117,

[5] ASM Metals Handbook 1992 vol 3 *Alloy Phase Diagrams*.

[6] Leyens C and Peters M 2003 *Titanium and Titanium Alloys Fundamentals and Applications* (WILEY-VCH Verlag GmbH & Co. KGaA, Weinheim) p 532,

[7] de Souza K A, Robin A 2007 Influence of concentration and temperature on the corrosion behavior of titanium, titanium-20 and 40% tantalum alloys and tantalum in sulfuric acid solutions *Materials Chemistry and Physics* **103** 351–60

[8] Zhukova Y S, Pustov Y A, Konopatsky A S, Filonov M R 2014 Characterization of electrochemical behavior and surface oxide films on superelastic biomedical Ti–Nb–Ta alloy in simulated physiological solutions *Journal of Alloys and Compounds* **586** S535–S538

[9] Karthikeyan T, Dasgupta A, Saroja S, Vijayalakshmi M, Raghunathan V 2004 Studies on weldability of Ti–5Ta–1.8Nb alloy *Journal of Nuclear Materials* **335** 299–301

[10] Salimov R A, Cherepkov V G, Golubenko J I, Krainov G S, Korabelnikov B M, Kuznetsov S A, Kuksanov N K, Malinin A B, Nemytov P I, Petrov S E, Prudnikov V V, Fadeev S N, Veis M E 2000 D.C. high power electron accelerators of ELV-series: status, development, applications *Radiation Physics and Chemistry* **57** 3–6 661–5

[11] Fadeev S N, Golkovski M G, Korchagin A I, Kuksanov N K, Lavruhin A V, Petrov S E, Salimov R A and Vaisman A F 2000 Technological applications of BINP industrial electron accelerators with focused beam extracted into atmosphere *Radiation Physics and Chemistry* **57** 3-6 653-5.

[12] *Recent Developments in the Study of Recrystallization* 2013 ed Peter Wilson, Chapter 4 Texturing Tendency in β-Type Ti-Alloys by Mohamed Abdel-Hady Gepreel DOI: 10.5772/53588

[13] Antonysamya A A, Meyer J, Prangnell P B 2013 Effect of build geometry on the β-grain structure and texture in additive manufacture of Ti-6Al-4V by selective electron beam melting *Materials characterization* **84** 153 – 68

[14] Ruktuev A A, Bataev I A, Golkovski M G, Bataev A A, Laptev I S, Junior A M J 2015 Structure, properties and texturing of Ti-Ta-Mo alloys produced by non-vacuum electron beam surface alloying of Ti *Applied Mechanics and Materials* **788** 230-6

IOP Conf. Series: Materials Science and Engineering **168** (2017) 012077 doi:10.1088/1757-899X/168/1/012077

The formation process of a bistable state in nanofluids

A I Livashvili, V V Krishtop, Y M Karpets and N M Kireyeva

Far Eastern State Transport University, Khabarovsk, Russia.

E-mail: krishtop@list.ru

Abstract. We study the theory of the dynamics of the concentration of nanoparticles in a liquid-phase environment under the influence of a light field. An exact solution for the nonlinear diffusion equation was found in the form of switching waves. It is shown that under the conditions of a stationary and nonlinear temperature coefficient of thermal conductivity of the medium, nanofluids become bistable.

1. Introduction

Nanoscale objects are a very popular subject for researchers because these systems have very specific properties [1–5]. Colloidal suspensions, or nanofluids as they are called today, are widely used in various fields of modern technology. For example, the magnetic fluid is used to polish optical components [6] because a slurry of silica particles in the liquid crystal substantially improves the characteristics of the optical drive [7]. Let us also note their use in chemical processes (catalysis) and the creation of new drugs, lubricants and so forth. With the increasing performance of electronic devices and the development of high technologies, it is necessary to create effective cooling systems and to manage large heat flows. One very promising development is related to molecular computers, which are based on switchable bistable molecules or their aggregates [8].

One of the ways to intensify heat transfer is to increase fluid thermal conductivity by adding solid particles with high thermal conductivity. Of particular interest in the creation of such suspensions are nanoparticles [9–12]. As shown by recent studies [8–15], the liquid phase of the medium, in which the components such as the dispersed nanoparticles are taken from wide bandgap semiconductors and dielectrics, are very effective for a number of nonlinear optical effects. In these media, unlike in homogeneous ones, the nonlinear optical response occurs due to light-wave induced changes in the refractive index and absorption coefficient, thermal diffusion phenomena due to electrostriction of particles. At the same time, the physical mechanisms involved, in particular the heat and mass transfer processes in these media, require further study in our view.

2. Theoretical model

The aim of our work is to conduct a theoretical study of the dynamics of nanoparticle concentration in the liquid-phase medium to be subjected to laser radiation of constant intensity, taking into account the thermal conductivity of the medium depending on its concentration. We believe that the size of the particles satisfies the condition: $a_0 << \hat{\lambda}$, where a_0 is its linear size and $\hat{\lambda}$ – is the wavelength of the light. Thus, we do not consider diffraction and scattering processes. We also assume that the density of the liquid and the particles are nearly the same, giving the option to exclude sedimentation processes from consideration.

We consider the liquid-phase medium with microparticles irradiated with a light beam evenly distributed over the cell intensity. As a result of the light field in the medium, temperature and concentration gradients occur, causing heat and mass transfer processes. These phenomena are described by the system of balance equations for temperature and particles [16]:

$$C_p\rho\frac{\partial T}{\partial t} = \lambda\nabla^2 T + \alpha I_0, \tag{1}$$

$$\frac{\partial C}{\partial t} = D\nabla^2 C + D_T\nabla\left[C(1-C)\nabla T\right]. \tag{2}$$

Note that in the heat equation (1), we have omitted the term responsible for the Dufour effect, because of its smallness, and in the diffusion equation (2) we do not consider a term corresponding to the action of gradient forces from the light field at this stage of research. The following notations are adopted: ∇ is the Laplace operator, T is the ambient temperature, $C = C(r,t) = \frac{m_0}{m}$ is the mass concentration of particulate matter (m_0 is the mass of the particles, m is the mass of the entire environment), C_p, ρ and λ are the permanent thermal fluid; I_0 - the intensity of light; α - optical absorption coefficient of the medium and D, D_T are the diffusion coefficients and thermal respectively.

In such a general statement of the system, (1) and (2) are unlikely to be solved. Therefore, we make some simplifying assumptions: we consider the one-dimensional case and exclude the contribution of the convective term, which occurs in equation (2). Then we take into account the fact that the processes of establishing the temperature goes much faster diffusion process. This makes it possible to study the diffusion processes at a fixed temperature: $\frac{\partial T}{\partial t} = 0$. In equation (1), we assume that the thermal conductivity depends on the concentration and that this dependence has the form [10, 11]:

$$\lambda(C) = \lambda_0 + \beta C = \lambda_0(1 + pC), \tag{3}$$

where $p = \frac{\lambda_0}{\beta}$ is the coefficient of proportionality. Note that this type of dependence was observed in a number of experiments [14]. Given the above, the diffusion equation (2) can be written as:

$$\frac{\partial C}{\partial t} = D\frac{\partial^2 C}{\partial x^2} - D_T\frac{\alpha I_0}{\lambda(C)}C(1-C). \tag{4}$$

Proceeding in this equation to the dimensionless variables, we obtain:

$\tau = \frac{S_T}{\lambda_0} I_0 \alpha D t$, $y = x\cdot\sqrt{\frac{S_T}{\lambda_0} I_0 \alpha}$, $S_T = \frac{D_T}{D}$ - Soret coefficient. If we also use the approach $\frac{1}{\lambda(C)} \approx \lambda_0(1-pC)$, we obtain the problem:

$$\frac{\partial C}{\partial \tau} = \frac{\partial^2 C}{\partial y^2} - pC^3 + (1+p)C^2 - C, \tag{5}$$

$$0 \le C \le 1, \quad -\infty < y < \infty, \quad 0 \le \tau < \infty. \tag{6}$$

Such parabolic equations with cubic nonlinearity were considered in [16, 17] for modelling dissipative media. First, we consider spatially homogeneous stationary states. Obviously, the zeros of the source function in equation (5): $C_1 = 0,\ C_2 = 1/p\ (p > 1),\ C_3 = 1$ correspond exactly to such states.

It is known that the kinetics of the dissipative system are highly dependent on the stability of the stationary states. In our case, the state $C = C_{1,3}$ is stable (they are derived from the source F'(C)<0 and the state $C = C_2$ is unstable. Thus, we found that the medium is bistable. Following [17], the solution of equation (3) will be found by substituting the Cole-Hopf:

$$C(y,\tau) = \frac{u_y \mu}{u(y,\tau) + u_0}, \tag{7}$$

where $u(y,\tau)$ is a new function and μ, u_0 are constant.

Substituting (7) into (5) and equating to zero the coefficients of the same powers of a system of linear equations to determine $u(y,\tau)$, we obtain:

$$u_\tau = 3u_{yy} - (1+p)u_y, \tag{8}$$

$$2u_{yyy} - (\operatorname{sgn}\mu)\cdot(1+p)u_{yy} + u_y = 0, \tag{9}$$

where $\mu = \pm\sqrt{2p}$. The characteristic equation corresponding to (9) can be written as:

$$k\left[2k^2 - (\operatorname{sgn}\mu)\cdot(1+p)k + 1\right] = 0, \tag{10}$$

its roots: $k_0 = 0,\ k_1 = \frac{1}{\sqrt{2p}},\ k_2 = \sqrt{\frac{p}{2}}.$

Further, given the symmetry of the equation (5) with respect to the replacement $y \to -y$, we restrict ourselves to a positive value of μ. Therefore, for the function we have:

$$\mu(y,\tau) = a_0 + a_1(\tau)\exp(k_1 y) + a_2(\tau)\exp(k_2 y) \tag{11}$$

Substituting (11) into (9) we find:

$$a_0 = A_0,\ a_i(\tau) = A_i \exp(\eta_i \tau),\ \eta_i = 3k_i^2 - k_i(1+p)\sqrt{\frac{p}{2}},\ i = 1,2. \tag{12}$$

where A_0 are the constants determined from the initial conditions.

Thus, the exact solution of the equation (5) will have the form:

$$C(y,\tau) = \sqrt{2p}\,\frac{N_1 k_1 \exp(k_1 y + \eta_1 \tau) + N_2 k_2 \exp(k_2 y + \eta_2 \tau)}{1 + N_1 \exp(k_1 y + \eta_1 \tau) + N_2 \exp(k_2 y + \eta_2 \tau)}, \tag{13}$$

where $N_i = \frac{A_i}{(a_0 + z_0)}$.

Note that such a solution was obtained by another method in [18].

3. Results and conclusions

Obviously, the solution (13) is continuous for any values and we note that solution (13) describes a bistable dynamics system, in which a wave state of the switching system is formed by a two-beam mechanism. Thus, phase 'partial' waves are linearly independent.

Passing from the exponent in (13) to the dimensional variables, we obtain the expression for the wave velocities:

$$\upsilon_{1,2} = \sqrt{\frac{S_T I_0 \alpha}{\lambda_0}} \left[(1+p)\sqrt{\frac{p}{2}} - 3k_{1,2} \right] D \qquad (14)$$

Given that $p > 1$, and the expression for $k_{1,2}$, it is evident that this 'two-phase' solution is a concentration of two plane waves with speeds of $\upsilon_{1,2}$. Thus, the exact solution of (14) describes the interaction of two concentration waves switching from the intermediate unstable state $(C_2 = 1/p)$ to the stable state $(C_1 = 0\ u\ C_3 = 1)$. Further, since $\eta_1 > \eta_2$, then we finally have: $C(y,\tau) \approx \sqrt{2p}\,\dfrac{N_1 k_1 \exp(k_1 y + \eta_1 \tau))}{1 + N_1 \exp(k_1 y + \eta_1 \tau)}$.

Note that if we consider the diffusion processes against a background of a fixed temperature with a constant coefficient of thermal conductivity, equation (3) is transformed into a well-known Kolmogorov-Petrovskii-Piskunov (KPP) equation [19], which, as we know, has only a single-wave solution.It is important, in this case, that the system loses its bistability (there are only two spatially homogeneous stationary states).

Of course, our approach is to some extent a model; nevertheless, the authors hope they have managed to discover under which conditions irradiated nanofluids acquire the properties of a dissipative bistable medium in which can propagate a switching wave.

We believe that not all of the issues have been exhaustively studied; for example, an inverse relationship between temperature and the concentration of nanoparticles in the dynamics of the system, taking into account the absorption coefficient of the concentration, was not considered. A detailed examination of this topic will be the subject of our further research.

References

[1] Apter B, Guilatt O and Efron U 2011 Ring-type plasmon resonance inmetallic nanoshells *Appl. Opt.* **50** 5457–64

[2] Krasnikov I, Popov A, Seteikin A and MyllyläR 2011 Influence of titanium dioxide nanoparticles on skin surface temperature at sunlight irradiation*Biomed. Opt. Express* **2** 3278–83

[3] Lee G J, Attri P, Choi E H, Kwon Y-W, Krasnikov I and Seteikin A 2014 Optical and structural properties of nanobiomaterials *J. Nanosci. Nanotechnol.* **14** 221–49

[4] Krasnikov I V, Seteikin A Yu and PopovA P 2015 Simulation of the effect of photoprotective titanium dioxide (TiO2) and zinc oxide (ZnO) nanoparticles on the thermal response and optical characteristics of skin *Opt. Spectrosc.* **118** 668–73

[5] Krishtop V, Doronin I and Okishev K 2012 Improvement of photon correlation spectroscopy method for measuring nanoparticle size by using attenuated total reflectance *Opt. Express* **20** 25693–99

[6] Bardakhanov S P, Novopashin S A andSerebrjakova M A 2012 Study conductivity nanofluids

nanoparticle alumina *Nanosystems: Phys. Chem. Math.* **3** 27–33

[7] Kreuzer M, Tschudi T, de Jeu WHand Eidenschink R 1993 New liquid crystal display with bistability and selective erasure using scattering in filled nematics *Appl. Phys. Lett.* **62** 17121–30

[8] Rambidi E 2007 Nanotechnology and molecular computers *M.: Fizmatlit* 457

[9] Hong T, Yang H and Choi CJ 2005 Study of the enhanced thermal conductivity of Fe nanofluids *J. Appl. Phys.* **97** 064311-1–064311-4

[10] Putnam S A, Cahill D G and Braun P V 2006 Thermal conductivity of nanoparticle suspensions *J. Appl. Phys.* **99**084308

[11] Rudyak V Y, Belkin A A and Tomilina E A 2010 On the thermal conductivity of nanofluids *Tech. Phys. Lett.* **36** 660–2

[12] Kul'chin Y N, Shcherbakov A V, Dzyuba V P, Voznesenskiy S S and Mikaelyan G T 2008 Nonlinear-optical properties of heterogeneous liquid nanophase composites based on high-energy-gap Al_2O_3 nanoparticles*Quant. Electron.* **38** 154–8

[13] Livashvili A I, Krishtop V V and Yakunina M I 2013 Electrostrictive self-action mechanism of radiation in nanofluids*Adv Condens Matter Phys* Article ID 591087

[14] Livashvili A I, Krishtop V V, Bryukhanova T N and Kostina G V 2015 Concentration dynamics of nanoparticles under a periodic light field *Phys. Procedia* **73** 156–8

[15] El-Ganainy E, Christodoulides D N, Rotschild C and Segev M 2007 Soliton dynamics and self-induced transparency in nonlinear nanosuspensions *Opt. Express* **15** 12207–18

[16] De Groot S R and Mazur P 1964*Non-equilibrium thermodynamics*(Moscow: Mir)

[17] Ognev A V, Petrovsky S V and Prostokishin VM 1995 The dynamics of wave formation in a dissipative bistable switching medium*ZHTF* **65** 1–8

[18] Danilov V G and Subochev P Yu 1991 Wave solutions of semilinear parabolic equations *TMF***89** 25–47

[19] Kolmogorov A N, Petrovskii I G and Piskunov N S 1937 A study ofthe diffusion equation with increase in the quantity of matter, and its application to a biological problem *Bull. Moscow Univ. Math.* **ser. A 1** 1–25

Modeling of electrostatic decapole-cylindrical mirror analyzer

Zh T Kambarova[1], A O Saulebekov[2] and D A Saulebekova[3]

[1] Ye.A. Buketov Karaganda State University, Universitetskaya str. 28, Karaganda 100028, Kazakhstan
[2] Lomonosov Moscow State University, Kazakhstan branch, Kazhimukan str.11, Astana 010010, Kazakhstan
[3]Nazarbayev University, Kabanbay Batyr Ave. 53, Astana 010000, Kazakhstan

E-mail: saulebekov@mail.ru

Abstract. Numerical modeling of electron-optical scheme of electrostatic decapole-cylindrical mirror energy analyzer has been carried out. Trajectory analysis of the charged particles motion in decapole-cylindrical field has been conducted. Two regimes of second-order angular focusing ("ring-ring" and "axis-ring" types) were found. Focusing properties of decapole-cylindrical analyzer in second-order angular focusing regime for different types of focusing were calculated. Instrument functions were obtained; energy resolution and luminosity of device were estimated.

1. Introduction

The mathematical method of construction a new class of multipole-cylindrical fields on the basis of the superposition of cylindrical and circular multipole fields, which proposed in [1-2], opens up the prospects to develop new effective systems of energy analysis. Based on this method, the wide range of various potential fields, of practical interest for development of new schemes analyzers with improved characteristics, was calculated. The calculation and analysis of potential properties of synthesized hexapole-cylindrical fields have allowed defining new schemes of mirror energy analyzers having broader functional capabilities compared with the cylindrical mirror [3-4].

The purpose of this work is a numerical modeling of electron-optical scheme of decapole-cylindrical energy analyzer of charged particles beams.

2. Description of decapole-cylindrical analyzer scheme

The scheme of the electrostatic mirror analyzer having the deflecting field which is the result of the superposition of cylindrical field and circular cylinder decapole, was studied in the present work.

$$U(\rho,z) = \mu\, Ln(1+\rho) + \gamma U_d(\rho,z), \qquad (1)$$

where $\mu\, Ln(1+\rho)$ is the cylindrical field, μ is the coefficient that specifies the weight contribution of the cylindrical field, $U_d(\rho,z) = \rho z^4 - 2\rho^3 z^2 + \frac{1}{2}\rho^2$ is the cylindrical decapole, γ is the weight contribution of the circular decapole.

The scheme of mirror energy analyzer with decapole-cylindrical field (1) is shown in Fig.1, and contributions of cylindrical field and decapole are respectively μ=1, $\gamma = -1/100$. It shows emerging of the axial trajectory of the particles beam from A ring source.

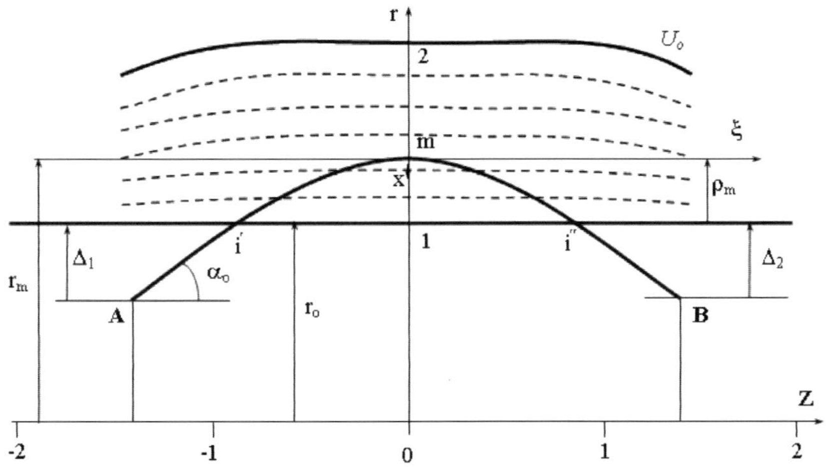

Figure 1. Scheme of energy analyzer based on the decapole-cylindrical field. A is the source of charged particles beam, B is the receiver , i' is the enter ring slit, i'' - the exit ring slit.

Because of the small contribution of circular field, the distribution of equipotential lines of decapole-cylindrical field, having small deviation from the straight line, is similar to one of equipotential lines of cylindrical field. The field is formed in the space between the two axial-symmetrical coaxial electrodes. The inner electrode of cylindrical shape (radius r_o) is at ground potential, the outer electrode having the curved profile is fed by deflecting potential U_o. At a certain ratio of geometrical and energy parameters of the analyzer, a charged particles beam coming from A ring source is reflected by the mirror field and focuses to B ring image.

3. Modeling of electron-optical scheme of mirror decapole-cylindrical energy analyzer

Numerical modeling of electron-optical scheme of energy analyzer based on the decapole-cylindrical field has been carried out in the work. The "Focus" numerical program of modeling axially-symmetrical corpuscular-optical systems was used to model the proposed energy analyzer [5]. Profile of outer electrode was defined on the basis of calculation equipotential lines in decapole-cylindrical field by using MathCAD program.

Fig.2 shows the scheme of second-order angular focusing regime of the "ring-ring" type in decapole-cylindrical analyzer with $\mu = 1$, $\gamma = -1/100$. The total length of the electron-optical system is 11.4. The potential of the outer electrode (2) having curved profile is 1. The radius of the inner cylindrical electrode (1) is equal to 4.2. The inner cylindrical electrode is at zero potential. The ratio of the energy of charged particle and the potential of the electrode is E /V = 1. The position of the source is x = 1.1; y = 2.9. The change step of the initial launch angle of particles change is 2^0. All dimensions are expressed in standard units.

IOP Conf. Series: Materials Science and Engineering **168** (2017) 012078 doi:10.1088/1757-899X/168/1/012078

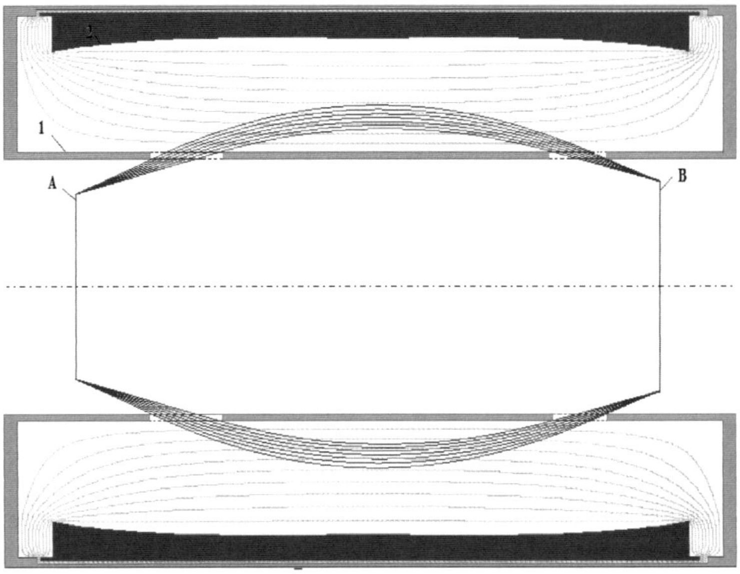

Figure 2. The scheme of the second-order angular focusing regime of the "ring-ring" type in decapole-cylindrical analyzer with $\mu = 1$, $\gamma = -1/100$. A is ring source, B is ring image, 1 - cylindrical electrode, 2 - outer deflecting electrode having curved profile.

According to Fig.2, the particles from A thin ring source in the range of polar angles from 30 to 42^0 are emitted and deflected from decapole-cylindrical field under the action of potential of external electrode having curved profile (2), and then focused to the B ring image. A ring source and its B image are located in the region of inner cylindrical electrode. Thus, the second-order angular focusing regime of the "ring-ring" type is performed in the system. From calculations of focusing properties it was found that the second-order angular focusing regime with the central angle of 34.885^0 was realized in the electron-optical system for a wide range of change of enter angles ($30\text{-}42^0$).

To calculate the instrumental function of decapole-cylindrical analyzer in the second-order angular focusing regime of the "ring-ring" type, the particles were launched from the ring source in the ranges of initial angles $30\text{-}42^0$ and initial energies 0.94-1.06. Fig.3 shows the instrumental function of decapole-cylindrical analyzer with μ=1 and $\gamma = -1/100$ for angular focusing regime of the "ring-ring" type. Relative energy resolution on half-height of instrumental function is 1% with $0{,}1R_{in}$ output aperture radius, and luminosity is $\Omega/2\pi$=12 %.

It also has been found that the second-order angular focusing regime of the "axis-ring" type was performed in the electron-optical system of decapole-cylindrical analyzer. In this case, the distance from the investigated sample to the analyzer is z = 3.6. This scheme of angular focusing provides the "sample-analyzer" long focal length.

Fig.4 shows the scheme of the second-order angular focusing regime of the "axis-ring" type in decapole-cylindrical analyzer. The interval of initial launched angles of charged particles is from 34^0 to 42^0. The ratio of the energy of a charged particle and the potential of the electrode is E / V = 1.3. The position of the point source is defined by x = 0; y = 0. The change step of the entry angle of particles is 1^0. The electron gun is located in the region of inner cylindrical electrode.

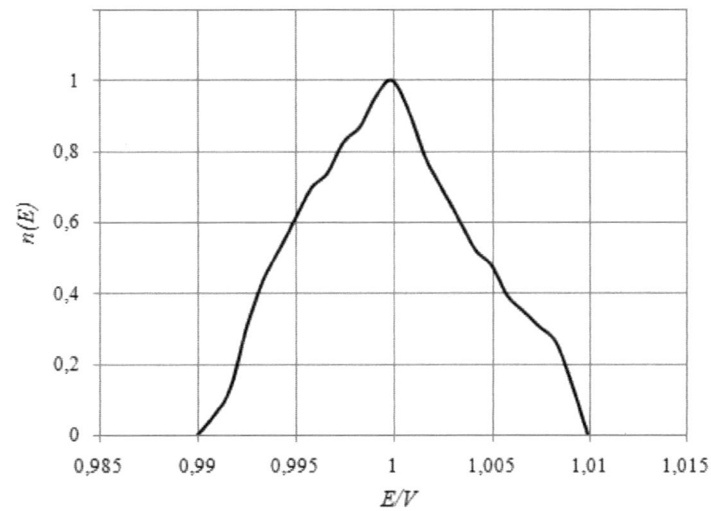

Figure 3. The instrumental function of decapole-cylindrical analyzer in angular focusing regime of the "ring-ring" type at $\mu = 1$, $\gamma = -1/100$.

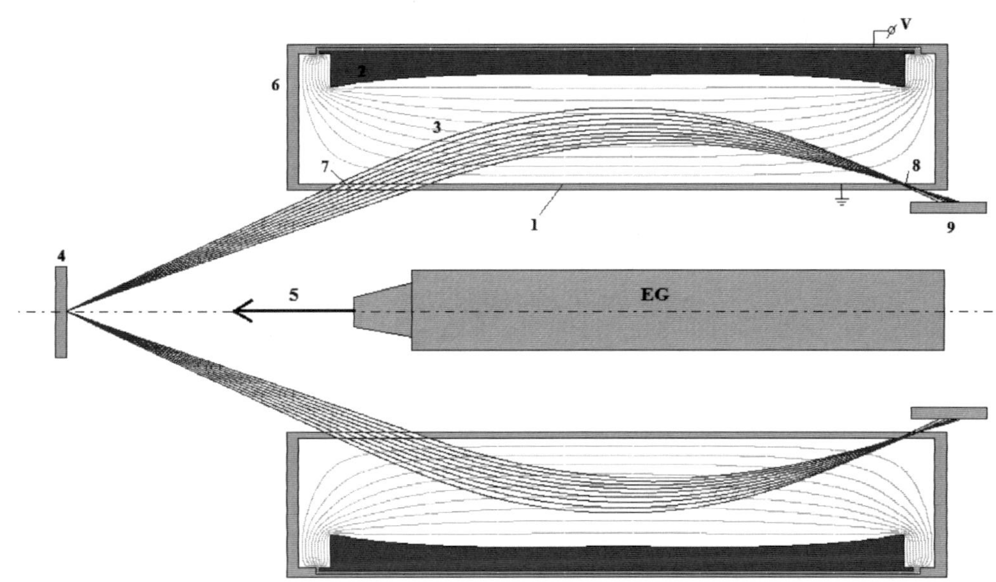

Figure 4. The scheme of the second-order angular focusing regime of the "axis-ring" type in decapole-cylindrical analyzer at $\mu = 1$, $\gamma = -1/100$. 1 - cylindrical electrode, 2 – outer electrode having curved profile, 3 - secondary electrons, 4 – investigated sample, 5 - primary electrons, 6 - magnetic screen, 7 and 8 – enter and exit slits, 9 – detector, EG - electron gun.

According to Fig.4, the secondary electrons (3) emitted by the action of the primary radiation (5) from the surface of the investigated sample (4), entered into the region of the analyzing decapole-cylindrical field through the enter slit (7), then reflected from the outer deflection electrode (2), and focused to the ring image on the surface of the inner cylindrical electrode (1). Next, they pass through the exit slit (8), and registered by the detector (9). This scheme with internal electron gun has a long focal distance from the investigated sample to the analyzer.

For instrumental function calculation a diaphragm is placed on the surface of the inner cylindrical electrode. The range of initial launch angle of particles is 34-42^0, the range of initial energies (or rather E /V) 1.287-1.313. Fig.5 shows the instrumental function of electron-optical scheme of decapole-cylindrical analyzer for the case of the angular focusing regime of the “axis-ring" type. Relative energy resolution on half-height of instrumental function of analyzer is 0.8% with 0.024R$_{in}$ output aperture radius, luminosity is $\Omega/2\pi$=8.6%.

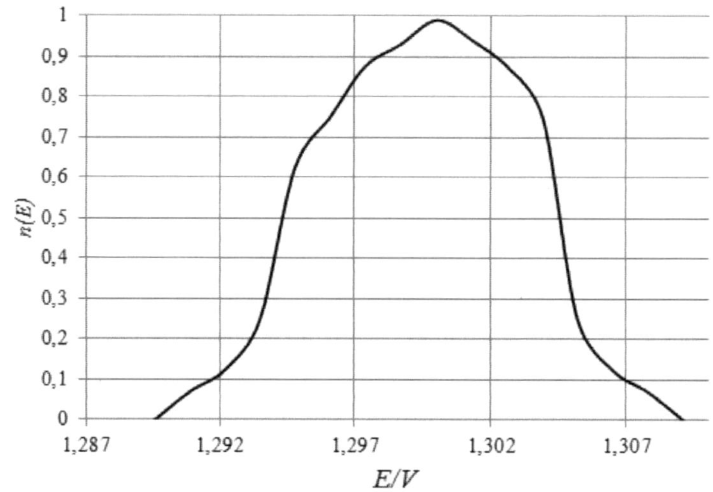

Figure 5. The instrumental function of decapole-cylindrical analyzer in angular focusing regime of the “axis-ring” type at $\mu = 1$, $\gamma = -1/100$.

The table 1 shows results of calculations of focusing properties of decapole-cylindrical analyzer at μ=1, $\gamma = -1/100$ for different types of second-order angular focusing.

Table 1. Focusing properties of decapole-cylindrical analyzer with μ =1, $\gamma = -1/100$ for different types of the second-order angular focusing.

Focusing type	«ring-ring»	«axis-ring»
Focusing order	2	2
Central angle of focusing	34.8850	34.70
X coordinate of focusing	10.41	17.415
Y coordinate of focusing	3.165	4
Reflection parameter	1	0.99

4. Conclusions

The numerical model of the electron-optical scheme of decapole-cylindrical analyzer with μ =1, $\gamma = -1/100$ has been obtained. Instrumental functions of analyzer for second-order angular focusing regimes of “ring-ring” and “axis-ring” types were calculated. Focusing properties of decapole-cylindrical analyzer for second-order angular focusing regimes of “ring-ring” and “axis-ring” types were determined by numerical calculations. It was established that on the basis of decapole-cylindrical field the high-aperture energy analyzers with second-order angular focusing can be built. Practical realizable scheme of the angular focusing regime of the “axis-ring” type with the long focusing distance makes it possible to use it as a dispersing element in the system for a local elementary analysis of micro- and nanometer scale regions.

Acknowledgements
This work was supported by the grant GF4-0815 of the Ministry of Education and Science of the Republic of Kazakhstan.

References

[1] Zashkvara V V and Tyndyk N N 1991 Axial-symmetrical multipoles, their application *Zh.Tech.Phys.* **61(4)** 148-157
[2] Zashkvara V V and Tyndyk N N 1996 Potential fields based on circular multipole series *Nuclear Instrum. and Meth.* **A370** 452-460
[3] Ashimbaeva B U, Chokin K Sh and Saulebekov A O 2005 Focusing Properties of a mirror analyzer with hexapole cylindrical field *J. of Electron Spectroscopy and Related Phenomena* **143** 29-32
[4] Ashimbaeva B U, Chokin K Sh, Saulebekov A O and Kambarova Zh T 2012 Modeling of electron-optical scheme of the analyzer with hexapole cylindrical field *Prikladnaya fizika* **2** 45-48
[5] Gurov V S, Saulebekov A O and Trubitsyn A A 2015 *Advances in imaging and electron physics. Analytical, Approximate-Analytical and Numerical Methods in the Design of Energy Analyzers* (Editor-in-chief Peter W. Hawkes CEMES-CNRS Toulouse, France. - Academic Press is an imprint of Elsevier) **192** 205

Investigation of the plasma processability of natural carbon bearing formations

V P Molchanov
Far East Geological Institute FEB RAS, 159 Prospekt 100-letiya Vladivostoka, Vladivostok, Russia, 690022

E-mail: vpmol@mail.ru

Abstract. In the south of the Russian Far East, a new perspective source of minerals was pioneered, which is the metal-bearing high carbon rocks of the Ruzhinskaya square. The rocks are rich in crystalline graphite, gold, platinum and carbon nanostructures (fullerene, nanotubes and diamond-like carbon). The technique of extraction of ultrapure (99.98%) crystalline graphite from these rocks has been developed using hydrometallugical methods. The obtained graphite was used as a raw material for plasma-chemical tests succeeded in the separation of nanodimensional carbon structures, part of which could be inherited from the natural graphite-bearing rocks. The results of investigation will be used in the development of resource-saving technology of minerals extraction.

1. Introduction

A new perspective source of minerals has been discovered in the south of the Russian Far East [1, 2]. The high carbonaceous rocks of the Ruzhinskaya square incorporate enormous resources of as crystalline graphite as carbon nanostructures, gold, platinum and rare earths. The quantity and quality of useful components concentrated in the graphite rocks of this area indicate that in the nearest future they will determine prospects of the Russian Far East mineral raw material base development.

The main task of the investigation was to study effects of various factors on a graphite-bearing rock in order to devise physicochemical principles of profitable and environmentally acceptable technology of mineral extraction. As is known, graphite of the Ruzhinskaya square represents several varieties [3] depending on formation conditions. The first variety is a result of gas-condensate crystallization from a deep reducing fluid and is characterized by the presence of diamond-like carbon. The second variety is a product of metamorphic recrystallization of sedimentary protolith and is distinguished by the presence of fullerenes and nanotubes. The plasma experiments for synthesis of carbon nanoindividuals from the high-clean graphite extracted from ore formations of the Ruzhinskaya square were carried out having regard to these facts.

2. Methodology of experiment

Commonly, the initial material for plasma chemical synthesis of carbon nanostructures is crystalline graphite [4]. This carbon modification corresponds to a condition of thermodynamic equilibrium, therefore production of the thermodynamically non-equilibrium (unstable) carbon modifications is possible only in conditions of substantial departure from the thermodynamic equilibrium. Such conditions are easily created under intensive energy supply causing sputtering and atomization of the condensed carbon. Subsequent dispersion of carbon vapor in the space of buffer gas goes with the vapor cooling that leads to occurrence of chemically unstable state of the carbon vapor in which content of free carbon atoms many times exceeds the equilibrium value defined by a value of gas temperature. The further condensation of carbon vapor in the buffer gas under the considered non-equilibrium conditions can bring into formation of the thermodynamically equilibrium graphite structure and multiple carbon nanostructures as well.

3. Results of experiment and their discussion

The initial material for the experiment was a sample of high carbon rock from the Ruzhinskaya square with gold content 0,1-0,2 g/t and the total REE concentration up to 100 g/t. The sample was separated into silicate and graphite components by flotation, with the pine oil as a foaming agent, the terpene alcohols as a basic reagent and the long-chain amines as a collector. During the flotation, a bulk of graphite as well as compounds of rare earth elements, silicon, aluminum, iron, magnesium, calcium, sodium, and titanium were concentrated in the foam, whereas ore minerals (gold, platinum) – in the slime [5]. According to X-ray fluorescent analysis, the flotation yielded industrial graphite with the ash content of 4-7 % that corresponds to 829-73 State Standard (the graphite available for making lubricants, coats of current-conducting rubber of the increased hardness, and so on.). However, graphite used in plasma chemistry as the initial raw material is required to have the lesser ach content. To lower the ach content, graphite is usually subjected to acid treatment or alloying with soda. In our case, the high ash content of the graphite is conditioned essentially by the presence of rare earths, silicon, aluminum, iron, magnesium, and calcium. In this connection, the treatment of the foam with ammonium hydrogen difluoride and subsequent acid leaching of the fluorinated product seemed to be the most appropriate decision for picking impurities. The physicochemical basis of fluoridation with ammonium hydrodifluoride is in the ability of oxygen-containing compounds of the transition and many non-transition elements to form easy-to-process ammonium fluometallates when interacting with NH4HF (Fig.1).

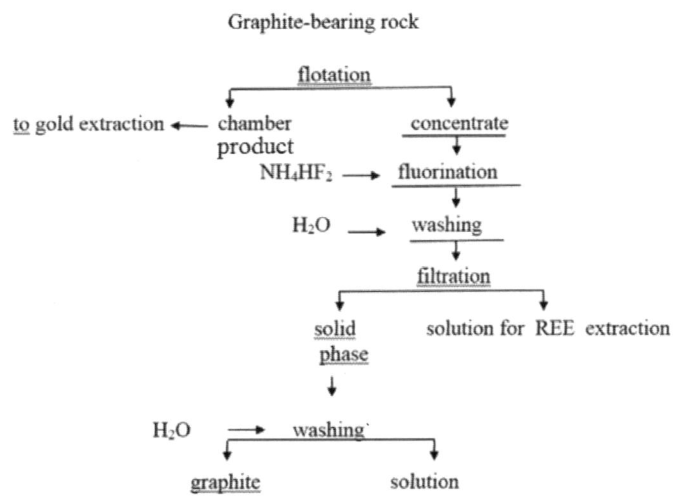

Figure 1. Scheme of graphite purification.

In the fluorination, 25 g of the graphite concentrate was mixed with 5 g of NH4HF2, dissolved in 5 ml of water. The dense paste was being slowly heated to 190-200o C; the temperature was then raised 400°C and remained for 2 hours. As a result of this standing, 2.8 g of (NH4)2SiF6 was desublimated in the nickel condenser and 0.8 g of a mixture of NH4HF2 and NH4F in the fluoroplastic one. The NH4HF2 - NH4F mixture can be further recycled. The sublimated ammonium fluorosilicate had high purity. The admixtures content did not exceed 10-4 mass. %. The non-volatile residue in the reaction container was remixed with 5 g of NH4HF2 dissolved in 5 ml of water. The obtained paste was heated to a temperature of 190-200°C and hold at this temperature for 2 hours. The weight of the fluorinated product was 24.7 g.

The X-ray diffraction data showed that the admixtures in the doubly fluorinated product represented water-soluble fluoammonium salts of aluminum, titanium, and iron and insoluble fluorite. The leaching process consisted in double dissolving of the fluorinated product in water at the liquid-to-solid ratio of 1:10, settling of the suspension and discharging of the transparent solution. The remained portion was filtered and the resulting filtrate was evaporated to dry salts. The mass of the yielded product was 1.44 g. According to the X-ray diffraction data, the product was a mixture of complex aluminum, iron and titanium fluoammonium salts. The ash-content of the graphite purified in this way was 0.56%.

The graphite purity can be raised by acid leaching because in this case the main admixtures are insoluble calcium and magnesium fluorides. The graphite obtained at this stage was subjected to water leaching and then treated with the 10% hydrochloric acid solution at the liquid-to-solid ratio of 1:8 or with the 15% nitric acid solution at the liquid-to-solid ratio of 1:5 for an hour at 700°C. The mixture was filtered, washed with water and dried. The purity of graphite in this case comprises 99.98% [6].

IOP Conf. Series: Materials Science and Engineering **168** (2017) 012079 doi:10.1088/1757-899X/168/1/012079

There is an assumption that the REE enter into the structure of natural graphite at nanolevel [5]. In this connection it was important to determine the composition of solution of the acid leaching of fluorinated product. For this purpose nitric solution was evaporated to dryness. In the residue obtained there have been discovered high concentrations of REE (Tab. 1) that clearly confirms our conclusions about the close bonds between the rare-earth components and natural graphite.

Table 1. Rare earth elements content in the evaporated filtrate.

Element	Content g/t	Element	Content g/t
Y	67.2	Gd	18.2
La	640.2	Tb	2.3
Ce	263.4	Dy	12.6
Pr	33.2	Ho	2.5
Nd	102.4	Er	7.0
Sm	16.8	Tm	0.9
E	3.75	Yb	5.3

Extra pure graphite can find a wide application as a constructive material in atomic power and heat engineering, as well as a starting material for preparing colloid graphite, graphitic oxide and extended graphite for use in microelectronics (production of electronic components), aircraft construction and automobile industry (fuel system parts, fuel pipes), defense industry.

The extra pure graphite was taken for making rods and a charge which were later used for producing the electric-arc plasma. The charge is a mixture of graphite powder and a collector looking like shavings of low carbon steel.

For experimental use, a plasma chamber was built. The hermetic right-angled chamber with internal dimensions 1000x800x80 mm is equipped with feed and evacuation system for buffer (and cooling, if necessary) gas, quartz windows for optical observations and measurements, and built-in thermo inlets for voltage supply on the plasmatron located inside the chamber. This laminar plasmatron worked as a generator of low temperature plasma. Coefficient of the plasma-generating gas (argon) compression consisted n=0.5-0.8. For electrodes, graphite rods of natural material were used. Electric arcing caused the charge melting.

Plasma chemical method application allowed obtaining the dispersive materials, which then have been studied with the help of Dron equipment, microprobe analyzer, atomic-force and scanning electron microscopes. The results of the study showed that the powder contained high amounts of fullerenes, carbon nano-tubes and diamond-like graphite.

The comparison of native carbon nanotubes discovered in the Ruzhinsky graphite (Fig. 2) and the synthetic nanoforms synthesized from the same material exhibits their great outward and inward resemblance. This similarity may be because of some carbon nano individuals (first of all diamond-like carbon) which could pass from the native graphite-bearing rocks. It seems that the study of native and synthetic carbon nano-objects generation should develop in two complementary ways. The developed technology for obtaining carbon nanomaterials with desired features will be in demand in various fields of science and technology. The designing of new techniques for the mass production of diamond-like carbon and carbonaceous nanotubes to be used, for example, in composite material [7], or nanotubes with given parameters for electronics and metrology [8] becomes perspective.

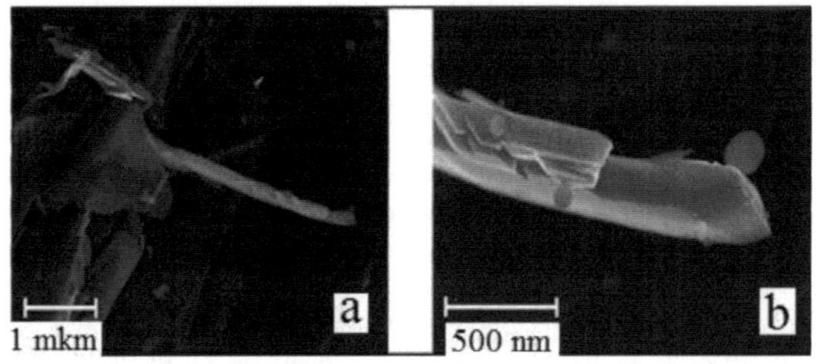

Figure 2. The image of carbonaceous nanostructure on native graphite – a, the enlarged image of the laminated carbonaceous nanotube with fullerite-like structures – b.

The fulfilled investigations made it possible to create a brand new integrated technology for a useful components extraction. At its first stage, flotation methods are needed to have noble metals and industrial graphite extracted. At the second stage the industrial graphite is treated by ammonium hydrofluoride to remove impurities, first of all rare earths. Then the obtained ultrapure graphite is subjected to plasma chemical procedure with the object of separating out the carbonaceous nanostructures (fullerenes, nanotubes, diamond-like carbon).

4. Conclusion

The fulfilled investigations made it possible to create a brand new integrated technology for a useful components extraction. At its first stage, flotation methods are needed to have noble metals and industrial graphite extracted. At the second stage the industrial graphite

Acknowledgments
The work was supported by RSF grant № 14-17-00420.

References:

[1] Khanchuk A I, Plyusnina L P, Molchanov V P and Medvedev E I 2007 *J. Pacific Geol.* **26** 1 70-80
[2] Khanchuk A I, Plyusnina L P, Molchanov V P and Medvedev E I 2010 *J. Geochemistry* **2** 115-25
[3] Khanchuk A I, Molchanov V P, Medkov M A, Gordienko P S and Dostovalov V A 2013 *Dokl. Earth Sciences* **452** 2 202-5
[4] Dubrovsy R and Bezmelnitsyn V N 2004 *Fullerens, nanotubes and cabon nanostructures* **101** 712
[5] Khanchuk A I, Molchanov V P, Blokhin M G, Elovsky E V, Medkov M A, Epov D G and Krysenko G F 2015 *Dokl. Chemistry* **460** 1 29-32
[6] Khanchuk A I, Krysenko G F, Epov D G, Voit A V, Molchanov V P and Medkov M A 2016 *J. Chem.Technol.* **7** 293-7
[7] Rakov E G 2013 *J. Chem.Technol.* **10** 2-7
[8] Soh H T, Morpurgo A and Kong J 1999 *J. Appl. Phys. Lett.* **75** 627-9

Study the spin configuration and the saturation magnetization of manganese-zinc ferrite nanoparticles by the Monte Carlo method

V A Rodionov and V A Zhuravlev
Tomsk State University, Lenina av., 36, Tomsk, 634050 Russia

E-mail: morridin70rus@mail.ru,

Abstract. In this work, the simulations of magnetic properties of nano-sized manganese ferrite particles with zinc replacement were performed. The percentage of replacement laid in range from 0% to 80%. The parameters of particles, including exchange integrals, were taken from experimental data received for $Mn_xZn_{1-x}Fe_2O_4$. The sizes of particles and thickness of defective surface layer were taken, taking into account real sizes distribution for manganese nanoparticles received by the way of mechanochemical synthesis. Simulations were performed using the Monte-Carlo methods, Metropolis algorithm.

1. Introduction

The interest in researching of magnetic properties of nanosized ferrites arose because of the wide use of ferrites in different practical areas, such as in medical purposes, in information storage devices, in absorbing coatings and in other different areas [1, 2]. In nanosized state the properties of ferrites may significantly differ from such properties for bulk material, this difference is especially significant for such properties as the total magnetic moment of particle, Curie temperature and magnetocrystalline anisotropy [3, 4]. In this regard, for predicting the properties of synthesized materials, simulations of behavior of nanosized ferrospinels while changing external parameters such as temperature and external magnetic field or while changing the composition of the substance, including the introduction of non-magnetic impurities in the material are needed.

To date, several different models for simulations of nanosized ferrimagnetic nanoparticles were suggested [5-12]. These models study particles of different dimensions, such as a three-dimensional [5-10] and quasi-one- dimensional [11, 12], moreover, models for three-dimensional particles differs from each other by count of layers including inner volume of particle, surface layer and some quantity of intermediate layers. Most of these models have antiferromagnetic interaction in surface layer [5-9], for inner volume the type of exchange interaction for some models is ferromagnetic [5, 6, 8] and for others is antiferromagnetic [7, 9]. In studied works, the model Hamiltonians for three-dimensional models take into account exchange between near neighbors in different layers, magnetocrystalline anisotropy for surface layer and for inner volume and influence of outer magnetic field. For quasi-one-dimensional models the Hamiltonians take into account exchange interactions between neighbors up to fourth coordination sphere, also in work [12] the model takes into account the possibility of nonmagnetic replacement. All models presented above lack binding with the concrete material and values of exchange integrals and anisotropy constants aren't presented, moreover, there are no three-dimensional models which take into account the possibility of non-magnetic replacement.

In all studied works the Monte-Carlo methods [13], Metropolis algorithm [5-9, 11, 12] or kinetic algorithm [10] are used for simulations.

In present work the three-dimensional model for nanosized manganese ferrospinel $Mn_xZn_{1-x}Fe_2O_4$ with non-magnetic replacement $0 \le x \le 0.8$ is studied. The simulations of spin configuration and specific saturation magnetization changes depending of temperature and non-magnetic replacement were performed. In the calculations we used the values of the exchange integrals given in [15] for the bulk samples of spinel $Mn_xZn_{1-x}Fe_2O_4$.

2. Description of the model and features of the algorithm

The structure of the model is represented by sphere, which is inscribed in cube with side equal to several elementary cells size, in that sphere there is another one, which represents inner volume, and difference between outer and inner spheres represents surface layer.

The peculiarity of the used model is phenomenological accounting dependency from the temperature. This model has two layers, surface layer and inner volume, but because of specifics of energy calculation there appears an implicitly defined intermediate layer, it happens because of interaction between atoms of surface layer with atoms of inner volume.

For simulation was used the Hamiltonian, which takes into account exchange interactions between near neighbors, magnetocrystalline anisotropy in surface layer and inner volume and influence of applied outer magnetic field. This Hamiltonian has the form [14]:

$$H = -J^c_{ab} m^c_a(T) m^c_b(T) \sum_{i \ne j \in c} \vec{S}_i \vec{S}_j - J^c_{bb} (m^c_b(T))^2 \sum_{i \ne j \in c} \vec{S}_i \vec{S}_j - J^s_{ab} m^s_a(T) m^s_b(T) \sum_{i \ne j \in s} \vec{S}_i \vec{S}_j -$$
$$-J^s_{bb} (m^s_b(T))^2 \sum_{i \ne j \in s} \vec{S}_i \vec{S}_j - K_c (m^c_a(T))^2 (m^c_b(T))^2 \sum_i (S^2_{ix} S^2_{iy} + S^2_{iy} S^2_{iz} + S^2_{ix} S^2_{iz}) \qquad (1)$$
$$-K_s m^s_a(T) m^s_b(T) \sum_k (S_k \vec{e}_k)^2 - g\mu_B \vec{H}_0 m(T) \sum_i \vec{S}_i,$$

where index c is used for ions of the inner volume of the particle, index s – is used for ions of the surface layer, J^x_{ij} – value of exchange integral in the corresponding sublattice, K_c – constant of the magnetocrystalline anisotropy in inner volume of the particle, K_s – constant of the magnetocrystalline anisotropy in surface layer, H_0 – the value of applied outer magnetic field, g – g-factor, μ_B – Bohr magneton. Unit vector e_k determines the relative location of the spins in the first coordination sphere of the considered ion in the surface layer and is defined by the following formula:

$$\vec{e}_k = \sum_i (\vec{p}_k - \vec{p}_i) / \left| \sum_i (\vec{p}_k - \vec{p}_i) \right|. \qquad (2)$$

Where vector p_k – unit radius-vector of the spatial orientation of the spin of selected metal ion, and vector p_i – unit radius-vector of the spin orientation for its nearest neighbors. $m_{a,b}{}^c$ and $m_{a,b}{}^s$ – normalized magnetizations of tetrahedral and octahedral sublattices internal volume and the surface layer of the particle, respectively. They are calculated by the following formula:

$$m^i_a(T) = m^i_a(0) * B\left(\frac{g\mu_b S_a H_0 + (\frac{S_a + S_b}{2})^2 z_{ab} J^i_{ab} + S^2_a z_{aa} J^i_{aa}}{kT}\right) \qquad (3)$$

$$m^i_b(T) = m^i_b(0) * B\left(\frac{g\mu_b S_b H_0 + (\frac{S_a + S_b}{2})^2 z_{ba} J^i_{ab} + S^2_b z_{bb} J^i_{bb}}{kT}\right). \qquad (4)$$

There $m^i_a(0)$ and $m^i_b(0)$ – magnetizations of the particle for corresponding sublattices at T = 0 K, where index i shows belonging to surface layer or to internal volume, $B(x)$ – Brillouin function, S_a and S_b – values of spins for magnetic ions of corresponding sublattices, H_0 – applied magnetic field, z_{aa}, zab, z_{ba} and z_{bb}, – the number of nearest neighbours for selected metal ions, where first index shows the type of the sublattice for selected metal ion, and second one shows the type of the sublattice for its neighbors, $J_{aa}{}^i$, $J_{ab}{}^i$ and $J_{bb}{}^i$ – exchange integrals for magnetic ions of corresponding sublattices and index i shows belonging to surface layer or to internal volume. In the formation of the arguments for the Brillouin function, the exchange interaction between atoms within the A sublattice is also taken into account, it gives insignificant contribution to the total energy of the particle, but noticeably affect the results obtained for the normalized magnetizations.

Also, there was developed the algorithm to simulate the presence of a non-magnetic replacement, it allows to relatively uniformly replace the randomly selected manganese ions by zinc ions. This algorithm is used at the stage of the model creation after the model is formed without replacement, and is as follows:

1) Calculating the quantity of atoms in A sublattice.
2) Calculating the total quantity of atoms for replacement

$$CA_{\text{rep}} = CA * Rep. \tag{5}$$

Where CA – total quantity of atoms in A sublattice, which lay in sphere, and Rep – percentage of replacement.

3) Calculating quantity of atoms for replacement in A sublattice, which lay in sphere, for each elementary cell of the particle.
   Calculations are made the fallowing way:
   a. Calculating CA^{i} - the quantity of atoms in A sublattice, which lay in sphere, for each elementary cell.
   b. For each elementary cell the minimal quantity of atoms for replacement is set:

$$CA^{\text{i}}_{\text{rep}} = Trunk(CA^{\text{i}} * Rep), \tag{6}$$

   where $Trunk(x)$ – the function of discarding the fractional part.
   c. Calculating difference between total quantity of atoms for replacement and the sum obtained by the formula (6) of atoms for replacement in all cells. This difference is the quantity of the remaining atoms for replacement.
   d. Then the random set of cells is selected, the number of selected cells in this set is equal to the difference obtained in the preceding paragraph, the quantity of atoms for replacement in these cells increases by one.
4) For each cell, the quantity of atoms, obtained in the previous step, is replaced. Atoms for replacement are randomly selected and their values of the exchange integrals, anisotropy constants and the influence of the contribution of the external magnetic field are set to zero.

3. The results of modeling ferrospinels $Mn_xZn_{1-x}Fe_2O_4$

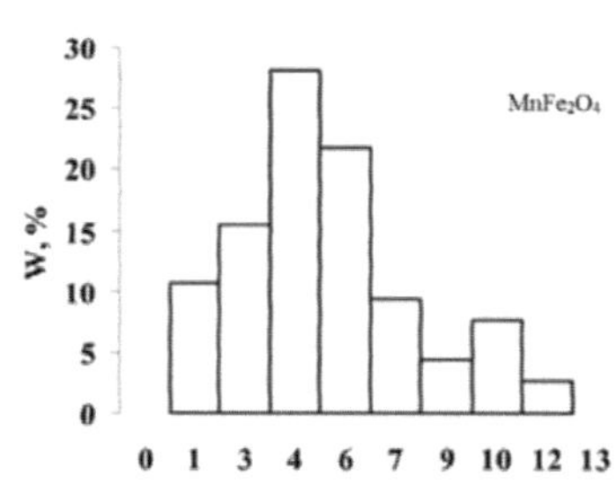

Figure 1. The distribution of nanoparticle sizes of manganese ferrite [3, 4].

Particles with size from 4 to 9 nm were selected for the simulation, that corresponds to the maximum of the size distribution for $MnFe_2O_4$ nanoparticles which was produced by mechanochemical synthesis method from salt systems [3, 4] (Figure 1).

The size range for particles from $5a$ to $9a$, with the thickness of the surface layer $1a$, where a - the size of the elementary cell of manganese ferrospinel, was selected for modeling, based on the data, presented in Figure 1. Such particles contain from 1.5 to 9 thousand magnetic ions [14].

For manganese spinel was discussed the situation with the substitution of the manganese atoms by the zinc atoms: $Mn_xZn_{1-x}Fe_2O_4$, the values of x = 0.2; 0.4; 0.6; 0.8 were considered. When substituting part of manganese atoms by zinc atoms, values of the exchange integrals differ, values exchange integrals for the internal volume for particles with different substitution percentage [15], are shown in Table 1, the exchange integrals for the surface layer are halved:

Table 1. The values of the exchange integrals for $Mn_xZn_{1-x}Fe_2O_4$ for different values of x.

Material	J_{AA}	J_{BB}	J_{AB}
$MnFe_2O_4$	-2K	-4K	-14K
$Mn_{0.8}Zn_{0.2}Fe_2O_4$	-2K	-4K	-14K
$Mn_{0.6}Zn_{0.4}Fe_2O_4$	-2K	-3.9K	-13.5K
$Mn_{0.4}Zn_{0.6}Fe_2O_4$	-2K	-3.8K	-13K
$Mn_{0.2}Zn_{0.8}Fe_2O_4$	-2.2K	-4.2K	-12.4K

For these materials were carried out calculations of the Curie temperature for quasi-bulk particles, i.e., particles without the defect surface layer. The obtained simulation results are very close to the experimental ones [15] (Figure 2.):

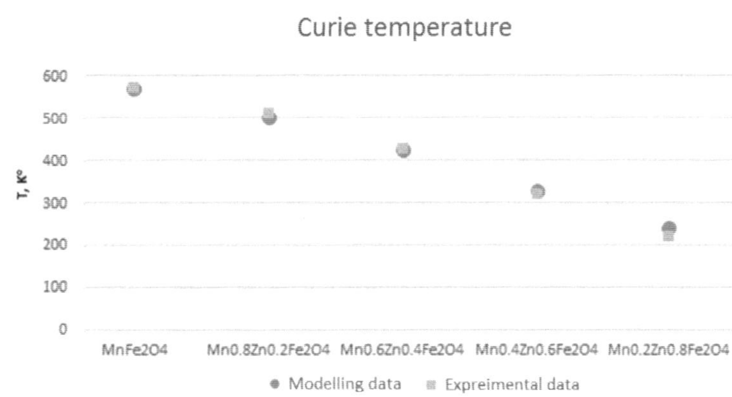

Figure 2. Dependency of Curie temperature T_c from ferrimagnetic particles composition.

Also the values of specific saturation magnetization for quasi-massive samples with different substitution values were performed (Table 2.):

Table 2. The values of specific saturation magnetization for $Mn_xZn_{1-x}Fe_2O_4$ for different values of x.

Concentration	x=1	x=0.8	x=0.6	x=0.4
Calculated	83	67	37	17
Experimental	80 [16]	69 [17]	41 [18]	23 [18]

According to the Table 2, the calculated values of the specific saturation magnetization are close to the experimental data [16-18].

The simulation results of change of the spin configuration when temperature changes for nanosized particles of manganese-zinc ferrospinel, show that the addition of nonmagnetic impurities significantly affects the speed of the particles transition to the paramagnetic state when heated. From Figure 3, which shows the dependence for $Mn_{0.8}Zn_{0.2}Fe_2O_4$ and $Mn_{0.6}Zn_{0.4}Fe_2O_4$, and figure 4, which shows the dependence of $Mn_{0.4}Zn_{0.6}Fe_2O_4$, is clear that with increasing percentage of atoms, which were replaced by non-magnetic ones, the Curie temperature decreases significantly. Graphics also reflect the fact that reducing the particle size in the case of nonmagnetic replacement in the nanoparticle of ferrospinel, similarly to the case without the replacement, results in a reduction of the specific magnetization of substitution and more rapid transition to the paramagnetic state. The bursts of magnetization, which are present on the graphs, are caused by a full transition of the surface layer in the paramagnetic state, at given temperatures, that may cause random ordering for surface atoms, which affect the atoms of inner layer, that is more important for the particles with small size.

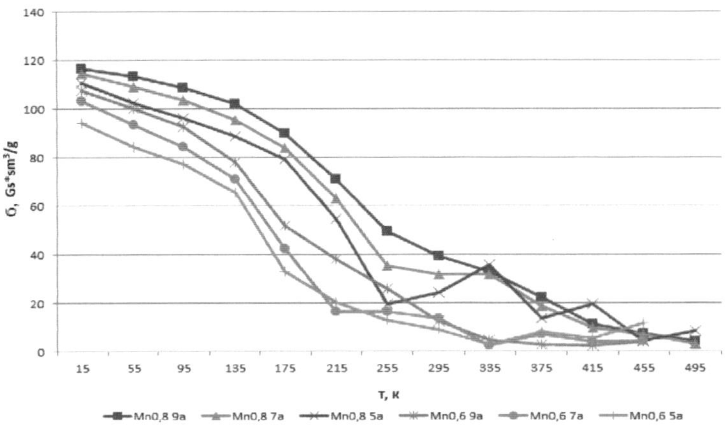

Figure 3. The temperature dependence of the specific saturation magnetization, for $Mn_{0.8}Zn_{0.2}Fe_2O_4$ and $Mn_{0.6}Zn_{0.4}Fe_2O_4$ particles, with sizes 5, 7 and 9 elementary cells.

By increasing the percentage of nonmagnetic replacement of up to 60%, for the nanosized particles, the transition to the paramagnetic state occurs before the reaching the room temperature (Figure 4.), which distinguishes this case from massive material, which has the Curie temperature of 315 K [15].

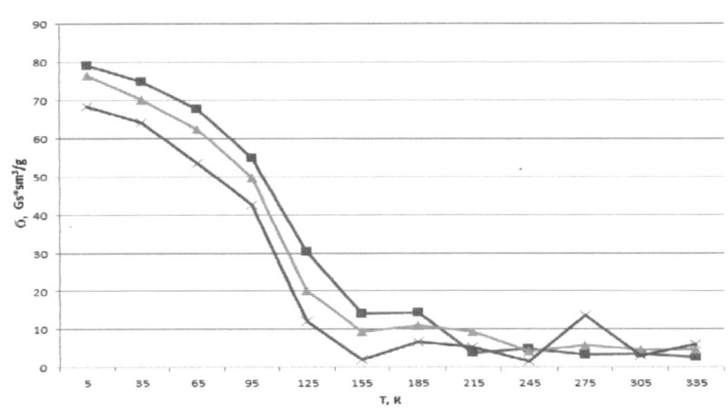

Figure 4. The temperature dependence of the specific saturation magnetization, for $Mn_{0.4}Zn_{0.6}Fe_2O_4$ particles with sizes 5, 7 and 9 elementary cells.

4. Conclusion

The paper described a technique for modeling spin configuration of nanoscale ferrimagnets with a spinel structure, and the results of the modeling of manganese ferrospinel, with the replacement of manganese ions by nonmagnetic ions of zinc in varying proportions were given. The simulation results showed that by increasing the proportion of non-magnetic replacement, the Curie temperature of both the bulk material and nanoparticles for manganese ferrite spinel significantly decreases, thus, transition to the paramagnetic state for nanoparticles is much faster due to the influence of the surface layer, nanoparticles with smaller size have a lower Curie temperature. Thus, to nanoparticles of manganese ferrospinel with nonmagnetic substitution of 60%, the Curie temperature falls below room temperature, while for the bulk samples it lies in the range above room temperature.

5. References

[1] Chinnasamy C N, Yang Aria, Yoon S D, and Hsu Kalin 2007 *J. Appl. Phys.* **101** 09M509
[2] Kinnari P and Upadhyay R V 2010 *J. Appl. Phys.* **107** 053907
[3] Naiden E P, Zhuravlev V A, Itin V I, Terechova O G, Magaeva A A, and Ivanov Y F 2008 *Solid State Physics* **50** 857
[4] Magaeva A A, Naiden E P, Terechova O G, Itin VI, Verchenkov K A, Stadnichenko A I and Boronin A I 2013 *M. Russian nanotechonologies* **8** 66
[5] Iglesias O, Batlle X and Labarta A 2005 *Condence Matter* **09** 53.
[6] Iglesias O, Labarta A and Batlle X 2008 *Journal of Nanoscience and Nanotechnology* **8** 2761
[7] Cabot A and Alivisatos A P 2008 *Funct. Mater* **11** 45
[8] Kalliopi T 2009 [Web resource] **http://www.ifj.edu.pl/conf/zakopane2009/lectures/Trohidou.pdf.**
[9] Adebayo K and Southern B W 2010 *Condence Matter* **03** 46
[10] Ruta S I 2013 [Web resource] **http://etheses.whiterose.ac.uk/6211/1/SergiuRutaThesisMSc_Final_version.pdf.**
[11] Spirin D V and Udodov V N 2007 **arXiv:cond-mat/0703639v1**
[12] Shabunina E V, Spirin D V, Popov A A, Udodov V N and Potekaev A I 2012 *Isv. Vusov. Physics* **55** 94
[13] Binder K 1982 *Monte-Carlo methods in statistical physics* (Moscow: Mir) 400
[14] Rodionov V A and Naiden E P 2014 *Advanced Materials Research* **1040** 70
[15] Zhilyakov S M and Naiden E P 1990 *The magnetic structure of diamagnetic-diluted cubic ferrimagnets* (Tomsk: Publ.TSU) 225
[16] Smit J and Wijn H P J 1962 *Ferrites*, (Moscow: Foreign Literature Publishing House) p. 504
[17] Mirshekari G R, Daee S S, Mohseni H, Torkian S, Ghasemi M, Ameriannejad M, Hoseinizade M, Pirnia M, Pourjafar D, Pourmahdavi M and Gheisari K 2012 *Advanced Materials Research* **409** 520
[18] Pissurlekar V J 2015 *International Journal of Science and Research* **4** www.ijsr.net Paper ID: NOV151205

Study of the magnetic anisotropy of the multiphase samples of the ferrimagnets with hexagonal crystal structure by the method of ferromagnetic resonance

V A Zhuravlev[1], R M Minin[2], V I Itin[2], Y M Lopushnyak[1], A V Zhuravlev[1] and E P Lilenko[1]

[1]National Research Tomsk State University, Lenina av., 36, Tomsk, 634050 Russia
[2] Tomsk Scientific Center SB RAS, pr. Academicheskii 10/4, Tomsk, 634021 Russia

E-mail: ptica@mail.tsu.ru

Abstract. The influence of machining conditions in a planetary ball mill on the phase composition, structural and magnetic parameters of hexaferrite powders composition $BaFe_{12}O_{19}$ was investigated. The properties of powders vary greatly depending on the power density and the time of machining. Magnetocrystalline anisotropy of multiphase powders was studied by the method of ferromagnetic resonance. The effective field of magnetic anisotropy is reduced by more than two times, with decreasing particle size of ~ 67 nm to ~ 10 nm when the processing time equal to 10 minutes. The flow of mechanochemical reactions during grinding leads to the disintegration of the hexagonal crystal phase and the formation of the magnetite phase with a small value of the magnetocrystalline anisotropy field.

1. Introduction

Currently, there is a marked increase in the number of publications devoted to the study of the physical properties and different aspects of the application of oxide ferrimagnets with hexagonal crystal structure (hexaferrites) [1]. The unique properties of hexaferrite are due to the large values of magnetocrystalline anisotropy fields and the rather large values saturation magnetization [2]. The surge of interest in this class of materials in recent years is due to the fact that a number of them are the multiferroics at room temperatures [1, 3, 4].

The traditional method of changing the properties of ferrimagnets is the chemical substitution of the iron ions by the other metal ions. However, there is an alternative way - by reducing its particle size to the submicron and nanoscale dimensions. This significantly increases the influence of a defective surface layer on the magnetic characteristics of the ferrimagnetic particles [5]. One of the methods to produce the nanostructured powders is machining in high energy ball mills [6 – 8]. Thus there is not only grinding but also takes place the mechanical activation (MA) of processed materials. According to [7, 8], the MA has little effect on the value of the effective anisotropy field of ultrafine powder of hexaferrite Zn_2Y with magnetic ordering of type the easy magnetization plane (EMP) [8], while the anisotropy field of powders of hexaferrite $BaFe_{12}O_{19}$ (Ba-M) with magnetic ordering of type the easy magnetization axis (EMA) significantly reduced [5, 6]. This decrease is due to the influence of the contribution of the anisotropy of the defective surface layer, which has in hexaferrites EMP type and, accordingly, reduces the effective anisotropy constant nanoparticles with EMA.

IOP Conf. Series: Materials Science and Engineering **168** (2017) 012081 doi:10.1088/1757-899X/168/1/012081

In this paper we investigated the phase composition, the structural parameters and the magnetic anisotropy of ultrafine powders of hexaferrite Ba-M. They were obtained by grinding in a high-energy planetary ball mill of the powders of industrial permanent magnets 19BA260 brand.

2. Preparation of samples and methods of research

The starting powder hexaferrite Ba-M was obtained from the ferrite permanent magnet 19BA260 brand. Magnet demagnetized by heating above the Curie temperature. It was annealed at a temperature 600 °C for 2 hours and cooled off with the oven. Then it was grinding in a ball mill. For the experiments the fraction of agglomerates of particles with a size less than 60 microns was used.

After preliminary grinding in a ball mill, powders were subjected to treatment in a high-energy planetary mill of the MVP type. Two regime were used, namely, a "soft" one with an energy intensity of 15 g and a "hard" one with an energy intensity of 60 g. The ratios of the ball mass to the powder mass were 5:1 and 20:1, respectively. Where g is the acceleration of free fall. The maximum processing time was 10 minutes for both modes.

The phase composition and the crystal lattice parameters of the powders thus prepared were studied by X-ray diffraction (SHIMADZU XRD-6000 polycrystalline diffractometer in the Bragg-Brentano geometry with a focusing pyrographite crystal monochromator inserted into the secondary gamma-quantum beam., *CuKa* radiation). The computer database of x-ray powder diffractometry PDF4+ of the International Center for Diffraction Data (ICDD, Denver, USA) was used for a qualitative analysis of the phase composition. Quantitative analysis of the phase composition, refinement of structural parameters of the detected phases and identification of coherent scattering regions (CSR) carried out by the program full-profile analysis Powder Cell 2.4. CSR data were used to estimate the average size of nanoparticles.

The magnetic anisotropy of multiphase hexaferrite powders was investigated by the ferromagnetic resonance (FMR) methods. The FMR spectrum was measured with using the passing through technique with standard waveguide in the frequency range 36 ÷ 53 GHz. To investigate the FMR, powders of the examined samples were put in the thin-walled quartz tubes with inner diameter of 0.7 mm and length of ~ 10 mm. The densities of powder samples were approximately identical. The tubes were arranged into a rectangular waveguide parallel to the wide wall of the waveguide in order that the alternating magnetic field was directed along the sample axis. A permanent magnetizing field was directed perpendicularly to the wide wall of the waveguide. Method of treatment of experimental data to determine magnetocrystalline anisotropy fields from FMR measurements is described in details in [9 - 11].

3. The results of X-ray diffraction analysis

Results of the study changes of the phase composition of the powders by mechanical activation time are shown in Figure 1 - "soft", in Figure 2 - "hard" mode. Curve 1 is initial phase content of hexaferrite $BaFe_{12}O_{19}$, curve 2 is the content of phase magnetite Fe_3O_4 and curve 3 is the content of phase hematite Fe_2O_3. The starting powder consists of 99.1 % of the hexagonal phase of Ba-M and 0.9 % Fe_3O_4 phase.

According to Figure 1, with the "soft" machining mode, a monotonic decrease of the initial phase Ba-M and an increase in the content of magnetite and hematite phases with increasing time of MA over 2 minutes are observed. Treated for 10 minutes, the powder contains 52.7 % $BaFe_{12}O_{19}$, 33.8 % Fe_2O_3 and 13.5 % Fe_3O_4.

A more complex pattern is observed in "hard" MA mode. The Ba-M phase content decreases more rapidly and at t = 10 min is only 8.4%. The content of the magnetite phase increased up to 91.3% at t = 10 min. The content of hematite changes not monotonically with increasing time of MA. First, it increases to about 4 minutes and then decreases to almost 0 % for 10 minutes dispersion time.

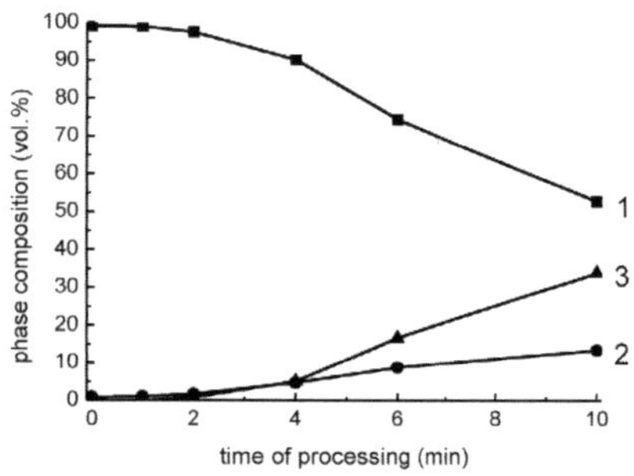

Figure 1. The dependence of the phase composition of the powders BaM on the activation time. "Soft" mode. 1 is $BaFe_{12}O_{19}$ phase, 2 is Fe_3O_4 phase, 3 is Fe_2O_3 phase.

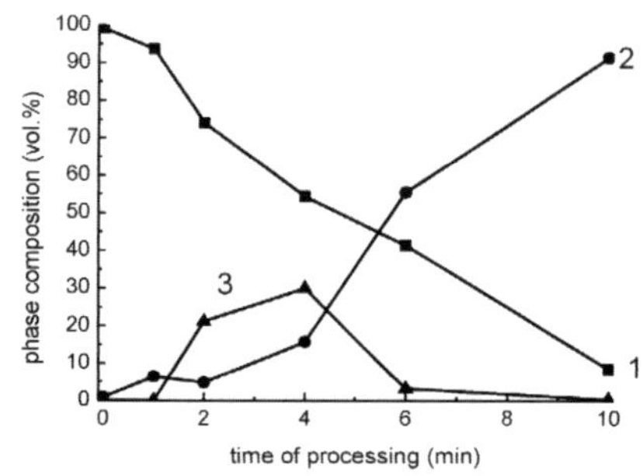

Figure 2. The dependence of the phase composition of the powders BaM on the activation time. "Hard" mode. 1 is $BaFe_{12}O_{19}$ phase, 2 is Fe_3O_4 phase, 3 is Fe_2O_3 phase.

Determined from X-ray analysis values of the lattice constants of the phase are close to known from the literature and are not presented here. The CSR values are shown in Table 1. The initial powder of a permanent magnet 19BA260 mark consists of nanostructured agglomerates less than 60 microns in size, consisting of particles with CSR $\approx$ 67 nm. Increasing the machining time leads to a substantial reduction of the initial particle size of the hexagonal phase in both processing modes to ~ 10 nm. The particle sizes of magnetite phase do not exceed $\approx$ 20 nm. At the "soft" mode the increase in the average size of the hematite particles with increasing time of MA to 32 nm at t = 10 min is observed.

Table 1. Dimensions of coherent scattering regions of investigated powders

Time, min	CSR, "soft" mode			CSR, "hard" mode		
	Ba-M, nm	Fe_3O_4, nm	Fe_2O_3, nm	Ba-M, nm	Fe_3O_4, nm	Fe_2O_3, nm
0	67	~16	-	67	~16	-
1	47	13	-	15	22	-
2	22	10	9	18	10	9
4	15	21	15	13	12	9
6	16	14	20	11	13	9
10	10	18	32	~10	17	-

4. The study of the magnetic anisotropy by the FMR method

A non-oriented polycrystalline ferrimagnets are the ***macroscopically*** isotropic medium. However, the shape of FMR line of polycrystals differs significantly from the FMR curves of isotropic media and of FMR curves of single-crystal samples. The FMR curves of polycrystals have substantially greater line width. At the resonance curves there are additional features: the maximums and (or) the steps. The main reasons for this are as follows: i) the magnetocrystalline anisotropy (MCA); ii) a different form of grains; iii) the dispersion of the values of MCA fields in various grains [12].

Investigations of the FMR spectra of the multi-phase samples of powders and polycrystalline ferrimagnets with a hexagonal crystal structure is the only ways to determine from the experiment important for practical applications the parameters of these materials [9 - 11]. These are: a) the values and sign of the magnetocrystalline anisotropy fields (H_{ai}); b) the value of the effective gyromagnetic ratio $\gamma = ge/2mc$, here g is effective g-factor, e is charge, m is mass of electrons, c is speed of light. The calculations of the FMR resonance curves [10] and a component of the permeability tensor [11] of uniaxial single-domain polycrystalline and powder materials have shown that on the form of the resonance curves one can determine what type of anisotropy: EMP or EMA has the material. Features on FMR curves (the maximums and (or) the steps) are observed near the magnetizing field's values

$H_{\parallel}$ and $H_{\perp}$ corresponding to the stationary directions on the angular dependence of the resonant field. The values of these resonance fields (or frequencies) are determined by the formulas [10]:

$$\omega_{\parallel} = \gamma_{\parallel}\left[H_{\parallel} + (\gamma_{\perp}/\gamma_{\parallel})H'_{a1}\right],\ \ \omega_{\perp} = \gamma_{\perp}\left[H_{\perp}(H_{\perp} - H_{\theta})\right]^{1/2}. \quad (1)$$

Here $\omega_{\parallel}$, $\gamma_{\parallel}$ and $\omega_{\perp}$, $\gamma_{\perp}$ are the resonant frequencies and gyromagnetic ratio's to the directions along the hexagonal axis c and in the basal plane, respectively; H'_{a1}, H_{θ} are the fields of magnetic anisotropy. These fields include the contributions from the magnetocrystalline anisotropy and shape anisotropy of the crystallites:

$$H_{\theta} = H'_{a1} + H_{a2} + H_{a3},\ H'_{a1} = H_{a1} + 4\pi M_S(N_{\perp} - (\gamma_{\perp}/\gamma_{\parallel})^2 N_{\parallel}). \quad (2)$$

In the formula (2) $H_{ai} = 2ik_i / M_S$ is the fields of magnetocrystalline anisotropy, $N_{\perp}$, $N_{\parallel}$ is the transverse and the longitudinal demagnetization factor of a particle having the shape of an ellipsoid of revolution, and $2N_{\perp} + N_{\parallel} = 1$. Here M_S is the saturation magnetization per unit volume. Thus, the study of the frequency dependence of the maximums and steps on FMR curves can estimate the values of $\gamma_{\parallel}$, $\gamma_{\perp}$, and obtain estimates for the anisotropy fields H'_{a1} and H_{θ}. Further, by a detailed comparison of the form of calculated and experimental resonance curves, you can clarify the values of the anisotropy fields H'_{a1} and H_{θ}.

The calculated imaginary part of the diagonal component of the permeability tensor (line) and experimental (points) curve of FMR of initial powder Ba-M at a frequency of 50 GHz is shown in Figure 5. The experimental curves in Fig. 5, 6 and 7 were normalized to the theoretical. The calculated FMR curve is estimated for the following parameters: M_S = 0.38 kGs; H_{a1} = 16.7 kOe; $\gamma/2\pi$ = 2.80 GHz/kOe and α = 0.1 is decay constants in the Landau-Lifshitz-Gilbert equation for a monocrystalline grain. Note that the value of H_{a1} is close to the value known from the literature (17 kOe) [2].

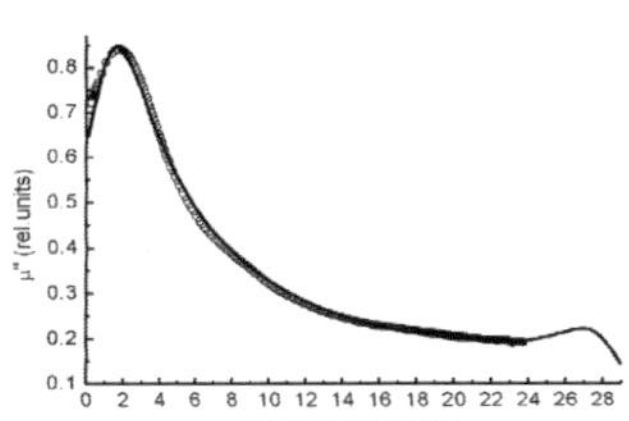

Figure 5. Experimental (points) and calculated (line) FMR curves Ba-M of initial powder. The frequency of 50 GHz.

The resonance curve in Figure 5 is a typical for the uniaxial polycrystalline and powder materials with the EMA and the large value of the anisotropy field. The low-field maximum corresponds to the resonance field for the crystallites in which the direction of the magnetic field closes to the direction of easy magnetization. The high-field feature corresponds to a resonant field of crystallites in which the field is oriented near the direction of hard magnetization. Because in the range of the magnetizing fields available for us on experimental curves FMR hexaferrite Ba-M was observed only one maximum in a field close to $H_{\parallel}$, the only experimentally determined values for us were the $\gamma_{\parallel} = \gamma$ and the H'_{a1}. Transformation of the form of powders FMR curves over time of MA is shown in Figure 6 ("soft") and Figure 7 ("hard") treatment regimes. The measurements frequency equals to 50 GHz.

It is evident that with increasing the time of MA the intensity of the peaks corresponding highly anisotropic phase Ba-M decreases and the line widths increases substantially for both processing modes. It is noteworthy that in the fields greater then 3 kOe, on the slopes of the curves there are additional features that can be interpreted as the disintegration of the initial phase of Ba-M into several fractions with different particle sizes and values of the anisotropy fields. Along with this process, the additional FMR peak corresponding to the phase with the low value of the anisotropy field appears and becomes more intense. This phase is the magnetite with the anisotropy field $H_a \approx 0.55$ kOe [2] in accordance with the data of X-ray analysis. It is evident that these processes occur more intensively with "hard" processing mode. From a comparison of the data in Figures 6 and 7 with the results of X-ray diffraction analysis (Figure 1, 2) it is seen that they are correlated. The content in the samples of highly anisotropic phase Ba-M with increasing t is decreases and the content of small anisotropic phase Fe_3O_4 is increases.

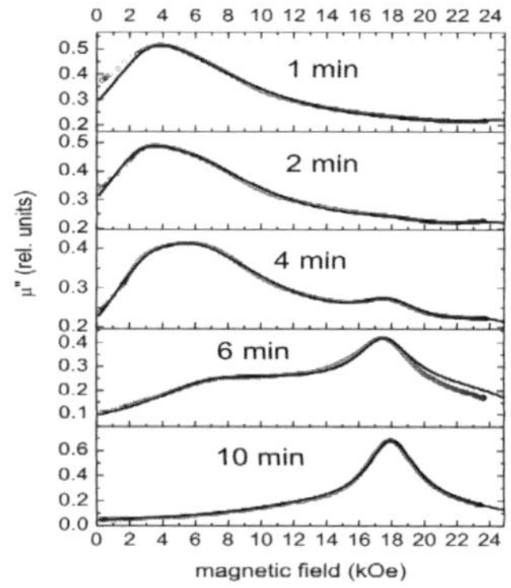

Figure 6. Experimental (points) and calculated (lines) FMR curves. "Soft" mode. MA time is shown in the figures.

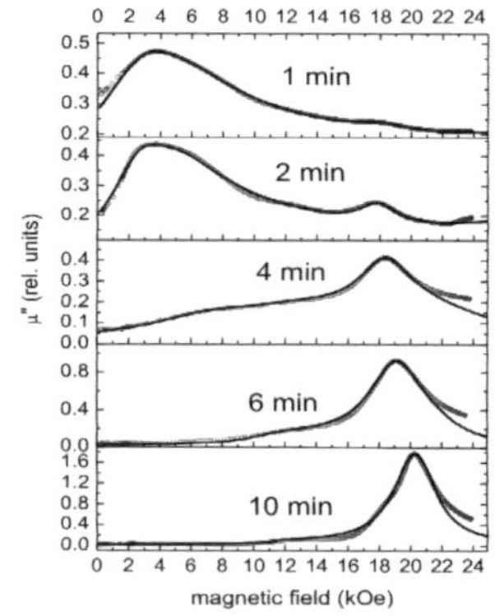

Figure 7. Experimental (points) and calculated (lines) FMR curves. "Hard" mode. MA time is shown in the figures.

The comparison of the FMR curves of the samples treated at 10 minutes of "soft" and "hard" mode shows that the main contribution to the total absorption gives the phase of Fe_3O_4, and the contribution of the phase Ba-M in both cases is much less. According to the data of X-ray analysis, the phase content of the Ba-M and Fe_3O_4 to t = 10 min is as follows: 52.7% and 13.5% of "soft" and 8.4% and 91.3% of "hard" modes, respectively. Thus 52.7% of the phase Ba-M for "soft" mode gives approximately the same contribution to the resonance curve as 8.4% phase Ba-M for "hard" mode. Since the common absorbance at FMR is proportional to the saturation magnetization, this fact can be explained by a significant decrease in the saturation magnetization of Ba-M phase during milling.

The theoretical interpretation of the experimental data was based on the results of X-ray diffraction (Figure 1, 2). The resonance curve of the sample was calculated as the sum of the contributions from the phase Ba-M and the magnetite phases with relevant content. The contribution from antiferromagnetic phase of hematite is not taken into account. We used data on the saturation magnetization of unit volume from [2] when calculate the imaginary parts of the diagonal components of the permeability tensor at the first stage. For Ba-M M_S = 0.38 kGs and for Fe_3O_4 M_S = 0.48 kGs. Further refinement of the M_S values was carried out by a detailed comparison of the shapes of calculated and experimental curves. The appearance of additional features on the resonance curves was seen as the result of the disintegration of the initial phase of Ba-M into fractions with different values of the anisotropy fields. To describe the resonance curves for a "soft" mode was sufficient to take into account only two fractions. In "hard" mode for processing times 1 and 2 minutes it must take into account the presence of the three fractions. The total content of Ba-M phase of the powder in such approach redistributed between the groups of fractions with appropriate relative volumes. The calculation results are represented by lines in Figure 6 and 7 and the parameters of calculated curves are summarized in Table 2. It is evident that at dispersing the particles of Ba-M there is a significant decrease in the magneto-crystalline anisotropy fields. According to modern concepts, such behavior is due to the increasing influence of the anisotropy of defective surface layer with decreasing particle size [5, 6, 8]. Effective anisotropy constant uniaxial magnetic nanoparticles can be written as: $k_{eff} = k_1 + k_f + k_S(V_S / V)$. Where $k_1 + k_f$ is the sum of the contributions from the magnetocrystalline anisotropy and shape anisotropy of the particles. These terms are written in the formula (2) as H'_{a1}. The value of $k_S = -1.9$ Erg/cm^3 is estimated in [5] contribution of the surface anisotropy, V_S is the volume of the defective surface layer, V is the volume of the particle. According to table 2, the reduction of particles size and the values of anisotropy fields H_a are going no monotonically because there are at least two fractions of Ba-M phase with markedly different values of H_a.

Table 2. Parameters of the calculated curves

Time of MA, min	$BaFe_{12}O_{19}$									Fe_3O_4	
	fraction 1			fraction 2			fraction 3				
	H_a, kOe	α	rel. volume	H_a, kOe	α	rel. volume	H_a, kOe	α	rel. volume	α	$\gamma/2\pi$, GHz/kOe
"soft" mode											
0	16.7	0.1	1	-	-	-	-	-	-	-	-
1	15.2	0.15	0.94	12.0	0.11	0.06	-	-	-	0.25	2.80
2	15.6	0.15	0.85	12.2	0.10	0.15	-	-	-	0.25	2.80
4	15.5	0.15	0.67	12.2	0.13	0.33	-	-	-	0.14	2.80
6	12.2	0.20	0.79	7.0	0.20	0.21	-	-	-	0.12	2.80
10	7.0	0.30	1	-	-	-	-	-	-	0.10	2.78
"hard" mode											
1	15.4	0.15	0.88	12.2	0.10	0.10	6	0.1	0.02	0.15	2.78
2	15.6	0.10	0.69	13.0	0.10	0.28	6	0.1	0.03	0.08	2.80
4	12.0	0.19	0.65	6.8	0.17	0.35	-	-	-	0.11	2.72
6	7.5	0.09	1	-	-	-	-	-	-	0.11	2.62
10	6.5	0.02	1	-	-	-	-	-	-	0.08	2.46

There is also a noticeable difference in the gyromagnetic ratios of magnetite phases with low anisotropy at various MA modes.

5. Conclusion

Thus, by dispersing in a high-energy planetary ball mill of the powders of industrial permanent magnets 19BA260 brand the values of the anisotropy field of Ba-M phase is decreased more than doubled and content of magnetite phase with low value of the anisotropy field is increased. Changing the properties of hexagonal phase is not monotonic and comes through fractions with markedly different values of the anisotropy fields.

Acknowledgments

This Research is supported by Tomsk State University Competitiveness Improvement Program reg. N 8.1.23.2015

6. References

[1] Pullar R C 2012 *Progress in Material Science* **57** 1191
[2] Smit J and Wijn H P J 1959 *Ferrite. Physical properties of ferrimagnetic oxides in relation to their technical applications* (Philips' Technical Library)
[3] Ebnabbasi K, Mohebbi M and Vittoria C 2013 *J. Appl. Phys.* **113** 17C703
[4] Ebnabbasi K, Mohebbi M and Vittoria C 2013 *J. Appl. Phys.* **113** 17C707
[5] Golubenko Z V, Ol'khovik L P, Popkov Yu A, Sizova Z I and Kamzin A S 1998 *Phys. Solid State* **40** 1718
[6] Naiden E P, Itin V I, Zhuravlev V A, Terechova O G, Lopushniak Yu M and Tcherbakov V E 2005 *Science of Sintering* **37** 107
[7] Zhuravlev V A and Naiden E P 2008 *Russian Physics Journal* **51** 902
[8] Zhuravlev V A and Naiden E P 2009 *Phys. Solid State* **51** 327
[9] Zhuravlev V A 1999 *Phys. Solid State* **41** 956
[10] Zhuravlev V A and Meshcheryakov V A 2014 *Russian Physics Journal* **56** 1387
[11] Zhuravlev V A and Naiden E P 2008 *Russian Physics Journal* **51** 38
[12] Gurevich A G and Melkov G A 1994 *Magnetization oscillations and Waves*, (Moscow: Fizmatlit) (in Russian).

Structure parameters and magnetic properties of nanosized strontium hexaferrite prepared by the sol-gel combustion method

V A Zhuravlev[1], R M Minin[2], V I Itin[2], M V Politov[1], I Yu Lilenko[1] and M R Ufimtsev[1]

[1]Tomsk State University, Lenin av., 36, Tomsk, 634050 Russia
[2] Tomsk Scientific Center SB RAS, pr. Academicheskii 10/4, Tomsk, 634021 Russia

E-mail: ptica@mail.tsu.ru

Abstract. The phase composition, structure parameters, and main magnetic characteristics of $SrFe_{12}O_{19}$ hexaferrite prepared by the sol-gel combustion method in combination with subsequent annealing at the temperature of 850°C for 6 hours are investigated. From the study of the magnetization curves in the pulsed magnetic fields the value of saturation magnetization of synthesized material was determined. The value of the magnetocrystalline anisotropy field was determined from the survey of the ferromagnetic resonance.

1. Introduction

Hexagonal ferrite of M-type $SrFe_{12}O_{19}$ is characterized by a large coercive force, rather large value of the saturation magnetization, high Curie temperature, large magnetocrystalline anisotropy, high electric resistance, high chemical stability and corrosion resistance. Due to the qualities listed and its low cost the ferrite is used in different spheres of science and technology as permanent magnets and bonded magnets, in various microwave devices (circulators, isolator, phase shifters, filters) and in magnetic and magneto-optic recorders of information with high density [1]. For the reasons listed above the materials which possess the controllable size and form of particles, narrow distribution of particles according to the size and with given magnetic characteristics are needed.

The traditional ceramic method of strontium hexaferrite synthesis is burning the mixture of $SrCO_3$ and Fe_2O_3 powders at a high temperature (> 1200 °C). The disadvantages of this method are high power consumption, uncontrolled morphology of particles, chemical heterogeneity and big size of particles and their agglomerates. Grinding of synthesized strontium hexaferrite powder leads to admixtures, elastic stresses and worsening of magnetic properties [2, 3].

To control the size and form of the particles as well as the magnetic properties of materials new energy-saving methods have been suggested. Such as sol-gel method and its modifications [4, 5], pyrolysis of aerosol, coprecipitation, hydrothermal method, SHS and etc. The full description of these methods can be found in the review [1]. As a rule, using these methods the nanosized and nanostructured particles are obtained which possess the properties different from the properties of massive polycrystalline materials due to the significant part of atoms present in the defective surface layer. By means of additional annealing one can change the size of particles, their morphology and the magnetic properties.

In the present work, the sol-gel combustion method is used to prepare nanosized strontium hexaferrite powders of M-type with the $SrFe_{12}O_{19}$ composition. The method is based on the preparation of initial colloid nanosystem that can react in the combustion mode. This method is the most suitable for the preparation of powder materials with good stoichiometric control, narrow particle size distribution and it requires relatively short processing time at lower temperatures [6]. Results of investigations the phase composition, structural parameters, and magnetic properties of the synthesized materials are presented.

Research of the magnetocrystalline anisotropy of the samples was carried out by the method of ferromagnetic resonance (FMR). FMR is the only method for the study of the magnetocrystalline anisotropy of the multi-phase powder samples consisting of a phases with significantly different values of the anisotropy fields [7].

2. Preparation of samples

As the reagents the aqueous solutions of analytical grade strontium nitrate $Ba(NO_3)_2$, analytical grade 9-aqueous iron nitrate $Fe(NO_3)_3 \cdot 9H_2O$, and lemon acid $C_6H_8O_7$ with concentration of 1 M were mixed in proportions

$$\{[Sr(NO_3)_2]:[Fe(NO_3)_3 \cdot 9H_2O]\}:C_6H_8O_7 = \{1:11.5\}:2.$$

The concentrated solution of NH_4OH ammonium hydroxide was added dropwise to the prepared mixture with constant stirring until pH of the solution became equal to 7. An HI98103 checker portable digital tester (HANNA Instruments) was used to measure pH of the solution mixture. The prepared sol was heated in an ES-6120 magnetic stirrer (EKROS) at temperatures from 80 to 90°C for 3–5 h. As a result of heating, the sol was transformed into a viscous gel of brown color which foamed. Further heating of the gel foam to temperatures 150–160°C caused its ignition and combustion that lasted for several minutes. After combustion, a friable powder was obtained which was easily grinded in a mortar converting into a fine powder. The $SrFe_{12}O_{19}$ hexaferrite nanoparticles were finally formed by annealing of the powder at the temperature of 850°C for 6 h.

3. The research techniques

The X-ray diffraction analysis was performed with a SHIMADZU XRD-6000 polycrystalline diffractometer in the Bragg–Brentano geometry with a focusing pyrographite crystal-monochromator inserted into the secondary gamma-quantum beam. For qualitative analysis of the phase composition, the PDF4+ computer database of x-ray powder diffractometry of the International Center of Diffraction Data (ICDD, Denver, USA) was used. The quantitative analysis of the phase composition and the refinement of the structural parameters of the detected phases were performed using the Powder Cell 2.4 program of full profile analysis.

The infrared spectrum was registered with a Nicolet 5700 infrared Fourier spectrometer (France) equipped with an attachment of diffuse reflection.

Magnetic measurements involved investigations of magnetization curves in pulsed magnetic fields up to 30 kOe using a setup described in [8] and measurements of ferromagnetic resonance (FMR) spectra by the standard waveguide passing through technique at frequencies in the range 37–53 GHz with an automated radio spectroscope. To investigate the FMR, powders of the examined samples were put in thin-walled quartz tubes with inner diameter of 0.7 mm and length of 10 mm. The tubes were arranged into a rectangular waveguide parallel to the wide wall of the waveguide in order that the alternating magnetic field was directed along the sample axis. A constant magnetizing field was directed perpendicularly to the wide wall of the waveguide.

4. The results of x-ray researches

The X-ray diffraction patterns of the synthesis product after ferritization at 850 °C for 6 hours are shown in Fig. 1. It can be seen that it contains basically diffraction peaks corresponding to $SrFe_{12}O_{19}$ hexaferrite. The traces of accompanying hematite phase (α-Fe_2O_3) are also present.

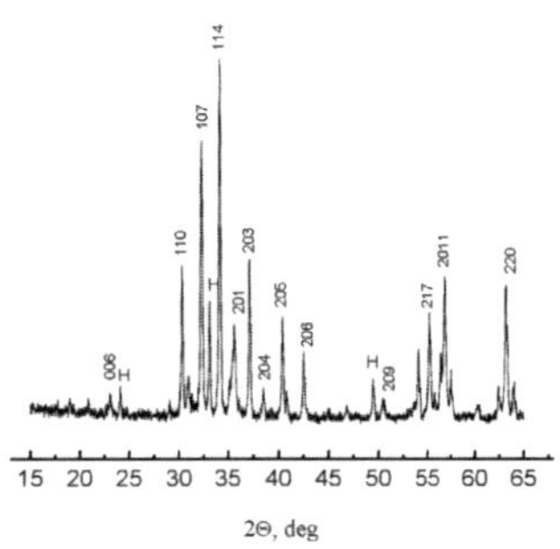

Figure 1. X-ray diffraction pattern of powder hexaferrite $SrFe_{12}O_{19}$ synthesized by sol-gel combustion method. The letters H show the traces of the hematite phase.

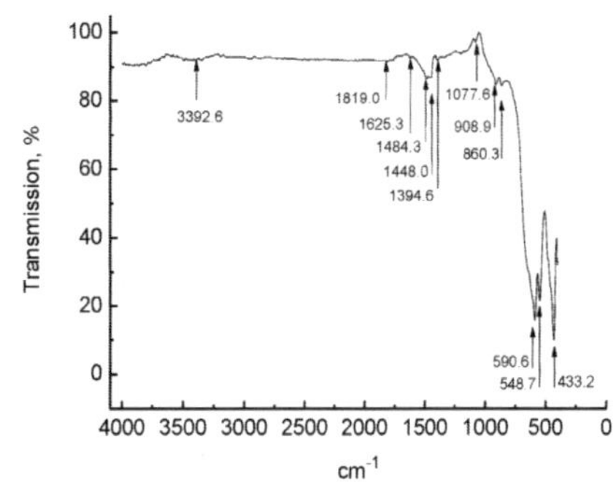

Figure 2. IR-spectrum of powder hexaferrite $SrFe_{12}O_{19}$.

Results of the x-ray diffraction analysis of the sample are given in Table 1.

Table 1. Phase composition and structural parameters of synthesized material after calcinations.

Phase constitution, %		Lattice constant, Å		CSR, nm	$\Delta d/d$*10^3
$SrFe_{12}O_{19}$	α-Fe_2O_3	*a*	*c*		
99.4	0.6	5.9175	23.3130	98	2.7

The content of the Sr–M main phase exceeded 99% and hematite α-Fe_2O_3 was the additional phase. The average crystallite sizes (~ 100 nm) were estimated based on an analysis of the physical broadening of diffraction lines from sizes of coherent scattering region (CSR). In the last column of the Table 1 shown the relative change in the interplanar distances $\Delta d/d$. The estimates of internal elastic microstrain $\sigma = E(\Delta d/d)$ can be carried out by that. Here E – Young's modulus.

5. FTIR spectrum of the sample

The Figure 2 shows the infrared spectrum of transmission coefficient of the obtained hexaferrite powder after annealing at T = 850 °C for 6 hours. At frequencies in the range 3500–3300 cm^{-1}, a very weak absorption band caused by strain-induced vibrations of O–H groups is observed, wide in comparison with other bands. It corresponds with a weak absorption band at 1625 cm^{-1} caused by H–O–H vibrations (bending vibrations of water molecules). The bands at 1484 - 1448 cm^{-1} and 860 cm^{-1} are due to the C–O vibrations due to the presence of lemon acid residues in the water.

The band localized at 909 cm^{-1} is probably attributed to the impurity phase α-Fe_2O_3 which corresponds with the data of X-ray diffraction analysis (Table 1). The bands observed at 433, 549 and 591cm^{-1} can be attributed to valence vibrations metal-oxygen of strontium hexaferrite. Similar results were obtained in [4] for strontium hexaferrite, synthesized by sol-gel combustion method with the usage of egg white.

Thus, the X-ray diffraction and spectrometric data given above prove that the synthesized material is hexaferrite $SrFe_{12}O_{19}$.

6. Research of the magnetization curves

A study of the magnetization curves of $SrFe_{12}O_{19}$ hexaferrite powder was conducted at the room temperature in the pulsed magnetic field by the method described in [8]. The results are presented in the Figure 3. The arrows show the direction of the magnetizing field alteration. The linear extrapolation of high-field part of the magnetization curves on the zero magnetizing fields gives the estimation for the specific saturation magnetization (σ_s). The specific residual magnetization (σ_r) can be determined from the point of intersection of the drop-down portion of the magnetization curve with the Y-axis.

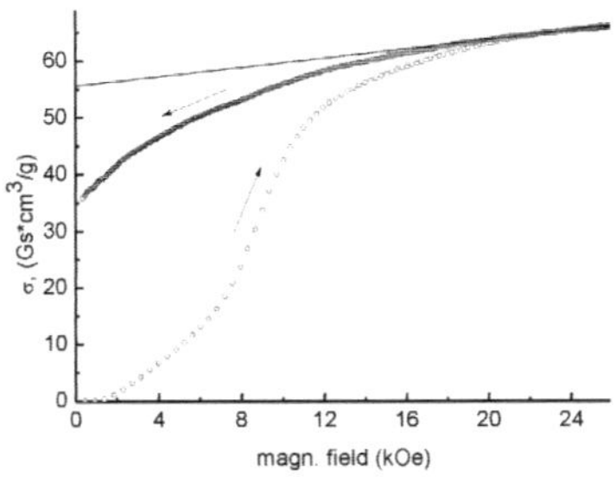

Figure 3. The magnetization curve of the synthesized strontium hexaferrite powder. The arrows show the direction of the magnetizing field alteration.

The values of the specific residual magnetization (σ_r) and the specific saturation magnetization (σ_s) are presented in the first row of the Table 2. In the last column of the table the saturation magnetization per unit volume calculated according to the formula $M_S = \sigma_S{\cdot}\rho$ are shown. Here $\rho = 5.11$ g/cm^3 is the density of strontium hexaferrite [1]. In the subsequent rows of the table for comparison are presented the results obtained in other studies.

Table 2. The research results of the magnetization curves

Sample	σ_r, Gs·cm^3/g	σ_s, Gs·cm^3/g	σ_r / σ_s	Size, nm	M_S, Gs
Sol-gel combustion, 850 °C	35.9	55.8 (0 kOe)	0.64	~ 100	285
Sol-gel [4], 1200 °C	-	56.2 (10 kOe)	-	≥ 1000	-
Sol-gel [5], 825 °C	-	62.6 (15 kOe) 70.8 (50 kOe)	0.49	135x50	-
Sol-gel [1], 1000-1200 °C	-	56 ÷ 70	-	-	-

In the paper [4] where the egg white was used as a fuel the value of σ_s measured in the field ~10 kOe is close to the one obtained in this study though the sizes of the particles differ significantly due to the different temperature of annealing. In the paper [5] the modified sol-gel synthesis is used and the hexaferrite particles were formed and annealed in the NaCl matrix. So a big value for the specific saturation magnetization was obtained. The data from the review [1] are given in the line 4 of the table. It is clear that the above values of σ_s are typical for strontium hexaferrites that synthesized by the sol-gel methods.

7. The study of the magnetic anisotropy by the FMR method

The study of ferromagnetic resonance spectrum for powder and polycrystalline hexaferrites gives an opportunity to experimentally determine the values and the sign of the fields of magnetocrystalline anisotropy (H_{ai}) as well as the value of the gyromagnetic ratio $\gamma = ge/2mc$, where g is the effective g-factor of the examined material, e is the electron charge, m is the electron mass, and c is the velocity of light. It has to be noted that in hexaferrites the γ parameter can be anisotropic [9]. The magnitudes and signs of the anisotropy fields H_{ai} define the type of magnetic ordering – the presence of either an easy magnetization plane (EMP) or an easy magnetization axis (EMA).

The method described in [9, 10] was used to calculate the resonance curves of FMR or components of permeability tensor of the single-domain polycrystalline and powder ferrimagnets with hexagonal crystal structure in the approach of non-interacting grains. The calculations showed:

- The shape of FMR curves makes it possible to define what type of anisotropy the material has: an EMA or an EMP.

- The resonance curves or field dependences of the imaginary parts of components the permeability tensor have two features: maximums and (or) steps. The low-field feature is caused by the resonance of crystallites in which the direction of the magnetic field ($\vec{H}$) is close to the direction of easy magnetization. The high-field feature corresponds to the resonance of the crystallites in which the field is oriented near the hard magnetization directions.

The features on the FMR curves are observed close to the values of the magnetizing fields which correspond to the stationary directions on the angular dependence of resonance field: $H_\parallel$ and $H_\perp$. The stationary directions are the maxima and minima on an angular dependence of the resonance field of monocrystalline grains. The values of these resonance fields (or frequencies) are determined by the formulas [10]:

$$\omega_\parallel = \gamma_\parallel\left[H_\parallel + (\gamma_\perp / \gamma_\parallel)H'_{a1}\right], \quad \omega_\perp = \gamma_\perp\left[H_\perp\left(H_\perp - H_\theta\right)\right]^{1/2}. \tag{1}$$

Here $\omega_\parallel$, $\gamma_\parallel$ and $\omega_\perp$, $\gamma_\perp$ are resonance frequencies and gyromagnetic ratios for directions along the hexagonal axis c of the crystalline lattice and in the base plane, H'_{a1}, H_θ are the magnetocrystalline anisotropy fields for these directions. These fields include contributions from magnetocrystalline anisotropy and anisotropy of the shape of crystallite:

$$H_\theta = H'_{a1} + H_{a2} + H_{a3}, \quad H'_{a1} = H_{a1} + 4\pi M_S(N_\perp - (\gamma_\perp / \gamma_\parallel)^2 N_\parallel). \tag{2}$$

In the formulas (2) $H_{ai} = 2ik_i / M_S$ are the fields of magnetocrystalline anisotropy (i = 1, 2, 3); $N_\perp$, $N_\parallel$ are demagnetizing factors of the particle that has a form of ellipsoid of revolution, $2N_\perp + N_\parallel = 1$.

Figure 4 shows the experimental resonance curves (points) of the sample Sr-M, measured in three frequencies. In the range of magnetizing fields to 23 kOe one maximum is observed in the field close to $H_\parallel$, which corresponds with the direction of EMA. With the decrease of frequency the maxima shift to the smaller fields. The increase of losses in zero fields with the decrease of frequency is defined by the approach of the high-frequency magnetic field to the frequency of natural ferromagnetic resonance which is defined by the formula:

$$\omega_{NFMR} = \gamma_\perp H'_{a1}\,. \tag{3}$$

According to the Figure 4, for the Sr-M hexaferrite this frequency is $\omega_{NFMR} \approx 2\pi$*48 GHz.

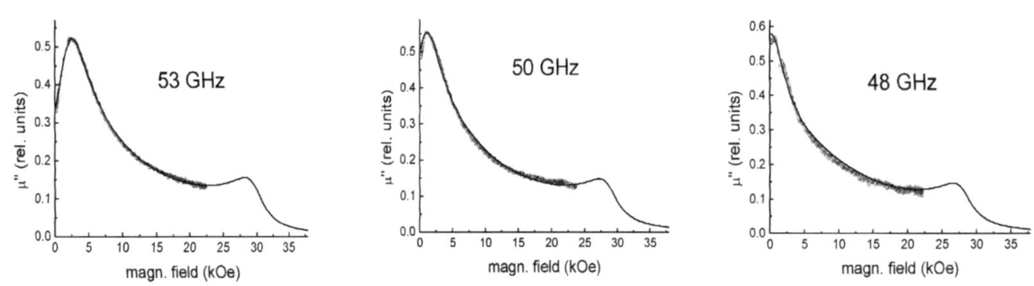

Figure 4. Curves of FMR Sr-M sample measured at different frequencies. The calculated curves are represented by lines, experimental data are represented by points.

The processing of experimental FMR spectra was carried out in two stages. In the first stage, the frequency dependences of the resonance fields were drawn. Then the estimations for magnetomechanical ratios γ and anisotropy fields (H'_{a1}) were performed by the least squares method for $\omega_\parallel$ according to the formula (1). At the second stage, more accurate values of these parameters were determined by a detailed comparison of the shapes of experimental and calculated resonance curves at different frequencies. In the resonance curves calculations the gyromagnetic ratio was considered to be isotropic $\gamma_{\shortparallel} = \gamma_{\perp} = \gamma$, and the value of anisotropy field $H_{\theta} = H'_{a1}$.

The calculated imaginary parts of the diagonal components of the permeability tensor are represented by lines in Figure 4. The calculated curves in the figure are shown in a wider range of magnetizing field than the experimental points. This is done in order to show a presence another peak near the field corresponding to the direction of hard magnetization ($H_{\perp}$). The resonance curves were calculated for values of γ, H'_{a1}, M_S and α, shown in the first row of Table 3. Here, α is a decay constant of monocrystalline grains in the Landau-Lifshitz-Gilbert equation of motion.

Table 3. Parameters of the calculated resonance curves

Sample	CSR, nm	$\gamma/2\pi$, GHz/kOe	H'_{a1}, kOe	M_S, Gs	M_r/M_S	α
$SrFe_{12}O_{19}$	98	2.78	17.25	285	0.64	0.12
$BaFe_{12}O_{19}$ [11]	87	2.80	15.40	366	0.58	-

The second row of Table 3 shows the characteristics of barium hexaferrite $BaFe_{12}O_{19}$. This sample was synthesized in [11] with the same procedure as in our work. According to the table, the particle size and the magnitude of the gyromagnetic ratio of strontium and barium hexaferrites of M-type differ slightly. The anisotropy field of hexaferrite Sr-M almost two kOe greater than anisotropy field of the BaM. This is consistent with the data for bulk materials [1]. Therefore, the ratio of M_r/M_S in SrM also higher, indicating the higher value of the coercive force.

8. Conclusion

Thus, the technology of synthesis of Sr-M hexaferrite by sol-gel combustion produces materials with magnetic properties close to the properties of the materials synthesized by conventional ceramic technology. The advantage of the proposed technology is a significant reduction in synthesis time and energy consuming process.

Acknowledgments

This Research is supported by Tomsk State University Competitiveness Improvement Program reg. N 8.1.23.2015

References

[1] Pullar R. C. 2012 *Progress in Material Science* **57** 1191
[2] Ding J, Maurice D, Miao W F, McCormick P G and Street R 1995 *J Magn. Magn. Mater.* **150** 417
[3] Mendoza-Suárez G, Matutes-Aquino J A, Escalante-García J I, Mancha-Molinar H, Ríos-Jara D and Johal K K 2001 *J. Magn. Magn. Mater.* **223** 55
[4] Li T, Li Y, Wu R, Zhou H, Fang X, Su S, Xia A, Jin C and Liu X. 2015 *J. Magn. Magn. Mater.* **393** 325
[5] Sapoletova N A, Kushnir S E, Li Y H, An S Y, Seo J W and Hur K H 2015 *J. Magn. Magn. Mater.* **389** 101
[6] Padampalle A S, Suryawanshi A D, Navarkhele V M and Birajdar D S 2013 *Int. J. of Recent Technology and Engineering* **2** 19
[7] Zhuravlev V A and Naiden E P 2009 *Physics of the Solid State* **51** 327
[8] Kreslin V Yu and Naiden E P 2002 *Prib. Tekh. Eksp.* **1** 63
[9] Zhuravlev V A 1999 *Physics of the Solid State* **41** 956
[10] Zhuravlev V A and Meshcheryakov V A 2014 *Russian Physics Journal* **56** 1387
[11] Naiden E P, Zhuravlev V A, Minin R V, Itin V I, Korovin E Yu 2015 *Russian Physics Journal* **58** 125

Luminescence of potassium sulphate crystals doped by Eu^{3+} ions

T Koketai, A Tussupbekova, A Baltabekov and E Mussenova

Karaganda State University named after academician Y.A. Buketov, Karaganda, Kazakhstan

E-mail: tussupbekova.ak@gmail.com

Abstract. This paper presents results of the study some spectral-luminescent properties of potassium sulphate crystals activated europium trivalent ions. The observed changes might be connected with the fact of crystals having Eu^{3+} ions and NO_3^- impurity ions. There was a proposition of the possibility of selective creation of impurity centers with the help of using various salts is normal for all the transition metal ions.

1. Introduction

K_2SO_4 crystals are considered as one of the most important construction materials in modern technology. The area of their usage keeps on expanding and the exploitation rules are becoming more and more complicated. The following problems which appear after this require a behavior forecast of construction materials in extreme conditions and modification of physical properties or creation of new materials with specific features. Transition metal trivalent europium ions are one of the traditional activators to study the spectral luminescence properties of crystals. Trivalent europium ions give rise to a narrow-band long-lived red luminescence that is not affected by incorporation of potassium ions. Divalent europium ions emit a UV–blue luminescence, consisting of a large spectral band centered at ca 430 nm. [1]. The photophysical properties of the Eu^{3+} ions oriented in K_2SO_4 crystal matrices were studied. The luminescence data for these doped crystals revealed them to be highly luminescent materials [2]. Rare earth nanomaterials with nanometer to submicronmeter crystalline sizes are an important class of materials displaying interesting physical and chemical properties especially in luminescence and catalysis fields. In the last decade, the synthetic chemistry and exploring of novel properties on rare earth nanomaterials have been developed rapidly. This review on the topic of controlled synthesis and properties of rare earth nanomaterials should be useful with these literatures mostly up to 2009 [3]. K_2SO_4 crystal, as well as potassium dihydrophosphate is a crystal with the composite anionic complex. Due to the practical application of these crystals in the last decades specific features of characteristic and impurity electronic exaltations are widely investigated [4]. Potassium sulphate is of interest for the study of luminescence, recombination and radiation processes and associated color centers. It is stable to radiation and the energy of activation is effectively transferred from the sulphate matrix to impurities [5]. The low symmetry of the lattice points of potassium sulphate gives confidence that we can avoid ambiguity in the interpretation of the nature of the electronic states of the color centers. This is the relevance of this study.

2. Experimental

The main investigation objects were K_2SO_4 doped by Eu^{3+} ions. K_2SO_4-Eu^{2+} crystals were grown from aqueous solutions. The impurity concentration was the same and equal to 0.1 mol%. Europium ions

IOP Conf. Series: Materials Science and Engineering **168** (2017) 012083 doi:10.1088/1757-899X/168/1/012083

included in the crystal lattice of K_2SO_4, during crystal growth from an aqueous solution. The thickness of the majority of the objects is approximately 1.5 mm.

The growth of K_2SO_4 crystals was produced of saturated aqueous solutions using isothermal evaporation of the solvent at 40^0C. After 10-12 days this process, there were 5 mm to 40 mm sized crystals distracted from a saturated K_2SO_4 solution. Sizes of the ready crystals radically vary depending on the components added into the final impurity solution. Activated single crystals were received after soluble metal salts were added to the final solution. Usually, concentration of the components added to the final solution is around 0.1 mol%.

Our work in particular included measurements of the absorption spectra of the luminescence excitation and emission at different temperatures which were used from the optical research methods. Absorption spectra of crystals and induced absorption spectra in the area of 200 – 800 nm were measured using photoelectron method on spectrophotometer SPh-16. Measuring optical density of the crystals was done relating to the air. Additional absorption spectra were measured towards non radiated crystals. Although, the certain number of situations involved using slightly radiated crystals. The last one took place to learn the influence of ionizing radiation on impurity absorption. Measurement of the absorption spectra was done at 80 – 320 K temperature range. The sources of light varied depending on the spectra range, starting with incandescent lamp ending with DDS-30 hydrogen lamp.

In all the optical measurement procedures, the research object was put into a nitrogen cryostat and then was cooled down to the temperature of liquid nitrogen and radiated with X-ray through beryllium window of URS-55a device. X-ray tube had a molybdenum anode. Measuring crooked thermo luminescence was done by supporting the constant speed of the heat in 9 K/min, which was controlled with the help of differential copper-constantan thermo steam.

To define the absorbed dose of the X-ray radiation we used Frike chemical dosimeter. Dosimeter readings do not depend on power of the absorbed radiation dose up to 10^5 $Gy{\cdot}s^{-1}$. Accuracy of Frike dosimeter measuring is approximately ±5%. As a result of the produced measurements it is defined that the power of radioactive dose from URS-55a X-ray device with X-ray tube and copper anode at the voltage of 35 kV and current of 10 mA makes about 150±10 $kGy{\cdot}s^{-1}$.

3. Results and Discussion

Figure 1 represents the typical thermally stimulated luminescence (TSL) curve for virgin potassium sulphate crystals with the radiated dose of X-ray quanta at the liquid nitrogen temperature 100 kGy (curve 1) and 200 kGy (curve 2). The image shows that there are two TSL peaks with the maximum points at 190 K and 280 - 300 K.

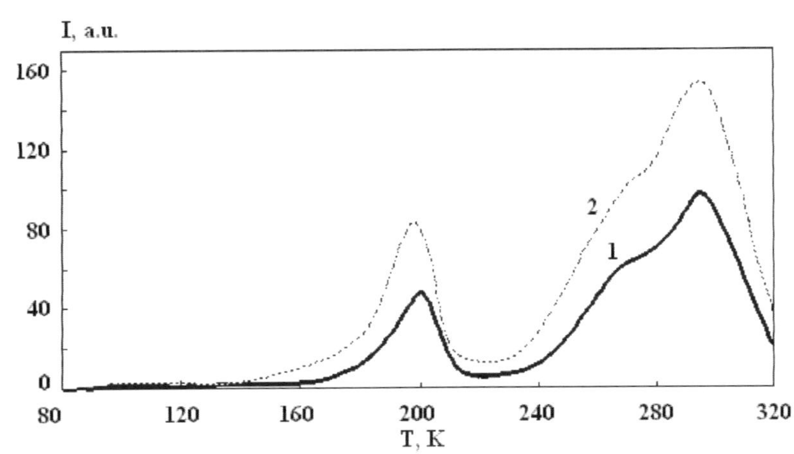

Figure 1. TSL curve for virgin potassium sulphate crystals

Figure 2 shows the accumulated lightsum (S) at the peak of the thermally stimulated luminescence in the 200K on the exposure dose of radiation. Dependence on time lightsum potassium sulphate irradiation with X-rays is linear. When exposure doses up to 90 minutes a tendency to saturation at the peak of the light sum in the recombination radiation were found at 200K. The presence of an energy barrier for light activation, and dependence of the accumulated light sum of the exposure dose of radiation are signs that the observed luminescence is the nature of recombination. The accumulation of defects responsible for the recombination process depends linearly on the radiation dose.

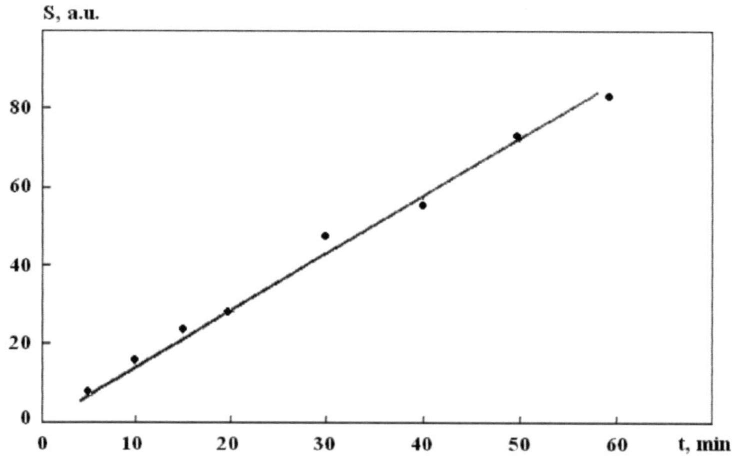

Figure 2. The dependence of the lightsum of the time of exposure dose of radiation TSL peak at 200K the crystal K_2SO_4

Figure 3 shows the TSL curve for K_2SO_4-$Eu(NO_3)_3$ (0.1 mol%) crystal. The dose of radiation X-rays at the temperature of liquid nitrogen was 4 kGy. Compare TSL curves in Figures 1 and 3 shows the high quality of their coincidence. Observed the same peaks of the recombination luminescence with peaks at 190K and 300K. The difference from the results shown in Figure 1 is a dramatic change in the attitude of the accumulated light sum in the peaks glow in the TSL curve.

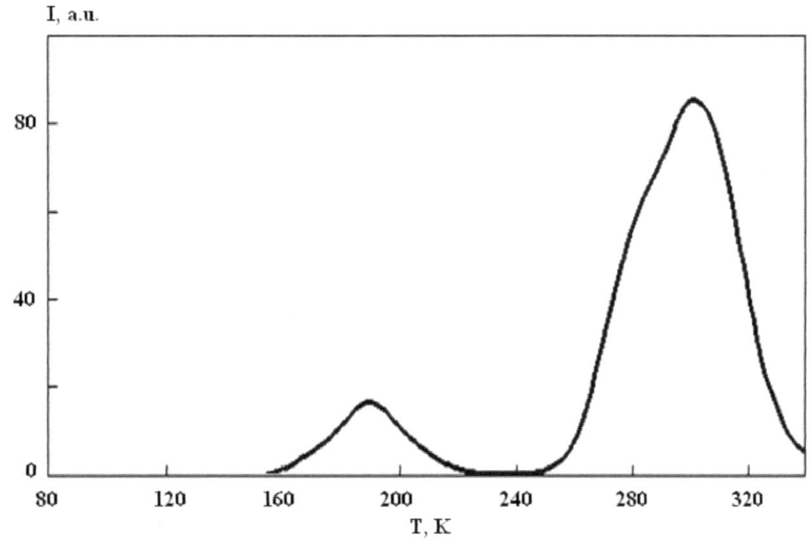

Figure 3. TSL curve of K_2SO_4-$Eu(NO_3)_3$ crystal (0.1 mol%)

The crystals doped with transition metal ions dominant recombination luminescence peak in the TSL curve is a glow with a maximum at 280K. The potassium sulphate activated by trivalent europium nitrate. The luminescence peak of research object does not stand out at 280K. The dominant is a glow with a maximum at 300K. New recombination luminescence peaks were not detected. To determine the effect of europium ions in the processes of formation, accumulation and recombination of radiation defects, TSL curves for crystals of K_2SO_4-KNO_3 were measured. TSL type curve for this sample is similar to that shown in Figure 3. The situation is complicated by the fact that the spectral composition of the TSL peaks for crystal K_2SO_4-$Eu(NO_3)_3$ with a maximum radiation were found at 3.5 eV.

The radiation in the optical band with a maximum at 3.1 eV is present in all the peaks of the TSL curve. However, this radiation is characteristic for non-activated potassium sulphate crystals. Therefore, to determine the impact of europium ions change the type of TSL curves and spectral composition of the radiation is not possible. The only thing that could set: at least up to doses of 50 kGy ratio lightsum low peak in the curve of the TSL and the group of high-temperature for K_2SO_4-$Eu(NO_3)_3$ crystals and K_2SO_4-KNO_3 does not depend on the dose of radiation and is 0.035 and 0.048, respectively.

It is known that the nitrate ions NO_3^- are acceptors for electrons [6]. Their presence in the matrix leads to a more rapid accumulation of hole centers. Also, europium ions increase the rate of accumulation of hole centers. The irradiation of ionizing radiation at room temperature, the optical density of a decrease in the absorption bands of 3.85 eV and 4.43 eV, and its increase in the band is 5.33 eV, which we associate with trivalent europium.

4. Conclusions

Thus, upon exposure to ionizing radiation divalent europium ions form with respect to the inner circle hole centers. TSL peaks in the region 280 - 300K associated with the electronic nature of the recombination process, i.e., the mobility of electrons or electron-rich center recombines with fixed hole. Reducing the ratio of the light sum in the TSL peaks in the sample doped with europium, it shows that radiation-induced impurity defects is the center of recombination. Thus, recombination processes in crystals of potassium sulphate ions europium activated involved impurity ions in the divalent state source. They are under the influence of ionizing radiation are transformed into venues hole nature, which occur at the recombination processes of thermal activation. It can be argued that the dissociative electron capture anion is sufficiently effective channel for the flow of free electrons in the crystals of sulfate and formation of radiation defects.

References

[1] Głowacki M., Solarz P., etc. 2016 Europium and potassium co-doped strontium metaborate single crystals grown by the Czochralski method *Journal of Crystal Growth.* Available online 9 July (in press)

[2] Osvaldo A Serra, Eduardo J Nassar, etc. 1998 Luminescent hourglass inclusion of europium complexes trapped in K_2SO_4 crystal *Spectrochimica Acta Part A: Molecular and Biomolecular Spectroscopy* **54** 13 2077–2080

[3] Chun-Hua Yan, Zheng-Guang Yan, etc. 2011 Handbook on the Physics and Chemistry of Rare Earths. Chapter 251–Controlled Synthesis and Properties of Rare Earth Nanomaterials **41** 275–472

[4] Baltabekov A, Koketaitegi T, Kim L 2010 *Eurasian Physical Technical J.* **7** 12

[5] Myrzakhmet M, Satayeva G, etc. 2015 *J. of Luminescence* **169** B 804

[6] Kim L, Kuketaev T 2003 Influence of phase transition in crystals on radiation-induced processes 12th international conference on radiation physics and chemistry of inorganic materials, Tomsk 45-46

IOP Conf. Series: Materials Science and Engineering **168** (2017) 012084 doi:10.1088/1757-899X/168/1/012084

Analyzers of charged particles flows with different functional capabilities

A O Saulebekov[1] and Zh T Kambarova[2]

[1] Lomonosov Moscow State University, Kazakhstan branch, Kazhimukan str.11, Astana 010010, Kazakhstan
[2]Ye.A. Buketov Karaganda State University, Universitetskaya str. 28, Karaganda 100028, Kazakhstan

E-mail: saulebekov@mail.ru

Abstract. Parameters of instruments developed on the basis of previously calculated different electron-optical systems for the analysis of charged particle flows were considered in the present work. Recommendations for their application in various fields of electron spectroscopy for the analysis of a solid surface were given.

1. Introduction

Classical cylindrical and spherical electrostatic mirrors and their possibilities are sufficiently well understood. However, the variable combinations at additional conditions open new functional capabilities of these devices. In the work it has been proposed the instruments, in which this approach was used. For example, the focal surface of the cylindrical energy analyzer of charged particles has a complex shape, which makes it impossible to use it in the spectrograph mode. It is overcome by imposing the conditions of straightening of the foci line.

2. Construction of Auger spectrometer for the analysis of very rough surface of a solid

In the base of developing is the idea of an Auger spectrometer creation, that being mounted on a standard flange DU-200 can be connected to ultrahigh-vacuum equipments for researches or technological needs.

For microanalysis of very rough surface of solid the system shown in Fig.1 can be used. The distinctive features of this system are the second-order angular focusing of charged particle beam in combination with a high linear dispersion on energy and small coefficient of longitudinal magnification. The longitudinal displacement of the source in this case is of less impact on energy resolution than in the classical cylindrical mirror (CM) [1-2], that is, the settings on the focus in a large range of cavities depths and heights of projections of a rough surface (e.g., fracture surface) is kept. Furthermore, by significantly reducing the entrance angle (to 10^0), the depth of probed cavities increases several times in comparison with CM. For CM the entrance angle is 42.3^0, i.e. depth of probed cavities approximately equal to their width. Increasing the depth of probing is shown in Fig.1 and is defined by the following formula

$$n = \frac{h_2}{h_1} = \frac{\tan 42.3^0}{\tan 10^0} = 5.16$$

Auger spectrometer for microprobing of very rough surfaces is shown in Fig.1. It consists of the side flange 1, on which all functional elements are collected , this flange is attached to the flange of the vacuum chamber 2; two cylindrical electrodes (inner 3 and outer 4); electrodes 5 for electrostatic protection of CM working field, the electron gun 6; outer 7 and inner 8 spherical electrodes [3]; exit aperture of diameter about 0.1-0.5 mm (typically used in spectrometers), receiver 9; sample 10, the investigated surface of which is perpendicular to CM axis and intersects it at the front focus; permalloy screen 11 for protection of analyzer from external electromagnetic fields; windows 12 for pass of charged particles beam through the inner cylindrical electrode that tightened by a one dimensional grid. The electron-optical characteristics of the given construction were calculated. The luminosity of energy analyzer is about 0.15% of 4π, and the energy resolution - 0.02%.

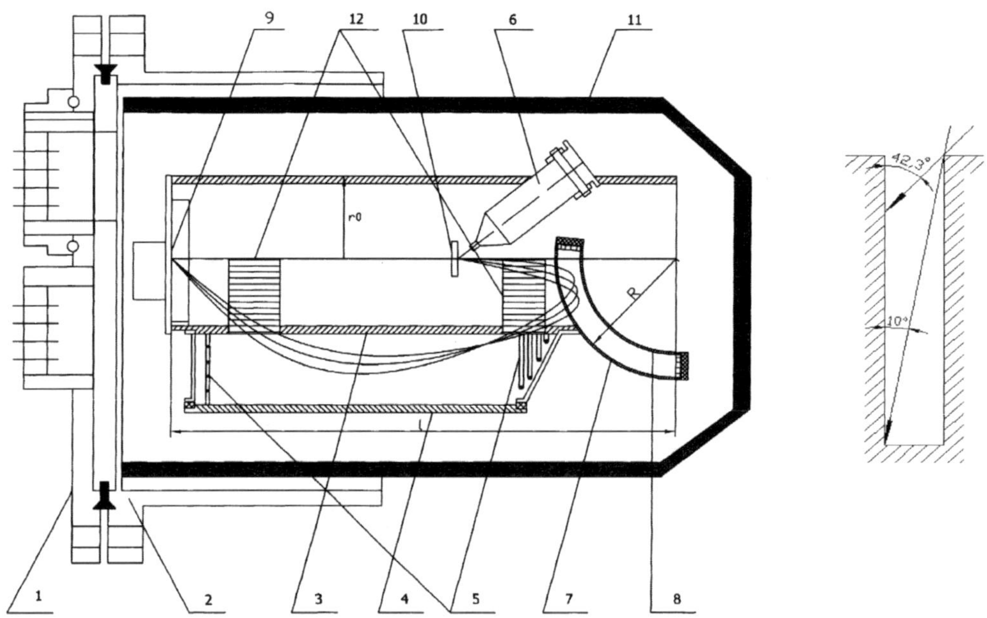

Figure 1. Auger spectrometer for the analysis of very rough surfaces.

3. Construction of the energy analyzer of the double filter type

Wide use of dispersionless electrostatic analyzers of double filter type is due to following properties. It is the reduced requirements to the angular focusing of the analyzed beam in the receiver area, which makes these devices less sensitive to disturbance electromagnetic fields and allows to work with sources of large sizes. Energy analyzer which works on the principle of double filter can be constructed on the basis of the scheme with the retro reflection of the charged particles beam. The design of the spectrometer is shown in Fig.2. Elements of construction (from 1 to 12) are the same as those of the Auger spectrometer for rough surfaces (Fig.1). The filter function of high-energy charged particles 13 performs a spherical field between two grids at the entrance of CM. At the pass of charged particles beam through the filter, the low-energy part of spectrum is cut off. After the beam enters CM it experiences the dispersing action of the cylindrical electrostatic field. The spherical mirror between two 7 and 8 electrodes (SM), performing the second filter [3], reflects charged particles with the energy from a small interval. This interval is adjacent to the low-energy edge of the spectrum separated by the first filter. As a result of two filters action, SM+CM+SM energy analyzer separates the narrow band of spectral interval.

In practice it is convenient that the axial trajectory of beam emerged from the SM field at the intersection point of a spherical electrode with symmetry axis. Fig.2 shows the scheme of the double filter, calculated for specific case. For the side trajectories with divergence of 10^0 values of maximum depth of penetration to field R_m are 0.84711 (+) and 0.84553 (-) respectively. Then $\Delta R_m / R_m =$ 0.1907% and $|(\Delta E)/E| = 1.37\%$. The luminosity of device is 6.1% of 4π.

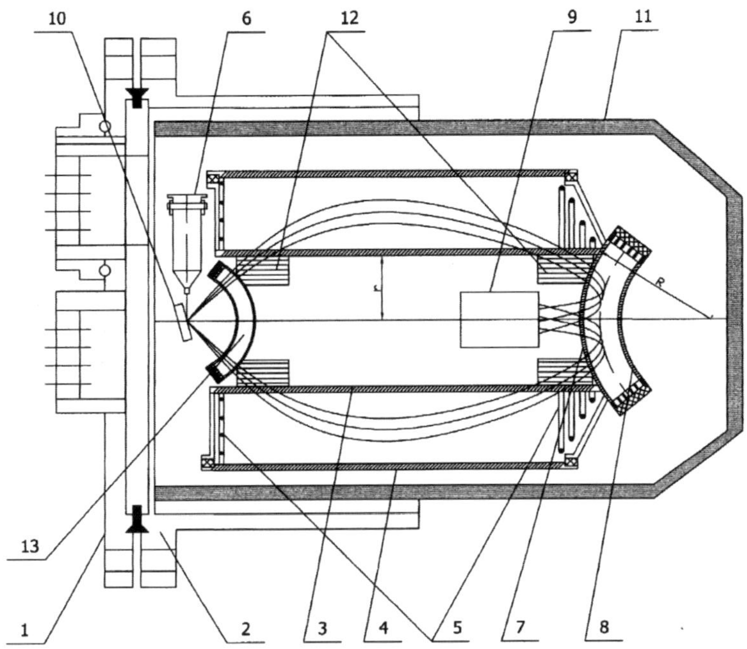

Figure 2. Construction of energy analyzer of double filter type.

4. Construction of the spectrograph on the based spherical and two cylindrical mirrors

The scheme for an energy- and angle-resolved spectrograph which provides high-quality angular focusing of charged particles beams of various energies along the focal surface shaped as a right circular cone was presented in Ref. [4]. An energy analyzer has been composed of successively arranged electrostatic spherical mirror and the cascades of internal and external reflection of a charged particles beam in the CM. In such case simultaneous registration of the finite intervals of charged particle distribution in both energy and angle by means of a position-sensitive detector of a relatively simple conic form, repeating a focal surface, is possible in practice. Express energy- and angle-analysis of the photoelectrons is possible only by means of high-luminosity energy analyzers. This requirement is met by a few known devices, including ESA-22 electrostatic electron spectrometer [5].

Theoretically developed construction of the spectrograph was described in this section; the spectrograph provides focal line straightening and is able to significantly reduce the measurement time of energy distributions in electron spectroscopy. General view of the spectrograph is shown in Fig.3. It consists of three agreed on parameters of electrostatic spherical and two cylindrical mirrors. The base component of spectrograph is a middle cylindrical electrode 1, which can be made of the pipe of certain radius (usually 20-30 mm, wall thickness is 0.5 mm, grade of steel is 12Ch18N10T). In the frame of electrode 1 three windows 2, 3 and 4 are necessary to be made. The windows are tightened by one-dimensional grid of nichrome wires with thickness of 0.15 mm arranged along generatrices of

IOP Conf. Series: Materials Science and Engineering **168** (2017) 012084 doi:10.1088/1757-899X/168/1/012084

the cylinder 1 with steps of 1 mm. Work spaces of electrostatic fields are placed: a) between the outer 6 and inner 5 spherical electrodes; b) between area of middle cylindrical electrode 1 and the inner cylindrical electrode 8; c) between area of middle cylindrical electrode 1 and the outer semi-cylindrical electrode 7; d) side cheeks 9, 10.

Let's consider the description of SM, operating in the internal reflection mode of charged particles beam. SM is consisted of two concentric spherical segments. The inner electrode 5 has a radius equal to twice the radius of the middle cylindrical electrode. The radius of the outer electrode 6 is determined from s reflection parameter [1]. For example, in construction of two-cascade cylindrical spectrograph, middle cylindrical electrode was made of a pipe of 21 mm radius. In this case, the inner spherical electrode will have radius of 42 mm, radius of outer electrode determined from the following formula

$$R_{outer} = \frac{R_{inner}}{1 - \frac{qu}{2Es}},$$

where q is the charge of analyzed particles, u is retarding potential difference between SM electrodes, E is kinetic energy of analyzed particles, R_{outer} and R_{inner} are radii of inner and outer spherical electrodes.

The inner electrode is usually pressed out of sheet of stainless steel with thickness of 0.5 mm, and is at ground potential, the outer electrode is carved and at retarding potential U_1. The inner electrode is provided with aperture windows 11 through which the charged particle beam enters the area of deflection spherical field and exits from it.

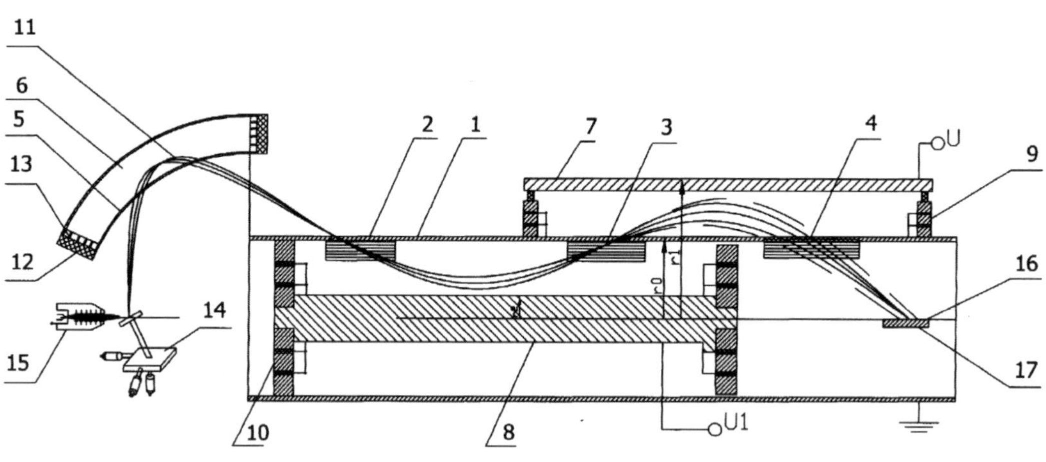

Figure 3. Construction of spectrograph on the based spherical and two cylindrical mirrors.

Windows are the slits in the form of a spherical zone disclosed by 180^0 in the azimuth angle. Each slit is covered by a one-dimensional grid of thin metal wires of 0.3 mm thickness in the form of circular arcs. Their radii equal to the radius of the inner spherical electrode. Wires are arranged at regular azimuthally interval in 2^0 in meridional planes, converging on the symmetry axis. As it is known, the transmission of the analyzer is defined as the ratio of intensity of the recorded charged particles flow and intensity of the monoenergetic flow entering analyzer. The scattering effect of the aperture grid for SM affects mainly to transmission of analyzer. At the same time focusing properties of SM are deteriorating insignificantly.

Spherical electrodes are centered by washers 12 made of an insulator, on which thin rings 13 are coaxially located. Potentials on the rings are distributed in accordance with the law 1/R in order to avoid the influence of the fringe field on the area of the charged particles beam pass . The sample 14 is fixed on the manipulator 15, allowed to move it in three mutually perpendicular directions with accuracy to 0.01 mm.

Let's now consider a cylindrical part of analyzer. 7, 8, 9 and 10 details are recommended for the construction from duralumin of D16T grade. The cylindrical electrodes 7 and 8 are supplied by U_1 and U_2 retarding potentials. On the sides cheeks 9 and 10 there are two protective ring electrodes which are at a potential in accordance with a logarithmic law of change of the cylindrical field from the radius. The value of the surface radius of the inner cylindrical electrode is selected based on requirements of equality deflecting potentials in cascades.

At the exit of analyzer, a plane diaphragm 16 with a rectilinear narrow are placed, its middle line coincides with symmetry axis of analyzer. Position-sensitive detector 17 in the form of a half-cylinder is located behind receiving aperture. Analyzer for protection from the influence of external electromagnetic fields was closed by a permalloy screen, which is not shown in Fig.3.

The value of luminosity of the spectrograph is about 1.7% of 4π, energy resolution of about 0.1%.

5. Conclusions

Schemes of devices and their geometrical and electron-optical parameters were shown. Ways of protection from the effect of fringe fields were listed. Devices can be used in fields such as electronic microprobe spectroscopy, photoelectron spectroscopy, Auger spectroscopy with increased scanning area, photoelectron spectroscopy with energy-angular resolution, Auger electron spectroscopy of rough surfaces of solid.

Acknowledgements
This work was supported by the grant GF4-0815 of the Ministry of Education and Science of the Republic of Kazakhstan.

References

[1] Trubitsyn A A 1995 A cylindrical mirror analyzer with high energy resolution *J. of Electron Spectroscopy and Rel. Phen.* **73** 305 – 310
[2] Zashkvara V V and Ashimbaeva B U 1998 Cylindrical mirror analyzer with adjustable angular focusing *J. of Electron Spectroscopy and Rel. Phen.* **94** 89 – 96
[3] Zashkvara V V, Saulebekov A O and oth. 1992 Electron-optical properties of electrostatic spherical mirror and systems based on it *Zh. Tech. Phys.* **62(6)** 189-204 [in Russian]
[4] Zashkvara V V and Ashimbaeva B U 1994 Theoretical basis for the scheme of an electrostatic energy and angle-resolved spectrograph *Nucl. Instr. and Meth.* **A340** 514-523
[5] Varga D, Tőkési K and Rajta I 1995 Design of an electrostatic electron spectrometer for simultaneous energy and angular distribution measurements *J. of Electron Spectroscopy and Rel. Phen.* **76** 433 – 436

Dependence the Integrated Energy of the Electromagnetic Response from Excitation Pulse Duration for Epoxy Samples With Sand Filler

V P Surzhikov[1] and A A Demikhova
Tomsk Polytechnic University, NDT Institute, 634050 Tomsk, Russia

E-mail: iva1954@mail.ru

Abstract. Results of research of influence of the excitation pulse duration on the parameters of the electromagnetic response of epoxy samples with filler the quartz sand presented in the paper. The electric component of a response was registered by the capacitive sensors using a differential amplifier. Measurements were carried out at two frequencies of the master generator of 65 kHz and 74 kHz. The pulse duration was changing from 10 to 100 microseconds. The stepped sort of dependence of the integrated oscillations energy in the response from duration of the excitation pulse was discovered. The conclusion was made about the determining role of the normal oscillations in formation of such dependence.

1. Introduction

Attraction of the phenomenon of compelled electromagnetic emission for development the non-destructive method of testing defects and the stress-strain state [1-7] determines the necessity of research and optimal conditions definition the excitation of responses of the mechanoelectrical transformations (MET), including as geometry of excitation and measurement and parameters of excitation pulse: frequency, duration.

In work [8] was studied the influence of a pulse duration of acoustic excitation on characteristics of the MET response of composite dielectrics. Shown difference of the compression pulses efficiency and stretching in responses formation of MET and significant contribution of own oscillations in response were suggested.

Below this assumption taking into account amplitude-frequency characteristics of an excitement pulse is considered.

2. Materials and methods

The description of a measurement technique is provided in works [9, 10]. Figure 1 shown block diagram of the measurement procedure.

[1]Corresponding author

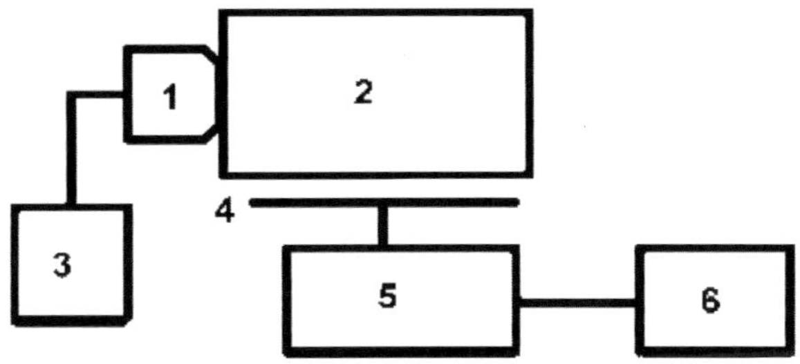

Figure 1. Block diagram of the measurement procedure: 1 – source of ultrasonic vibrations; 2 – sample; 3 – master oscillator; 4 – capacitive sensor; 5 – differential amplifier; 6 – I/O board NI PCI – 6133.

In our experiments the electric component of response was recorded by a capacitive sensor. Experiments were made on the epoxy samples with size 60.5 × 80× 91.5 mm. Quartz sand with a particle size not more than 0.5 mm was used as filler. The filler volume amounted to 0.5 epoxy volume. The piezoelectric transducer on the basis of ceramics of CTS 19 was a source of ultrasonic oscillation. It located on the verge of the sample with size of 60.5 × 80 × 91.5 mm.
The acoustic pulse was initiated by an electric harmonic pulse from the master generator with fixed frequency of 65 kHz and 74 kHz. The pulse duration varied from 10 μs to 100 μs with a step equal 2 μs. The capacitive sensors were installed parallel to the sample verges with size of 91.5 × 80 mm at the distance of 1 mm from a surface. Signals from the capacitive sensor were amplified by the differential amplifier In experiments using the I/O board NI PCI - 6133 was registered the temporal realization of responses from the MET sources at their activation by 140 acoustic pulses with a period of 4096 μs. In the analysis were used the responses averaged by entire temporal realization. The experiments error did not exceed 5%.

3. Results and discussion

Figure 2 illustrates the dependence of integrated energy of the MET responses from duration of the excitation pulses for two values of the master generator frequencies of 65 kHz and 74 kHz.
Dependence for 65 kHz has a step-like sort with duration of steps equal to the period of the generator increasing the duration of the excitation pulse leads to change in amplitude-frequency characteristics of the MET responses both at 65 kHz (Fig. 3) and at 74 kHz (Fig. 4).

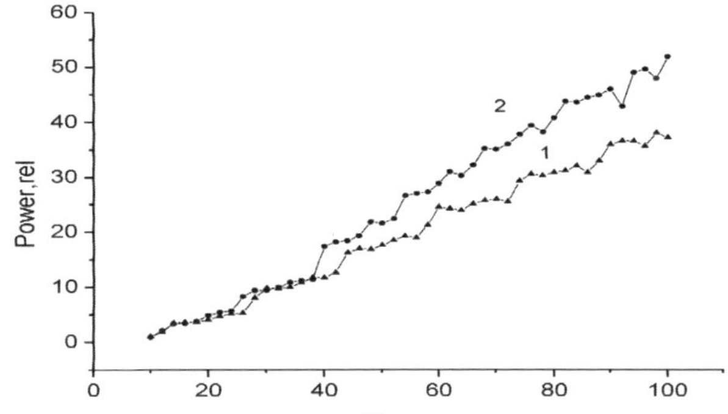

Figure 2. Dependence the integrated energy of the mechanoelectrical transformations response from the excitation pulse duration at 65 kHz (curve 1) and 74 kHz (curve 2).

As seen from the figures a consequence of increasing duration at 65 kHz is a significant decrease of amplitude in the frequency range 69 kHz - 74 kHz. At frequency 74 kHz the pulse duration increase is compliant the amplitude decrease in the frequency range 62 kHz - 67 kHz, and a power increase in the strip of 72.5 kHz.

The calculated amplitude-frequency characteristics of an excitement pulse for two frequencies used in experiment are given in figures 5 and 6.

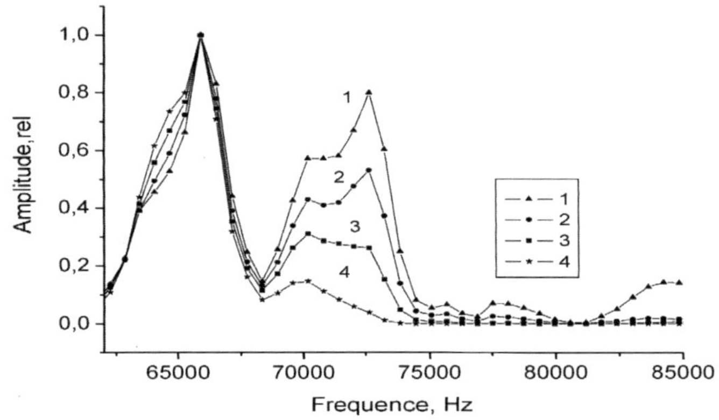

Figure 3. Power spectrum of the electromagnetic response at the excitation pulse frequency of 65 kHz with different pulse duration: 1 – 12 μs, 2 – 33 μs, 3 – 66 μs and 4 – 100 μs.

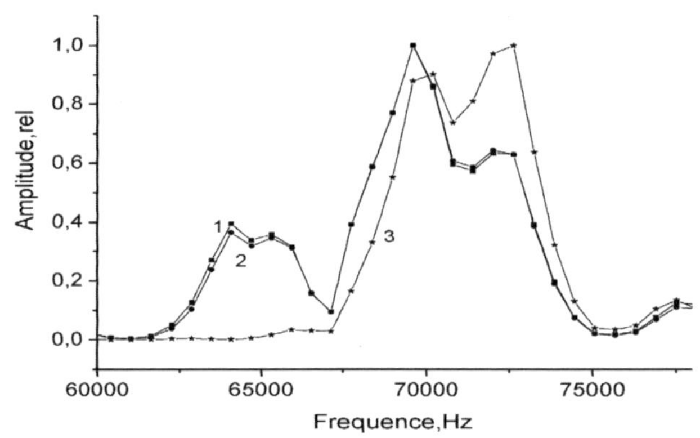

Figure 4. Power spectrum of the electromagnetic response at the excitation pulse frequency of 74 kHz with different pulse duration: 1 – 12 μs, 2 – 33 μs and 3 – 100 μs.

It is seen that with increase a pulse duration the spectrum band excitation is narrowed from 170 kHz to 10 kHz at 100 μs with maxima on the master generator frequencies.
In discussing the experiments results, we adhered to the following logical sequence.

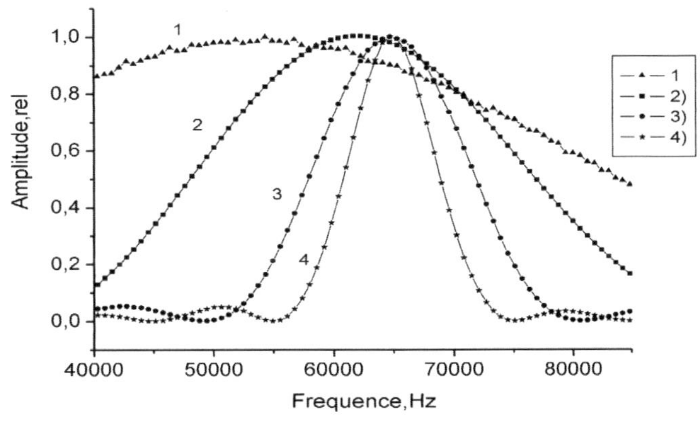

Figure 5. Normalized the power spectrum of excitation pulse for frequency of 65 kHz with different pulse duration: 1 – 12 μs, 2 – 33 μs, 3 – 66 μs and 4 – 100 μs.

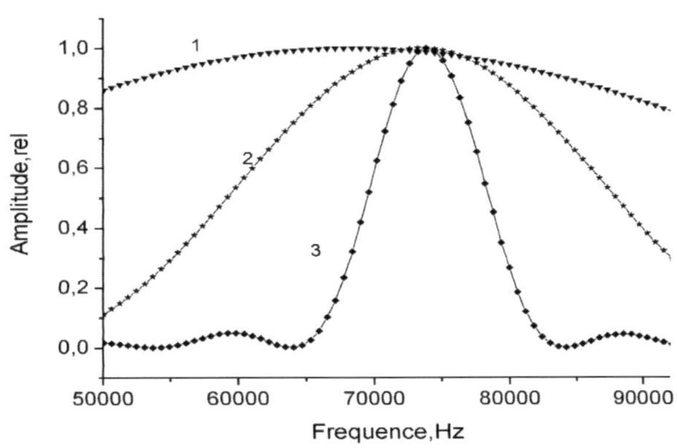

Figure 6. Normalized the power spectrum of excitation pulse for frequency of 74 kHz with different pulse duration: 1 – 12 μs, 2 – 33 μs, 3 – 100 μs.

A voltage pulse from the master generator forces the piezoelectric transducer to radiate ultrasonic oscillations in the certain frequency range. The type of amplitude-frequency characteristics of ultrasound in this range is determined by piezoelectric transducer parameters and emitter design. And as a result, the type of the amplitude-frequency characteristics of a response is determined by frequency range which given by the duration of excitation pulse, the parameters of amplitude-frequency characteristics of the ultrasound, the set of allowed frequencies in sample and the used measurement geometry.

Thus, changes in the amplitude-frequency characteristics of response are determined by decrease the width of an excitation spectral band. In this case at the frequency of 65 kHz (Fig. 3) the excitation spectrum narrowing leads to an apparent bandwidth allocation in response spectrum in the frequency range of 63 kHz - 67 kHz. At the frequency of 74 kHz (Fig. 4) the limitation of the excitation spectrum excludes from spectrum response the frequencies in the range of 63 kHz - 67 kHz. Therefore, appearance of the "steps" on dependence (Fig. 2) naturally connected with acoustic vibrations in the frequency range of 63 kHz - 67 kHz.

For the purpose of identify the oscillations type perpetrated with a frequency of 65 kHz was measured the longitudinal acoustic waves velocity (2585 m/s) and knowing the oscillation frequency was determined the wavelength equal to 1/2 the sample length. This fact, and also results of work [9-12] allowed to assume the main role of normal oscillations in the "steps" formation.

4. Conclusion

The materials of researches given above led to the suggestion that determining share in energy of the MET response and respectively in the formation of uneven dependence of integrated energy of a response are contribute the normal oscillations.

Acknowledgements

This work was supported by the Ministry of Education and Science of the Russian Federation in part of the "Science" program.

References

[1] Yavorovich L V., Bespal`ko A A, Fedotov P I, Popelyaev A I 2015 IOP Conf. Ser.: Mater. Sci. Eng. **81** 012055
[2] Bespal'ko A A, Surzhikov A P, Yavorovich L V, Fedotov P I 2012 Russ J Nondestruct+ **48** 4
[3] Bespal'ko A A, Yavorovich L V, Fedotov P I 2011 Russ J Nondestruct+ **47** 10
[4] Bespal'ko A A, Yavorovich L V, Viitman E E, Fedotov P I 2010 J Min Sci+ **46** 2
[5] Bespal'ko A A, Yavorovich L V, Fedotov P I 2012 J Min Sci+ **43** 5
[6] Fursa T V, Asanov V A, Saveljev A V 2003 Russ J Nondestruct+ **8** 8
[7] Fursa T V, Demikhova A A, Vlasov V A 2014 Russ J Nondestruct+ **50** 5
[8] Surzhikov V P, Yavorovich L V, Zhanchipov B D 2016 IOP Conf. Ser.: Mater. Sci. Eng. **110** 012104
[9] Surzhikov V P, Khorsov N N, Khorsov P N Russ J Nondestruct+ **48** 2
[10] Surzhikov V P, Fedotov P I, Khorsov N N 2015 Tech Phys+ **60** 3
[11] Bespal'ko A A, Yavorovich L V, Fedotov P, Pomishin E, Vedyashkin M V 2016 *MATEC Web Conf.* **79** 01039 DOI: http://dx.doi.org/10.1051/matecconf/20167901039
[12] Plotnikova I V, Vedyashkin M V, Mustafina R M, Plotnikov I A, Tchaikovskaya O N, Bastida J, Verpeta M 2016 *MATEC Web Conf.* **79** 01019 DOI: http://dx.doi.org/10.1051/matecconf/20167901019

IOP Conf. Series: Materials Science and Engineering **168** (2017) 012086 doi:10.1088/1757-899X/168/1/012086

Complex defects in crystal scintillation materials and phosphors

V Lisitsyn[1], **L Lisitsyna**[2] **and E Polisadova**[1]
[1]National Research Tomsk Polytechnic University, Lenin Avenue, 30, Tomsk, Russian Federation
[2]Tomsk State University of Architecture and Building, Solyanaya sq., 2, Tomsk, Russian Federation

E-mail: lisitsyn@tpu.ru

Abstract. The possibility of the existence of complex defects in pure and doped crystal phosphor discussed in work. The luminescent properties of mono- and nanocrystals of zinc tungstate, the powders of YAG with various compositions are studied. It is shown that the intrinsic defects, impurities, oxygen vacancies, the hydroxyl groups may be present in the structure of the complex defects (nanodefects). Nanodefects form during synthesis and have high efficiency of the transfer of excitation energy to the emission centres.

1. Introduction

The existing models assume uniform distribution of the lattice defects in the crystal. A defect, a dopant, is an elementary unit that possesses specific properties which do not depend on the matrix it is located in. This assumption is confirmed by the studies of radiation-induces processes in the crystal with a perfect structure and a low dopant concentration. This approach was used to establish the structure of defects of different types (emission centers in particular), the processes associated with defect migration, transformation, and transfer of excitation energy and excitation relaxation. However, crystal phosphors frequently contain large concentrations of defects, dopants; the crystal structure is complex and imperfect. Therefore, it can be assumed that defects enter these crystals in the form of complex systems during synthesis. The complexes are supposed to contain a dopant, co-dopants and intrinsic lattice defects to compensate the difference in the charge of the matrix and dopant ions, and elastic tension in the region of the dopant location. This complex is expected to be a nano-sized formation as compared to a point defect and it can be referred to as a nanodefect [1,2]. Crystal phosphors with high concentration of defects which form nanodefects can be referred to as highly imperfect phosphors. Formation of these complex defects with emission and absorption centers during synthesis is supposed to significantly affect the processes of the excitation energy transfer to emission and relaxation centers.

The key information on the formation of nanodefects and their structure available to date was obtained from the study of LiF crystals doped with polyvalent Fe, Ti, U and W ions. The

[1] To whom any correspondence should be addressed.

charge of multivalent ions differs from the charge of Li replaced by these ions. To compensate the charge and elastic stresses, intrinsic defects are introduced in the vicinity of the dopant ion: interstitial ions and vacancies. Crystals containing multivalent ions are grown from the mix with ions incorporated in the mix in the form of oxides, which means that oxygen ions enter the crystal. High multivalent ion concentrations can be incorporated in the crystal in case the crystal is grown in air in the presence of LiOH in the mix. During synthesis, multivalent ions introduced in the LiF crystal with oxygen ions, OH, and intrinsic defects form nanodefects. It was experimentally shown that nanodefects contain oxygen ions [3–6], OH [7–8], interstitial ions and vacancies [2]. A nanodefect contains an emission center with O^{2-} ion in its structure and, probably, an anion vacancy [1, 9–11].

A nanodefect has a large capture cross section for charge and energy carriers. Electronic excitations captured by nanodefects are effectively transferred to the emission center. Therefore, the presence of nanodefects makes the LiF crystal containing multivalent ions an efficient converter of radiation energy into luminescence. It can be assumed that in scintillators and phosphors, which typically contain a high concentration of defects (impurity and intrinsic defects), nanodefects formed during synthesis can significantly affect radiative processes. This paper presents experimentally obtained information on the presence of nanodefects in scintillation crystals and phosphors.

2. Research Methods

The time-resolved luminescence spectroscopy were used to find out the presence of nanodefects in highly imperfect scintillator and phosphor crystals. The luminescence in the samples was excited by high-energy pulsed electron fluxes, laser radiation with λ_{exit} of 337, 266 and 355 nm and LED chip with λ_{exit}=452 nm. The density of the electron flux energy was 70 mJ/cm^2, the average energy of the accelerated electrons was ~250 keV and the duration of the excitation pulse was 10 ns. The pulse duration of the laser flux was 2 ns. Phosphors were excited using LED chips with an emission band at 360, 395 and 452 nm. To measure kinetic characteristics, pulse voltage was applied to the chip to generate a 70 ns flash.

In excitation by electron fluxes and laser radiation, the spectra and kinetics of luminescence decay were recorded using the PMT-84-6 photomultiplier with MDR-3 monochromator and LECROY digital oscilloscope 6030 (350 MHz). The time resolution of the measuring system was 7 ns. When excited by chip, the luminescence decay kinetics was recorded using the MDR-204 monochromator PMT "Hamamatsu 10720-20" and Tetronix DPO-3033 oscilloscope with 2 ns time resolution.

In steady-state measurements, the luminescence and excitation spectra were measured using the Agilent Cary Eclipse spectrofluorimeter, and a pulsed xenon lamp was used as a light source. The integrated luminescence spectra were recorded using the Avantes AvaSpec-3648 and AvaSpec-2048 fiber optic spectrometers in the range of 200–1100 nm.

The elemental phosphor composition was determined using the Quanta3D 200i scanning electron microscope with an energy dispersive X-ray analysis system (EDAX). The morphology of the phosphor powder was investigated with the Leica DM 6000 M optical microscope and Quanta3D 200i electron microscope.

The scintillation material samples were provided by the Institute for Scintillation Materials NAS of Ukraine. The phosphors chosen for the study were commercial samples.

3. Experimental results and discussion

3.1. Existence of nanodefects in metal tungstate crystals

It has been found that radioluminescence of $PbMoO_4$, $CaMoO_4$, $CaWO_4$, $PbWO_4$, $CdWO_4$; $CdWO_4$-Li_2CO_3; $CdWO_4$- Li_2CO_3-Bi_2O_3, LiF:WO_3 and $Li_2W_{0.95}Mo_{0.05}O_4$ crystals and other crystals of this type have similar spectral characteristics. The luminescence spectra of the crystals contain a band with a maximum of about 2.6 eV and half-width ~0.6 eV. Similar spectral-kinetic parameter values of the excited emission in oxygen-containing materials different both in the crystal lattice type and in the type of the co-dopant indicate some common features of the radiative center structure [12–14].

Tungstate and molybdate crystals are typically grown by the Czochralski method from the mix prepared by mixing Zn, Pb and Cd metal oxides with tungsten and molybdenum oxides in appropriate proportions to obtain a stoichiometric composition. The crystal grown by this method is supposed to contain numerous lattice defects, and this should affect the material characteristics. In [13], the authors suggest that highly imperfect crystals, tungstates, molybdates and polyvalent ion doped LiF crystals, contain complex defects (nanodefects) which represent a set of intrinsic lattice defects incorporated during the growth of impurity ions, including oxygen ions, OH^-.

Introduction of Li and Bi impurities in $CdWO_4$ crystal leads to uncontrolled changes in the light output. This suggests an indirect role of Li and Bi in the scintillation process. Doping should have an effect on the degree of intrinsic imperfection due to mutual charge compensation and, consequently, on nanodefect formation. Therefore, incorporation of the dopant impurity results in changed light output, but this change depends on the nanodefect formation associated processes during crystal growth [15,16].

The spectral-kinetic characteristics of the luminescence were studied for $ZnWO_4$ and $ZnWO_4$-Eu prepared in the form of nano- and bulk crystals. Bulk crystals were grown in air by the Czochralski method, and nanocrystals were grown under microwave activation at 120° C for 30 minutes by the hydrothermal method. The size of the grain nanocrystals was about 25 nm.

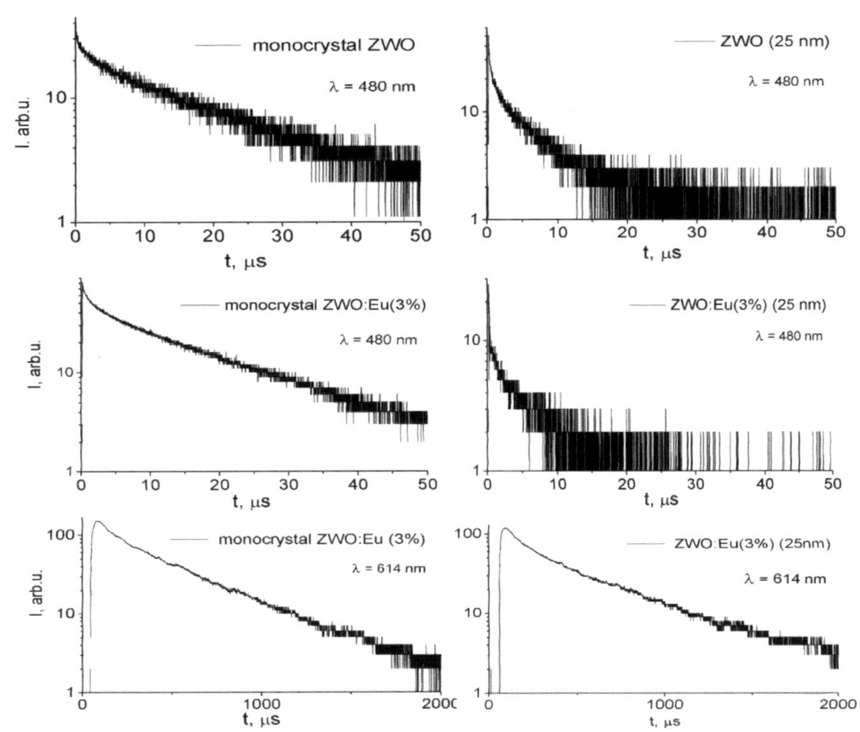

Figure 1. Kinetic curves of PCL decay for ZnWO4 single crystal, single crystal doped 3% Eu ions, powder crystals with size grains 250 nm

In work [17] it was found that the spectra of the macro- and nanocrystal luminescence are similar in shape and position. In the $ZnWO_4$ crystal spectrum, a characteristic scintillation band can be observed with a maximum at 490 nm. In the $ZnWO_4$-Eu crystal spectrum, apart from this band, a narrow band caused by europium can be observed at 614 nm in the long wavelength region [18]. However, after pulse excitation, the kinetic curves of the luminescence decay in $ZnWO_4$ nano- and bulk crystals are found to be distinctly different.

The kinetic curves of the luminescence decay for the studied samples are shown in Fig. 1. The investigation of the spectral-kinetic curves of the PCL decay of undoped $ZnWO_4$ crystals of different dispersion indicated the following results. The characteristic luminescence decay time of single crystals is 26 μs. The luminescence decay time of grain-like nanocrystals with a size of 25 nm is about ~ 7 μs. The kinetic characteristics in the main scintillation band (host luminescence) in monocrystals and nanocrystals ZnWO4:Eu change in a similar manner (Fig.1). However, the decay time in the band 614 nm caused by europium ions is weakly dependent on the morphology and size of the crystal of zinc tungstate. The decay time at band of luminescence 614 nm for mono- and nanocrystals it is about 0.4 ms.

These findings indicate that in a nanometer-sized crystal, the surface of the crystal significantly affects its radiative characteristics. When the crystal size is greater than one micrometer in diameter, the radiative characteristics remain unchanged. This occurs if the sizes of the emission complexes are comparable to those of the crystals. This assumption is supported by a weak effect of crystal sizes on the radiative characteristics of the europium centers.

3.2. The possibility of existence of nanodefects in white LED phosphors

LEDs with blue light converted by phosphor to visible one are the most effective sources of white light. Most commonly used phosphors are those based on yttrium aluminum garnet doped with rare earths. The phosphors are synthesized in air from the mix, a mixture of powders of at least four oxides. The crystal produced contains a high concentration of crystal defects. Therefore, it can be expected that nanodefects are formed in white LED phosphors during synthesis that determines high efficiency of blue light to white light conversion.

A set of commercial YAG phosphors taken for the study differed in the ratio of the basic elements of their composition. The studied phosphors were solid solutions of Y, Al, Gd and Ga oxides doped with Ce. All the phosphors represented powders of microcrystals or microfluxes less than 30 μm in size.

In the phosphors, the deviation from the average aluminum content does not exceed 20%, the content of yttrium changes by 4 times, and that of gadolinium changes from 0 to 12%. According to the specification, all the samples are doped with cerium. In a number of phosphors, cerium was not detected as its weak band is often overlapped by the spectral background caused by other elements present in the phosphor.

The excitation and luminescence spectra of all the phosphors were measured. Table 1 shows the results of the characteristics obtained for the excitation and luminescence spectra measured by conventional method using the Cary Eclipse spectrofluorimeter. In the table, λ_{em}, λ_{ex}, λ_{emm}, $\lambda_{ex\text{-}m}$ indicate the wavelengths of the luminescence and excitation spectra, and FWHM of the corresponding bands in electron-volt.

The kinetic characteristics of photoluminescence decay in phosphors were measured under excitation by radiation from the chip with $\lambda_{ex\text{-}m}$=452 nm. The values of the characteristic luminescence decay times are given in Table 2. The characteristic decay times τ of all the investigated phosphors are found within 60–66 ns. An exception is the SOM-5000 phosphor; its decay time is 53 ns. The measured decay times coincide with those obtained in [19–20].

Table 1. Characteristics of the luminescence and excitation spectra of the studied phosphors.

Phosphors	The spectral characteristics							
	Luminescence				Excitation λ_{emm}			
	λ_{ex}=340 nm		λ_{ex}=440 nm					
	λ_{emm}, nm	ΔE, eV	λ_{emm}, nm	ΔE, eV	$\lambda_{ex\text{-}m}$ nm	ΔE, eV	$\lambda_{ex\text{-}m}$ nm	ΔE, eV
SDL5000	515	0.396	515	0.393	344	0.257	446	0.441
SDL 4100	532	0.413	533	0.403	341	0.322	446	0.457
SDL 3000	-	-	-	-	341	0.273	453	0.421
SDL1000	532	0.416	530	0.440	340	0.273	446	0.419
SD3300	532	0.390	532	0.384	340	0.261	448	0.447
MG 531 W S800	533	0.398	533	0.399	340	0.313	445	0.447
MG 633 1 W S800	533	0.401	533	0.405	340	0.322	448	0.447
L-2083-2-+L2085-1-S1000	533	0.408	533	0.415	340	0.332	446	0.409
AWS5-90818-1	533	0.398	533	0.406	340	0.332	448	0.447
EDL5100	511	0.420	511	0.411	340	0.322	448	0.447
FL-4255	512	0.411	512	0.419	345	0.331	450	0.494
FL-5049	560	-	564	0.502	340	-	480	0.448
FL-6040	500	-	510	0.273	-	-	510	0.542

Table 2. Characteristics of the photoluminescence bands and decay times in the maxima of the luminescence bands for phosphors.

Phosphors	Characteristics of band. $\lambda_{ex\text{-}m}$ = 452 nm		
	$\lambda_{em\text{-}m}$, nm	FWHM, eV	τ, ns
AWS-5/90818-1	535±2	0.474	61
L-2086/2W	550±2	0.457	62
MG-397/2W	545±2	0.438	64
MG-531/W	532±2	0.447	62
MG-558/2W	537±2	0.457	65
MG-571/W	535±2	0.447	62
SDL3300	553±2	0.443	64
SDL3000	556±2	0.475	66
SDL3400	550±2	0.455	65
SDL4000	532±2	0.468	61
SDL4100	530±2	0.462	60
SDL5000	517±2	~0.450	53

The comparison of the results obtained in the study of the radiative characteristics of phosphors and their elemental composition allows the following conclusion. No direct relation of the radiation characteristics can be observed, i.e. spectral and kinetic characteristics are not related to the composition of the studied phosphors. This implies that the emission characteristics of phosphors depend not only on the elemental composition, but on the lattice defects introduced into the crystal during synthesis. Variation in the elemental composition causes changes in the defect structure of the crystal, which can result in the formation of similar type of defect and nanodefect complexes.

4. Conclusion

The results obtained in the study of zinc tungstate based scintillation crystals imply that complex nanometer-sized defects are formed in these crystals during synthesis. The kinetic characteristics of the luminescence of the investigated $ZnWO_4$ and $ZnWO_4$:Eu crystals dramatically change for crystals which are 10 nm and less in size.

In the studied white LED phosphors, the spectral and kinetic characteristics of the luminescence in the main bands are similar irrespective of the great difference in the elemental composition of the phosphors. This suggests that the emission centers and relaxation of these centers in the studied phosphors are similar. Apparently, the radiative characteristics of the phosphor depend not only on the composition but also on the imperfection of the medium related primarily to the intrinsic defects. It may be assumed that during synthesis complex defects (nanodefects) with emission centers containing intrinsic defects are formed in the phosphor. Nanodefects cause a sharp increase in the efficiency of the transfer of excitation energy to the emission centers.

References

[1] Lisitsyna L, Oleshko V, Putintseva S and Lisitsyn V 2008 *Optics and Spectroscopy* **105** 531
[2] Lisitsyna L, Lisitsyn V. 2013 *Physics of the Solid State* **55** 2297
[3] Nepomnyashchikh A, Radzhabov E, Egranov A 1984 *Color luminescence centers and LiF crystals* (Novosibirsk: Nauka)
[4] Lisitsyna L A, Putintseva S N, Lisitsyn V M, Oleshko V I. 2008 *Inorganic Materials* **44** 659
[5] Shvarts K K, Kristapson Ya Zh, Lusis D Yu, Podin A V 1967 *Radiation Physics* V Zinatne, Riga 179
[6] Korzhik V 2003 *Physics of scintillators on the basis of oxygen monocrystals* (Minsk BSU)
[7] Stoebe G 1970 *J. Phys. Chem. Solids*. **31** 1291
[8] Wedding B, Klein M 1969 *Phys. Rev*. **177** 1274
[9] Egranov A V, Radjabov E A 1992 *Spectroscopy of the oxygen and hydrogen impurity centers in alkaline and haloid crystals* (Novosibirsk: Nauka)
[10] JainV K 1986 *J. Phys. D: Appl. Phys.* **19** 1791
[11] Rebane L A 1968 Trudy IFA AN ESSR **37** 14
[12] Kolobanov V N, Kamenskikh I A, Mikhailin V V et al. 2002 *Nuclear Instruments and Methods in Physics Research A* **486** 496
[13] Lisitsyna L A, Korepanov V I, Lisitsyn V M, Eliseev A E, Timoshenko N N, Dauletbekova A K 2011 *Optics and Spectroscopy* **110** 529
[14] Spassky D, Vasil'Ev A, Kamenskikh I et al. 2009 *Physica Status Solidi (A) Applications and Materials Science* **206** 1579
[15] Kornulo A, Jankowska-Frydel A, Kuklinski B, Grinberg M, Krutiak N, Moroz Z, Pashkowsky M 2004 *Radiation Measurements* **38** 707
[16] Lisitsyn V M, Valiev D T, Lisitsyna L A, Tupitsyna I, Polisadova E F, Oleshko V I 2013 *Journal of Applied Spectroscopy* **80** 361
[17] Lisitsyn V M, Valiev D T, Tupitsyna I A, Polisadova E F et al 2014 *J. of Luminescence* **153** 130
[18] Lisitsyn V M, Valiev D T, Tupitsyna I A, Polisadova E F et al 2014 *Advanced Materials Research* **872** 128
[19] Xinbo Yang, Hogjun Li, Qunyu Bi, Liangbi Su, Jun Xu 2009 *J. of Crystal Growth* **311** 3692
[20] Zorenko Y V, Savchin V P, Gorbenko V I et. al 2011 *Physics of the Solid State Volume* **53** 1620

Physico-chemical Modification of the Fibrous Filter Nozzles for Purification Processes of Water and Air

S V Bordunov[1], O V Galtseva[2], N M Natalinova[2], A A Rogachev[3] and Ruizhi Zhang[4]
[1]Superviser, Public Corporation Scientific - Innovation Enterprise "Echteh", Tomsk, Russia
[2]Associate Professor, National Research Tomsk Polytechnic University, Tomsk, Russia
[3]Senior Staff Scientist, Associate Professor, Francisk Skorina Gomel State University, Gomel, Belarus
[4]Student, National Research Tomsk Polytechnic University, Tomsk, Russia

E-mail: rabota2013tpu@mail.ru

Abstract. A set of experiments to study physical and chemical modification of the surface of fibers is conducted to expand the area of their application for purification of water, gas and air (including that in conditions of space). The possibility of modification of filter nozzles in the process of fiber formation by particles of coal of BAU type, copper sulfide and silver chloride is experimentally shown. The fraction of the copper sulfide powder less than 50 microns in size was crushed in a spherical mill; it was deposited on fiber at air temperature of 50^0 C and powder consumption of 0.5 g/l of air. The resulting material contained 6–18 CuS particles per 1 cm of the fiber length. An effective bactericidal fibrous material can be produced using rather cheap material – CuS and relatively cheap natural compounds of sulphides and oxides of heavy metals.

1. Introduction
The development of various industries requires new approaches to purification of ventilation air and industrial gases. Filtering through porous filter medium is one of the most advanced methods of purification of water and gas from natural and anthropogenic pollutants. Filter nozzles made from fibers are most preferable for these purposes.

2. Experimental
The technology of processing recyclable materials and thermoplastics waste (polyethylene, polypropylene, polystyrene, polyethylene terephthalate and their compositions) to produce fibrous materials has been developed, in particular, for filtration purification of liquids and gases.

The technology [1–3] under varying parameters of the technological process allows production of fibers with a wide range of diameters (1–500 μm) with simultaneous coating of the surface of the resulting fibers with powders for production of filters for different purpose. Sedimentation of active materials adsorbed on the fiber surface in water or water-organic solutions and interaction of these materials with the surface of the carrier-material with the following formation of the adsorptive material layer are the main methods of modification of fibrous materials based on polyolefins which are highly resistant to the majority of chemical reagents and oxidizers [4–10].

Fixation of the reagent modifier on the fiber surface through adsorption, electrostatic interaction, formation of hydrogen communications or other types of interactions including adsorption of chemical compound vapors on the fibrous material surface with subsequent increase in its adhesion energy, for example, by hydrolysis, is reported in [11].

The method of producing fibers from waste polymers is based on spraying the streams of the molten polymer which are coming out from the rotating reactor intersecting the airflow [1–3].

Earlier we published a number of papers [12–14] about the use of fibers which are produced by the technology [1–3] for filtration purification of water from oil products and other pollutants [15-18] with the methane-butane fraction being separated from natural gas before transportation [19-22] by the characteristics of the fibers with the surface modified by anti-bactericidal additives.

These additives can work a sufficiently long time in the process of gas purification; however, they are washed off from the fiber surface in continuous filtration of liquids through filter with a linear speed of 0.1 m/min or more within 5–6 days. In this regard, the possibility of applying different multifunctional additives to the fiber surface was studied during the process of production of these fibers so that the powdery additives were introduced into the polymer fiber at the stage of the liquid melt solidification. Only this technique can provide reliable bond of the powder particles with the basis of the produced fibers.

Figure 1 shows the schematic diagram of the installation, where some elements are added into the system of the feed of intersecting airflows which allows introduction of the powders of various materials into the air to produce fibers with special properties [2]. We used particles of BAU coal and copper sulfide for modification of the surface of fibrous sorbents which are used for purification of water and air from hydrocarbons and disinfection through filtration, respectively.

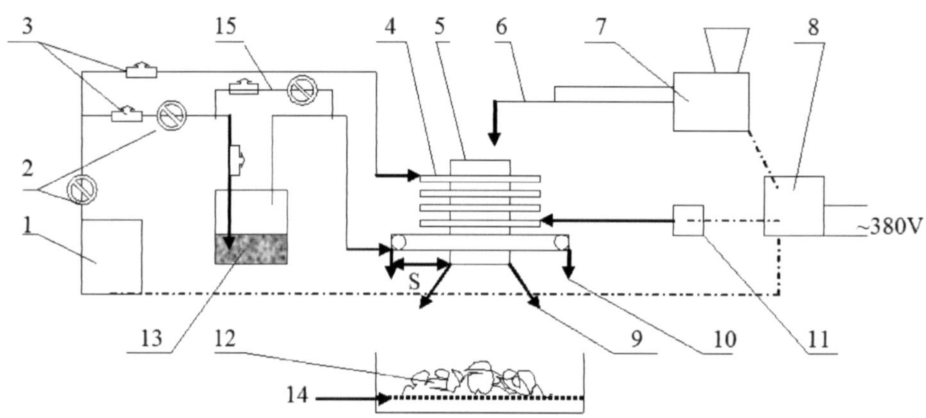

Figure 1. The schematic diagram of the installation for producing fibers with simultaneous coating of their surface with powders, where S is the distance from the edge of the reactor to the air duct, 1 – the compressor, 2 – the flowmeters, 3 – the valves; 4 – the inductor, 5 – the rotary reactor, 6 – the feed line of the polymer melt, 7 – the extruder, 8 – the control panel, 9 – the direction of movement of the resulting fiber, 10 – the direction of movement of air with the sprayed powders from the annular duct, 11 – the power source of inductor, 12 – the resulting fiber, 13 – the capacity of the source powders, 14 – the receiver of the resulting fiber, 15 – the bypass line of the air.

During fiber production, the viscosity of the polymer will vary depending on the distance between the point of melt emergence from the reactor to the point of its meeting with the air stream. At the first stage of the process, the streams of the molten polymer flow from the reactor nozzles under the effect of centrifugal force. When the streams of melt and air flowing from the inlet holes meet, the diameter of the polymer streams decreases and the melt is cooled. During the experiments, valves regulated the mass rate of the powder, BAU coal and copper sulfide feeding.

Despite high density of the melt streams which are removed away from the reactor through the Archimedes's spiral, part of the powder will not get into the melt streams and will be caught in the box and in the filter. Dusty air is sucked away through this filter by the fan.

The distance from the edge of the reactor to the air duct (S) and the temperature of the air determine two important characteristics of the produced fibers: the mechanical strength (rapid cooling of the melt provides high crystallinity and fragility) and density of the powder coating. The density of the powder coating is visually determined with a microscope as a quantity of particles per 1 cm of length of fibre.

At the first stage of the experiments on modification of polypropylene fibers, the consumption of the BAU powder was 20 g/min, and the consumption of air was 200 l/min, i.e. the ratio of the components was 0.1 g of the BAU powder per 1 liter of air. The average diameter of the resulting fibers ranged from 70 to 150 μm depending on the speed of the reactor rotation.

Only relative elongation of the fibers changed sufficiently since they break due to the decreased degree of crystallinity when S and air temperature increase. The increased number of the particles of the BAU powder on the fibers was achieved by increased consumption of powder. This dependence was linear in the range from 0.1 g/l to 0.5 g/l of air. The number of particles per 1 cm of the fiber increased up to 14÷16. However, due to losses of the powder caught by the filter these losses increased from 80 to 95%.

The microscopic analysis of the fiber samples was conducted with the microscope MBS–9 in the transmitted light with magnification up to 100 times to determine the location of individual threads, their shape and surface structure to calculate the specific surface of the test samples. The structure of the obtained fibrous materials represented a combination of adhesive bonded fibers which had a pronounced anisotropy in the distribution of the fibers in the volume. The average diameter and the specific surface of the fibres were determined visually using the MBS-9 optical microscope (20x lens, 5x ocular and the scale interval of the stage micrometer of 0.0154 μm).

The tested fiber was placed on the microscope stage of the MBS-9 microscope; the most frequently occurring diameter of the fibers was determined.
The specific surface of the fiber was calculated by equation [1] using the obtained value of the fiber diameter:

$$S_{SP} = \frac{4}{D\rho} \cdot \frac{\sum \nu_i \sigma_i^2}{\sum \nu_i \sigma_i^3}, \tag{1}$$

where D is the most frequently occurring diameter of the fibers; ρ is the adsorbent density; ν_i is the portion of the fibers of this diameter out of the total number of fibers; σ_i is the fraction of the diameter of this size out of the most frequently occurring diameter.

The bulk density of the samples was determined at the delivery condition in free packing by the pycnometric method according to the GOST 18995.1-73 using VLR-200 scales and carbon tetrachloride of "pure grade" type as pycnometric liquid. The porosity of the sample (the ratio of the volume of voids ("pores") to the entire volume of the material) was calculated using the obtained data.

The parameters of the fibrous materials used were as follows:

- the bulk density of 138–145 g/m^3;
- the porosity of 80 %;
- the diameter of the fiber of 25÷75 μm.

Figure 2 shows the image of the fiber produced from polypropylene with a diameter of 120 μm with the BAU coal particle, which was made with the electron microscope at the Nano-Center, Tomsk Polytechnic University.

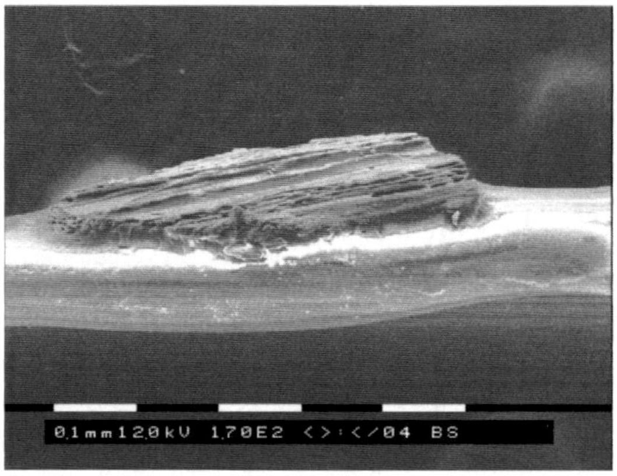

Figure 2. The image of the fiber produced from polypropylene with a diameter of 120 μm with the BAU coal particle.

3. Results and considerations

The fibers were produced from recycled polypropylene with a diameter of 100÷120 μm with powders of copper sulfide and silver chloride deposited on the surface using the installation (Figure 1). Previously, the powder was sifted through the sieve (50 μm); the fraction with the size less than 50 μm was used for modification. The calculation of the number of the copper sulfide powder particles on the first three samples of fibers, which were received at different modes, under a microscope showed that 6, 14 and 18 particles of the modifying additive fall within 1 cm of the fiber length, on average (Table 1). The fourth sample of the fiber, which was modified by silver chloride, contained 11 particles of powder per 1 cm of length, on average. The average value was calculated using the results obtained in twenty parallel experiments; the variation of the data did not exceed 25% relative to the mean value. The fifth sample of the fiber modified by copper sulfide consisted of a mixture of the first three samples with the ratio 1:1:1 (this sample was not tested). The fiber samples were packed into a glass column with an inner diameter of 20 mm to a height of 200 mm, and these samples were compacted under pressure of 1.5 kg/cm^2.

The anti-bacterial characteristics of the samples were studied in the laboratory of the Centre of the GosSanEpidemNadzor (Seversk, Russia) by the methodology according to the order M3 No. 720 and methodical instructions No. 287-113. For studies, the daily culture of the test strain Escherichia coli GCPM 240418 M–17 (Research Institute of Viral Infections, Sverdlovsk, Russia) was used; this corresponded to a concentration of 0.93·109 cells/ml for bacteria of the intestinal group at a temperature of 10–30 ^{0}C. A suspension of bacteria was prepared; its optical density corresponded to 10 units according to the turbidity standard (OSO–42–28–85 P) of the State Scientific-Research Institute for Standardization and Control (Moscow, Russia).

The comparison of the degree of turbidity of the experimental and reference samples was carried out visually. The solutions of the required concentration were prepared by initial dilution through consecutive transfer of 1 ml of the suspension in test tubes filled with 9 ml of sterile distilled water. Testing of the initial suspension and water after filtration was carried out in accordance with MUK 4.2.1018-01 "Sanitary-microbiological analysis of drinking water". Carefully mixed 1 ml of each sample was placed in sterile Petri dishes. After that, 20 ml of the melted and cooled to 45÷49 ^{0}C nutritious agar were poured into each cup and quickly mixed. Simultaneously, 0.1 ml of the infected water was seeded on the environment for intestinal bacteria (Endo, Levin). The colony counting was performed after 24 hours of incubation at 37 ^{0}C.

The research results (the average for three parallel filters) are presented in Table 1.

Regeneration of filters was carried out by 70% solution of ethyl alcohol with subsequent washing with distilled water.

Table 1. Antibacterial properties of fibers modified by powders of copper sulfide and silver chloride.

	Filter	The number of microorganisms in 1 ml				
		50	150	280	780	1280
1	6 particles of CuS per 1 cm of length of fiber	There is no growth	There is no growth	There is no growth	There is no growth	There is no growth
2	14 particles of CuS per 1 cm of length of fiber	There is no growth	There is no growth	There is no growth	There is no growth	There is no growth
3	18 particles of CuS per 1 cm of length of fiber	There is no growth	There is no growth	There is no growth	There is no growth	There is no growth
4	11 particles of CuS per 1 cm of length of fiber	There is no growth	There is no growth	There is no growth	There is no growth	There is no growth

4. Summary

The data presented in Table 1 shows that all the samples of the modified fibers are active to this kind of microorganisms which belong to typical coliform bacteria. These samples are sanitary-indicative microorganisms. A filter with a minimum number of particles per 1 cm fiber length effectively destroys microorganisms within the whole studied range of concentrations even after 15 cycles of regeneration.

Thus, the conventionally used compounds of chromium, cadmium, zirconium, natural sulfur-containing minerals and organic compounds (sulfadimethoxine, chlortetracycline, etc.) can be employed for production of anti-bacterial fiber materials according to the above technology.

References

[1] Bordunov V V *et al* 2002 RF Patent 2179600 *Byull. Izobret.* **5**
[2] Bordunov S V 2010 RF Patent 2388854 *Byull. Izobret.* **13**
[3] Galtseva O V *and* Bordunov S V 2014 *Testing. Diagnostics* **13**
[4] Tolonen H *and* Sjölind S 1996 *Mechanics of Composite Materials* **31** 4
[5] Surzhikov A P *et al* 2006 *Rus. Phys. J.* **49** 5 506-10
[6] Surzhikov A P *et al 2002 Rus. Phys. J.* **45** 12 1190-94
[7] Ivanov Y *et al* 2001 *Polymer Composite Materials — Interface Phenomena & Processes* (Kluwer Academic Publishers, Bulgaria)
[8] Murphy J *Additives for Plastics* (Elsevier Ltd, 2001)
[9] Watt J P *et al* 1976 *Rev. Geophys. Space Phys.* **14** 4
[10] Chen L A *et al* 1998 *Braz. J. Chem. Eng.* **15** 4 DOI:10.1590/S0104-66321998000400005
[11] Uzhov V N *and* Valdberg A Y, *Purification of industrial gases from dust* (Chemistry, Russia, 1981)
[12] Mustafina R M *et al* 2016 *Journal of Physics: Conference Series* **671** (1) 012046 DOI: 10.1088/1742-6596/671/1/012046
[13] Bordunov V V *et al* 2001 *KORUS'2001* **2**
[14] Bordunov S V *and* Bordunov V V 2009 *Non-ferrous metals* **12**
[15] Natalinova N M *et al* 2016 *SIBCON'2016* **1** 7491866 DOI: 10.1109/SIBCON.2016.7491866

[16] Galtseva O V *and* Bordunov S V 2015 *IOP Conference Series: Materials Science and Engineering* **81** 112010 DOI: 10.1088/1757-899X/81/1/012062
[17] Gavrilin A N, Moizes B B 2013 *Testing. Diagnostics* **9** 81
[18] Nizhegorodov A I *et al* 2016 *Journal of Physics: Conference Series* **671** 012037
[19] Vasendina E A *et al* 2016 *IOP Conf. Series: Materials Science and Engineering* **110** 012070
[20] Rogachev A A 2004 *Russian Journal of Applied Chemistry* **77** **(2)** 281 DOI: 10.1023/B:RJAC.0000030367.63884.89
[21] Rogachev A A, Yarmolenko M A, Xiaohong Jiang, Shen Ruiqi, Luchnikov P A, Rogachev A V 2015 *Applied Surface Science* **351** 811-18
[22] Luchnikov P A, Rogachev A V, Yarmolenko M A, Rogachev A A, Bespal'ko A A Dynamics of combined electron beam and laser dispersion of polymers in vacuum 2016 *IOP Conf. Series: Materials Science and Engineering* **110** 012081

IOP Conf. Series: Materials Science and Engineering **168** (2017) 012088 doi:10.1088/1757-899X/168/1/012088

Control System of Parameters of the Azimuthal Module

I V Plotnikova[1], O V Galtseva[1], O N Tchaikovskaya[2] and S A Tchekarova[3]

[1]Associate Professor, National Research Tomsk Polytechnic University, Tomsk, Russia
[2]Professor, National Research Tomsk State University, Tomsk, Russia
[3]Student, National Research Tomsk Polytechnic University, Tomsk, Russia

E-mail: rabota2013tpu@mail.ru

Abstract. Analytical and experimental studies of the azimuthal module of two-component vibrational micromechanical gyroscope were conducted. It is shown that the micromechanical gyroscope is a system with distributed parameters. The frequency analysis is performed using software T-Flex. The influence of mechanical disturbances on the movement of azimuthal module in the form of translational and angular oscillations is shown; the natural frequencies of the azimuth are defined.

1. Introduction

Micromechanical inertial sensors belong to the field of measuring technique and integrated electronics. They are used for measuring of two components of angular speed of object by one micromechanical component. Micromechanical gyroscopes (MMG) can be applied in production of digital camcorders and cameras, mobile phones, intellectual toys, as the measuring instruments of angular speeds in control systems of small-sized aircrafts, in updatable inertial strapdown systems, for systems of navigation and control of cars and other land vehicles and in the gyroscopic inclinometers which determine the direction of drilling. The use of the MMG allows reducing of weight and size characteristics and self-cost of these systems [1].

Features of micromechanical inertial sensors are minor dimensions, cost and power consumption [2-4]. The measured signal is very small; at present the MMG have enough low precision [5]. The efforts of developers are focused on creating of ways of improving of the precision of micromechanical gyroscopes [6–9]. Most of the MMG measures only one component of angular speed of the object.

2. Experimental

Development of the modern systems of orientation (SO) and movement control of objects for various purposes is associated with their miniaturization and reducing of energy and operating expenses.

One of the promising directions in the development of SO is application of micromechanical gyroscopes; they are made from the modern engineering materials, according to the new technologies in the micromechanics [1]. These devices have minor dimensions, cost and power consumption.

Currently, the MMG are inferior to other types of gyroscopes for precision, but they are exceeds them in weight and size characteristics, cost and power consumption.

With due regard for the high rate of improvement of accuracy parameters in this paper the dynamic characteristics of two-component micromechanical gyroscope and automatic setting of system parameters of MMG are considered at which are provided the best characteristics of SO, based on these gyroscopes.

SO are used for determining of position of devices and moving objects relatively of given basic directions.

Basic directions are at orientation in near-earth space:

- the direction of the midday line;
- the direction of the local vertical.

Geographically oriented coordinate system $o\xi\eta\varsigma$ is often used as a directional reference (Figure 1). Origin of this system is related to the object and moves with it.

Axis $o\xi$ and $o\eta$ lie in the horizon plane, the axis is directed on tangent to meridian to the north, the axis $o\varsigma$ is directed on vertical to the Zenith, the axis $o\eta$ is directed to the west.

The vector of angular speed of rotation of the Earth Ω_E is decomposed into two orthogonal components:

$$\Omega_H = \Omega_\xi = \Omega_E cos\varphi; \quad \Omega_V = \Omega_\eta = \Omega_E \sin\varphi,$$

where φ, λ are geographic latitude and longitude of the location of the object.

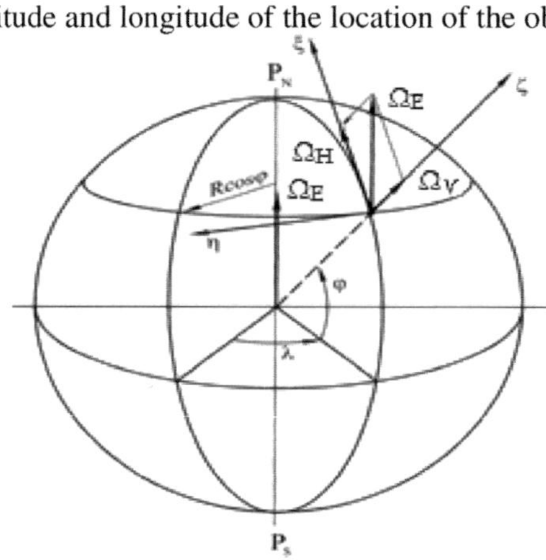

Figure 1. Geographically oriented earth axis system.

Orientation of object relatively to the reference coordinate system is shown in Figure 2.

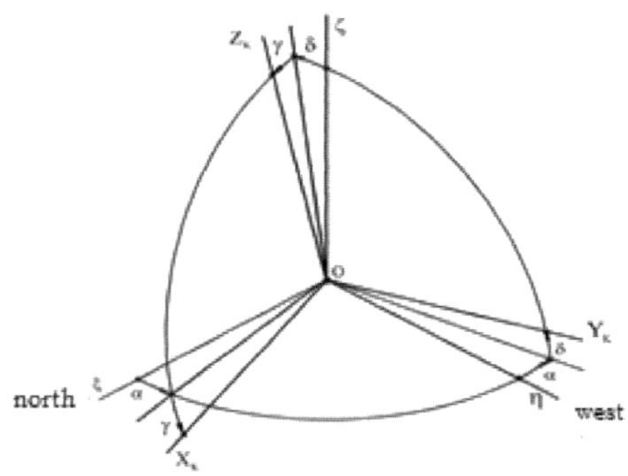

Figure 2. Coordinate systems for defining of orientation of the object.

According to Figure 2 the orientation of the object relative to the reference coordinate system is determined by the angles α, γ, δ, where α is azimuth of the object; γ, δ are angles of coverage of the object from the plane of the horizon; $OX_K Y_K Z_K$ is the coordinate system which associated with the object rigidly.

Principle scheme of micromechanical SO is presented in Figure 3. It contains the two-axis pendulum 2, which is fixed in the gimbal subweight 3. The pendulum performs a function of the geometrical vertical reference.

The sensing element (SE) of the azimuth module 1 is located on the pendulum. The signals x(t), y(t) are withdraw from the pendulum for determination of the azimuthal angle α.

Angle sensors 4 and 5 are on the axes of subweight of the pendulum; information about angles $\bar{\gamma}(t)$, $\bar{\delta}(t)$ is determined by these sensors.

Object deviations from the horizon plane are passed through the analog-digital converter (ADC) 6 to the computer 7 which provides:

- calculation of the orientation angles α, γ, δ and geographical latitude φ;
- filtering of signals;
- determining the ending of the transition process of the pendulum;
- generation of control signals by exciter;
- formation of control signals of the system of resonant settings of SE.

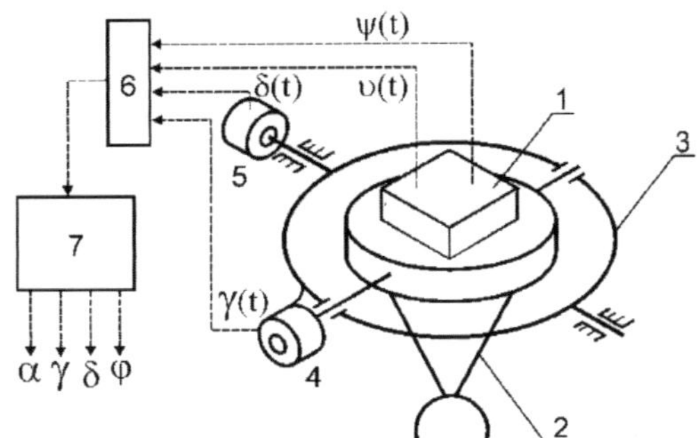

Figure 3: Principle scheme of orientation system.

Researched system has two independent channels:

- the azimuth channel (for measuring of azimuth of the object α);
- the horizontal channel (for measuring of angles of deviation of the object from the plane of the horizon γ, δ).

The sensing element of the azimuth channel is a two-component oscillatory vibratory micromechanical gyroscope [10].

The oscillations of the gyroscope are described by equations

$$\begin{aligned}&(m+m_p)\ddot{x}+\mu\dot{x}+(2k_2-(m+m_p)(\Omega_z^2+\Omega_y^2))x-\\&-2m\Omega_z\dot{y}+2(m+m_p)\Omega_y\dot{z}+m\Omega_x\Omega_y y+\\&+(m+m_p)\Omega_z\Omega_x z=0;\\&m\ddot{y}+\mu\dot{y}+(2k_1-m(\Omega_z^2+\Omega_x^2))y+2m\Omega_z\dot{x}-\\&-2m\Omega_x\dot{z}+m\Omega_x\Omega_y x+m\Omega_z\Omega_y z=0,\end{aligned}$$

where m, m_p are inertial mass of the body and internal frame; x, y are motion of the inertial mass along the axes X, Y; μ_1, μ_2 are coefficients of forces of viscous friction; k_1, k_2 are stiffness of elastic elements of suspension, Ω_x, Ω_y, Ω_z are angular speeds of turning of the basis.

Sensibility and pass band are the main dynamic characteristics of the speed sensor. The amplitude on the one coordinate reaches the maximum values and it on the other coordinate reaches the values which are lower almost on two orders at the excitation of oscillations of inertial body on the one of the partial frequencies of the system [11–16].

$$\omega_{01} = \sqrt{\frac{k_1}{m}},\ \omega_{02} = \sqrt{\frac{k_2}{(m+m_p)}}.$$

The maximum sensitivity of the gyroscope is achieved by resonance tuning which corresponds to the equality of partial frequencies to the oscillation frequency of the exciter (Figure 4).

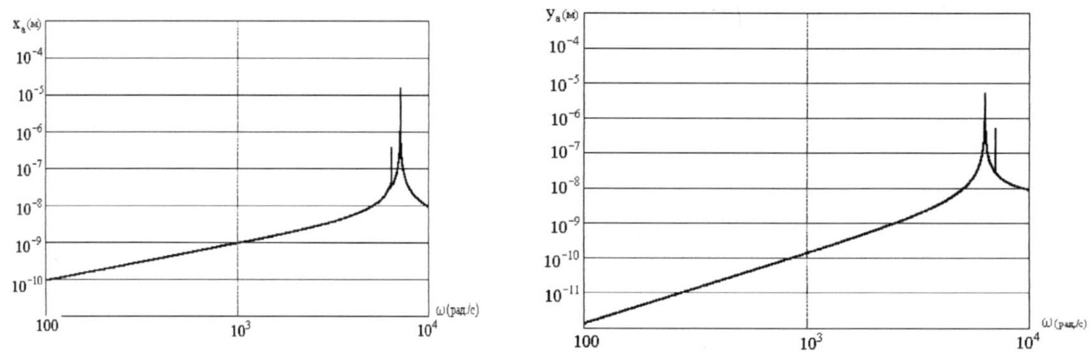

Figure 4. Amplitude - frequency characteristics of MMG.

Figure 5 shows the form of natural oscillations of MMG.

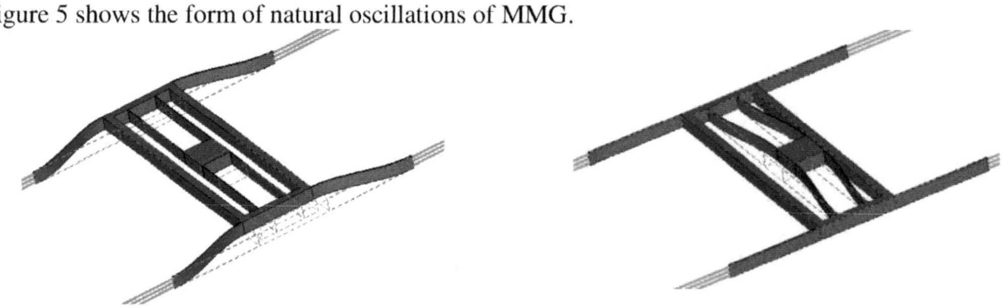

1106.8 Hz 1174.0 Hz

Figure 5. Form of natural oscillations of MMG.

The sensing element, which is an inertial body, converts the value of the current angular speed into the relative movement of construction elements. The electronic part is used for transformation of the measured capacity into voltage, for formation of control of primary oscillations of inertial body, for formation of control of secondary oscillations, for formation of output signal of gyroscope.

The change of temperature of surroundings causes change of the mechanical properties of the elastic subweight; it leads to change of natural frequencies. In addition, the natural frequencies depend on speeds and the design parameters of the gyroscope Ω_x, Ω_y. It is a cause of infringement of resonant setting of the device and emergence of additional errors in working.

Given the acute nature of resonance due to the high quality factor of silicon MMG, precise resonant tuning is required. The AFC system is used; this system can be realized through regulation of "stiffness" of the suspension by means of the vibrodrive. As the voltage of the drive electrodes varies, "stiffness" of the system changes; therefore, the natural frequency of the gyroscope changes as well.

3. Results and considerations

The algorithm of the system of locked loop of frequency is designed in such a way that the initial value of the stiffness k_0 is set on the first step. The value of this stiffness should be such that the system worked obviously not in resonance.

After then the current value of the stiffness k begins to strive for resonant value k_r, and it increases on size Δk at each step. If the amplitude did not increase at some step as at resonance, but it decreased, it means that the resonance is passed and it is necessary take a step back.

The work of system of locked loop of frequency is simulated using software Simulink MATLAB; the results are presented in Figure 6.

The first two graphs show the amplitudes of output signals of the gyroscope along the axes *OX* and *OY*. The third graph represents the dependence of the output signal of Microcontroller block, which implements the algorithm of work of system of the auto setting, from time.

On these graphs show, system of the auto setting of the frequency gradually increases stiffness, bringing the system to resonance.

Due to the last increasing of stiffness the amplitude of the output signal decreases. It suggests that the resonance is passed; the value of additive to the stiffness is exceeds the required value. On the basis of this information the system of auto-adjust of frequency decides to return on a step backwards and to stop the work. The considered system of auto-adjust of frequency is realized by a microcontroller.

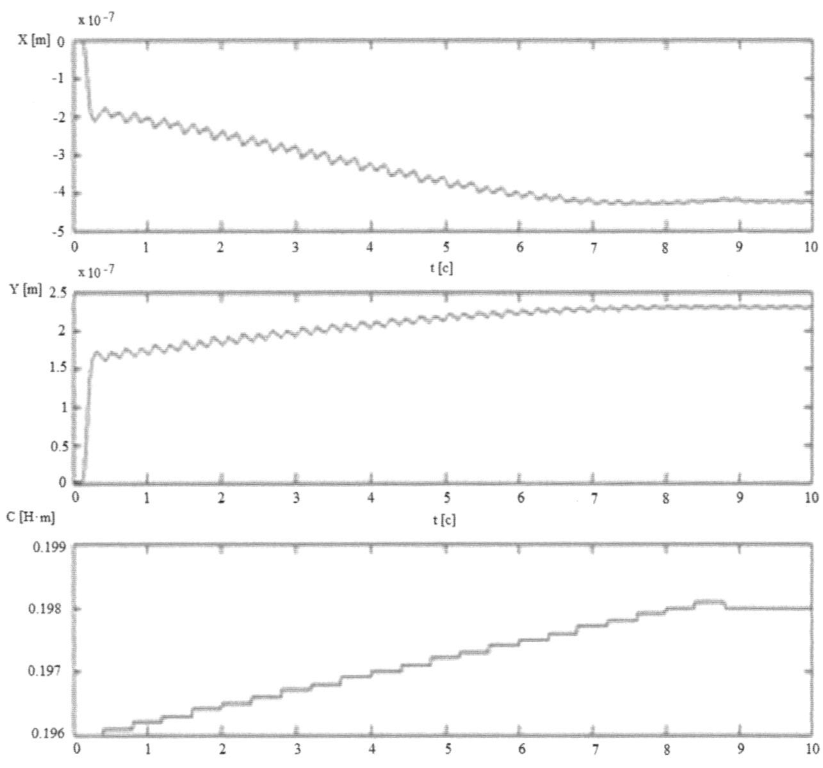

Figure 6. The output signals of MMG at auto-adjust of its own frequencies.

4. Conclusion

Thus, important characteristics of the inertial sensor are accuracy, measuring range, pass band, indication setting time, resistance to vibration, compatibility with GPS/ GLONASS, operability at low and high temperatures, analog and digital output signals.

Automatic resonance setting of the azimuthal module allows obtaining maximum sensitivity, which is especially important in micromechanical systems, where the amplitudes of the output signals are small.

References

[1] Parent A *et al* 2007 *Proc. of IEEE SENSORS* **72850** 4388541 DOI: 10.1109/ICSENS.2007.4388541
[2] Doronin V P *et al* 2015 *Proc. of ICINS 2015* **112822** 323–326
[3] Nesterenko T G *et al* 2007 *Proc. of ICINS 2007* **109057** 52–53
[4] Luchnikov P A 2015 *Russian Physics Journal* 57 9 1239-45
[5] Tao Y *et al* 2011 *Sens. Actuators, A* **168(2)** 286–299 DOI: 10.1016/j.sna.2011.04.039
[6] Kempe V 2011 *Magnetoelastic torque sensors* (England: Cambridge University Press)
[7] Surzhikov A P *et al* 2016 *IOP Conf. Series: Materials Science and Engineering* **110** 012002 DOI: 10.1088/1757-899X/110/1/012002
[8] Natalinova N M *et al* 2016 *Proc. of* EECEA2016 **1** 52 DOI: 10.1109/EECEA.2016.7470765
[9] Lawrence A., Modern inertial technology – navigation, guidance and control, *Springer-Verlag*, 1998, 278 p.
[10] Vasileva O V *et al* 2015 *Proc. of Int. Conf. MEACS 2015* 7414881 DOI: 10.1109/MEACS.2015.7414881
[11] Nesterenko V P *et al* 2001 *Proc. of KORUS 2001* **116335** 975048 39–42 DOI: 10.1109/KORUS.2001.975048
[12] Aleutdinov A D, Ghygazov S A, Mylnikova T S, Luchnikov P A 2015 *IOP Conf. Series: Materials Science and Engineering* **81** 012069
[13] Galtseva O V *et al* 2015 *Proc. of SENSIG '15* 91–95
[14] Apostolyuk V 2006 *Theory and design of micromechanical vibratory gyroscopes* (US: Springer) 173–195 DOI: 10.1007/0-387-25786-1_6
[15] Surzhikov A P *et al* 2006 *Rus. Phys. J.* **49** 5 506-10
[16] Vasendina E A *et al* 2016 *IOP Conf. Series: Materials Science and Engineering* **110** 012070 DOI: 10.1088/1757-899X/110/1/012070

IOP Conf. Series: Materials Science and Engineering **168** (2017) 012089 doi:10.1088/1757-899X/168/1/012089

Pneumatic device of the preload and dynamic loads balancing to reduce the intensity of thermal processes in the metal cutting process

A N Gavrilin[1,5], B B Moyzes[1,6], A I Cherkasov[2,7], I A Ditenberg[3,8], A R Khairullin[4], E V Glinenko[4]
[1] Associate professor, Tomsk Polytechnic University, Tomsk, Russia
[2] Senior lecturer, Tomsk Polytechnic University, Tomsk, Russia
[3] Associate professor, Tomsk State University, Tomsk, Russia
[4] Student, Tomsk Polytechnic University, Tomsk, Russia

E-mail: [5]tom-gawral@list.ru, [6]mbb@tpu.ru, [7]acherkasov@tpu.ru, [8]ditenberg_i@mail.ru

Abstract. Improved reliability of the technological system «machine-tool-instrument-detail» is an important current task. Backlashes and insufficient stiffness of technological system lead to intensive wear of the cutting tool, increasing the heat in the cutting zone. Due to high temperature in the thin surface layers of the workpiece and tool thermal processes may occur which are similar to release and can cause the structural changes of the material. The current article presents the final design of the device which has been developed to reduce the intensity of thermal processes in metal cutting.

1. Introduction

Improving the quality of metalworking and reliability of Technological System «Machine-Tool-Instrument-Detail» (TS MTID) is always an important current task. There are some methods to solve this problem: to change the machining process mode (feed, spindle rate speed, depth of cut), installation of special devices to enhance the stiffness, vibration level reducing [1–5] etc. For example, due to the stiffness enhancing the durability of technological system elements significantly increases.

Insufficient stiffness of process system parts leads to overheating in the cutting zone which impacts directly on the resistance of the tool to weariness , the quality of process material and the cutting efficiency.

Impact of pressure and high temperature causes thermal processes (chemical reaction between the tool and workpiece materials), that changes the chemical composition and physical and chemical properties of the tool surface and the workpiece.

It is known that stiffness of the technological system is almost always the dominating factor determining precision at all types of mechanical machine processing which do not involve implementation of size or profile cutting tool [6].

One of the most effective measures to improve the stiffness of the machine-tool, is the elimination of backlashes [7]. The main method of elimination of backlashes is to provide a preload, which is included while designing the technological system by using the particular units design or using special mechanisms for the tension force generation F_t. In both the first and second cases, tension force F_t

results in a shift in ratio of deformation to force $\frac{\Delta_1}{\Delta F_1}$ downward (Figure 1), so that when the technological system operates, two zones of stiffness can be marked:

- «low» – without F_t;
- «high» – with F_t.

In the area of «small» stiffness a backlash in the joints influence the technological system operation which downgrades the parts processing. There is no backlash in the area of «high» stiffness. In such a case, significant increase of precision and quality of the parts processing is attained as well as the increase of time for life of instrument operation and the reliability of the technological system operation [8-11].

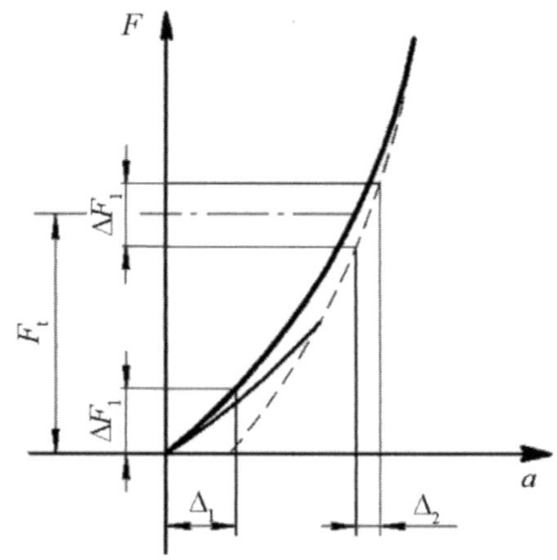

Figure 1. Graph of dependency between the applied force and deformation TS.

2. Design of the pneumatic device

There known devices for tension implementation by load suspension. But this design increases the dynamic loads in the technological system by increasing the system inertia. There are solutions to create a preload by means of pneumatic springs however, due to the limited spring stroke and the inability to change the power characteristics of these pneumatic springs when required, but the implementation of such springs is partly neutral [12, 13].

The developed pneumatic device with a simple design and reliability for operation increases the stiffness of any TS MTID when these shortcomings are absent as well as dampens occurring vibrations.

Figure 2 shows the basic diagram of a pneumatic device.

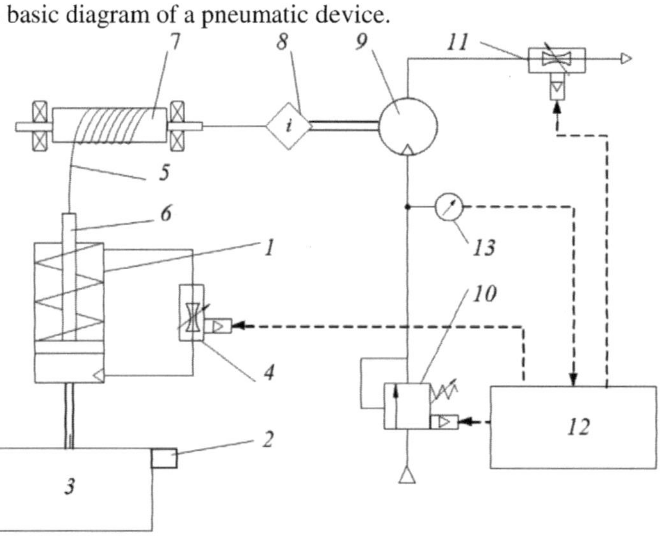

Figure 2. Structure diagram of the preload device and balancing of the dynamic loads in the drive unit of the movable part of the machine-tool.

Pneumatic device for a movable unit of the machine-tool contains a damper *1*, build in a form of a single stock pneumatic cylinder. There is a return spring fitted in the hydro-cylinder end. Vibration sensor *2* is fixed on the piston end of the hydro-cylinder which is jointed to a movable unit of the machine-tool *3*, for example with a table.

Stock end and head end of the pneumatic cylinder are interconnected by tubes by means of controlled throttle *4*. One end of the wire *5* is fixed to the rod of the pneumatic cylinder *6*. The other end of the wire is coiled on the wheel *7*, and fixed thereon.

The wheel *7* is fixed to a stationary part of the machine-tool and through a planetary gear reducer *8* is connected to the drive shaft of the pneumatic motor *9* which is connected with a pressure source, such as a pump by the tube through a reducing valve *10*. There is a pneumatic accumulator (on the structural diagram are not shown) before reducing valve *10*. The regulated throttle *11* connecting to the pneumatic system is mounted on the outlet of pneumatic motor *9*. Controlled throttle *4* and the vibration sensor *2* are connected to the numeric control device *12*.

Gravity force G and the cutting force F_c are in charge when machining the details which carry out the movement of the movable machine units (screw gear, the gear of rack bar - tooth gear, etc.):

$$F_{c} = F_{const} \pm dF \approx F_{const} \pm \sum_{i=1}^{n} \left(dF_{i} \cdot \sin \omega_{i} t + \varphi_{i}\right),$$

where F_{const} – constant cutting force component;
$\pm dF$ – variable cutting force component;
dF_i, ω_i φ_i – amplitude, frequency, phase of a i harmonic.

The component force $\pm dF$ is caused by variable cross section of cut or intermittent nature of the cutting process (milling). And when contour machining on the numeric control device there is a significant dynamic force dF_{in} related to the presence of the inertial mass of the movable unit – actuator of machine tool:

$$dF_{in} = m_{a} \frac{d^{2}z}{dt^{2}},$$

where m_a – movable unit mass, H;
$\frac{d^2 z}{dt^2}$ –movable unit acceleration, m/s^2.

These variable components of the forces have a negative impact on the accuracy of machining.

3. Calculation method and operating principle of the device

Pneumatic device operates as follows. When operating, these forces act on the damper body *1*. The effect of the variable component of the forces $\pm dF$ gives rise to a pulsing air flow dQ between the stock end and head end of the damper *1* cavities, which passes through a controlled throttle *4* generates the damping force dF_d

$$dF_{d} = S \frac{dQ^{2}}{K_{th}^{2}} S_{th}^{2},$$

where S – stock area of damper, m^2;
dQ – pulsating gas flow rate, m^3/s;
K_{th} – throttle ratio (determined by the design), m^3/N·s;
S_{th} – through area of controlled throttle, m^2.

It's necessary to do the ratio for optimal damping:

$$\frac{C_{d}}{K_{d}} = f_{nf} = \sqrt{\frac{C_{a}}{m_{a}}}, \tag{1}$$

where f_{nf} – the natural frequency oscillations of the movable unit of the machine, Hz;
C_a – stiffness of the movable unit of the machine, N/m;
m_a – mass of the movable unit of the machine, kg.

C_d – damper stiffness, N/m:

$$C_d = \frac{E_{rm} \cdot S_{pr}^2}{W_C} + C_s,$$

where E_{rm} – reduced modulus of the bulk elasticity for damper cavities, N/m^2;
W_C – the total volume of air under pressure, m^3;
C_s – the stiffness of the return spring damper, Nm;
K_d – damping coefficient, Ns/m:

$$K_d = \frac{dF_d}{V_{pr}},$$

V_{pr} – the rate of movement of the damper piston relatively to its body, m/s.

The equation below is based on the assumption that non-extensible rope (or other rod) or it's reduced value of deformation in comparison with the deformation and backlashes of the movable machine parts.

$$V_{pr}=V_a \text{ and } a_{pr}=a_a,$$

where V_a – conveying speed;
a_a, a_{pr} – accelerations of the movable operating element and the damper piston.

When the condition (1) is met the pneumatic device will perform as a high-frequencies filter, which leads to a significant reduction of the dynamic and impact forces on the movable machine unit by damping. Rotational moment on the wheel *7* is generated by pneumatic motor *9*. The required force F_r on the wire *5*, is:

$$F_r = \frac{q}{2\pi}(p_0 - p_{th}),$$

where q – the work volume of pneumatic motors *9*, m^3/v
p_0 – the pressure on pneumatic accumulator, H/m^2;
p_{th} – pressure before the adjustable throttle *11*, H/m^2

The generated force is greater than the weight of the movable unit of the machine G

$$F_r > G = m_a \cdot g, \quad (2)$$

where g –gravitational acceleration m/s^2

When the condition (2) is met the force F_d on the wire *5* is directed coincidently the same way as the cutting force F_c. This will create a permanent tension Δ of the joints of movable unit of the machine-tool as well as compression on the magnitude

$$\Delta = \frac{F_c + F_r}{C_a},$$

thus, neutralizing the backlashes magnitude in mechanical transmission which perform the movable units of machine-tool.

A certain pressure in the stock end of the hydraulic cylinder of damper *1* and the compression of the return spring is attained. This provides the normal operation of the damper *1* at a large stroke magnitude and increases the stiffness of elements of the movable unit of the machine-tool. The control valve *10* and throttle *11* are used to calibrate the required torque and rotational speed of the pneumatic motor shaft *9*.

The digital controller *12* controls the value f_{th} of the controlled throttle *4* area, depending on the movement and vibrations rates level occurring in the mobile unit of machine-tool and measured by the vibration sensor *2*. The digital controller passes control signals to the controlled throttle *4* of the damper *1*.

The differential equation that describes the operation of the pneumatic damper, is the follows:

$$m_a a_a + \alpha V_a + C_d x_a = F, \tag{3}$$

Where the variation low F takes values from:

$$F = \begin{cases} F_c, \text{ in the zone of low stiffness} \\ F_t + F_c, \text{ in the zone of higt stiffness} \end{cases}.$$

The mathematical model (simulation) made it possible to conduct theoretical study on the pneumatic device operation.

Structural diagram of the pneumatic device by which calculations were carried out in the Lab View National instruments software product, is shown below (Figure 3).

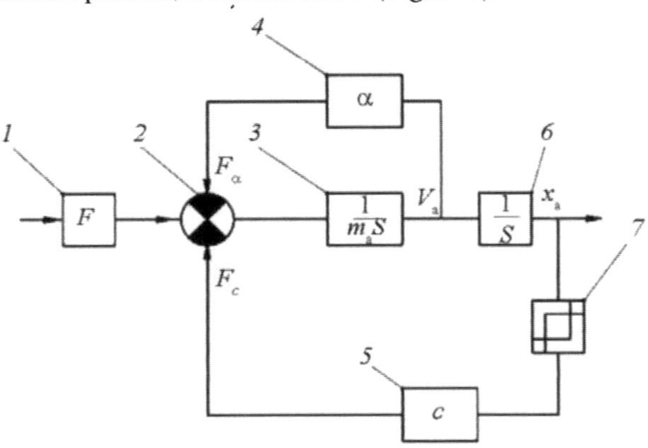

Figure 3. Structural diagram.
1 – tension; *2* –summation unit; *3* – 1/*S* Laplace operator and mass of the technological system unit; *4* – friction ratio; *5* – stiffness of the technological system unit; *6* – Laplace operator; *7* – element, describing the backlash of the mechanical unit.

The obtained results of the research are presented in the form of spectrograms (Figure 4).

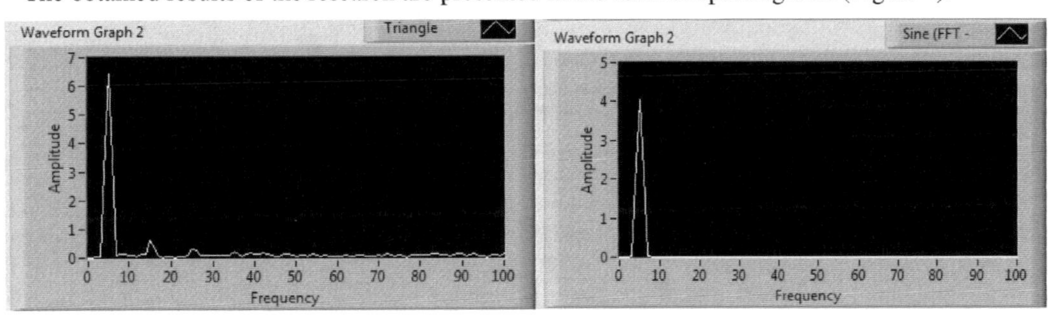

a *b*

Figure 4. Spectrograms of the process obtained by simulating the operation of the damper
a – at a «low» stiffness of the system and backlash presence;
b – at a «high» stiffness of the system and backlash absence.

The first diagram (Figure 4*a*) shows the presence of background noise, which can be explained by the presence of backlash in the joint.

The second diagram (Figure 4*b*) built in the presence of a damping device (at backlash absence) and which shows the absence of noise.

4. Conclusion

The results of the simulation conducted in the Lab View National instruments, confirmed the adequacy of the structural diagram (Figure 3) and the promising use of the damping device.

Basing on further comparison of spectrograms we made a conclusion that the presence of backlash between the operating body and the machine-tool leads to an increased oscillations amplitude that negatively affected the processing quality of the parts [14-16].

As a result of applying preload, the reduction in vibration and noise in the technological system is attained, the quality of treatment increases, tool life increases by reducing the intensity of thermal processes in metal cutting. It is confirmed by experimental data.

References

[1]. Gavrilin A N, Moyzes B B 2013 *Kontrol'. Diagnostika (Testing. Diagnostics)* **9** 81-84
[2]. Gavrilin A N Moyzes B B Cherkasov A I 2015 *Applied Mechanics and Materials* **756** 35-41
[3]. Gavrilin A N 2013 *Kontrol'. Diagnostika (Testing. Diagnostics)* **11** 23-26
[4]. Gavrilin A, Moyzes B, Zharkevich O 2015 *Journal of Vibroengineering* **17 (7)** 3495-3504
[5]. Korotaev A D et al. 2015 *Russian Physics Journal* **57** (10) 1301
[6]. Korablev P A 1962 *Machining accuracy in the instrument engineering* (Moscow, Mashgis)
[7]. Avrutin S V 1962 *Rational work of a Mille* (Moscow, Mashgis)
[8]. Surzhikov A P, Frangulyan T S, Ghyngazov S A, Lisenko E N, Galtseva O V 2006 *Russian Physics Journal* **49** (5) 506-10
[9]. Plotnikova I V *et al* 2016 *MATEC Web Conf.* **79** 01019 DOI: http://dx.doi.org/10.1051/matecconf/20167901019
[10]. Bespal'ko A A, Yavorovich L V, Fedotov P, PomishinE, Vedyashkin M V 2016 *MATEC Web Conf.* **79** 01039 DOI: http://dx.doi.org/10.1051/matecconf/20167901039
[11]. Galtseva O V and Bordunov S V 2015 *IOP IOP Conf. Ser.: Mater. Sci. Eng.* **81** 012062
[12]. Vasendina E A *et al* 2016 *IOP Conf. Ser.: Mater. Sci. Eng.* **110** 012070
[13]. Mustafina R M *et al* 2016 *Journal of Physics: Conference Series* **671** (1) 012046
[14]. Boldin L A Machine Tools (Operational Issues). Moscow, Mashgiz, 1957 260
[15]. Nizhegorodov A I, Gavrilin A N, Moyzes B B 2016 *Key Engineering Materials* **685** 320-324 DOI: 10.4028/www.scientific.net/KEM.685.320
[16]. Nizhegorodov A I, Gavrilin A N, Moyzes B B, Ditenberg I A, Zharkevich O M, Zhetessova G S, Muravyov O P, Bets M 2015 *Journal of Vibroengineering* **18** (6) 3734

Effect of powder compaction on radiation-thermal synthesis of lithium-titanium ferrites

A P Surzhikov[1], E N Lysenko[1], V A Vlasov[1], A V Malyshev[1], M V Korobeynikov[2] and M A Mikhailenko[3]
[1] National Research Tomsk Polytechnic University, Tomsk, Russia
[2] Budker Institute of Nuclear Physics SB RAS, Novosibirsk, Russia
[3] Institute of Solid State Chemistry and Mechanochemistry SB RAS, Novosibirsk, Russia

E-Mail: malyshev@tpu.ru

Abstract. Effect of powder compaction on the efficiency of thermal and radiation-thermal synthesis of lithium-substituted ferrites was investigated by X-Ray diffraction and specific magnetization analysis. It was shown that the radiation-thermal heating of compacted powder reagents mixture leads to an increase in efficiency of lithium-titanium ferrites synthesis.

1. Introduction

It is known that the titanium doped lithium ferrites are widely used in microwave engineering [1-5]. The ions substitutions significantly improve the properties of ferrites, since the Ti^{4+} introduction in the lithium spinel reduces the probability of Fe^{2+} formation increasing the electrical resistivity of ferrite [6]. These ferrites are characterized by low values of dielectric losses and high temperature stability.

Radiation-thermal (RT) synthesis is one method that allows to synthesize the materials and modify their properties effectively. In [7-12] RT method was successfully used for synthesis of lithium-substituted ferrites under conditions of high-energy electron beams heating. It was found that RT heating of the initial reagents mixture significantly increases the reactivity of the solid-phase system that leads to a decrease in a synthesis temperature and increase in homogeneity of the end product.

Among the most important factors in ferrites preparation, determined the efficiency of solid-phase synthesis, is an operation of compacting the reaction mixtures. This operation forms the initial conditions for the diffusion mass transfer and therefore the study of the compaction effect is necessary in the development of RT methods for ferrites production.

Previously, it was found that the compaction of reaction mixture increases the efficiency of the thermal synthesis of $Li_{0.5}Fe_{2.5}O_4$ lithium ferrites [13]. In the present work we studied the effect of compaction of reagents powder mixture in lithium-substituted ferrites synthesis by thermal heating and RT heating via pulsed electron beam.

2. Experimental techniques

Lithium-titanium ferrite with the general formula of $Li_{0.5(1+x)}Fe_{2.5-1.5x}Ti_xO_4$ (x = 0.2) was synthesized from Li_2CO_3-TiO_2-Fe_2O_3 mixture. Reagents mixture was obtained by weighing the required amounts of pre-dried components and then dry mixed in an agate mortar with 10 fold mixing through a mesh with a cell of 80 μm. Two types of samples were investigated: powder with a density of 0.95 g/cm^3 (powder samples) and powder compacted by the single-action compacting under a pressure of 200 MPa into tablets with diameter of 15 mm and thickness of 2 mm (compacted samples). Average density of compacted samples was 2.6 g/cm^3.

Thermal (T) synthesis of samples was carried out in a laboratory resistance furnace. Radiation-thermal (RT) synthesis of the samples was performed on the pulsed electron accelerator ILU-6 at the Institute of Nuclear Physics SB RAS (Novosibirsk) [14]. The electron energy was 2.4 MeV, beam current in the pulse was 400 mA, pulse duration is 500 μs, and pulse frequency is 7-15 Hz. The dose within the range of one pulse was 800 kGy/s. Average radiation dose was ca. 5 kGy/s and in the mode of isothermal synthesis was ca. 3 kGy/s. Heating of the samples was carried out by electrons energy, without external sources of heat. The desired temperature was kept using certain values of pulse frequency. The duration of heating and cooling stages does not exceed 3 min. The both T and RT synthesis was carried out in air. We used two temperatures (600°C and 750 °C) and times (10 min and 120 min) for each type of heating.

Analysis of the phase composition for investigated samples was conducted by X-ray diffractometer ARL X'TRA with semiconductor Si (Li) Peltier detector with Cuk$_\alpha$ radiation. A diffractogram was measured in the angle range 2θ = (20–70)° at a scanning speed of 0.02 °/s. Phase identification was performed using PDF-4+ powder database of International Centre for Diffraction Data (ICDD). X-ray diffraction patterns were processed by the method of full-profile Rietveld analysis using complex software Powder Cell 2.4.

In addition, the control of synthesis degree by measurements of specific magnetization in synthesized samples was carried out using the automated complex for magnetic properties study in pulsed magnetic fields [15].

3. Result and discussion

Figure 1 presents X-ray diffraction patterns for powder samples (figure 1a, c) and compacted samples (figure 1b, d) after T (figure 1a, b) and RT (figure 1c, d) synthesis of Li_2CO_3-TiO_2-Fe_2O_3 mixture. These reflections are a superposition from the phases of the initial components and phases of $LiFeO_2$, $LiFe_5O_8$ and spinel phases of lithium-titanium ferrites of $Li_{0.5(1+x)}Fe_{2.5-1.5x}Ti_xO_4$ composition, including the $Li_{0.6}Fe_{2.2}Ti_{0.2}O_4$ final composition. The data of XRD analysis for registered spinel phases in synthesized samples are summarized in table 1.

Table 1. Content of spinel phases in $Li_{0.5(1+x)}Fe_{2.5-1.5x}Ti_xO_4$ (weight %).

Synthesis mode	Type of synthesis	Powder samples	Compacted samples
600 °C –10 min	T	0	0
	RT	9.3	29.6
600 °C – 120 min	T	11.5	2.6
	RT	22.4	49.8
750 °C – 10 min	T	28.3	45.6
	RT	43.6	93.6
750 °C – 120 min	T	59	91.9
	RT	77	100

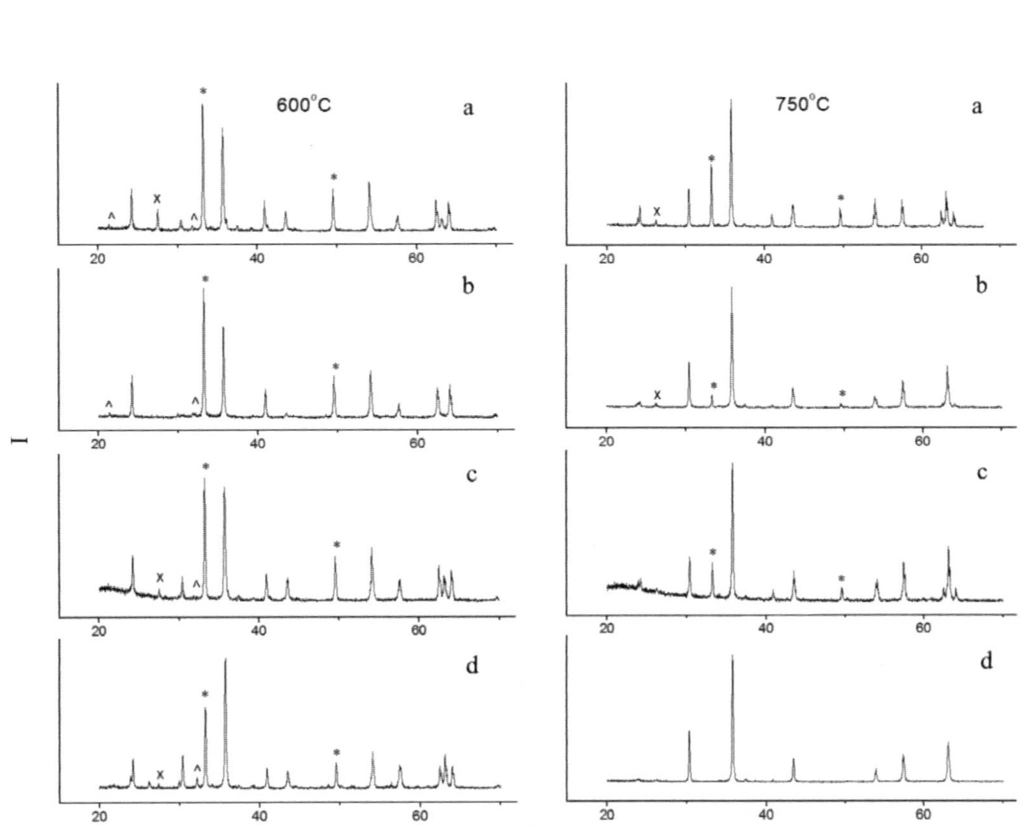

Figure 1. X-ray diffraction pattern of Li_2CO_3-TiO_2-Fe_2O_3 powder mixture (a, d) and compacted mixture (b, d) after T (a, b) и PT (c, d) synthesis at temperatures of 600 °C and 750 °C during 120 min. Reflections of initial components: (*) - Fe_2O_3; (×) – TiO_2; (^) - Li_2CO_3.

For thermal synthesized samples, $Li_{0.6}Fe_{2.2}Ti_{0.2}O_4$ lithium-titanium ferrite is formed by all modes except heating at 600 ° C for 10 min, in which only the initial components phases are observed. The effect of the electron beam heating even more accelerates this process. For RT synthesized samples, there is a formation of the spinel phases under all conditions of synthesis. It should be noted that 51.8 wt. % of $Li_{0.6}Fe_{2.2}Ti_{0.2}O_4$ out of 77 wt. % spinel phases is observed for powder samples in condition of RT synthesis at 750 °C for 120 min, while 100 wt. % content of spinel phase is formed in compacted samples at the same condition (table 1). The results showed that the synthesis reaction rate of lithium-titanium ferrite is higher in the compacted samples. This is expressed in faster transformation of the initial reagents in the spinel phases.

The data of specific magnetization of Li_2CO_3-TiO_2-Fe_2O_3 mixture are shown in table 2. The results showed an increase in the specific magnetization in compacted samples compared to powder samples. It also indicates a higher content of magnetic spinel phases in the synthesized compacts.

4. Conclusion

Use of non-compacted powder mixture of the initial reagents decreases the efficiency of solid-phase synthesis of $Li_{0.5(1+x)}Fe_{2.5-1.5x}Ti_xO_4$ ferrites. The including a compaction operation of powder reagents mixture in ferrites preparation leads to an increase in concentration of the spinel phases and decrease in content of initial components in a lithium-titanium ferrites, synthesized by thermal and radiation-thermal heating. In the latter case, a radiation-thermal heating is the most effective method to synthesize lithium-titanium ferrites.

Table 2. Specific magnetization of Li_2CO_3-TiO_2-Fe_2O_3 mixture (emu/g).

Synthesis mode	Type of synthesis	Powder samples	Compacted samples
600 °C –10 min	T	1.1	0.3
	RT	5.9	10
600 °C – 120 min	T	1.3	6.6
	RT	14.9	34
750 °C – 10 min	T	11.9	29.4
	RT	22.9	47.8
750 °C – 120 min	T	27.9	53.1
	RT	37.6	53.5

5. Acknowledgements
The research was supported by the Ministry of Education and Science of the Russian Federation in part of the "Science" program.

References
[1] Smit J and Wijn H P J 1959 *Ferrites Physical Properties of Ferrimagnetic Oxides in Relation to their Technical Applications* (Eindhoven: Philips Technical Library) p 299
[2] Baba P D, Argentina G M, Courtney W E, Dionne G F and Temme D H 1972 Fabrication and properties of microwave lithium ferrites *IEEE Trans. Magn.* **8** 83–94
[3] Jiang X N, Lan Z W, Yu Z, Liu P Y, Chen D Z and Liu C Y 2009 *J. Magn. Mater.* **321** 52–55
[4] Malyshev A V and Peshev V V 2007 *Russ. Phys. J.* **50** 2 161–164
[5] White G O and Patton C E 1978 Magnetic properties of lithium ferrite microwave materials *J. Magn. Magn. Mater.* **9** 4 299–317
[6] Manjula R, Murthy V R K and Sobhanadri J 1986 Electrical conductivity and thermoelectric power measurements of some lithium-titanium ferrites *J. Appl. Phys.* **59** 2929–32
[7] Surzhikov A P, Lysenko, Malyshev A V, Vlasov V A, Suslyaev V I, Zhuravlev V A, Korovin E Y and Dotsenko O A 2014 Study of the radio-wave absorbing properties of a lithium-zinc ferrite based composite *Russ. Phys. J.* **57** 5 621–626
[8] Surzhikov A P, Pritulov A M, Lysenko E N, Sokolovskii A N, Vlasov V A and Vasendina E A 2012 Influence of solid-phase ferritization method on phase composition of lithium-zinc ferrites with various concentration of zinc *J. Therm. Anal. Calorim.* **109** 63–67
[9] Surzhikov A P, Lysenko E N, Vlasov V A, Malyshev A V and Nikolaev E V 2013 Investigation of the process of ferrite formation in the Li_2CO_3-ZnO-Fe_2O_3 system under high-energy actions *Russ. Phys. J.* **56** 6 681–685
[10] Malyshev A V, Lysenko E N and Vlasov V A 2015 *Ceramics International* **41** 10 13671-13675
[11] Surzhikov A P, Vasendina E A, Lysenko E N and Nikolaev E V 2014 *Inorganic Materials: Applied Research* **5** 2 102–106
[12] Aleutdinov A D, Ghygazov S A, Mylnikova T S, Luchnikov P A 2015 *IOP Conf. Series: Materials Science and Engineering* **81** 012069
[13] Pritulov A M, Usmanov R U, Galtseva O V, Kondratyuk A A, Bezuglov V V and Serbin V I 2007 *Russ. Phys. J.* **50** 2 187–192
[14] Auslender V L 1994 ILU-type electron accelerator for industrial technologies *Nuclear Instruments and Methods in Physical research* B 89 46–48
[15] Kreslin V Yu and Naiden E P 2002 *Instruments and Experimental Techniques* **45** 1 55–57

IOP Conf. Series: Materials Science and Engineering **168** (2017) 012091 doi:10.1088/1757-899X/168/1/012091

Conformer of the peroxynitrite ion formed under photolysis of crystalline alkali nitrates – cis or trans?

V Kh Pak[1], V A Anan'ev[1], E P Dyagileva[2], S Yu Lyrshchikov[3], M B Miklin[1] and M A Rezvova[1]

[1]Kemerovo State University, Krasnaya str., 6, Kemerovo, 650043, Russia. E-mail: eprlab@kemsu.ru

[2]Kemerovo State Medical Academy, Voroshilova str., 22[a], Kemerovo, 650029, Russia. E-mail: dyagilevaep@mail.ru

[3]National Research Tomsk Polytechnic University. Energetic Institute. Lenina avenue, 30. Tomsk. 634050. Russia. E-mail: serstud@mail.ru

e-mail:eprlab@kemsu.ru Phone:+73842588286

Abstract. The optical and infrared reflectance spectra of the crystalline powders prepared by co-crystallization of caesium nitrate, nitrite, and peroxynitrite from alkali solution have been studied. We find that the trans conformer forms under photolysis of crystalline pure caesium nitrate. Under its dissolution the trans conformer transforms to the cis conformer.

1. Introduction

The peroxynitrite ion ($ONOO^-$) plays an important role in a variety of physiological processes [1-4] and in photochemistry of the nitrate ion [5-7]. It can exist in two geometric forms, as cis and trans conformer, due to the partial double bond between the nitrogen atom and the first peroxide oxygen atom [1]. On the basis of quantum chemical calculations it was concluded that in gas phase cis conformer is more stable than trans conformer [1,8]. In tetramethylammonium peroxynitrite crystal the peroxynitrite ion was observed in cis form [9]. Pfeiffer and Mayer suggested that in vivo peroxynitrite is formed in trans form [10]. Both forms were produced under UV irradiation (193 nm) of potassium nitrate in an argon matrix [11]. In some cases the geometric form of the peroxynitrite ion is not established [6,7]. So, the geometric form of the peroxynitrite ion in condense phase depends on the synthesis method. To describe the reactions involving the peroxynitrite ion need to know its geometric structure because it may play a major role in peroxynitrite reactivity [12].

The maximum position of the optical absorption band due to cis conformer of the peroxynitrite ion in aqueous solution should be observed at 310.4 nm while for the trans conformer at 338.8 nm based on Krauss' calculations [13]. Other calculations indicate that the maximum position should be observed at 303 and 374 nm for the cis and the trans conformer, respectively [14]. The absorption band due to peroxynitrite, synthesized from nitrite and hydrogen peroxide, observed in alkali solution has maximum position at ~302 nm [15]. The absorption maxima for KOONO, formed by photolysis of KNO_3 in solid argon matrix, were observed at 325 and 375 nm. These bands were attributed to absorption of cis and trans conformer, respectively [11]. Photolysis of crystalline potassium and caesium nitrates results in optical band at 350 and 355 nm, respectively, due to peroxynitrite [6,16]. It has been suggested that this band corresponds to absorption of trans conformer [17].

For various quantum chemical methods calculated harmonic fundamentals for cis and for trans conformer of the peroxynitrite ion are dissimilar up to 250 cm^{-1} [18]. Moreover maxima of bands observed in IR spectra of the peroxynitrite ion in solids differ from the calculated data for cis and for trans conformer [11,18,19]. At the same time maxima of bands observed in laser Raman spectrum of the peroxynitrite ion synthesized from nitrite and hydrogen peroxide in alkali solution are close to the calculated data for cis conformer [1].

The goal of present work is to study optical and IR reflectance spectra of peroxynitrite, synthesized by different methods in alkali solution, introduced into caesium nitrate crystals by co-crystallization. Comparison of these spectra with corresponding spectra of peroxynitrite, obtained by photolysis of pure nitrate salt, will determine geometric form of it for the latter case.

2. Experimental

All chemicals used were analytical reagent grade or better. Saturated alkaline solutions of caesium nitrate (pH 13) were prepared by dissolution of appropriative amount of solid NaOH and solid

i) untreated salt;

ii) untreated salt and a small amount of caesium nitrite salt;

iii) photolyzed salt

(solutions 1-3, respectively). All the solutions were placed in Petri dishes. In addition, solution 4 was prepared by photolysis of Petri dish with solution 1. Caesium nitrate crystalline powders were grown by slow evaporation of the solutions during 24 h at room temperature. The powders before using were washed with saturated alkali solution of caesium nitrate and then they were stored at room temperature for 1 h.

Photolysis was carried out with a low-pressure mercury lamp with uviol glass at room temperature. Further in the text we denote used samples as follows:

- powder obtained from solution 1 is sample 1;
- powder obtained by photolysis of sample 1 during 1 h is sample 2;
- powders obtained from solutions 2-4 are samples 3-5, respectively.

The procedures for the chemical analysis of peroxynitrite and nitrite were described earlier [20]. Concentration of peroxynitrite in sample 2 and nitrite in sample 3 was $\sim 10^{-5}$ mol/g and $\sim 5\times 10^{-3}$ mol/g, respectively. Concentration of peroxynitrite in samples 4,5 was approximately 50 and 100 times smaller than in sample 2, respectively. Concentration of nitrite in samples 4, 5 was approximately a two hundred times smaller than in sample 3.

The optical and infrared reflectance spectra of the samples were recorded at room temperature by means of a "Shimadzu UV-2450" and Tenzor 27 spectrophotometers, respectively. To achieve good signal-to-noise ratio for IR reflectance spectra the results of 250 scans were accumulated and averaged.

3. Experimental results

Fig. 1 displays optical reflectance spectra of samples 1-5. The single band due to nitrate with maximum position at 306 nm for sample 1 was observed. As seen, spectra 2-5 differ from spectrum 1. Fig. 2 displays difference spectra calculated as the difference between spectra 2-5 and spectrum 1 (hereinafter, spectra 2'-5').

The broad band with maximum position at 359 nm and the narrow band with maximum position at 370 nm are observed in the spectra 2' and 3', respectively. The parameters of the former band are similar to the one for optical absorption band due to peroxynitrite in photolyzed caesium nitrate crystal [16], while the parameters of the latter band and its vibrational structure (indicated by arrows) are similar to the one of the optical absorption band due to nitrite introduce into caesium nitrate crystal by co-crystallization [21]. Based on these facts and the composition of samples 2 and 3 we can conclude that the band at 359 nm is due to peroxynitrite, and the band at 370 nm is due to nitrite.

The single weakly band with maximum position at ~310 nm is observed in spectrum 4'. Because sample 4 contains both nitrite and peroxynitrite, we can conclude that this band is due to the latter.

The spectra of samples 1-3 are identical. So, both nitrite, introduced into the powder by co-crystallization, and peroxynitrite, introduced into the powders by UV irradiation, has not band in this region of the infrared reflectance spectra. As seen, in the infrared reflectance spectra of samples 4 and 5 two bands at 630 and 640 cm^{-1} were observed. These bands unstable and they are not registered in the infrared reflectance spectra after storage for 24 hours at room temperature. Simultaneously peroxynitrite completely decay, based on the data of chemical analysis. So, it can be concluding that bands at 630 and 640 cm^{-1} are due to peroxynitrite.

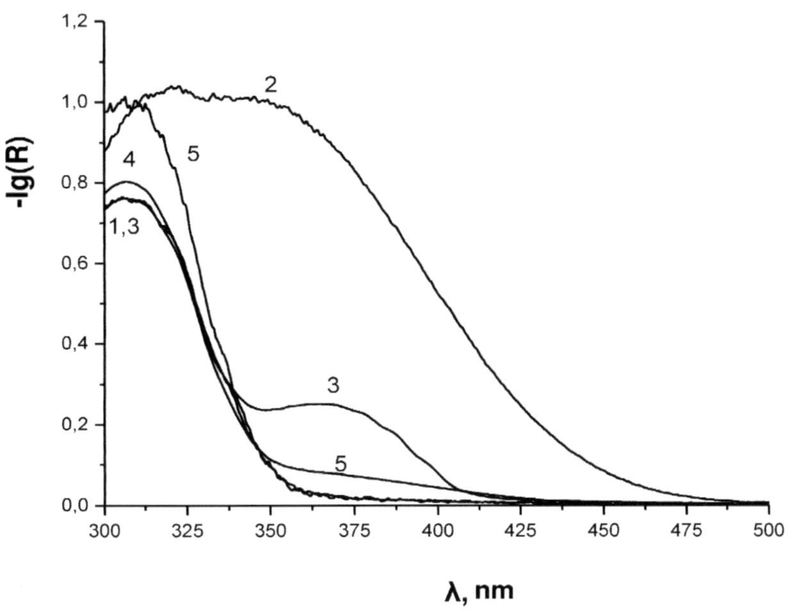

Figure 1. Optical reflectance spectra of the samples 1-5 (1-5, respectively).

The spectrum 5' consists of two bands. The band in the long-wavelength region with maximum position at ~370 nm can be represented as a superposition of the bands in the spectra 2' and 3'. The shape of the band in the short-wavelength region with maximum position at ~310 nm is similar to the shape of the band in the spectrum 4'. So, spectrum 5' can be described as superposition of spectra 2'-4', and it is due to nitrite and peroxynitrite.

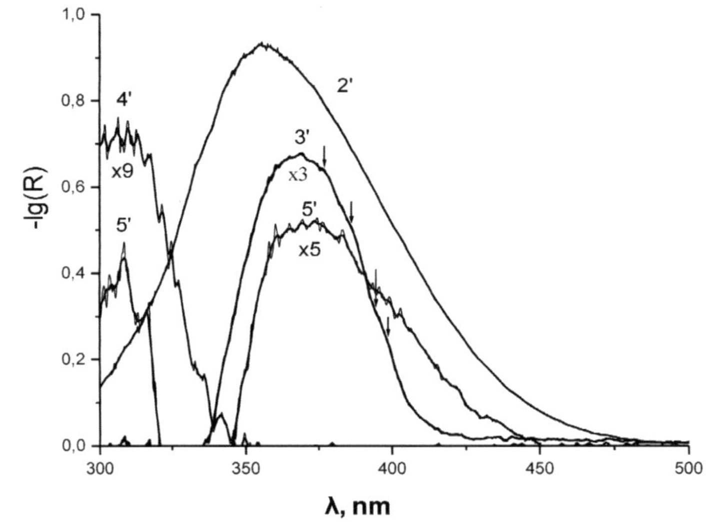

Figure 2. Difference spectra of the samples 2-5 (2'-5', respectively).

Fig. 3 displays the infrared reflectance spectra of samples 1-5.

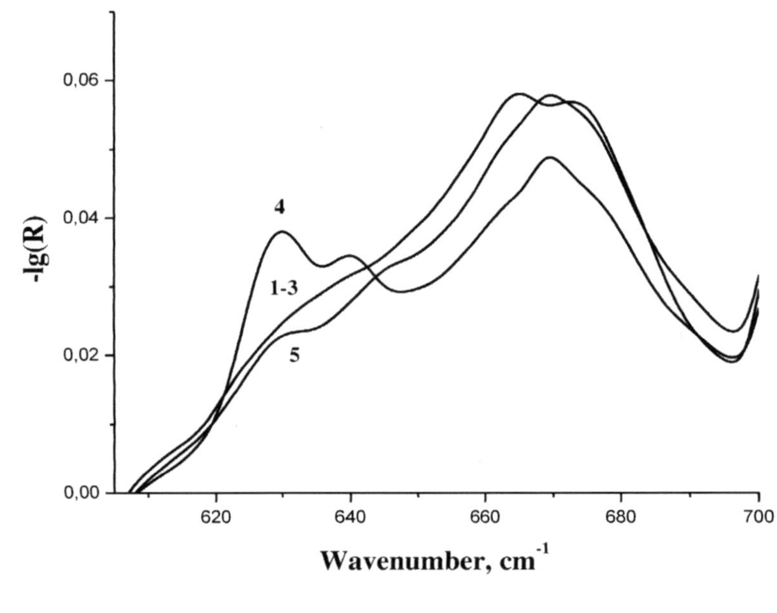

Figure 3. IR spectra of the samples 1-5 (1-5, respectively).

4. Discussion

Comparison of the optical and infrared reflectance spectra of UV irradiated samples 2 with the spectra of the sample 4, contained peroxynitrte introduced by co-crystallization, data of chemical analysis on peroxynitrite, and stability of it in the corresponding samples allow us to suggest that spectroscopic parameters of peroxynitrite in sample 2 differ from the one in the sample 4. Since, peroxynitrite in aqueous alkali solution is in cis form [14,15] then in sample 4 it should be in this form. So, peroxynitrite in sample 2 must be in trans form and its formation under photolysis can be represented as follows

$$NO_3^-(solid) \xrightarrow{h\nu(253.7nm)} trans-ONOO^-(solid).$$

Trans conformer transform to more stable in aqueous alkali solution cis conformer under dissolution

$$trans-ONOO^-(solid) \rightarrow cis-ONOO^-(aq)$$

and this conformer introduce to crystal lattice under co-crystallization

$$cis-ONOO^-(aq) \rightarrow cis-ONOO^-(solid).$$

From our experimental data it follows that peroxynitrite in sample 5 is in cis and trans forms. Because peroxynitrite in aqueous alkali solution is stable as cis conformer then trans conformer in sample 5 can be form under photolysis of powder which lies at the bottom of a Petri dish. It could be argued that in crystalline caesium nitrate cis conformer in contrast with trans conformer is unstable and it isomerizes into the nitrate ion during storage.

5. Conclusion

Dissolving of UV irradiated samples generate in a solution trans conformer of the peroxynitrite ion. So, it allows us to study its chemical reactions and it would be useful to explain a variety of physiological processes involving the trans conformer of the peroxynitrite ion.

Acknowledgements

This works was supported state assignment of Ministry of Education and Science of Russian Federation.

References

[1] Tsai H H, Hamilton T P, Tsai J-H M, Woerd M van der, Harrison J G, Jablonsky M J, Beckman J S, Koppenol W H 1996 *J. Phys. Chem.* **100** 15087
[2] Ferrer-Sueta G, Radi R 2009 *ACS Chem. Biol.* **4** 161
[3] Trujillo M, Naviliat M, Alvarez M N, Peluffo G, Radi R 2000 *Analusis* **28** 518
[4] Koppenol W H 1998 *Química nova* **21** 326
[5] Mack J, Bolton J R 1999 *J. Photochem. Photobiol.* A **128** 1
[6] Plumb R C, Edwards J O 1992 *J. Phys. Chem.* **96** 3245
[7] Anan'ev V, Miklin M J. 2005 *Photochem. Photobiol.* A **172** 289
[8] Olson L P, Bartberger M D, Houk K N 2003 *J. Am. Chem. Soc.* **125** 3999
[9] Wörle M, Latal P, Kissner R, Nesper R, Koppenol W H 1999 *Chem. Res. Tox.* **12** 305
[10] Pfeiffer S, Mayer B 1998 *J. Biol. Chem.* **273** 27280
[11] Lo W-J, Lee Y-P, Tsai J-H M, Beckman J S 1995 *Chem. Phys. Lett.* **242** 147
[12] Beckman J S, Carson M, Smith C D, Koppenol W H 1993 *Nature* **364** 584
[13] Krauss M *Chem. Phys. Lett.* 1994 **222** 513
[14] Kirsch M, Groot H de 2002 *J. Biol. Chem.* **277** 13379
[15] Drexler C, Elias H, Fecher B, Wannowius K J, Fresenius 1991 *J. Anal. Chem.* **340** 605
[16] Anan'ev V, Miklin M, Nelyubina N, Poroshina M J. 2004 *Photochem. Photobiol.* A **162** 67
[17] Anan'ev V 2010 *IOP Conf. Ser.: Mater. Sci. Eng.* **15** 012078
[18] Relph R A, Bopp J C, Johnson M A, Viggiano A A 2008 *J. Chem. Phys.* **129** 064305
[19] Liang B Y, Andrews J 2001 *J. Am. Chem. Soc.* **123** 9848
[20] Kriger L D, Miklin M B, Dyagileva E P, Anan'ev V A 2013 *Rus. J. Phys. Chem.* A **87** 319
[21] Anan'ev V, Seliverstov M 2005 *Ceramics – Silikáty* **49** 68

Mechanisms of change of superficial properties of polymeric materials in discharge plasma

P A Luchnikov[1], O A Sarkisov[2], A A Rogachev[2,3], E Ph Pevtsov[1] and T A Demenkova[1]

[1] Moscow Technological University (MIREA), 78, Vernadsky av., Moscow, 119454, Russia
[2] F. Skoryna Gomel State University, 104, Sovetskaya Street, Gomel 246019, Belarus
[3] Belarusian State University of Transport, 34, Kirov St., Gomel, 246653, Belarus

E-mail: xamdex@gmail.com

Abstract. In work analytical expressions for assessment of the tribochemical changes and definition of the most effective direction of regulation of contact interactions are received. These results can be used for the analysis of experimental data at a research of mechanisms of change of polymeric materials superficial properties.

1. Introduction

In practice for improvement of physical and chemical properties of materials of the electronic equipment, the creation of the set superficial nanostructures by etching and thermal annealing, the ion-beam and thermal methods of activation and modification of materials are widely used [1–6]. At the same time the offered recommendations, technical solutions in this area have empirical character as systematic researches of the nature of the proceeding processes, their influences, for example, on morphology, the adsorptive, electrophysical properties of materials are practically absent. Under certain conditions of such activation the manifestation of specific radiation and stimulated physical and chemical processes which are capable essential to change structure of polymeric materials and their operational properties [4, 7-8].

2. Technique of calculations

Within the relaxation and diffusive theory of interphase processes [9] the opportunity of the analytical description of the processes proceeding in the radiation modified polymeric materials when using various schemes of their processing is given. At the same time the main attention is paid to consideration of influence of a dose of radiation and processing time on kinetics of adhesive interaction, value of friction force, wear intensity of materials at operation in friction couple.

At activation (electromagnetic or corpuscular) impact on materials [3, 7] in superficial layers of polymers the following main processes proceed: – heating, activation of diffusive processes; – excitement and ionization of macromolecules; – a rupture of macromolecules, formation of low-molecular fragments, free radicals and ions, the subsequent processes of interaction with their participation; – the subsequent chemical transformations of active fragments, ions and radicals, their interaction among themselves and with atoms of the contacting environment (stitching, oxidation, etc. processes); – dispersion (etching) of the modified superficial layer under the influence of ions, high-vigorous particles, activation influence; – structural and morphological changes as a result of decrease of molecular weight, increase of segmental mobility, polarity of macromolecules etc.

As a result after processing depending on the nature of polymer, a dose of radiation and temperature formation of polar groups, for example, =OH, –O–C–, =N–O–, is possible, low-molecular products of thermal degradation, radiolysis, which are localized on the surface of material and interact with molecular fragments, exerting corresponding impact on their superficial properties [1].

In the relaxation-diffusive theory superficial properties of material are determined by value of density of the active centers (AC), their distribution by layer thickness. Generation of the active centers, the nature of their influence on processes of adhesive interaction depend on conditions of carrying out activation processing. At the same time realization of two main characteristic of technological receptions is possible: carrying out preliminary activation processing of material; a combination of ionic activation influence with other processes, for example, sedimentation of covering, at adhesive interaction, friction, etc.

At assessment of efficiency of preliminary modifying of polymeric material it is necessary to consider the nature of the proceeding post-radiation processes which kinetics is defined by concentration and stability of the formed active particles (radicals, ions, etc.), temperature and the oxidizing environment.

When carrying out activation processing of a surface the external environment can play a role of chemically active element of system and in it as a result of influence factors (ions, electrons, electromagnetic radiation) the corresponding processes influencing their properties also proceed.

In [10, 11] it is shown that superficial properties of polymeric materials enough with difficulty depend from nature substances, intensity of influence, and also temperature. Increase of the adsorptive activity of a surface at rather low doses of activation processing is connected with emergence in superficial layers of the acyl and peroxide radicals and their subsequent oxidation. The polar groups which are on a surface carry out a role of the active centers and increase adhesion interaction. In some cases change of properties in the course of processing is explained also with generation of the low-molecular products which are formed in polymer at influence of radiation, ions, their subsequent ion- or thermostimulated desorption.

Taking into account the representations stated above when determining energy of interphases interaction E at adhesive contact with other surface corresponds to expression [9]:

$$E = N_a^0 \varphi(t)\bar{u}(t) + \int_0^t v_a(\tau)\varphi(t-\tau)\bar{u}(t-\tau)d\tau \quad , \tag{1}$$

where N_a^0 – density of the communications which are formed on a surface in a contact zone in initial moment; $v_a(\tau)$ - speed of generation of adhesive communications in the course of contact; $\varphi(t-\tau)$ и $\bar{u}(t-\tau)$ – the functions describing change of density of adhesive communications and their average energy in the course of interaction respectively.

It is necessary to consider a number of the physical and chemical processes proceeding in superficial layers of the radiation-modified materials:

– radiation and the ion-stimulating destruction of macromolecules. In this case changes of AC can be defined by a ratio:

$$dN_o = n_0(\alpha_1 q_1 + \alpha_2 q_2)dt\,, \tag{2}$$

where: n_0 – density of macromolecules in superficial layer; q_1, q_2 – power of the radiation and ionic components of the activating stream; α_1, α_2 – the coefficients defining resistance of macromolecules to destruction at impact of radiation and a stream of ions respectively;

– change of the chemical composition of macromolecules as a result of interaction with active molecules of the environment as oxidation;

$$dN_x = k_1 C dt\,, \tag{3}$$

where: C– concentration of reactive atoms of the environment;

IOP Conf. Series: Materials Science and Engineering **168** (2017) 012092 doi:10.1088/1757-899X/168/1/012092

– etching of in superficial layer by high-vigorous corpuscular stream.

The kinetics of ionic dispersion will register:

$$dN_{mp} = -k_2 N q_2 dt, \tag{4}$$

where k_2– etching coefficient (ionic dispersion).

Density of the active centers of macromolecules N, its change in processing influences the size of density of adhesive communications N_a. In case of contact of polymer with the surface of other material having the sum of the areas of active microsites of S_a on the planimetric surface of contact by square S_0, according to [4]:

$$dN_a = aWdN., \tag{5}$$

The surfaces of substrate fixed on the active centers of macromolecules don't participate in further processes of generation of adhesive communications therefore change of number of the free active centers N of macromolecules on a surface taking into account ratios (2), (3) is described by the relaxation equation:

$$\frac{dN}{dt} = \frac{N_0 - N}{\tau_p} + n_0(\alpha_1 q_1 + \alpha_2 q_2) + k_1 C - k_2 N q_2, \tag{6}$$

where: N_0 – number of the active centers which are in superficial layer and capable to come on a surface; τ_p – characteristic time of diffusion relaxation of the active centers.

The solution of the equation (6) under an entry condition $N(t=0)=N_1$ is expression:

$$N = \left(N_1 - \frac{c}{b}\right)\exp(-bt) + \frac{c}{b}, \qquad 7)$$

where $c = \frac{N_0}{\tau_p} + n_0(\alpha_1 q_1 + \alpha_2 q_2) + k_1 C$; $b = \frac{1}{\tau_p} + k_2 q_2$.

On the basis of (5) taking into account (7) we will receive expression for density of adhesive communications N_a in the form:

$$N_a = \frac{S^0}{S_0}(C - N_0)\left[\left(N_1 - \frac{c}{b}\right)\cdot\frac{1}{(-2b)}(\exp(-2bt) - 1) - \frac{c}{b^2}\left(e^{-bt} - 1\right)\right], \tag{8}$$

At initial stage in processing $t << 1/b$ density of the active centers linearly increases under the law $N = N_1 + N_{st}(bt-1)$, where N_{st} – the stationary density of AC.

The analysis (7) shows that quasistationary value of density of AC is defined by the power of active processing and a ratio of its electromagnetic and ionic components.

$$N_{st} = \frac{N_0 + n_0(\alpha_1 q_1 + \alpha_2 q_2)\tau_p + \tau_p k_1 C}{1 + k_2 q_2 \tau_p}, \tag{9}$$

At ion-beam activation it is possible to accept that $q_1=0$ and if mobility of AC is very low ($1/\tau_p$ is close to 0) and there is no change of the chemical composition of a polymer superficial layer ($k_1 = 0$) stationary value of density of AC is equal:

$$N_{st,u} = n_0 \frac{\alpha_2}{k_2}, \tag{10}$$

At such processing the activation effect which is shown in change of number of AC doesn't depend on power and dose of ionic processing. Here change of parameters of adhesive interaction is reached only by etching of surface.

When processing in the smoldering category of plasma for which it is characteristic the combination of impact on the processed surface of the accelerated ions and UF-radiations, size of concentration $N_{st.tl}$ AC will be equal:

$$N_{st,tl} = \frac{n_0}{k_2}\left(\alpha_1 \frac{q_1}{q_2} + \alpha_2\right), \qquad (11)$$

At comparable values of parameters of ion-plasma processing of a surface of polymeric material change of concentration of AC is defined as:

$$\Delta N = N_{st,tl} - N_{st,u} = \frac{n_0}{k_2}\alpha_1 \frac{q_1}{q_2}, \qquad (12)$$

Thus, on the basis of (12) it is possible to conclude that processing in the smoldering category is more effective, and the maximum activation influence is reached in the presence in a stream, influencing a surface, of intensive electromagnetic radiation. Let's note that the parameters k_1, k_2, a_1, a_2 in the ratios given above depend from nature of polymeric material, temperatures and their values substantially define activation effect.

3. Conclusions

The received analytical expressions for special cases give the chance to estimate extent of influence on superficial processes of these or that radiation or the tribochemical changes, to define the most effective directions of regulation of contact interactions, and also to use them for the analysis of experimental data. So, the analysis of ratios (7), (9) – (11) allows to note features of kinetics of the adsorptive influence, the most effective methods of decrease of parameters of contact interaction, for example, as a result of decrease of segmental mobility of macromolecules, generation of low-molecular products.

Acknowledgments
This work was supported by the Ministry of Education and Science of Russian Federation.

References

[1] Kestelman V N Physical methods of modification of polymeric materials. M.: Chemistry 1980. 223 (in Russian)
[2] Ghyngazov S A, Vasil'ev I P, Surzhikov A P, Frangulyan T S, Chernyavskii A V 2015 *Technical Physics* **60** 1 128-32
[3] Ivanovsky G F, Petrov V I Ion-plasma processing of materials. M.: Radio and communication, 1986 232 (in Russian)
[4] Luchnikov P A, Rogachev A V, Yarmolenko M A, Rogachev A A, Bespal'ko A A 2016 *IOP Conf. Ser.: Mater. Sci. Eng.* **110** 012081
[5] Surzhikov A P, Frangylyan T S, Ghyngazov S A 2012 *Russian Physics Journal* **55** 4 345-52
[6] Luchnikov P A Structure 2015 *Russian Physics Journal* **57** 9 1239-45
[7] Aleutdinov A D, Ghygazov S A, Mylnikova T S, Luchnikov P A 2015 *IOP Conf. Ser.: Mater. Sci. Eng.* **81** 012069
[8] Rogachev A A, Yarmolenko M A, Xiaohong Jiang, Shen Ruiqi, Luchnikov P A, Rogachev A V 2015 *Applied Surface Science* **351** 811-818
[9] Bui M V, Rogachev A V 1997 Relaxation-diffusion theory of interfacial processes (Gomel: BelSUT)
[10] Berlin A A, Basin V B Bases of adhesion of polymers. M.: Chemistry, 1969 320 (in Russian)
[11] Puchkin Yu N, Baydarovtsev Yu P, Vasilets V N, Ponomarev A N 1983 *HVE* **17** 4 368-371 (in Russian)

IOP Conf. Series: Materials Science and Engineering **168** (2017) 012093 doi:10.1088/1757-899X/168/1/012093

Influence of reagents mixture density on the radiation-thermal synthesis of lithium-zinc ferrites

A P Surzhikov[1], E N Lysenko[1], V A Vlasov[1], A V Malyshev[1], M V Korobeynikov[2] and M A Mikhailenko[3]

[1]Tomsk Polytechnic University, Tomsk, Russia
[2]Budker Institute of Nuclear Physics SB RAS, Novosibirsk, Russia
[3]Institute of Solid State Chemistry and Mechanochemistry SB RAS, Novosibirsk, Russia

E-mail: vlvitan@tpu.ru

Abstract. Influence of Li_2CO_3-ZnO-Fe_2O_3 powder reagents mixture density on the synthesis efficiency of lithium-zinc ferrites in the conditions of thermal heating or pulsed electron beam heating was studied by X-Ray diffraction and magnetization analysis. The results showed that the including a compaction of powder reagents mixture in ferrite synthesis leads to an increase in concentration of the spinel phase and decrease in initial components content in lithium-substituted ferrites synthesized by thermal or radiation-thermal heating.

1. Introduction

Zinc doped lithium ferrospinels are the basic ferrite materials that widely used in microwave engineering [1-6]. The introduction of Zn^{2+} ions in $Li_{0.5}Fe_{2.5}O_4$ lithium spinel increases the saturation magnetization (M_s) of ferrites, however, a decrease in M_s is observed with increasing Zn^{2+} content at $x \geq 0.4$.

There are several methods that increase the efficiency of ferrite synthesis and change their properties. One such method, named radiation-thermal (RT), is becoming increasingly important and widespread in the present time. The RT method is successfully tested in the synthesis of some oxide systems, including the synthesis of lithium-titanium [7], and lithium-zinc [8-10] ferrites under high-energy electron heating.

Compaction of reacting mixture is among the most important operations determining the efficiency of RT synthesis, since a compaction degree affects the initial conditions for the diffusion mass transfer. In [11], it was found that an increase in degree of reaction mixture compaction improve the efficiency of $Li_{0.5}Fe_{2.5}O_4$ lithium ferrite thermal (T) synthesis. In the present work, the influence of powder mixture reagents density on an efficiency of lithium-zinc ferrite synthesis under T and RT heating was investigated by XRD and specific magnetization analysis.

2. Experimental techniques

The object of the study was $Li_{0.5(1-x)}Zn_xFe_{2.5-0.5x}O_4$ lithium-zinc ferrite (x=0.2) synthesized from Li_2CO_3-ZnO-Fe_2O_3 mixture. Reagents mixture was obtained by weighing the required amounts of pre-dried components and then dry mixed in agate mortar with 10 fold mixed through a mesh with a cell of 80 μm. We investigated two types of samples: not compacted powder with 0.95 g/cm^3 density (further, powder samples) and powder with 2.6 g/cm^3 density (further, compacted samples) that compacted by unilateral cold compaction under 200 MPa pressure into tablets with a 15 mm diameter and a 2 mm thickness.

Thermal heating of samples was carried out in a resistance furnace. Radiation-thermal heating of samples was performed on the pulse electron accelerator ILU-6 at Institute of Nuclear Physics SB RAS (Novosibirsk) [12]. The electron energy was 2.4 MeV, beam current pulse – 400 mA, pulse duration –500 µs, pulse repetition frequency – 7-15 Hz. Average radiation dose for isothermal heating mode was ca. 3 kGy/s. In limits of one pulse, a dose was 800 kGy/s. Samples heating and the preset temperature were carried out by decelerated electrons energy, without recourse to heat external sources. Duration of nonisothermal stages (heating and cooling) did not exceed 3 min. Both T and RT synthesis were realized in air. The modes with temperatures of 600 °C and 750 °C and times of 10 min and 120 min were used for both heating.

Analysis of sample phase composition was conducted by X-ray diffractometer ARL X'TRA with Peltier semiconductor Si (Li) detector with Cuk_α radiation. To analyse phase composition and materials structure, we used a symmetric Bragg diffraction geometry [13-14]. Measurements of diffraction patterns were spent in the range 2θ = (20-70)° at 0.02 °/s scanning speed. Phase identification was carried out using the PDF-4+ International Centre for Diffraction Data (ICDD) powder database. Received XRD patterns were processed by full-profile analysis method using Powder Cell 2.4 software.

Moreover, an additional control of synthesized samples was performed by saturation specific magnetization measurements using the automatic complex for study of magnetic properties in pulsed magnetic fields [15].

3. Result and discussion

Figure 1 shows X-ray diffraction patterns for Li_2CO_3-ZnO-Fe_2O_3 samples after T and RT synthesis. The marked reflections correspond to oxides or carbonate phases. Unmarked reflections belong to both α-Fe_2O_3 and spinel phases, from which the reflections merge into single-peak reflections. XRD analysis data, showing the content of $Li_{0.5(1-x)}Zn_xFe_{2.5-0.5x}O_4$ spinel phases in synthesized samples, are summarized in Table 1.

XRD patterns for thermal synthesized samples indicate the presence of initial components phases, lithium orthoferrite, lithium pentaferrite, as well as $Li_{0.5(1-x)}Zn_xFe_{2.5-0.5x}O_4$ lithium-zinc ferrites phases including the $Li_{0.4}Zn_{0.2}Fe_{2.4}O_4$ final composition. $Li_{0.4}Zn_{0.2}Fe_{2.4}O_4$ phase is formed in all modes except 600 °C – 10 min, in which there is only initial components phases and lithium orthoferrite.

According to the XRD data for RT synthesized samples, the reflections from the initial components phases and lithium ferrites spinel phases are observed. $Li_{0.4}Zn_{0.2}Fe_{2.4}O_4$ phase is formed in all modes including 600 °C – 10 min mode in which 9.7 wt.% this ferrite out of 13.7 wt.% spinel phases is observed for powder samples. Lithium-zinc ferrite formation is increased by increasing temperature and time of synthesis and reaches 88.9 wt.% out of 100 wt.% spinel phases in compacted samples at 750 °C – 120 min RT mode.

Overall, the results confirmed previous findings that the rate of ferrites formation is higher in conditions of heating reagents mixture by high-energy electron beam. In additional, the results show that the rate of lithium-zinc ferrite formation is higher in compacted samples, and so, the compaction of reacting mixture is a necessary process in ferrite production.

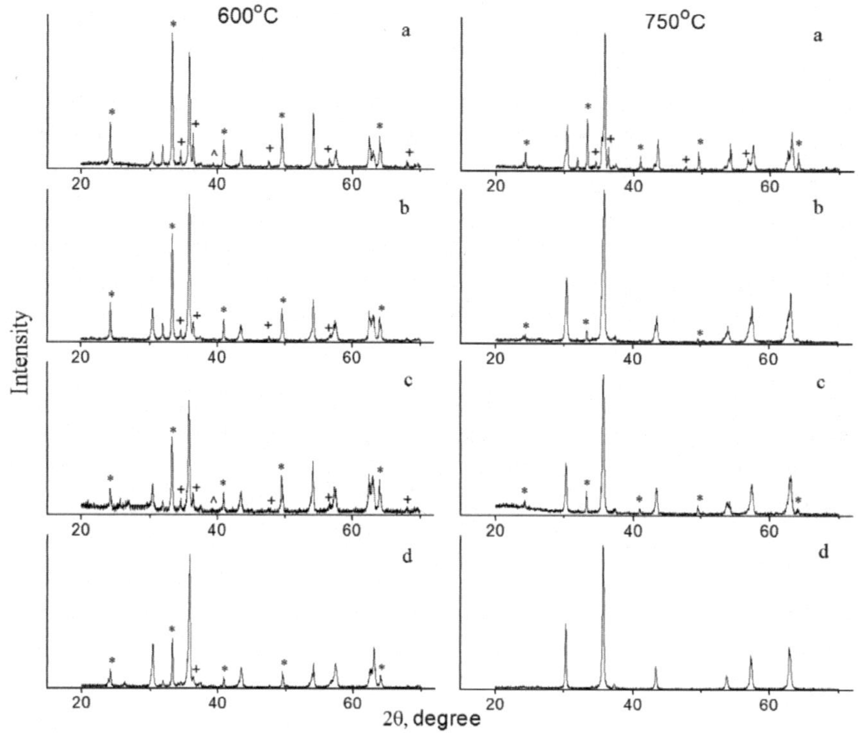

Figure 1. XRD pattern of Li_2CO_3-ZnO-Fe_2O_3 powder samples (a, c) and compacted samples (b, d) after T (a, b) and RT (c, d) heating at 600 and 750 °C for 120 min. Reflections of initial components: (*) - Fe_2O_3; (+) – ZnO; (^) - Li_2CO_3

Table 1. Content of spinel phases in $Li_{0.5(1-x)}Zn_xFe_{2.5-0.5x}O_4$ (weight %)

Synthesis mode	Type of heating	Powder samples	Compacted samples
600 °C –10 min	T	0	0.7
	RT	13.7	67.4
600 °C – 120 min	T	16	45.9
	RT	45.1	75.2
750 °C – 10 min	T	25.4	58.5
	RT	58.1	91.1
750 °C – 120 min	T	65.4	93.9
	RT	88.9	100

Magnetization data of Li_2CO_3-ZnO-Fe_2O_3 mixtures are given in table 2. The results showed an increase in magnetization of compacted samples compared to the powder samples, which also indicates a higher content of magnetic spinel phases in the synthesized compacts.

Table 2. Saturation magnetization of Li_2CO_3-ZnO-Fe_2O_3 mixture (emu/g)

Synthesis mode	Type of heating	Powder samples	Compacted samples
600 °C –10 min	T	1.2	-
	RT	8.9	43.5
600 °C – 120 min	T	10.9	31.5
	RT	20	48.7
750 °C – 10 min	T	13.4	34.2
	RT	40.4	66
750 °C – 120 min	T	36.5	55.9
	RT	59	68.7

4. Conclusions

• Use of uncompacted powder of initial reagents decreases the efficiency of solid-state synthesis of $Li_{0.5(1-x)}Zn_xFe_{2.5-0.5x}O_4$ ferrites.

• The application of compaction operation of powder reagents mixture leads to an increase in spinel phase concentration and decrease initial components content in lithium-substituted ferrites synthesized in thermal and radiation-thermal heating at the same temperature-time modes of synthesis.

5. Acknowledgements

This work was supported by the Ministry of Education and Science of the Russian Federation in part of the "Science" program.

References

[1] Levin B E, Tret'yakov Yu D and Letyuk L M 1979 Physical and Chemical Principles of Synthesis, Properties, and Application of Ferrites [in Russian]. (Moscow: Metallurgiya)
[2] Jiang X N, Lan Z W, Yu Z, Liu P Y, Chen D Z and Liu C Y 2009 *J. Magn. Mater.* **321** 52–55
[3] Mazen S A and Abu-Elsaad N I 2012 *J. of Magnetism and Magnetic Materials* **324** 20 3366-3373
[4] Smit J and Wijn H P J 1959 *Ferrites. Physical properties of ferrimagnetic oxides in relation to their technical applications* (Eindhoven: Philips Technical Library) p 299
[5] Baba P D, Argentina G M, Courtney W E, Dionne G F and Temme D H 1972 *IEEE Trans. Magn.* **8** 83–94
[6] White G O and Patton C E 1978 *J. Magn. Magn. Mater.* **9** 4 299–317
[7] Surzhikov A P, Lysenko E N, Vlasov V A, Malyshev A V and Vasendina E A 2016 *Materials Chemistry and Physics* **176** 110-114
[8] Malyshev A V, Lysenko E N and Vlasov V A 2015 *Ceramics International* **41** 10 13671-13675
[9] Surzhikov A P, Lysenko E N, Malyshev A V, Vlasov V A, Suslyaev V I, Zhuravlev V A, Korovin E Y and Dotsenko O A 2014 *Russian Physics Journal* **57** 5 621-626
[10] Malyshev A V, Lysenko E N, Vlasov V A and Nikolaeva S A 2016 *Ceramics International* **42** 16180-16183
[11] Pritulov A M, Usmanov R U, Galtseva O V, Kondratyuk A A, Bezuglov V V and Serbin V I 2007 *Russian Physics Journal* **50** 2 187-192
[12] Auslender V L 1994 *Nuclear Instruments and Methods in Physical research* **B 89** 46–48
[13] Mittemeijer E J and Welzel U 2013 Modern Diffraction Methods. (Weinheim: Wiley VCH)
[14] Meisner L L, Lotkov A I, Ostapenko M G, Gudimova E Y and Arysheva G V 2012 *Inorganic Materials: Applied Research* **3** 5 401-407
[15] Kreslin V Yu and Naiden E P 2002 *Instruments and Experimental Techniques* **45** 1 55–57

IOP Conf. Series: Materials Science and Engineering **168** (2017) 012094 doi:10.1088/1757-899X/168/1/012094

On positrons irradiation in the defects of vacancy type

AI Kupchishin [1,2]
[1] Abay Kazakh National Pedagogical University, 13 Dostyk ave., Almaty, Kazakhstan
[2] al-Farabi Kazakh National University. 71 al-Farabi ave., Almaty, Kazakhstan

E-mail: ankupchishin@mail.ru

Abstract. It was predicted the radiation of low energy positrons at their interaction with the defects of the vacancy type. Models of positron traps have been proposed. The energy transitions of positrons in the defects were calculated.

Introduction

It is known that a change in the physical properties of the irradiated and deformed metals, alloys and other materials is not only due to a change in the crystal structure of a solid, but also the electronic structure of defects. Investigation of the influence of defects on the electronic properties of materials allows to identify common patterns of electrons kinetic properties in imperfect crystals (using the method of positron annihilation) and to obtain information about the electronic structure of the defects. This is primarily due to the unique ability of positron to localize in areas with excess electronic charge in crystals, which are usually inherent to the defects in the crystal structure allowing to solve link issues of electronic and atomic structures [1 – 3]. The impact of defects on the parameters of positron annihilation in metals was first discovered by I.Y. Dekhtyar et al. [4, 5] for nickel and its alloys with iron. It is shown that plastic deformation narrows the angular correlation curves, and count rate increases at the maximum. This is due to the presence of excess negative charge in the defect area during the formation of stress field around the dislocations. The resulting local dipoles cause the polarization of S- and d-bands and positrons are captured effectively by dislocation cores, which was confirmed by experimental data on the angular correlation and positron lifetime [6, 7]. The best capture occurs on polymers [8]. In this case, the energy spectrum of electrons in the defect is undergoing significant changes, the probability of positrons annihilation with conductivity electrons increases by reducing their annihilation in the atomic core and due to the high lifetime of positrons in these areas. Various models of positron capture are currently used to obtain quantitative results on the change in electronic structure of metals and alloys, as well as analytical relations between the intensity of photons annihilation and pulsed electron spectrum [9, 10]. We propose a model of positrons capture by vacancy traps and model of positron emission in the defects.

1. Physical models

Probability of positrons capture by trap can be provided when positron falling to defect area. Corresponding expressions for the probability can be obtained from quantum-mechanical calculations. Here are some model calculations.

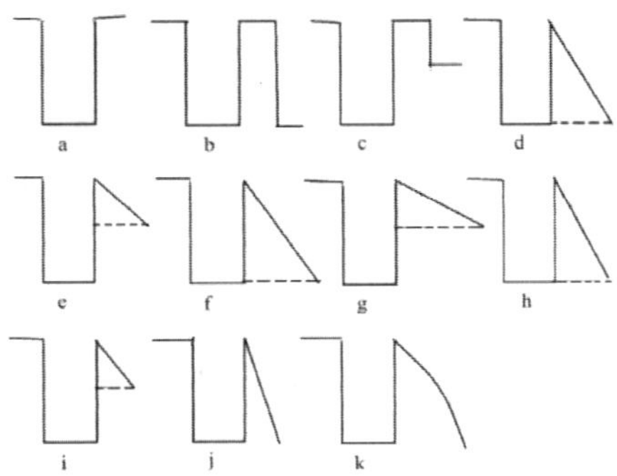

a – one-dimensional potential hole; b, c – rectangular potential barrier; d, e – Coulomb barrier; f, g – inverse-square barrier; h, i – parabolic barrier; j – linear barrier

Figure 1. Models of potential barriers of positron traps.

1.1. One-dimensional potential hole

In the case of one-dimensional hole, the probability of positron escape from the defect can be represented as follows (Figure 1 a)

$$\beta_i(\mathrm{E}) = \begin{cases} 0, when \mathrm{E} > U_0, \\ 1, when \mathrm{E} > U_0. \end{cases}$$

where U_0 is depth of the potential hole of the trap.

1.2. Rectangular potential barrier model

In the first approximation, the probability of positrons escape from the traps is determined by Wentzel-Kramers-Brillouin method. For rectangular potential barrier required probability is represented in the form of (Figure 1 b):

$$\beta_i(\mathrm{E}) = \begin{cases} 1, if \mathrm{E} \geq V_0, \\ \exp\left\{ -\frac{2\sqrt{2m_0}}{\hbar} \sqrt{V_0 - \mathrm{E}} (r_2 - r_1) \right\}, if \mathrm{E} < V_0. \end{cases} \quad (1)$$

The probability for other models can be found similarly.

1.3. Model of positrons emission in vacancy defects

Considering the simplest Schrödinger equation it can be assumed that a positron-vacancy (cluster) system is characterized by certain values of the positron energy. In this case, transition of the positron from one state to another is carried out with a discrete loss of energy by emission. In the defects, positron emits low-energy photons, wavelength of which can be determined from the following relationship:

$$\lambda = \frac{hc}{E_\gamma}.$$

Setting the values of E_γ energy in the range of 0.001 – 10 eV, and assuming that $h = 6.62 \cdot 10^{-34}$ J/s, and $c = 3 \cdot 10^8$ m/s, $1 \leq \lambda \leq 10^4$ nm range can be obtained. Thus, the photon spectrum is above the visible region (by energy).

Positrons motion in rectangular potential hole with U_0 depth and a width can be considered as follows. It is obvious that the spectrum of doubly degenerate levels is continuous when $E > U_0$, and is discrete when $E < U_0$. Schrödinger equation in the range of $0 < x < a$ is written as:

$$\varphi'' + \frac{2m}{\hbar^2}(E - U_0)\varphi = 0.$$

Solution of this equation has the following form:

$$\varphi = A_2 \sin(hx + y), K = \sqrt{\frac{2m}{\hbar^2}(E - U_0)}. \tag{2}$$

It is easy to fund graphically and results in at least one discrete energy level E_1. The number of levels increases with the size of the hole and its depth.

2. Results and Discussion

β_i (E) probability for various values of U_0 and $r_2 - r_1$ was calculated using the formula (1) in relation to the difference of the potential barrier height and positron energy E (Figure 2). It can be seen that the probability is characterized as follows: the higher positron energy or the lower potential barrier height, the higher probability of positron escape from the trap. With increasing width of the potential barrier β_i (E) value is reduced. The target probability behaves similarly for other capture models.

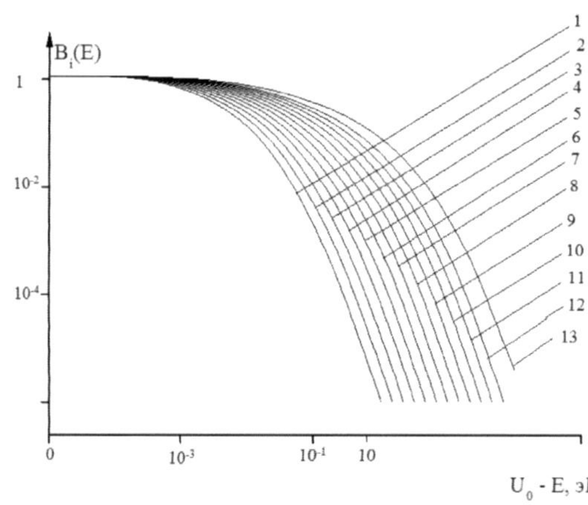

1–10; 2 – 9.3; 3 – 8.4; 4 –7.0; 5 – 6.0; 6 – 5.0; 7 – 4; 8 – 3.0; 9 – 2.0; 10 –1.5; 11 – 1.0; 12 – 0.75; 13 – 0.53

Figure 2. Dependence of the escape probability for positrons on the $(U_0 - E)$ value for various values of the potential barrier width $(r_2 - r_1) \cdot 10^{-10}$ m.

As can be seen from the calculations, the interaction potentials of positrons with vacancies, divacancies and vacancy complexes have the value of a few electron volts. In the first approximation, the positron-vacancy interaction or a complex can be represented in the form of the potential hole (or the potential hole with a barrier). The vacancy estimate for molybdenum at a ≈ 0.32 nm and U_0 = 1.2 eV gives E_l = 0.72 eV; for the vacancy of (a ≈ 0.64 nm) there are two levels: E_l = 0.37 eV; E_2 = 1.13 eV. From (2) it can be seen that the positron emission is the most possible for large vacancy complexes, the R size of which is several interatomic distances. For the energy of 10^{-3} – 10 eV the range is 2Å≤ R≤ 200Å. Consequently, using the positron (by emissions) it is possible to study the defects by the above energy range and approximate size and concentration. In real crystals, the interaction of positrons will essentially vary from the model, although the physical nature of the process is not changed.

Conclusion

1. The models of the potential barriers of the positron traps is proposed, as well as the model of positrons emission in vacancy defects. The probability of positrons escape from the traps is calculated and it is shown that the higher positron energy or the lower potential barrier height, the higher probability of positron escape from the trap. With increasing width of the potential barrier β_i (E) value is reduced.

2. From the results obtained it follows that the positron emission is the most possible for large vacancy complexes, the R size of which is several interatomic distances. For the energy of 10^{-3} – 10 eV the range is 2Å≤ R≤ 200Å. Using the positron (by emissions) it is possible to study the defects (its size and concentration).

References

[1] Dekhtyar I Y 1975 *Metals, electrons, lattice* Electronic structure of defects in metals 228 – 252
[2] Goldanskiy B I 1968 *Physical Chemistry of positron and positronium* (M.: Nauka) 174
[3] Brande W, 1967 *Positron annihilation. – N.Y.,L.: Academ. Press* Positron annihilation in molecular substances and ionic crystals 155 – 182
[4] Dekhtyar I Y, Levina D A, Mikhalenkov V S 1964 *Proceedings of AS of the USSR* Annihilation of positrons with electrons in plastically deformed metals **156** 4 795 – 798
[5] Dekhtyar I Y, 1974 *Physics Reports* The use of positrons for the study of solid **9** 5 243 – 353
[6] Doyama M, Hasiguti R R, I973 *Crystal Lattice Defects* Studies of Lattice defects by means of positron annihilation **4** 3 139 – 163
[7] West R N, 1973 *Advan. Physics* Positron studies in condensed matters **22** 3 263 – 283
[8] Kupchishin A I, Nauryzbaev M K, Kupchishin A A, Komarov F F, Remnev G E, Voronova N A, Kaykanov M 2015 *IOP Conf. Ser.: Mater. Sci. Eng.* **81** 011002
[9] Boos E G, Shmygalev Y V, Shmygaleva T A et al. 2015 *Cascade-probability method. Solution of radiation-physical problems, Boltzmann equations. Relation with Markov chains* (Almaty: Abay KazNPU, SRINCTM of al-Farabi KazNU, Kama LLP) 388
[10] Kupchishin A I, Kupchishin A A, Voronova N A, Kirdyashkin V I 2016 *IOP Conf. Ser.: Mater. Sci. Eng.* **110** 012040

IOP Conf. Series: Materials Science and Engineering **168** (2017) 012095 doi:10.1088/1757-899X/168/1/012095

Analysis of the thermal modes of Focal Plane Arrays

E Ph Pevtsov[1], T A Demenkova[1], P A Luchnikov[1] and V V Vetrova[1]

[1]Integrated Circuits, Nanoelectronics Devices and Microsystems Design Center, Moscow Technological University (MIREA), 119454, Vernadsky 78, Moscow, Russia

E-mail: pevtsov@mirea.ru

Abstract. Techniques and programs for calculation and modeling of thermal characteristics of MEMS-designs of radiation thermal receivers are offered. Comparative estimates of characteristics of radiation receivers with heatsensitive thin films are executed. Results of researches can be used for development of original programs for modeling of thermal receivers of radiation.

1. Introduction

As it was predicted, in reviews [1, 2], in process of development of opportunities of micron technologies developers of IR-systems began to understand that uncooled thermal detectors, though yield photon and quantum in the extreme parameters, possess a number of characteristics which do them irreplaceable in the extensive field of applications. In particular, it belongs to tasks of observation and recognition of objects at small distances (to 2500 m), infrared microscopy, medical and industrial diagnostics, etc. The main benefit of thermovision systems on multielement thermal detectors in relation to systems with photon and quantum devices, is that their work doesn't require chilling up to the cryogenic temperatures. In English-speaking literature these devices call "uncooled matrix detectors in the focal plane» UFPA (Uncooled Focal Plane Arrays). Calculations of thermal balance form a basis of optimization of a design and execution of thermal detectors of radiation of any types [3-7]. The task of optimization of a design of the detector consists in making heating process by radiation stationary during the time corresponding to shot time.

2. Technique of calculations

As a rule, the corresponding physical models come down to the solution of the linear parabolic differential equations in private derivatives. However simple assessment of time constant of heating of the thermal detector radiation can be received for the following reasons. Let's consider the idealized case in which the receiver with a reference temperature T_0 represents the thin plate located in vacuum between two planes from which lower is a heatdrain with a fixed temperature of T_0 and upper is a radiation source from background which temperature jump changed from T_0 to T_b (scene temperature), and we will put that the plate is at the room temperature and scene temperature not much more (for example, by one degree) exceeds initial, i.e. $T_b \approx T_0 \approx 300$ K.

Believing that distribution of temperature on amount of a thin plate can be neglected, i.e. that its temperature is identical on all its amount, we determine how the average temperature of a plate T rather initial T_0 values changes over time.

The equation of thermal balance describing radiating transfer of heat from more heated plate with a temperature T_b (or T) to more colder plate with a temperature T (or T_0), has an appearance:

$$c_V ShdT = S\left[4\sigma T_b^3 (T_b - T) - 4\sigma T_0^3 (T - T_0)\right]dt \quad , \tag{1}$$

where c_V – volume thermal capacity of plate material, S – its area, h – thickness.

Here the left part represents the energy acquired by the plate with area S and thickness h for dt, and in the right part the first member corresponds to radiation heat exchange between a background and the top surface of a plate, and the second member – to radiation heat exchange with heat drain. In view of $T_b \approx T_0 \approx 300$ K and doing substitution $\theta(t) = T(t) - T_0$, taking into account an entry condition $\theta(0) = T(0) - T_0 = 0$, we receive:

$$\theta = \frac{(T_b - T_0)}{2}\left(1 - e^{-\frac{t}{2\tau}}\right), \tag{2}$$

where $\tau = \dfrac{c_V h}{4\sigma T_0^3}$ – the thermal constant of time of radiation exchange process depending on thickness of plate and its heat properties.

At $T_0 = 300$ K thermal conductivity of the single site of plate surface (integrated coefficient of radiating heat exchange) $G = 4\sigma T_0^3 = 6{,}1$ W/(m^2·K) and the value of time constant corresponding to our task in seconds, where $[c_V]$ = J/(m^3·K) and [h] = m.

In real thermal detectors it is necessary to consider outflow of heat through design elements at the expense of heat transfer that significantly reduces the average temperature of the detector and, therefore, his sensitivity. The time constant is generally proportional to the attitude of a total thermal capacity H_Σ of the detector and elements of design towards total heat conductivity of G_Σ of elements through which there is a leakage of heat:

$$\tau_\Sigma = \frac{H_\Sigma}{G_\Sigma}, \tag{3}$$

For the design of the detector represented in Figure 1 in which thickness of each detector is supposed a lot of smaller than its radius R, for definition of distribution transitional function of temperature in the case under consideration it is possible to consider not depending on coordinate x (i.e. to consider that the plate evenly heats up on thickness). Then the corresponding thermal task can be reduced to cylindrical, namely to the solution of task on heating by the radiation of a thin edge with radius R and thickness constant b on the heat-conducting cylinder with radius.

The equation of heat conductivity satisfying this model has an appearance:

$$K_p\left(\frac{\partial^2\theta}{\partial\rho^2} + \frac{1}{\rho}\frac{\partial\theta}{\partial\rho} - \frac{1}{\chi_p}\frac{\partial\theta}{\partial t}\right) = \frac{G(T_b + T_0 - 2\theta)}{b}, \tag{2}$$

where ρ – radius in system of cylindrical coordinates; $\theta(\rho)$ – edge temperature in a point with the coordinate of ρ, K_p – coefficient of heat conductivity of material of the detector χ_p – heat diffusivity coefficient ($\chi_p = K_p / c_V$), c_V – a volume thermal capacity of detector material.

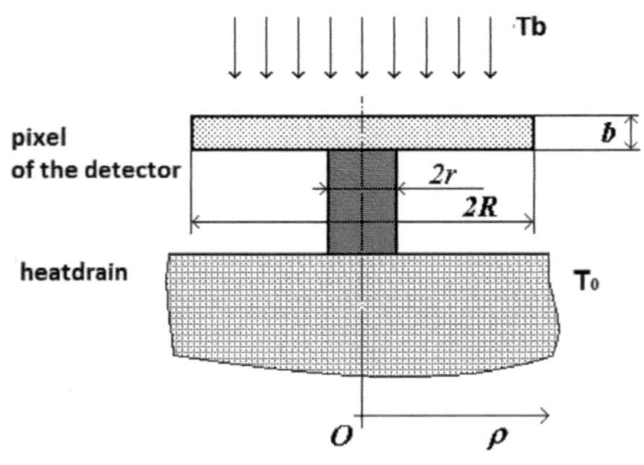

Figure 1. To calculation of the thermal detector with heatdrain in the center.

Passing to dimensionless coordinates and parameters we receive a classical regional task with the mixed Dirichlet and Neumann boundary conditions:

$$\frac{\partial^2\theta}{\partial\xi^2}+\frac{1}{\xi}\frac{\partial\theta}{\partial\xi}-\frac{\partial\theta}{\partial\tau}+\beta^2\theta=\frac{\beta^2}{2}(T_b+T_0), \tag{3}$$

$$\theta(\rho,0)=T_0$$

$$\theta(1,\tau)=T_0$$

$$\left.\frac{\partial\theta}{\partial\xi}\right|_{\left(\frac{R}{r},\tau\right)}=0$$

where: $\xi=\frac{\rho}{r}$ – dimensionless radius; $1\le\xi\le\frac{R}{r}$; $\tau=\frac{\chi_p}{r^2}t$ – dimensionless time; $\beta^2=\frac{2Gr^2}{K_p b}$ – the parameter characterizing radiating heat exchange and design.

The nonstationary solution of the equation (2), i.e. calculation of temperature transitional function can be received by the method of final differences leading to the system of the linear equations describing changes of temperature in splitting grid knots at different timepoints. The example of such calculation of profiles of the specified temperature for this task of the program written in language C ++ is given in Figure 2. At the same time the following typical values of parameters have been accepted: heatsensitive material – polivinilidenftorid film (K_p = 0.13 w/m·K, c_v = 2.3·10^6 J/m^3 K), the sizes of reception element 1×100×100 um^3, diameter of bump contact is 20 um.

Criterion of correctness of the executed calculations is comparison of the stationary decision received by a numerical method with the same decision received analytically. In [3] it is shown that at the analytical decision has an appearance:

$$\theta_\infty=\frac{1}{2}\left[1-\frac{K_1\left(\beta\cdot\frac{R}{r}\right)I_0(\beta\xi)+K_0(\beta\xi)I_1\left(\beta\cdot\frac{R}{r}\right)}{K_1\left(\beta\cdot\frac{R}{r}\right)I_0(\beta)+K_0(\beta)I_1\left(\beta\cdot\frac{R}{r}\right)}\right], \tag{10}$$

where I_0, I_1, K_0 and K_1 – the modified Bessel functions of zero and first order. This dependence for the same parameter values is provided in Figure 2 by the shaped line. The stationary decision received as a result of calculations by method of grids matches it within 5 %.

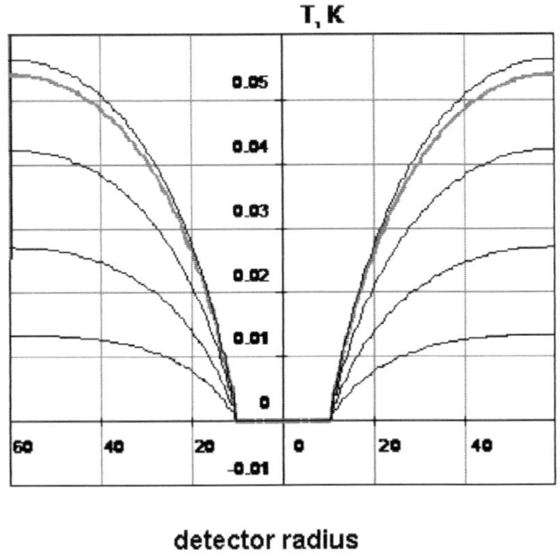

Figure 2. Change of temperature after inclusion of radiation in heatsensitive layer in the detector on bump contact with heat drain in the center

The comparison of different options of detector design given in Figure 3. is an example of application of the developed method of calculation. The specified average temperatures calculated for different types of designs with the heat sensitive polymeric film are given in this drawing. Apparently, the most effective heating happens for the design with the thermal drain in the center. In time about 0.02 s with values of the specified average temperatures of a layer about $10^{-3}...10^{-2}$ K one degree of a scene by, and optimum values of its thickness constitute 1...2 um are reached. The optimum relation of radius of the separate single detector to radius of the maintaining element of the contact design lies within 5...10. In general these conclusions match the results given in reviews [8, 9].

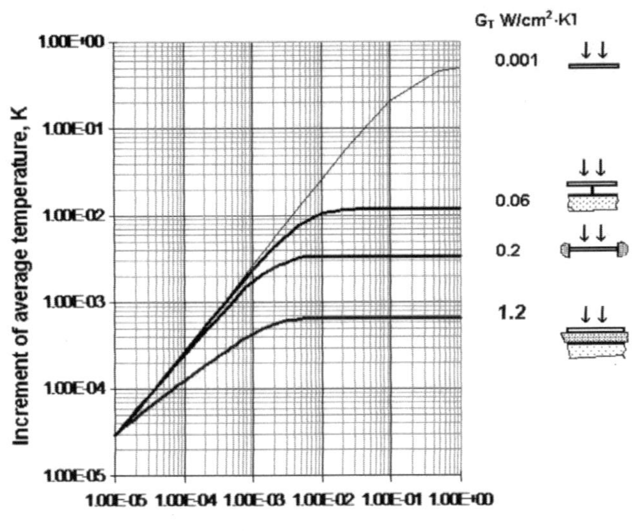

Figure 3. The specified average temperatures and heat conductivity per unit area for different types of designs with the heat sensitive film.

3. Conclusions

The techniques and programs for calculation and modeling of thermal characteristics of MEMS-designs of thermal detectors of radiation based on determination of the transitional functions of average temperature of heat sensitive layers and ratios signal/noise are developed. The offered techniques allow to carry out comparative estimates of characteristics of detectors of radiation on heat sensitive (for example, pyroelectric) thin films. On the basis of developed techniques the analysis of standard options of MEMS-designs is made. It is shown that the design with the film raised over the plate with low heat conductivity, in particular for pyroelectric detectors – films from a poliniviniledenfotrid or the organic pyroelectricic appears the most effective. General heat conductivity in this case doesn't exceed 0.1...0.05 W/cm^2·K in case of the sizes of elements of the detector $1\times100\times100$ um^3 и $1\times50\times50$ um^3 Development of the offered techniques consists in development of original programs of modeling of the thermal detectors of radiation on the basis of universal tools for designing and modeling of MEMS, for example, on the basis of the CoventorWare.

Acknowledgments
This work was supported by the Ministry of Education and Science of Russian Federation

References

[1] Pevtsov E F, Chernokozin V V Matrix IR-receivers for small-size thermal imaging cameras 2001 *Electronic Components* **1 2 3**. (in Russian)
[2] Skidmore G D Uncooled 10um FPA Development at DRS Proc. SPIE 9819, Infrared Technology and Applications XLII, 98191O (20 May 2016) doi: 10.1117/12.2229079
[3] Luchnikov P A, Rogachev A V, Yarmolenko M A, Rogachev A A, Bespal'ko A A 2016 *IOP Conf. Ser.: Mater. Sci. Eng.* **110** 012081
[4] Luchnikov P A 2015 R*ussian Physics Journal* **57** 9 1239-45
[5] Aleutdinov A D, Ghygazov S A, Mylnikova T S, Luchnikov P A 2015 *IOP Conf. Ser.: Mater. Sci. Eng.* **81** 012069
[6] Rogachev A A, Yarmolenko M A, Xiaohong Jiang, Shen Ruiqi, Luchnikov P A, Rogachev A V 2015 *Applied Surface Science* **351** 811-18
[7] Surzhikov A P, Frangulyan T S, Ghyngazov S A, Lisenko E N, Galtseva O V 2006 *Russian Physics Journal* **49** 5 506-10
[8] Carslaw H S, Jaeger J C *Conduction of Heat in Solids* 2-nd ed. Oxford University Press. New York (1959) (ISBN 0198533683)
[9] Pevtsov E F, Tchuiko A V Software and hardware for researches of electrophisical properties of structures on base of active dielectrics. MSTU MIREA HERALD. 2015 **2(7)** 262-71

Radiation Stability of Metal Nanowires

S A Bedin[1], F F Makhin'ko[1], V V Ovchinnikov[1,2], N N Gerasimenko[1,3] and D L Zagorskiy[1,4]

[1]Institute of Electrophysics, Ural Branch, Russian Academy of Sciences, ul. Amundsena 106, Yekaterinburg, 620016, Russia
[2]El'tsin Ural Federal University, ul. Mira 19, Yekaterinburg, 620002 Russia
[3]National Research University of Electronic Technology (MIET), pl. Shokina 1, Zelenograd, Moscow 124498 Russia
[4]Shubnikov Institute of Crystallography, Russian Academy of Sciences, Leninskii pr. 59, Moscow, 119333 Russia

E-mail: bserg5@gmail.com

Abstract. The aim of this work is to investigate the radiation stability of pure nickel and iron–nickel $Fe_{0.56}Ni_{0.44}$ alloy nanowires fabricated by matrix synthesis using polymer track membranes and Ar^+ and Xe^+ (E = 20 keV, j = 300 μA/cm^2) beam irradiation. The dependence of the stability of nanowires on their diameter, fluence, and type of implanted ions is investigated. The assumption that the thermalized regions of dense cascades of atomic displacements (thermal spikes) play an important role in the nanowire structure change is made. These regions are nanosized zones of explosive energy release and heated to several thousands of degrees.

1. Introduction

In recent years, scientists have been increasingly interested in the production, investigation, and application of nanosized materials. Among these materials is a promising type, namely arrays of one-dimensional metal structures, in other words, nanowires (NWs). Arrays of NWs made of copper and silver are interesting as ion and electron emitters (cold cathodes), whereas NWs made of iron are promising as magnetic materials (sensors, materials for high-density magnetic recording).

Among the important characteristics of the NW arrays is their stability. It is known [1] that the specific geometry and the developed surface lead to their lower (compared with volume analogues) chemical and thermal stability. According to preliminary results, the radiation stability of these arrays is different from that of volume materials. The study of the radiation stability is of interest because of at least two factors.

First, radiation treatment is a kind of modification and control of the properties of such NWs.

Second, radiation resistance, the ability to retain properties under different radiation conditions, in many cases is one of the most important service parameters.

It should also be noted that, in addition to the practical aspect, the issue of structural changes under irradiation is of independent theoretical interest. In contrast to the studies of radiation defects in volume metals, there is a small number of works dedicated to nanostructures of this kind. The main part of these studies is devoted to materials with nanosized grains. For example, it is noticed in [1] that among features of nanostructures can be the inability to accumulate radiation defects due to the fact that such defects can immediately break the surface.

Among several works actually dedicated to the study of NWs, articles on the modeling of the impact of ion beams on metal NWs using molecular dynamics methods [2, 3, 4] should be noted.

In addition to the works modeling of radiation effects, there are works devoted to the study of the radiation impact on the metal NWs. For example, a method for the ion beam modification of Ni and Co NWs 200 nm in diameter is described in [5]. These NWs fabricated by template synthesis using a porous alumina oxide were irradiated with 3-MeV Cl^{2+} ions at a fluence of 10^{15} cm^{-2}. As a result, an oxidation layer on the surface of the NWs was reduced, which was supported by high-resolution TEM. The features of the surface structure changes were investigated under such irradiation.

In [6], silver NWs 120 nm in diameter and 20 μm in length were irradiated at room temperature with 5-MeV He^+ ions at a beam current of 50 nA. Insignificant changes at a fluence of $5 \cdot 10^{15}$ cm^{-2} were observed, whereas the initially crystal structure of NWs was transformed into the amorphous structure at a fluence of more than $8 \cdot 10^{16}$ cm^{-2}.

In [7], silver NWs were irradiated with 3-MeV protons (beam current of 50 nA) at room temperature. Under irradiation with fluence of 10^{14}-10^{15} cm^{-2}, NWs were welded in contact areas to form a kind of grid of nanowires.

In [8], 5-MeV carbon ion irradiation-induced changes in the electrical conductivity of silver NWs were studied at different ion fluences ranging from 10^{12} to 10^{16} cm^{-2}. The electrical conductivity increases and, then, decreases with increase in fluence. The first behavior was explained by the formation of contacts between the wires; the second, by an increased defectiveness in some wires.

In our work we investigated some points of radiation stability of NWs made of 3d-metals, namely pure nickel and iron-nickel $Fe_{0.56}Ni_{0.44}$ alloy. The effects of the fluence and the temperature of a target on the stability of NW arrays were investigated upon irradiation with Ar^+ and Xe^+ ions (E = 20 keV). The used energy relates to the energy range of the most commonly used to modify the structure and properties of materials.

2. Experiment

The investigated NWs were fabricated by the template synthesis. Commercial track membranes with a pore diameter from 0.03 to 0.1 μm (Joint Institute for Nuclear Research, Dubna) were used as template matrices. This deposition process is described in detail in [9, 10].

A Watts electrolyte that contains 300 g/lof $NiSO_4 \cdot 7\ H_2O$, 45 g/l of $NiCl_2 \cdot 6\ H_2O$, and 38 g/l of H_3BO_2 was used to prepare nickel NWs, at an electrolyte temperature of 50–60°C.

The $Fe_{0.56}Ni_{0.44}$ alloy was deposited from the electrolyte of the following composition: 16 g/l of $NiSO_4 \cdot 7H_2O$, 40 g/l of $NiCl_2 \cdot 6H_2O$; 16 g/l of $FeSO_4 \cdot 7H_2O$; 1-2 g/l of antiscorbutin ($C_6H_8O_6$); and less than 1 g/l of sodium laurel sulfate ($NaC_{12}H_{25}SO_4$).

NW arrays separated from the polymer template were irradiated with continuous Ar^+ ion beams using an ILM-1 implanter equipped with a PULSAR-1M technological ion source based on a low-pressure glow discharge with a hollow cold cathode, which can operate in the continious and pulsed-periodic modes (S_{beam} ~ 100 cm^2; the energy and ion current density in the continuous mode were varied in the following ranges: E = 5-50 keV, j = 10-500 $\mu A/cm^2$) [11].

NWs were irradiated with Ar^+ and Xe^+ (Figure 1) onto a copper substrate with a rotated holder (angle between the ion beam axis and a normal to the substrate surface was 45°; E = 20 keV, j = 300 $\mu A/cm^2$, a rotation rate of 8 rpm) at various fluences and in several temperature conditions:

(i) irradiation with Ar^+ ions, $F = 10^{16}$-10^{18} cm^{-2}, beam heating below 150°C (fluences were achieved in several stages to avoid sample overheating);

(ii) irradiation with Ar^+ ions, $F = 10^{16}$-10^{18} cm^{-2}, beam heating to 550°C (fluence was continuously accumulated);

(iii) irradiation with Xe^+ ions, $F = 10^{16}$-10^{17} cm^{-2}, beam heating to 550°C (fluence was continuously accumulated).

Initial and irradiated samples were investigated using an FEI Scios scanning electron microscope.

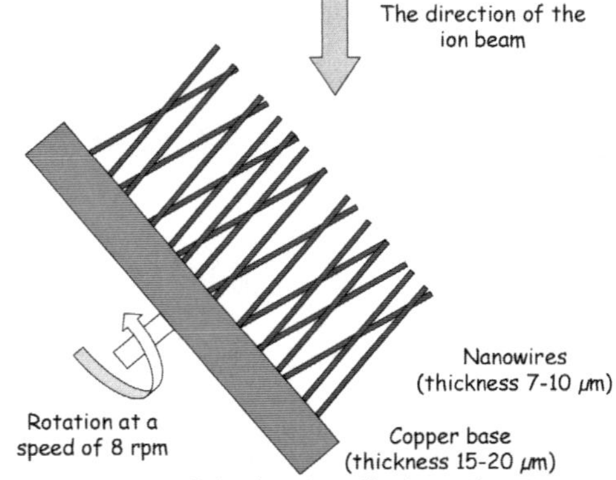

Figure 1. Scheme of the ion irradiation of NW assemblies.

3. Results and discussions

Figure 2 shows scanning electron microscopy images of initial (before irradiation) NWs made of Ni with a diameter of 60 nm (Figure 2 a) and NWs made of $Fe_{0.56}Ni_{0.44}$ alloy with a diameter of 100 nm (Figure 2 b).

Figure 2. Scanning electron microscopy images of (a) initial Ni NWs 60 nm in diameter and (b) FeNi NWs 100 nm in diameter.

Below, Figure 3 a demonstrates Ni NWs 60 nm in diameter after Ar^+ ion irradiation. The arrow indicates the irradiation direction. At a fluence of 10^{17} cm^{-2} and an irradiation time of 1 min (Figure 3 a), melted NW tops (see the left top angle in the image) are observed. No noticeable changes are observed at a lower fluence of 10^{16} cm^{-2}. An increase in the fluence to $5 \cdot 10^{17}$ cm^{-2} (irradiation time of 5 min with a step of 1 min) results in melting of a NW bundle up to its middle, which is clearly observed in the scanning electron microscopy images (Figure 3 b). A further increase in the fluence (Figure 3 c) (irradiation time of 10 min with a step of 1 min) causes subsequent melting and deformation of NWs. Continuous irradiation with heating (Figure 3 d) (5 min) leads to the melting of the all surface of a NW bundle. The wires are less melted inside the bundle.

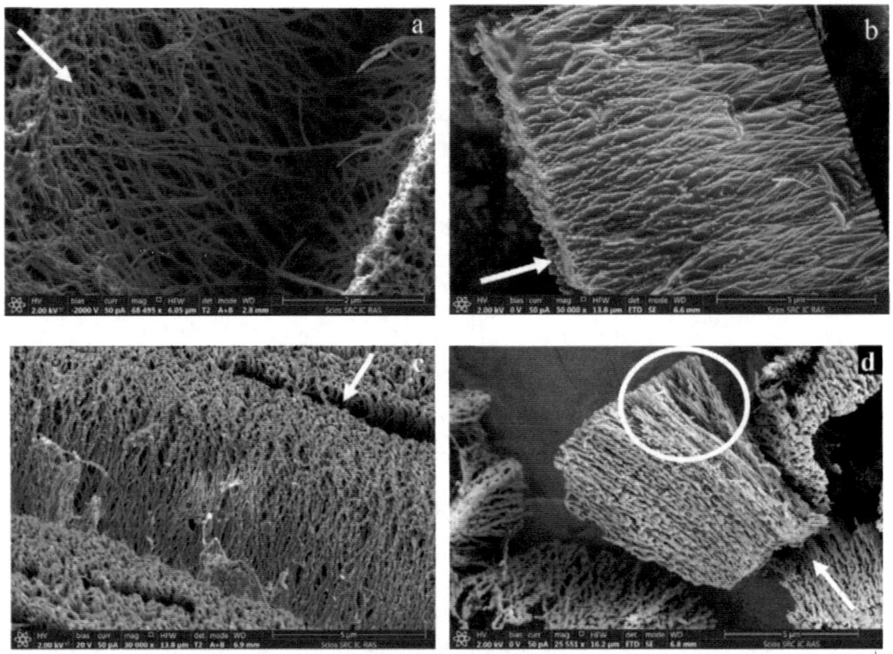

Figure 3. Scanning electron microscopy image of Ni NWs 60 nm in diameter after Ar^+ ion irradiation with different fluences, heating to not higher than 150°C (accumulated fluences were achieved in several stages to avoid overheating): (a) F=10^{17}cm^{-2}; (b) F=5·10^{17}cm^{-2}; (c) F=10^{18} cm^{-2}; and (d) F=5·10^{17}cm^{-2}, heating to a stationary temperature of 550°C.

Figure 4 shows $Fe_{0.56}Ni_{0.44}$ NWs 100 nm in diameter after Ar^+ ion irradiation. In this case, wires have a greater diameter; therefore, this seems to be the reason for their less deformation after irradiation. It is noticeable that tops of the wires are "sharpened" from the direction of beam irradiation at the increased fluence.

Figure 5 shows NWs irradiated with Ar^+ and Xe^+ ions for comparison purposes. In both cases, irradiation conditions were the same: E = 20 keV, j = 300 μA/cm^2, a rotation rate of 8 rpm, an angle of 45°, heating to 550°C. Irradiation was performed in a continuous mode for 5 min. It can be seen from the figures that wires are noticeably deformed after Xe^+ irradiation and are different from those irradiated with Ar^+.

Taking into account the temperatures (150 and 550°C), which are insufficient for melting pure nickel and iron-nickel alloy, the observed melting of nanowires can be explained by the formation of *thermal spikes* under irradiation with heavy ions [12, 13]. They are almost spherical (in the case of the implantation of heavy ions into heavy matrices) regions of dense cascades of atomic displacements with a typical diameter of 5-10 nm (about 9 nm for Ar^+ and 6 nm for Xe^+ [15] implanted in Ni with the used energy). The mentioned regions thermalized for 10^{-12} s (cooling time of ~10^{-11}), reaching temperatures of 3000-5000 K and above. The average depth at which these regions are formed is about 5-10 nm at an ion energy of 20 keV.

Since the energy release per atom of the cascade is significantly higher in the case of Xe^+ than that of Ar^+ [15], Figure 5 shows more significant bending of nanowires with the formation of knots (buds) and even branches under Xe^+ ion irradiation due to the possible splashes of the thermal peak regions out of the nanowires. Covering of adjacent nanowires with molten metal is possible.

In view that the number of atoms per unit of the surface area of metals is ~10^{15} cm^{-2}, we can conclude that the NW surface (which is not in the field of geometric shadow) has melted repeatedly at fluences of 10^{16}-10^{18} cm^{-2} due to the formation of thermal spikes [13, 14]. We put forward this mechanism as an explanation for the observed effect of melting of the NWs at low total temperatures.

Figure 4. Scanning electron microscopy image of $Fe_{0.56}Ni_{0.44}$ alloy NWs 100 nm in diameter after Ar^+ ion irradiation with different fluences, with heating to 150°C: (a) F=10^{17}cm^{-2}; (b) F=5·10^{17}cm^{-2}; (c) F=10^{18} cm^{-2}; and (d) F=5·10^{17}cm^{-2}, with heating to 550°C.

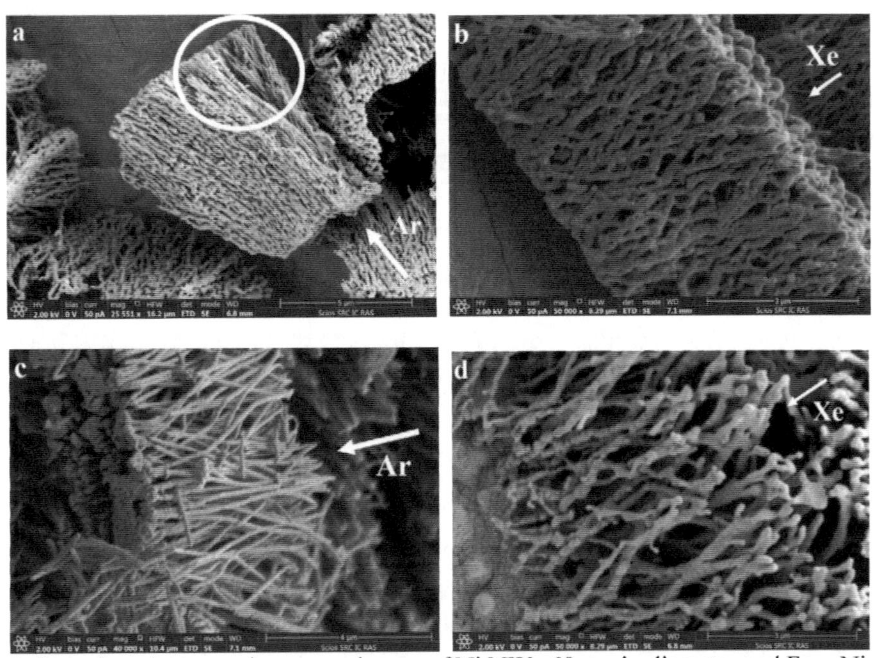

Figure 5. Scanning electron microscopy image of Ni NWs 60 nm in diameter and $Fe_{0.56}Ni_{0.44}$ alloy NWs 100 nm in diameter after Ar^+ and Xe^+ ion irradiation with F = 5·10^{17}cm^{-2} (heating to 550°C).

It should be noted that damaged near-surface volume is comparable with the volume of nanowires (taking into account their rotation). There is an opinion that some decrease in the melting temperature may be caused by a sharp increase in the number of nonequilibrium defects in the NW volume. Although, it is evident that this mechanism cannot be the only explanation for the observed melting effect, but it can ease the melting process.

Note that damageable subsurface volume is comparable to a volume of NWs (in view of their rotation). Because of this, a marked decrease in the melting temperature may occur due to a sharp increase of the number of non-equilibrium defects in the volume of NWs. Although only this mechanism can hardly explain the observed melting effect, but it is also likely to contribute.

4. Conclusions

As a result, the study showed that under both Ar^+ and Xe^+ ion irradiation conditions (E = 20 keV, j = 300 μA/cm^2, fluences of 10^{16}-10^{18} cm^{-2}), NWs are deformed and melted even upon slight general (not local) beam heating (to 150°C). The effect of Xe^+ ion irradiation is more pronounced.

As an explanation for the observed effect of melting of NW arrays at decreased temperatures we propose to consider the formation of thermal spikes under ion irradiation.

The conclusion about the low radiation stability of the NWs under used Ar^+ and Xe^+ ion irradiation conditions can be drawn. It is further interesting to study the effect of the chemical composition, the geometric characteristics of nanowires, the ion energy, and the kind of implanted ions on the radiation stability in detail. X-ray diffraction analysis and Mössbauer spectroscopy are required to obtain the information about possible amorphization of the NW surface by rapid quenching of thermal spikes and about other structural features of nanowires.

Anknowledgments

We thank V.V. Artemov (Shubnikov Institute of Crystallography, Russian Academy of Sciences) for the assistance in electron microscopy studies.

This work was supported by the Russian Scientific Foundation, project no. 15-19-10054.

References

[1] Andrievski R A 2014 *Physics-Uspekhi* **57** 10 945
[2] ZhenYu Yang et al. 2013 *Science China Physics, Mechanics and Astronomy March* **56** 3 498
[3] Weina Lia et al. 2013 *Nuclear Instruments and Methods in Physics Research Section B: Beam Interactions with Materials and Atoms* **307** 15 158
[4] Zou Xue-Qing et al. 2010 *Chinese Physics B* **19** 3 036102
[5] Soung Kyu Park et al. 2009 *Current Applied Physics* **9** 847
[6] Shehla H, Ali A. et al. 2015 *Chin. Phys. Lett.* **32** 9 096101
[7] Honey S et al. 2016 *Chinese Physics B* **25** 4 045105-1
[8] Bushra Bari et al. 2015 *Current Applied Physics* **15** 642-647
[9] Bedin S A, Rybalko O G, Polyakov N B, Zagorskiy D L, Razumovskaya I V, Bondarenko G G and Oleinikov V A 2010 *Inorganic Materials: Applied Research* **1** 4 359
[10] Zagorskiy D L, Korotkov V V, Frolov K V, Sulyanov S N, Kudryavtsev V N, Kruglikov S S and Bedin S A 2015 *Physics Procedia* **80** 144
[11] Gavrilov N V, Mesyats G A, Nikulin S P, Radkovskii G V, Eklind A and Perry A J 1996 *J. Vac. Sci. Technol. A* **14** 1050
[12] Ovchinnikov V V 2008 *Phys. Usp.* **51** 955
[13] Ovchinnikov V V, Makhin'ko F F, Solomonov V I, Gushchina N V and Kaigorodova O A 2012 *Technical Physics Letters* **38** 86
[14] Ovchinnikov V V Makhin'ko F F and Solomonov V I 2015 *Journal of Physics: Conference Series* **652** 012070
[15] Biersack J P and Haggmark L G 1980 *Nucl. Instr. & Meth.* **174** 257

Modeling of radiative - conductive heat transfer in compositing materials

P A Luchnikov[1], V I Nefedov[1], N A Trefilov[1], A N Dementiev[1] and A P Surzhikov[2]
[1]Moscow Technological University (MIREA), 119454, Vernadsky 78, Moscow, Russia
[2]Institute of Non-Destructive Testing, National Research Tomsk Polytechnic University, 30, Lenina Avenue, 634050, Tomsk, Russia.

E-mail: xamdex@gmail.com

Abstract. A layer of composite material is investigated, which is heated one-sidedly with one-dimensional energy transfer accounting for thermal conductivity and radiation. A mathematical model is suggested for non-stationary coefficient thermophysical problem under radiative-conductive heat transfer in a material layer. Temperature dependencies of thermal capacity and thermal conductivity coefficient of composite radio-transparent material have been determined through numerical modeling by solving the coefficient reverse problem of thermal conductivity.

1. Introduction

Composite radio-transparent materials are widely utilized as protection devices for aircraft antennas [1, 2]. In the process of exploitation, these composite materials experience influence of various environment factors, such as discharge plasma [3–5], electrization and optical radiation [6], locally increased temperature and others. The crucial factor determining composite's properties is the increased temperature, which alters such primary dielectric parameters of composite materials as complex relative dielectric permittivity [7, 8]. In order to measure temperature change within the sample one uses high-temperature thermocouples with small conductor cross-section. The thermocouples are embedded into the sample's volume during its formation stage. The coordinates of thermocouple's joint in the sample are defined via X-ray imaging. Thus, several thermocouples with different setting depths are placed in various cross-sections of the sample.

2. The samples and research method

In order to calculate the temperature in spaces between the thermocouples of the sample a specific mathematical apparatus is used applying numerical mathematical methods. One uses solutions for non-standard direct problems of thermal conductivity and inverse coefficient ones under radiative-conductive heat transfer in a flat layer of the absorbing, emitting and scattering medium. One-dimensional problem is commonly considered for the purposes of measurement. In the mathematical setting of the thermophysical problem of determining temperature distribution in the thickness of a flat dielectric sample one considers a flat layer of a thermophysically isotropic material.

When solving thermal conductivity problems, first-, second- or third-type boundary conditions may be used in arbitrary combinations for each of the surfaces. As for the radiation transfer, within each of the spectral ranges the sample boundaries can be either semi-transparent or non-transparent for self-radiation and radiation of external heat sources. The diversity of boundary condition types allows modeling arbitrary heat response conditions and processing the experiment results, which have been conducted on installations varying in design and sample heating conditions [8, 9]. When solving coefficient reverse problems, both separate and integrated determination of temperature dependencies of volumetric heat capacity cq (T) and thermal conductivity coefficient λ (T) are valid.

Thermophysical properties of the sample an $d\lambda(T)$ arenon-linear and temeprature-dependant, whileitsoptical-physical properties, such as absorption and scattering factors, refraction index and scattering indicatrix, depend on temperature and wavelength of the heat-associated radiation.

We consider a layer of composite material which is in condition of one-sided heating with energy transferred one-dimensionally via thermal conductivity and radiation. In the case initial temperature distribution within the sample and heat exchange conditions at its boundaries are known. When implementing numerically, the composition of heat exchange parameters set is determined by a concrete type of heat response. Thus, in case of radiative microwave heating or infrared heating, one specifies densitiesof radiation fluxes that are brought to the sample surfaces, their spectral and angular distributions. Under the experimental conditions with convective heating created by gas burners or arc plasma sources, we specify gas environment temperature and temperature dependency of the heat-transfer coefficient, Moreover, in all the cases of thermal response optical properties of the boundaries are considered established.

The temperature dependencies cq (T) and$\lambda(T)$ of the material are determined by solving a reverse problem of thermal conductivity or radiative-conductive heat transfer. When coefficient calculation methods are applied, thermophysical properties of the material are among causal characteristics whereas temperature patterns and radiation fields are consecutive. In the process of the experiment temperatures at separate points of the sample are meausred. In this case measuring temperature at one inner point of the sample allows to restore only one temperature dependency – either cq (T) or $\lambda(T)$. In order to restore both dependencies one should measure the temperatures in at least two points within the sample thickness and heat flux on one of its surfaces.

3. Mathematical modeling

We shall now take into consideration a mathematical model of the non-stationary coefficient thermophysical problem. The mathematical model of radiative-conductive heat transfer in a layer of absorbing, emitting and scattering medium includes thermal conductivity equation as well as radiation transfer equation transformed using a binomial approximation of the moments method [10, 11]. The equations for inner heat sources and initial condition for the thermal conductivity equation as well as boundary conditions for the thermal conductivity and radiation transfer equations also make up the mathematical model.

$$cq(T)\cdot\frac{\partial T(t,r)}{\partial t}=\frac{1}{r^n}\cdot\frac{\partial}{\partial r}\cdot\left(r^n\cdot\lambda(T)\cdot\frac{\partial T(t,r)}{\partial r}\right)+q_V; \tag{1}$$

$$R_1<r<R_2;\ R_1<r<R_2;\ T(0,r)=f_0(r); \tag{2}$$

$$-K_1\cdot\lambda(T)\cdot\frac{\partial T(t,R_1)}{\partial r}=\alpha_1(T)\cdot T(t,R_1)+\beta_1(\tau)+\gamma_1(T); \tag{3}$$

$$K_2\cdot\lambda(T)\cdot\frac{\partial T(t,R_2)}{\partial r}=\alpha_2(T)\cdot T(t,R_2)+\beta_2(t)+\gamma_2(T); \tag{4}$$

$$R_1<r<R_2$$

$$\frac{1}{r^n}\cdot\frac{\partial}{\partial r}\cdot\left(\frac{r^n}{\varpi_v}\cdot\frac{dM_{0_v}(r)}{dr}\right)-3\cdot\alpha_v\cdot M_{0_v}(r)=-3\cdot\alpha_v\cdot B_v^*(T); \tag{5}$$

$$-\frac{1}{2\cdot\varpi_v}\cdot\left(\frac{1}{3}+R_{1_{W1_v}}\right)\cdot\frac{dM_{0_v}(R_1)}{dr}+\frac{1}{2}\cdot\left(\frac{1}{2}-R_{0_{W1_v}}\right)\cdot M_{0_v}(R_1)=$$
$$=q_{0_{W1_v}}\cdot\left(2\cdot\eta_{m_{W1}}\cdot Q_{0_{W1_v}}+\eta_{\delta_{W1}}\cdot Q_{1_{W1_v}}\right)+2\cdot E_{W1_v}\cdot n^2\cdot B_{W1_v}(T_{W1}); \tag{6}$$

$$\frac{1}{2\cdot\varpi_\nu}\cdot\left(\frac{1}{3}+R_{11_{W2_\nu}}\right)\cdot\frac{dM_{0_\nu}(R_2)}{dr}+\frac{1}{2}\cdot\left(\frac{1}{2}-R_{01_{W2_\nu}}\right)\cdot M_{0_\nu}(R_2)=$$
$$=q_{0_{W2_\nu}}\cdot\left(2\cdot\eta_{m_{W2}}\cdot Q_{01_{W2_\nu}}+\eta_{\delta_{W2}}\cdot Q_{1_{W2_\nu}}\right)+2\cdot E_{W2_\nu}\cdot n_\nu^2\cdot B_{W2_\nu}(T_{W2}); \tag{7}$$

$$q_V=\int_{\bar{\nu}_1}\alpha_\nu\cdot\left(M_{0_\nu}(r)-B_\nu^*(T)\right)\cdot d\nu, \tag{8}$$

$$\varpi=\alpha+\beta(1-p),$$

where: T is the temperature, K; t is time; r – stand for coordinate; cq – volumetric thermal capacity, J/m^3; λ – thermal conductivity coefficient, W/m·K; α – absorption coefficient, m^{-1}; β – scattering coefficient, m^{-1}; p – scattering indicatrix parameter; f_0 – initial temperature distribution; $K_{1,2}$ – the variable indicating the boundary condition type.

With $K_{1,2}=0$ first-type boundary conditions are set. In this case

$$\alpha_{1,2}(T)=-1;\qquad \gamma_{1,2}(T)=0;\qquad \beta_{1,2}(\tau)=f_{1,2}(\tau);$$

$f_{1,2}(\tau)$ is the surface temperature dependency in time.

With $K_{1,2}=1$ second- and third-type boundary conditions are set. And in this case

$$\alpha_{1,2}(T)=-\alpha_{f1,2};\ \beta_{1,2}(\tau)=q_{TW1,W2};$$

$$\gamma_{1,2}(T)=\int_{\bar{\nu}_1+\bar{\nu}_2}\varepsilon_{W1,2_\nu}\cdot q_{\Pi_{W1,W2_\nu}}d\nu-\int_{\bar{\nu}_1+\bar{\nu}_2}\varepsilon_{W1,2_\nu}\cdot B_{W1,2_\nu}d\nu+\alpha_{f1,2}\cdot T_{f1,2};$$

where: $\alpha_{f1,2}$ is the heat transfer coefficient, W/m^2; ε – is the emissivity of the boundary surface; q_Π – stands for the density of the radiation flux incident, W/m^2; q_T – density of the flux absorbed by the surface, W/m^2; T_f – environment temperature, K; B – is the Planck function; n – is refraction index; $B^*=4\cdot n^2B$; R_{mn}, Q_{mn}, Q_m, E are integral optical characteristics of the boundary surfaces; η_m – portion of diffuse radiation in the incedent radiation flux; η_δ – portion of the directed radiation in the incident radiation flux; q_0 – radiation flux incident to the semi-transparency area of the material, W/m^2; M_0 – intensity moment of the zero-order radiation, W/m^2. The ν index indicated frequency dependence, and $W1$, $W2$ are related to the front and rear surfaces accordingly.

Solving the system (1)–(8) is carried out by the method of finite differences with implicit difference scheme applied. The coefficient problem of radiative-conductive heat transfer is usually formulated in extreme setting [10]. The algorithm for determining the coefficients of equation (1) – λ (T), cq (T) – is realized via minimizing the functional linking the experimental and estimated temperatures

$$S=\sum_{i=1}^{N}\int_0^{\tau^e}\left(T^e(\tau,r_i)-T(\tau,r_i)\right)^2d\tau=\min, \tag{9}$$

where: N – is the number of temperature sensors; $T^e(t,r_i)$ are experimental temperature values; $T(t,r_i)$ – estimated temperature values, determined from the solution of (1)–(8); t^e – experiment duration.

Solving the radiative-conductive heat transfer is usually carried out applying the principle of iterative regulation [12, 13]. In this case the condition of residual is used to stop the process of minimization:

$$S\approx\delta^2. \tag{10}$$

$$\delta^2=\sum_{i=1}^{N}\int_0^{\tau^e}\Delta T^e(t,r_i)^2dr, \tag{11}$$

where $\Delta T^e(t,r_i)$ is the error of experimental temperature values.

The sought-for temperature dependencies are expanded into series by the number of basis functions:

$$\lambda(T)=\sum_{m=1}^{M_\lambda}\lambda_m\cdot\psi_{\lambda_m}(T); \tag{12}$$

$$cq(T)=\sum_{m=1}^{M_C}C_m\cdot\psi_{C_m}(T), \tag{13}$$

where: λ_m, C_m are expansion factors; ψ_λ, ψ_C are basis functions; M_C, M_λ stand for the number of members in the expansion.

Cubic and linear *b*-splines are often used as basis functions [13]. So the sought-for spline coefficients appear to be the values of thermophysical parameters of the sample material in the spline-approximation vertices.

Expansion of the sought-for dependencies cq (*T*) andλ (*T*) to basis functions allows one to bring problem (1) to the problem of determining the $M_C + M_\lambda$ function minimum of the variables:

$$S = S(\lambda_1, \lambda_2, ..., \lambda_{M_\lambda}, c_1, c_2, ..., c_{M_c}) = \min. \tag{14}$$

In order to solve this problem we can use, for instance, a gradient minimization method – the Davidon-Fletcher-Powell method [12]. The valuesof $\partial S/\partial \lambda_i$ gradient tvector; $i = \overline{1, M_\lambda}$; $\partial S/\partial C_i$; $i = \overline{1, M_C}$ are determined from the dual problem. The Davidon-Fletcher-Powell method involves determining the minimum of function *S*along the chosen search direction.

Acknowledgments
This work was supported by the Ministry of Education and Science of Russian Federation.

References

[1] Prigoda B A, Kokunjko B C Radomes of aircraft M.: Mashinostroenie 1970 340 p (in Russian)
[2] Walton J D Radome Engineering Handbook N.V.: Marcel Dekker, 1970 520 p
[3] Luchnikov P A, Rogachev A V, Yarmolenko M A, Rogachev A A, Bespal'ko A A Dynamics of combined electron beam and laser dispersion of polymers in vacuum 2016 *IOP Conf. Ser.: Mater. Sci. Eng.* **110** 012081
[4] Luchnikov P A Structure Formation Mechanisms and Electrical Properties of PVD Fluoropolymer Films 2015 *Russian Physics Journal* **57** 9 1239–45
[5] Rogachev A A, Yarmolenko M A, Xiaohong Jiang, Shen Ruiqi, Luchnikov P A, Rogachev A V Molecular structure, optical, electrical and sensing properties of PANI-based coatings with silver nanoparticles deposited from the active gas phase 2015 *Applied Surface Science* **351** 811–18
[6] Aleutdinov A D, Ghygazov S A, Mylnikova T S, Luchnikov P A Thermal shock removal of defective glass-enamel coating from cast-iron products 2015 *IOP Conf. Ser.: Mater. Sci. Eng.* **81** 012069
[7] Luchnikov P A, Sigov A S, Trefilov D N, Trefilov N A Wave dielcometric materials: monograph under the editorship A.S. Sigov. M.: Scientific world, 2014 300 p (in Russian)
[8] Alifanov O M Identification of heat transfer processes in aircraft. M.: Mashinostroenie, 1979. 216 p (in Russian)
[9] Prosuntsov P V, Reznik S V Determination of thermophysical properties of translucent materials 1985 *Engineering-physical journal* **496** 971–76 (in Russian)
[10] Prosuntsov P, Reznik S Mathematical model of the inverse problem of radiative-conductive heat transfer in semitransparent scattering medium Proceedings of the Siberian Department. USSR Academy of Sciences. Series technique Sciences. 1986 **2** 10 3–9 (in Russian)
[11] Surzikov A P, Lysenko E N, Ghyngazov S A, Frangulyan T S 2002 *Russian Physics Journal* **45** 10 989-96
[12] Himmelblau D. Applied nonlinear programming. M.: Mir, 1975. 534 p (in Russian)
[13] Forsyth J, Malcolm M, Moler K Computer methods for mathematical computations. M.: Mir, 1980 279 p (in Russian)

IOP Conf. Series: Materials Science and Engineering **168** (2017) 012098 doi:10.1088/1757-899X/168/1/012098

Sintering of Ceramics with Microwave Radiation

V I Nefedov[1], P A Luchnikov[1], N A Trefilov[1] and K A Selenya[2]

[1]Moscow Technological University (MIREA), 119454, Vernadsky 78, Moscow, Russia

[2]Scientific Center of Applied Electrodynamics, 190103, Riga Avenue, 26, St. Petersburg, Russia

E-mail: xamdex@gmail.com

Abstract. The method of ceramics sintering in the EM-field of UHF band radiation (microwave heating) has been investigated. The paper presents analytical expressions to estimate temperature distribution in the material treated in normalized heating time and pulse duration with Bouguer and Fourier criteria applied. An installation for drying and sintering ceramic samples with microwave radiation (microwave-heating chamber of 30 dm^3 and a continuous wave magnetron of 3 kW) has been described.

1. Introduction

Microwave heating of the materials treated is carried out through the EM-wave energy absorption in the substance. Like radiation heating microwave heating is characterized by electromagnetic waves penetrating the material treated and providing its heating due to molecular friction of dielectric losses [1–4]. This method is highly distinguished from convective heating or heat-transferring ones [5-7]. Dielectrics, such as fluorine-polymers [8], have extremely small dielectric losses and are not actually heated as compared with organic semiconductors [9].

Microwave heating of materials has high efficiency in the technologies of microwave synthesis, continuous rubber vulcanization, ceramics and nano-ceramics sintering. The resulting effect of this leads to a considerable increase of chemical reaction rate as compared with the conventional ways of heating [10], thus providing a tenfold decrease of the technological process time [11].

2. Microwave heating model

The simplest microwave heating model is the plain-wave power absorption by semi-infinite dielectric lossy medium. In this case the volume density of heating power can be specified by the following:

$$D = \frac{1}{2} E_0^2 \exp(-2k_0 z \sqrt{\varepsilon\mu\sqrt{1+tg^2\delta}\,\sin^2\frac{\delta}{2}}),$$

where: E_0 – is the electric field intensity amplitude on the medium surface; z – is distance from the medium boundaries; k_0 – stands for the wave number for free space; ε,μ – are relative dielectric and magnetic permeabilities of the medium; $tg\delta$ – is the tangent of the medium dielectric losses' angle.

Fig. 1 represents the graphs of heating volume density P/P_{max} depending on the normalized coordinate z/z_0 at various $tg\delta$ values of the medium.

It is evident from the graph that microwave heating takes place unevenly in medium depth and the magnitude of the non-uniformity increases as the medium thickness and dielectric losses angle tangent increase. If the uniform temperature field in a stationary mode is required, one should determine the heating timeline taking into consideration the sample thickness. To a first approximation, the average complex relative dielectric permeability ε_m of the units threated under the sparse packing is determined by the Lichtenecker equation

$$\varepsilon_m = \varepsilon_{ou}^{\eta},$$

where: ε_{ou} – is the complex relative permeability of the treated units; η – is relative portion of the microwave-heating chamber volume taken up by the material being treated. When using sparse packing the average dielectric losses angle tangent decreases.

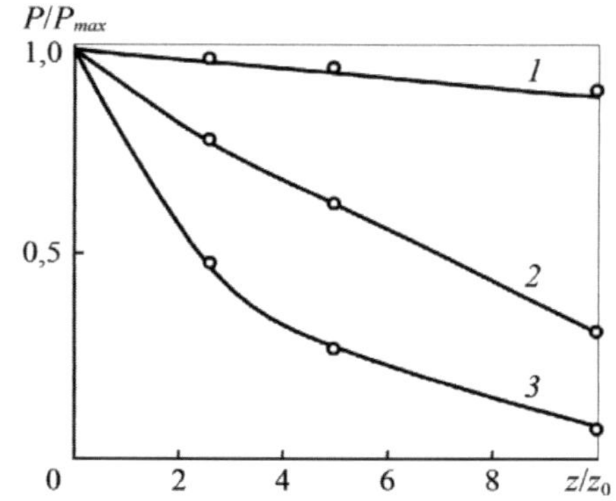

Figure 1. Heating relative volume density dependence P/P_{max} of z/z_0 normalized coordinate with medium $tg\delta$ values: 1 – 0,001; 2 – 0,1; 3 – 1,0.

The electromagnetic field power density carried by the wave inside the medium decreases as it travels away from the boundary and is described by the following ratio:

$$\frac{P}{S} = \frac{E_0^2}{2|W_c|}\exp(-2\alpha z)\cos\frac{\delta}{2},$$

where W_c – is the medium wave impedance.

The temperature distribution in the depth of the medium on microwave heating in the linear approximation satisfies the equation of thermal conductivity at radiation absorption according to the Beer–Lambert–Bouguerlaw which reads [12]:

$$\frac{\partial \vartheta}{\partial t} = a\frac{\partial^2 \vartheta}{\partial z^2} + \Im\frac{2\alpha}{c\gamma}\exp(-2\alpha z),$$

with the second-type boundary conditions

$$\frac{\partial \vartheta}{\partial z}(0,t) = \frac{\partial \vartheta}{\partial z}(l,t) = 0$$

where: ϑ – is the excessive temperature; $\Im$ – stands for the irradiance within the medium; a – is the medium thermometric conductivity factor; c – is specific heat capacity of the medium; γ – is the specific weight of the medium; l – stands for the thickness of the heated layer.

$$\vartheta(z,t) = T^0(z,t) - T_0^0, \quad \Im = \frac{P}{S}.$$

The solution to the equation depends on the irradiance time function type. At microwave heating, rectangular functions are commonly used, and for them the solution to the time interval $t < 1$ and $t > 1$ equation has the following form [12]:

$$\vartheta(z,t\le 1) = t + \frac{2}{Fo[1-\exp(-Bu)]} \times \sum_{n=1}^{\infty} \frac{[1-(-1^n)\exp(-Bu)]}{\pi^2 n^2\left[1+\frac{\pi^2 n^2}{Bu^2}\right]}[1-\exp(-\pi^2 n^2 Fot)]\cos(\pi n z),$$

$$\vartheta(z,t\ge 1) = 1 + \frac{2}{Fo[1-\exp(-Bu)]} \times$$

$$\sum_{n=1}^{\infty} \frac{[1-(-1^n)\exp(-Bu)]}{\left[1+\frac{\pi^2 n^2}{Bu^2}\right]}\{\exp[-\pi^2 n^2 Fo(t-1)] - \exp(-\pi^2 n^2 Fot\}\cos(\pi n z),$$

Here we use the normalized time t/t^* to the heating pulse duration t^* as well as Bouguer (Bu) and Fourier (Fo) criteria.

$$Bu = 2\alpha l,\quad F_0 = \frac{at}{l^2}.$$

Table 1 gives the calculated dependencies $\vartheta(z/l)$ for the time point $t = 1$, and Table 2 lists the surface heated temperature dependencies $\vartheta(z = 0,t)/\,\vartheta_{av}$ for different time points after heating completion in case of one-sided heating.

Table 1 - The temperature changes in the medium after microwave heating completion ($Fo = 10^{-2}$, $t = 1$).

	z/l	0	0,3	0,7	1,0
	10^{-2}	1,004	1,002	0,998	0,996
Bu	10^{-1}	1,043	1,02	0,9799	0,958
	$5{*}10^{-1}$	1,224	1,095	0,8967	0,8007
	∞	11,28	0,1725	<0,001	<0,001

Table 2 - Surface temperature change after microwave heating completion ($Fo = 10^{-2}$).

	t/t^*	1	2	3	4	6	8	10
	10^{-2}	1,004	1,004	1,003	1,003	1,002	1,002	1,002
Bu	10^{-1}	1,043	1,037	1,032	1,029	1,024	1,019	1,016
	$5{*}10^{-1}$	1,224	1,188	1,165	1,147	1,119	1,097	1,079

t – Time, t – Time of sample heating*

The data given in Table 1 and Table 2 are obtained for a linear model with the values of Fourier and Bouguer criteria typical for ceramics. One should account for the real materials having non-linear thermophysical properties, thus the thermal conductivity factor can vary in magnitude during the heating process [13]. Moreover, the calculated dependencies in Table 1 and Table 2 do not take into account chemical transformations and palasticizers and moisture gasification.

In order to compare the microwave heating and the convection one the last line of Table 1 shows the temperature change in the medium when applying heat to the surface only with the heating time being the same.

It should be emphasized that from the electrodynamics standpoint, determining field structures in microwave heating chambers appears to be difficult for a variety of reasons [1–4]:

– the size of heating chambers is substantially more than the wavelength of the electromagnetic oscillation, which implies that numerous coherent eigenmodes with unknown amplitudes exist in the chamber volume;

– the spectrum of the MW-heating chamber eigenmodes depends on the way the material treated fills the chamber. The material characteristics (its geometric form, its layout inside the chamber, complex relative dielectric permeability, etc.) are mainly defined by accidental factors and may alter during the treatment.

3. Experimental research

When considering simplified models of the microwave-heating chambers it is reasonable to use approximate calculation methods.

So as to define fields in the closed-type chambers one usually applies the chamber model of a cavity resonator. Provided walls of the resonator have a simple form, coincide with the coordinate surface and have infinite conductivity, the electromagnetic field in the cavity at the moment of resonance is the solution of the Helmholtz equation, satisfies simple boundary conditions and is therefore determinable. Possible field types in the heating chamber represent the eigenfunctions of the corresponding electrodynamic problem and are known for the parallelepiped and cylinder chambers. Therefore, the full-field in heating chambers of the forms mentioned above can be presented as a series expansion by the eigenfunctions. The expansion coefficients depend on the type and location of the exciter in the heating chamber.

It is supposed that if even distribution of the microwave heating power is achieved in an empty chamber, it will remain when the chamber is filled with the material treated. Such analysis approach is an eigenfunction method.

A lab installation has been designed for drying and sintering ceramics with microwave radiation. The installation comprises the microwave heating chamber of 30 dm^3 volume, which is excited by a continuous wave magnetron with maximum output oscillation power of 3 kW.

The material treated was set on applicators — thin ceramic tube segments.. Chamber volume filling coefficient is equal to ~ 0.1…0.2. To control the technological process a witness sample disk made of the material treated was placed inside the chamber. A tungsten-rhenium thermocouple was embedded in the sample with its outputs screened from the microwave field exposure. A radio-wave phasing tester of the factor of the probing EM-wave transmitting through the thicknessof the witness sample was used as an additional control system for the material treatment process. For this the sample was placed in the gap of the dielectric wave-guide, created with the ceramic tubes embedded in the measuring radio-wave channel. The measurements were carried out at the microwave field wavelength of 3 cm. Microwave heating and sintering of small-sized ceramic units have been demonstrated to be promising.

Acknowledgments
This work was supported by the Ministry of Education and Science of Russian Federation.

4. References

[1] Introduction to Microwave Sample Preparation: Theory and Practice Eds. H M Kingston, L B Jassie American Chemical Society, Washington DC, 1988
[2] Sample preparation in microwave ovens: Theory and practice. ed. by G M Kingston, L B Jesse. M.: Mir, 1991. 336 p (in Russian)
[3] Arkhangelsky Yu S, Devyatkin I I Microwave heating units for intensification of technological processes. Saratov: Saratov state University 1983 140 p (in Russian)
[4] Voronin M I, Karmanov I N, Koshelev A V Physics of heating of microwave dielectrics for aircraft and their protection. Under the General editorship of M. Y. Voronin. Novosibirsk: SSGA, 2008 156 p (in Russian)
[5] Aleutdinov A D, Ghygazov S A, Mylnikova T S, Luchnikov P A Thermal shock removal of defective glass-enamel coating from cast-iron products 2015 *IOP Conf. Ser.: Mater. Sci. Eng.* **81** 012069
[6] Luchnikov P A, Rogachev A V, Yarmolenko M A, Rogachev A A, Bespal'ko A A Dynamics of combined electron beam and laser dispersion of polymers in vacuum 2016 *IOP Conf. Ser.: Mater. Sci. Eng.* **110** 012081
[7] Surzhikov A P, Frangulyan T S, Ghyngazov S A, Lisenko E N, Galtseva O V 2006 *Russian Physics Journal* 49 5 506-10
[8] Luchnikov P A Structure formation mechanisms and electrical properties of PVD fluoropolymer films 2015 *Russian Physics Journal* **57** 9 1239–1245
[9] Rogachev A A, Yarmolenko M A, Xiaohong Jiang, Shen Ruiqi, Luchnikov P A, Rogachev A V Molecular structure, optical, electrical and sensing properties of PANI-based coatings with silver nanoparticles deposited from the active gas phase 2015 *Applied Surface Science* **351** 811–818
[10] Katz J D, Blake R D Microwave sintering of multipli alumina and composite components 1991 *Amer. Ceram. Soc. Bull* **70** 8 1304–1308
[11] Kubrakova I V, Myasoedova G V, Eremin S A Smple Preparation in the conditions of microwave heating 2006 *Methods and objects of chemical analysis* 1 27–34 (in Russian)
[12] Grigoriev B A, Need A V, Shibanov V B Tables for calculating transient temperatures of flat bodies under heating by radiation. M.: Nauka, 1971 708 p (in Russian)
[13] Krzhizhanovsky R E, Stern Z Y Thermophysical properties of non-metallic materials. L.: Energy, 1973. 334 p (in Russian)

Electromagnetic response of the three-layer construction on the basis of barium hexaferrite and a foam glass

V Suslyaev[1], O Kazmina[2] ,G Kuleshov[3], E Korovin[4] K Dorozhkin[5] and E Lebedeva[6]

[1] Assistant Professors,National Research Tomsk State University, Tomsk, Russia,
[2] Professors,National Research Tomsk Polytechnic University, Tomsk, Russia,
[3] Assistant Professors, National Research Tomsk State University, Tomsk, Russia,
[4] Assistant Professors, National Research Tomsk State University, Tomsk, Russia,
[5] Engineer, National Research Tomsk State University, Tomsk, Russia.
[6] Postgraduate, National Research Tomsk Polytechnic University, Tomsk, Russia

E-mail: susl@mail.tsu.ru

Abstract. This paper contains results of study of the frequency dependence of reflection coefficient of the ceramic surface coated with the construction consisting of following layers: metal, composite on the basis of ferrite with hexagonal structure, and foam glass. It is shown that foam glass layer reduces significantly the reflecting characteristics of the construction.

1. Introduction

In the recent years considerable attention is paid to absorbing materials which are widely applied to solve problem of electromagnetic compatibility at production of elements of high-frequency electronics, in the field of equipment of multifunctional screened rooms and defensive industry [1–5]. Given weight and dimensional characteristics, absorbing materials differ on broadbandness, efficiency of interaction with electromagnetic radiation. The main requirements to absorbers along with the maximal absorption of the microwave energy and its minimum reflection are mechanical strength of material, ecological stability and low density, in order to avoid increase of the weight of the protected object.

It is known that magnetic absorbing materials have the least thickness but range of their application is limited to conducting surfaces. However, the problems of protection of dielectric surfaces, for example, of ceramic ones are often met in practice. At the same time change of ceramics structure as result of introduction of additional fillers into structure is undesirable as it can change consumer properties of the product in whole. Such problem can be solved by applying conducting layer on the ceramic surface coated with composite materials on the basis of magnetic material. Carbonyl iron [6, 7] and ferrite with structure of spinel and garnet are the most widely used magnetic material. Unfortunately, the majority of magnetic materials work well in relatively low-frequency range while it is required to widen frequency area up to terahertz. Ferrites with hexagonal structure which have the greatest values of fields of magnetic crystallographic anisotropy, that determines frequencies of natural ferromagnetic resonance, are the most promising materials for application at high frequencies [8, 9]. To ensure efficiency of interaction with electromagnetic radiation it is necessary to match complex wave resistance of coating and air environment. As rule, lamination constructions with decrease of concentration of active phases are used for this purpose. As far as ferrites possess the larger specific weight, total weight of the construction increases. Application of tapered or wedge-like constructions sharply increases coating sizes. These shortcomings can be excluded, using foam glass which is light, long-lived, fireproof material as the matching layer [10, 11]. Variant of multilayer construction consisting of three materials including foam glass, ferrite composite and metalized ceramics is considered in this paper.

The work purpose is to define reflection coefficient of the three layer construction in wide frequency range from 20 to 260 GHz.

2. Raw materials and methods of investigations

The traditional method to obtain foam glass is the powder method based on expansion of foam forming mixture prepared from glass powder with gas forming agent. Use of waste glass as source raw materials solves important ecological problem of utilization of the part of household wastes. Broken glass of the most widespread types of glass such as tare glass and sheet one is the most suitable. Researches on utilization of broken electron-beam tubes, computer monitors and TVs screens are actively carried out, too [12]. One of the main requirements to source raw materials is constancy of chemical composition of glass since deviations from it result in instability of foam glass properties. In this work we have used broken glass presented by sub-standard production of glass-works for manufacturing of LED lamps. Glass has stable composition given in tab. 1 and differs from sheet glass and tare one by presence of barium oxide which is entered into batch by barium sulfate to increase expansion ability of glass melt.

From glass of given composition we have prepared fine ground glass powder with particle size less than 60 μm with addition of carbon-containing gas developing agent in the form of highly active soot with specific surface area of 16 m2/g. According to results of the works which have been carried out early, it has been established that to organize expansion process it is enough 0.5 wt. % of soot. In this experiment soot fulfills the role not only gas developing agent, but also electrically conductive filler which is not completely removed from material at expansion. Combination of porous structure of foam glass with presence of carbonaceous particles provides efficient absorption of electromagnetic wave. Therefore the content of soot in foam forming mixture has been increased up to 1 wt. %.

Table 1. The chemical composition of the glass.

Material	Oxides Content, wt. %							
	SiO_2	Al_2O_3	Fe_2O_3	CaO	MgO	Na_2O	K_2O	BaO
Glass	71,9	0,60	0,10	5,50	3,20	16,10	0,80	1,80
Oxides swing	± 1,5	–	–	± 0,5	± 0,3	± 0,1	± 0,2	± 0,2

The electromagnetic characteristics were investigated using the equipment of the «Center of Radio Measurements at Tomsk State University». Spectra reflection coefficient were measured by the method of free space using a radio spectroscope built around an E8363B Vector Network Analyzer (Agilent Technologies) in the range 20–40 GHz and an STD-21 teraherz spectrometer in the range 115–260 GHz.

3. Experimental results

We have used for measurements the flat samples of ceramics coated with copper foil layer with thickness of 0.5 mm. Reflection coefficient of such construction is practically equal to R= 0 dB – it is the total reflection. Applying of foam glass with thickness of 6.2 mm on this construction (fig. 1.1.) has decreased reflection coefficient almost twice (R= –3 dB) in the range of 20 - 40 GHz (graph.1, fig 2). It is connected with absorbing properties of foam glass which have been revealed earlier. It is connected with absorbing properties of foam glass which have been revealed earlier [11].

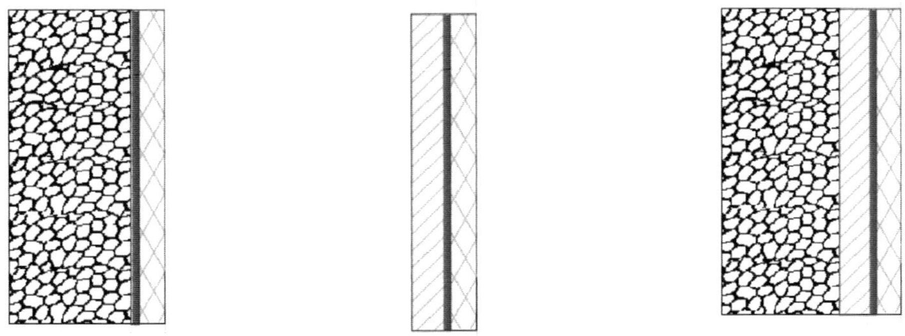

1) Foam Glass on Metallized Ceramics

2) Ferrite Composite on Metallized Ceramics

3) Foam Glass + Ferrite Composite on Metallized Ceramics

Figure 1. The principle structure of the investigated multilayered materials.

The curve 2 in the figure 2 corresponds to the construction shown in the figure 1.2 when the layer of magnetic absorber is applied on the metal layer. The absorber represents the composite in which active phase has been presented with ferrite powder of hexagonal structure – $BaFe_{12}O_{19}$ (60 % of total mass), and binding agent has been presented with water dispersion paint. The thickness of composite layer was 1.5 mm. Composite has been polymerized on air in 4 hours. The size of ferrite particles was less than 80 μm. The mixture has been stirred with the "Alena" ultrasonic disperser with action power of 20 W and with frequency of 22 kHz. The reflection coefficient is presented by oscillating function which reflects interference properties of such absorber which are defined with frequency dependences of magnetic and dielectric permittivity and thickness of the particular layer.

On average the reflection coefficient of the sample of ferrite composite on metalized ceramics differs from the coefficient for the sample of foam glass on metalized ceramics a little that is connected with high contrast of complex wave resistance of magnetic coating with free space, while the reflecting properties of foam glass are defined only by the absorbing properties. The reflection coefficient of the construction has been decreased considerably when foam glass layer has been applied on the layer of magnetic composite (fig. 1.3). In low-frequency area reflection has decreased in 10 times and at high frequencies in hundred and more times (graph 3, fig. 2). At that it should be noted the broadbandness of the created construction.

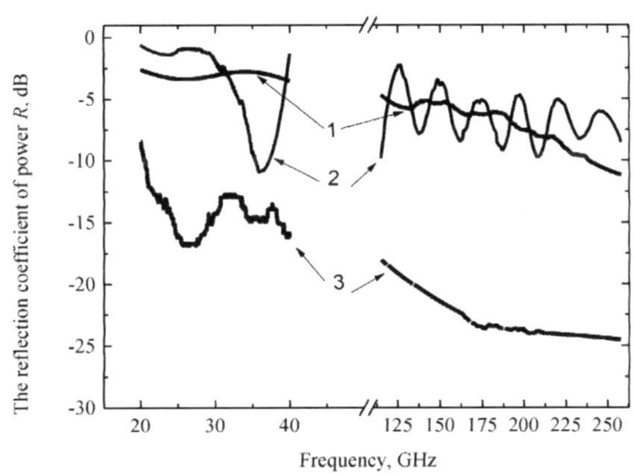

Figure 2. Spectra reflection coefficient of the samples.

4. Conclusions

The conducted research showed that ceramic coatings can be protected from influence of electromagnetic radiation in a wide frequency range using composition: metal, magnetic absorber and foam glass. This construction has relatively small weight, high performance of interaction with microwave, and low weight and dimensional characteristics.

Acknowledgments

Synthesis of materials is supported by Tomsk State University Competitiveness Improvement Program. Measurements of electromagnetic parameters are supported by Russian Federation Presidents grant MK-6957.2015.8.

References

[1] He Y, Gong 2009 *A Lett J Exp.* **85** 58003
[2] Qin F and Brosseau C 2012 *J. Appl. Phys.* **111** 061301
[3] Zhang X F, Dong X L, Huang H, Liu Y Y, Wang W N, Zhu X G, Lv B, Lei J, P Lee C G 2006 *Appl. Phys. Lett.* **89** 053115
[4] Suslyaev V, Kuznetsov V, Zhuravlev V, Mazov I, Korovin E, Moseenkov S, Dorozhkin K 2013 An investigation of electromagnetic response of composite polymer materials containing carbon nanostructures within the range of frequencies 10 MHz-1.1 THz *Russ. Phys. J.* **55** 8
[5] Suslyaev V I, Malinovskaya T D, Melentyev S V, Dorozkin K V *Int. Conf.* (Tomsk) vol 1040 (*Adv. Mater. Res.*) p 137
[6] Zhuravlev V, Suslyaev V, Dotsenko O, Babinovich, A 2011 Composite radio-absorbing material based on carbonyl iron for millimeter wavelength range *Russ. Phys. J.* **53** 8
[7] Zhang B, Feng Y, Xiong J, Yang Y, Lu H *IEEE Transactions on Magnetics* 2006 **42** 1778
[8] Kwon H J, Shin J Y, Oh J H 1994 *J. Appl. Phys.* **75** 6109
[9] Zhuravlev V, Suslyaev V 2006 Analysis and correction of the magnetic permeability spectra of $Ba_3Co_2Fe_{24}O_{41}$ hexaferrite by using Cramers-Kronig relations *Russ. Phys. J.* **49** 8
[10] Chen M, Zhu Y, Pan Y, Xu H 2011Guo J *Mater. Des.* **32** 3013
[11] Kazmina O, Dushkina M, Suslyaev V, Semukhin B 2014 Porous material for protection from electromagnetic radiation *AIP Conf. Proc.* **1623** 241
[12] He Y, Gong R 2009 *EPL* **85** 58003

IOP Conf. Series: Materials Science and Engineering **168** (2017) 012100 doi:10.1088/1757-899X/168/1/012100

Degradation of emission parameters in LiF-MeO crystal scintillators

L A Lisitsyna[1] and R N Suleimen[2]
[1] State University of Architecture and Building, Sq. Solijanaja 2, Tomsk, Russia
[2] L.N. Gumilyov Eurasian National University, St. Satpayev 2, Astana, Kazakhstan

E-mail: lisitsyna@mail.ru

Abstract. The paper presents the results of the research in spectral and kinetic parameters of the luminescence in phosphors based on LiF-MeO crystals. The potential of these phosphors used as scintillators in the ionizing radiation field is discussed.

1. Introduction

Scintillating materials based on oxygen compounds may be divided into two types: scintillators of a doped type and natural scintillators. LiF crystal is the scintillator of a doped type. The prospect of widespread using of this material based on the features of doped crystals structure making light output dopant emission comparable to that in natural scintillators [1, 2]. Application of oxygen-containing crystal phosphors as radiation detectors in medical diagnosis, customs control, detectors for fundamental science is being discussed in [3-5, 6, 7]. However, the potential of these materials as scintillators in the ionizing radiation field cannot be assessed since no data of the effect of radiation-induced defects in lattice crystals on scintillation characteristics of the material is available. (It is known that LiF lattice is very sensitive to ionizing radiation). Furthermore, information on radiation modification of the oxygen-containing impurity in the composition of the luminescence centers especially important for the understanding of the aging process in doped crystals placed in radiation field.

2. Object and Method

We used pulse spectroscopy methods to investigate LiF crystals doped with oxides of various metals (Me: W, Ti, Fe, Li). Imperfection of the crystals (initial and induced by ionizing radiation) was estimated through the analysis of the infrared absorption spectra in the range of 4000–1000 cm^{-1} at 300 K, pulsed photoluminescence (PPL) and pulsed cathodoluminescence (PCL) spectra in the spectral region of 6–1 eV and in the time interval of 10^{-8}–10^{-1} s after the action of exciting pulses. PPL was excited at 300 K by laser pulses with photon energy of 4.66 eV and 10 ns pulse duration. PCL was excited by single electron pulse (EP) with the following parameters: pulse width of 10 ns, the average electron energy of 250 keV. The magnitude of the electron fluence per pulse was varied in the range of 10^{10} – 10^{13} cm^{-2}.

Irradiation of the crystal was carried out by electron pulses with repetition rate of $2 \cdot 10^{-2}$ Hz in the temperature range of 100–300 K.

The temporal and spectral resolutions of the measurement path were equal to 20 ns and 2 nm, respectively. The transparency cutoff of undoped LiF crystal and all doped crystals LiF was found in the UV region near 12 and 6 eV, respectively. The IR spectra of the LiF crystals doped with oxides of W, Ti or Fe contained a band at 3730 cm^{-1} that belongs to hydroxide, and a number of narrow bands in the region of 3500–3600 cm^{-1}, which belong to Me-OH complexes. The LiF-Li_2O crystals were transparent in the IR region up to 1000 cm^{-1}.

Under exposure to laser radiation with photon energy of 4.66 eV, in all the as-grown crystals the PPL at 3.1 eV occurred 'figure 1'. It was inertialess relative to the excitation pulse.

The kinetics of the dopant PPL decay at 300 K is described as a sum of two exponential functions:

$$J(E,t) = J_{01}\exp(-\frac{t}{\tau_1}) + J_{02j}\exp(-\frac{t}{\tau_{2j}}), \quad (1)$$

where: J_{01}, J_{02j}, and τ_1, τ_{2j} are the peak values of the emission components intensity and the magnitudes of the characteristic decay times, respectively 'figure 1b'; j is type cation-dopant (Li, Ti, W, Fe); E is the energy of the emitted photon. The inequality observed for all the crystals are $\tau_1 << \tau_{2j}$ and $J_{01} > J_{02j}$.

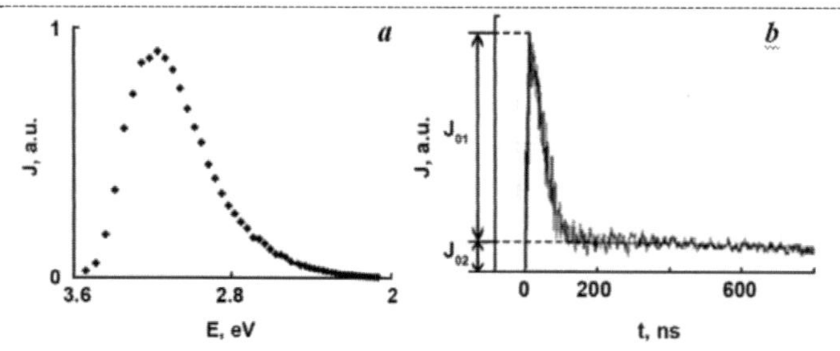

Figure 1. PPL spectrum at 300 K of the as-grown LiF-WO_3 crystal (a), and the kinetics of PPL decay at 2.8 eV (b).

The contribution of the PPL long-living decay component in amplitude value of emission intensity at 3.1 eV does not exceed 20% at 300 K.

The spectra of the PPL decay components are found to be different 'figure 2'. The peak position of the short-living dopant emission band (3.1 eV) with FWHM equal 0.5 eV 'figure 2a' and the emission duration ($\tau_1 \leq 20ns$) do not depend on the type of the cation-dopant. The peak position of the long-living dopant emission band depends on the type of the dopant and occurs at 2.7 (LiF-WO_3), 2.8 (LiF-TiO_2), 2.9 (LiF-Fe_2O_3) and 3.1 эВ (LiF-Li_2O) at 300 K 'figure 2b'.

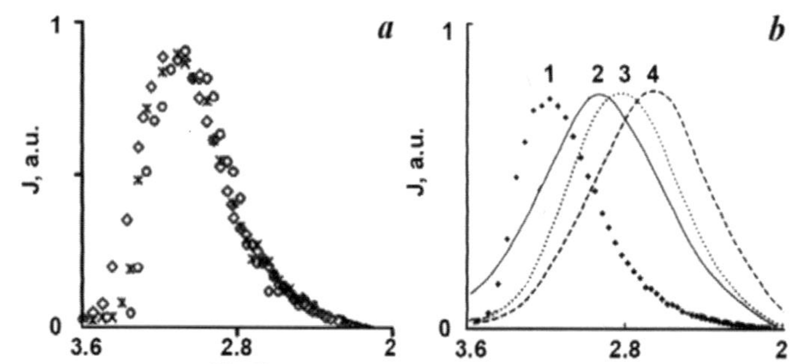

Figure 2. PPL spectra of short-living (a) and long-living components (b) at 300 K measured with 10 ns (a) and 10 μs time delay with respect to the end of the laser pulse action in as-grown LiF crystals doped with Li_2O (1), Fe_2O_3 (2), TiO_2 (3) and WO_3 (4). The spectra are normalized.

The luminescence spectra of the decay components were measured as $J_{0i} = f(E)$, where J_{0i} is amplitude value of the intensity of the respective component at the emission energy E.

The duration of the long-living dopant emission component PPL decay depends on the cation dopant type. It is shown that when its mass reduces, the dopant emission decay time increases 'figure 3'.

3. Results and Discussion

1.1. Luminescence spectra of the preliminary irradiated doped crystals

Exposure to ionizing radiation (flux of 250 keV electrons) of the crystals causes formation of intrinsic primary electron and hole color centers (CCs) F_l and F_i^o, where F_l is an electron localized at anion vacancy (V_a), F_i^o is a fluorine atom in interstitial state. At $T \geq 250K$ addition to primary CCs complex CCs are created (F_2, F_2^+, F_3^+...) under ionizing radiation of crystal. In PPL spectrum in addition to dopant emission bands new bands at 2.33 and 1.85 eV belonging to radiative transitions in the F_3^+ and F_2 CCs, respectively, occur. The characteristic decay time of the additional bands coincides with the laser pulse duration. Figure 4 shows the PPL spectrum of the LiF-Li_2O crystal irradiated by electrons beam at 300 K and measured with time delay of 10 ns (curve 1) and 10 μs (curve 2) with respect to the end of the laser pulse action. (In both spectra, the bands at 3.1 eV belong to short-living and long-living components of the dopant center PPL. F_3^+ and F_2 emission bands positions are marked with arrows).

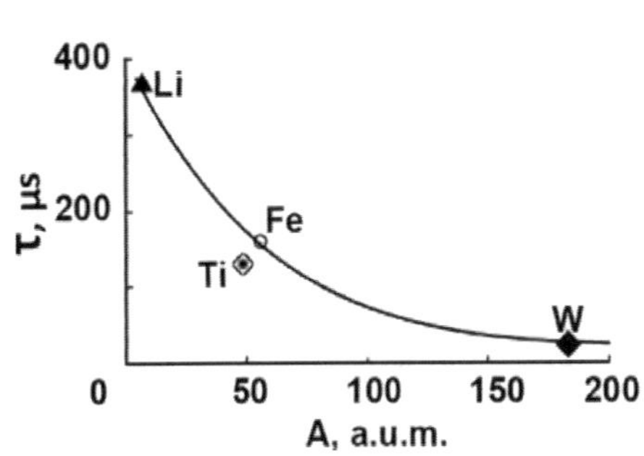

Figure 3. Dependence of the decay time of the long-living dopant PPL emission at 300 K on the cation-dopant mass.

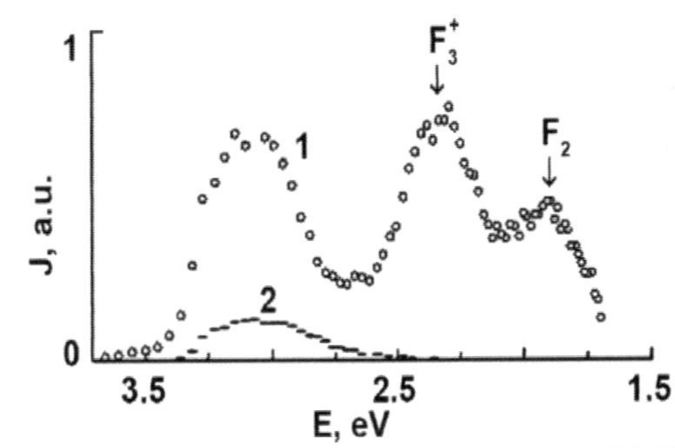

Figure 4. PPL spectra of the irradiated LiF-Li_2O crystal at 300 K measured with a time delay of 10 ns (curve 1) and 10 μs (curve 2) with respect to the end of the laser pulse action. D=10^3Gy. (F_3^+ and F_2 emission bands are marked with arrows).

It was found that the absorption band of F_2 and F_3^+ CCs (2.75 eV) and emission long-living band (at 3–2.7 eV in different crystals 'figure 2b') occur in the same spectral region. The dopant emission is revealed to be reabsorbed by CCs that causes degradation of the dopant emission band 'figure 5'. The reabsorption efficiency depends on CCs concentration and the degree of the CCs absorption band and dopant emission band overlapping (minimum overlapping of these bands occurs in LiF-Li_2O crystal).

It is evident from figure 5 that emission intensity due to degradation of the dopant emission band decreases in the region of 2.75 eV, i.e., in the region of maximum sensitivity of phosphor to radiation. The effect increases if concentration of F_2 (F_3^+) CCs increases with increasing absorbed dose by crystals.

To avoid a decrease in the sensitivity of the scintillator due to the reabsorption of the dopant emission by radiation defects it is necessary to decrease the temperature of operation of the scintillator (at $T < 250K$ the efficiency of creating F_2 and F_3^+ CCs is low), or enter a limited amount absorbed by the scintillator dose D_0, below which no dose dependence of emission intensity occurs. The following condition will be satisfied: if $D < D_0$, $J \neq f(D)$.

In addition to CCs formation, exposure to ionizing radiation of LiF-MeO (Me: W, Ti, Fe) crystals results in the destruction of hydroxide ions followed by creation of molecular O_2^- ions in an emission state with the absorbed dose-dependent efficiency. In the LiF-Li_2O crystals transparent in the IR spectral region, ionizing radiation does not induce creation of molecular oxygen ions.

The PCL spectrum of the molecular oxygen ions is excited within the nanosecond EP at $T \geq 100K$ and consists of ten equidistant bands with 0.15 eV band spacing 'figure 6'. PCL of O_2^- ions is short-living and the value of the characteristic decay time at 300 K is 40 ns. The O_2^- ion emission cannot be optically excited with quantum energy of 4.66 eV in LiF-MeO crystals. Figure 6 shows, for example, the PPL and PCL spectra for irradiated LiF crystal doped with iron oxide measured in a nanosecond time interval after the end of the excitation pulse action.

Result presented in figure 6 shows that in one the same irradiated crystal spectra PPL and PCL are different. PPL spectrum consists of dopant band at 3.1 eV and the bands at 2.33.and 1.85 eV belonging to F_3^+ and F_2 CCs, respectively. PCL spectrum consists of the CCs bands at 2.33 and 1.85 eV and electron-vibronic bands belonging to O_2^- ions. Week dopant bands at 3.1 and 2.8 eV are coved with emission of O_2^- ions.

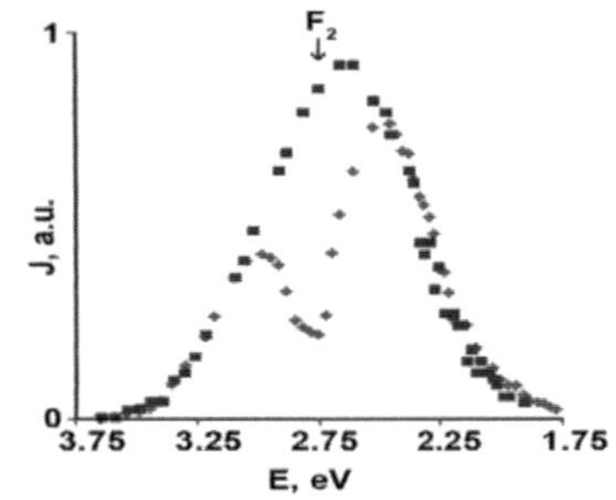

Figure 5. PPL spectra of the LiF-WO_3 crystal before (upper curve) and after irradiation at 300 K (lower curve). D=4.10^{4}Gy. (F_2 absorption band position is marked with arrow).

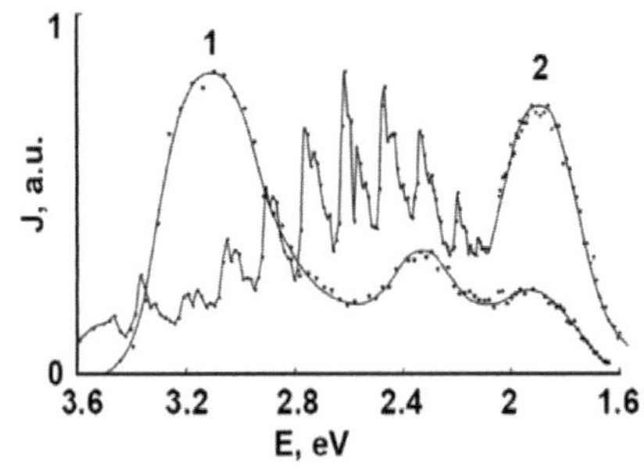

Figure 6. PPL (1) and PCL (2) spectra at 300 K for irradiated LiF-Fe_2O_3 crystal measured with 10 ns time delay after the end of the excitation pulse action. D=6.10^{4}Gy.

1.2. Effect of irradiation temperature on PCL kinetic parameters

The temperature dependence of the characteristic time of the short-living dopant emission with the spectrum in the form of the band at 3.1 eV is shown in figure 7a. The character of dependence and the decay constant values recorded at different temperatures do not depend on the cation dopant type, and these are identical for all the investigated crystals. The temperature dependences of the characteristic decay time of the long-living dopant emission in LiF crystals doped with tungsten or titanium oxide are shown in figure 7b.

It is found that the temperature of the crystal under irradiation affects the kinetics of the long-living dopant luminescence buildup. If $T < 250K$, for all crystals the PCL buildup occurs within a nanosecond electron pulse action 'figure 8a'. In the range 270–300 K, an additional post-radiation buildup stage is observed in all the studied crystals 'figure 8b'. The common feature of the crystals is their non-transparency in UV energy ≥6 eV, which is largely associated with the presence of oxygen- containing impurities.

So, post-radiation dopant emission buildup is observed only in the PCL spectra and absent under optical excitation of the same irradiated crystals. The PCL kinetics at the long-living dopant band at, $T \geq 270K$, compared with PPL decay kinetics described in equation (1), can be described as follows:

$$J(E,t) = J_{01}\exp(-\frac{t}{\tau}) + \left[J_{02j} + J_{03j}(1-\exp(-\frac{t}{\tau_{3j}}) \right] \exp(-\frac{t}{\tau_{2j}}) \quad (2)$$

where τ is emission decay time (ns) of three types of short-living centers in excited state (F_3^+ CCs, O_2^- and O^{2-} ions) which emission bands overlap with long-living dopant emission band; τ_{3j} τ_{2j} are the characteristic time of long-living dopant emission buildup and decay, respectively.

Thus, inertialess buildup of dopant emission occurs in the excitation of the crystals as both laser and electron pulses 'figure 1b' and 'figure 8a'. However, stage of post-radiation buildup dopant emission occurs only when crystals are excited by EP at high temperature. It means that post-radiation buildup is not an inherent property of the dopant and it occurs due to energy transfer from lattice to the dopant with participation of radiation defects which are mobile in temperature region of ≥270 K.

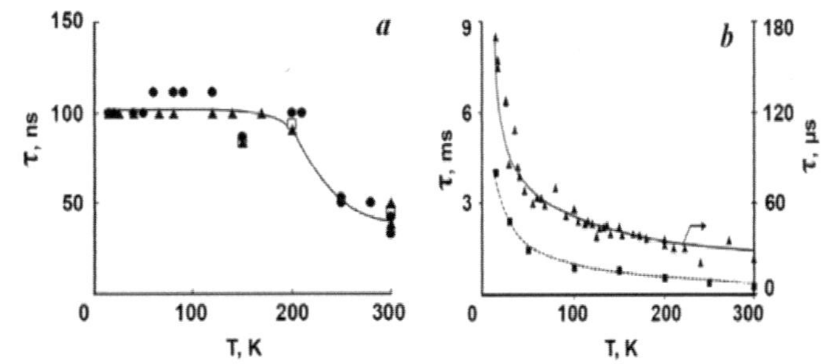

Figure 7. Temperature dependence of the characteristic decay time for short-living dopant emission at 3.1 (a) and long-living dopant emission (b) in LiF crystals doped with WO_3 (upper curve) and TiO_2 (lower curve).

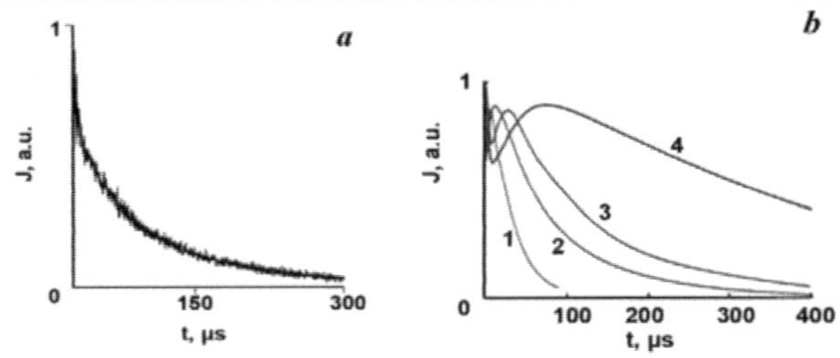

Figure 8. PCL dopant kinetics at 250 K for LiF doped with TiO_2 (a) and at 300 K in LiF crystals doped with WO_3 (1), Fe_2O_3 (2), TiO_2 (3) and Li_2O (4) (b).

A superposition of delay emission process, realizing with delay parameter τ_{2j}, and process of the post-radiation buildup, realizing with buildup parameter τ_{3j}, leads to degradation of the emission pulse form 'figure 8b'and the increasing of duration PCL pulse. In one and the same irradiated crystal the duration of PCL dopant emission pulse longer than the duration of PPL dopant emission pulse due to stage of post radiation buildup of the dopant cathodoluminescence 'figure 8'.

The nature of the doped crystal emission can be interpreted as follows.

1. The band in the region of 3.1 eV belongs to allowed intracenter transition on free O^{2-} ions being present in all the studied crystals can be described by equation (3):

$$O^{2-} + E \rightarrow \left(O^{2-}\right)^{*} \xrightarrow{\tau_1} O^{2-} + h\nu_0, \qquad (3)$$

(In equation (3) $h\nu_0 = 3.1\text{eV}$, τ_1 is time decay allowed emission transition in free O^{2-} ion).

2. The long-living dopant emission with spectral-kinetic parameters, which depend on the type of the cation-dopant, is the result of allowed intracenter transition on O^{2-} ion being located in the $Me^{n+}mO^{2-}$ complex, where n is charge state of dopant in crystal lattice, m is the number of oxygen ions coordinated with Me^{n+}. The structure of the impurity complex defines both the value of the energy emitted by oxygen ion and the probability of the radiative transition. Long duration of the allowed emission transition on O^{2-} ions is determined by the duration of the intermediate reversible transfer of the charge from $(O^{2-})^*$ ion to Me^{n+} ion and can be described by equation (4):

$$Me_j^{\,n+} mO^{2-} + E \rightarrow Me_j^{\,(n-1)+}(m-1)O^{2-}O^{-} \xrightarrow{\tau_{2j}} Me_j^{\,n+}(m-1)O^{2-}(O^{2-})^{*} \xrightarrow{\tau_1} Me_j^{\,n+} mO^{2-} + h\nu_j \,, \qquad (4)$$

where: Me_j and n are the type cation-dopant and its charge state in lattice, respectively (Me_j: W, Ti, Fe or Li and n: 6, 4, 3, 1, respectively).

For Me_j: W, Fe, Ti the observed inequality is $h\nu_j < h\nu_0$. Apparently the value of ΔE_{em} equal to $(h\nu_0 - h\nu_j)$ is the characteristic of an energy loss in the process of lattice relaxation to create the appropriate type impurity complex MeO during LiF crystal growing. According to results presented in figure 2b, value ΔE_{em} is equal to 0 (in LiF-Li_2O), 0.2 (in LiF-Fe_2O_3), 0.3 (in LiF-TiO_2), 0.4 eV (in LiF-WO_3). The value of ΔE_{em} is determined not only by type on cation-dopant but also the number of oxygen ions coordinated with it.

In equation (4) the observed inequality is $\tau_{2j} >> \tau_1$. The value of τ_{2j} is determined by original valence of the dopant; the dopant mass; the degree of overlapping of the wave functions of the cation-dopant and the oxygen ion in the excited state; duration of the transfer of hole (O^-) between equivalent (or nonequivalent) O^{2-} ions coordinated with Me^{n+}; a degree of lattice relaxation during the charge transfer to the cation-dopant. (The fact that the electron of the excited O^{2-} ion is easily transferred to the neighboring defect having an effective positive charge and that this reaction is reversible are proved in [8]).

4. Conclusions

In the process operation of phosphorus based on crystals LiF-MeO in the field of ionizing radiation a decrease of light output of the scintillators and a change in the parameters of the scintillation pulse occur with increasing absorbed dose.

Reducing the light output due to the decrease in the intensity of the long-living dopant emission is caused by the reabsorption of the dopant emission by intrinsic radiation-induced defects, which concentration increases when the absorbed dose grows. The external parameters such as the value of the absorbed dose or the temperature of the crystal by irradiation may lead to an invariance of the dopant emission. So, if $T \leq 250K$, ionizing radiation virtually does not lead to creation of complex electron CCs, and hence, it does not cause dopant emission reabsorption. In the operation of the scintillator in the field of ionizing radiation at $T \geq 250K$ post-radiation dopant emission buildup of the PCL occurs. Overlapping emission decay and emission buildup results in changing the shape of the scintillation pulse in PCL and, as a result, the increase in the duration of the scintillation pulse.

The duration of the scintillation pulse can be maintained stable through the introduction of restrictions on the temperature of crystal operating in field of ionizing radiation. At $T \leq 250K$ no emission buildup process can be observed. The radiation fields with photon energy less than the band-gap are considered to be most favorable for operation of scintillators based on LiF-MeO. This provides the stability of the emission pulse duration and high light output.

References

[1] Lisitsyna L, Lisitsyn V 2013 *J. Physics of the Solid State* 55 2297
[2] Lisitsyna L A, Tupitsyna I A, Trefilova L N 2015 *Proc. IOP Conf. Series: Materials Science and Engineering* **81** 012024
[3] Ronda C 2008 *Luminescence. From Theory to Applications* (Weinheim: WILEY-VCH Verlag GmbH & Co. KGaA)
[4] Nikl M, Laguta V, Vedda A 2008 *J. Phys. Stat. Sol. b* **245** 1701
[5] Korzhik M V 2003 *Physics of scintillators based oxygen-containing single crystals* (Minsk: BSU) p 263
[6] Nowotny R 2004 *J. Phys. Med. Biol. b* **49** 2599
[7] Atroschenko L, Burachas S, Galchinetsky L, Grinyov B, Ryzhikov V and Starzhinsky N 1998 Scintillator crystals and detectors of ionizing radiation based on them (Kiev: Naukova Dumka) p 310
[8] Egranov A V and Radzhabov E A 1992 Spectroscopy of Oxygen and Hydrogen Centers in Alkali_Halide Crystals (Kiev: Naukova Dumka) p 160

Impact of nanosecond proton beam processing on nanoblocks of copper

Y V Borodin[1], A Y Mantina[1], V Pak[2] and X X Zhang[3]
[1]Tomsk Polytechnic University, 30 Lenina Av., Tomsk, 634050, Russia,
[2] Kemerovo State University, 6 Krasnaya St., Kemerovo, 650043, Russia,
[3]Harbin Institute of Technology, 11 Siling St., Nangang District, Harbin, 150001, China

E-mail: uryborodin@tpu.ru

Abstract. X-ray studies in conjunction with the method of recoil nuclei and electron microscopy of irradiated plates polycrystalline Cu by nanosecond high power density proton beams (E = 120 keV; I = 80 A/cm^2, t = 50 ns) showed nano block nature of the formation of structure in the surface layer target and condensed-formed film.

1. Introduction

Analysis of the X-ray diffraction and electron microscopy pictures of the crystals' electron density distribution in the surface layer includes establishing the causes of the contrasting line [1, 2], nano blockiness [3], modulating the structure [4] and other phenomena caused by various influences [5, 6]. A number of technical difficulties obstruct the direct use of formulas to estimate the scattering intensity of X-ray and electron beams at a certain point of space [7]. We calculate only the integral intensities of the diffraction maxima caused by the small size of the crystallites, the ultimate highs wide, non-periodical scattering centers, a large beam divergence and etc. [7-9]. When we calculate the theoretical intensity of the polycrystalline substances that consist of nano blocks (≤1000), we use crystal mosaic model. However, a good matching is observed for weak reflexes that were calculated by the kinematic theory. Differences are associated with mismatched shapes and sizes of coherent scattering and nano blocks. The integral intensities calculation of the diffracted crystal layers doped with protons, which the authors conducted [10], showed a decrease in these differences. As in the model of crystal mosaic, nano blocks in protonated layers are slightly disoriented, similar in size of coherent scattering regions and have few defects. It seems necessary to study the features of X-ray diffraction by nano block structure of the surface layers doped with protons and condensed films.

2. Experimental methods

Irradiation of polycrystalline Cu plates by nanosecond high power density proton beams (E = 120 keV; I = 80 A / cm^2, t = 50 ns) led to increasing proton concentration to a depth of 10 microns, depending on the resulting number of pulses. The nuclear-recoil method found that in the layer depth ≤ 1.2 m at 5 irradiating pulses the concentration of hydrogen nuclei ^{1}H reaches $8{\cdot}10^{19}$ cm^{-3}. X-ray measurements were performed on a DRON-3 diffractometer in monochromatic Cu-Kα eradiation by sliding-beam method.

3. Results and Discussion

Except reflexes CuO, Cu_2O and face-centered cubic-Cu strong oscillations of the background and the redistribution of the intensity of reflection (111) Cu are observed on the radiogram of X-ray irradiated plates (Figure 1). It is taken into account that the integrated intensity of the diffraction reflections is proportional to the thickness of the film and the penetration depth of X-rays at a target. Depending on the number of irradiated pulses (N) and the observed values of X-ray modulation periods (ν), calculated values of ribs of nano blocks Cu (α) which are based on the Fibonacci sequence of steps (F_n) are arranged in series (Table 1).

Table 1. The sizes of nano blocks after radiation of plates of copper.

N, pcs		1			2		5		11	31
ν, Å	118	400	501	133	800	71.4	400	48.7	43.8	200
F_n	7	9	10	7	11	6	9	5	5	8
a, Å	110	289	468	110	757	68.3	289	42.1	42.1	179

In the series (Table 1) fluctuations in the calculated values of the size of nano blocks based on the step of Fibonacci sequence from periods of the observed X-ray modulation ranges from 2 to 15%, which may be due to the presence of nano blocks CuO, Cu_2O and inaccurate selection of modulation oscillations. Despite the differences in the experimental and calculated values of nano blocks' sizes, we can take a structural element (SE) with the volume V as the minimum volume of scattering X-ray diffraction

$$V = V_{e.c.}\sqrt{r}\ , \qquad (1)$$

where $V_{e.c}$ – elementary cell volume;
r = 1,618 - the golden ratio.

Use of scattering volume of the SE in the expression of the kinematic theory leads to the coincidence of the calculated and experimental values and intensities. That is indicated by the authors in [10] for polycrystalline Al. Applying experimentally determined value of the interplanar distance of the most densely packed plane with a maximum intensity line (dmax) we can determine the size of the SE from

$$A = d_{max} N^{1/2} \tau^{1/6} K^{-1/3}\ , \qquad (2)$$

where $N=H^2+K^2+L^2$ – is the sum of square indexes of high packed plane;
K - is a shape coefficient of nanoparticles (K_{cub} = 1; K_{oct} = 0.4714; K_{tetr} = 0.1222179).

X-ray diffraction analysis of condensed films generated by spraying Cu on K-8 glass substrates were carried out by the method of the moving beam. We analyzed the intensity distribution of the diffracted beam in the reflection area of (111) Cu.

Depending on the number of pulses irradiating a target, changes in the modulation period values are observed on the X-ray films. The calculated value of the edge lengths with a = 5.64 Å , that are included in the cubic-shaped nano blocks of the SE varies for the series (Table 2).

Table 2. The sizes of nano blocks after radiation of the condensed copper films on substrates from K-8 glass.

N, pcs	1			2		5		11
v, Å	71.4	133	200	71.4	200	48.8	160	501
F_n	6	7	8	6	8	5	8	10
a, Å	68.3	110	179	68.3	179	42.1	179	468

The series (Table 2) shows that some percentage of the nano blocks from the SE maintained their sizes when transferring substance to a substrate. The amount of substance transferred for 31 pulses forms typical film structure of polycrystalline copper. The first layer (single-pulse irradiation of the target) is divided into small parts when it is separated from the substrate. This points at a poor adhesion of deposited and periodically connected nano blocks of the SE with the substrate. Maximum continuity of condensable films is achieved when a target is sprayed with two pulses. Modulation with a period of 71.4 Å observed in the single-layer and double-layer films can be attributed to the presence of the densest hexagonal packings of atomic layers in the transition between the nano blocks and the substrate areas. Hexagonal structure of these transitional layers' material with the substrate is observed at the electron- microscopic picture. Long-wave modulation of the film lattice increases in the multi-pulse mode and voids and cracks appear. When target is sprayed with five pulses, growth of crystal Cu, with size of the SE nano blocks is 199.8 Å begins in the film. A further increase in the number of pulses provides formation of polycrystalline copper with a modulation period 499.5 Å in the film.

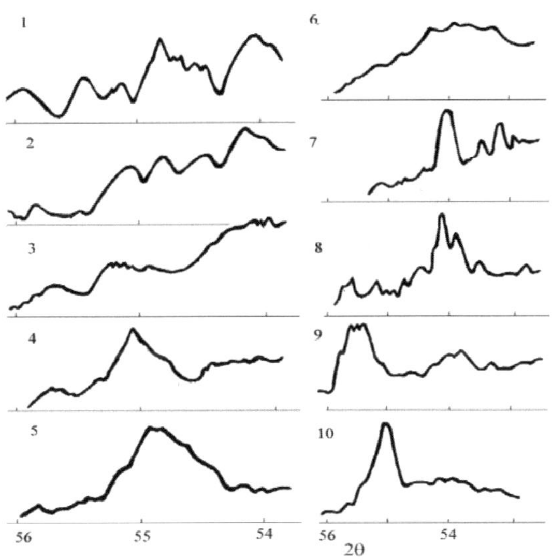

Figure 1. A portion of the Cu X-ray picture near reflex (111) of the subsurface layer of the target (5.1) and condensed film (6-10) during irradiation with nanosecond pulses of protons (pieces) 1 (1,6) 2 (2,7) 5 (3,8), 11(4,9), 31(5,10).

4. Summary

Thus, the conducted radiographic study in conjunction with the method of recoil nuclei and electron microscopy showed the nano block nature of the structure formation in the surface layer of the target and in the formed condensed film.

Due to the high concentration of protons, it can be noted with sufficient approximation that micro blocks mostly are formed from the structural elements.

References

[1] Cowley J M 1995 *Diffraction Physics, third revised edition* (Elsevier) DOI:10.1016/B978-0-444-82218-5.50024-9

[2] Chang S L 1998 *Multiple Diffraction of X-Rays in Crystals* (Moscow: Peace) p 234

[3] Borodin Y 2011 *Proceedings of the 6th International Forum on Strategic Technology, IFOST 2011* **1** 218–21

[4] Iziumov I A 1987 *Neutron diffraction on long periodic structures* (Moscow: Energoizdat) p 200

[5] Ghyngazov S A, Vasil'ev I P, Surzhikov A P, Frangulyan T S, Chernyavskii A V 2015 *Technical Physics* **60** 1 128-32

[6] Surzhikov A P, Frangulyan T S, Ghyngazov S A 2014 *Journal of Thermal Analysis and Calorimetry* **115** 2 1439-45

[7] Iveronova V I and Revkevich G P 1978 *Theory of X-ray Scattering* (Moscow: Moscow State University Press) p 278

[8] Rezvova M A, Zhevnyk V D, Pak V, Borodin Y V and Kachina E V 2016 *IOP Conf. Ser.: Mater. Sci. Eng.* **110** 012073

[9] Borodin Y V, Ermolaev D S, Pak V and Zhang K 2016 *IOP Conf. Ser.: Mater. Sci. Eng.* **110** 012072

[10] Borodin Y V and Sergeev A N 2008 *Proceedings of the 3rd International Forum on Strategic Technology, IFOST 2008* 174-76

The effect of temperature on the properties of grapheme oxide langmuir films

E Seliverstova[1], **N Ibrayev, R Dzhanabekova, V Gladkova**
Institute of Molecular Nanophotonics, Karaganda 100028, Kazakhstan

E-mail: genia_sv@mail.ru

Abstract. The effect of temperature on the optical properties and structure of graphene oxide langmuir films was studied. It is shown that the decrease of optical density of the films in the absorption band of graphene oxide take place when the film was heated above 100 °C. Increasing the temperature of the films over 200 °C leads to the growth of its optical density. As it was shown with the scanning electron microscopy the structure of grapheme oxide films annealed at different temperatures is practically unchanged. Measured the Raman spectra show that in the films there are a partial reducing of disordered and high-defective graphene oxide LB films occurs under heat treatment. This process accompanied by ordering of amorphous sp^3 clusters and its transition in sp^2-bonded carbon clusters.

1. Introduction

Many studies demonstrate that the optical and structural properties of graphene and coatings based on it are not only depending on acidity of medium but on its temperature too. For example, in the work of [1] it has shown that high temperature and chemical treatment of graphene films is not only leads to a decrease in their resistance, but also to increase of their transparency. As it was demonstrated in works of [2, 3] this is due to partial reduction of graphene oxide. Since LB film based on graphene will be subjected to not only of light but also to heat exposure in solar cells, it is necessary to examine how these factors will affect their optical and structural properties. Unfortunately, at present there are practically no studies on the effect of temperature on the structural and optical properties of graphene oxide Langmuir-Blodgett (LB) films SLGO.

2. Experiment

Single layer graphene oxide was used for dispersion preparation (SLGO) (Cheaptubes, USA). The suspensions of graphene oxide were prepared in dimethyl formamide (DMF, Sigma Aldrich) by ultrasonication. and used without purification. Solid films were obtained by Langmuir–Blodgett (LB) technology. Monolayers were formed on the surface of the subphase by spreading of solution in the LB through (KSV NIMA Medium). The deionized water was cleaned by AquaMax and was used as subphase. The resistivity of the deionized water was equal to 18.2 $M\Omega \cdot cm^{-1}$. The surface tension of water was equal to 72.8 mN/m at pH=5.6 and temperature 22 °C.

Obtained films were heated to various temperatures from 50 to 250℃ in the muffle furnace during 1 hour/ Temperature was changed in increments of 25℃. Absorbance and transmittance of LB films after annealing was measured with using of Cary-300 spectrophotometer (Agilent).Measurements of

[1] To whom any correspondence should be addressed.

Raman spectra were carried out with the spectrometer Solver Spectrum under laser excitation with wavelength of generation equal to 473 nm in the National Nanotechnology Laboratory of the Al-Farabi Kazakh National University

3. Results and discussion

The dependence of optical density and transmission via temperature of graphene oxide LB films is shown on the Figure 1.

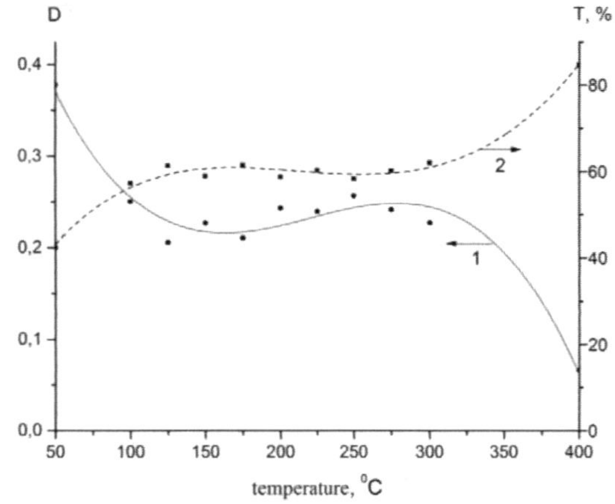

Figure 1. Dependence of optical density (curve 1) and transmittance (curve 2) of SLGO LB film via temperature (λ_{reg} = 226 nm).

As it can be seen, when the SLGO LB films was heated up to 100 ° C there is a twice reduction of graphene oxide absorption. Further annealing of films leads to the decreasing of D values by 10 – 15% for every 25 °C. At temperatures above 200 °C the changes in optical density was not registered. In this case, as it was seen with the naked eye, the color of the film was changed from dark brown to black and gray, indicating on partial reduction of graphene oxide to graphene. With this process a slight increase in the optical density of the film in the temperature range from 225 to 275 °C is connected, apparently.

Film heating up to 400 °C leads to a significant decrease in the absorbance of SLGO, which is associated with the destruction of the LB film. The transmittance of the film behaves in the opposite way. Thus, the annealing improves the transparency of the film from 40 up to 60%.

SEM studies reveals that the structure of film SLGO heated at different temperatures is almost constant (Figure 2).

In order to explain the decrease in absorbance of LB films of graphene oxide the Raman spectra were recorded. The measurement results are shown in Figure 3 and Table 1.

Table 1. Position and intensity of Raman bands of SLGO in LB films after heating.

Temperature, °C	D, cm^{-1}	I, a.u.	G, cm^{-1}	I, a.u.	I_D/I_G	2D, cm^{-1}	I, a.u.	I_{2D}/I_G
0	1360	5279	1602	6979	0,76	2740	1767	0,25
125	1360	3900	1602	4920	0,78	2720	1469	0,29
200	1360	5751	1602	7372	0,79	2720	1916	0,26

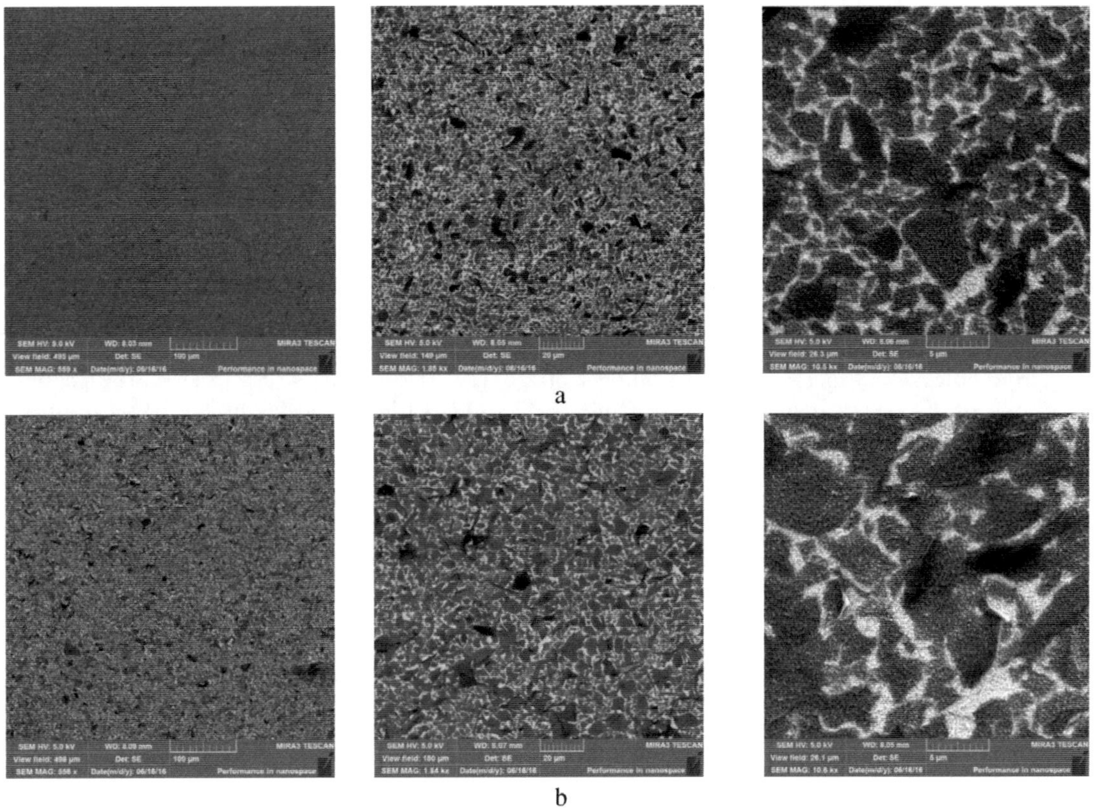

Figure 2. SEM images of SLGO LB films after heating at temperature of: (a) 50 ℃, (b) 200 ℃.

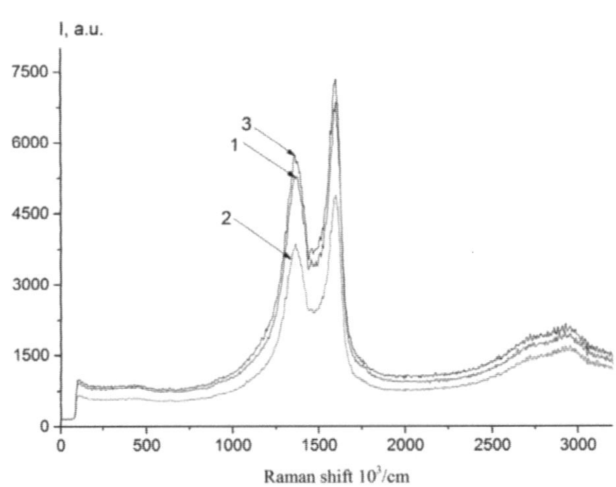

Figure 3. Raman spectra of SLGO LB films before (1) and after heating at temperature of: 2 – 125 ℃, 3 – 200 ℃.

There are several bands can be clearly distinguished in the Raman spectra of the synthesized LB films. As a rule, the most informative are G and D-bands [4, 5]. The table show that the G-band of graphene oxide is appears at 1602 cm^{-1}. This position is shifted to higher frequencies compared to the position of the G-band of graphite (1581 cm^{-1}) and is the result of the presence of certain double bonds which resonate at higher frequencies [6, 7]. After a complete reduction of graphene oxide to graphene this band usually shifts to the lower frequencies [4].

D-band of SLGO in LB film exhibits at 1360 cm^{-1}, which characterizes the degree of imperfection of graphene and it is active only if the defects are involved in the double resonance scattering near the Brillouin zone [8]. Therefore commonly intensity ratio of I_D/I_G is used to establish the dimensions of the sp^2-domains in carbon materials. The obtained data indicate that the ratio of the intensities of G and D-bands was slightly increased with increasing of temperature. Typically, different authors show that this ratio should be reduced in the case of reduction of graphene oxide [5]. However, as shown in works of [9-14], the increase of I_D/I_G is associated with an increased formation of bonds defects or large number of small bonded sp^2-domains of carbon during the process of reduction. Also in disordered media, such as amorphous and crystalline graphite, increase of I_D/I_G ratio reveals the ordering process and increasing in the number of aromatic rings in the structure [15]. The possible presence of disordered regions in the structure of SLGO LB films indicates by the presence of 2D band in the Raman spectra. It is known that the position of this band and the ratio I_{2D}/I_G, are directly related to the number of layers of graphene nanostructures [15, 16]. As it is seen from the table, the annealing of the film at a temperature of 200 °C lead to a shift of 2D band toward lower frequencies. This means that during the heating process the thickness of the graphene oxide LB films of were decreased [16].

Thus, it can be concluded that the heat treatment leads to the partial reduction of disordered and high defective graphene oxide in the LB films, accompanied by ordering of amorphous sp^3-regions and their transition to the sp^2-bonded carbon clusters.

4. Conclusion

The optical and structural properties of graphene oxide LB films at temperature influence were studied. It is shown that heating of the SLGO LB films over 100℃ leads to the decrease in the optical density of the samples. The film structure is not changed. As shown by measurements of Raman spectra, there is partial reduction of disordered and high defective graphene oxide in the LB films, accompanied by ordering of amorphous sp^3-regions and their transition to the sp^2-bonded carbon clusters as well as a decrease in the thickness of the LB films.

References

[1] Zheng Q B, Gudarzi M M, Wang Sh J, Genga Ya and Li Sh 2011 *Carbon* **49** 2905
[2] Zheng Q, Hing Ip W, Lin X, Yousefi N, Yeung K K, Li Zh and Kim J K 2011 *ACS NANO* **7** 6039
[3] Qing Zh, Li Sh and Jun Y 2012 *Trans. Nonferrous Met. Soc. China*. **22** 2504
[4] Sutar D S, Narayanam P K, Singh G, Botcha D V, Talwar S S, Srinivasa R S and Major S S 2012 *Thin Solid Films* **520** 5991
[5] Jorio A, Dresselhans M, Saito R and Dresselhaus G F 2011 *Raman spectroscopy in graphene related systems* (Verlag: Wiley-VCH) p 356
[6] Yang D Q and Sacher E 2006 *Langmuir* **22** 860
[7] Paredes J I, Villar R S, Solis F P, Martinez A A and Tascon J M 2009 *Langmuir* **25** 5957
[8] Su C Y, Xu Y, Zhang W, Zhao J, Tang X, Tsai C H and Li L J 2009 *Chem. Mater.* **21** 5674
[9] Stankovich S, Dikin D A, Piner R D, Kohlhaas K A, Kleinhammes A, Jia Y, Wu Y, Nguyen S T and Ruoff R S 2007 *Carbon* **45** 1558
[10] Xu Y, Bai H, Lu G, Li C and Shi G 2008 *J Am Chem Soc* **130** 5856
[11] Lambert T N, Luhrs C C, Chavez C A, Wakeland S, Brumbach M T and Alam T M 2010 *Carbon* **48** 4081

[12] Choi E Y, Han T H, Hong J, Kim J E, Lee S H, Kim H W and Kim S O J 2010 *Mater Chem* **20** 1907
[13] Shen J, Hu Y, Shi M, Lu X, Qin C, Li C and Ye M 2009 *Chem Mater* **21** 3514
[14] Fan Z, Wang K, Wei T, Yan J, Song L and Shao B 2010 *Carbon* **48** 1686
[15] Ferrari A C and Robertson J 2006 *Phys Rev Lett* **97** 187401
[16] Park J S, Reina A, Saito R, Kong Jand Dresselhaus G 2009 *Carbon* **47** 1303

XPS study of the structure of nitrogen doped a-C film

D G Piliptsou[1], A S Rudenkov[1], A V Rogachev[1,2], Xiaohong Jiang[1], P A Lychnikov[3] and V A Emel'yanov[2]

[1]International Chinese-Belarusian scientific laboratory on vacuum-plasma technology, Nanjing University of Science and Technology, Xiao Ling Wei 200, Nanjing 210094, China,
[2]Francisk Skorina Gomel State University, Sovetskaya, 104, Gomel 246019, Belarus;
[3]Moscow Technological University (MIREA), 119454, Vernadsky 78, Moscow, Russia

E-mail: pdg_@mail.ru

Abstract. It was found change of phase composition and structure of the carbon coating, subjected to ion nitriding. It is shown that ion nitriding is in the surface layers of increasing the share of sp2-phase with a decrease in the size of the clusters. It was found that the ion nitriding of carbon coatings are formed CNx connections to increase the number of the type of N-Csp2 bonds.

1. Introduction

Amorphous diamond like films (a-C) have been arousing scientific interest for more than 30 years due to their unique properties. As it is known [1, 2] that the properties of a-C films are defined by the ratio of carbon atoms with sp3 and sp3of electronic hybridization. The efficient technological management method of structure and properties of such films is their doping by various chemical elements and compounds. The doping of a-C films is carried out by nitrogen for the purpose of synthesis of C3N4 carbon nitride which has theoretically been expected in the research possessing high hardness, wear resistance, and also good heat conductivity and optical transparence. As it is mentioned in [3], the introduction of nitrogen into the structure of a-C leads to the essential structural change and properties of a film. At the same time, the influence nature depends on the method and the mode of deposition. In order to produce nitrogen containing a-C, various methods of nitrogen introduction into the structure of a carbon matrix are applied, as follows:

1) the deposition of carbonic films in the environment of molecular nitrogen (a-C:Nx);

2) the deposition of carbonic films with simultaneous treatment of the growing film by nitrogen ions with energy to 300 eV (a-C:N+low);

3) the subsequent ionic treatment of earlier formed carbonic films by nitrogen ions with energy to 1,5 keV (a-C:N+high).

We are studied how nitrogen ions with energy to 1,5 keV change the structure of carbon films.

2. Experimental

The deposition of a carbon film is produced by the evaporation of a graphite target with the influence of an impulse discharge. The carbon films formed with the frequency of impulse discharge of 20 Hz a-C:N+high are subjected to the subsequent treatment by nitrogen ions with energies up to 1,5 keV (Fig. 1). Voltage discharge is 350 V. The polished wafers of silicon monocrystal (100) and quartz serve as a substrate. In order to clean the substrate, the treatment of its surface is produced for 15 min by argon ions with energy 400 eV and with ionic current density of 25 A/m^2.

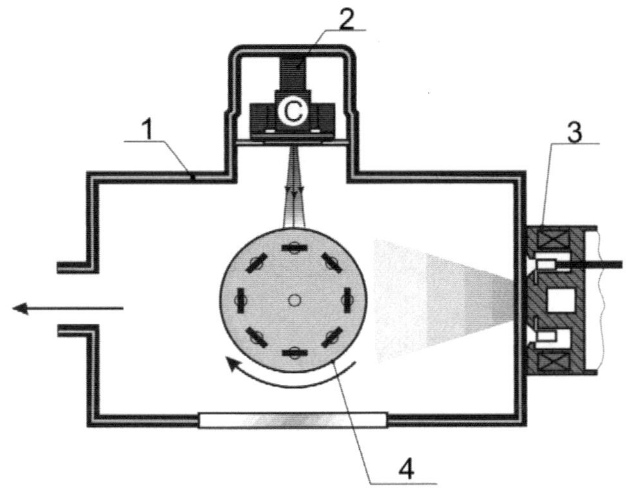

1 - vacuum chamber; 2 - pulse carbon plasma source; 3 - ion source; 4 - substrates

Figure 1. The sheme of the vacuum unit intended for the deposition of carbon films in conditions of ion bombing and the subsequent ion treatment of alloyed carbon films.

The microstructure of the carbon films is determined by Raman spectroscopy methods (the length of the exciting radiation wavelength is 532 nm and power is 5 mW) (Senterra, Bruker).
The chemical composition and structure of the bonds are determined by X-ray photoelectron spectroscopy (XPS, PHI Quantera II) by initiating the substance by aluminum Kα-radiation with photon energy 1486,6 eV and total power of 250 W. Before analyzing the carbon film is etched by the flow of argon ions with 5 keV energy. This energy is sufficient to remove the surface hydroxyl and hydrocarbon contaminants adsorbed while the films stay in the air. The scale of energy bonds in the XPS spectrum is calibrated by C1s (284,6 eV). XPS spectrum of C1s and N1s is estimated by XPSPEAK software. The spectrum of the background radiation is modeled with Shirley function.

3. The results and their analysis

It is stated that further treatment of an already formed carbon film a-C:N+high results in partial implantation of nitrogen ions, their content in the film does not exceed 2,5...3,5 at. % according to the energy dispersive analysis. It should be noted that the formation of carbon films in ion assisting conditions (regardless of assisting ion energy) exerts a significant influence on the phase composition. When assisting the ratio ID/IG increases from 0,5 to 0,9 and the G-peak width decreases to 174,6 cm-1. After subsequent ion nitriding the ratio ID/IG equals 0,58 and the G-peak width is equal to 189,2 cm^{-1}. These changes are explained by different mechanisms of phase transitions sp3→sp2 of carbon films. At ion assisting the main cause of phase composition change are inelastic collisions of nitrogen ions and carbon in a plasma, while ion nitriding is superficial, and etching processes and local heating of the upper film layers predominate [3–5].

The further treatment with nitrogen ions causes significant phase transitions in single-component films of carbon, which is proved by XPS. The presence of N1s peak in XPS spectra indicates the process of implantation of nitrogen ions into the surface layers during subsequent processing [6, 7].

Table 1 shows that the subsequent processing of single-component ionic carbon a-C:N+high films leads to an increase in the content of the sp2-hybridized carbon atoms, which is partly confirmed by Raman spectroscopy, introducing a slight increase in ID/IG ratio. The changes in the content of sp3-phases are difficult to judge, because with the decomposition of a component with a binding energy of 285,4 eV ~ we mean Csp3 and C – N, C = N bonds.

However, even with the contribution to the intensity of the component of compounds such as C and C = N–N type, it is seen that the proportion of integral area slightly decreases in all cases. It is clear from the analysis of N1s that at the ion treatment the bonds of CNx type are formed with predominant content of N– Csp2N- type bonds.It is found that the subsequent ion nitriding increases the microhardness, regardless of the type of carbon film.

Ion nitriding of carbon films leads to the formation of compounds of nitrogen and carbon with a predominant content of N–Csp2 bonds. In addition, there is a slight increase of sp2-hybridized carbon atoms, which is confirmed by XPS and Raman spectroscopy. On the basis of what can be assumed that during the bombing occurs the implantation of nitrogen ions into the surface layers, followed by the formation of chemical compounds of carbon and nitrogen.

Table 1 - Analysis of XPS of carbon films after ion nitriding

Film	Peaks of XPS spectrum	Bond type	Peak, eV ± 0,2 eV	Width, eV	Area share, %
a-C	C1s	Csp^2	284,4	1,21	34
		Csp^3	285,3	1,34	65
		C–O	286,5	0,59	1
	O1s	O–C	531,9	3,42	–
a-C:N^{+high}	C1s	Csp^2	284,5	1,07	50
		Csp^3, C=N, C–N	285,4	1,65	48
		C–O	286,7	1,02	2
	N1s	N–sp^3C	398,2	1,37	38
		N–sp^2C	399,9	1,96	57
		N–O	401,7	0,78	3
	O1s	O–C	531,9	2,39	–

It is stated that at ion nitriding (1,5 keV) of carbon films the relative content of N–Csp3 bonds 3,5…3,7 times higher than in carbon films formed in the presence of molecular nitrogen or in assisting with energy nitrogen ions to 0,3 keV. This fact can be explained by a higher energy of implanted nitrogen ions as well as lower content of sp2-clusters in the initial film [7, 8].

In practice for improvement of physical and chemical properties of materials of the electronic equipment, the creation of the set superficial nanostructures by etching and thermal annealing, the ion-beam and thermal methods of activation and modification of materials are widely used [9–12].

4. Conclusion

In conclusion, as a result of analysis of the microstructure of film a-C:N+high films the mechanism of influence of ion nitriding on the structure, processes of chemical interactions and properties of carbon films are defined. Increased nitrogen activity and decrease in the kinetic energy of carbon ions in the plasma are favorable for the formation of chemically related structures in N-alloyed carbon films. It is defined that at ion nitriding (up to 1,5 keV) the relative content of carbon films of N- Csp3 bonds is 3,5…3,7 times higher than in carbon films formed in the presence of molecular nitrogen or in the assisting conditions with energy nitrogen ions to 0,3 keV. This fact can be explained both by a higher energy of implant nitrogen ions and lower content of sp2-clusters in the initial film.

Acknowledgements.
This work was supported by the Ministry of Education and Science of Russian Federation, Belarusian Republican Foundation for Fundamental Research (project No. T16M-015, for 2016-2018), National Natural Science Foundation of China (project No. 51373077, for 2015–2016), and Intergovernmental Cooperation Projects in Science and Technology of the Ministry of Science and Technology of PRC (projects No. CB11-09 & CB11-11).

References

[1] Majumdar A, Das S C, Bogdanowicz R, Shripathi T, Langel W. Rainer Hippler Role of nitrogen in evolution of sp2/sp3 bonding and optical band gap in hydrogenated carbon nitride 2013 *Vibrational Spectroscopy* **66** 63–68

[2] Aleutdinov A D, Ghygazov S A, Mylnikova T S, Luchnikov P A Thermal shock removal of defective glass-enamel coating from cast-iron products 2015 *IOP Conf. Ser.: Mater. Sci. Eng.* **81** 012069

[3] Liu D G, Tu J P, Zhang H, Chen R, Gu C D Microstructure and mechanical properties of carbon nitride multilayer films deposited by DC magnetron sputtering 2011 *Surf. Coat. Technol.* **205** 3080–3086

[4] Chen R, Tu J P, Liu D G, Yu Y L, Qu S X, Gu C D Structural and mechanical properties of TaN/a-CN_x multilayer films 2012 *Surf. Coat. Technol.* **206** 2242–2248

[5] Voevodin A A Comparative study of wear-resistant DLC and fullerene-like CN_x films produced by pulsed laser and filtered cathodic arc depositions 2005 *Surface &Films Technology* **197** 116–125

[6] Broitman E Mechanical and tribological properties of CNx films deposited by reactive magnetron sputtering 2001 *Wear.* **248** 55–64

[7] Cohen M L Structural, electronic and optical properties of carbon nitride 1995 *Material Science Engineering* **209** 1–4

[8] Gammon W J Experimental comparison of N(1s) X-ray photoelectron spectroscopy binding energies of hard and elastic amorphous carbon nitride films with reference organic compounds 2003 Carbon **41** 1917–1923

[9] Luchnikov P A, Rogachev A V, Yarmolenko M A, Rogachev A A, Bespal'ko A A Dynamics of combined electron beam and laser dispersion of polymers in vacuum 2016 *IOP Conf. Ser.: Mater. Sci. Eng.* **110** 012081

[10] Luchnikov P A Structure Formation Mechanisms and Electrical Properties of PVD Fluoropolymer Films 2015 Russian Physics Journal **57** 9 1239–1245

[11] Rogachev A A, Yarmolenko M A Xiaohong Jiang, Shen Ruiqi, Luchnikov P A, Rogachev A V Molecular structure, optical, electrical and sensing properties of PANI-based coatings with silver nanoparticles deposited from the active gas phase 2015 Applied Surface Science **351** 811–818

[12] Surzhikov A P, Frangulyan T S, Ghyngazov S A, Lisenko E N, Galtseva O V 2006 *Russian Physics Journal* **49** 5 506-10

Diffusion mass transfer in ionic materials under intense electron irradiation

I G Bochkarev[1], S A Ghyngazov[2], T S Frangulyan[2], A B Petrova[2] and A V Chernyavskii[2]

[1] Institute of Solid State Chemistry and Mechanochemistry of the Siberian Branch of the RAS, Novosibirsk, Russia
[2] National Research Tomsk Polytechnic University, Tomsk, Russia

E-mail: chernyavski@tpu.ru

Abstract. The results of studies on the impact of an electron beam with the energy of 1-2 MeV on diffusion processes in materials with ionic bonds are presented in the paper. Used electron beam intensity is allowed to provide heating of the material to temperatures of 1600 K. Diffusion of Na, Mg, Al ions into single crystals KBr in the temperature range 573-883 K, Al ions in the NiO-AlO system at 1373-1573 K, was studied. Diffusion annealing carried out under thermal and radiation-thermal heating of the samples. Then diffusion coefficients were determined. It was found stimulating action of irradiation on diffusion processes of Mg, Al ions in Kbr and Al ions in the NiO-Al_2O_3 system, which consists in increasing the diffusion coefficients at radiation-thermal annealing. The observed effect is achieved by increasing the effective rate of diffusion jumps.

1. Introduction

Concentrated streams of energy can be used for processing organic and inorganic materials [1-4]. Several authors [5-9] proposed a method based on the use of high-intensity beam of the high-energy electrons to heat and to activate the process of sintering of ceramic materials. A beam of the high-energy electrons (1.4-2.0 MeV) transfers its energy to the entire volume of irradiated material at the same time. Thereby, the thermal conductivity factor limiting the process rate is eliminated. Apart from the purely thermal effect of heating speed, it may have a phenomenon of the radiation-thermal activation of the diffusion processes underlying ceramic production. Therefore, research directed on revealing the impact of the irradiation on mass transfer, are of great interest.

This report summarizes the studies of high-temperature diffusion carried out by us in several ionic compounds under irradiation of accelerated electrons beam, providing both a high rate of radiation defects, and heating of the samples. Here we studied the effect of the irradiation on different types of diffusion processes.

Diffusion crystals doping of the impurities examined by the example of the most simple in composition and structure of alkali halide crystals (AHC). Diffusion limiting solid phase synthesis was investigated by conducting the reaction leading to the formation of spinel $NiAl_2O_4$ by the interaction between the nickel oxide and alumina.

2. Experimental results and discussion

2.1. Research of heterodiffusion under intense electron irradiation in AHC

Single crystals of potassium bromide were used in the experiments. It was studied the diffusion of impurity ions with different charge state: Na^+, Mg^{2+}, Al^{3+}. Diffusion annealing carried out under conditions of thermal and electron beam heating of the crystals. We used electron accelerator ELV-6 (INP SB RAS, Novosibirsk) for radiative heating of samples with the electron energy of 1.4 MeV and the continuous beam current density of 2.5 mA/cm^2.

The depth profiles of impurity ions were investigated by secondary ion mass spectroscopy (SIMS). The technique of the profiles measurement, excluding the effects of the crater and the charging surface, is described in [10-12].

2.1.1. Diffusion of ions Na^+. Diffusion carried from the thin film of NaCl (200-400 nm thickness) deposited on the surface KBr single crystal by thermal evaporation under vacuum. Diffusion annealing of the different samples was performed in the temperature range of 573-883 K for 20 min. In the semi-logarithmic coordinates the temperature dependence of the coefficients of thermal (D_T) and radiation-thermal (D_{RT}) diffusion of sodium ions in KBr lattice approximated by two temperature portions with different activation energy of the process (the activation energy is given in eV):

$$D_T = 1.3 \cdot 10^{-7} \cdot \exp(-\frac{0.51}{kT}) + 7.3 \cdot \exp(-\frac{1.68}{kT}), \text{ cm}^2/\text{s};$$

$$D_{RT} = 2.3 \cdot 10^{-7} \cdot \exp(-\frac{0.55}{kT}) + 3.8 \cdot \exp(-\frac{1.67}{kT}), \text{ cm}^2/\text{s}.$$

Irradiation did not affect on the diffusion process of isovalent impurity over the entire temperature range.

Mechanisms of impurities thermal diffusion in AHC analyzed at the literature [13]. It is found, that the impurities homologues diffusion, as well as self-diffusion is performed by the vacancy mechanism. In accordance with these views the low-temperature portion is likely caused by the migration of sodium on the biographical origin cation vacancies, the concentration of which is determined primarily crystal contamination of multivalent cationic impurities. The concentration of these vacancies (N_{imp}) was estimated by us on the basis of the ion conduction measurements of used crystals. According to the calculations the measured diffusion coefficients were provided by the original defectiveness of the cation sub-lattice at 10^{-7} mol. fractions in the low-temperature area. The activation energy of diffusion in this area is close to the activation energy for migration of the cation vacancies, which is consistent with the considered model of the process. Diffusion limited thermodynamically equilibrium vacancies at higher temperatures.

In accordance with the mechanism of Dienes-Damask [14], the main condition for the stimulation of diffusion processes in radiation fields is the excess of the stationary concentration of radiation vacancies (N_{irr}) over the thermodynamically equilibrium concentration (N_{th}) and N_{imp}. The calculations of these values show that it does not hold in the realized experimental conditions. N_{th} was determined according to the formula proposed by Lydiard [15], and N_{irr} in accordance with the kinetic model Dienes-Damask. The rate of radiation defects was set equal to 10^{-3}-10^{-9} s^{-1}. Thus, the results of the study suggest that the Dienes-Damask mechanism of the radiation diffusion stimulation is ineffective at high temperatures.

2.1.2. Diffusion of heterovalent impurities in the KBr. The thin metallic films of aluminum or magnesium were deposited on the crystal surface by thermal evaporation under vacuum in this series of experiments. The diffusion profiles were measured by SIMS. The results in Table 1 indicate the intensification of heterovalent impurities diffusion in potassium bromide under electron irradiation.

Temperature dependence of ion diffusion coefficients of Al^{3+} and Mg^{2+} for thermal and radiation-thermal annealing are described by expressions (the activation energy is given in eV):

$$D_T^{Al}(T) = 5.58 \cdot 10^{-7} \exp(-\frac{1.03}{kT}),\ \text{cm}^2/\text{s};\quad D_{RT}^{Al}(T) = 1.18 \cdot 10^{-2} \exp(-\frac{1.68}{kT}),\quad \text{cm}^2/\text{s} \qquad (1)$$

$$D_T^{Mg}(T) = 4.18 \cdot 10^{-6} \cdot \exp(-\frac{1.2}{kT}),\ \text{cm}^2/\text{s};\quad D_{RT}^{Mg}(T) = 3.81 \cdot 10^{-3} \cdot \exp(-\frac{1.6}{kT}),\ \text{cm}^2/\text{s}. \qquad (2)$$

Table 1. The diffusion coefficients of heterovalent cations in KBr.

Temperature, K	Mg^{+2} ions		Al^{+3} ions	
	D_T, cm^2/s	D_{RT}, cm^2/s	D_T, cm^2/s	D_{RT}, cm^2/s
673	$5.4 \cdot 10^{-14}$	$4.8 \cdot 10^{-14}$	$2.65 \cdot 10^{-14}$	$3.0 \cdot 10^{-14}$
763	$1.0 \cdot 10^{-13}$	$1.3 \cdot 10^{-13}$	$9.3 \cdot 10^{-14}$	$9.8 \cdot 10^{-14}$
813	$1.3 \cdot 10^{-13}$	$3.8 \cdot 10^{-13}$	$2.1 \cdot 10^{-13}$	$4.9 \cdot 10^{-13}$
853	$3.2 \cdot 10^{-13}$	$1.4 \cdot 10^{-12}$	$4.45 \cdot 10^{-13}$	$1.2 \cdot 10^{-12}$
883	$6.2 \cdot 10^{-13}$	$3.6 \cdot 10^{-12}$	$8.3 \cdot 10^{-13}$	$3.5 \cdot 10^{-12}$

Equations (1) - (2) show that radiation exposure causes an increase in the activation energy of the diffusion process and the pre-exponential factor D_0. In this respect, the situation is radically different from the characteristic of the impurity monovalent ions diffusion where these parameters are insensitive to the effects of the irradiation. Obviously, the observed effect of the stimulation of the magnesium and aluminum ions diffusion under electron irradiation is associated with their charge state different from the charge matrix ions. Multiply charged cations interact effectively with the generated electron-hole pairs in the lattice of AHC in the field of radiation. However, they can change their charge at the electron localization and make the diffusion jumps in different charge states. Perhaps for this reason there is an increase the activation energy of diffusion. The growth D_0 shows an increase in the frequency of jumps under irradiation, which is naturally related to the concentration of vacancy defects. As shown above, the Dienes-Damask mechanism implementation is not able to significantly affect this value. Theoretically, one can assume the existence of spatial correlations between diffusing ion and the cation vacancies accumulated in the surrounding area, which may increase the probability of ion hopping. However, quantifying the contribution of this process to increase the parameter D_0 is not yet possible.

2.2. Investigation of the diffusion controlling the solid-phase synthesis under electron-beam heating.
To study the nickel aluminate spinel synthesis reaction was chosen. This reaction is characterized by a number of advantages that make it suitable for use in diffusion experiments. $NiAl_2O_4$ is the only product of the reaction between the starting oxides, the production rate constant $NiAl_2O_4$ is relatively large and there is practically no mutual solubility between the α-Al_2O_3 and NiO. Experiments were carried out as follows. A thin nickel film with thickness 0.25-0.30 microns was deposited by thermal evaporation in vacuum on a polished surface of the aluminum oxide crystals. Subsequently nickel film was oxidized to NiO in air at the temperature of 1073 K insufficient for spinel formation. Then, solid-phase synthesis of the compound was carried out at higher temperatures by thermal and radiation-thermal heating. The irradiation was made by the accelerator ILU-6 (INP SB RAS, Novosibirsk) with the electron energy of 2.0 MeV. Absorbed dose rate is 20-60 kGy/s. The layer thickness of the spinel formed on the side of NiO was measured with a micro-interferometer MII-11 after removing from sample of unreacted NiO by hydrogen reduction at 823K for three hours and then selectively dissolving the metal in 10% nitric acid solution. The reduction temperature was chosen to 80-90 K below start temperature of the recovery $NiAl_2O_4$.

The growth kinetics of the spinel layer under irradiation as well as in the thermal heating obeyed the parabolic law $x^2 = k \cdot t$, where x is the thickness of the product layer; k - parabolic constant of formation rate of $NiAl_2O_4$; t - time. Based on this law, it was calculated the values of k, and set their temperature dependence for the compared methods of sample heating (the activation energy is given in eV):

$$k_T = 2.7\times10^5 \exp\left(-\frac{5}{kT}\right),\ \text{cm}^2/\text{s};\quad k_{RT} = 1.8\times10^6 \exp\left(-\frac{5}{kT}\right),\ \text{cm}^2/\text{s}.$$

It is seen that the irradiation does not alter the activation energy of the process. Radiation-stimulated growth constant k is caused by an increase in pre-exponential factor, suggesting an increase in the frequency of cations jumps.

To quantify the diffusion process, the thermodynamic theory of solid-phase reactions [16, 17] should be used, which is based on the following provisions:

1. The reaction rate is limited by diffusion of ions through the product layer.
2. The product layer is compact and its non-equilibrium defects (dislocations, grain boundaries) do not make a decisive contribution to the ion mobility.
3. The reactions at the phase boundaries occur more rapidly than diffusion processes. Hence, local thermodynamic equilibrium is established at the phase boundaries.
4. Individual ions move independently. There is electroneutrality at any cross section of the product.

According to [18] the rate of $NiAl_2O_4$ formation is limited by Al^{3+} ions diffusion through the product. Parabolic k and rational ($k_1 = k/2V$, where V - equivalent volume of spinel) reaction constant directly related to diffusion coefficient of the limiting reaction ions. In the case of oppositely directed diffusion fluxes of positive ions, subject to the conditions of $D_A^{2+} \gg D_B^{3+} \gg D_O^{2-}$ the theory gives the following expression relating the average value of the diffusion coefficient of the limiting reaction ions with a rational reaction rate constant:

$k_1 = z_i \cdot C_i \cdot \overline{D}_i \cdot \dfrac{4\Delta G}{3RT}$,where Z_i; C_i; D_i - ion charge, its equivalent concentration and the average value of the diffusion coefficient, respectively; $\Delta G°$ - standard Gibbs free energy of formation $NiAl_2O_4$, equal 12.6±1.7 kJ/mol. The calculation results showing an increase in the diffusion coefficient of ions in the radiation field are presented in table 2.

Table 2. The diffusion coefficients Al^{3+} ions in $NiAl_2O_4$.

T, K	1373	1423	1473	1523	1573
D_T, cm²/s	$2{,}4\cdot10^{-15}$	$1{,}0\cdot10^{-14}$	$4{,}2\cdot10^{-14}$	$1{,}7\cdot10^{-13}$	$3{,}1\cdot10^{-13}$
D_{RT}, cm²/s	$1{,}3\cdot10^{-14}$	$7{,}1\cdot10^{-14}$	$2{,}3\cdot10^{-13}$	$9{,}6\cdot10^{-13}$	$3{,}8\cdot10^{-12}$

3. Conclusions

Thus, a plurality of the above results is evidence of the stimulating action of intense irradiation on different types of diffusion processes in dielectric materials. This effect is achieved by increasing the effective rate of diffusion jumps.

Acknowledgements
This work was supported by the Ministry of Education and Science of the Russian Federation within the framework of the government "Science" program.

References

[1] Luchnikov P A, Rogachev A V, Yarmolenko M A, Rogachev A A, Bespal'ko A A 2016 *IOP Conf. Series: Materials Science and Engineering* **110** 012081
[2] Luchnikov P A 2015 *Russian Physics Journal* **57** 9 1239-45
[3] Rogachev A A, Yarmolenko M A, Xiaohong Jiang, Shen Ruiqi, Luchnikov P A, Rogachev A V

2015 *Applied Surface Science* **351** 811-18

[4] Aleutdinov A D, Ghygazov S A, Mylnikova T S, Luchnikov P A 2015 *IOP Conf. Series: Materials Science and Engineering* **81** 012069

[5] Neronov V A, Melekhova T F, Tatarintseva M I and Voronin A P 1988 *Soviet Powder Metallurgy and Metal Ceramics* **27** 2 111–114

[6] Neronov V A, Voronin A P, Tatarintseva M I, Melekhova T F and Auslender V L 1986 *J. of the Less – Common Metalls* **117** 1-2 391-394

[7] Surzhikov A P, Frangylyan T S, Ghyngazov S A 2012 *Russian Physics Journal* **55** 4 345-52

[8] Malyshev A V, Lysenko E N and Vlasov V A 2015 *Ceramics International* **41** 10 13671-13675

[9] Surzhikov A P, Frangulyan T S, Ghyngazov S A, Vasil'ev I P and Chernyavskii A V 2016 *Ceramics International* **42** 12 13888-13892

[10] Gyngazov S A, Surzhikov A P, Frangulyan T S and Chernyavskii A V 2002 *Russian Physics Journal* **45** 8 753-758

[11] Surzhikov A P, Chernyavskii A V, Ghyngazov S A and Frangulyan T S 2002 *Russian Physics Journal* **45** 12 1190-1194

[12] Borodin Y 2011 *Proceedings of the 6th International Forum on Strategic Technology, IFOST 2011* **1** 218–21

[13] Beniere F 1972 Diffusion in ionic crystals *Physics of Electrolyte*s vol 1 ed J Hladik (London-New York:Academic Press) chapter 6

[14] Dienes G J and Damask A C 1958 *J. Appl. Phys.* **29** 12 1713-1721

[15] Lidiard A 1957 *Ionic conductivity* (Berlin u. a., Springer-Verl. Handburch der Physik. Hrsg. von S. Flügge. Bd. XX Teil 2)

[16] Schmalzried H and Wagner C 1962 *Z. phys. Chem. N.F.* **31** 202-221

[17] Schmalzried H 1962 *Z. phys. Chem.N.F.* **33** 111-128

[18] Pettit F S, Randklev E H and Felten E J 1966 *J. Amer. Ceramic Soc.* **49** 4 199-203

Increase the strength characteristics of polymer films by radiation graft polymerization

M A Rezvova[1], V D Zhevnyk[1], V Pak[1], V Ya Denisov[1], Y V Borodin[2]
[1]Kemerovo State University, Krasnaya st. 6, 650 043, Kemerovo, Russia
[2]Tomsk Polytechnic University, Lenina st. 30, 634 050, Tomsk, Russia

E-mail: rezvovamaria@mail.ru

Abstract. The possibility of increasing the strength characteristics of polymer films based on polyethylene and polyamide by radiation graft polymerization was investigated. Two methods of graft polymerization (direct method and the method of polymer mixtures cross-linking) on the PMMA films surface were studied. The possibility of increasing the strength and elasticity of polymeric films by radiation modification of polymethyl methacrylate was shown.

1. Introduction

The work is devoted to studying the radiation graft polymerization of polymethyl methacrylate polymer films used for the medical devices manufacture, in order to increase their strength characteristics. The studies were conducted with using of ionizing γ-radiation Co^{60} at the facility RHM - γ - 20.

The graft polymerization is one of the most promising methods of modern modifications of various polymers and polymeric materials. The principal difference between the graft polymerization from the conventional type polymerization is a reaction initiation nature. The initiation of graft polymerization is carried out by active sites generated one way or another in the starting polymer macromolecules and not to monomer molecules or specially introduced into the initiator system [1, 2].

The radiation graft polymerization is based on the generation of active sites radical and ionic nature in the various polymers under the ionizing radiation influence followed by graft polymerization of various monomers (acrylates, acrylamide, N-isopropylacrylamide, N-vinylpyrrolidone, etc.). The radiation graft polymerization solves the problem of obtaining modified materials with a predetermined set of processing and performance properties. The radiation output and the graft polymerization rate depends on both nature of the substrate and nature of the monomer. The radiation effects is affected by many factors such as the energy spectrum and intensity, exposure time during irradiation and thereafter, the temperature at which the irradiation and temperature, which arises in the material as a result of interaction with radiation, manufacturing technology of the samples and their subsequent treatment, the level of internal stress therein, the content and composition of impurity.

An advantage of radiation graft polymerization is its versatility, enabling to modify the polymers of virtually any nature, size and shape (film, fibers, powders, etc.). Each graft polymerization method involves processes, which are characterized by a variety state of polymer - monomer system [2-7].

The objects of study in this paper are selected polymeric films: polyamide film (nylon 6, h = 0,015 mm) and a polyethylene terephthalate film (PET, h = 0,012 mm) and high density polyethylene (HDPE, h = 0,06 mm).

Nylon is a synthetic polyamide obtained by polycondensation of adipic acid and hexamethylenediamine. It has properties such as heat resistance, flexibility, elasticity, high strength, dimensional stability and chemical agents. Nylon is cross-linked by radiation. Tendency to crosslink was shown under such changes, as a reduction in the solubility, increasing temperature resistance, etc. [8]. Nylon is not soluble in most organic

IOP Conf. Series: Materials Science and Engineering **168** (2017) 012105 doi:10.1088/1757-899X/168/1/012105

solvents, it does not give the effects of mild acids, alkalis and salt water, which allows its use in medical devices.

Polyethylene is a plastic physiologically neutral material. It has low vapor and gas permeability, does not react with alkali of any concentration and solutions of any salt, resistant to gasoline, water, oil. Different branching degree of the polymer chain polyethylenes of high (LDPE) and low pressure (HDPE) determines the difference between the properties of these materials. Exposure to radiation on polyethylene leads to increase its tensile strength at room temperature until the irradiation dose approximately equal 10 Mrad; with further increasing of the radiation dose on the tensile strength starts to decrease slowly. Elongation at break also increases slightly and then decreases sharply. Such decrease of the elongation at break due to increased crosslinking density are also found in many other polymers [9-11].

Polyethylene terephthalate is a polyester, formed by condensation of ethylene glycol and terephthalic acid. It is characterized by high strength, resistance to abrasion and multiple strains in tension and bending and are relatively resistant to the action of light, x-ray, gamma - rays [12, 13]. The predominant process during irradiation is stitching.

Solving the problem of the polymeric materials hardening is related to the search for ways to improve the performance of products based on them, and, above all, health care products.

1. Results and Discussion

In the first stage of the work we investigated the possibility of polymethyl methacrylate (PMMA) grafting on the films surface. PMMA is approved for use in products for implantation into the human body. As is known from the literature, PMMA grafting to polymer fibers (films) under certain conditions of synthesis leads to raising their strength characteristics [14-16].

Two methods of radiation graft polymerization was investigated in this work: method of polymer blends cross-linking and direct method of graft polymerization [17].

The graft polymerization process was performed in two stages: in the first stage polymer film was placed in a 1% solution of polymethyl methacrylate in methyl methacrylate (MMA) for 20 minutes, then extracted, dried from monomer and exposed γ-irradiation Co^{60} in an inert atmosphere with absorbed dose 25 - 157 kGy at 40 ° C. After irradiation, the film was washed in MMA and dried. The film tensile strength (σ, MPa) and elongation at puncture (deformability - ε, %) was determined on a tensile testing machine RFI - 30.

The calculation of the deformation-strength characteristics was conducted by the formulas:

$$\sigma = k_\sigma \frac{P}{hd}\text{, MPa} \tag{1}$$

$$\varepsilon = k_\varepsilon \frac{l}{d} \times 100,\% \tag{2}$$

where: k_σ = 1,1 and k_ε = 2,2 – normalization coefficients, depending on the diameter of the indenter (lancing device); P - load at which the sample breaks, kg / cm^2; h - sample thickness, in millimeters; d - diameter of the working area (4 mm); l - elongation to break (puncture), mm.

The PMMA re-polymerization was carried out at once grafted films. 7 steps PMMA grafting were performed for each film. The test results of radiation-modified PMMA (by the method of crosslinking polymer blends) the polymer films are shown in Figures 1 and 2.

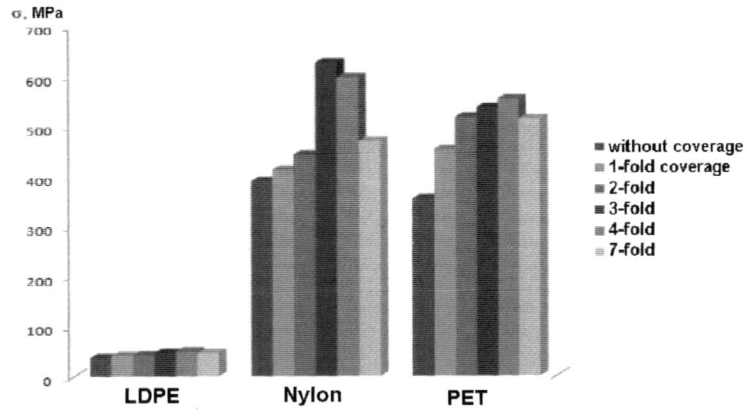

Figure 1. Tensile Strength of polymer films, modified by radiation crosslinking PMMA polymer blends at break when tested at a puncture.

The results show increase film strength in selected conditions of modifying. As it is seen in Figure 1, with triple vaccination PMMA polyamide film on its tensile strength increased by over 50%. The same increase was also noted for the strength of the PET film but quadruple grafting. Changes in film strength of LDPE under these conditions of the grafting polymerization insignificant. Increasing the multiplicity of the covering to 7 reduces the strength characteristics of the films, which may be related:

1 – with increasing thickness of the grafted PMMA layer, properties which at some point begin to dominate vaccination properties of the film itself;

2 – a polymethylmethacrylate destruction at high doses (PMMA polymer is advantageously degraded);

3 – with an increase in crosslink density in the substrate under the action of radiation which is to "overcure" stage leads to structural changes in the film itself.

Change in the deformability of the films studied by grafting PMMA by crosslinking polymer blends is shown in Figure 2.

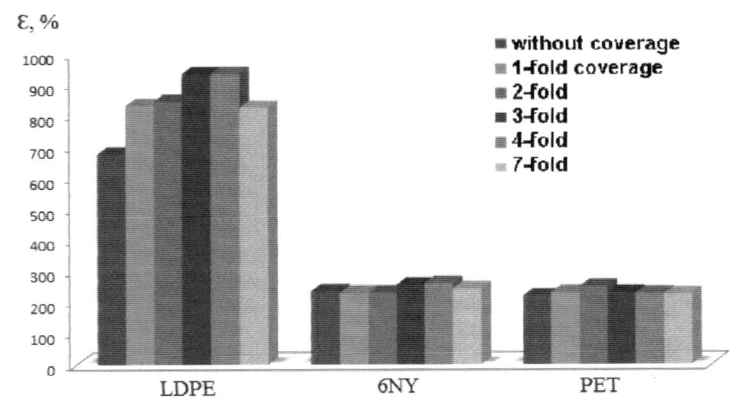

Figure 2. The deformability of polymer films modified PMMA by radiation crosslinking polymer blends during the test to rupture at a puncture.

With this method of grafting a polyethylene film for an increase in elasticity of up to 38% at 4.3 times of the coating. For polyamide and polyethylene terephthalate films elasticity remains virtually unchanged.

Modifying polyamide film "Nylon 6" and PET film held by direct graft polymerization method using an excess of methyl methacrylate monomer (MMA). LDPE film is unstable in MMA under the influence of ionizing γ - radiation. The test film images were exposed in MMA monomer at 40 ° C at a total absorbed dose of radiation 18 kGy, with the suppression of the homopolymerization process. After reaction, the samples gain (ΔP, %) and the strength characteristics of the films was measured. The test results obtained films grafted "Nylon 6" and PET are shown in Table 1.

Table 1. Changes in the weight and strength of "Nylon 6" films and PET modified by the direct method of radiation graft polymerization.

Film	Absorbed Dose, kGy (Mrad)	Film weight gain ΔP, %	σ,MPa	ε,%	Δ σ, %	Δ ε,%
	0.0	-	449±31	286±39	-	-
Nylon 6	18 (1.8)	4	699±54	347±82	56	21
	0,0	-	387±57	330±1	-	-
PET	18 (1.8)	2.5	475±23	275±1	22	-15

The data in Table 1 shows that carrying out a direct reaction in the monomer graft polymerization at an absorbed dose of 18 kGy results in increased strength both test films. The most significant hardening obtained for the film "Nylon 6" (vaccinated 4% PMMA). Formability increases with rising the strength for the polyimide film, while the PET film strength decreases with increasing the elasticity. In addition, the increase in strength to PET is almost twice less than that of the polyamide film 6NY. This can be explained by the fact that the graft to polyethylene terephthalate require higher absorbed doses than for grafting to the polyamide. This is due to the dissipation of radiation energy polyester benzene ring. A further increase of the absorbed radiation dose in the system of a polymer film - the monomer leads to deformation of the film due to the strong growth of the grafting to the surface layer of PMMA films.

Modification of study PMMA polymer films in the process of radiation grafting confirmed changes in the IR - spectra: in the area of 1730 cm-1 band appears, characteristic of an ester group of methyl methacrylate [18]. As an example of Figure 3 shows the spectra of LDPE film precursor film and the film with a single or double coated PMMA polymer blends by crosslinking.

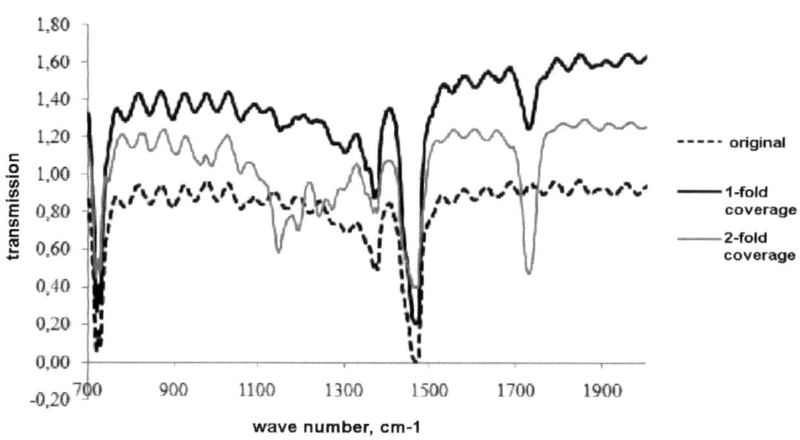

Figure 3. IR spectra of LDPE films, original and modified by PMMA.

2. Conclusion

Comparing the results obtained in this study during the radiation-chemical grafting to the polymer PMMA film, it can be concluded that the method for crosslinking polymer blends of polymeric materials studied most versatile and effective in terms of increasing their strength properties. In addition, each substrate requires an individual approach when choosing a radiation graft polymerization conditions.

References

[1] Ivanov V S 1988 Radiation chemistry of polymers p. 320
[2] Shur A M 1969 *Macromolecular compounds* p. 656
[3] Krul' L P 1990 *J. Russian Chemical Reviews* **59** 807-26
[4] Pikaev A K 1987 *Modern Radiation Chemistry* (Moscow: Science) p. 448
[5] Cetlin B L 1973 *Radiation graft polymerization* (Moscow: Science) p. 454
[6] Ahad I Ul, Bartnik A, Fiedorowicz H, Kostecki J, Korczyc B, Ciach T and Brabazon D 2014 *Journal of Biomedical Materials Research Part A* **102** 3298–3310
[7] Luchnikov P A 2015 *Russian Physics Journal* **57** 9 1239-45
[8] Kabanov V Ya, Fel'dman V I, Yershov B G, Polikarpov A I, Kiruhin D P and Apel' P Yu 2009 *J. High Energy Chemistry* **43** 5-21
[9] Kabakchi A M, Pyankov G N, Yarmilko E G, Panasenko V V and Fedchishin I F 1972 *J. Radiation chemistry* **2** 373-80
[10] Madani M, El-Bayoumi A S, Aly S S, Dayem A M and Hammam M 2012 *Journal of physics* **1** 10-16
[11] Madani M 2010 *J. Bull. Mater. Sci.* **33** 1 65–73
[12] Gotoh K, Yasukawa A and Kobayashi Ya 2011 *Polymer Journal* **43** 545–551
[13] Abdolahifard M, Bahrami S H and. Malek R M A 2011 ISRN Organic Chemistry 2011 8 pages
[14] Mondal I H 2013 *Journal of Engineered Fibers and Fabrics* **8** 42-50
[15] Jaganathan S K, Balaji A, Vellayappan M V and Subramanian A P 2015 *Journal of Materials Science* **50** 2007–18
[16] Penga J, Wanga M, Qiaob J and Weia G 2005 *J. Radiation Physics and Chemistry* **72** 739–43
[17] Sehgal T and Rattan S 2010 *J. Radioanal Nucl. Chem.* **286** 71–80
[18] Dehant I 1976 Infrared spectroscopy of polymers (Moscow: Chemistry) p. 471

Development the Adjusted Mathematical Model of Responses in the Nondestructive Testing Defectiveness Method Based on the Mechanoelectrical Transformations.

P N Khorsov[1], V P Surzhikov and A A Demikhova
Tomsk Polytechnic University, NDT Institute, 634050 Tomsk, Russia

E-mail: horsov@tpu.ru

Abstract. The improving version the responses mathematical model when using the method of mechanoelectrical transformations for nondestructive testing of defects in composite dielectric materials was examined. The refinement is conditioned by necessity of account of the frequency-dependent attenuation. For this has been used spectral approach: decomposition of excitation pulse in range and the formation of the amount each attenuate spectral component with the sample pulse response.

1. Introduction

An informative parameter in active nondestructive testing methods is interaction of excitation acoustic wave with testing object defects. Testing difficulties arise, because often defects (for example, cracks) have sizes considerably smaller, than excitation waves lengths. Therefore actual is to provide repeated reflection of excitation waves through testing object. At the same time the wave front repeatedly crosses defective zones and small deformations of front gradually accumulate.

As a result of it the registered response may differ significantly from a response from testing object with other defectiveness degree. To realize this, is necessary to provide the reverb mode, in which occur multiple reflection acoustic waves from the testing object borders. The reverb phenomenon is used in some acoustic testing methods [1-3]. On the same principle the mechanoelectrical transformations method (MET) was built [4].

In the MET method a testing object from composite dielectric material is excited by a short pulse of set form. The excitation acoustic wave impacts on mechanoelectrical converters that can be as double electric layers on dissimilar materials boundaries and also inclusions possessing piezoelectric properties. Due to changes the dipole moments of converters is formed the alternating electromagnetic field which may be registered by using external capacitance receivers. Using the MET method shows its high sensitivity to defectiveness and stress-strain state degree of dielectric materials [5-7].

For the study the MET method possibilities a mathematical model of responses in the ray approximation was developed [8]. On the model was shown that response parameters depend on system geometry: the excitation device – a sample – the signal receiver, from a sample parties ratio in the parallelepiped form from the location of the mechanoelectrical transformations sources.

In a narrow frequency strip the spectral and temporary responses characteristics of model and real testing objects with small defectiveness in the experiment identical geometry conditions were similar

[1]Corresponding author

that testifies applicability of the ray approximation when researching on mechanoelectrical transformations method and possibility the mathematical model use as a standard for comparison of defectiveness and stress-strain state degree for real testing objects. However in a wide excitation spectral band the rather big difference between really obtained data and calculated ones by the mathematical model is observed. The reason for this is that the mathematical model does not take into account the spectral components frequency-dependent attenuation of excitation wave at its passing through the sample.

2. Modeling

A pulse mechanical excitation has rather wide range even in the conditions a radio pulse with harmonic filling. The high-frequency components of an excitation acoustic wave in the passing process by a sample because of internal friction attenuate faster than low-frequency inverse proportion to a square of frequency [9].

The harmonious signal attenuation from time t at a given frequency f0 is described by the following formula:

$$A(t) = A_0 \cdot \sin(2 \cdot \pi \cdot f_0 \cdot t) \cdot \exp(\frac{-t}{\tau(f_0)}), \quad (1)$$

where A_0 –a harmonic wave amplitude in an excitement point; $\tau(f_0)$ – an attenuation time constant for a given frequency.

$$\tau(f_0) = \frac{1}{f_0^2 \cdot k}$$

k – constant multiplier;

Changing the frequency spectrum of the excitation pulse from the passage of time on the model requires a modification of the mathematical model, which would make it an adequate by the real of attenuation processes.

The response of sig_i is formed by convolution to the sample pulse characteristic h_i and temporary realization of a excitement signal s_i:

$$sig_i = \sum_{j=0}^{i} h_j \cdot s_{i-j}, \quad (2)$$

The following procedure of the accounting of spectral components attenuation was considered:

1) decomposition a form of an mechanical excitation pulse in a row of Fourier;
2) each harmonic component is represented in the form of attenuated harmonica;
3) its convolution with impulse response characteristic of the mathematical model is performed;
4) summarize all harmonious responses and to receive a full response in which frequency-dependent attenuation will be considered.

The problem is that in case of the attenuating harmonious signal its range expands according to a formula:

$$S_f = \frac{A_0 \cdot i}{\sqrt{2 \cdot \pi} \cdot \left(2 \cdot \pi \cdot (f - f_0) - \frac{-i}{\tau(f) \cdot k} \right)}, \quad (3)$$

where i – imaginary unit.

Consequently, appear the lateral spectral bands which are added to values of other harmonious components.

On the other hand, as seen in fig. 1 with frequency increase the amplitude of a specified harmonica decreases due to redistribution of energy in lateral frequencies. I.e. the competing processes take place: increases in amplitude of this harmonica due to replenishment from ranges of the next harmonicas and its reduction due to energy redistribution.

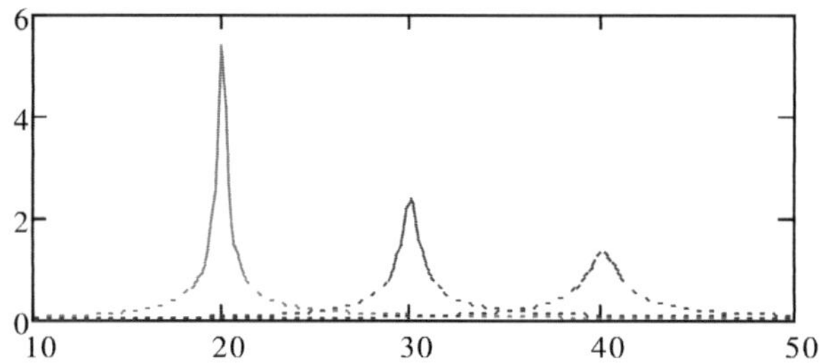

Figure 1. AFC of attenuated harmonic signals on the frequencies of 20, 30 and 40 kHz.

Let's consider a range of a single impulse. Its feature is that its amplitude-frequency characteristic (AFC) does not depend on frequency and, therefore, represents the straight line parallel to a frequencies axis. When passing a single impulse through a solid body its harmonious components will fade according to the equation (1).

Process of pulse signal attenuation in a solid body from time was simulated. The sum of the harmonious components attenuation of pulse excitation sig(t) on a formula (1) was calculated using the Mathcad:

$$sig(t) = \sum_{j=0}^{n} A_j \cdot \cos(2 \cdot \pi \cdot f_j \cdot t + \varphi_j) \cdot \exp\left[-t \cdot (f_j)^2 \cdot k\right] \quad , \tag{4}$$

where the A_j - the amplitude of the j harmonic expansions in the Fourier series; φ_j - the phase of the harmonic

Then the spectrum sig (t) was calculated using a Fourier transform. Fig. 2 shows the AFC of the (synthesized) spectrum.

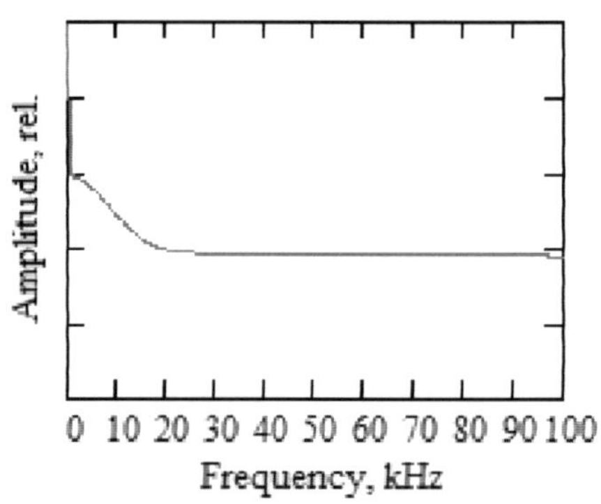

Figure 2 AFC of the synthesized range of a single pulse signal.

As appears from fig. 2 a total response of the spectrum attenuation harmonics of excitation single pulse it is also almost frequency independent at frequencies above 20 kHz. Raise of AFC at frequencies below 20 kHz is caused by limitation of selection length. It can be explained with the fact that the attenuated harmonious signal does not decrease to zero on the end of temporary realization, i.e. becomes discontinuous. Therefore there is Gibbs's effect distorting Fourier's decomposition. With increase in length of realization the error due to Gibbs's effect is displaced to the area of lower frequencies, showed calculation for mathematical model. Taking into account the specified circumstance it is possible to draw a conclusion on the acceptability of the algorithm of the accounting of attenuation frequency dependence for calculation by the mathematical ray model offered above [1]. Experiment by receiving a response from a concrete sample was carried out and calculation with use of the system same geometrical arrangement the impact device – a sample – the receiver of a signal and the same form of the excitation pulse was performed.

The electromagnetic response from the concrete sample for a given geometry of the experiment was received, as well as was calculated by the mathematical model using the same geometrical parameters and the same shape excitation pulse. The sample sizes - 100*80*60 mm3. The dot source of pulse mechanical excitation was located in the center of a lateral side, and a plate of capacitive receiver with size of 10*10 mm2 at the opposite side across from source. Frequency of digitization was 1 MHz (Δt=1μs); duration of temporary realization - 3 ms.

3. Results and discussion

In fig. 3 the temporary characteristic of excitation pulse displayed. In fig. 4 – it's AFC (the continuous line). In the same figure the dashed line showed AFC of the excitation pulse synthesized taking into frequency attenuation.

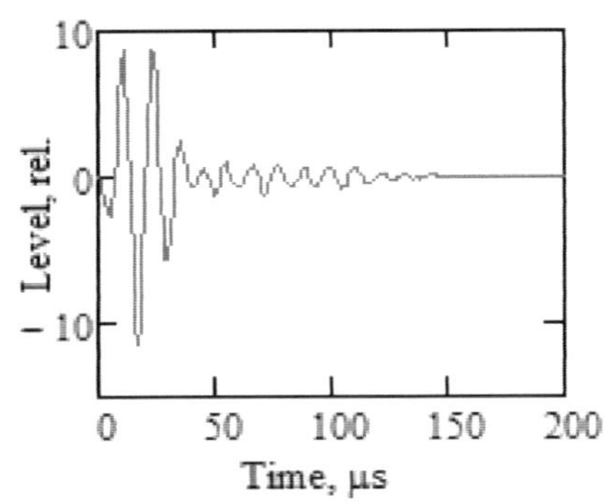

Figure 3. Temporary realization of excitation pulse.

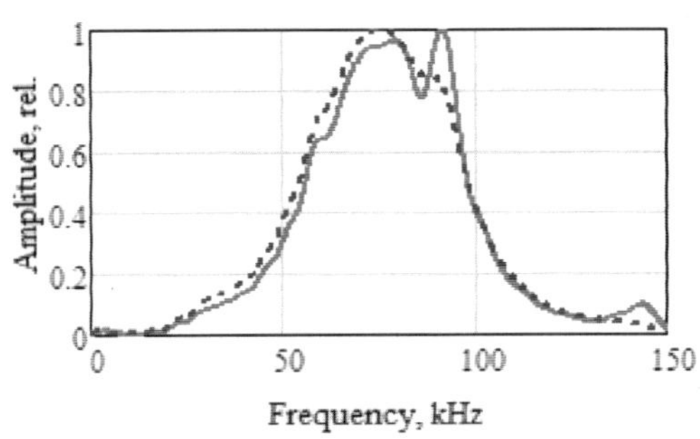

Figure 4. AFC of excitation pulse (the continuous line) and AFC of synthesized taking into frequency attenuation (the dotted line).

Apparently from fig. 4 real and synthesized AFC practically coincide.

For response calculation by the mathematical model was calculated the pulse characteristic of h(t) by the parameters of real experiment shown in fig. 5.

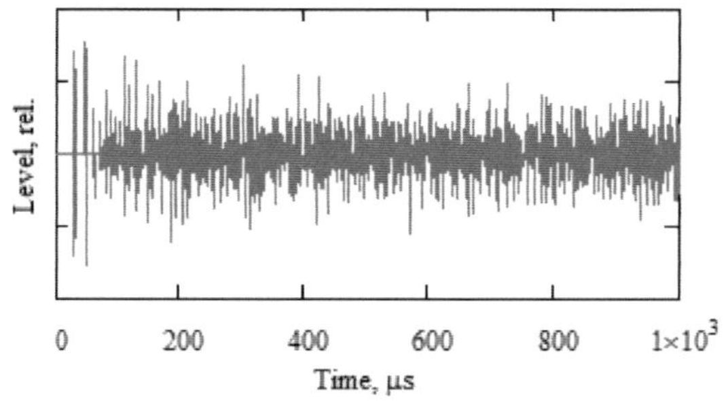

Figure 5. The pulse characteristic calculated by the mathematical model.

The pulse characteristic was calculated at model excitation with signal duration in one interval digitization time (1 μсек).

As seen from figure 5 in spite of the fact that the sample is excited by a spherical acoustic wave which level decreases in inverse proportion to distance from an excitement point, the pulse characteristic carries not decreasing character because energy of a signal does not dissipate in space, external concerning a sample, and is reflected from its borders and again participates in process of interaction with sources of mechanoelectrical transformations. I.e. in the specified model we neglect loss of acoustic energy at the expense of its exit from the sample boundaries.

The pulse characteristic represents imposing of a straight line and reflected from borders of a pulse wave with amplitudes inversely proportional to the passed way. Calculation of a response of res (t) was carried out on the following formula:

$$res(t) = \sum_{k=0}^{n} \exp(-f_k^{\ 2} \cdot k \cdot i \cdot \Delta t) \cdot \sum_{j=0}^{i} a_k \cdot \sin(2 \cdot \pi \cdot f_k \cdot (i-j) \cdot \Delta t) \cdot h_j + \varphi_k) \quad , \tag{5}$$

where n – number of the foldable harmonicas, which was chosen equal 450 that allowed to capture the range of frequencies from 0 to 150 kHz.

In fig. 6 temporary realization of the calculated response, and in fig. 7 – its AFC is displayed. In the same figure also shown excitation pulse AFC in the form of a dotted line.

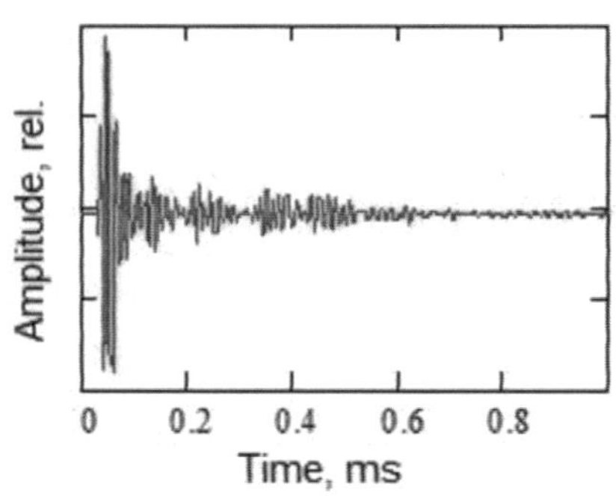

Figure 6. Temporary realization of the calculated response.

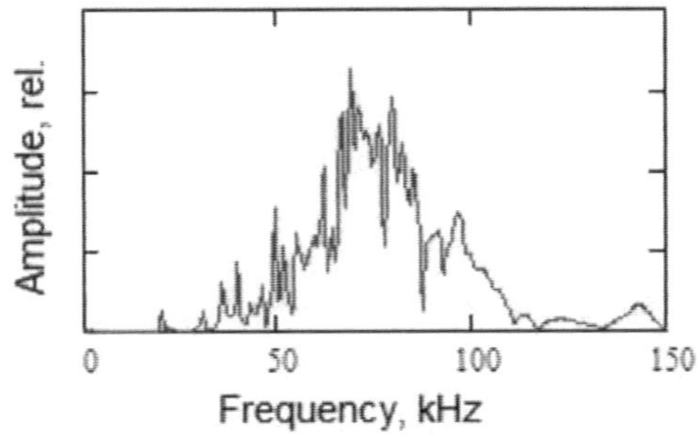

Figure 7. AFC of the temporary realization of the calculated response.

The comparative analysis of calculation results for model with a real response showed their qualitative similarity. Difference is caused, mainly, by distinction of internal structure. In model the sample is without defects and structureless whereas the real sample has a certain concentration of defects and has difficult multi-scale structure. Nevertheless the settlement response can be quite used as a reference point concerning which the assessment of change of defectiveness and structure under influences of external impact, for example, of mechanical loading, change of temperature and humidity in the conditions of monitoring of testing object.

4. Conclusion

The approach used to create the adjusted mathematical model, allows calculating responses that can serve as reference point for evaluation the defectiveness materials degree on experimental data.

Acknowledgments
This work was financially supported by The Ministry of Education and Science of the Russian Federation in part of the science activity program.

References

[1] Slasten M I 2012 *Izvestiya YUFU* **19** 134
[2] Feng-Hua Li, Jian-Jun L 2002 *Chinese Physics Letters* **19** 8
[3] Couvreur L., Couvreur Ch 2004 *The Journal of VLSI Signal Processing* **36** 2–3
[4] Freemantle R J, Alper T, Challis R E 1993 *Measurement Science and Technology* **4** 10
[5] Surzhikov V P, Khorsov N N, Khorsov P N 2011 *Control. Diagnostics* **11**
[6] Surzhikov V P, Khorsov N N, Khorsov P N 2012 *Russ J Nondestruct+* **48** 2
[7] Surzhikov V P, Khorsov N N, Khorsov P N 2012 7th International Forum on Strategic Technology (IFOST - 2012) **2**
[8] Khorsov N N 2007 *Russ Phys J+* **50** 2

IOP Conf. Series: Materials Science and Engineering **168** (2017) 012107 doi:10.1088/1757-899X/168/1/012107

Thermal effects of carbonated hydroxyapatite modified by glycine and albumin

S A Gerk[1], O A Golovanova[1] and M V Kuimova[2]

[1]F.M. Dostoevsky State University, Omsk, Russia
[2]National Research Tomsk Polytechnic University, Tomsk, Russia

E-mail: gerksa_11@mail.ru

Abstract. In this work calcium phosphate powders were obtained by precipitation method from simulated solutions of synovial fluid containing glycine and albumin. X-ray diffraction and IR spectroscopy determined that all samples are single-phase and are presented by carbonate containing hydroxyapatite (CHA). The thermograms of solid phases of CHA were obtained and analyzed; five stages of transformation in the temperature range of 25-1000°C were marked. It is shown that in this temperature range dehydration, decarboxylation and thermal degradation of amino acid and protein connected to the surface of solid phase occur. The tendency of temperature lowering of the decomposition of powders synthesized from a medium containing organic substances was determined. Results demonstrate a direct dependence between the concentration of the amino acid in a model solution and its content in the solid phase.

1. Introduction

Among a wide range of modern biomaterials used for the restoration of bone defects, composites, based on carbonated hydroxyapatite (CHA) and organic components such as collagen and non-collagenous proteins, polysaccharides, polymers, etc., are in wide demand [1-5]. Crystal-chemical similarity of CHA with inorganic basis of bone mineral provides a high biocompatibility of such materials used in implantation; namely, it induces biochemical reactions similar to those of bone remodeling [6, 7]. The use of organic compounds in the quality of dopants approximates the conditions of the synthesis of materials to bone mineralization *in vivo*, taking place on a collagen matrix and with biological fluids [1-3, 8]. Therefore, it is possible to obtain CHA similar in crystallinity and morphological characteristics to the native bone that also improves the osteointegration of the implant with the surrounding living tissues. Despite a large number of works on the preparation of biocomposites [1-8], the influence of organic materials on the structure and properties of CHA, the mineral component of bone tissue, are not adequately covered. Thus, the role of amino acids of the polypeptide chain of collagen, composing the crystalline segments -Gly-X-Y-, (Gly – glycine, proline, etc.) and amorphous regions (lysine, histidine and others) are not fully described [9, 10]. The impact of organic components of biofluids (albumins, globulins, lipids, carbohydrates, etc.) on the characteristics of the solid phase is not fully considered [11]. Therefore, the present work is devoted to the preparation of CHA powders from model solutions of human synovial fluid containing glycine and albumin, study of their composition and thermal effects.

2. Material and methods

CHA was produced by "wet synthesis" from the model solution approximated by electrolyte composition to the synovial fluid of an average healthy person at values pH = 7.40±0.05 and ionic strength - 0.174±0.001 [8]. This CHA was used as a control sample. Two series of model experiments were carried out. In the first series of syntheses, in the initial solution containing anions (HPO_4^{2-}, HCO_3^-, Cl^-, SO_4^{2-}) glycine was added (CH_2NH_2COOH, α- amino acetic acid of CP brand, the range of concentrations from 0.04 to 0.24 mol/l), the second – albumin (bovine serum albumin (BSA) of A brand; the range of concentration from 5 to 10 g/l). When selecting the concentrations of glycine and albumin, we were guided by their content in the collagen of the bone tissue and biological fluids (synovia, blood plasma), respectively [12, 13]. The solution containing the anions (HPO_4^{2-}, HCO_3^-, Cl^-, SO_4^{2-}) and the organic substance was spilled to the solution with calcium and magnesium cations at a rate of 5 ml/min. Further, the acidity of the model system was adjusted by adding a 10% solution of HCl until the value of pH = 7.40±0.05 was reached. Crystallization of the solid phase was carried out at room temperature (20-22 °C) for 7 days. At the end of this period, the precipitate was separated from the solution by filtration, washed three times with water, dried at 80 °C to completely remove the chemically not of the connected water, until a constant weight was achieved.

A study of the phase composition was conducted using the X-ray powder diffractometer D8 Advance (Bruker) under the following conditions: Cu-kα radiation (wavelength 0.15406 nm) scanning step – 0.05°, time of accumulation of a signal – 4 seconds/point, div. slit=0.5, voltage and filament current 40 kV and 40 mA, respectively; scan region 2θ – 5-80°. The transcripts of the diffractograms were obtained from the databases of powder diffraction ICDD PDF-2 in the program EVA (Bruker). IR spectra of the precipitates were recorded on a spectrophotometer "FSM-2202". The scanning range was 400-5000 cm^{-1}, the resolution was 8 cm^{-1}. The samples were prepared by compression into tablets with KBr. The thermal analysis was carried out on a STA-449 C synchronous thermal analyzer "NETZSCH". The samples were calcined in platinum crucibles in the atmosphere of the air from 25 to 1000 °C at the rate of 10 °C/min. The mass of the sample varied depending on the values of the recorded signals of the mass loss effect and the thermal effect. The mass was 10-20 mg. The quantitative data of mass loss during annealing were obtained using software Proteus 7.10. By obtained thermogravimetric (TG) and differential thermal (DTG) curves, the mass losses of the components of the samples were determined by heating. A differential gravimetric curve (DTA) was used for the characteristics of the observed thermal effects (changes in energy, °C/min). The specific surface areas of the samples were studied by the method of single-point nitrogen adsorption at 77.4 K on the adsorption apparatus "Sorbtometr", produced by Katakon LLC, Russia. The calculation of the obtained values of S_{BET} (m^2/g) was done using the BET method.

3. Results and Discussion

Using XRD and IR-spectroscopy, it was found that all precipitates are single-phase and are presented by CHA (figure 1a, b), regardless of the presence of organic substance in the initial model solution. On IR spectrums there are absorption of the bonds of inorganic groups and molecules: PO_4^{3-}, CO_3^{2-}, -OH и H_2O. The presence of modes of vibrations of the organic groups $-NH_2$, $-CH_n$, C=O, -N-H indicates that the glycine and albumin are in the composition of the precipitates.

However, according to data of the thermal analysis, the presence of organic substance in the initial medium reflects on the form of the curves TG, DTG and DTA (figure 2), which indicates different mass losses and thermal effects of these powders. Comparing the obtained thermal curves of the control sample (figure 1a) with the published data [14-20], we can distinguish five stages of thermo-transformation in the temperature range 25-1000 °C.

In the first stage (*I, 25-280 °C*, endothermic effect there is the removal of chemically free water and decomposition of volatile impurities (such as adsorbed CO_2).

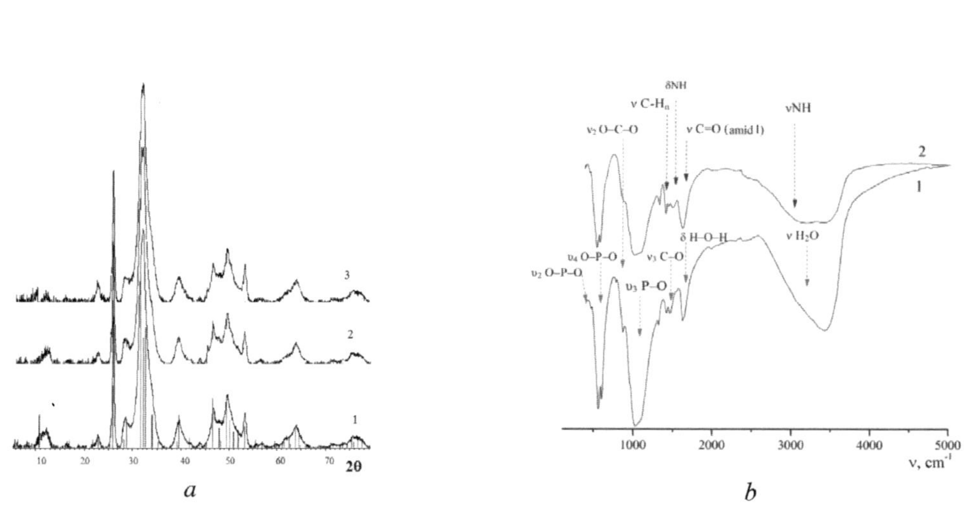

Figure 1. Pattern (a) and IR spectra (b) CHA: 1) pure; 2) 0.16 mol/l glycine; 3) 5 g/l albumin.

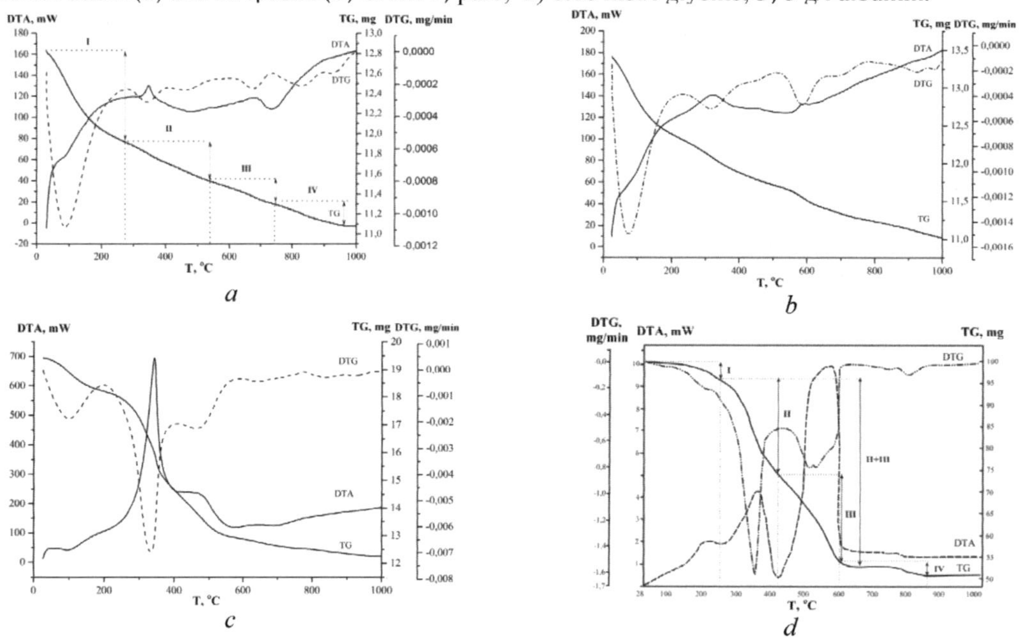

Figure 2. Thermograms CHA: a) pure; b) 0.04 mol/l glycine; c) 5 g/l albumin; d) human bone tissue «in the norm» [20].

The second and third stages (exothermic effects) are followed by the loss of water associated in various ways with the solid phase: physically adsorbed (*II, 280-470 °C*), chemisorbed and crystallization (*III, 470-750 °C*). Schematically the processes of water removal on the data can be represented as:

$$[\text{КГА} \cdot nH_2O] \cdot mH_2O_{(sol.)} \rightarrow \text{КГА} \cdot nH_2O_{(sol.)} + mH_2O_{(g.)} + Q; \ \text{КГА} \cdot nH_2O_{(sol.)} \rightarrow \text{КГА}_{(sol.)} + H_2O_{(g.)} + Q.$$

At higher temperatures, at the fourth and fifth stages (*IV and V, 750-900 °C and 900-1000 °C*) there are two endothermic peaks at 830 and 940 °C on DTA curves. In these temperature ranges, the transformation of CHA occurs, namely, the removal of carbonate ions from its structure in the form of CO_2 and transition of non-stoichiometric carbonated hydroxyapatite to the stoichiometric phase or β-$Ca_3(PO_4)_2$:

$$Ca_{10}(PO_4,CO_3)_6(OH)_{2(sol.)} \rightarrow Ca_{10}(PO_4)_6(OH)_{2(sol.)} + 6\ CO_{2(g.)} - Q;$$

$$Ca_9(PO_4)_{6-x-y}(HPO_4)_y(CO_3)_x(OH)_{2-y(sol.)} \xrightarrow{800\ ^\circ C} 3\ \beta\text{-}Ca_3(PO_4)_{2\ (sol.)} + x\ CO_{2(g.)} + 2\ H_2O_{(g.)} - Q\ [22].$$

In works [1, 23] it is marked that at a temperature of 850 °C, the transformation of GA hexagonal syngony into monoclinic form (hexagonal P63/m → monoclinic P21/b) is possible due to the reorientation of the dipolar groups OH^- in the triangular channels of Ca^{2+} ions of its crystal lattice. At 900 ^{0}C the formation of oxyhydroxyapatite from stoichiometric HA is possible according to the reaction [24]:

$$Ca_{10}(PO_4)_6(OH)_{2(sol.)} \rightarrow Ca_{10}(PO_4)_6(OH)_{2-2x}O_x n_{x(sol.)} + xH_2O_{(g.)},$$ где n_x – vacancy, x<1.

It was found that the powders, synthesized in the presence of an organic component, are characterized by the highest total losses of mass, and their transformation takes place at lower temperatures (I-IV, Table 1, figure 2 and 3a). Perhaps this is due to the presence of organic substance in the solid phase, which is consistent with the data of IR-spectroscopy.

Table 1. Temperatures of transformation and mass losses (Δm) of the components of powders.

Stages	T beginning of transformation (°C)			Δm (mg)							
					In the presence of glycine (mol/l)				In the presence of albumin (g/l)		
	Control sample	In the presence of glycine	In the presence of albumin	Control sample	0.04	0.08	0.16	0.24	5	7	10
I	25	25	25	7.098	7.367	7.300	8.352	10.647	6.282	5.381	5.430
II	280	240	200	2.340	3.302	3.408	3.728	4.095	18.280	18.130	20.440
III	470	450	400	2.496	2.964	3.015	3.057	3.744	9.166	9.195	12.449
IV	750	670	600	1.248	1.524	1.700	1.417	1.872	1.905	2.090	1.741
V	900	850	800	0.390	1.354	1.268	1.268	1.395	1.287	1.097	1.076
I+V		25-1000		13,572	16,511	16.733	17.822	21.753	35.920	35.893	41.136

It is known that thermal decomposition of aliphatic amino acids, to which belongs glycine, as well as peptides occurs at temperatures not exceeding 400 °C (T_m glycine = 232 - 262 °C; T_m albumin <90 °C.) [15, 16, 25]. Thus, the thermal degradation of glycine occurs in two steps. At the initial stage (fast stage, $\Delta H > 0$) at 200-270 °C, oligopeptides and diketopiperazine (cyclic dipeptide) are formed. Methylamine, formamide, acetamide, propionamide, N-atsetilamid and acetic acid, water, carbon dioxide are by-products. Further, at 270-370 °C slow stage ($\Delta H < 0$) of complete oxidation of degradation products proceeds. From data of Table 2 it follows that the thermal transformation of glycine, which is present in the obtained powders, occurs in stage III and corresponds at 2000 seconds to the exothermic peak on the DTA curve (321 °C). Exactly on the third stage with the increasing content of amino acids in the initial medium, the mass losses of glycine increase linearly. However, less energy is consumed for the destruction of the samples containing larger amounts of amino acids (figure 3*b*). Possibly at a lower concentration of glycine in an aqueous solution, components of zwitterions (NH_3^+- $C^{\alpha}H_2$ - COO^-) bond with the ions of calcium and PO_4^{3-} CHA more tightly and have a high adsorption activity in relation to their surface. This correlates well with the data of the specific surface area (Table 2). It is seen that the samples obtained from the model medium with low content of amino acid have the lowest specific surface.

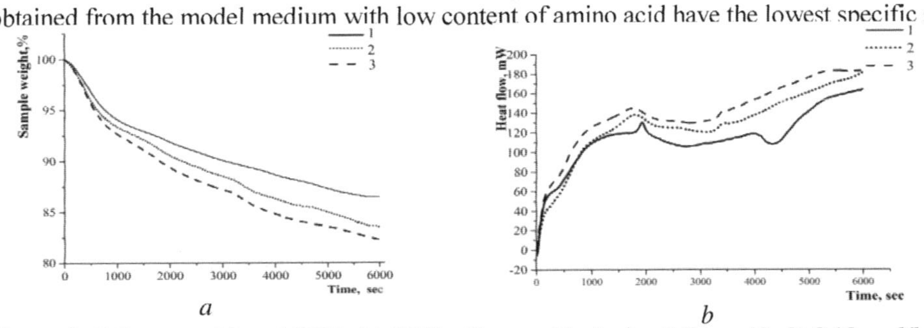

Figure 3. TG curves (a) and DTA (b) CHA: 1) pure; 2) glycine 0.04 mol/l; 3) 0.12 mol/l.

Table 2. The specific surface area of the powders.

CHA	Control sample	In the presence of glycine (mol/l)			In the presence of albumin (g/l)		
		0.04	0.08	0.016	5	7	10
S_{BET}, m^2/g	130±7	70±1	52±2	75±3	44±2	48±2	25±2

Interesting results were obtained in the analysis of the thermograms of the samples synthesized in the presence of albumin (figure 4, Table 1). It was shown that the presence of protein leads to higher mass loss at temperatures above 200 – 240 °C and does not depend on its mass in the model solution.

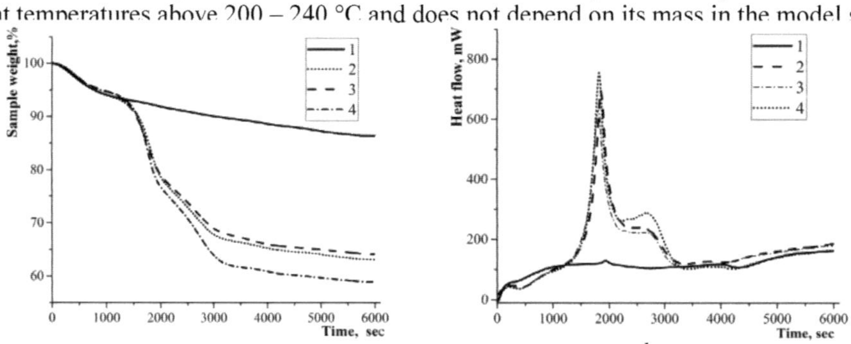

a *b*

Figure 4. TG curves (a) and DTA (b) CHA: 1) pure; albumin – 2) 5 g/l; 3) 7 g/l; 4) 10 g/l.

We believe that, despite the low temperature of denaturation of pure albumin, the destruction of the polypeptide chains bounded with the surface of CHA occurs in the temperature interval 200-400 °C ... 450 (stage II, DTA maximum at 340 ... 250 °C and 2000 seconds figure 3a). We obtained similar results previously while studying the thermal curves of bone tissue (step II, figure 2d). In [21] it is shown that at this stage (DTA maximum/peak at 360 °C) there is a thermal oxidation of organic substances with less molecular mass than a collagen; for example, non-collagen proteins (albumin). In addition, high temperatures of thermal deterioration of albumin in the solid phase can be explained by a complex set of amino acids. Human and serum albumin are globular proteins, whose primary structure includes 585 and 582 residues respectively. These residues can be different amino acids: neutral, basic and acidic; their side radicals can be various functional groups (e.g., tryptophan, phenylalanine, tyrosine, etc.) [16]. It is known that the destruction of such amino acids may consist of four or five stages, and proceed at temperatures of 300 ... 400 °C and 600 ... 700 °C [26]. The presence of charged functional groups and zwitterions of amino acids provides a stronger interaction of albumin with the surface of CHA, which is in agreement with data from works about the chemisorption mechanism of their interaction [27]. The powders synthesized from the medium containing the protein, unlike the samples prepared in the presence of glycine, are characterized by the lowest specific surface area of powders. At the maximum content of albumin (10 g/l), the sample is formed, for which S_{micro} value is 5 times lower than the specific surface area of the control sample (Table 2).

In all samples, for which the precipitate was held in the presence of organic substance, a significant loss of mass at temperatures of more than 600 ... 670 °C (stage IV and V, Table 2) was recorded. At these temperatures, the removal of carbonate ions from CHA structure occurs. This fact indirectly indicates that in the crystalline grid of CHA of these precipitates, there are numerous carbonate ions; therefore, the samples with glycine characterized by low crystallinity and regularity of structure. At IR spectra of these powders, the reduction of intensity of mode υ_3(P–O) и υ_4(O–P–O) of phosphate groups (figure 1b) is indicated.

4. Conclusion

Thus, CHA powders were synthesized in the presence of albumin and glycine from prototypes of synovial fluid. It was shown that organic substance is present in the solid phase of the samples.

Investigated thermal effects and mass loss of the samples CHA obtained in the presence of glycine and albumin. It was found that the thermal decomposition of organic substance associated with surface area of CHA mainly takes place at temperatures of 200-400 ... 450°C. It was shown that with increasing content of amino acid in the initial medium, mass losses of glycine increased linearly. The mass loss for the powders obtained in albumin environment does not depend on its concentration in a model solution. For all samples obtained in the medium containing organic substance, unlike the control sample, the smallest specific surface area and crystallinity are characteristic.

Acknowledgments
The research was partially supported by the Russian Foundation for Basic Research (grant mol_a No. 16-33-00535 and No. SP-933.2015.4).

Reference

[1] Starikova V V and Rudchenko S O 2010 *Vest. Kharkov Un-ty* **14** 35
[2] Assumption W A, Zakharov V N and Villager M A 2014 *Patent.* 22534789
[3] Moysenovich M M, Arhipova A Yu, Orlova A A, Drutskaya M S, Volkova S V, Zaharov S E, Agapov I I and Kirpichnikov M P 2014 *Acta Naturae* **1(20)** 103
[4] Yanovska A A, Kuznetsov V N, Stanislaus A S, Gander E V, Pogorelov M V and Danilchenko S N 2015 *Chem., phys. and tech. of surface* **4** 535
[5] Samchenko Y M, Boldeskul I E, Sukhodub L B, Danilchenko S N, Birch L I and Ullberg Z R 2009 *Nanostruct. Materials* **1** 81
[6] Lukin Yu C 2010 *Dis. ... Cand. Techn. Sciences* (Moscow: Russian University of chemical technology) p 161
[7] Dorozhkin S V 2009 *Mater.* **2** 399
[8] Golovanova O A, Lemesheva (Gerk) S A and Ismailov R R 2013 *Patent.* 2496150(379).
[9] Green D W, Goto T K, Kim K S and Jung H S 2015 *J. Roy. Soc.* **3** 1
[10] Torbenko V P and Karsavina B S 1977 *Functional biochemistry of bone* (Moscow: Medicine) p 272
[11] Martin R B 1999 *Mater. Sci. Forum.* **1** 5
[12] Matveeva E L 2007 *Abstract. Dis. Doctor. Bio. Sciences* (Tyumen: Tyumen State University) 24
[13] Krylova N N 1954 *Meat Biochemistry* (Moscow: Pieprasit) p 320
[14] Goloshchapov Y N, Kashkarov V M, Rumyantsev N A, Seredin P B, Lenshin A C, Agapov B L and Domashevskaya E P 2011 *Condens. matter and interphase boundaries* **4** 427
[15] Smeltsova I L 2013 *Dis. ...Can. Chem. Sciences* (Nizhniy-Novgorod: Nizhny Novgorod State University of architecture and construction) 104
[16] Badelin V G, Tyunina E Y and Mezhevoy I N 2014 *Liq. Cryst. and their Appl.* **3** 43
[17] Nedoseko V B, Gorbunova I L and Drozdov V A 2004 *Dent.* **4** 13
[18] Sudakova T V, Litvinov S D, Baev E E and Gusev V V 2009 *Act. Prob. of new science* **10** 7
[19] Safronova T V, Putlyaev V I, Kurbatov S A, Shatalov T B, Larionov D S, Kozlov D A and Evdokimov P V 2015 *J. Inorg. Materials* **1** 1269
[20] Guzeeva T I, Guzeev V V, Leonova L A, Lelyuk O A, Krikunenko A S and Shatokhina Y M 2009 *Vest. Tomsk Polytech. Un-ty* **3** 47
[21] Lemesheva S A and Golovanova O A 2009 *Chem. for sustainable development* **3** 327
[22] Bogdanova E A, Sabirzyanov N A 2014 *Mater.* **10**
[23] Zakharov N A, Klyuev V A, Sentsov M Y and Toporov Y P 2012 *J. Techn. Physics.* **4** 153
[24] Barinov S M 2005 *Bioceramics of calcium phosphate* (Moscow: Science) p 204
[25] Sarnatskaya V V, Yushko LA, Korneeva L N, Sahno LA, Maslenny V N, Snezhkova E A, Mihalovskiy S V and Nikolaev V G 2005 *Effer. therapy* **11** 10
[26] Pshonkina N N 2011 *Fundament. Research* **12** 1067
[27] Morgues P A, Serro A P and Saramago B J 2003 *Biomaterials* **24** 451

Study of the effects of heat-treatment of hydroxyapatite synthesized in gelatin matrix

A V Zaits[1], O A Golovanova[1], M V Kuimova[2]
[1]Department of Inorganic Chemistry Omsk F. M. Dostoevsky State University, Russia
[2]National Research Tomsk Polytechnic University, Tomsk, Russia

E-mail: Golovanoa2000@mail.ru

Abstract In the study, the isothermal thermogravimetric analysis (TGA) of hydroxyapatite synthesized in gelatin matrix (HAG) has been performed. 3 wt.% HAG samples were synthesized from the solution simulating the human extracellular fluid (SBF). X-ray diffraction and IR spectroscopy were used to determine the composition. During the experiment, increase in the calcination temperature up to 200°C–800°C was found to cause weight loss. The study of phase composition revealed that heat treatment does not affect the phase composition of the solid phase, which is composed of hydroxylapatite (HA). The prepared HAG (3 wt.% gelatin) samples are shown to have low thermal stability; the degradation of the samples occurs at 400°C.

1. Introduction

The global bioceramics demand on the market currently reaches ~40 billion euros per year, the estimated annual growth rate is of 7–12% and the required amount of the material to meet the demand is estimated at tens of tons [1–3]. The number of patients in need of surgery to restore the integrity of the bone is increasing every year: in the US, the annual number of patients exceeds more than 1 million people (of which over 300 thousand people need hip and knee prosthetics, the same holds true for dental implants).

Materials based on hydroxyapatite $Ca_{10}(PO_4)_6(OH)_2$ are analogs of the mineral component of bone, and they are considered the most promising biocomposites to be used as substitutes for bone defect repair due to their high bioactivity [4–9].

It is known that the main components of bone are type I collagen (~ 20%), a mineral phase (~ 60%), water (~ 9%) and non-collagenous proteins (~ 3%), the rest being polysaccharides and lipids [10, 11]. The organic part of bone is composed of collagen fibers and proteins, such as osteocalcin, osteonectin and fibronectin [12]. Gelatin is a natural polymer formed through hydrolysis of collagen. In contrast to the latter, it is more stable and exhibits less antigenicity, thereby can be effectively used as a calcium phosphate based organic matrix biomaterial [13]. Gelatin contains biologically active functional amino acid groups, and it is an advanced material for bone regeneration, including its combination with hydroxyapatite [14]. Since gelatin is a denatured form of collagen and it contains significant amounts of biologically functional amino acid groups, the investigation of properties of hydroxyapatite synthesized in the gelatin matrix (HAG) under near-physiological conditions is a challenging and relevant physicochemical problem.

2. Materials and methods

The SBF was used as a model solution of extracellular fluid. A solid phase was prepared by precipitation from aqueous solutions; the starting components of the system were $CaCl_2$, $MgCl_2$, K_2HPO_4, $NaHCO_3$, Na_2SO_4, NaCl [15] solutions and 3 wt.% gelatin (figure 1).

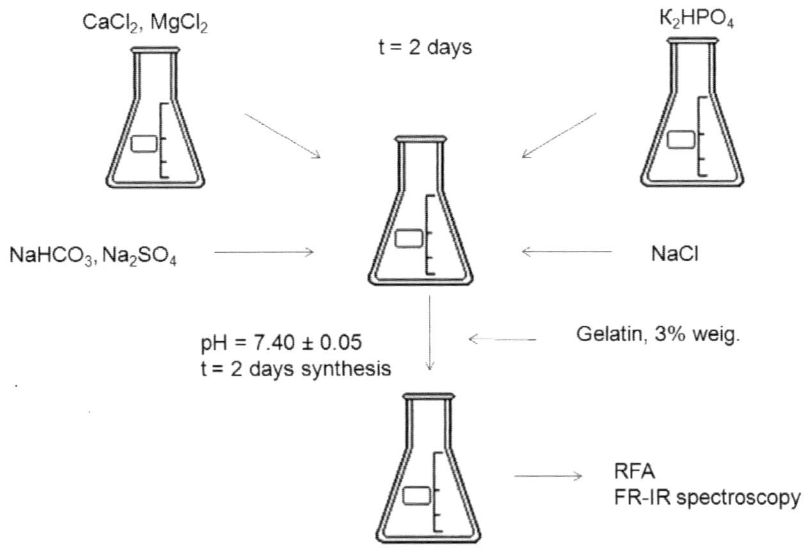

Figure 1. Scheme for HAG synthesis from SBF.

Thermal effects were studied by isothermal thermogravimetric analysis. Hydroxyapatite samples synthesized in the presence of gelatin were placed in dry ceramic crucibles. Then the samples were placed in a muffle furnace and kept for 2 h. The calcination temperature was varied from 200 to 800°C by increment of 200°C. After that, the crucibles were allowed to cool to room temperature, and then weighed on an analytical balance.

The solid precipitates were analyzed by FT-IR spectroscopy and X-ray diffraction (XRD) – qualitative and quantitative phase analysis, measurement of the crystallite sizes (coherent scattering regions – CSR). The IR spectra were recorded using a FT-801 spectrometer (the samples were prepared in the form of KBr pallets). The XRD of the powder samples was performed with the Bruker D8 Advance X-ray diffractometer, and the diffraction patterns were identified using the software package EVA (developed by Bruker AXS).

3. Results and discussion

The XRD results showed that the samples synthesized in SBF under varying concentration of gelatin were single-phase and represented hydroxyapatite. The diffraction patterns indicated HA reflexes at 25.8° (002), 31.7° (211), 32.2° (112) and 32.9° (300) (figure 1).

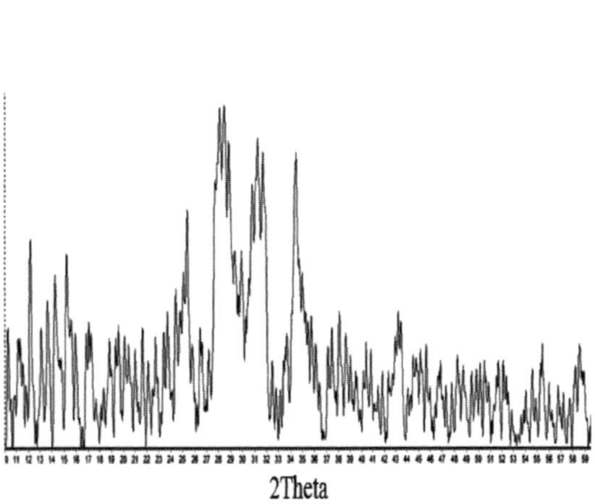

Figure 2. Diffraction pattern of the hydroxyapatite powder crystallized in the presence of 3% gelatin solution.

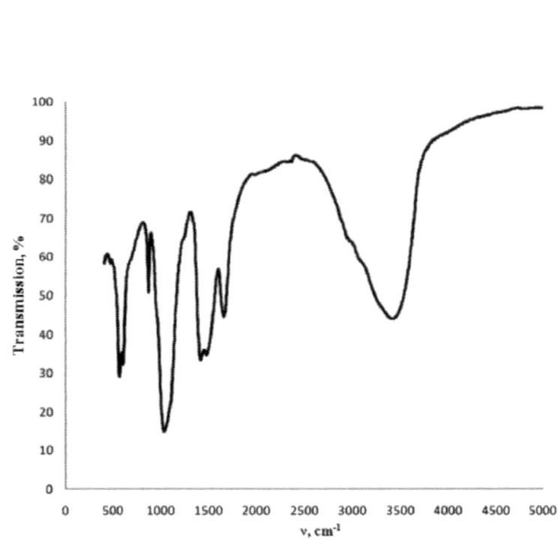

Figure 3. Results of FT-IR spectroscopy for the hydroxyapatite sample synthesized in a gelatin matrix.

The lattice parameters of the prepared solids are increased (Table. 1) as compared to those of the unmodified hydroxyapatite that can be attributed to possible HA sorption in the gelatin matrix.

Table 1. Lattice parameters.

Sample	Initial gelatin concentration, %	Lattice constants, Å	
		a	*c*
HA	-	9.414	6.865
Gelatin	3	9.432	6.881

The analysis of the IR-spectra (figure 3) identified the bands that determine the HA structure: 1040–1080 (ν_3), 960, 840 (ν_2), 602, 574 (ν4) and 473 (ν_2) cm^{-1} corresponding to the phosphate group vibrations, a broad band at 3440–3570 cm^{-1} related to deformation vibrations of the OH^--groups and valent vibrations of the adsorbed water. Carbonate ions are observed to localize in the phosphate ion lattice position with the corresponding absorption bands of the C–O deformation vibrations of the CO_3^{2-} ion bonds at 875 cm^{-1} and valent vibrations at 1415–1481 cm^{-1} in the IR-spectra [16]. A corresponding band of the deformation vibrations of the C=C bond in the region of 1658 cm^{-1}.

The study of the prepared hydroxyapatite samples synthesized in the presence of gelatin showed continuous weight loss (figure 4).

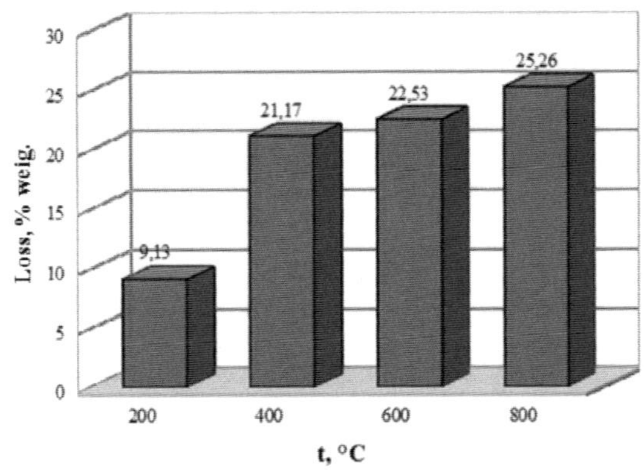

Figure 4. Diagram of TGA for 3 wt.% HAG samples.

The TGA diagram demonstrates the highest weight loss at 200°C and 400°C that can be attributed to adsorption and removal of the chemically bound water from the HAG samples. An increased weight loss at 400°C can be caused by burn out of the organic phase (gelatin) in the HA samples. At 600°C, the weight decreases insignificantly, while at 800°C, the weight loss increases by 2 times (compared to that at 600°C), which indicates the removal of the carbonate groups from the HA structure. These data is consistent with the results obtained by FT-IR spectroscopy (figure 5).

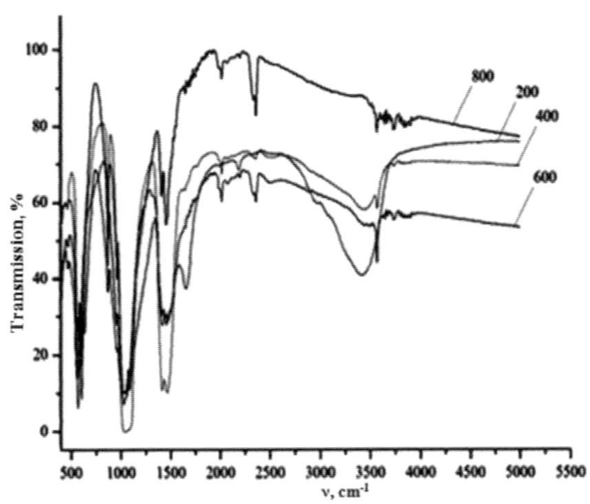

Figure 5. Results of FT-IR spectroscopy for 3% HAG samples subjected to heat treatment at 200, 400, 600 and 800°C.

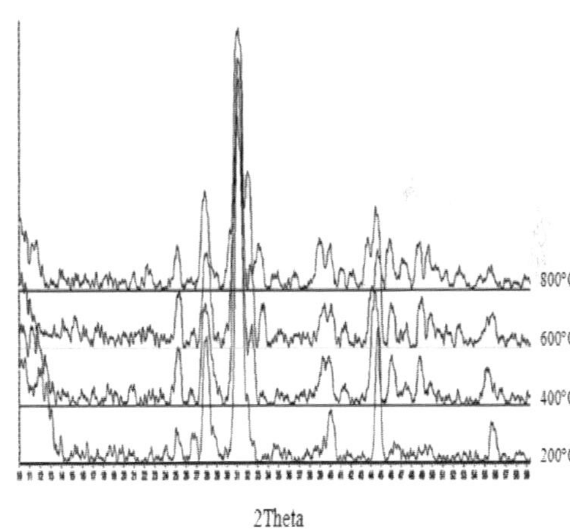

Figure 6. X-ray diffraction patterns for 3% HAG after heat treatment.

The X-ray diffraction patterns (figure 6) of the HAG samples subjected to heat treatment show peak splitting at 400°C. In our opinion, this is caused by degradation of the gelatin components (amino acids), and thereby release of the carbonate component that has been partially released in the form of carbon dioxide, and partially entered the calcium phosphate structure. At 600°C and 800°C, the intensity of peak splitting did not increase and remained at the same level. This is most likely due to the fact that no orbitals are left vacant to bind carbonate ions in calcium phosphate as compared to that in case of unsubstituted hydroxyapatite.

The analysis of the IR-spectra (figure 3) identified the bands that determine the HA structure: 1040–1080 (v_3), 960, 840 (v_2), 602, 574 (v4) and 473 (v_2) cm^{-1} corresponding to the phosphate group vibrations, a broad band at 3440–3570 cm^{-1} related to deformation vibrations of the OH^--groups and valent vibrations of the adsorbed water. Carbonate ions are observed to localize in the phosphate ion lattice position with the corresponding absorption bands of the C–O deformation vibrations of the CO_3^{2-} ion bonds at 875 cm^{-1} and valent vibrations at 1415–1481 cm^{-1} in the IR-spectra [16]. A corresponding band of the deformation vibrations of the C=C bond in the region of 1658 cm^{-1}.

The results of FT-IR spectroscopy showed that the spectrogram of the samples not subjected to heat treatment (figure 3) is identical to that of the samples heated at 200°C (figure 5). This suggests that no significant change in the HAG structure occurs at 200°C. During the analysis, several overlapping spectra were observed: in the region from 2700 cm^{-1} to 3800 cm^{-1}, which, apart from the -OH water vibrations, include vibrations of the amino groups, and methyl and methylene groups. In addition, $–H_2$ groups overlap the carboxyl groups in the region of 1400 cm^{-1}.

As can be seen in Fig 4, at 400°C, the peak significantly decreases in the region of 3500 cm^{-1} that indicates removal of water from the structure and degradation of amino acids, and methyl and methylene groups. The shape of the carboxyl group doublet in the region of 1415–1466 cm^{-1} changed as well that can be confirmed by the removal of the $–CH_2$ group from the prepared sample.

When the temperature rises to 600°C, the intensity of the C=C band (1658 cm^{-1}) is not observed, and the intensity of the C-O band (1420–1466 cm^{-1}) decreases, which indicates the start of the carbonate component destruction. In addition, the O-H band (3569 cm^{-1}) reduces, which is also indicative of the start of the hydroxyapatite destruction.

At 800°C (figure 6), the spectrogram shows further reduction of the O-H band (3569 cm^{-1}) and C-O band (1420–1466 cm^{-1}).

At 200°C, a corresponding deformatio band of the C=C bond is observed in the region of 1658 cm^{-1}. At temperature increase up to 400°C, the peak intensity falla, and at temperature up to 600°C, it completely disappears, which is consistent with the data in [17].

4. Conclusion

The isothermal thermogravimetric analysis revealed four stages of thermal decomposition of composite materials: 1) removal of water and easily volatile impurities (t=25°C–280°C); 2) gelatin pyrolysis (t=280°C–400°C); 3) calcium pyrophosphate phase formation (t=400°C–600°C); 4) removal of carbonate ions from the HA structure and its decomposition (t > 600°C).

Acknowledgements
This research was supported in part by the Russian Foundation for Basic Research (grant No. 15-29-04839 ofi_m).

References

[1] Hench L, Johns D 2007 *Technosphere* **1** 301
[2] Suchanek W, Yashimura M 1998 *J. Mater. Res.* **13** 94
[3] Zethraeus N, Borgstorm F, Strom O et al. 2007 *Osteoporosis International* **18** 9
[4] Astakhov I N 2008 *Surgery.* **12** 34
[5] Putlyaev V I 2004 *Soros Educational J.* **1** 44
[6] Fomin A S, Komlev V S, Barinov S M, et al. 2006 *Advanced Mater.* **2** 51
[7] Kosyakov A N 2010 *Orthopedics, Traumatology and Prosthetics* **4** 105
[8] Gnedenkov S V 2011 *Advanced Mater.* **2** 49
[9] Sidelnikov A I 2000 *Info-Dent.* **5** 10
[10] Orlovskii V P, Komlev V S, Barinov S M 2002 *Neorg. mater.* **38** 973

[11] Murugan R, Ramakrishna S 2005 *Cryst Growth Des.* **5** 111
[12] Li J, Chen Y, Yin Y, Yao F, Yao K 2007 *Biomaterials* **28** 781
[13] Zhou Y, Zhao Y, Wang L, et al. 2012 *Radiat. Phys. Chem.* **81** 553
[14] Golovan A P, Rugal A A, Gun'ko V M, et al. 2010 *Surface* **2**. 244
[15] Golovanova O A, Berdinskaya M V, Zaits A V, et al. 2014 *J. of Structural Chem.* **5**. 992
[16] Khlusov I A, Surmeneva M A, Surmenev R A, et al. 2012 *Bulletin of Siberian medicine* **4** 78
[17] Solonenko A P, Golovanova O A 2013 *Rus. J. of Inor. Chem.* **58** 1420

IOP Conf. Series: Materials Science and Engineering **168** (2017) 012109 doi:10.1088/1757-899X/168/1/012109

Production of silver nanoparticles in water solution by radiation treatment

M A Mikhailenko[1], M V Korobeinikov[2,3], A A Bryazgin[2,3], B P Tolochko[1]

[1]Institute of Solid State Chemistry and Mechanochemistry SB RAS, Kutateladze, 18, 630128 Novosibirsk, Russia
[2]Budker Institute of Nuclear Physics SB RAS, Lavrentieva, 11, 630090 Novosibirsk, Russia
[3]Novosibirsk National State Research University, Pirogova, 2, 630090 Novosibirsk, Russia

E-mail: M.V.Korobeynikov@inp.nsk.su

Abstract. Radiation-chemical synthesis of silver nanoparticles was studied. The silver nanoparticles in arabinogalactan (AG) water solution are stabilized in conglomerates, it is fixed by rise of additional bands in the optical absorption spectra. Pre-radiation treatment of AG causes crosslinking and oxidation. Pretreated AG solution increases the stability of conglomerates containing silver nanoparticles in case of dilution.

1. Introduction

The aqueous solutions containing silver nanoparticles are now used in medical practice, and their applications are widening Production of stable water solutions containing silver nanoparticles is important both for theoretical and practical considerations. It is witnessed by numerous works [1]. Water-soluble polymers, including natural biopolymers, added into water are efficient stabilizers of silver nanoparticles [2]. Natural biopolymers can favor to form nanoparticles having narrow size distribution perhaps due to solution structuring and formation of permolecular structure. Natural polymers can have some advantages, for example, they can simultaneously be deoxidizers for silver ions and stabilizers for resulting dispersions. Other important properties of natural biopolymers are biocompatibility and low toxicity.

Arabinogalactan (AG) is biocompatible branched natural polysaccharide that can act both as deoxidizer and stabilizer. Beside these AG is capable to structure on it's own into globules of various sizes depending on concentration in solution [4].

Publication [5] describes deoxidizing properties of the AG, a highly alkaline medium was used to activate it. Silver or silver hydroxide precipitation in alkaline medium is inevitable, and a reaction localization (in solution or heterogeneously on sediments' surface) is not known. Alkaline infusion for activation of AG deoxidizing properties can also be regarded as chemical deoxidizing.

Chemical deoxidizing methods for silver nanoparticles formation lead to product pollution. It restricts application of reaction product or requires additional stages to remove intermediate products. And presence of chemical deoxidizer influences on stabilization and self-organization processes. A work [6] shows that silver nanoparticles formed in photochemical deoxizing process are ordered on nanocarbon tubes functionalized by sodium dodecylulphate, and collective photon effects are displayed. A chemical deoxidizing resulted in chaotic nanoparticles arrangement on carbon nanofibers.

Another way to produce metal nanoparticles without chemical deoxidizers is radiation-chemical process [7]. Powerful ionizing radiation required for this process can be generated either by unstable isotopes or by electron accelerators. Advantages of modern electron accelerators as radiation sources are high radiation intensity, possibility of operative process control and high reproducibility [8]. Radiation treatment can be done by electron beam or by X-ray generated by special converter [9]. Metal ions deoxidization during water solutions irradiation can be done by water radiolysis products [10] and perhaps by free electrons.

Aim of our work was to study electron beam treatment possibilities for production of stable water solutions containing silver nanoparticles and AG as a stabilizer. Dependence of solution properties on preliminary AG solution electron beam treatment was studied, it was compared with AG solution without preliminary electron beam treatment.

2. Experiment

Arabinogalactan (Mr = 17,000 by GPC) was separated out of larch wood (Lárix sibírica Ledeb.) [11]. 0.01 M $AgClO_4$ water solution in 0.05M $HClO_4$ aqueous solution was produced by silver oxide powder (product of extra pure grade $AgNO_3$ thermal decomposition) dissolving.

Reaction solutions electron-beam treatment was carried out by pulse linear electron accelerator ILU-6 (BINP SB RAS). The electron energy was 2.4 MeV, beam pulse current was 320 mA, pulse repetition rate was 5 Hz, pulse duration was 0.5 ms and average beam current was 0.45 mA.

Solutions were treated in polytetrafluoroethylene containers. The solution layer thickness was up to 5 mm. Containers with solutions accumulated required doses by multiple passing under accelerator beam window at a speed of 2 cm/sec, dose per pass was 5 kGy. Intervals between passes were about 30 seconds. 5 minute time breaks were inserted after every 5 passages (dose of 25 kGy) to prevent excessive solutions heating up.

Preliminary AG electron beam treatment was done in 5% water solution, doses were 25, 50 and 75 kGy.

Silver nanoparticles were produced in water solutions containing 2.5% AG and 0.005 M of $AgClO_4$ (pure silver concentration was 0.535 g/l). This solution was treated with dose of 100 kGy.

UV-spectra were obtained by using a Cary-50 spectrophotometer (Varian). The absorption spectra were recorded within 1 day after the electron-beam treatment, solutions were placed in quartz cuvettes with absorbing layer of 1 cm. The solutions were diluted in 20, 40 and 320 times, see Figure 1.

The microstructure, phase structure, elemental and chemical composition of local sites were fixed by microscopic observations using

- TEM - JEM-2100 (Jeol, Japan) with addition of energy dispersive X-ray analysis (EDXA) on analyzer Inca-250,
- SEM - S-3400M (Hitachi, Japan) with the possibility of a local elementary analysis.

Solutions containing nanoparticles were placed on aluminum substrate and dried at ambient conditions. The final water removal from samples was carried out under reduced pressure.

Molecular weight distribution of the samples was determined by gel-permeation chromatography(GPC) on an Agilent 1200 chromatograph with a 1260 Infinity refractive-index detector (30°C, PL aquagel-OH 40, 300x7.5 mm, 0.1 M $LiNO_3$, 1 ml/min). The column was calibrated using standard dextran samples (Sigma-Aldrich). The samples were studied as 0.2% weight solutions in 0.1 M $LiNO_3$ prepared using ultrasound.

3. Results and discussion

Just after electron beam treatment the treated solutions acquired dark red color that after storage during some hours changed into orange-red. The formed solutions contacting silver nanoparticles were stable during many months – no nanoparticles precipitation was fixed.

IOP Conf. Series: Materials Science and Engineering **168** (2017) 012109 doi:10.1088/1757-899X/168/1/012109

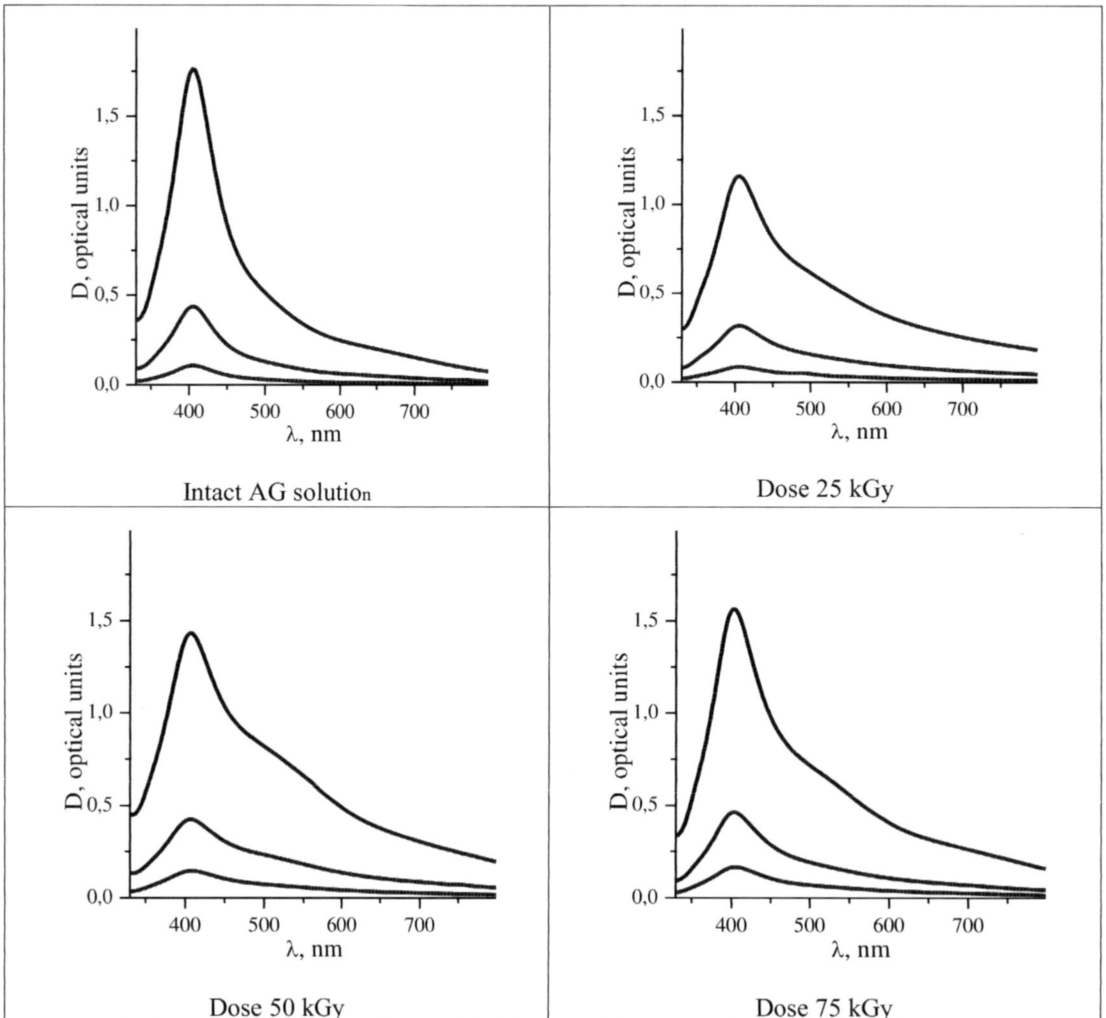

Figure 1. UV-spectra of Ag nanoparticles stabilized in aqueous AG solutions – intact and preliminary irradiated, before and after 2 dilutions with rates of 20 times.

Figure 1 shows optic absorption spectra of the silver nanoparticles in intact AG aqueous solution and in preliminary treated solutions with doses 25, 50 and 75 kGy before and after 2 dilutions having rates of 20 times. Absorption spectrum peaks in all cases is in region 403-407 nm. The spectra's forms are practically the same for all samples.

All spectra after dilution in 20 times show sufficient absorption in the region of more than 500 nm. Spectral factorization in coordinates optical density – frequency (reciprocal number, cm^{-1}) using Gauss method gave us 3 main spectral components with peaks in 400-410, 515-520 and 640-660 nm ranges.

Presence of some longitudinal plasmon resonances (LPR) in absorption spectra can be stipulated by some reasons [2,12]:

- presence of particles having sizes of more than 50 nm that leads to sufficient increase in quadruple component input into LPR;
- presence of faceting or great anisotropy in crystals' forms;
- particles' self-organization effects.

Last variant is most probable in our case as the dilution of AG solution without preliminary radiation treatment leads to practical disappearance of absorption in the region of higher than 500 nm.

The dilution of AG solution preliminary treated by radiation did not result in absorption disappearance in the range of 510-520 nm even after dilution in 1680 times.

Electron microscopical study of films formed by solutions evaporation has shown that their surfaces are homogeneous. Particles of more dense material appeared after partial polymer matrix burning out by electron beam, element analysis confirmed that they are silver particles. Some particles with sizes less than 100 nm have low aspect correlation and form clusters with sizes up to 500 nm.

Nanoparticles' vicinity in the clusters engenders additional plasmon resonance lines like in [6]. Separate nanoparticles are practically not fixed conforming hypothesis about ordering.

High resolution electron microscopy images shown in Figure 2 witness presence of circular silver nanoparticle clusters. Big nanoparticles are located mainly on surface. These pictures correspond with earlier published data about AG water solutions gained by terahertz radiation. It was fixed that water solution AG concentration increase up to 1% results in globule size increase up to 75-80 nm. Concentration increase over 1% results in agglomeration of the formed globules [13].

Comparative experiments with water and dimethyl sulfoxide as solvent demonstrated that hydrogen bonds plays a decisive role in AG agglomeration [5].

So we consider the radiation-chemical silver nanoparticle synthesis results in nanoparticles stabilization due to fastening on the AG agglomerates' surfaces.

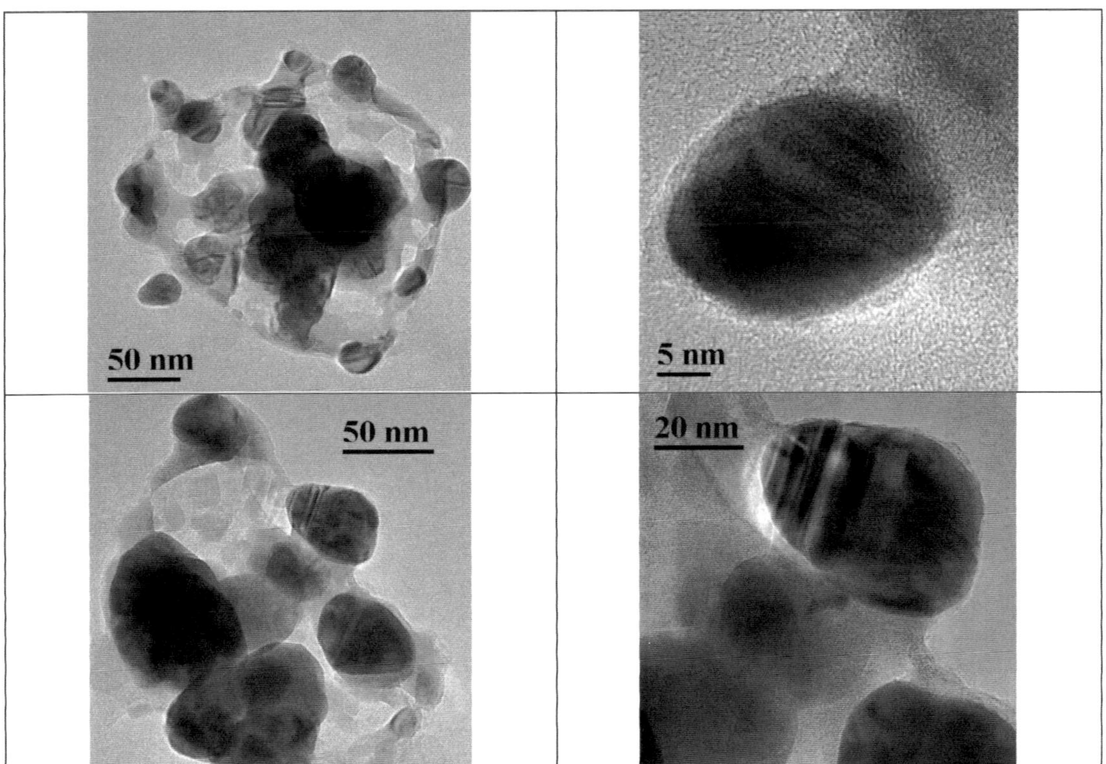

Figure 2. Electron microphotographs of composite formed in AG aqueous solution preliminary treated with the dose of 75 kGy.

AG practically did not manifested its deoxidizing properties in chosen conditions of radiation-chemical silver reduction. Radiolysis reduction rate is higher than chemical reduction rate.

Nevertheless there is possibility that nanoparticles nuclei were formed and fastened on globules' surfaces during the process of AG and $AgClO_4$ reaction solutions preparation.

The AG aqueous solution electron beam treatment results in partial AG oxidation and crosslinking. 3-dimensional crosslinked gels formation was not observed after treatment of 2.5-10% aqueous AG solutions. AG is oxidized during irradiation by water radiolysis products, but differently than by classical peroxide oxidation [14]. Sedimentation by acetone carried out in 2.5-10% AG solutions after electron beam treatment with doses up to 150 kGy similarly as in [14] demonstrated low (less than 1%) low molecular weight fraction share. The classic peroxide oxidation gives low molecular weight fraction share up to 10%.

Probably AG is oxidized during irradiation like dextrans [15], and in classic dextran oxidation by potassium permanganate [16, 17] C-C bond breakages are going in pyranose and furanose rings without sufficient polymer chain cracking.

Figure 3 shows mass-molecular distributions of AG isolated from intact and irradiated 5% aqueous solutions.

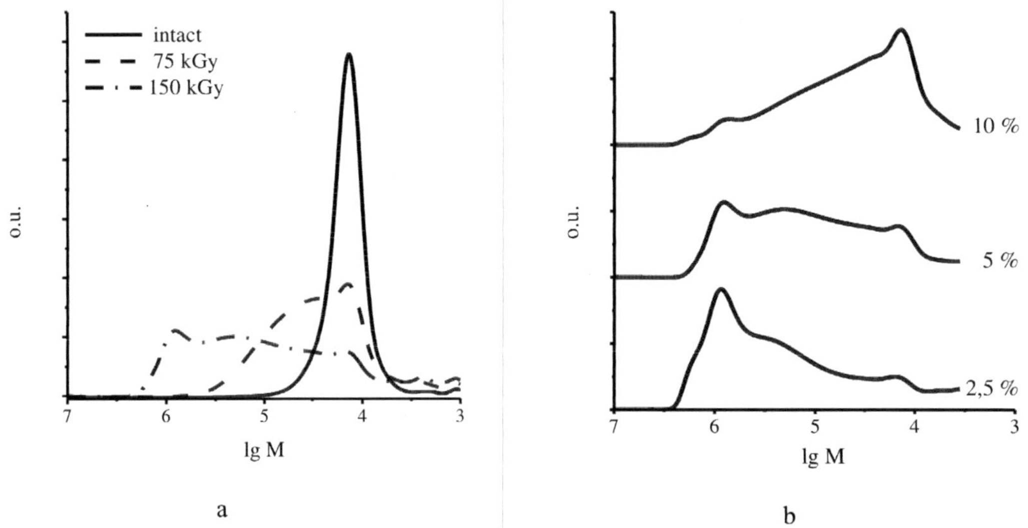

Figure 3. GPC-cartoon of initial AG and after electron beam treatment.

GPC studied had shown that AG solutions electron beam treatment resulted in MMP dispersion increase and sufficient molecular weight growth up to values of more than 1000000 (lg M = 6). Maximum crosslinking rate was observed in low concentration AG solutions (Figure 3b, dose 150 kGy). Chromatograms of irradiated AG 2.5-10% solutions had initial AG peak witnessing reaction incompleteness. It can be caused by heterogeneity of system.

4. Conclusion

The formed solutions contacting silver nanoparticles were stable during many months – no nanoparticles precipitation was fixed. The following dilutions up to 1680 times does not cause the silver nanoparticles precipitation.

The data get earlier and our studies results show that AG solutions can be regarded as heterogeneous systems with ordered globules. The ordered nanoparticles clusters with small distances between them can be formed on their surfaces and it is stimulating cooperative effects. Perhaps the nuclei of the nanoparticles are forming during the reaction mixtures preparation.

Preliminary AG solutions irradiation results in formation of greater molecules and probably in formation of groups capable of strong hydrogen bonds formation. It helps to preserve AG conglomerates in solutions after dilutions.

The radiation-chemical reduction of silver aqueous AG solution leads to silver nanoparticles ordering. The appearance of additional absorption band in UV-spectra is due to the close proximity of the silver nanoparticles in these structures. The AG cross-linking and oxidation are going during electron-beam treatment. Preliminary electron beam treatment of aqueous AG solution improves conglomerated silver nanoparticles stability, and dilution in 1680 times does not cause the silver nanoparticles precipitation.

The authors are grateful to Dr. A.G Zelinskii (ISSC SB RAS) for providing a silver salt solution and Mr. V.I. Kondratyev (BINP SB RAS) for assistance in scanning electron microscopy studies.

5. Acknowledgement

X-ray diffraction experiments were carried out with involvement of equipment belonging to the shared research center SSTRC, BINP SB RAS.

References

[1] Quang Huy Tran, Van Quy Nguyen, Anh-Tuan Le 2013 *Adv. Nat. Sci.: Nanosci. Nanotechnol.* **4** 033001

[2] Krutyakov Yu A, Kudrinskiy A A, Olenin A Yu, Lisichkin G V 2008 *Russian Chemical Reviews* **77** 3 233-257

[3] Jegatha Christy and M Umadevi 2012 *Adv. Nat. Sci.: Nanosci. Nanotechnol.* **3** 035013

[4] Mikhailenko M A, Shakhtshneider T P, Eltsov I V b, Kozlov A S, Kuznetsova S A, Karacharov A A, Boldyrev V V 2016 *Carbohydrate Polymers* **138** 5 1-7

[5] Sukhov B G, Aleksandrova G P, Grishchenko L A, Medvedeva S A, Trofimov B A, Feoktistova L P, Sapozhnikov A N, Proidakova O A, T'Kov A V 2007 *Journal of Structural Chemistry* **48** 5 922-927

[6] Jean-Christophe Valmalette, Zhenquan Tan, Hiroya Abe & Satoshi Ohara 2014 *Scientific Reports* **4** 5238

[7] Ershov B G, Abkhalimov E V, Sukhov N L 2005 *High Energy Chemistry* **39** 2 55-59

[8] Auslender V L, Bryazgin A A, Faktorovich B L, Gorbunov V A, Kokin E N, Korobeinikov M V, Krainov G S, Lukin A N, Maximov S A, Nekhaev V E, Panfilov A D, Radchenko V N, Tkachenko V O, Tuvik A A, Voronin L A 2002 *Radiation Physics and Chemistry* **63** 3-6 613-615

[9] Auslender V L, Bukin A D, Voronin L A, Kokin E N, Korobeinikov M V, Krainov G S, Lukin A N, Radchenko V M, Sidorov A V, Tkachenko V O 2004 *Radiation Physics and Chemistry* **71** 1-2 295-297

[10] Ershov B G 2004 *Russian Chemical Reviews* **73** 1 101–113

[11] Kuznetsova S A, Kuznetsov B N, Mikhailov A G, Skvortsova G P 2006 Russian Federation patent RU 2273646

[12] Mikhailenko M A, Shakhtshneider T P, Eltsov I V b, Kozlov A S, Kuznetsova S A, Karacharov A A, Boldyrev V V 2016 *Carbohydrate Polymers* **138** 5 1-7

[13] Jaysen Nelayah, Mathieu Kociak, Odile Stephan, F Javier Garcia de Abajo, Marcel Tenci, Luc Henrard, Dario Taverna, Isabel Pastoriza-Santos, Luis M Liz-Marzin & Christian Colliex 2007 *Nature Physics* **3** 348 – 353

[14] Borisov I M, Zimin Yu S, Monakov Yu B, Shirokova E N, Mudarisova R Kh, Muslukhov R R, Medvedeva S A, Tolstikov G A 2004 *Russian Chemical Bulletin* **53** 2 318-324

[15] Shkurupiy V A, Krasnov V A, Gulyaeva E P, Troitskiy A V, Bogdanova L A, Machneva T V, Grishin O V, Auslender V L, Korobeinikov M V 2002 *Radiation Physics and Chemistry* **63** 3-6 691-695

[16] Shkurupiy V A, Troitskiy A V, Potapova O V, Luzgina N G 2008 Russian Federation patent RU 011718

Structure and properties of radiation modified polyethylene

U V Ancharova[1,3*], M A Mikhailenko[1], M R Sharafutdinov[1], B P Tolochko[1], K B Gerasimov[1], M V Korobeynikov[2,3] and A A Bryazgin[2,3]

[1]Institute of Solid State Chemistry and Mechanochemistry SB RAS, Kutateladze, 18, 630128 Novosibirsk, Russia
[2]Budker Institute of Nuclear Physics SB RAS, Lavrentieva, 11, 630090 Novosibirsk, Russia
[3]Novosibirsk National State Research University, Pirogova, 2, 630090 Novosibirsk, Russia

E-mail: ancharova@gmail.com

Abstract. Polyethylene was irradiated with intensive electron beam with different doses. Structure changes after irradiation *ex situ* and *in situ* during heating up to its melting point were studied using synchrotron radiation X-Ray diffraction and thermal analysis.

1. Introduction

The intensive electron beam can modify structure and properties of inorganic and organic materials, crystal or amorphous. Ionizing radiation impact on polymer materials in different ways, such as introduction of defects into the crystal structure, breaking bonds in polymers, that in turn can lead to formation of other bonds in the structure and changes in mechanical or physical-chemical properties of polymer materials [1].

Polyethylene structure $-(-CH_2-CH_2-)n-$ looks like zigzag chains of carbon atoms, each carbon atom of the chain is connected with two hydrogen atoms. Usually chains tend to form crystalline domains with amorphous surrounding. Polyethylene crystal structure [2] has orthorhombic symmetry with the space group Pnam (#62), with lattice parameters a=7.41Å; b=4.94Å; c=2.55Å. Polymer chains extend along *c*-axis in the crystal and each unit cell consists of one formula unit $-(-CH_2-CH_2-)_n-$ and contain two perpendicularly zigzagged chains (see the Figure 1).

The irradiation causes two oppositely going processes in polymers – scissioning (cleavage) and crosslinking of the chains [3]. Crosslinking is the formation of a three-dimensional polymer network. In subjecting any polymer to electron beams or X-rays, the amount of energy input and multitude of hemolytic bond cleavages can result in either polymer chain extension and then crosslinking or polymer breakdown and scissioning.

The crosslinking prevails in polyethylene resulting in mechanical properties changing.

To understand the nature of polymer properties changing one ought to comprehend the changes in local structure after irradiation. The most direct way is to investigate the structure by X-ray diffraction method, especially synchrotron radiation XRD, as the X-Ray scattering on light elements is not intense enough. Thermal analysis helps to connect structure changes with changing in properties of polyethylene [4].

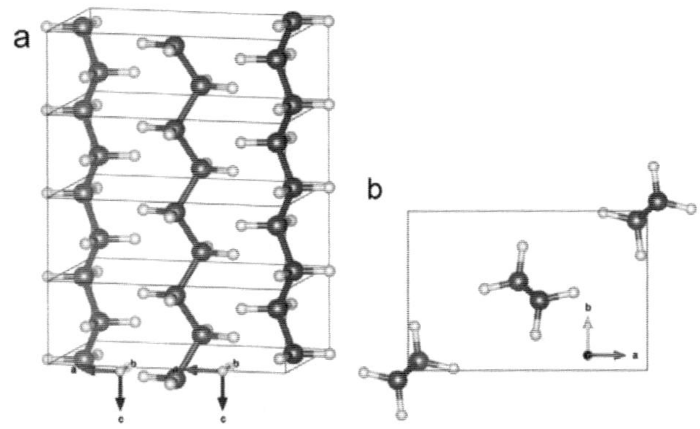

Figure 1. Polyethylene crystal structure (a) general view; (b) projection on (001) plane.

2. Experiment

LDPE - Low Density Polyethylene samples having thicknesses of ~1 mm were tested. The samples were subjected to electron beam treatment with doses of 2, 4, 6, 8, 16 and 40 kGy at room ambient conditions (temperature and pressure). The electron beam was generated by ILU-6 electron accelerator (located in BINP SB RAS, Novosibirsk, Russia), beam parameters were as follows: electron energy of 2.4 MeV, pulse beam current of 320 mA, pulse duration of 0.6 ms, pulse repetition rate of 2 Hz, underbeam transportation velocity of 2 cm/s. The absorbed dose was determined by number of underbeam passages, during one passage sample got the dose 2 kGy during approximately 4-5 seconds under beam window. The samples temperature was kept below 35°C to avoid structural changes caused by temperature increase. Time between passages during radiation treatment was not less than 30-40 seconds.

Ex situ investigations of polymers structure changes in XRD experiments were carried out at the experimental station on the 4th beamline of VEPP-3 storage ring in BINP SB RAS. X-rays had a wavelength of 0.3685 Å. In situ investigations during heating of polyethylene have been made at 5b SR experimental station of VEPP-3 storage ring, λ=1.516 Å. Heating has been made in opened furnace at air up to the temperature of polyethylene melting point ~120°C.

Samples for thermal analysis were prepared as follows. For calorimetric investigations round tablets with diameter 3 mm were cut of the plates of polyethylene. To remove residual stresses in material after that process they were annealed at 70°C during 2 hours. After that samples were irradiated at the same conditions as that for XRD investigations.

Calorimetric investigations were carried out with using DSC 200F3 (NETZSCH) calorimeter in sealed crucibles in the temperature range from -160°C to 140°C with a heating rate of 10°C/min.

3. Results and discussion

Initial polyethylene samples and samples were irradiated by electron beam with doses of 2, 4, 6, 8, 16 and 40 kGy and then were investigated in the bulk state by synchrotron radiation X-ray diffraction method. Figure 2 shows resulting XRD patterns, demonstrating certain changes in polyethylene crystalline structure after irradiation. One can see that for initial state width of peaks were significantly higher than for irradiated samples, it points to increase in crystalline degree (size of domains and/or perfection in crystallite structure). It can be seen that the positions of the peaks tend to change for some peaks: they shift in different directions depending on crystallographic indexes. It means that crystal structure changes: the XRD analysis shows that the crystalline lattice expands along *a* and *b* axes and shrinks along *c* axis (see Figure 1a). That phenomenon can be connected with polymer cross-linking in transversal direction of the main chains in crystal, that also can cause changes in mechanical properties after irradiation, because cross-linking takes place not only in crystalline (visible in XRD) but also in amorphous part of the material.

Structure changing in polyethylene (PE) during heating up to melting point were studied. The irradiated and initial samples were heated up to 120°C with a rate of 5°C/min and their XRD were investigated in continuous mode with a frame rate of one frame per second. The changes in position and width of the (110) peak for PE during heating after irradiation with different doses are presented in Figure 3.

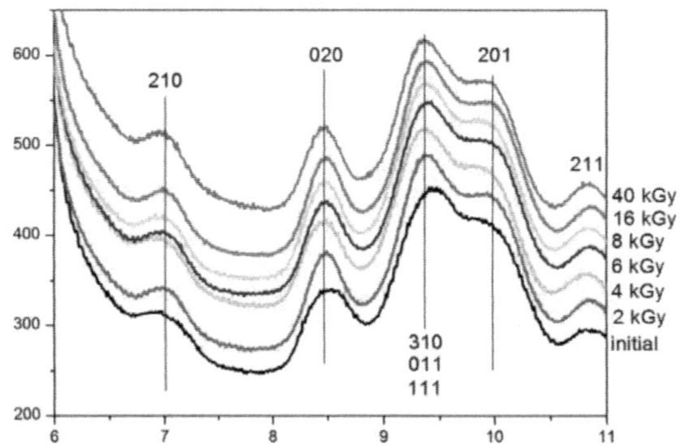

Figure 2. XRD scattering intensity $I(2\theta)$ in arbitrary units for polyethylene after irradiation with high energy electrons of different doses (0 -40kGy); λ=0.3685Å.

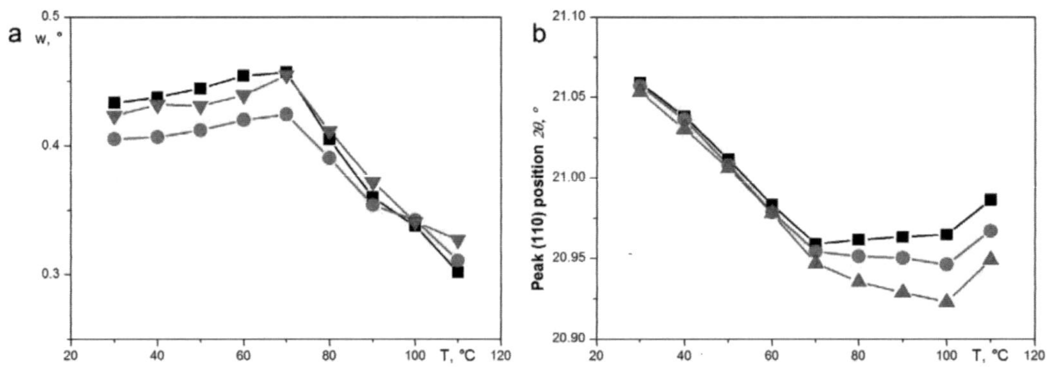

Figure 3. Changes in (a) widths; (b) positions of (110) XRD reflection of PE during heating of irradiated polyethylene after irradiation with doses 0 kGy (■), 4 kGy (•), 40 kGy (▲), λ=1.516Å.

Widths of (000) peak (small angle scattering) for any considered irradiation doses gradually increase and reach maximum at the temperature of 100°C right before the melting point of PE and then gradually decrease, while widths of (110) peak also gradually increase to the point 70°C and then gradually drop down (fig. 3*a*). At the same temperatures the tendency in changes of peak position (fig. 3*b*) changes: below 70°C it gradually decrease, then behavior for different doses differs but in general does not change and after 100°C tends to increasing. Changes in intensity of (110) peak depending on temperature (fig. 5a) also correlate. *That means that that temperatures are singular points in changes in structure and properties of PE.* Probably the first of them is connected to the temperature of activation of collective oscillations in chain, that can initiate changes in structure: the process of crosslinking in crystalline fraction of polymer; and the second one is the temperature of pre-melting. Main differences for samples irradiated with different dozes are observed in a final peak position, it has the lower values the higher was irradiation doze (fig. 3b). It means that after irradiation structure of PE becomes to be more loose and stretched along transversal directions towards polymer chains.

Figure 4 shows thermoanalysis results for irradiated and initial polyethylene samples.

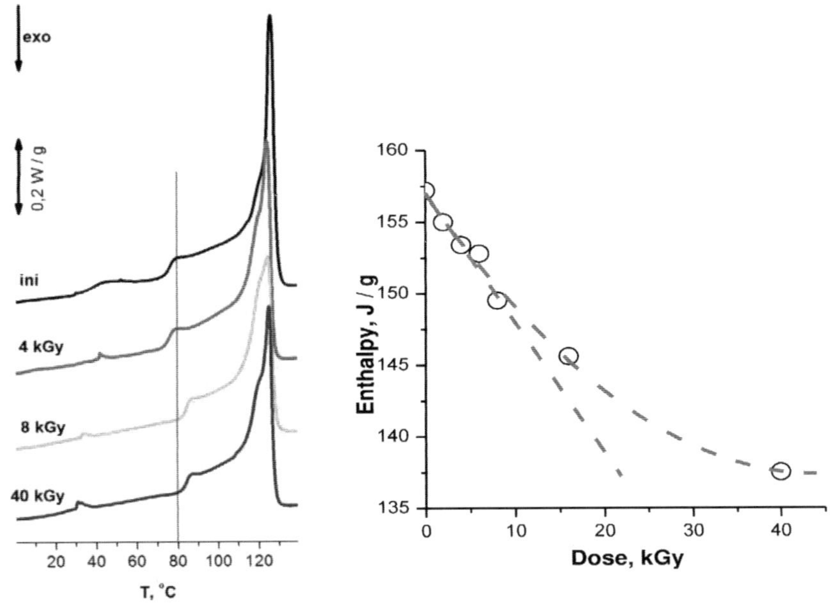

Figure 4. (left) absorbed energy during heating of irradiated polyethylene having doses from top to bottom: 0: 4; 8; 40 kGy; (right) thermal effect depending on irradiation dose.

The most part of energy absorbed by the material during irradiation with low doses is to increase the internal energy of the sample and the variation of heat can be described by a linear dependence Y=157.1-0.88X by points below 10 kGy. During calculations with using points above 10 kGy the dependence of the thermal effect on the absorbed dose of ionizing radiation can be described by the second-order polynomial $Y=156.96-0.89X+0.01016X^2$, here X is the irradiation dose. Thermal effects were calculated using a sigmoidal baseline from 0 to 150°C using Proteus Analysis software.

The values of melting points change without any clear dependence and lay in the range of 124-127°C. It disappears the pronounced thermal effect of low-temperature softening at 45°C (existing for initial sample), which is usually referred by properties to amorphous phase of polyethylene. Also it changes the temperature of viscous fluid transition (plastic) appearing at 75°C for initial sample, which can be referred by properties to the ordered phase of polyethylene. Thermal effects of glass transition unfortunately failed to be reliably obtained.

Obviously at the beginning stage of irradiation it occurs mostly transformations of not ordered part of PE. As mobility of their chains allows to achieve the maximal efficiency of irradiation. Then after crosslinking the mobility of polymer network of not-ordered fraction approach to that for ordered part of polyethylene and efficiency of irradiation drops down.

4. Conclusion

Structure investigation show that polyethylene after irradiation tend to form more 'loosening' structure, it points on creating bonds during crosslinking in transversal to chains direction, that affect also the crystal structure properties and changes in lattice parameters. Comparing XRD and thermoanalysis results one can conclude that the singular point of ~70-75 °C can be connected to crosslinking of ordered fraction of PE, occurring during heating of the sample as a result of increasing mobility in chain network.

The described XRD method is rather fast not nondestructive analysis method and can be used for gel fraction estimation and measurement.

IOP Conf. Series: Materials Science and Engineering **168** (2017) 012110 doi:10.1088/1757-899X/168/1/012110

5. Acknowledgement

X-ray diffraction experiments were carried out with involvement of equipment belonging to the shared research center SSTRC, BINP SB RAS.

References

[1] Forgács P, Tolochko B P, Sheromov M A 1981 *Polymer Bulletin* **6** 127
[2] Bunn C W 1945 Chemical Crystallography, Oxford University Press 233
[3] Industrial radiation Processing With Electron beams and X-rays 201. International Atomic energy Agency technical document, http://www.cirms.org/pdf/Industrial%20Radiation%20Processing%20-%20May%202011%20-%20Revision%206.pdf
[4] Minkova L, Nikolova M, Nedkov E 1988 *Journal of Macromolecular Science B: Physics* **1** 99

Mixed feed and its ingredients electron beam decontamination

V V Bezuglov[1,3], A A Bryazgin [1,3], A Yu Vlasov [1,3], L A Voronin [1,3], Yu V Ites[2], M V Korobeynikov [1,2,3], S V Leonov [2], M A Leonova [2], V O Tkachenko [1,3], E A Shtarklev [1,3] and Yu G Yuskov [2]

[1]Budker Institute of Nuclear Physics SB RAS, Lavrentieva, 11, 630090 Novosibirsk, Russia
[2]Institute of Experimental Veterinary of Siberia and Far East SB RAS, 630501 Krasnoobsk, Novosibirsk region, Russia
[3]Novosibirsk National State Research University, Pirogova, 2, 630090 Novosibirsk, Russia
E-mail: M.V.Korjbeynikov@inp.nsk.su

Abstract. Electron beam treatment is used for food processing for decades to prevent or minimize food losses and prolong storage time. This process is also named cold pasteurization. Mixed feed ingredients supplied in Russia regularly occur to be contaminated. To reduce contamination level the contaminated mixed feed ingredients samples were treated by electron beam with doses from 2 to 12 kGy. The contamination levels were decreased to the level that ensuring storage time up to 1 year.

1. Introduction

Russia has the vast areas with cold severe climate. The goods are delivered to the people living and working in these Northern Territories only during summer. The poultry factories are also supplied with mixed feed only in summer time, and the decontamination of the mixed feed will decrease the feed losses during the year.

Mixed feed ingredients supplied in Russia regularly occur to be contaminated. The main problems are caused by fish flour and soya. The fish flour is produced on the ship far away in the ocean and then pass the long way to the final consumers by ships and then more than 5 thousand kilometers by train and often by river boats. Contaminated soya is supplied from Far East and its way is on trains and river boats and barges.

When the train (tens of wagons) or barge with mixed feed ingredients arrived to poultry factory the products are assayed. The bacterial contamination tests last 2 weeks, and only after that ingredients can be used for mixed feed preparation. If the test results show that contamination level exceeds the admissible limits there is no way to send the products back to seller – too long and too expensive way back, or no possibility to deliver back because of short navigation period on northern rivers.

So usually the poultry farm officials have choice of 3 possibilities – to discard the contaminated ingredients, to "dilute" contaminated ingredient by not contaminated components to decrease contamination level to admissible values (rarely happening possibility) or to put it in production and to add great amount of antibiotic preparations into mixed feed to prevent poultry loss. The antibiotics prevent poultry losses but accumulate in poultry meat, and it will not meet sanitary requirements – again financial losses. The contaminated mixed feed can contaminate feed kitchen, and it will be hard to make it contamination free.

Thermal treatment is not suitable to reduce contamination level because it will spoil the ingredients' nutritional properties.

Electron beam treatment can be the problem solution – it reduces the contamination level without heating (so it is also named "cold pasteurization") and without spoiling nutritional properties.

The radiation sterilizing abilities were discovered in the beginning of XXth century. First patent describing radiation usage for food preservation is dated July 17 1930 [1].

The radiation treatment of food really started in the end of 40-s in XX century. Food irradiation became normal practice in the end of XXth century due to technical development of electron accelerators, increase of their energy and beam power resulting also in irradiation cost decrease and welfare growth – all these factors make the food radiation treatment economically viable process.

Accelerators development and radiation technologies are described in [2]. Electron beam treatment is used for food products to prevent spoiling and prolong the storage time. Main facts about food irradiation and research results are given in [3-7]. Food irradiation practices are described in [8-13].

Budker Institute of Nuclear Physics (BINP) is one of the world leaders in accelerator development and production [2]. More than 140 industrial electron accelerators were supplied and installed by BINP all over the world.

2 electron accelerators produced by BINP were installed in 1982 in Odessa port elevator to decontaminate the wheat imported from Canada if the contamination levels exceed admissible values. Later 2 electron accelerators produced by BINP were installed in port elevator in Guangdong province.

We have not find information that anyone used electron beam treatment for decontamination of mixed feed or its ingredients. We have carried out research work to estimate the possibilities of electron beam treatment for mixed feed ingredients processing. It was assumed that it is better to treat the ingredients with high contamination level than ready mixed feed.

The main aim of the work was to determine the minimal dose allowing to prevent the mixed feed from spoiling during the transportation to the Northern Territories and following storage during the period up to the year – till the next year navigation. The second aim was to estimate nutrition properties degradation after electron beam treatment.

2. Experimental

A pulse radio frequency electron accelerator ILU-6 installed in BINP is used for experimental works – development of radiation technologies and development of accelerator systems. Many radiation technologies were elaborated on this machine.

Numerous researches have shown that the pulse electron beam better suppresses microflora than continuous electron beam due to higher impact during the pulse. [14-18]

Figure 1 shows dose-depth curve for electron beam treatment. Beam penetration ability R_{opt} at energy of 2.5 MeV is 0.8 g/cm^2.

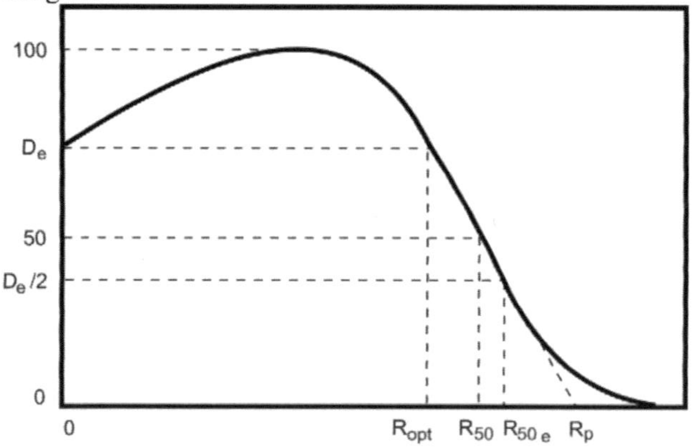

Figure 1. Dose-depth curve for electron beam treatment.

ILU-6 parameters are listed below:
Electron energy – up 2.4 MeV
Pulse beam current – up to 420 mA
Pulse duration – 0,5 ms
Pulse repetition rate – up to 50 Hz
Scanning width – 90 cm
Sample table underbeam transportation velocity – up to 8 cm/s

Electron beam and X-rays generate free radicals and, in the presence of water, a constituent of all living matter, highly reactive short-lived hydroxyl radicals. Both can lead to single and double-strand breaks, intra-strand cross-links and thus damage DNA, including breaking the hydrogen bonding between the nucleotides in the dyad. With but a few double strand breaks, DNA is no longer able to repair or replicate itself. Without the ability to correctly replicate their own DNA structure, cells die. Pathogens in food or on medical products exposed to sufficient ionizing radiation are then died and bio-burdens eliminated [2].

The more complicated is the organism the less is the lethal dose. Simplest microbes (especially in spore form) and mold fungi possess highest resistance to irradiation.

Typical doses for various irradiation processes:

- Sprouting inhibition 0.1 – 0.2 kGy
- Disinsection 0.3 – 0.5 kGy
- Parasites development inhibition 0.3 – 0.5 kGy
- Fruits maturing delay 0.5 – 1.0 kGy
- Fungi inhibition 1.5 – 3.0 kGy
- Bacteria inhibition 1.5 – 3.0 kGy
- Sterilization 15 – 30 kGy
- Polymerization 25 – 50 kGy

For most bacteria the dose of 2 kGy is so called D_{10} – the dose that results in 10 times reduction in survived bacterial population. So we have taken 2 kGy as minimal dose and dose step.

3. Electron beam treatment of packed samples and its results

The mixed feed and its ingredients samples were subjected to electron beam treatment by ILU-6 pulse electron accelerator. Electron beam treatment parameters were:
Electron energy – 2.4 MeV
Beam penetration ability – 0.8 g/cm^2
Pulse beam current – 320 mA
Pulse duration – 0.5 ms
Pulse repetition rate – 5 Hz
Scanning width – 90 cm
Underbeam transportation velocity – 2 cm/s
Treatment doses in out experiments varied from 2 to 12 kGy.

In experiments we use the feed contaminated by coliform (1.5*10^6 cfu/g), spore forms (1.2*10^6 cfu/g) with total bacterial count of 3.0*10^6. The study of microbial contamination of feed samples was carried out by methods provided by state regulations of Russia (GOST 51426-99, ISO 7251-2005, ISO 7218-2008, ISO 11133-1-2008) for microbiological testing of feed.

The mixed feed samples were placed into the hermetically sealed packs. The packs with the mixed feed were placed on the moving table and were transported under the electron beam that was scanning across the table moving direction overlapping all table surface.

Figure 2 shows the moving table with the mixed feed samples and its 9 ingredients inside the packs. The treatment doses varied from 2 to 12 kGy. The packing was done so that the product mass thickness was about 0.3-0.4 g/cm^2 that is 2-3 times less than electron beam penetration depth R_{opt} at energy of 2.4 MeV that nearly equal to 0.8 g/cm^2 to ensure good dose uniformity.

Figure 2. Samples of mixed feed and of 9 its ingredients on moving table prepared for electron beam treatment.

Treated samples were tested using the standard microbiological technique. The usual and selective mediums were used for incubation during 24 hours at temperature of 37°C. The bacterial colonies amount, rates of coliforms and spores were counted.

Total microbial count was reduced to an acceptable concentration after electron beam treatment with doses of 4 kGy and higher.

Coliforms were not fixed after EB treatment with the dose of 12 kGy.

The spore forms of micro-organisms survive after treatment dose of 12 kGy.

Figure 2 represents the treatment results. The diagram shows amount of colony-forming units per gram dependence on the treatment doses. Red line is the legal limit that is allowed in poultry feed in Russians Federation – $5*10^5$ cfu/g. Coliform bacteria are not allowed in poultry feed.

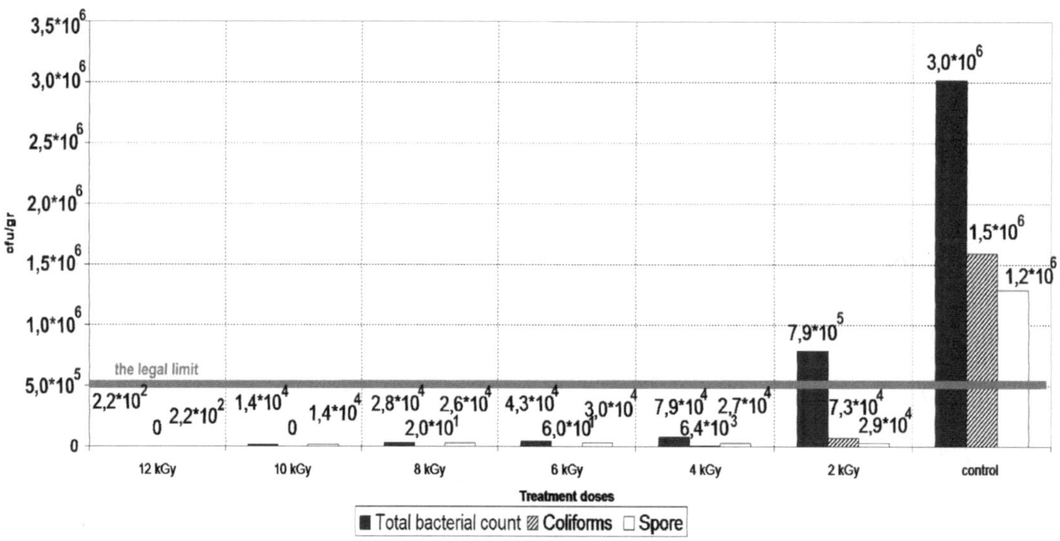

Figure 3. Electron beam treatment results - amount of colony-forming units per gram dependence on treatment doses.

4. Conclusion

Total microbial burden in mixed feed and its ingredients samples was reduced to an acceptable level after electron beam treatment with doses of 4 kGy and higher.

Coliforms were not fixed after EB treatment with the dose of 12 kGy.

The spore forms of micro-organisms survive after dose of 12 kGy.

The EB treatment can decrease the mixed feed losses and prolong the storage time. This method can be potentially applicable for treatment of mixed food lots with contamination level exceeding the admissible values.

References

[1] Wüst O 1930 Procédé pour la conservation d'aliments en tous genres *Brevet d'invention* 701302 (July 17)

[2] Industrial radiation Processing With Electron beams and X-rays, 2011. International Atomic energy Agency technical document, http://www.cirms.org/pdf/Industrial%20Radiation%20Processing%20-20May%202011%20-%20Revision%206.pdf

[3] Facts about food irradiation. Document of Food and Environmental Protection Section. Joint FAO/IAEA Division of Nuclear Techniques in Food and Agriculture, http://wayback.archive.org/web/20060316042916/http://www.iaea.org/programmes/nafa/d5/public/foodirradiation.pdf

[4] GAO, Food Irradiation: Available Research Indicates That Benefits Outweigh Risks, GAO/RCED-00- 217 (Washington, D.C.: Aug. 24, 2000). http://www.gao.gov/archive/2000/rc00217.pdf

[5] GAO-10-309R Federal Oversight of Food Irradiation http://www.gao.gov/assets/100/96545.pdf

[6] Food Safety News/ Texas Researchers Devise New Irradiation Method http://www.foodsafetynews.com/2010/12/texas-research-devise-new-irradiation-method/

[7] Irradiated food/ Global development&Irradiotion as a quarantine treatment Irradiation technology industrial application forum 19-20 july 2011 http://www.latu.org.uy/docs/7-Carl-Blackburn-Alimentos-irradiados-medida-cuarentenaria.pdf

[8] ASTM F1355 - 06(2014) Standard Guide for Irradiation of Fresh Agricultural Produce as a Phytosanitary Treatment

[9] ASTM F1356 - 08 Standard Practice for Irradiation of Fresh and Frozen Red Meat and Poultry to Control Pathogens and Other Microorganisms

[10] ASTM F1736 - 09(2016) Standard Guide for Irradiation of Finfish and Aquatic Invertebrates Used as Food to Control Pathogens and Spoilage Microorganisms

[11] ASTM F1885 - 04(2010) Standard Guide for Irradiation of Dried Spices, Herbs, and Vegetable Seasonings to Control Pathogens and Other Microorganisms

[12] ASTM E2449-05(2013) Standard Guide for Irradiation of Pre-packaged Processed Meat and Poultry Products to Control Pathogens and Other Microorganisms

[13] ASTM F1640 - 09 Standard Guide for Packaging Materials for Foods to Be Irradiated

[14] Tallentire A, Purdie J W and Ebert M 1974 Increased response of anoxic Bacillus megaterium spores to radiation at high dose-rates *International Journal of Radiation Biology* **26** 5 435-443

[15] Tallentire A and Barber D J W 1978 The Interrelationship between Pulse Length and Dose-rate within the Pulse in the Enhancement of Atomic Damage in Wet Bacillus megaterium Spores Irradiated with Pulses of Electrons. Sixth Symposium on Microdosimetry (Booz and Ebert, edt.) Harwood Academic Publishers Ltd. 1061-1069

[16] Barber D J W and Tallentire A 1980 Modification by Chemical and Physical Means of the Pulse Dose Effect in Irradiated Bacteria Spores. Seventh Symposium on Microdosimetry (Booz, Ebert and Hartford, edt.). Harwood Academic Publishers Ltd. 1145-1155

[17] Tallentire A 1983 Oxygen effect, hydrogen peroxide yields, and time scale of interation of potentially damaging species in electron pulse irradiated bacterial spores *Advances in space*

research **3** 8 105-112
[18] DeVaux L C, et al. Effect of Electron Beam Dose Rate on Microbial Survival. Proceedings AccApp '07, Pocatello, ID (July 29 – August 2, 2007) 388-393

Institute of Physics
Dirac House, Temple Back
Bristol BS1 6BE UK

ISSN: 1757-8981
ISBN 978-1-5108-3672-3

9 781510 836723